Harnessing Indigenous Epistemologies for Sustainable Progress

Patrick Ngulube
University of South Africa, South Africa

IGI Global
Scientific Publishing
Publishing Tomorrow's Research Today

Vice President of Editorial Melissa Wagner
Managing Editor of Acquisitions Mikaela Felty
Managing Editor of Book Development Jocelynn Hessler
Production Manager Mike Brehm
Cover Design Phillip Shickler

Published in the United States of America by
IGI Global Scientific Publishing
701 East Chocolate Avenue
Hershey, PA, 17033, USA
Tel: 717-533-8845
Fax: 717-533-8661
Website: https://www.igi-global.com E-mail: cust@igi-global.com

Library of Congress Cataloging-in-Publication Data

Names: Ngulube, Patrick, editor.
Title: Harnessing indigenous epistemologies for sustainable progress / edited by Patrick Ngulube.
Description: Hershey, PA : IGI Global Scientific Publishing, [2025] | Includes bibliographical references and index. | Summary: "An approach to knowledge development and transmission that is more pluralistic and inclusive is supported by the incorporation of Indigenous viewpoints and wisdom into academia, which challenges the predominance of Western epistemologies"-- Provided by publisher.
Identifiers: LCCN 2024051504 (print) | LCCN 2024051505 (ebook) | ISBN 9798369384473 (hardcover) | ISBN 9798369384480 (paperback) | ISBN 9798369384497 (ebook)
Subjects: LCSH: Ethnoscience--Research. | Ethnoscience--Study and teaching. | Sustainability. | Education, Higher.
Classification: LCC GN476 .H37 2025 (print) | LCC GN476 (ebook) | DDC 001.089--dc23/eng/20241220
LC record available at https://lccn.loc.gov/2024051504
LC ebook record available at https://lccn.loc.gov/2024051505

British Cataloguing in Publication Data
A Cataloguing in Publication record for this book is available from the British Library.

Editorial Advisory Board

Table of Contents

Detailed Table of Contents

This chapter develops a framework integrating Indigenous spirituality with ethical structures to support environmental values, focusing on the Khoisan's culture and knowledge systems. Utilising deep ecology aligns Khoisan spirituality with sustainable development. The study addresses the need for frameworks combining Indigenous and Western values in social and economic models. The inductive content analysis of interviews, observations, and literature generates themes to support this integration. The framework contributes to Indigenous Knowledge Systems by offering alternative models for coexistence, promoting holistic development that balances economic, cultural, and environmental sustainability. The study challenges contemporary development paradigms, emphasising the importance of Indigenous Knowledge in fostering sustainable practices and a harmonious relationship with nature. By incorporating interconnectedness and deep ecology, it advocates for social and economic systems that respect and integrate Indigenous spirituality and Western values.

Fourth World knowledge systems, often overlooked in discussions of human sustainability, focus on the dynamic relationships between people, land, and the cosmos. These systems are typically associated with non-state nations and people, whose interactions with state entities often involve conflict over ancestral territories. Pursuing balance and harmony for life sustainability is not a primary consideration in these interactions. The author uses the Four Directions metaphor to symbolise the relational connection between human experience, land, and the cosmos, blending qualitative, quantitative, and relational reasoning. This framework has local, regional, and global implications for future sustainability. A conceptual framework is presented integrating Indigenous scientific knowledge with conventional sciences to explain various phenomena. Additionally, ALDMEM is introduced, a mechanism for mediating knowledge system applications, ensuring the consent of Fourth World people and state populations, and aligning with the principle of free, prior, and informed consent.

This chapter outlines the interpretation of the contours of Indigenous onto-epistemologies (existing prior to colonization by the West) as proffered by various sages and scholars, including the relevance of these for the current era. It focuses on the axiological commitments towards strengthening the relational balance across human and more-than-human existence. It indicates implications for enabling learning exchanges between Indigenous and Western ways of doing science, but with the proviso that we reflect critically on the values that the science and its products are serving (wittingly or unwittingly). It contends, drawing on and extending the work of a host of Indigenous authors, that the quest for supposed value freedom in the practice of science is a chimera and that our knowing processes are necessarily value-invested. This starting point prompts us to experiment further with ways of rendering the process of world-making (locally and globally) more inclusively beneficial and sustainable.

The Indigenous paradigm provides a framework to guide methodological decisions through its axiological, ontological, and epistemological assumptions. This paradigm should be prioritized in research conducted in Indigenous communities because it encompasses beliefs related to spiritual connectedness and decolonization that are not found in other research paradigms. Its permeable borders are illustrated by an example of an Indigenous/transformative study from Africa that puts the Indigenous and transformative paradigms in conversation with each other to advance justice.

This chapter originates from the author's research thesis, which investigated Indigenous dignity from prehistoric epochs through the concept of time, space and knowledge, Moreover, Indigenous constitutive rules, that is, peace, ecology, inclusive development and Indigenous leadership, transforming Indigenous dignity through triangulating rethinking thinking, transdisciplinarity, and cognitive justice into Indigenous people's dignity. Meta-synthesis and genealogy analysis guided by these constitutive rules when applied to qualitative studies data from prehistoric this is prior to the historic advent of and contemporary epochs during systemic negation of Indigenous knowledge. Thus, leading discussions through restorative action by integrating UNSDGs and AU Agenda 2063 into transforming Indigenous people's dignity into Indigenous knowledge epistemologies. Towards futuristic imperatives for integrating within modernity in research and public policy for implementation in the Indigenous Knowledge Hub.

This chapter argues that there is a sustained onslaught on epistemologies and identities of the inhabitants of the Global South through the hegemony of Western ways of knowing through the promotion of globalization. It posits that peddling this myth sustains Western epistemological hegemonies that have perpetuated and sustained the extirpation of those of the Global South. Through these subtle weapons of mass destruction, the researchers observed that most formerly colonized Indigenous communities have lost their identities and cultures. They further contend that scholarships by the countries in the Global North, including the new geo-political arrivants China and Russia continuously pummel and destroy indigenous identities under the guise of building relationships. Through these observations, the chapter accentuates the observation that while people are not being decimated physically, this occurs subtly. The chapter concludes by emphasizing the importance of embracing and harnessing all epistemologies to build better more sustainable and resilient communities.

This chapter provides a theoretical study through exploring the concept of cognitive empire alongside Afrocentricity, highlighting how the European construction of the modern world system has perpetuated a triumphalist narrative. Since the 15th century, Europe has systematically undermined African intellectual traditions, disrupting Indigenous cognitive processes and facilitating conquest. This historical trajectory has led to the dominance of European epistemologies while marginalising non-European contributions. In response, decolonial movements in Africa aim to reclaim oppressed humanity and challenge the moral decline associated with Eurocentric paradigms. The chapter argues that Indigenous knowledge systems are crucial for envisioning alternatives to Western-centric views and advocating for cognitive justice. Ultimately, it positions Afrocentricity as vital for reimagining knowledge production in the African context.

This systematic literature review focuses on using Indigenous theories and theorising in African scholarship. African scholars have tended to use Western and other theories without critically evaluating them for their relevance and ability to give a full explanation of the dynamics of their Indigenous phenomena. A clear conceptualisation of these terms can assist African scholars in theorising and using theories that explain their knowledge claims appropriately and contextually. This systematic review aimed to track the use of Indigenous theories in African scholarship and to understand the extent to which Indigenous theories have been employed. A thematic synthesis of the studies showed limited use of Indigenous theories and theorisation in African scholarship. The chapter mapped and provided insights into the use of theory in African scholarship and provided strategies on how African scholars can theorise and use Indigenous theories in their research in the quest for epistemic freedom.

This chapter theorises cultural communities' alienation from their cultural roots due to the effect of deterritorialization, colonization, globalization and westernization, and this has also given rise to cultural change or acculturation. The chapter examined conversations concerning the reclamation of misappropriated heritage in cultural communities in Southern Africa. The study employed an Afronographic approach, which is an Afrocentric paradigm that places the interests of the African people at the centre of the analysis of the deterritorialization and reterritorialization of knowledge intending to reclaim ways of knowing and culture. Afrocentricity theory was employed to underpin this study because it is a novel research methodology that is deeply rooted in the African people's experiences, unlike Eurocentric theories that disorder Africans in the sideline of epistemology.

Preserving and disseminating indigenous knowledge in the digital age presents unique ethical challenges. There is a need to use frameworks such as the CARE and FAIR principles in the digital preservation of Indigenous knowledge. A literature review was conducted to unpack how the principles can be integrated into the digital preservation of Indigenous knowledge. The databases searched were Scopus, Web of Science, and Google Scholar. The findings show that integrating both principles require community-led governance models, culturally appropriate metadata systems, and collaborative partnerships between indigenous communities, researchers, and technologists. These approaches ensure that indigenous groups retain authority over sharing knowledge while benefiting from digital advancements. This, emphasising that there is a need for a paradigm shift that respects Indigenous epistemologies and the development of technology that prioritises Indigenous communities by all stakeholders.

Indigenous African epistemologies have contributed to the preservation and sustainable use of African heritage and resources for centuries. Indigenous African knowledge systems (IK) were transmitted orally by Indigenous knowledge-practising elders of the community who also served as IK custodians of the knowledge system. However, the marginalisation and displacement of African Indigenous epistemologies by Western knowledge systems have threatened their existence. This chapter argues that promoting the documentation and preservation of African Indigenous knowledge is crucial, not only for posterity but to protect it from exploitation, distortion and extinction. This chapter adopts the Afrocentric theoretical framework on African Indigenous epistemologies and calls for more research and documentation of African Indigenous knowledge systems to be conducted to promote the preservation and sustainable use of local heritage and resources. In addition, the rights to African Indigenous knowledge systems need to be protected by law, and patents should be given to local products.

This study aims to establish how music and dance can be used to drive IKS promotion to preserve national heritage. The study was undertaken to answer questions about the type of music and the views of both the musicians and listeners. Southern Africa has a rich history of indigenous knowledge systems (IKS) grounded in a strong oral tradition. IKS of Southern Africa can be traced from the arts, agriculture, traditional medicine, music, education, and so on. As the social-cultural values are predominantly expressed in the linguistic and musical interactions of the cultural practitioners and custodians, language, dance, and music may be seen as repositories and organic inventory systems for the Indigenous knowledge legacy. The musicians tell the stories of their lives: their identity, experiences, encounters, attachments, emotions, and sentiments. They portray the interconnectedness of their cosmology and spirituality. Making one question the understanding of indigenous identity.

The chapter examines the integration of Indigenous Knowledge Systems (IKS) in the context of the mopane worm (Imbrasia belina) economy amongst the Kalanga ethnic group in Mangwe District, southwestern Zimbabwe. The mopane worm, a vital resource in southwestern Zimbabwe, plays a crucial role in subsistence and a source of income for the Kalanga. In the context of this book, IKS, passed down from generation to generation, governs the harvesting, processing, and consumption of mopane worms. This chapter examines the role of Kalanga IKS in sustainably managing the mopane ecosystem, emphasising Kalanga ecological practices, harvesting patterns, preparation, and preservation techniques. It also analyses the socio-economic impact of the Mopane economy, looking at how IKS facilitates local trade and markets, provides livelihoods, and enhances food security in a semi-arid environment. By researching and documenting these, the study accentuates and illuminates our understanding of the importance of IKS in promoting sustainable development while preserving cultural heritage.

This chapter explores the harnessing of indigenous entrepreneurship (IE) in Botswana for sustainable development in the 21st century. It examines two case studies of Indigenous entrepreneurs in Botswana, highlighting the use of traditional and modern knowledge. One of these Indigenous entrepreneurs sells donkey milk products, and another deals with fashion apparels. Data were collected through in-depth

Preface

INTRODUCTION

The pursuit of cognitive fairness in academic institutions necessitates a deliberate and sustained emphasis on indigenous epistemologies and epistemic freedom. Historically, knowledge production and dissemination have been dominated by Western paradigms, often marginalizing indigenous ways of knowing. The process of "indigenizing the academy" seeks to challenge this imbalance by integrating indigenous perspectives, methodologies, and wisdom into scholarly discourse. This integration not only ensures epistemic justice for Indigenous peoples but also enriches academic inquiry by fostering a more pluralistic and inclusive intellectual landscape. By acknowledging the validity and relevance of indigenous knowledge systems, academia can create a more just and equitable learning environment where cognitive diversity is recognized and celebrated.

Indigenous knowledge systems, deeply rooted in cultural traditions, historical narratives, and environmental interconnections, offer invaluable insights into sustainable development, ecological stewardship, and community well-being. These knowledge systems are not only repositories of historical wisdom but also dynamic and evolving frameworks that continue to contribute meaningfully to contemporary global challenges. Recognizing and validating these epistemologies fosters a more comprehensive understanding of African civilizations and their multifaceted contributions to the global knowledge base. The promotion and preservation of African heritage depend on the conscious acknowledgment and application of indigenous knowledge in academic and policy-making spheres.

This volume, *Harnessing Indigenous Epistemologies for Sustainable Progress*, is a collective effort to foreground indigenous knowledge as a legitimate and vital contributor to academia and sustainable development. By bringing together scholars, researchers, and practitioners from diverse fields, this book provides a platform to explore how indigenous epistemologies can inform and transform contemporary academic and policy discourses. The chapters within this volume challenge the prevailing epistemic hegemony and advocate for a more equitable recognition of indigenous ways of knowing in the quest for sustainable futures.

The primary objectives of this book are to:

- Promote cognitive fairness in academic institutions by emphasizing indigenous knowledge and epistemic freedom.
- Challenge the dominance of Western epistemologies by incorporating indigenous viewpoints and wisdom into academia, fostering a more pluralistic and inclusive approach to knowledge development and transmission.

- Uphold the epistemic rights of Indigenous peoples and enrich scholarly discourse with diverse methodologies and perspectives through the "indigenizing of the academy."
- Create a more equitable and just learning environment that benefits all academics and students by celebrating cognitive diversity and elevating epistemic freedom in the academy.
- Preserve and promote African heritage by acknowledging and valuing the web of indigenous knowledge passed down through generations.
- Enhance understanding of African civilizations and their contributions to the global knowledge base through the adoption of indigenous epistemologies.
- Provide distinctive perspectives on sustainable living, community health, and environmental stewardship, reflecting the close ties of indigenous knowledge systems to the history and ecosystems of the African continent.

CHAPTER OVERVIEW

Chapter 1: A Deeper Look into Indigenous People's Spirituality Epistemologies and Creating New Social and Economic Systems for Sustainability

Author: Mosiuoa Ngaka, Independent Consultant, South Africa

This chapter presents a framework integrating Indigenous spirituality with ethical structures to promote environmental values, drawing on the Khoisan's cultural knowledge systems. It explores deep ecology as a means to align Khoisan spirituality with sustainable development. The study highlights the need for frameworks that merge Indigenous and Western values in social and economic models. By employing inductive content analysis, it develops themes supporting this integration, ultimately contributing to Indigenous Knowledge Systems. The chapter challenges contemporary development paradigms, advocating for holistic sustainability that respects Indigenous spirituality and Western values.

Chapter 2: The Spiral of Knowledge: The Fourth World and Sustainability

Author: Rudolph C. Ryser, Center for World Indigenous Studies, United States

This chapter explores Fourth World knowledge systems, emphasizing the dynamic relationships between people, land, and the cosmos. Using the Four Directions metaphor, it explains how human experience intertwines with the natural and cosmic world. The chapter presents a conceptual framework that integrates Indigenous scientific knowledge with conventional sciences. Additionally, it introduces ALDMEM, a mechanism for applying and mediating knowledge systems while ensuring the consent of Fourth World peoples. The work underscores the necessity of balancing Indigenous and state knowledge for sustainability.

Chapter 3: Harvesting Indigenous Onto-Epistemologies for Contributing to a More Balanced and Sustainable Future

Author: Norma. R. A. Romm, University of South Africa, South Africa

This chapter examines Indigenous onto-epistemologies and their relevance in contemporary times, focusing on strengthening relational balance across human and more-than-human existence. It calls for critical reflection on scientific values and their impact on society. The chapter argues that the notion of value-free science is a misconception, emphasizing that all knowledge systems are value-invested. By integrating Indigenous and Western scientific methods, it proposes ways to create more inclusive, sustainable world-making processes.

Chapter 4: Indigenous and Transformative Paradigms: Permeable Borders for a Just and Sustainable Future

Author: Donna Mertens, Independent Consultant, United States

This chapter highlights the Indigenous paradigm as a guiding framework for research in Indigenous communities. It emphasizes the paradigm's axiological, ontological, and epistemological foundations, particularly in relation to spiritual connectedness and decolonization. Using an African study as an example, the chapter illustrates how Indigenous and transformative paradigms can be integrated to advance justice in research methodologies.

Chapter 5: Indigenous Knowledge Epistemologies as Futuristic Imperatives

Author: Joy Ndwandwe, Indigenous Knowledge Hub Mountain View, Eswatini

This chapter explores Indigenous dignity through the lenses of time, space, and knowledge. It delves into Indigenous constitutive rules, including peace, ecology, and inclusive development, while advocating for rethinking cognitive justice. The chapter examines how Indigenous knowledge epistemologies can be incorporated into modernity and policy frameworks, aligning with global sustainability goals such as UNSDGs and AU Agenda 2063.

Chapter 6: Global Higher Education: The Sustenance and Hushed Annihilation of Indigenous Knowledges and Identities

Authors: Jacob Mapara, Ethel Siamena, Crispen Mazodze, Chinhoyi University of Technology, Zimbabwe

This chapter critiques the continued erosion of Indigenous epistemologies and identities due to the dominance of Western knowledge systems. It argues that globalization perpetuates the epistemological hegemony of the Global North, undermining the identities of formerly colonized Indigenous communities. The chapter highlights the need to embrace and preserve diverse knowledge systems to create resilient communities.

Chapter 7: Dismantling the Eurocentric Cognitive Empire and Re-Centering African Knowledges in Pursuit of Cognitive Justice

Author: Moloi Lehasa, University of South Africa, South Africa

This chapter examines the concept of cognitive empire and Afrocentricity, exploring how European epistemologies have dominated the modern world system since the 15th century. It discusses how African decolonial movements seek to reclaim Indigenous intellectual traditions and advocate for cognitive justice. The chapter argues that Indigenous knowledge systems are essential for challenging Eurocentric perspectives and fostering alternative worldviews.

Chapter 8: Indigenous Theories and Theorisation in African Scholarship in the Quest for Epistemic Freedom

Author: Patrick Ngulube, University of South Africa, South Africa

This chapter reviews the use of Indigenous theories in African scholarship, noting the prevalent reliance on Western theories. Through thematic synthesis, it identifies the limited application of Indigenous theories and calls for their broader incorporation. The chapter provides strategies for African scholars to theorize using Indigenous frameworks, thereby advancing epistemic freedom.

Chapter 9: Deterritorialization and Reterritorialization of Indigenous Knowledge

Authors: Tlou Maggie Masenya, Durban University of Technology, South Africa; and Collence Chisita, University of South Africa, South Africa

This chapter examines the alienation of cultural communities from their Indigenous roots due to colonization, globalization, and Westernization. It explores the reclamation of misappropriated Indigenous heritage in Southern Africa, employing an Afronographic approach grounded in Afrocentricity theory. The chapter emphasizes the need to reclaim Indigenous ways of knowing and preserve cultural identity.

Chapter 10: Integrating CARE And FAIR Principles in the Digital Preservation of Indigenous Knowledge

Authors: Josiline Chigwada, and Patrick Ngulube, University of South Africa, South Africa

This chapter addresses ethical challenges in preserving Indigenous knowledge in the digital age. It advocates for integrating CARE and FAIR principles to ensure community-led governance, culturally appropriate metadata, and collaborative partnerships. The chapter stresses the importance of prioritizing Indigenous communities in digital preservation efforts.

Chapter 11: Promoting the Documentation and Preservation of African Indigenous Epistemologies for Posterity

Author: Stanley Nil Murwira, Belvedere Technical Teachers College, Zimbabwe

This chapter discusses the importance of documenting and preserving African Indigenous epistemologies. It critiques the marginalization of Indigenous knowledge and calls for legal protections and patents to safeguard local intellectual traditions. The chapter adopts an Afrocentric framework to promote the sustainable use of African heritage.

Chapter 12: Driving the Indigenous Knowledge System Quest Through Music and Dance

Author: Khatija Bibi Khan, University of South Africa, South Africa

This chapter explores how music and dance are vital Indigenous knowledge and heritage repositories. It highlights the role of artistic expression in preserving cultural identity and Indigenous cosmology. The chapter underscores the interconnectedness of language, spirituality, and music in sustaining Indigenous knowledge systems.

Chapter 13: Knowledge System and the Mopane Worm Economy Amongst the Kalanga of Mangwe in Southwestern Zimbabwe

Author: Innocent Nyathi, Ministry of Primary and Secondary Education, Zimbabwe

This chapter explores the integration of Indigenous Knowledge Systems (IKS) within the mopane worm (*Imbrasia belina*) economy among the Kalanga people in Mangwe District, southwestern Zimbabwe. The mopane worm is a vital resource for subsistence and income generation. The chapter delves into Kalanga ecological practices, harvesting patterns, preparation, and preservation techniques that ensure sustainable management of the mopane ecosystem. It also examines the socio-economic impact of the mopane worm trade, emphasizing the role of IKS in facilitating local markets, improving livelihoods, and enhancing food security. Ultimately, the chapter highlights how preserving IKS contributes to sustainable development and cultural heritage.

Chapter 14: Harnessing Indigenous Entrepreneurship in the 21st Century for Sustainable Development

Author: Bonny Ngakane, University of South Africa, South Africa

This chapter investigates the potential of Indigenous entrepreneurship in Botswana for promoting sustainable development. Through two case studies, the research identifies how traditional and modern knowledge systems contribute to entrepreneurial innovation. The study highlights the lack of dedicated structures and policies supporting Indigenous entrepreneurship, revealing that government assistance remains generalized. The findings underscore the importance of IKS as capital for start-ups and the role of digital technology in expanding Indigenous markets. The chapter introduces the Ngakane model, offering policy recommendations to create a sustainable and innovative Indigenous entrepreneurial ecosystem.

Chapter 15: Locating the Doma Indigenous Knowledge Systems in Changing Environments: An Exposition of the Doma Way of Life in The Zambezi Valley, Zimbabwe

Author: Kwashirai Zvokuomba, University of Johannesburg, South Africa

This chapter examines the resilience of the Doma people in Zimbabwe's Zambezi Valley as they maintain their Indigenous knowledge systems in an evolving environment. Utilizing an ethnographic approach, the study applies Antonio Gramsci's cultural hegemony theory to assess how the Doma preserve traditional practices despite increasing external pressures. The research highlights their semi-nomadic lifestyle, coping strategies, and the vulnerabilities they face. The chapter ultimately seeks to understand the retention of the Doma knowledge system and the challenges threatening its sustainability.

Chapter 16: Harnessing Language, Adult Education, and Indigenous Knowledge for Sustainable Urban Agriculture in Lagos

Author: Amidu Owolabi Ayeni, Adeola Funmilola Afonja and John Olubunmi Faloju, University of Lagos, Nigeria

This chapter explores the role of language, adult education, and Indigenous knowledge in promoting sustainable urban agriculture in Lagos, Nigeria. Based on structured interviews with 114 adult urban farmers, the study identifies these elements as essential for modernizing agricultural practices. Findings suggest that effective integration of Indigenous knowledge with contemporary science and technology can enhance agricultural sustainability. The chapter advocates for policies that incorporate language and adult education into agricultural training programs, ensuring the effective transmission and use of improved agro-technologies.

Chapter 17: Pedagogy of Ecological Kinships: Community Teaching-Learning Practices of the Siddis in Gujarat

Author: Sayan Dey, Bayan College (affiliated with Purdue University Northwest), Oman and Niharika Sinha, Alliance University, India

This chapter investigates the community-based pedagogical practices of the Siddi people in Gujarat, India. Tracing their Afro-Indian heritage, the study highlights how music, dance, and spiritual traditions such as Zikr songs, Dhamaal dances, and Urs celebrations serve as alternative educational tools. These cultural expressions facilitate the preservation and transmission of ancestral knowledge, creating an ecologically and emotionally sustainable learning environment. The chapter discusses how these practices offer radical pedagogical alternatives rooted in ecological kinship and cultural memory.

Chapter 18: Fostering Cultural and Cognitive Justice through Inclusive Learning: Promotion of Indigenous Knowledge Systems for Sustainable Learning

Author: Edward Shizha, Wilfrid Laurier University, Canada

This theoretical chapter argues for the integration of Indigenous African knowledge into formal education to achieve sustainable learning. It critiques the exclusion of Indigenous epistemologies from mainstream schooling and advocates for cultural and cognitive justice in knowledge production. The study emphasizes that meaningful education must reflect students' lived experiences and social realities. It concludes that African school curricula should be grounded in Indigenous knowledge systems to promote sustainability and equity in learning.

Chapter 19: Answering the Call to Integrate Indigenous Education into the Nursing Curriculum

Author: Muhammad Salman Alam, Farabi College of Nursing and Health Sciences Charsadda, Pakistan, and Abel Jacobus Pienaar, Boitekanalo College, Botswana

This chapter examines the urgent need to incorporate Indigenous education into nursing curricula, particularly in low- and middle-income countries. Using qualitative research methods, including educational workshops and interviews, the study identifies key themes: the characterization of Indigenous knowledge, its impact on cultural and psychological well-being, and the role of nurses in advocating for Indigenous health systems. The chapter proposes strategies for integrating Indigenous education into nursing, fostering cultural inclusivity, and enhancing evidence-based decision-making in healthcare.

Chapter 20: Transforming Research in Postgraduate Education: An African Indigenous Perspective

Author: Mishack Thiza Gumbo, University of South Africa, South Africa

This chapter critiques the dominance of Western paradigms in postgraduate education and advocates for an Indigenous research framework. It outlines three key areas for transformation: reshaping institutional culture, redefining supervisory roles, and incorporating Indigenous worldviews into academic discourse. The chapter proposes a postcolonial Indigenous paradigm and contributes to the ongoing efforts to Africanize higher education curricula and research methodologies.

Chapter 21: The Body as a Divination Bone: An Application of the Bone Oracle as a Methodology

Author: Tebogo Victoria Kgope, University of South Africa, South Africa

This chapter explores Indigenous African epistemologies through the conceptual framework of divination bones, drawing from the works of Credo Vusamazulu Mutwa, Dani Nabudere, and Ngugi wa Thiong'o. It examines Indigenous African methodologies of knowledge transmission, using metaphorical interpretations of divination bones to analyze historical and contemporary societal structures. The study applies traditional concepts such as motlhakola (dispossession), thwagadima (danger), and mpherefere (chaos) to Indigenous African philosophical thought, offering a unique methodological approach to understanding cultural knowledge systems.

CONCLUSION

This book is intended for a diverse audience, including policymakers, academics, postgraduate students, researchers, and professional organizations committed to the advancement of indigenous knowledge systems. Policymakers will find insights into the practical integration of Indigenous epistemologies into sustainable development frameworks. Academics and postgraduate students will benefit from the rigorous discussions on cognitive justice and epistemic freedom. Researchers and scholars will discover innovative methodologies and perspectives that broaden the scope of intellectual inquiry.

The thematic sections of this volume encompass critical discussions on re-imagining Indigenous epistemologies in academia for a sustainable future, exploring the intersection of Indigenous knowledge and epistemic freedom in support of cognitive justice, and harnessing Indigenous knowledge for the preservation and promotion of African heritage. Through these discussions, we aim to illuminate the transformative potential of indigenous epistemologies in addressing contemporary global challenges.

The integration of Indigenous epistemologies into mainstream academic discourse is not merely an act of inclusion; it is a necessary step toward achieving genuine epistemic justice and sustainability. As we navigate the complexities of modern knowledge production, it is imperative to recognize that wisdom is not confined to any single tradition or framework. The richness of indigenous knowledge, with its deep ecological, social, and spiritual dimensions, offers profound lessons that can guide us toward more harmonious and sustainable ways of living.

We hope that this book serves as a catalyst for further discourse and action in advancing indigenous epistemologies within academic and policy frameworks. The journey toward cognitive justice requires the collective efforts of scholars, policymakers, and communities committed to dismantling epistemic hierarchies and fostering a more inclusive knowledge ecosystem. By embracing the wealth of indigenous wisdom, we move closer to a world where knowledge is truly diverse, equitable, and transformative.

Acknowledgements

I have been influenced by a diverse group of people in my growing interest in indigenous knowledge systems, far too many to name them all. I would like to express my sincere gratitude to the members of the editorial board whose encouragement was instrumental in bringing this book to fruition. Thank you all for believing in and supporting this project.

I would like to express my gratitude to the reviewers who acted as gatekeepers, ensuring the quality and coherence of the chapters. Many of the authors also served as referees, and I sincerely appreciate their dual roles. I am also grateful to the authors whose work is included in this collection. Their commitment to their work and this project was sterling. Colleagues at IGI Global Scientific Publishing, including Mrs Jocelynn Hessler, Director of Book Development, maintained a steadying hand throughout the process of developing this book. I would like to thank Dr. Leslie for her work on the review reports and her wonderful job attending to the reviewers' comments on behalf of her late husband and our Indigenous brother. Finally, I acknowledge Professor Jacob Mapara for his wonderful editorial support. All this would not have been possible without the support of my family and close indigenous friends.

In Memoriam

The Editorial Board and I would like to express our deepest sympathy following the recent passing of Dr. Rudolph C. Rÿser, a member of our Editorial Board and the author of Chapter 2. His passing is a significant loss to our community dedicated to indigenous knowledge systems. We will honour his memory not only through this book but also through the lectures, papers, and research that will continue to inspire others.

Chapter 1
A Deeper Look Into Indigenous People's Spirituality Epistemologies and Creating New Social and Economic Systems for Sustainability

Mosiuoa Nehemiah Ngaka
Independent Researcher, South Africa

ABSTRACT

This chapter develops a framework integrating Indigenous spirituality with ethical structures to support environmental values, focusing on the Khoisan's culture and knowledge systems. Utilising deep ecology aligns Khoisan spirituality with sustainable development. The study addresses the need for frameworks combining Indigenous and Western values in social and economic models. The inductive content analysis of interviews, observations, and literature generates themes to support this integration. The framework contributes to Indigenous Knowledge Systems by offering alternative models for coexistence, promoting holistic development that balances economic, cultural, and environmental sustainability. The study challenges contemporary development paradigms, emphasising the importance of Indigenous Knowledge in fostering sustainable practices and a harmonious relationship with nature. By incorporating interconnectedness and deep ecology, it advocates for social and economic systems that respect and integrate Indigenous spirituality and Western values.

INTRODUCTION

The disconnect between humanity, the environment, and nature is vividly reflected in South Africa through environmental degradation, pollution, poverty, high crime rates, and reduced social cohesion. Ikeke and Ukutsemuya (2024) highlight that this environmental chaos severely damages ecosystems and erodes community cohesion, leading to economic instability. Issues such as oil spills, deforestation, and terrorism disrupt livelihoods and infrastructure, while the encroachment of the Sahara Desert exacerbates

DOI: 10.4018/979-8-3693-8447-3.ch001

these challenges, thus threatening both environmental and economic sustainability. The influences of global consumerism, colonisation, and apartheid have diminished respect for nature, which was once central to ancient African values (Mampilly, 2015).

To address these challenges, South Africa must rectify historical injustices by combating poverty and creating jobs to foster a more inclusive socio-economic system. Policymakers are tasked with prioritising the needs of marginalised communities, thus ensuring equitable access to essential resources such as land, clean water, energy, and housing despite the decision-making challenges this presents.

Indigenous Spirituality is fundamental in establishing sustainable social and economic systems. It emphasises deep connections to the land, ancestors, and the natural world. This spirituality encompasses diverse beliefs and practices that underscore the interdependence of all living beings, fostering stewardship that informs environmental relationships. Many Indigenous communities perceive the Earth as a living entity that promotes sustainable practices focusing on ecological balance and community well-being.

Moreover, integrating Indigenous Spiritual modalities into healthcare systems offers pathways to a more inclusive and sustainable health approach. Valuing diverse healing practices enriches the healthcare landscape and supports community resilience and cultural identity amidst dominant biomedical paradigms. Collaborative efforts must prioritise respect and dialogue to harness the strengths of all medical systems (Josyula et al., 2015).

This chapter explores Indigenous peoples' spirituality and epistemologies, aiming to integrate these perspectives into developing sustainable social and economic systems. By examining the intersection of Indigenous Spiritual practices and sustainable resource management, this study seeks to enhance the understanding of how these frameworks can foster resilient and equitable socio-economic systems. The chapter is structured to include an introduction, a literature review on Indigenous Spirituality and Sustainable Systems, sections on methodology, results and discussion, and a concluding summary.

Aims and Objectives

The study explores the integration of Indigenous spirituality and practices into new social and economic systems to promote sustainability.

The objectives were to:

i. Explore the integration of Indigenous spirituality and practices into new social and economic systems to promote sustainability.
ii. Assess the impact of Indigenous knowledge on local decision-making and resource management.
iii. Determine the effectiveness of deep ecology theory in addressing ecological issues by integrating Indigenous values.
iv. Promote the recognition and utilisation of Indigenous knowledge in global development initiatives.
v. Develop a framework for social and economic systems that incorporates ancient spiritual traditions, ethical principles, and sustainable practices.

LITERATURE REVIEW

Indigenous Spirituality as a Framework for Sustainability

Indigenous spirituality is foundational for creating sustainable systems. Recognising spirituality as a manifestation of a divine presence that sanctifies all living and non-living entities highlights how these beliefs can lead to ethical frameworks conducive to sustainability. The literature emphasises the strong socio-economic ties between nature and Indigenous life, fostering new perspectives on well-being that incorporate natural values into daily life (Sangha et al., 2015, p. 198).

The Impact of Indigenous Spirituality Practices and Knowledge

Herbal medicine is a significant example of Indigenous knowledge that has influenced global populations. For instance, the Nepali people have been using traditional medicine for generations, with about 80% of the rural population depending on it for primary healthcare (Prasad et al., 2020). Although a singular definition remains elusive, distinctive characteristics set Indigenous Knowledge apart. It is culturally specific and informs local decision-making across various domains, including agriculture, health, and resource management. Community life in Indigenous societies is deeply rooted in communal rituals and relationships, making the codification of implicit knowledge inherently challenging. Notable examples include adopting modern bean varieties in Rwanda and Colombia (Virk et al., 2021).

Indigenous knowledge provides essential problem-solving tools for local communities, particularly marginalised groups, and significantly contributes to global discussions on development. However, this knowledge should be utilised more in development processes globally (Magni, 2017, p. 442). Understanding what local communities already know can make initiatives more contextualised and responsive to their needs.

Responses to Environmental Crises

The literature highlights the importance of Indigenous spirituality in addressing modern environmental challenges, underscoring the urgent need for resilient systems to confront issues like industrial pollution and resource depletion. According to Omoyajowo et al. (2024), Indigenous spirituality is vital for developing sustainable social and economic systems, focusing on conservation and a deep bond with the ecosystem. Grounded in various cultural traditions, it offers valuable perspectives to tackle crises stemming from materialism and secularisation. By integrating Indigenous spiritual viewpoints, we can foster a comprehensive approach to environmental stewardship that promotes respect for nature and acknowledges the interconnectedness of all living things. This blend of multicultural spiritual perspectives enhances our understanding of conservation and strengthens our relationship with the environment. Ultimately, it facilitates more effective ecosystem protection strategies that benefit the planet and the communities dependent on it.

Non-Anthropocentric Nature of Indigenous Spirituality

African spiritual perspectives emphasise a deep respect for nature, promoting an ecological spirituality that fosters harmonious socio-economic systems. This mutual development between humanity and the environment is called meliorism, which suggests that people inherently strive to improve their socio-economic circumstances while effectively utilising the Earth's resources (Hossain, 2021). Mwiti (2014) demonstrates this concept through the Meru people's veneration for the environment, which is evident in their naming of sacred lakes and rituals during times of need. This viewpoint supports the discussion on incorporating spirituality into socio-economic frameworks.

CREATING SUSTAINABLE SYSTEMS THAT ALIGN WITH INDIGENOUS PRACTICES

Indigenous communities employ various traditional responses to adverse climate conditions, including resource diversification and changes in species used (Magni, 2017, p. 442). Even acknowledging Indigenous knowledge as crucial for environmental sustainability, international recognition has not significantly increased Indigenous participation in decision-making processes (Magni, 2017, p. 442).

Buen Vivir as an Alternative for Progress

The increasing interest in Indigenous knowledge situates these systems as viable alternatives for progress. The notion of *Buen Vivir*, derived from the reactions of Latin American Indigenous groups to harmful development strategies, encompasses values and norms shared across various communities. It represents a new conceptualisation of development that includes rights to land, culture, and self-determination (Magni, 2017, p. 438). However, more effort is needed to integrate these principles into broader societal frameworks.

The Integration of Knowledge Systems and the Implications for the 2030 Agenda

Research has shown that participatory approaches are vital for a balanced representation of different knowledge systems, mainly traditional and scientific knowledge. These approaches promote cross-cultural understanding and have proven effective in climate change adaptation, natural hazard prevention, and biodiversity conservation. For instance, frameworks integrating Indigenous and scientific knowledge have succeeded in various contexts, including hydro-meteorological hazards in Southeast Asia (Magni, 2017, p. 443).

This literature review comprehensively examines Indigenous spirituality and its potential to inform sustainable social and economic systems. It emphasises the significance of Indigenous Epistemologies in addressing contemporary ecological challenges and advocates for frameworks that prioritise the intrinsic value of all life. Through this alignment, the review establishes a robust theoretical foundation for understanding how Indigenous practices can reshape socio-economic structures toward sustainability.

PHILOSOPHICAL ASSUMPTIONS

The philosophical foundations of this study are rooted in a critical realism ontological and epistemological stance. This philosophical approach posits a superlative intelligence that exists independently of sensory observation. The inductive content analysis approach was deemed the most suitable method for gathering data from diverse sources, while data unification was employed to integrate these sources. Thematic content analysis was utilised to extract themes from the data, ensuring participants' anonymity and confidentiality in their responses.

Ontology is a framework that accurately represents the complexities of the real world. It is designed to organise and manage knowledge about entities and their interrelationships. Consequently, ontologies not only organise raw data but also encapsulate the meaning behind that data. Until now, developers of ontologies have had to manually define the semantics of the objects and the relationships between them (Adamo, 2020; Guizzardi, 2020; Hnatkowska et al., 2020; Machado, 2021; Panayiotou, 2024; Wei, 2020). This concept addresses the 'what' of existence regarding entities and the 'how' of categorising them based on their similarities and differences. Ontology aligns with critical realism, which posits that reality exists independently of human cognition or understanding. Bhaskar (2008) contends that experiences perceived through the five senses do not encompass the entirety of existence, suggesting that a more profound understanding lies beyond mere physical sensations. He argues that reality is constructed through understanding the social processes that shape it.

A phenomenological approach was particularly fitting for this chapter as it is context sensitive. The ontological perspective asserts that Indigenous spirituality diverges from Western interpretations and critiques the unethical abandonment of primordial spiritual practices. In line with the notion of deep ecology, Kaunda (2021, p. 92) asserts that African Indigenous religions emphasise human responsibility towards nature and that the relationship between humans and the natural world was viewed as symbiotic before the arrival of Western missionaries.

For centuries, scientists and philosophers have explored the fundamental question of how knowledge is obtained. Knowing involves knowing information and its accuracy and utilising various processes and sources to collect it. Epistemology, a branch of philosophy, examines how knowledge is acquired and validated. This field can encompass a spectrum of approaches, from accepting beliefs based on authority to engaging in empirical research or drawing insights from personal experiences. A key component of epistemology involves evaluating the justification for a belief, suggesting that we are justified if we have a solid basis for that belief (Rupp et al., 2024, p. 253).

The reality concerning the Indigenous spiritual structures, values, and knowledge systems of the Khoisan cannot be directly observed or comprehended through positivist methods. Consequently, Interpretivism emerges as a suitable philosophical assumption, with a phenomenological methodological approach that aims to understand reality from the perspectives of social actors. The core tenet of critical realism is that reality exists beyond what is immediately observable, and this reality can be constructed and understood through the experiences of social actors.

RESEARCH METHODOLOGY

This study employed a qualitative approach to ethnography. Ethnographic research gives rich insights into human, social, and organisational activities (Myers, 1999). While diverse ethnographic approaches allow researchers to comprehensively understand various cultural, social, and economic aspects, including naturalistic ethnography, this study used collaborative ethnography, ensuring that the research process and outcomes were relevant and beneficial to the community (Lassiter, 2005a). This approach emphasises mutual respect and shared authority in the research process (Rappaport, 2008). The benefits of collaborative ethnography include the trustworthiness in ethical research practices (Lassiter, 2005a), empowerment of participants (Lassiter, 2005b), rich and diverse data (Lassiter, 2021) and sustainability (Menezes & Costa, 2024).

Research Design

To develop a social cohesion framework, this study focused on the life of the Khoisan, aiming to identify current practices compared to those lost. Primary data gathered through interviews and observations of surviving San's spiritual practices informed what was practised. In contrast, secondary data, including literature on the ancient Khoisan, revealed what had been lost. The study adopted a critical realism ontological and epistemological stance, recognising an objective reality beyond sensory perception that is aligned with primordial spirituality.

Critical realism necessitates knowledge construction from diverse sources, including transcribed interviews, observations, and published material about the Khoisan way of life. Following the guidance of Elo et al. (2014, p. 4), purposive sampling was utilised to select knowledgeable informants for Inductive content analysis. Data collection occurred in various locations, including Askham in the Northern Cape and George in the Western Cape.

Description of Participants and Their Contexts

The study focused on the Khoisan of the Kalahari and Western Cape, situated in the Mier area in the Northern Cape, bordering Botswana and Namibia. They were divided into two groups: the settled population, whose subsistence depended on livestock, government grants and gardening, and the hunter-gatherers, whose lifestyle was mainly characterised by a nomadic culture. According to Pakendorf and Stoneking (2021), the Khoisan of Kalahari are said to be the offshoot of the distinct race called Khoisanoid, whose place of abode was East, Central and Southern Africa.

Ten study participants were male, aged 27 to 71, including two Khoisan activists, two Khoisan chiefs, and one descendant of the Khoisan royal blood. One of the two female participants was a Khoisan chief in the 41-50 age range, and the second was a community leader in the 31-40 age range, as illustrated in Table 1.

Table 1. Summary of demographic details of participants

Participant	Age	Gender	Race	Status in the Khoisan Community
1	40-31	Male	Coloured	Khoisan & Activist
2	50-41	Male	Coloured	Khoisan & Activist
3	50-41	Female	Coloured	Khoisan/ Khoisan Chief
4	70-61	Male	Coloured	Khoisan/ Khoisan Chief
5	60-51	Male	Coloured	Descendant of the Khoisan Royal Blood
6	40-31	Female	Coloured	Community Member/Leader
7	70-61	Male	Coloured	Khoisan/ Khoisan Chief
8	40-31	Male	Coloured	Community Member
9	60-51	Male	Coloured	Community Member
10	30-21	Male	Coloured	Community Member
11	80-71	Male	Coloured	Community Member
12	50-41	Male	Coloured	Community Member

(Field Data 2016 and 2017)

Content Analysis

The inductive content analysis, detailed in Figure 1, was employed to analyse three types of data: (1) transcribed interviews, (2) field notes from observations, and (3) secondary literature. Utilising data unification (White & Marsh, 2006), these data sources were merged into a single open coding frame, enabling the emergence of themes to build the primordial spiritual framework. This approach aligns with critical realism, emphasising the construction and validation of knowledge through multiple perspectives.

Designing Content Analysis

The analysis involved detailed thematic content analysis, beginning with the open coding of all notes, transcribed interviews, and literature. Following this, information was then classified. A comprehensive list of codes was generated, ensuring that all participant responses were covered by at least one code. Responses that did not fit existing codes were re-examined, recoded, or removed. The coded responses were then scrutinised for conceptual relevance, leading to the generation and analysis of subcategories and categories, ultimately resulting in a framework for the primordial spiritual practices of the Khoisan. See Figure 1.

Figure 1. Content analysis methodology procedure framework

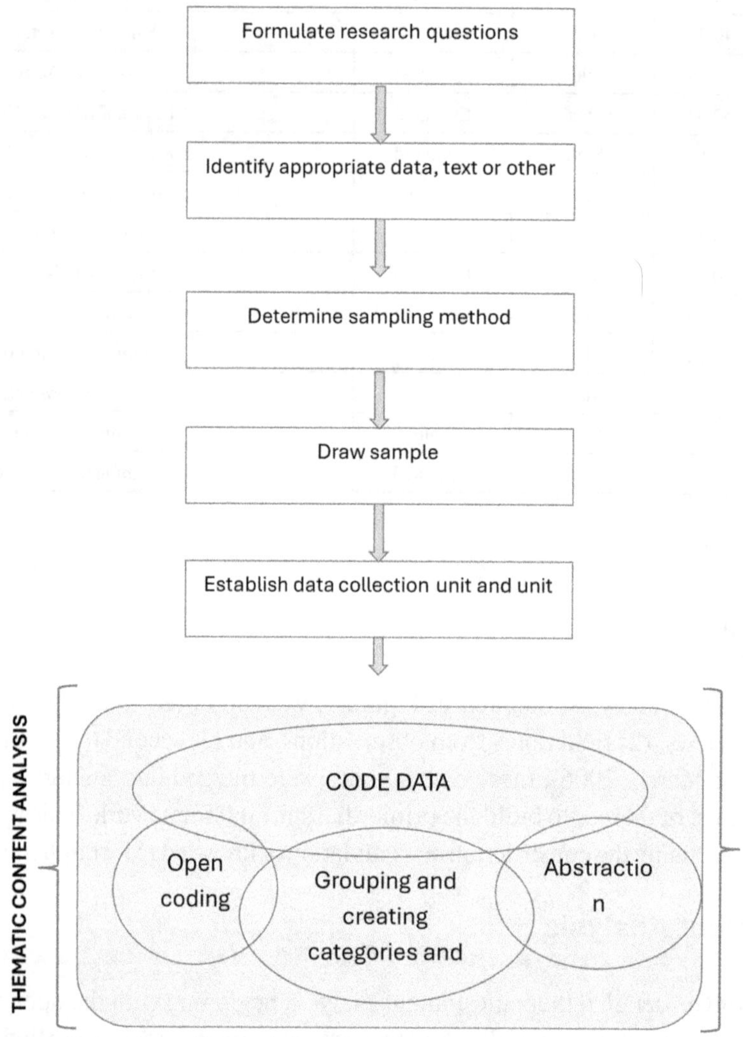

(Developed by the author and adapted from White and Marsh, 2006)

Content Analysis Methodology

The qualitative content analysis aims not to achieve generalisability but to foster transferability (Lim, 2024). According to White and Marsh (2006), data collection units are defined as units of information derived from interview questions or themes that arise during the data unitisation process, which involves observations, interview transcripts, or texts from various published sources. Following data collection, Thematic Content Analysis started with open coding, wherein phrases and text segments were tagged in correspondence to observation notes, literature, and questions posed throughout the fieldwork. Through abstraction, categories and subcategories were created to enhance comprehension and generate new insights.

Unit of Analysis and Data for the Study

The data sources identified were interviews, transcripts, observations, and Primordial Spiritual practices of the ancient Khoisan, which were used to address the research questions.

Sampling

Sampling is a method for selecting a specific number of data representatives from the research population (Gentles et al., 2015). This study utilised purposive sampling, which facilitates the selection of information-rich cases for detailed investigation (Gentles et al., 2015). Participant recruitment was based on willingness, knowledge of the Khoisan way of life, seniority, age, gender, and maturity. The geographical location of participants was also a critical selection criterion, given that the Khoisan are primarily found in specific regions of South Africa.

Participants were drawn from the Northern and Western Cape. Roy et al. (2015) define the unit of analysis in terms of three questions: Who?, What? and For which? which relate to the analysis of information and the conclusions that stem from it. At the time of the study, the Khoisan population was estimated to be around 100,000 (Prigg, 2014). In the Northern Cape, four participants consented to partake in the study, while eight participants from the Western Cape also agreed to participate. Among the 12 participants, three were Chiefs, and their ages ranged from 27 to 71 years.

Credibility and Transferability in Qualitative Research

The concepts of validity, reliability, and generalisability in qualitative research have been the focus of considerable debate, as noted by (Noble & Smith, 2015a). Trustworthiness, data quality, and transparency guide qualitative research according to (Lim, 2024). In qualitative contexts, trustworthiness signifies the consistency between the researcher's interpretations and the actual phenomena being studied (Kakar et al., 2023) and, thus, the credibility and transferability of the study. To establish credibility and transferability, this study ensured the representation of diverse realities from various participants in different locations, which were corroborated by insights from existing literature. The measures adopted were consistent with recommendations by Noble and Smith (2015b). Credibility and transferability were further corroborated by identifying similar findings reported by other researchers cited within this study.

Ethical Considerations

This study was conducted across two provinces: the Northern and the Western Cape. To safeguard the interests of the participants, the researcher, and the university, a consent letter was signed by all participants following a comprehensive description of the study. Participants were informed of their right to withdraw from the study at any point. In the Northern Cape, the gatekeeper was the initial point of contact and established the study rules. In the Western Cape, the gatekeeper permitted the researcher to proceed after signing a consent letter on behalf of the participants. Participants were assured that their identities would remain confidential, with pseudonyms employed throughout the study. They were also informed that the contents of the study would be treated as confidential, and they signed a letter of consent to participate. Fieldwork was conducted in March and November 2016 and February 2017, respectively.

Participants were informed that their participation was voluntary and that they could withdraw at any time. They were assured of their anonymity, and upon adhering to all ethical protocols, they signed consent letters, after which fieldwork commenced.

Data Analysis

At first, the researchers considered using NVivo for data analysis but decided against it due to the smaller datasets, which would be better suited for a manual approach. They noted that NVivo could be time-consuming and demand careful planning to maintain data integrity and accuracy (Kraiwanit et al., 2023, p. 6).

Braun and Clarke (2023) delineate two thematic analysis approaches: interpretative themes, which delve into deeper meanings, and superficial topic summaries. They emphasise that authentic interpretative themes should emerge organically during the analysis, underscoring the necessity for depth and creativity.

Two studies exemplify this distinction: Fraser et al. (2021) explored themes in transgender adults' healthcare experiences, while Frohard-Dourlent et al. (2020) identified significant themes related to surgical readiness assessments in Canada. Both studies underscored the value of nuanced understanding over mere summaries.

The text contends that reliance on topic summaries within the reflexive thematic analysis undermines its objective of revealing deeper meanings and narratives inherent in the data. It advocates for an analytical approach that fosters thorough understanding rather than adherence to predefined categories that may restrict the analysis's efficacy.

The analysis process began with open coding of notes, transcribed interviews, and literature, as illustrated in Figure 1. This was succeeded by further open coding to categorise the information into categories and subcategories. After compiling a comprehensive list of codes based on participants' responses, each response was re-evaluated to ensure alignment with at least one code. New codes were formulated for responses that did not correspond with existing ones, while irrelevant responses were recorded or excluded. Upon confirmation of all relevant codes, the coded responses underwent a subsequent review, culminating in establishing subcategories and categories. This iterative process ultimately resulted in a framework for understanding the primordial spiritual practices of the San and their role in fostering sustainable social and economic systems.

RESULTS AND DISCUSSION

The study illustrates the intrinsic link between Indigenous spirituality and deep ecology, reinforcing that recognising the divine presence in all entities can cultivate ethical frameworks that are conducive to sustainable practices (Sangha et al., 2015). This supports integrating these spiritual modalities into contemporary systems for enhanced sustainability.

The thematic content analysis underscores how Indigenous Spirituality informs ethical decision-making, suggesting that incorporating Indigenous Knowledge Systems can have a positive impact on local resource management. This is aligned with evaluating the role of such knowledge in local governance.

The study's conceptual framework derived from deep ecology served as a foundation for developing new social and economic systems that integrate ancient spiritual traditions. This directly relates to the objective of creating frameworks that emphasise sustainability through ethical principles rooted in Indigenous spirituality.

By critically evaluating existing literature surrounding Indigenous spirituality through the lens of deep ecology, the study reinforced the effectiveness of this theoretical approach in addressing ecological issues. This supports the objective of using Deep Ecology to integrate Indigenous values for sustainable solutions. This study outlines its epistemological and ontological stance and the content analysis methodology, which employs thematic content analysis for data analysis.

Figure 2. An expanded illustration of thematic content analysis

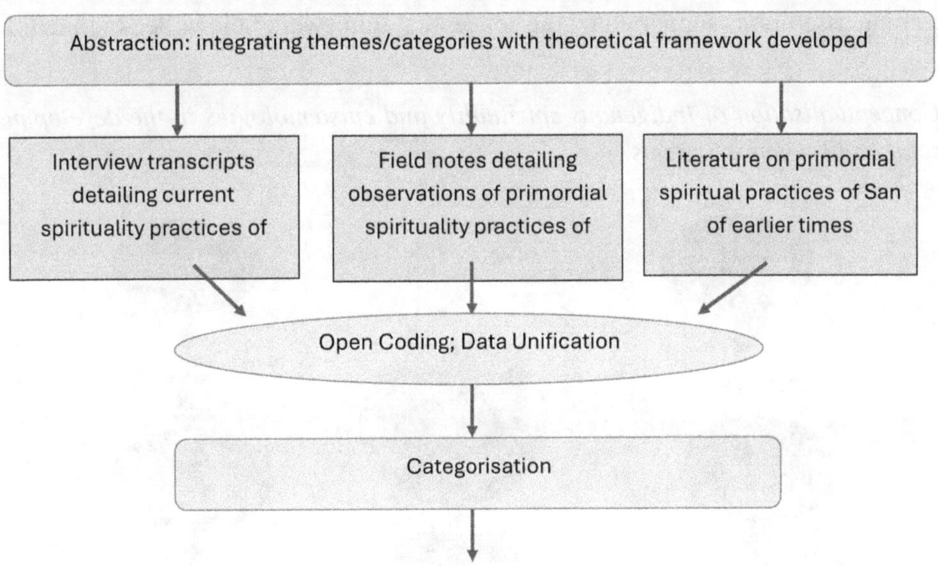

(Author's initiative)

As outlined in Figure 2, the findings highlight Indigenous spirituality's non-anthropocentric nature and emphasis on ecological balance. This aligns with the aim of integrating these practices into new socio-economic systems to foster sustainability. The mention of sacred sites and rituals among the Meru people (Mwiti, 2014) illustrates how these practices can promote environmental stewardship and sustainable living.

The results emphasise the culturally specific nature of Indigenous knowledge and its role in shaping local decision-making, particularly in Tharu agricultural practices. By integrating Indigenous and scientific knowledge, farmers are making informed decisions that enhance resilience to climate change (Chaudhary, Erskine, & Acciaioli, 2022, p. 13). This clearly aligns with the objective of assessing Indigenous Knowledge's impact on local resource management.

The emergence of the concept of *Buen Vivir* as an alternative development framework in the findings reflects the need for integrating ancient spiritual traditions and ethical principles into broader socio-economic systems. The research underscores the importance of participatory approaches that blend

Indigenous and scientific knowledge (Magni, 2017, p. 443), aligning with the objective of developing frameworks that promote sustainability.

The analysis confirms the theoretical framework of deep ecology, showcasing how Indigenous spirituality offers insights into resilience against environmental crises (Ikeke & Ukutsemuya, 2024). The findings advocate for ecological spirituality and highlight the necessity of recognising the intrinsic value of all life. This aligns with the objective of evaluating Deep Ecology's effectiveness in integrating Indigenous values to address ecological challenges.

The findings point to a growing recognition of Indigenous knowledge as vital for sustainable progress, reinforcing its importance in global development initiatives. However, they also highlight the challenges in integrating these principles into broader societal frameworks, emphasising the need for participatory approaches that facilitate the incorporation of Indigenous knowledge into development practices (Magni, 2017, p. 443).

The foregoing assertion is supported by the conceptual framework below: See Figure 3.

Figure 3. Conceptualisation of Indigenous spirituality and epistemologies in the development of sustainable social and economic systems

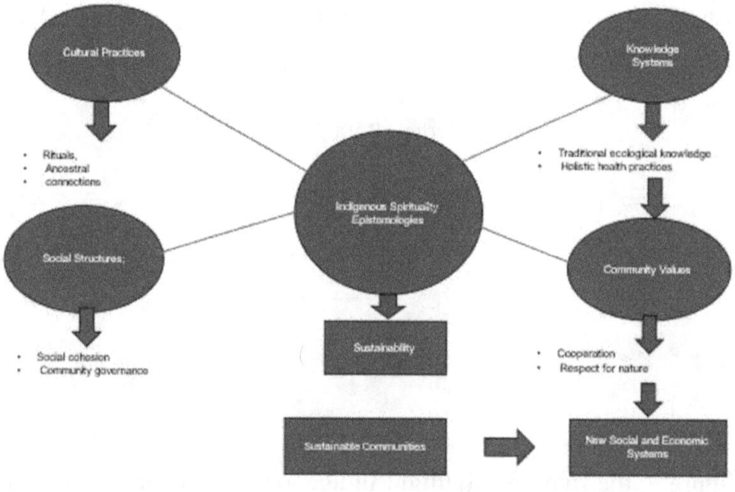

(Author's initiative)

THEMES AND CONTENTS EMERGING FROM INTERVIEW RESPONSES

Belief in God

The Khoisan hold a deep conservation ethic, perceiving nature as an embodiment of the spiritual realm. Their belief in God, the life creator, is demonstrated through rituals that call upon healing spirits and signify important life changes. This blend of spirituality with sustainability highlights the necessity for new economic frameworks that honour these values, thus emphasising a spiritual bond with the environment. Participants emphasised that their faith in God was central, viewing all aspects of nature as reflections of the divine.

Rituals, Culture, and Socialisation

The Khoisan's dedication to their spiritual beliefs was evident in rituals such as the full moon and harvesting ceremonies. Despite experiencing persecution that restricted their practices, these rituals remained vital to their cultural identity. Elendu (2024) points out that rituals are essential in Indigenous cultures, reinforcing the notion that incorporating such practices can enhance sustainable development and social cohesion.

Traditional Medicine and Healing

The Khoisan employ traditional medicine grounded in their spiritual beliefs, using local herbs to address various health concerns. This practice exemplifies how Indigenous knowledge shapes health and resource management, showcasing a comprehensive approach to well-being that fuses physical and spiritual health. The emphasis on medicinal practices underscores the pressing need to document this knowledge, particularly amid worries about cultural extinction (Boezak, 2015).

The idea that technology acts as a voice for marginalised cultures stems from the historical connections between ancient nomadic tribes and the spiral tribe, expressed through music and dance over millennia. Farber (2015) notes that Indigenous technology is represented by the outlawed Indigenous dagga plant, utilised to treat multiple ailments. Hemp seed oil is also highlighted for its potential benefits in addressing constipation, cancer, skin conditions, and high blood pressure.

Farber (2015) affirms that hemp seed oil is an effective remedy for atopic dermatitis and antiretroviral treatment (ARV). This is supported by research from New York University's Comprehensive Epilepsy Centre, which found a 45% reduction in seizures within three months using a purified cannabidiol compound. The oil has a low peroxide value, is non-toxic, and lacks psychotropic properties, providing a calming effect on the body.

Imam et al. (2024) suggest that hemp seed oil, tea tree oil, and ginseng oil are beneficial omega fatty acid supplements that can address various ailments, including cancer and skin issues, and promote hair growth. Mwiti (2014) argues, however, that scientific knowledge should be assessed through universally accepted theoretical frameworks before being recognised as legitimate.

According to Suchitra and Parthasarathy (2020), ginger extracts have been widely utilised for their anti-inflammatory, digestive, and anti-tumour properties, with fresh ginger exhibiting antiviral effects against COVID-19.

Even today, many nations that have legalised cannabis struggle to create regulations that effectively tackle their socio-economic challenges, identifying cultivation and commercialisation as pivotal issues. Safety concerns among cannabis users were highlighted compared to other inhaled, medical, or recreational substances. Legalisation has been recognised as crucial for ensuring safety and facilitating the commercialisation of cannabis (Gwala, 2023).

Ancestors as Mediators Between the Living and Non-Living

In Khoisan spirituality, ancestors play a vital role in guiding decisions, linking the living to the spiritual world. This ancestral connection provides a framework for resource management, emphasising the importance of incorporating Indigenous beliefs into local governance. As noted by Đorđević Belić (2024), ancestors act as intermediaries, shaping the community's spiritual and ecological choices.

Ethical and Social Structures

The egalitarian character of the Khoisan society is reflected in their rituals, which promote respect and collaboration among community members. These ethical principles lay the groundwork for social structures that encourage collective well-being and sustainable practices. Participants stressed that their rituals not only signified life transitions but also strengthened community bonds, aligning with the objective of fostering inclusive social systems (Takada & Noguchi, 2024).

Collaboration

The Khoisan's emphasis on teamwork and collective effort is essential to their social structure. This focus on collaboration fosters social systems that prioritise community welfare and environmental stewardship. Their deep respect for all beings, including non-human entities, underscores the interconnectedness vital for sustainability.

Conservation Ethic, Nature, and the Universe

The Khoisan's spiritual beliefs highlight a profound respect for the environment, viewing it as a sacred entity. This perspective aligns with deep ecology principles that advocate for the interconnectedness of all living things. Their practices, such as ascribing sacred significance to natural occurrences, demonstrate how Indigenous values can address current ecological challenges (Hossain, 2021).

This environmental ethic is reflected in the Meru people's reverence for their surroundings, as they conserve swamplands and forests, bestowing holy names upon lakes and swamps, such as Iria ria Thai, meaning *God's Lake* (Mwiti, 2014). This philosophy resonates with Naess' concept of Deep Ecology, which promotes a non-anthropocentric approach emphasising unity and mutual respect among all entities, living and non-living (Hossain, 2021). This happens alongside a call for equity in ecological justice that governs human interactions with the universe (Haddad & Solomon, 2024).

Equality, Visions, Dreams, and Trance Dance

The Khoisan's belief in dreams as a means of communication with ancestors reflects a rich cultural legacy that can inform global development efforts. Their egalitarian values and respect for spiritual communication highlight the importance of integrating Indigenous Knowledge into broader societal frameworks, as suggested by (Đorđević Belić, 2024). This narrative intricately weaves together the themes with relevant aims and objectives, illustrating how Khoisan beliefs and practices can guide sustainable social, economic, and ecological systems.

OBSERVATIONS OF KHOISAN SOCIAL AND ECONOMIC ACTIVITIES

Observations Made in Kalahari

On 21 to 25 March 2016 at the commencement of the fieldwork, the participants in Kalahari showed me how nature was a source of the Khoisan's artistic expressions. An illustration below shows a family of Khoisan at work making ethnic jewellery pieces.

The Khoisan use nature as a source of artistic expression. See Figure 4 below:

Figure 4. Using natural substances to make ethnic jewellery pieces

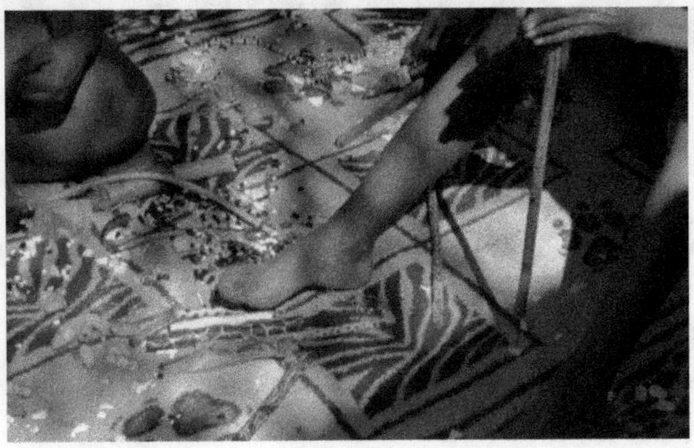

(Photograph by the author)

They use white shells that grow from the tree under whose shade we were sitting, and it is called the *camethon peel* and it has small seeds inside, and the hole is then opened from each of the seeds. Other materials were the ostrich bone and another animal bone which is in circular shape, and a *katool* (a plant that resembles a stick). The illustration of the ***tree*** that was captured during the researcher's fieldwork is shown below.

Camethon peel, *ostrich* bone, and *katool* were used to create a hole in a tree, which was captured during fieldwork. See Figure 5 below:

Figure 5. The tree whose seeds are used to make shackle for dance and other body adornments

(A photograph by the author)

Materials from different natural sources are used for making necklaces. For example, *steenbok* skin is used for making strings after it has been put in a plastic bag to rot; this is to ultimately hold the ostrich bone that makes the necklace. They also used the porcupine quill as part of making a necklace. The product of what I observed being prepared and crafted is below:

Materials from natural sources are used to make necklaces, such as steenbok skin and porcupine quill. See Figure 6 below:

Figure 6. Some of the products made while watching

(A photograph by the author)

The Khoisan make use of a stone called! *Nau Klip stone or other stone* for drawing is only found in Elandsrand in Mpumalanga/Limpopo. They also use natural resources, such as dancing shackles from a white shell with an edible living worm. They also drew pictures that depicted the culture and spirituality of the San, using a hard wire that had been put on fire until it was red. This activity is represented in Figure 7:

Figure 7. Indigenous fine art - process of drawing using hot wire

(A photograph by the author)

Observations Made at Goedverwacht, Piketberg in the Western Cape

The Khoisan in the Western Cape had a special event in February 2017 where the new chiefs would be inaugurated. The ceremony was delayed due to a funeral in the community and was led by the regional chief, senior chief, chiefs and elders. The ceremony was characterised by a procession, with the Chieftaincy singing, dancing, and chanting prayer. On arrival at the actual place, there was a small hut made of grass, so everyone who entered had to kneel. See Figure 8 below:

Figure 8. Small hut made of grass where inauguration rituals were conducted

(Photograph by the author)

The house was encircled with stones to form the boundary of the hut, and the procession of chiefs and elders entered through a passage. The regional chief welcomed the procession and highlighted the rules of the ceremony, such as not going beyond the boundary, and only the anointed could enter. The senior chief then took over and began the inauguration process. The 11 chiefs were inaugurated on the day, with the presiding chief calling them to take off their shoes and elders summoned into the yard of the hut. The shaman and other designated chiefs were dancing, praying and singing Khoisan praise songs (Table 2), such as:

Table 2. Khoisan praise songs

Khoisan language	English translation
Sida Sida	It could be an invocation, possibly expressing reverence or calling on a deity.
Gera Tsoa Tsoa	might emphasise a request or command to "go" or "move," possibly in a spiritual sense of urging action or progress.
Gans Gans Tsi	Could relate to a call for divine attention or intervention, with "Tsi" emphasising urgency or immediacy.
!Kho Kha Tsi	Might reference an appeal to a spiritual force or deity, invoking a connection with the divine.
Ti Elotse Ti Elotse	Could be a repetitive chant, invoking action or movement, possibly in praise or petition.
Gan Gan	The repetition of "Gan" could indicate a strong or emphatic call, possibly addressing a higher power or deity. "Gan" in some Khoisan languages can mean "to call" or "to invoke."
!Khutse	This could refer to something related to "god" or "divinity," such as calling upon a higher spiritual entity or asking for divine presence. The exclamation mark (which signifies a click sound) indicates the use of Khoisan's distinctive phonetic feature, which can be part of spiritual chants or invocations.
O Rte Dats Ge	O re: "O" might refer to a pronoun (like "you" or "we"), and "re"** can often mean "are" or "to do" in many Southern African languages. Dats ge: This could be a form of praise or acknowledgement, potentially meaning "you are holy" or "you are great," as "data" could relate to greatness or reverence and "ge" could be a form of a possessive or connecting particle.
Ke Is Aroma	Ke: In many Southern African languages, "ke" can mean "I am" or "I am the one." This could mean "the" or "this", depending on the context. Aroma: In this case, "aroma" could refer to a pleasant smell. Aroma is often used metaphorically in religious contexts to refer to offerings or the presence of the divine, as in "the sweet smell of incense," which represents the connection to the divine or to a state of holiness.
'!Kho Tse // ie Ila	'!Kho tse //ie ila '!Kho: This refers to "the spirit" or "the divine presence," often used to invoke a sacred power or higher being. In many Khoisan cultures, the term "Kho" can refer to a spiritual or ancestral force. tse: This could refer to something like "all," "the people," or "the living," depending on the context. //ie: This is a form of the verb "to be" or "to exist" in certain Khoisan languages, particularly the San languages. ila: This can mean "to lift," "to raise," or "to praise," suggesting an act of spiritual elevation or honour. Interpretation: The phrase may be translated as "The spirit (or divine presence) raises all" or "The divine presence lifts up the people" in a spiritual or praising sense. It is likely referring to the uplifting power of the divine in a ceremonial context.
Hoaba Lumisa'	Hoaba: This word could refer to a high or exalted state, often used in praise contexts. It may also refer to something being elevated or honoured. lumisa: This verb often means "to praise," "to glorify," or "to honour." Interpretation: "Hoaba lumisa" could be translated as "Exalt and praise" or "Honour and glorify." It seems to be an invocation or command to raise praise or to acknowledge and honour the divine or ancestral spirits.

FURTHER RESEARCH DIRECTIONS

Research on Indigenous spirituality epistemologies and sustainable social and economic systems can delve into several key areas, including epistemological foundations of Indigenous spirituality, Indigenous Knowledge Systems and sustainability, new social systems inspired by Indigenous epistemologies, economic systems based on Indigenous principles, cross-cultural and interdisciplinary approaches, policy implications and advocacy, case studies and practical applications:

- Study and compare different Indigenous spiritual epistemologies across various cultures. Look into how these belief systems inform understanding of the world, nature, and community.
- Investigate how colonisation and modernity have impacted Indigenous Spirituality and Knowledge Systems. Consider the resilience and adaptation of these epistemologies over time.
- Explore the ontological and cosmological views inherent in Indigenous spiritual practices, focusing on how they perceive the interconnectedness of all life forms and the natural world.
- Research how Indigenous Spiritual beliefs contribute to sustainable environmental practices. Focus on case studies where these practices have led to successful conservation efforts.
- Examine how Traditional Ecological Knowledge (TEK) can be integrated into modern environmental management systems. Look at examples of co-management between Indigenous communities and governments or non-governmental organisations (NGOs).
- Investigate Indigenous communities' strategies to adapt to environmental changes while maintaining cultural integrity.
- Explore models of governance rooted in Indigenous spiritual principles, such as consensus-building, community ownership, and resource stewardship.
- Study Indigenous approaches to justice that focus on restoration and healing rather than punishment. Compare these with contemporary justice systems and evaluate their potential for broader application.
- Investigate how Indigenous spirituality fosters social cohesion and a sense of identity within communities. Consider how these principles can inform new social systems that prioritise community well-being.
- Research Indigenous economic models (IEM) that emphasise sustainability, reciprocity, and balance. Focus on alternative economic systems such as barter, communal ownership, and resource sharing.
- Study the challenges and opportunities for Indigenous economies in the context of globalisation. Look at how Indigenous communities navigate global markets while preserving their economic practices.
- Explore how Indigenous communities are innovating within their traditional economic frameworks to create new opportunities for sustainability and growth.
- Investigate ways to bridge Indigenous Knowledge Systems with Western scientific and economic paradigms to create hybrid sustainability models.
- Collaborate with experts in Anthropology, Sociology, Environmental Science, and Economics to enrich your understanding and develop comprehensive models for new social and economic systems.
- Explore how research on Indigenous spirituality and sustainability (IKSS) can inform public policy. Focus on areas like environmental legislation, land rights, and cultural preservation.

- Research the role of Indigenous leadership in developing new social and economic systems and how policy and advocacy can support these.
- Identify and analyse successful case studies where Indigenous epistemologies have led to sustainable social or economic systems. Document the processes, challenges, and outcomes.
- Consider designing pilot projects that apply Indigenous principles to create new social or economic systems. Measure their impact and scalability for broader applications.

Research focusing on the preceding areas may contribute to a deeper understanding of Indigenous spirituality and its potential to inspire sustainable social and economic systems.

CONCLUSION

This study explored Indigenous spirituality, practices, and nature into new social and economic systems for sustainability. The findings highlight the importance of Indigenous knowledge in promoting a holistic development approach that is economically, culturally, and environmentally sustainable. Indigenous Spirituality and practices offer valuable insights into sustainable practices by emphasising the interconnectedness of all life and a harmonious relationship with nature. This perspective contrasts with many contemporary development paradigms prioritising economic growth over environmental and social well-being.

The discussion suggests pathways for collaboration between policymakers, practitioners, and scholars with Indigenous communities to co-create sustainable systems. It addresses challenges like bridging the gap between traditional knowledge and modern scientific approaches and protecting Indigenous rights and autonomy.

In conclusion, integrating Indigenous spirituality, practices, and nature into new social and economic systems offers a promising path to sustainability. This approach honours Indigenous wisdom and provides practical solutions to pressing challenges. Genuine partnerships with Indigenous communities are essential to ensure their knowledge and practices are respected, protected, and integrated into broader development agendas. This can create sustainable, just, and inclusive systems, paving the way for a more harmonious coexistence with our planet.

ACKNOWLEDGEMENT

This chapter is partly based on a Doctor of Philosophy thesis of Mosiuoa Nehemiah Ngaka, completed in 2018 at Durban University of Technology under the supervision of Prof P Msweli (Promotor) titled: investigating primordial spirituality practices, social capital, ethic structures and processes in the South African environment. Department of Public Administration in the Faculty of Management Sciences.

REFERENCES

Adamo, G. (2020). *Investigating business process elements: A journey from business process management to ontological analysis, and back* (Doctoral dissertation, University of Genoa, Italy).

Bhaskar, R. (2008). *Dialectic: the pulse of freedom, with new introduction*. Routledge. DOI: 10.4324/9780203892633

Boezak, W. (2015). *Struggle of an ancient faith – the Khoisan of South Africa*. [Self-published].

Braun, V., & Clarke, V. (2023). Toward good practice in thematic analysis: Avoiding common problems and be (com)ing a knowing researcher. *International Journal of Transgender Health*, 24(1), 1–6. DOI: 10.1080/26895269.2022.2129597 PMID: 36713144

Chaudhary, B. R., Erskine, W., & Acciaioli, G. (2022). Hybrid knowledge and climate-resilient agriculture practices of the Tharu in the western Tarai, Nepal. *Frontiers in Political Science*, 4, 969835. DOI: 10.3389/fpos.2022.969835

Dhir, R. K. (2016). *Sustainable development goals: Indigenous peoples in focus*. International Labour Organization (ILO).

Đorđević Belić, S. (2024). Communication with the deceased in dreams: Overcoming the boundary between this world and the otherworld or its conceptualisation strategy? *Religions*, 15(7), 828. DOI: 10.3390/rel15070828

Elendu, C. (2024). The evolution of ancient healing practices: from shamanism to Hippocratic medicine: a review. *Medicine*, 103(28), e39005. DOI: 10.1097/MD.0000000000039005 PMID: 38996102

Elo, S., Kääriäinen, M., Kanste, O., Pölkki, T., Utriainen, K., & Kyngäs, H. (2014). Qualitative content analysis: A focus on trustworthiness. *SAGE Open*, 4(1), 2158244014522633. DOI: 10.1177/2158244014522633

Farber, T. (2015, December 13). The ugly truth behind the plunder of nature. *Sunday Times*, •••, 15.

Fraser, G., Brady, A., & Wilson, M. S. (2021). "What if I'm not trans enough? What if I'm not man enough?": Transgender young adults' experiences of gender-affirming healthcare readiness assessments in Aotearoa, New Zealand. *International Journal of Transgender Health*, 22(4), 454–467. DOI: 10.1080/26895269.2021.1933669 PMID: 37808530

Frohard-Dourlent, H., MacAulay, M., & Shannon, M. (2020). Experiences of surgery readiness assessments in British Columbia. *International Journal of Transgender Health*, 21(2), 147–162. DOI: 10.1080/26895269.2020.1742842 PMID: 33015666

Gentles, S. J., Charles, C., Ploeg, J., & McKibbon, K. A. (2015). Sampling in qualitative research: Insights from an overview of the methods literature. *The Qualitative Report*, 20(11), 1772–1789. DOI: 10.46743/2160-3715/2015.2373

Guizzardi, G. (2020). Ontology, ontologies, and the "I" of FAIR. *Data Intelligence*, 2(1-2), 181–191. DOI: 10.1162/dint_a_00040

Gwala, R. S. (2023). The legalisation of cannabis in South Africa: proposing an economic value chain model for South Africa. In *Rapid innovation and development in the global cannabis market* (pp. 98–122). IGI Global. DOI: 10.4018/978-1-6684-6346-8.ch006

Haddad, B. M., & Solomon, B. D. (2024). Ecological economics as the science of sustainability and transformation: Integrating entropy, sustainable scale, and justice. *PLOS Sustainability and Transformation*, 3(2), e0000098. DOI: 10.1371/journal.pstr.0000098

Hnatkowska, B., Kozierkiewicz, A., & Pietranik, M. (2020). Semi-automatic definition of attribute semantics for the purpose of ontology integration. *IEEE Access : Practical Innovations, Open Solutions*, 8, 107272–107284. DOI: 10.1109/ACCESS.2020.3000035

Hossain, M. F. (2021). Anthropocentric approach to the environment: An overview. *Journal of Social Sciences*, 1711, 43.

Ikeke, M., & Ukutsemuya, J. (2024). The Afrocentric imperative in tackling African environmental challenges: Afroecosolidarity perspectives. *East African Journal of Environment and Natural Resources*, 7(1), 17–27. DOI: 10.37284/eajenr.7.1.1675

Imam, S. S., Mehdi, S., & Mansuri, A. (2024). A comprehensive review on the pharmacological potential of red ginseng oil, tea tree oil, and hemp seed oil. *European Journal of Biomedical*, 11(4), 167–177.

Josyula, K. L., Nambiar, D., Narayan, V., Sathyanarayana, T. N., Porter, J., & Sheikh, K. (2015). Cultural consonance, constructions of science and co-existence: A review of the integration of traditional, complementary, and alternative medicine in low- and middle-income countries. *Health Policy and Planning*, 30(8), 1067–1077. DOI: 10.1093/heapol/czu096 PMID: 25171821

Kakar, Z. U. H., Rasheed, R., Rashid, A., & Akhter, S. (2023). Criteria for assessing and ensuring the trustworthiness in qualitative research. *International Journal of Business Reflections*, 4(2), 148–173. DOI: 10.56249/ijbr.03.01.44

Kaunda, M. K. (2021). Mother earth, mother Africa and African women's role in Indigenous religions. *African Journal of Gender and Religion*, 27(1), 91–117. DOI: 10.14426/ajgr.v27i1.89

Kraiwanit, T., Limna, P., & Siripipatthanakul, S. (2023). NVivo for social sciences and management studies: A systematic review. *Advance Knowledge for Executives*, 2(3), 1–11.

Lassiter, L. E. (2005a). *The Chicago guide to collaborative ethnography*. University of Chicago Press. DOI: 10.7208/chicago/9780226467016.001.0001

Lassiter, L. E. (2005b). Collaborative ethnography and public anthropology. *Current Anthropology*, 46(1), 83–106. DOI: 10.1086/425658

Lassiter, L. E. (2021). Collaborative ethnography: trends, developments, and opportunities. In B. Diamond & S. El-Shawan Castelo-Branco (eds), *Transforming ethnomusicology volume i: methodologies, institutional structures, and policies*. New York. DOI: 10.1093/oso/9780197517604.003.0004

Lim, W. M. (2024). What is qualitative research? An overview and guidelines. *Australasian Marketing Journal*, •••, 14413582241264619. DOI: 10.1177/14413582241264619

Lundquist, C., Harhash, K. A., Armenteras, D., Chettri, N., Mwamodenyi, J. M., Prydatko, V., & Acebey Quiroga, S. (2016). Building capacity for developing, interpreting and using scenarios and models. In *Methodological assessment of scenarios and models of biodiversity and ecosystem services* (pp. 225–254). Secretariat of the Intergovernmental Platform for Biodiversity and Ecosystem Services.

Machado, L. M. O. (2021). Ontologies in knowledge organization. *Encyclopedia*, 1(1), 144–151. DOI: 10.3390/encyclopedia1010015

Magni, G. (2017). Indigenous knowledge and implications for the sustainable development agenda. *European Journal of Education*, 52(4), 437–447. DOI: 10.1111/ejed.12238

Mampilly, Z. (2015). Decolonizing the United States: Lessons from Africa. *Ufahamu. Journal of African Studies*, 38(3), 1–22.

Menezes, M., & Costa, C. S. (2024). The role of collaborative ethnography in placemaking. *Humans*, 4(3), 284–297. DOI: 10.3390/humans4030018

Mwiti, G. (2014). African Indigenous psychotherapy. *Journal of Psychology and Christianity*, 33(2), 142–153.

Myers, M. (1999). Investigating information systems with ethnographic research. *Communications of the Association for Information Systems*, 2. Advance online publication. DOI: 10.17705/1CAIS.00223

Noble, H., & Smith, J. (2015a). Issues about reliability and validity. *Evidence-Based Nursing Research*, 18(2), 163. PMID: 25653237

Noble, H., & Smith, J. (2015b). Issues of validity and reliability in qualitative research. *Research Made Simple. Evidence-Based Nursing*, 18(2), 34–35. DOI: 10.1136/eb-2015-102054 PMID: 25653237

Omoyajowo, K., Danjin, M., Omoyajowo, K., Odipe, O., Mwadi, B., May, A., & Rabie, M. (2024). Exploring the interplay of environmental conservation within spirituality and multicultural perspective: Insights from a cross-sectional study. *Environment, Development and Sustainability*, 26(7), 16957–16985. DOI: 10.1007/s10668-023-03319-5 PMID: 37363007

Pakendorf, B., & Stoneking, M. (2021). The genomic prehistory of peoples speaking Khoisan languages. *Human Molecular Genetics*, 30(R1), R49–R55. DOI: 10.1093/hmg/ddaa221 PMID: 33075813

Panayiotou, C. (2024). Ontological foundation for representing metaphors as part of LOD. *IEEE Access : Practical Innovations, Open Solutions*, 12, 73810–73838. DOI: 10.1109/ACCESS.2024.3396871

Prasad, S. M., Patel, B. D. P., Tiwari, N. N., Sapkota, S., & Shah, B. (2020). Traditional healers of Nepal: Their knowledge, skill, practices, and technology. *Journal of Ayurveda Campus*, 1(1), 7–15. DOI: 10.51648/jac.3

Prigg, M. (2014, December 5). The dwindling South African tribe who number just 100,000 but were once the most common humans on earth. *MailOnline*. Retrieved on 10 October from https://www.dailymail.co.uk/news/article-2868526/The-dwindling-South-African-tribe-number-just-100-000-common-humans-earth.html

Rappaport, J. (2008). Beyond participant observation: Collaborative ethnography as theoretical innovation. *Collaborative Anthropologies*, 1(1), 1–31. DOI: 10.1353/cla.0.0014

Roy, K., Zvonkovic, A., Goldberg, A., Sharp, E., & LaRossa, R. (2015). Sampling richness and qualitative integrity: Challenges for research with families. *Journal of Marriage and Family*, 77(1), 243–260. DOI: 10.1111/jomf.12147

Rupp, D. E., Thornton, G. C.III, Bisbey, T. M., Hoover, A. N., Salas, E., & Murphy, K. R. (2024). An epistemology for assessment and development: How do we know what we know? *Industrial and Organizational Psychology: Perspectives on Science and Practice*, 17(2), 252–268. DOI: 10.1017/iop.2024.3

Sangha, K. K., Le Brocque, A., Costanza, R., & Cadet-James, Y. (2015). Ecosystems and Indigenous well-being: An integrated framework. *Global Ecology and Conservation*, 4, 197–206. DOI: 10.1016/j.gecco.2015.06.008

Suchitra, M. R. & Parthasarathy, S. (2020). *International Journal of Research in Pharmaceutical Sciences*.

Takada, A., & Noguchi, T. (2024). Diversity and transition of stepfamilies among the G|ui and G||ana. *Hunter Gatherer Research*, •••, 1–11. DOI: 10.3828/hgr.2024.37

Virk, P. S., Andersson, M. S., Arcos, J., Govindaraj, M., & Pfeiffer, W. H. (2021). Transition from targeted breeding to mainstreaming of biofortification traits in crop improvement programs. *Frontiers in Plant Science*, 12, 703990. DOI: 10.3389/fpls.2021.703990 PMID: 34594348

Wei, T. (2020). *Terminology and ontology for cultural heritage: application to Chinese ceramic vessels. Formal Languages and Automata Theory*. Université Grenoble Alpes.

White, M. D., & Marsh, E. E. (2006). Content analysis: A flexible methodology. *Library Trends*, 55(1), 22–45. DOI: 10.1353/lib.2006.0053

KEY TERMS AND DEFINITIONS

Social and Economic Systems: A social science and a branch of economics that focuses on the relationship between social behaviour and economics.

Sustainability: The ability to maintain a process continuously over time.

Epistemology: The philosophical study of nature, origin, and limits of human knowledge.

Indigenous Knowledge Systems: The knowledge and practices that local people develop over generations of living in a particular environment.

Primordial Spirituality: Primeval early spirituality.

Chapter 2
The Spiral of Knowledge:
The Fourth World and Sustainability

Rudolph C. Ryser
Center for World Indigenous Studies, USA

ABSTRACT

Fourth World knowledge systems, often overlooked in discussions of human sustainability, focus on the dynamic relationships between people, land, and the cosmos. These systems are typically associated with non-state nations and people, whose interactions with state entities often involve conflict over ancestral territories. Pursuing balance and harmony for life sustainability is not a primary consideration in these interactions. The author uses the Four Directions metaphor to symbolise the relational connection between human experience, land, and the cosmos, blending qualitative, quantitative, and relational reasoning. This framework has local, regional, and global implications for future sustainability. A conceptual framework is presented integrating Indigenous scientific knowledge with conventional sciences to explain various phenomena. Additionally, ALDMEM is introduced, a mechanism for mediating knowledge system applications, ensuring the consent of Fourth World people and state populations, and aligning with the principle of free, prior, and informed consent.

INTRODUCTION

There is a widely accepted tendency in political, academic, and social institutions to take for granted what we mean when using the word "Indigenous." "Indigenous" or "Native" are terms of political art and academic convention often used interchangeably as an adjective describing "people" or "people" (Fenstad et al., 2002; Agrawal, 2004; Atleo, 2004; Ryser, 2022). The words Fourth World people use for their collective identity commonly translate to mean "people" or a people of a place, a feature of the Earth or homeland. The author applies the descriptive expression "Fourth World", commonly understood in the knowledge systems of people globally, unless specifically indicating the word used to identify the people.

The "Fourth World" concept in 20th-century discourse was initially introduced in policy and academic discourse in the early 1970s at the beginning of the global Indigenous "engagement" movement and the height of the North American Indian land rights movement led by Secwepemc Grand Chief George Manuel. The book titled *The Fourth World: An Indian Reality* (Manuel & Posluns, 1974) not only led the emerging global land rights movement but placed the Fourth World imprint on the collective identity of

DOI: 10.4018/979-8-3693-8447-3.ch002

what the United Nations would come to identify as 1.3 billion "Indigenous people." Manuel advanced the idea that Fourth World people are not just nations within states, but they are nations occupying ancestral territories and political space with international states within larger geopolitical processes. Struggling to maintain their ancestral territories and cultural and political identities, they existed simultaneously within and beyond the conceptual limits of the state. The more than 6000 nations, Manuel and Posluns (1974) asserted, were political equals to states, and their knowledge of the ancestral territories was essential to peaceful relations and human sustainability. Our modern usage of the "Fourth World" is derived from the book, but their usage of "Fourth World" is rooted in the Hopi origin story.

In the Hopi oral tradition, humanity emerged over time in four worlds. The Hopi Knowledge Keepers (Kikmongwi), who have the responsibility for preserving and passing on the Hopi cultural and spiritual knowledge to future generations, have, through their traditions, informed us of the worlds of emergence. The First World, called *Tokpela* or "endless space," Tokpela was the world the "Sun Spirit" Tawa made with the first four men and four women, each speaking different languages. According to Hopi legend, men and women soon forget the Creator Tawa. Tawa had intended that humans live in harmony, but they became corrupt and unbalanced, failing to live in accord with the Creator's laws and balance with nature and all of life. Tawa saved just a few people who understood the need for harmony and took shelter with the ants underground. Tawa then destroyed the world with fire.

The Second World, *Tokpa, meaning "dark midnight", was created by Tawa and his nephew Sotuknang, characterized by* the survivors of the Tokpela and their offspring multiplied and migrated across the land. With so much natural life, these people became greedy and selfish and hoarded goods for themselves. Tawa destroyed this world with ice, freezing everything except a few people loyal to the teachings of harmony and balance. These people were sheltered and protected by ants underground.

The Third World, the Kikmongwi explanation, was the Kuzkurza or "Cold World", also created by Tawa and his nephew Sotuknang. The survivors of the Second World would inhabit this new world—Tokpa. The descendants of these people would become potters and cloth weavers, and they would grow crops for food. As makers of things, the people of Kuzkurza began making weapons and violently attacking each other to steal and take possession of land and resources. Once again, Tawa decided this was not good, so destroying the Kuzkurza with water became the answer, and once again, Tawa saved only a few people who followed the law of harmony and balance. The people took shelter with Spider-Woman in a hollow reed as the waters destroyed the world.

The Fourth World, *Tuwaquchi, meaning the "world complete", was created by Tawa and Sotuknang,* and it was peopled by the survivors of the third world and their descendants. The survivors emerged from the *sipapu*, the place of emergence and return," a sacred place in the Grand Canyon in what is now Arizona, United States. The people, as the story is retold, were guided by Spider Woman and *Masauwu*, the Spirit of Death and Earth God, to migrate across the land until they reached their permanent home between the Colorado River and the Rio Grande become the people we now know as the Hopi.

It was this story Grand Chief George Manuel carried to different people he visited around the world and learned that the concept of "world complete" that many people commonly understand the Hopi called Tuwaquchi or Fourth World. He decided that all these different people were the people of the Fourth World[2]-- the ***world complete***.

Using the terms in this way draws distinctions between people who are members of an ancestral community with ancestral ties to a territory. Settler populations and their descendants cannot claim such ancestral ties to a territory as they create an identity. Imperialism or colonialism is implicit in naming these distinctions. "Fourth World people" are distinct cultural and political societies. These people are colonized, for example, by European kingdoms, Chinese imperial dynasties, and Arabic Emirates from the 13th to the 20th centuries, and they are usually declared to be minority populations within a larger population under a newly proclaimed political jurisdiction comprised of settler populations or sometimes dominated by a ruling nation.[3] The term "Indigenous," though meaning "original," is a category borne of *statism*. The use of "Indigenous" by the states presumes that the people identified as Indigenous are a minority population under the centralized control of the state. This control is assumed since the state is defined as having the authority to apply and enforce universal laws within a bounded territory by exercising police power through the state's political apparatus. It has become a political term of art adopted by the United Nations, International Labor Organization, states' governments, and many Fourth World nations to designate the people colonized and re-colonized[4] by newly formed states. Indeed, neo-colonial processes are at the root of sub-regional and regional conflicts that are widely referred to as "civil wars" when, in fact, they are wars of self-determination or land control between Fourth World people occupying their ancestral territories and the state (Ryser, 1985). The neo-colonial process incorporated people into the new states established on top of Fourth World ancestral territories, sometimes dividing Fourth World territories between the new states.

COLONIALISM AND THE INTERRUPTION OF BALANCE AND HARMONY

Colonialism is only one of at least four expressions of European imperialism since the fifteenth century (Smith, 1986, 1987; Smith, 1999, 2012), with one of those expressions originating from the perspective of Indigenous societies or original nations. European expansionism occurred within Europe for centuries before the fifteenth century and was not limited to European Kingdoms. Waves of imperialism also swept across Asia from China and Manchuria, the Middle East from Saudi Arabia, and Moors that had come from the 7th century to dominate Northern Africa moved into southern Europe hundreds of years before the fifteenth century. After the British and Ottoman Empires collapsed in the early 20th century, the decolonization movement left behind complex political disruption among Fourth World nations, frequently resulting in what is today called "civil war." These "wars" were most often resistance movements by Fourth World nations to state domination.

Imperialism involves the invasion of the territories of pre-existing nations, deemed as "tribes," "primitives," "nomads", or similar terms signifying inferior otherness. Long before the emergence of the modern state in the 17th century, nations struggled to maintain their cultural and political identities against invading and assimilative forces. Those assimilative forces were military victors and invasive migrating people who came to control the historical narratives that eventually became the origin stories of today's states, with the nations (or their remnants) becoming subsumed discursively into those narratives. Some of those nations are recognizable to us today; aside from the original nations of the Americas (Méxica, Maya, Anishinabe, Lakota, etc.), Scotland, Frizland, Waripiri, Euskadi, Maasai, Palestinians, Dinka, Bedouins, and Catalonia are examples of countries and people who maintained their national identities but have been incorporated into states. Many more are far less well known—such as the Kurds of Kurdistan, the Rohingya of Myanmar, the Naga of India, and the Otomi of Mexico—but they can be found on all the world's continents except Antarctica.[5]

These historical circumstances and political/cultural clashes created a need to understand how the world and relations between people were changing. There had been no conceptual framework to explain what had happened between nations and emerging states or what changes began to unfold in the 20th century when nations that had been captives in empires acted to extricate themselves to become internally autonomous, independent states. The need for a theoretical framework that provides concepts aiding in elaborating social, economic, political, geopolitical, and cultural understanding of the nation remaining as newly distinguishable national identities distinct from states became palpable by the mid-twentieth century. These nations had become a strategic and geopolitical fact as state after state convulsed with political and military conflicts over land rights, collective rights, self-determination, and cultural determinism. In Burma, Russia, Somalia, South Africa, Canada, Nicaragua, the Philippines, Rwanda, Mexico, Indonesia, the United States of America, India, Chile, Bangladesh, Japan, Norway, Nigeria, and Israel, the nations captured within the struggle against state assimilation and claims and encroachments made by the state and corporations into nations' territories.

In this broader context of developing political lexicon, national and international movements, regional and sub-regional conflicts, and emerging international multi-state inquiries "into the situation of Indigenous people", scholars began to formulate a theoretical framework within which the disparate influences might be better understood and explained.

FOURTH WORLD KNOWLEDGE ROOTS

Fourth World leadership evolved in the 1970s as nations' representatives began to engage international mechanisms concerned[6] with "Indigenous rights" and "human rights." As nations' representatives engaged each other, it became apparent that a common vocabulary—a Fourth World socio-cultural-political lexicon—would contribute to a better understanding of what increasingly recognized as a shared experience. At the same time, it also became apparent that a commonly understood conceptual framework would be necessary to engage in cross-cultural dialogue (nation-to-nation and nation-to-state) on various topics. Such a framework would not only bridge the thinking of nations but aid in structuring dialogues between nations and states (Cobo, 1982; Martinez, 1997). Scholars and policymakers among Fourth World people recognized that this theoretical framework must provide concepts helpful in understanding nation-to-nation relations, nation-to-state relations, and particularities such as problems of climate change, human sustainability, land rights, education, collective rights, social change, self-determination, political organization, territorial and natural resource use, organization of society, the arts, horticulture, health, and food accessibility.

While international relations theorists[7] continued to assert that nations and states are to be considered the same and that Fourth World nations are to be considered ethnic groups, the emergence of stronger political identities in the Fourth World necessitated a separation of these terms to provide geopolitical clarity. Through the efforts of a few political leaders and scholars, the shape and content of Fourth World knowledge systems emerged in the 1970s and 1980s, moving past much of conventional 19th and 20th-century political theory.

Since Manuel's seminal book was published a half-century ago, a body of scholarly work built on Fourth World knowledge systems has begun to develop. Applied in the fields of Fourth World Geopolitics (Ryser, 2012), studies in human development, change, and environmental sustainability, developing Fourth World scholarship bridges international relations theory, environmentalism, human sustainability, and studies about Fourth World people, offering a conceptual framework that recognizes the international political character of dominated people while de-naturalizing the modern state system. It also recognizes the struggles for political self-determination of nations far older than 500 years. From this perspective,

Indigenous people in the western hemisphere would be only one subset of the taxonomy we call Fourth World nations. Even though Fourth World scholars conceived their scholarship in ancestral North America, it is not widely known within Indigenous academic studies. The goal of this chapter is to advance Fourth World knowledge systems as a research methodology appropriately suited to understanding the particularities of sustainability in the context of the geopolitical/neoliberal world order of consumerism and development while demonstrating the conceptual utility of the Fourth World knowledge systems in their application to modern knowledge construction.

FOURTH WORLD KNOWLEDGE SYSTEMS ROOTED IN INTERDEPENDENCE AND SUSTAINABILITY

Fourth World political activists frequently call for diplomatic representatives, scholars, states' governments, and international institutions to respect Indigenous knowledge. References are made to the importance of Indigenous knowledge in understanding how to deal with the adverse effects of climate change, ensure the sustainability of populations, maximize the benefits of food and water to rapidly growing populations, and resolve conflicts between political adversaries in domestic and international arenas.[8]

Little is said about the structure, principles, rules, and concepts of Fourth World knowledge systems rooted in the culture of balance, harmony, interdependence and sustainability of each distinct people and the environment and how any of these Indigenous knowledge systems may relate to each other and can be applied to vexing social, economic, and political problems. Indeed, some of the most determined advocates demand respect for Indigenous people and their knowledge, and it is hard to explain the very system of knowledge they say must be respected—much less how that knowledge will be applied. The absence of such clarity leaves states' representatives and academics promising to respect such knowledge without having the slightest idea of what knowledge they are respecting.

LOCUS OF KNOWLEDGE

There are different types of knowledge that knowledge function differently when "owned" by an individual, a family, a community, or within a trans-community[9] environment. Having an understanding or familiarity with a fact, subject, place, situation, or experience is knowledge. Access to "personal knowledge"—as in knowledge owned by an individual or a family—is strictly prohibited in some societies. Other forms of knowledge associated with the community or trans-community environment may have virtually no access restrictions.

One type of knowledge answers the question of "how" something is done, a second type of knowledge is "knowing" that something is true, while a third kind of knowledge answers the question of "that" is true.

Knowledge of "how" to do something explains or asserts a process. Knowledge-based on "how" is a distinct kind that may be expressed independently of facts or may depend on demonstrable facts. While knowing how to do something depends on factual evidence about how a thing is done or explained, applying "how" knowledge does not necessarily require implicit or explicit facts.

A second type of knowledge answers a question by knowing something exists or is true. One of India's most distinguished twentieth-century scholars, Sarvepalli Radhakrishnan (1888 – 1975), wrote cogently about a second type of knowledge he interpreted as "knowing" or "intuition." Intuitive knowledge or "knowing" that something is true can be achieved through "immediate apprehension"- as in, "This flower is beautiful!" Such knowledge is not dependent on facts or reasoning. Knowing something is true relies on inspiration and immediate cognition (Radhakrishnan, 1929; 2008).

A third type of knowledge answers the question "if," resulting in a conscious action to produce a result. This knowledge confirms the reality of a form of knowledge that may be a fixed or transformational reality—which is to say knowledge "if" is actualized or it takes a new or different form that must be considered new knowledge.

Ageless knowledge systems constructed by Fourth World people and their earlier ancestors remain very much in evidence in medicine, architecture, water engineering, music, food security, governance, sculpture, population control, mental health, horticulture, animal husbandry, agriculture, history, literature, economics, and international relations. It would be folly to suggest that these many endeavours are not rooted in the earliest knowledge systems of humankind. Scholarly understanding of the application of knowledge systems in these and many other aspects of life is the task repeatedly enshrouded in mystery and intellectual bigotry. Fourth World knowledge offers a perspective, a set of concepts, and a worldview. It also points to methods for applying ancient knowledge from different Fourth World perspectives to shared problems of communities and humanity.

FOURTH WORLD THOUGHT EPISTEMOLOGICAL FOUNDATIONS

Fourth World knowledge systems state that the concepts of comparison, relational reasoning, a balance between contending forces, and equality of kind (that human beings are part of all living things and not the dominant living thing) will—when applied in life and thought—ensure comity between people, between people and living nature, and with the forces of the cosmos. Fourth World knowledge systems also assert that human cultures—as with the cultures of other animals, plants, rivers, mountains, and lands—are defined by this evolving and dynamic relationship between people (animals, plants), the land, and the cosmos.

Fourth World knowledge systems assert that history, memory, and thought processes are multi-dimensional—where two-dimensional thought (linear past progressing to the future, fatalistic, cyclical, or providential) is seven-dimensional. Tracing thought diagrammatically; the process is more like a spiral[1] where motion and change move in all directions simultaneously in time and space (UNGA, 2007).

Four Directions

The knowledge systems of the Fourth World are captured in the visual and tangible metaphors depicting the directions from which the Sun and the Moon traverse the sky and the places where people, foods, and other animals can reside together. The *four directions* are symbolically rendered as horizontal and vertical lines intersecting at their midpoints (See Figure 1) from the basic Kongo people's[10] Four Directions cosmogram) demarcate space pointing to the planet's polar north and south, the rising sun and moon, and the "tree of life" (the stars making up the Milky Way Galaxy). The vertical, polar line is in dynamic motion, moving slightly back and forth, mirroring the earth's wobbles in space. The horizontal line points to the east and west and moves up and down on its axes, reflecting the seasonal changes owing to the earth's changing angle relative to the sun as it follows its orbit around the sun.

Figure 1. Nkisi: Kongo Four Directions cosmogram from South-Central Africa

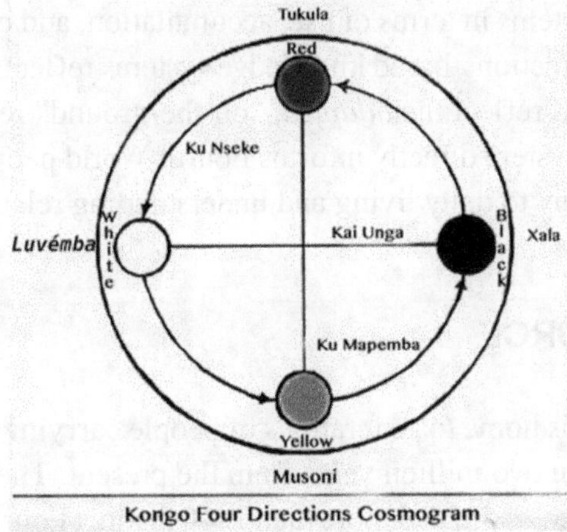

(Author)

The Kongo's *Nkisi* is a traditional symbol from South Central Africa, specifically from the Kongo people who live in the Democratic Republic of Congo, Angola, and Congo-Brazzaville. The *nkizi* is a circular diagram representing the Kongo worldview and their understanding of the universe. Among the Kongo this four directions symbol has a deep symbolic meaning, representing the interconnectedness of all things in the universe and between all things on the land. The central cross represents the unity of the physical and spiritual worlds, and the four quadrants represent the four elements and the four directions. The colours used in the cosmogram are also highly symbolic, representing different aspects of the Kongo worldview and harmonizing life. The *nkisi* serves as a powerful reminder of the importance of respecting and honouring the natural world and the interconnectedness of all things in the universe.

As with the Kongo many other Fourth World people symbolize the four directions, for example, as a medicine wheel,[11] calendar,[12] four distinct symbols,[13] and prayer ceremonies in the four directions toward the rising sun, setting moon, and the polar north and the south. The four directions dynamic symbol not only operates on a single two-dimensional plain but is further amplified by three additional directions demonstrated by passing a line through the middle axis of both lines extending toward the centre of the Earth, the middle of the intersecting lines and outward toward the centre of the galaxy. In its simplest form, this remarkable metaphor reflects the relationships between fixed physical points in space on the planet's surface and with the galactic points in motion inside and outside the planet.

The Toltecs assigned *Four Texcatlipocas*[14] to the four directions; they assigned every aspect of life to all the four directions where human beings live—the planet's surface—and their relation to the cosmos. Toltecs assigned metaphysical attributes of life to the direction below (the underworld or centre of the Earth), the centre axis around which all things exist and the direction above where all things originate. While all seven directions essentially define and position the human experience and the experience of all living reality, the four directions provide the most tangible guidance for human beings standing on the ground. Each of the four directions has a colour, a sound, a plant, an animal, a dance, medicine, a shape, a climate, an environment, soil, water, and a name suited to each direction.

The Fourth World is embedded in the knowledge systems of Indigenous people worldwide who share these "four directions." Rendered in its totality, one must come to terms with the actual use of knowledge to understand the knowledge systems in terms of use, accumulation, and construction (Yeo et al., 2013). Despite their origins, the four directions-based knowledge systems reflect similar themes and structures, and differentiated micro-systems reflect the *adapted,* "on-the-ground" realities of the different people. Each Fourth World knowledge system directly informs Fourth World people' experiences and thoughts, thus linking balance and harmony to daily living and understanding relationships.

BWANGA: THE VITAL FORCE

The vast continent of Africa is home to generations of people carrying the genetic markers of human beings extending back more than two million years from the present. The Center for World Indigenous Studies Fourth World Atlas Project estimated more than 1000 distinct Fourth World nations located in the 54 African states. This pervasive "vital force" profoundly influences many of these nations particularly south of the Sahara Desert; it is the force that animates, strengthens, and weakens all life—unifying all things. It is Bwanga. As Temples (1959) observes:

> The vital force can be intensified and compacted and can exteriorise itself at what we may call a nodal point or vital centre. This vital centre, this nodal point, this particularised manifestation or sign of the vital force, is called "kijimba" by the Baluba. A wild beast may be pierced by ten arrows without dying, while another beast succumbs to the first shot. This is because the one arrow has touched the vital centre, or one of its vital centres (p 39).

It is the *vital force* that forms relationships between all things, and its *kijimba* confirms that relationship. Knowledge of the force is confirmed by kijimba. However, there are two kinds of knowledge: absolute and immutable and relative and speculative. Tempels (1959) further notes:

> ... if anyone wishes to take for himself, or to make use of, the vital force of an inferior being, he should try to procure for himself a like "kijimba" which signifies and materialises the vital relation between the other being and himself. It is, moreover, the "kijimba" that one finds as the chief element, the active principle, the source of energy in every "bwanga" (p. 39).

Fourth World Africans store, retrieve, and communicate knowledge orally embedded in slogans, ceremonial or spiritually derived language, poetry, leadership lists of reigning monarchs, narratives or tales, commentaries. The "four directions" defined in the image of the Kongo Cross incorporates a directional red pyramid pointing to the north (toward the Sahara Desert), and black pyramid pointing to the south (toward the southern part of the African continent), and coloured circles (See Figure 1) contain deep meaning connecting the people, the land and the cosmos.[15]

The knowledge systems briefly discussed here share several commonalities: the intimate relationship between the personal and impersonal, comprehension of universal life force, the unity of all things physical and metaphysical, the four directions metaphor, and the origins of all things ultimately centering on the founding of land on water—an island. By comparing the four systems we see that each spring from "local" (individual, family, and community) conceptions, and that the local radiates to trans-local environments to continental systems. There has for centuries been a sharing between cultures, a cross-cultural conversation that built continental-wide systems of knowledge while energizing the evolution of individual, family, community, and sub-regional knowledge complexes facilitating human adaptation to localized environmental and social changes. While observers and scholars generally refer to "Indigenous knowledge" as "localized," limited, and functional, the reality is that all knowledge systems function in a dynamic human cultural environment where exchange and sharing is not only practiced, but also necessary to achieve continuity. Cross-cultural fertilization of ideas and experiences is a normal dynamic for any successful culture, and it is notable that the expression of knowledge will have localized attributes, but such knowledge sits within a broader knowledge paradigm.

INTEGRATING FOURTH WORLD KNOWLEDGE FOR SUSTAINABILITY

Shared living among different people and their environment depends on a shared understanding of knowledge systems. While human reasoning is universal (Goel & Dolan, 2001; Haskell, 2008; Stenning &Lambalgen, 2012), it takes different forms and may be influenced by different experiences. Inquiring into the nature of material and immaterial realities has occupied human beings from the beginning of time—peering out into the forest, at a mountain, or into the sea while also casting a gaze on the sky. The sense of wonder and the search for explanations caused the mind and body to reach for methods to answer the rapidly forming questions as events and phenomena that came into vision. For some people, formalized religions and theologies provided conceptual and practical structures to explain the mystical. For other Fourth World people, relying on natural mental powers, dreams, and analogies is an essential method of inquiry. Still, others turned to syllogisms to reason answers to questions, and some chose to employ methods to determine cause and effect, concluding that if one experiments frequently.

In addition to the rationalist methods just discussed, dreaming, waiting, comparing multiple variable relationships, counting, watching, stressing (as in physical stress), visioning, taking entheogens, analogical reasoning, and relational reasoning are some of the key methods for testing and applying Fourth World thought (Lyman & O'Brien, 2001; Hummel & Holyoak, 2005; Crone et al., 2009).

Relational Reasoning and Expression/Narrative Analysis

Indigenous evaluation is relational and multidimensional. It is a rational process expressed in a phrase, "relational reasoning", relying on the more familiar concept of induction. The relationship between multiple domains (i.e., events, locations, objects, movement, time, sound, colour) is iteratively assessed while working toward a predicted goal (Ryser, 2012). Where a factor is not productive, it is set aside, leaving the remaining "domains" that increasingly acquire meaning about each other and together contribute to the explanation, description or affirmation of a particular *truth* (Landy & Goldstone, 1991; Goel & Dolan, 2001; Hummel & Holyoak, 2005; Crone et al., 2009; Son et al., 2010).

Fourth-world scholars and positivist scientists deal with facts but often express them in stories or narratives. Stories circulate throughout cultures, providing meaning for both personal and external phenomena. Identity is closely associated with the narrative, and the narrative reveals perceptions and meaning. Within individuals and cultures, various ways of knowing and borders of knowing are permeable (Ryser, 1997).

Storytelling (History, Present, Future - Simultaneity)

Story imposes, eliminates, and controls time and space so that the mind—the Spirit—can learn, explore, remember, alter, expand, and contract ideas, experiences, and perceptions. Through this miraculous process, it is possible to transport oneself or a group of people to another land, another time, and another dimension, rendering what may be impossible in the material world wholly and completely possible and accessible. Using the "story medium" in Fourth World society serves as a learning process, remembering history and foretelling the future as much as explaining the present. A researcher employs storytelling in two circumstances:

1. To imagine or contemplate how a problem can be understood and solutions tested
2. To create a simulated environment for a group of individuals serving as sources of information to reveal their perceptions, experiences, memories, predictions, and contemporaneous perceptions.

In the first circumstance, a researcher may wish to examine how to traverse a great distance over land and sea to achieve a particular goal. Developing a story that places either the researcher or an alter ego in the position of carrying out the mission allows the researcher to consider the aids and obstacles on the path to the goal. By so doing, it becomes possible to consider, examine and otherwise experience how a problem can be solved.

A researcher can structure an approach to solving a problem by following the story to its logical conclusion. In the second circumstance, a researcher may employ a story to help community members develop a common understanding of a problem they jointly seek to solve. Participants in a group may be presented with a story that offers many options for solving a problem considered by the story protagonists, allowing each participant to identify with one or more story members. By doing so, each participant does not risk anything personally but can "experiment" with solutions without risk.

Storytelling has many different functions in society, but in research, it serves as an important method for understanding and explaining material and immaterial phenomena such that a researcher can consider remote intergalactic problems without leaving earth, solve structural problems confronted by architects and engineers and examine the mysteries of the mind. The story has played an important role in research and society for millennia. A good storyteller can make dramatic and successful contributions to human understanding and meeting human needs (Hart, 2010; McDonough, 2010; Fleischhacker at al., 2011; Kaya & Seleti, 2013, Sinclair, 2013).

Dreaming, Meditation, Waiting

Dreaming, meditation, and waiting perform a function similar to storytelling in that these methods of obtaining information and understanding bridge the gap between the material and immaterial realities (physical and metaphysical) essential to reality as a whole (Stone, 2004; Zauditu-Selassie, 2007; Hart, 2010; Sinclair, 2013). When practised skillfully to acquire knowledge, dreaming, meditation, and waiting collapses time and space such that the past, present and future exist simultaneously. Under this condition, the "dreamer" may access knowledge not otherwise available in the material world. This can be a powerful research method when considering health, emotional, and social problems such as psychological breakdown or social upheaval.

Dreaming can also contribute to solving problems such as finding food and water in a barren location or understanding how best to find a plant or capture an animal that may give itself for food. Dreaming can also play an important role in understanding how to respond to a natural calamity such as a flood, drought, sudden freeze, or other climate change. Finally, a dreamer may contribute to defining and explaining how best to defend against an enemy or define the best approach to resolving a dispute without resorting to violence.

Waiting is associated with inspiration or a moment of realization. A researcher may not readily comprehend the answer to a question, but by consciously and intentionally "waiting" the answer may become apparent. This method is particularly appropriate in studies involving complex relationships between individuals or events that may require compression of time and space to fully understand the implications of interactions.

Stressing

When stressed, the body adjusts to maximize the benefits of core support. In many societies, these core supports contribute to what is understood as "visioning" or intensified contemplation that allows consciousness to extend well beyond the body. Stressing may be accomplished by running for a long duration, running up a hill until exhausted, swimming vigorously for a long time, enduring heat, cold fasting or experiencing pain for a period. Stressing has the ability to concentrate the conscious self-such that a problem or question can become more clearly conceived and comprehended, thus allowing for the possibility of solving a problem or obtaining an answer (Logsdon et al., 2009; Korn, 2023).

Stressing the body increases natural hormones that prompt the body to act in an urgent situation or what is commonly called "fight, flight or freeze." This result can be turned into the process of discovering new knowledge. In traditional societies, the "sweat lodge", with its intense humidity and heat, can clean the mind and spirit, producing mental clarity and spiritual transparency that changes the unknown or obscure into a known and apparent. Vision quests where an individual may become isolated without the comforts afforded by community and home may have similar mind cleansing and spiritual clarity. Similarly, the Sundance ceremony consists of daily practice, a sacred pipe, monthly prayer ceremonies and an annual ritual of stressing the body through fasting and prolonged exposure to the elements. All of this contributes to healing the body and sharing healing powers with others who are sick. Stressing is a method for cleansing and revealing powers that cross from the Spirit world to become available for ordinary people mediated by a specially designated group member.

Narrative Analysis, Portraiture, Metaphor

The spoken word, sculpted object, pictogram, or painted image constitutes symbolic or metaphoric devices containing meaning. Such expressions in the research constitute data sources that can be structured to become subject to analysis. Narrative analysis is employed in Fourth World research as well as conventional research. The object of analysis is frequently the "story," shared everyday experiences, oral histories, and these sources may be aided by a stone marked with memory gashes or a piece of wood with marks on it, as in the *alchuringa* used by the Aboriginals in Australia to tell stories that repeat a theme or particular body of knowledge or pictograms used the Mexíca, Egyptians, and other people on walls or forms of paper or bark containing drawings that symbolize elements of a story, history, or knowledge to be retold.

Comparing

The act of comparing involves recognizing whether there are similarities and dissimilarities between objects or factors upon which one focuses attention. Comparing allows the discovery of truths that are not apparent when considering an object or factor alone. The comparison may give rise to understanding contrasts that richly enhance similarities that were not otherwise apparent. The researcher must use this approach cautiously but always recognize the idea, object, factor or changing circumstance that is not otherwise apparent without comparison.

Comparisons are useful, for example, when a researcher wishes to evaluate characteristics or a pattern of factors that may explain the effects of social or cultural changes in a society. To conduct a useful comparison, one must establish criteria for sameness, difference and irrelevance. This evaluation method can explain the political experience of a Fourth World nation seeking to prevent the destruction of their forest by comparing the experience of one or more nations with this problem to the situation of a nation that is the research focus. A comparison may reveal decisions or actions that could or should not be taken to solve the problem. As with other rationally derived conclusions, those resulting from comparison rely upon persuasion for others to use the results. The comparison relies on additional methods of inquiry to confirm observations and other tests based on the senses.

Comparison is equally useful in architecture, where useful information may be gleaned by comparing structural or design practices from ancient structures to develop a modern structure. Similarly, a comparison may amplify otherwise invisible reasons for medical conditions, agricultural failures, or successes. Simple comparisons involving only a few variables are common, but comparisons involving many are less common.

Retrodiction

This is a process of taking systematic steps back in time, considering, comparing or establishing relations between pairs or multiple factors, objects, ideas, events, stories, arts and other manifestation of physical or metaphysical manifestations to establish or predict what may happen in the future, define what is occurring in the present or explain the meaning of an event or what actually occurred in the past. Physical, economic, political, and cultural fractures from a period much before the predicted outcome can be defined by examining and drawing relationships between apparently unrelated historical, documentary, artefact, symbolic, social, and sources of information and working back in time to discover how or what may have existed that can be used to predict what can occur in the future if similar circumstances exist.

Retrodiction may be compared with "reverse engineering" or reconstructive reasoning. Understanding what occurred in the past requires careful reconstruction based on known factors that can help explain earlier knowledge. Retrodiction can be used as a single method or in combination with other methods to consider what a people may respond to climate change based on how earlier generations responded to changing climate. Similarly, food security may be examined for future policy considerations based on successful or failed experiences from the past. Retrodiction may also be used in research focusing on historical sociology, archaeology, evolutionary biology, psychology, and political change.

Analogizing, Analogic Reasoning

Analogy is an approach to thinking and explaining by comparing different factors to discover similarities as the best source of evidence to estimate unknown aspects of a new alternative. In other words, the ordinary process of associating an object with qualities "A" and B" that are shared by a second object, and the first object also has a quality "D", the second object will also share the qualities of "D." By this we mean that one may draw an analogy as a conclusion from multiple case examples organized by a question to form an answer. When there is closer agreement between alternative factors, there is a likelihood of stronger evidence supporting a conclusion. This approach is not much different from reasoning by logic (Sowa & Majumdar, 2003).

Remembering, Sharing

Memory is a powerful tool in research that may seem self-evident, but the approach is often discounted. Remembering can be triggered by an event, symbol, or social interaction where individuals share their thoughts. The capacity to remember is an essential part of research when a researcher must contend with many factors that are to be relationally evaluated or retrodicted by considering variables from the past to predict what may exist or come into being in the future. Remembering is perhaps the oldest method of providing reliable information about the past, present, and future. The method for distributing outcomes from remembering involves sharing that may be carried out through oral storytelling, dance, sculpture, drawing, and painting.

Talking Circles, Community Participation

Community is an essential ingredient in research. The best means for reconstructing or structuring social interaction (community) is to involve participants in a process that directly causes collective thought; actions and inspiration of a group can be captured. The Talking Circle is much like a focus group, but questions of a facilitator don't necessarily guide it as in a focus group. Using story and group or individual simulation recognizable to members of Talking Circles can trigger and encourage inspired observations, thoughtful analysis, production of new ideas and effective personal responses. Analysis of words and expressions generated by participants then reveals new meanings useful to the Talking Circle members and the researcher.

Dance, Song, Image Rendering

Various forms of expression need not solely depend on words (symbolically rendered or orally rendered). Observing or participating in dance, song, and construction of images can demonstrate new knowledge or explanations that may otherwise be inaccessible to the researcher. The researcher may learn much more as a participant than as an observer, so it is important to seek opportunities to become a primary participant and not merely a secondary observer.

Dance is frequently a method for revealing qualities, behaviours, and stories when oral storytelling is insufficient. Song performs the same function, as does the making of images. These methods demonstrate information and reveal the results of inquiries by lending body, voice, sounds and images as approaches to "directly" explain or show the results of what is already known. While dance, song and images play important ritualistic roles in society, they also teach and shape society's and its individuals' social and intellectual character.

Methods of Learning

Group discovery, observing, practice, relational reasoning, trial and error, mirroring (shadowing), oral literature practices, ceremonies, and institutional structures (political, religious, social) are instruments of learning within Fourth World knowledge. These approaches use the Four Directions metaphor to remember the new knowledge acquired. Investigations into learning necessarily require the involvement of the community that benefits from the learning outcomes. Consequently, conducting group learning through Talking Circles (allowing for community participation), in addition to reviews of historical information, can reveal patterns that indicate aspects of learning that are not understood. Employing comparisons, counting, and observations that can be examined in a group where critical assessments become possible beyond the understanding of a single person.

Oral expressions presented in rituals that follow a recognizable pattern can then provide the learner or learners with expansive understanding, especially when questions and counter-comments become part of the process. Learning in the ways we have described follows a natural path of discovery benefiting from the social and cultural connections of community members or small groups of learners.

Fields of Application

Fourth World thought, and the knowledge systems that undergird the ideas can be applied to stories, dance, ethics, philosophy, athletics, horticulture, agriculture, literary expressions, architecture, planning, construction, design, environment, diplomacy, governance, community and family relations, war, trade, economics, education, health, social organization, entertainment, design, music, development of technologies, and the organization and conduct of business, government, diplomacy and scientific inquiry. The applied theory and knowledge systems may be better understood with this example of applied Fourth World thought and *realpolitik* as applied to the field of Fourth World Geopolitics (Griggs, 1992; Nietschmann, 1994; Atleo, 2004; Ryser, 2012).

POLITICAL: FOURTH WORLD NATIONS

Fourth, World knowledge systems view people as self-defining. Nietschmann (1994) writes, "A people is distinguished by a common history, a common geographical location and homeland, cultural or linguistic links, religious or ideological links, racial or ethnic ties, a common economic base, and an adequate number of individuals asserting common identity." Quoting Sills (1993) and Smith (1986), Seton advances the argument that when people become politically mobilized, claiming the right to self-determination, they appear to form nations. When faced with threats from common enemies, people politically organize and mobilize against these threats. For example, the Dine (Navajo) were a people who traditionally lived in a "dispersed, non-national pattern, organized in clans as basic operational units, until they were faced with a need to form a nation under military leaders who united them to defend their lives against encroachment and genocidal attacks by Euro-American invaders" (Seton, 1999). Today, the "Navajo Nation" is officially recognized by the United States, though not as a political equal.

One field in which Fourth World thought has significant application is Fourth World Geopolitics—the study and understanding of relations between Indigenous (Fourth World) nations and the world's 194 states. In this context, early scholarship sought a conceptual framework that could aid the process of understanding rapidly changing relations between Fourth World people and the states.

Fourth World ideas are evident in diplomacy applied to the construction of relations between nations and between nations and states,[16] architectural designs, and even in the development of artificial intelligence in computers where relational reasoning is an essential component of developing program language that emulates human thought processes. The underlying conceptual framework of Fourth World Theory and its application may be better understood by considering the metaphoric organizing symbol that links human beings to the earth, to each other and the cosmos.

Nations occupying ecological niches within externally imposed state boundaries saw early demands for using a theoretical framework at the United Nations Stockholm Environmental Conference in 1972. Through collective recognition of shared realities, under the leadership of Secwepemc Chief George Manuel, the initial contours of the Fourth World were born.

Fourth World thought provides a framework that examines the relationship between Fourth World nations, states, transnational corporations, and non-governmental organizations. Rudolph C. Ryser, Bernard Nietschmann, John H. Bodley, David Hyndman, Kathy Seton, Danielle Elford, and Richard Griggs—preeminent students of Fourth World theories—used it as an analytical lens through which to test the existence of social phenomena where political and cultural changes are rapidly occurring in Fourth World communities. Seton (1999) describes Fourth World ideas as "scrutinizing how colonial empires and modern states invaded and now encapsulate most of the world's enduring nations and people." Quoting Nietschmann, she adds, "It also explores how this destructive expansion jeopardizes the world's biological and cultural diversity and ultimately rebounds to break down and break up states" (Nietschmann, 1994, pp. 225-226). Through ground-level analysis, writings, and maps, Seton (1999) continues that Fourth World rectifies the distorting and obscuring of Indigenous nations' identities, geographies and histories; "that make up the usually hidden "other side" in the invasions and occupations that produce most of the world's wars, refugees, genocide, human rights violations, and environmental destruction" (Nietschmann, 1994, p. 227).

Economy

In its most basic form, the economy is concerned with sustainability in distributing goods and services and perpetual growth in the ordinary sense. The wealth of people necessarily involves the value and benefit of land and natural life, as well as human activity's health and education capabilities. Fourth World thought allows researchers to go well beyond conventional numeric numbers to measure economy (recognizing its unsustainability in terms of human need and nature's capacity) by incorporating the total human, land, and natural life and goods and services value as a long-term sustainability emphasis. An economy can be evaluated in terms of its effective balance between human, land and natural life and the production of goods and services. Where raw materials are extracted, the economy must, under Fourth World thought, balance those extractions as reductions against increased human capacity or production of goods and services. If extractions of raw materials result in a lower quality of life (health, education), then extraction constitutes a net loss to the economy. Fourth World perspective requires consideration of all factors to support decisions that advance the quality of life.

Traditional Healing Arts and Sciences

Traditional Healing Arts and Sciences is a broad field, and Fourth World ideas are defined in the field. Studies of nutrition, community health, education, social order, and mental and physical health are all within the frame of Fourth World theoretical inquiry.

Traumatic experiences and other unexpected catastrophes affect the functioning of the whole being physically, emotionally, mentally, and spiritually. Natural rhythms are underlying the integrated functions of mind/body/spirit in humans and animals alike that pulsate, oscillate, and vibrate (Korn, 2013). Chronic disease (cancers, hepatitis, arthritis, heart disease, autoimmune disorders, and emotional imbalances) comprises more than 80% of all human health conditions. They are accessible when research is rooted in the cultural context of the chronic disease. Evident symptoms may indeed be a function of multiple layers of relationships involving individuals, their environment, their family, their friends, community structures, the air, water, and the land. Treating only the symptoms of a chronic disease can only delay the inevitable: decline and death over time. Applying multiple factors to produce a plan for prevention as well as treatment and considering the many variables ultimately allows the researcher, the medical practitioner, or the traditional medicine practitioner to assist the person struck down by chronic disease to recover. In many instances, it may become necessary to heal a community and remove the multiple variables that produce chronic conditions. Understanding the relationships between different and unrelated factors becomes essential for understanding and addressing human health questions.

ALDMEM[17] MEDIATING KNOWLEDGE AND NEGOTIATED CONSENT

Academic and research institutions are appropriate mechanisms between different societies to blend epistemological frameworks to achieve a sustainable future for all living things on the planet. That process is underway, as demonstrated by scholarly contributions to this anthology. However, there are other practical and immediate means by which Fourth World spiralist knowledge systems and the linear, rationalist/empirical knowledge system can be mediated. Fourth World scholars have developed a plan to establish an international negotiating organization called ALDMEM that can directly influence efforts to achieve a sustainable future.

The plan was developed recognizing the need to implement the United Nations principle of free, prior, and informed consent (FPIC) of Fourth World people when a people's interests may be adversely affected non-Fourth World policies and practices. Direct engagement by practitioners of different knowledge systems in the FPIC process can make sustainability in the future a reality.

According to international decisions taken by states' governments and agreed to by Fourth World nations, the process must be carried out when states or any of their corporations, academic institutions, non-governmental organizations, or multi-lateral organizations wish to take actions that affect Indigenous nations' interests: wellbeing, culture, economy, environment, or society or exercise of self-determination. States' governments acted in the World Conference on Indigenous People high-level plenary meeting of the General Assembly 2014 Outcome Statement stating that UN member states commit to implementing FPIC. In their commitments, UN member states voted to (para 3) *"cooperate in good faith with Indigenous people ... through their ... representative institutions" and to secure "approval of any project affecting their lands or territories and other resources ... through their free, prior, and informed consent."* These statements assert the intention of states to implement the UN Declaration on the Rights of Indigenous People (2007), the ILO Convention 169 (1989), and other international agreements. In addition, Indigenous nations have adopted other international laws to implement FPIC, including the 1994 International Convention on the Rights of Indigenous Nations and the Alta Outcome Document of 2013. FPIC is a necessary concession by States' Governments and Indigenous Nations to freely negotiate the consent of Indigenous nations based on "political equality." The laws agreed to by states and nations require that engagements go beyond "consultations" and directly involve the negotiation of consent in the form of agreements.

Negotiated Consent Agreements applying blended knowledge systems can involve the introduction of state-originated health programs, education institutions, economic policies, environmental policies, cultural practices, and access to ancestral territories and the resources in those territories historically relied on by nations to sustain life—matter of importance to sustainability. As important as the FPIC process is, the reality is that there is no authorized mechanism to implement the process to facilitate the identification of problems, negotiate consent, and enforce agreements. However, the ALDMEM plan offers a realistic approach to bringing all the different knowledge systems together to achieve a negotiated pathway for sustainability that benefits all of humanity.

As political equals to states, Fourth World people must sit at the same table with states, corporations, and other entities to establish a new international mechanism to promote and carry out the FPIC process—the only alternative to state-based organizations that alone have great difficulty envisaging effect sustainability. The Board Chair of the Center for World Indigenous Studies, working with the President of the National Congress of American Indians, jointly drafted the following plan for a new international institution called ALDMEM (See Figure 2). The plan is shown in the diagram below, illustrating the key elements and actions the political parties must take to implement FPIC.

Figure 2. ALDMEM for sustainable future

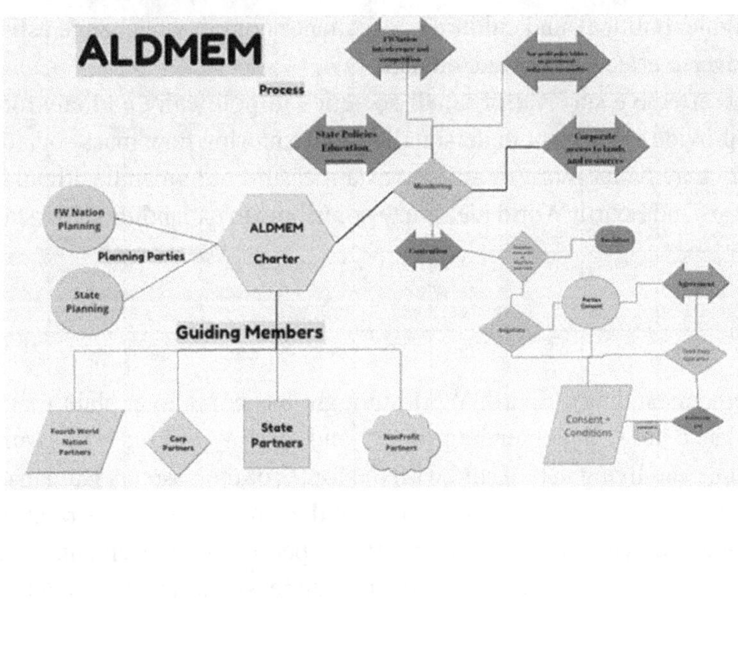

(Author)

The purpose, mission, and goals are the focus of dialogue and implementation by Fourth World nations, states' governments, corporations, academic institutions, and non-governmental organizations invited to the planning table. Recognizing the knowledge systems on which nations and states/corporations operate is essential for mediating the application of knowledge for a sustainable future. Some aspects of the epistemological frameworks between systems of human knowledge can work together and serve as the means for defining effective approaches to a mutually agreed, sustainable future.

FUTURE RESEARCH DIRECTIONS

Researchers employing Fourth World knowledge systems as a conceptual framework for economic, political, educational, health and environmental research have benefited from blending research methods to explain complex problems. Given the success of this research model where conventional sciences and Fourth World sciences work in parallel and a complementary fashion there seem to be no limits to the new directions of research.

Energy and the environmental research are two obvious problem areas that respond to human need. Understanding the relationship between the land, the people and the cosmos may open new areas of theoretical physics as well as methods for drawing on solar and thermal energy. Research in architecture may reveal new and more appropriate habitations developed before the industrial age that will have new applications in the present and future ensuring minimal adverse effects on the environment and maximum energy efficiencies.

New and extensive research in community trauma and individual trauma health can benefit significantly from Fourth World knowledge. Communities that have been torn apart by war, environmental collapse, and social upheaval must be better understood to achieve healing. Traumas in communities may contribute significantly to individual mental health stresses and traumas that may only be effectively treated within a healthy community context. Research into community trauma and into the relationship of individual trauma will respond to a profound human need.

Fourth World knowledge systems may serve the essential conceptual role where researchers examine the social, economic, political, and cultural consequences of climate change especially focusing on refugees from the adverse effects of climate change.

Research into the effective survival of small societies in politically and environmentally stressed circumstances may provide significant understanding for exploring how mass societies may be reorganized. Size of society may be far more significant as a measure of human health and productivity than nearly any other factors and Fourth World ideas may contribute to expanding understanding in this field.

CONCLUSION

Based on these understandings, Fourth World thought has come to explain that comparison, relational reasoning, balance between contending forces, and equality of kind will assure comity between people, between people and living nature, and with the forces of the cosmos. Human culture, as with the cultures of other animals, plants, rivers, mountains, and the lands, forms the structure of life and living defined by the evolving and dynamic relationship between people (animals, plants, etc.), the land and the cosmos. History, memory and thought process are multi-dimensional where instead of two-dimensional (linear past progressing to the future, fatalist, cyclical or providential thinking) thought is seven or more dimensions and successful comprehension of reality demands employing all seven dimensions.

Fourth World knowledge systems offer conceptual frameworks that embrace the intuitive and the rational or perhaps more specifically, a blend of rational reductionism and inductionism. This blending of methods may permit researchers to understand relationships of elements within an environment of chaos as well as logical and mythological problems. Researchers are encouraged by Fourth World thought as a framework to design studies that are multi-dimensional to present and explain complex problems in ways manageable at the human scale. That is the object of learning—to explore realms not immediately accessible, but to interpret and make understandable those domains and compound domains that define the physical and metaphysical realities. Ultimately the Fourth World working with other systems of knowledge will sustain life well into the future.

REFERENCES

Agrawal, A. (2004). Indigenous and scientific knowledge: Some critical comments. *Indigenous Knowledge and Development Monitor*, 3(3), 7–8.

Atleo, E. R. (2004). *Tsawalk, a Nuu-chah-nulth worldview*. UBC Press.

Bohaker, H. (2006). The significance of Algonquian kinship networks in the Eastern Great Lakes Regions. *The William and Mary Quarterly*, 63(1), 23–52. DOI: 10.2307/3491724

Bruce, H. G. (2012). *Northern lights: Fourth World Nations and the geopolitical dance in the Arctic*. SIT Graduate Institute.

Cobo, J. R. M. (1982). *Study of the problem of discrimination against Indigenous populations*. Final Report. United Nations.

Crone, E. A., Wendelken, C., van Leijenhorst, L., Honomichl, R. D., Christoff, K., & Bunge, S. A. (2009). Neurocognitive development of relational reasoning. *Developmental Science*, 12(1), 55–66. DOI: 10.1111/j.1467-7687.2008.00743.x PMID: 19120413

Elford, D. (2002). Conservation by Self-Determination in Central America. *Fourth World Journal*, 5(1), 98–149.

Emeagwali, G. (2003). *African Indigenous knowledge systems (AIK): implications for the curriculum*. African World Press.

Fenstad, J. E., Hoyningen, P., Hu, Q., Kokwaro, J., Salick, J., Shrum, W., & Subbarayappa, B. V. (2002). *Report from the ICSU Study Group on Science and Traditional Knowledge*. International Council for Science.

Fleischhacker, S., Ries, M., Vu, A., & McPhail, A. (2011). Engaging tribal leaders in an American Indian healthy eating project through modified talking circles. *Family & Community Health*, 34(3), 202–210. DOI: 10.1097/FCH.0b013e31821960bb PMID: 21633212

Fowler, J. (2011). The concept of indigeneity: Can the Declaration on the Rights of Indigenous People be understood within western liberal philosophy? *Queensland Law Student Review*, 4(1), 35–46.

Gehl, L. (2012). Debwewin journey: A methodology and model of knowing. *Alternative*, 8(1), 53–65. DOI: 10.1177/117718011200800105

Goel, V., & Dolan, R. J. (2001). Functional neuroanatomy of three-term relational reasoning. *Neuropsychologia*, 39(9), 901–909. DOI: 10.1016/S0028-3932(01)00024-0 PMID: 11516443

Griggs, R. (1992). *Background on the Term "Fourth World"*. DayKeeper Press. Olympia, WA USA, Center for World Indigenous Studies. Occasional Paper #18.

Griggs, R., & Hocknell, P. (1996). *The geography and geopolitics of Europe's Fourth World*. IBRU, Boundary and Security Bulletin.

Hall, A. (2003). *The American empire and the Fourth World: the bowl with one spoon*. McGill Queen's University Press.

Hart, M. A. (2010). Indigenous worldviews, knowledge, and research: The development of an Indigenous research paradigm. *Journal of Indigenous Voices in Social Work*, 1(1), 1–16.

Haskell, D. L. (2008). Tarascan kingship the production of heirarcy in the Prehispanic Patzcuaro Basin, Mexico. PhD Doctoral, University of Florida.

Hipwell, W. T. L. (1997). Industria, The Fourth World, and the question of territory. *Middle States Geographer*, 30, 1–10.

Hummel, J. E., & Holyoak, K. J. (2005). Relational reasoning in a neurally plausible cognitive architecture an overview of the LISA Project. *Current Directions in Psychological Science*, 14(3), 153–157. DOI: 10.1111/j.0963-7214.2005.00350.x

Kaya, H. O., & Seleti, Y. N. (2013). African Indigenous knowledge systems and relevance of higher educaiton in South Africa. *The International Education Journal: Comparative Perspectives*, 12(1), 30–44.

Korn, L., Logsdon, R. G., Polissar, N. L., Gomez-Beloz, A., Waters, T., & Ryser, R. (2009). A randomized trial of a CAM therapy for stress reduction in American Indian and Alaskan Native family caregivers. *The Gerontologist*, 49(3), 368–377. DOI: 10.1093/geront/gnp032 PMID: 19377083

Korn, L. E. (2023). *Rhythms of recovery: integrative medicine for PTSD and complex trauma*. Routledge. DOI: 10.4324/9781003231455

Landy, D., & Goldstone, R. L. (1991). Relational reasoning is in the eyes of the beholder: how global perceptual groups aid and impair algebraic evaluations. *Proceedings of the Annual Meeting of the Cognitive Science Society*, 27, 2509.

Lyman, R. L., & O'Brien, M. J. (2001). The direct historical approach, analogical reasoning, and theory in Americanist archaeology. *Journal of Archaeological Method and Theory*, 8(4), 303–342. DOI: 10.1023/A:1013736416067

Manuel, G., & Posluns, M. (1974). *The Fourth World: an Indian reality*. Con Mills, Collier-Macmillan.

Martinez, M. A. (1997). *Study on treaties, agreements and other constructive arrangements between states and Indigenous populations: Final Report*. United Nations.

McDonough, K. S. (2010). *Indigenous experience in Mexico: readings in the Nahua intellectual tradition. PhD Doctoral*. University of Minnesota.

McGuire, P. D. (2010). Exploring resilience and Indigenous ways of knowing. *A Journal of Aboriginal and Indigenous Community Health*, 8(2), 117-131.

Nietschmann, B. Q. (1990). Conservation by conflict in Nicaragua. *Natural History*, (November), 42–49.

Nietschmann, B. Q. (1991). Nietschmann rejoings: Conservation by self-determination. *Research & Exploration*, 7(3), 373.

Nietschmann, B. Q. (1994). The Fourth World: nations versus states. In Wood, D. J. D. A. W. B. (Ed.), *Reordering the world, geopolitical perspectives on the twenty-first century* (pp. 226–242). Westview Press.

Nsamenang, A. B. (2006). Human ontogenesis: An Indigenous African view on development and intelligence. *International Journal of Psychology*, 41(4), 293–297. DOI: 10.1080/00207590544000077

Owusu-Ansah, F. E. & Mji, G. (2013). African Indigenous knowledge and research. *African Journal of Disability*, 2(1), Art. #30 5 pages. .DOI: 10.4102/ajod.v2i1.30

Paredez, D. M. (1964). *Hunab Ku síntesis del pensamiento filosófico maya*. Orion.

Poonwassie, A., & Charter, A. (2001). An Aboriginal world view of helping: Empowering approaches. *Canadian Journal of Counselling/Revue. Canadian Counsellor*, 35(1), 63–73.

Radhakrishnan, S. (1929, 2008). *An idealist view of life*. Aristophanes Press - archINFORM.

Rodriguez, R. C. (2010). *Amoxtli- The Xodex in Lak Ech-Panche Be and Hunab Ku and the forgotten 1524 debate*. Eagle Feather Research Institute.

Ryser, R. (1985). Fourth World Wars: Indigenous nationalism and the emerging new international political order. In Boldt, M., & Long, J. A. (Eds.), *The quest for justice: Aboriginal people and Aboriginal rights*. University of Toronto Press. DOI: 10.3138/9781442657762-028

Ryser, R. (1994). *International covenant on the rights of Indigenous nations*.

Ryser, R. (1998). Observations on 'self' and 'knowing. In Wautischer, H. (Ed.), *Tribal epistemologies* (pp. 17–29). Ashgate.

Ryser, R. (2011). Fourth World geopolitics. H. G. Bruce. Unpublished.

Ryser, R. (2012). *Indigenous and traditional knowledge. ecosystem management and sustainability*. Berkshire Publishing Group.

Ryser, R. (2012). *Indigenous nations and modern states: the political emergence of nations challenging state power*. Routledge. DOI: 10.4324/9780203139882

Ryser, R. (2015). The Anáhuac knowledge system: A dialogue between Toltecs and Descartes. *Fourth World Journal*, 14(1), 31–47.

Ryser, R. (2022). We are people first. E. Barrett Ristroph (ed.), *Who is Indigenous: perspective on indigeneity, identity, dispossession, and transition*. Ristroph Law, Planning, and Research.

Seton, K. (1999). Fourth World Nations in the era of globalisation an introduction to contemporary theorizing posed by Indigenous nations. *Fourth World Journal*, 4(1), 49–69.

Sinclair, N. J. (2013). *Nindoodemag Bagijiganan: a history of Anishinaabeg Narrative. PhD*. The University of British Columbia.

Smith, A. D. (1986, 1987). *The ethnic orgins of nations*. B. Blackwell.

Smith, L. T. (1999, 2012). *Decolonizing methodologies: research and indigenous people*. Zed Books.

Son, J. Y., Smith, L. B., & Goldstone, R. L. (2010). Connecting instances to promote children's relational reasoning. *Journal of Experimental Child Psychology*, 108(2), 260–277. DOI: 10.1016/j.jecp.2010.08.011 PMID: 20926099

Sowa, J. F., & Majumdar, A. K. (2003). Analogical reasoning. *International Conference on Conceptual Structures,* Dresden, Germany. Springer-Verlag.

Stenning, K., & Lambalgen, M. V. (2012). *Human reasoning and cognitive science*. MIT Press.

Stone, C. L. (2004). *In place of gods and kings: authorship and identity in the relación de Michoacán*. University of Oklahoma Press.

Tempels, P. (1952, 1959). *Bantu philosophy*. Présence Africaine, Victor Petrus Keuppens.

UNGA. (2007). United Nations Declaration on the Rights of Indigenous People. GA 61/295 (A/61/I.67 and Add. 1). U. N. G. Assembly. United Nations.

Vela, R. C. (2015). Fourth World People: Continuity, philosophies and sources. *Fourth World Journal*, 13(2), 21–28.

Yeo, A. W., Zaman, T., & Kulathuramaiyer, N. (2013). Indigenous knowledge management in the Kelabit Community in Eastern Malaysia. *International Journal of Sociotechnology and Knowledge Development*, 5(1), 23–36. DOI: 10.4018/jskd.2013010103

Zauditu-Selassie, K. (2007). Women who know things: African epistemologies, ecocriticism, and female spiritual authority in the novels of Toni Morrison. *The Journal of Pan African Studies*, 1(7), 38–57.

KEY TERMS AND DEFINITIONS

Blended Research: Systematically employing Fourth World research methods with positivist quantitative and qualitative methods.

Cosmogram: A geometric figure depicting a cosmology - a study of the universe's beginnings and its changes.

Culture: The dynamic and evolving relationship between a people, the land, and the cosmos.

Four Directions: A concept noting the rising and setting sun and the earth's poles connecting celestial events with events in human experience and the earth.

Knowledge System: A conceptually constructed body of ideas, observations, and methods for comprehending through understanding and intuition.

Metaphor: A form of speech, observation or other symbol that combines unrelated objects, practices and/or concepts into a singular image (as in graphic), expression (as in story) or physical action (as in dance).

Science: A systematic construction, ordering and expression of knowledge.

ENDNOTES

1. The Maya-speaking people of the Yucatan Peninsula and the region of what is now southern Mexico and Guatemala characterized their way of thinking, conceiving knowledge as the Hunab Ku. This mode of thought comes from the infinite repetition of four spiraling arcs, indicating the intercon-

nectedness of all life, the four cardinal directions, the four arcs of time and perpetual movement in all directions through space. The Mayan conception is reflected in the knowledge systems of people worldwide. For more on this subject, see Ryser (1998).

2. Author's personal communications with Grand Chief George Manuel.

3. So-called "decolonized" populations throughout Asia, Africa, the Pacific, Melanesia, East Asia, Eastern Mediterranean, Caribbean, and elsewhere were essentially "recolonized" by the leadership of an "installed" (by the former colonizer) nation or by the most powerful nation resulting in a single nation or faction of a nation ruling over other nations now held within the jurisdiction of the new state. Burma, Syria, India, Nigeria, Namibia, Rwanda, Iraq, Afghanistan, the former Yugoslavia, and Indonesia are decolonized states that essentially recolonized nations within newly established state boundaries.

4. To denote that during the decolonization period following 1948 many states were formed with an Indigenous nation controlling the central government that subsequently continued colonial policies over other nations within state boundaries. Burma/Myanmar, Nigeria, Namibia, Afghanistan, Iraq are notable examples of "recolonization".

5. The International Working Group on Indigenous Affairs (IWGIA) in Denmark, Survival International in the United Kingdom and Center for World Indigenous Studies (CWIS) in the United States of America have documented more than 5000 distinct people in the world though the popular press does not regularly report about them.

6. Fourth World nation-oriented NGOs and representatives of Fourth World communities and nations began participating in international meetings of states concerning the Bio-Diversity Convention, Intellectual Property Rights, the UN Working Group on Indigenous Populations, United Nations Convention on Climate Change, International Labor Organization and other regional meetings of states on a wide range of topics. These initiatives called on Indigenous interlocutors to prepare language explaining their policy recommendations and sometimes detailed explanations of the recommendations. Representatives began adopting the language of these meetings and searching for a language that better explained their understanding of problems and solutions from their community or nation's perspective.

7. Notably Edward Hallett Carr (1892 – 1982), who saw his historiographic approach to the theory as providing the foundation for a new international order, Hans J. Morgenthau (1904 – 1980), with his theories rooted in realism asserted that "statism" describing "nation states" as the main characters in international relations excluding roles for non-state actors. Robert O. Keohane (1941 -) is associated with neoliberalism and advocates the principle that states must be primarily concerned with absolute gains over the interests of other states.

8. Representatives of Indigenous advocacy organizations such as the Philippines-based Indigenous People International Centre for Policy Research and Education (TEBTEBBA), Forest People Programme in the United Kingdom, Inuit Circumpolar Council (ICC) with offices in Russia, United States of America, Canada and Greenland joined by Indigenous scholars and political leaders have actively pursued a policy of installing *respect for Indigenous knowledge* language in the Climate Change Treaty and Convention on Bio-Diversity for example.

9. This expression is meant to suggest movement between communities.

10. The Kongo people are a Fourth World nation spread across several Central African countries. In the Democratic Republic of Congo, the Kongo people primarily inhabit the western and southwestern regions of the country, including the provinces of Kwilu, Kwango, and Bas-Congo. In Angola, the Kongo people primarily inhabit the northern and western regions of the country, including the provinces of Zaire, Uige, and Bengo. In Congo-Brazzaville, the Kongo people primarily inhabit the southern and western regions of the country, including the departments of Kouilou and Pointe-Noire.

11. Assiniboine, Anishinabe, Lakota, Shawnee, Haudenosaunee, Xochicuautla, Nu-chal-nulth, Sámi, Igbo, Kongo, Yoruba, Lithuanian, Gaels, Kurds, Bedouin, Euskal, Mapuché, Purépecha, Dani, Ainu, and the Nur are among the many nations around the world where the four directions or the encircled symbol constitutes a valued mnemonic. One's memory can be triggered by symbolic information in the four directions concerning stages of life (birth, adulthood, death,); the seasons of spring, summer, autumn and winter (represented as well by the equinox); the qualities of life including intellectual, physical, emotional, spiritual, and relational; natural elements of fire, air, water and earth; the kinship animals of turtle, eagle, bear, deer, buffalo, whale, among others; and plant life, corn (in all its varieties), camote, chia, sweetgrass, sage, cedar, tobacco, and various grains.

12. The Mayan Calendar (Tzolk'in) is one example of the four directions embedded and defining a calendar. The calendar defines the east, north, west, and south with colours identifying four colours of corn, directions and elements (red – North - air, yellow – East - water, blue – South – earth, and white – West - air).

13. In Chinese and neighboring people' symbology these are the Dragon (East), Vermillion Bird (South), White Tiger (West) and Black Turtle (North) representing the direction and season with specific and individualized characteristics that have material and immaterial significance. The symbols may vary from people to people in China, Korea, Vietnam, and Japan and there may be added symbols (for the center providing a fifth element) as in the Yellow Dragon of the Center (earth).

14. Texcatlipocas is a plural form of Tezcatlipoca, a central deity in Aztec religion. He is associated with a variety of concepts, such as the night sky, hurricanes, obsidian, and conflict. He was considered one of the four sons of Ometecuhtli and Omecihuatl, the primordial dual deity https://en.wikipedia.org/wiki/Tezcatlipocahttps://www.worldhistory.org/Tezcatlipoca/.

Tezcatlipoca is often translated from the Nahuatl as "smoking mirror." It alludes to his connection to obsidian, the material from which mirrors were made in Mesoamerica. They were used for shamanic rituals and prophecy, and as such Tezcatlipoca is additionally associated with divination. Some possible synonyms for Tezcatlipoca are: • Smoking Mirror: a literal translation of his name that emphasizes his link to obsidian and vision. • Black Tezcatlipoca: one of his manifestations as the god of the north and of Texcocohttps://www.worldhistory.org/Tezcatlipoca/. • Hummingbird Sorcerer: another name for his manifestation as the god of Tenochtitlan and the patron of Huitzilopochtli

15. In LandofKam's Blog Derric "Rau Khu" Moore discusses this in "Kongo Cross/Cosmology on Colonoware (retrieved from https://landofkam.wordpress.com/tag/kongo-cross/page/2/ December 2014).

16. This is demonstrated by the application of principles from the Haudenosaunee Great Law. Based in the Great Law the Haudenosaunee, Abenaki, Lenape, Shawnee, Wyandot and Anishinabek among many other nations observed three basic principles of diplomacy: Gáiwoh (righteousness), Skénon (health), and Gashasdénshaa (power). These are applied in direct dialogue and negotiations. The first principle of Gáiwoh when practiced properly requires that human beings in society and between nations be just and their relations must be balanced and just. The second principle of Skénon requires a soundness of mind and body and the establishment of peace (becoming one heart, one heart, one body and becoming one people). Finally, the third principle of Gashasdénshaa affirms the centrality of law and custom backed by such force as is found to be necessary to ensure that Gáiwoh prevails. This is the highest and most practical form of diplomacy and means to bring about political equality."

17. Ancestral Lands Decolonization, Monitoring, negotiating and Enforcement Mechanism

Chapter 3
Harvesting Indigenous Onto-Epistemologies for Contributing to a More Balanced and Sustainable Future:
Exploring Options for Performing Better Worlds

Norma Ruth Arlene Romm
https://orcid.org/0000-0002-1722-9720
University of South Africa, South Africa

ABSTRACT

This chapter outlines the interpretation of the contours of Indigenous onto-epistemologies (existing prior to colonization by the West) as proffered by various sages and scholars, including the relevance of these for the current era. It focuses on the axiological commitments towards strengthening the relational balance across human and more-than-human existence. It indicates implications for enabling learning exchanges between Indigenous and Western ways of doing science, but with the proviso that we reflect critically on the values that the science and its products are serving (wittingly or unwittingly). It contends, drawing on and extending the work of a host of Indigenous authors, that the quest for supposed value freedom in the practice of science is a chimera and that our knowing processes are necessarily value-invested. This starting point prompts us to experiment further with ways of rendering the process of world-making (locally and globally) more inclusively beneficial and sustainable.

INTRODUCTION

Before I begin my discussion in this chapter, I should specify that as far as my positionality is concerned, I am English speaking South African – classified as White in terms of Apartheid South Africa's racialized categorizations. These classifications were introduced officially in 1948 by the Nationalist government to formally structure our social, economic and political relations along the lines of racialized

DOI: 10.4018/979-8-3693-8447-3.ch003

groupings. However, in my orientation, I sympathize with an Indigenous approach to understanding our human-to-human and other connections in the web of life, as expressed in the African concept of *Ubuntu* (translated as I am because we are), as well as in other concepts, such as the Andean concept of *Sumak Kaway* or *Buen Vivir* (translated as living well) or the Native American (Lakota) expression of *Mitákuye Oyás'iŋ*, which points to our connection with "all our relations".

In terms of my Indigenous-oriented self-identification, I espouse the values which I detect from numerous authors' writings about Indigenous understandings of relationality (as outlined in, for example, Romm 2015, 2017, 2018, 2020, 2021, 2024). These include values which I draw out from a myriad of in-person interactions with people Indigenous to Africa in particular, in and outside of academia, as well as from intercultural contact with various Native Americans and Latin Americans.

With reference to Africa, I concur with interpretations of Ubuntu expressed by sages and scholars offering an ontology and epistemology (onto-epistemology) which is grounded in an understanding (and sense) of our being part of a web of relations that comprise human and more-than-human worlds – which cannot be separated as they are interdependent. When people's being-in-the-world is infused with Ubuntu, the strengthening of their interrelations is sought. As Molefe (2014, p. 29) puts it:

> The fact that all being and reality is interrelated in the African traditions is not merely a descriptive fact but also one which entails a normative load. The normative package flowing from this view of reality is that the interrelatedness of being is a state of being we ought to strive to maintain (my emphasis).

The African conception of connectedness thus contains an axiology – a value orientation to create an increasing relationality and to resist non-relational dynamics built on dominance and the attempt to exercise control (instead of working towards a relationship of reciprocity). In the course of this chapter, I explain in detail what this implies in terms of resisting the hegemony of "modern" (Western) science with its way of relating to, rather than with, the world (human and more-than-human) which it helps to form. I use the phrase more-than-human in keeping with a non-anthropocentric view that does not centralize humans in relation to the rest of nature. (e.g., Chilisa, 2020, p. 24; Inoue, Ribeiro, Goncalves, Basso & Moreira, 2023, p. 3; Mabunda & McKay, 2021, p. 376; McIntyre-Mills, Lethole, Makaulule, Wirawan, Widianingsih & Romm, 2023, p. 781.) As Addae and McIntyre-Mills elucidate, a non-anthropocentric viewpoint implies that we "extend solidarity with nature and recognize that people are people through connecting with others and the environment" (2022, p. 57). This resonates with the African philosophy of Ubuntu and has implications for generating more sustainable connections with "nature", of which indeed we are a part.

Ngara, referring to the concept of Ubuntu, points out that it is "found in diverse forms in many societies throughout Africa. She points to the etymology of the word, where NTU refers to the "dynamic life force or power responsible for all of creation" (2017, p. 343).

In similar fashion to Molefe, she notes that:

> From a NTU perspective, the overriding focus of life is to be in harmony with the forces of life. Harmony implies living life – not [a quest to] controlling life. It means steering life while recognizing that there are other forces that will, in part, determine the vehicle and the direction of travel. When one is in harmony [developing a reciprocal relationship] they are NTU. (2017, p. 344)

Ngara interprets that "a person in Indigenous African cosmology is perceived as interdependent, interrelated and interconnected with all of creation" (2017, p. 344). Like Molefe, she notes that this cosmology goes hand in hand with a value system: it "calls on us to constantly repair, renew and transform the world, making it more beautiful and beneficial than when we inherited it" (2017, p 345). More importantly, we cannot separate ourselves, as supposedly distant knowers, from *influencing the web of relations in the course of our "knowing" processes.* As she states (citing Capra, 1975), "these relations always involve the observer" (p. 349). That is, the relations in the complicated web of existence include ourselves as knowers, who need to be acknowledged as *participants in world shaping as we engage with the world which we help to form.* I draw out the implications of this participative, performative, view of knowing in this chapter. (The use of the term performative implies that we perform (shape) the world as part of our involvement therein – cf. Romm, 2021.)

When discussing Ubuntu as a worldview, some scholars (Indigenous and non-Indigenous alike) focus more on the interpretation of Ubuntu as implying a relationship between people as *humans*, without stressing our relationships with *all of creation*. In Romm and Lethole, we argue that although the human focus of Ubuntu can be invoked by the expression "I am because we are", the "we" can be widened to include the whole community of interdependent relations (extending beyond geographical and national boundaries), of which nature too is seen as part of the community. We explain how the African view of totems, which suggests that people share an identity with a totem (such as an animal, a river, a lake, a tree, etc.) implies a non-anthropocentric approach (Romm & Lethole, 2021, p. 87). It is this non-anthropocentric approach to feeling ourselves as closely knitted with other life forms that ties together the numerous Indigenous worldviews as advanced across the globe. Hence, it can be said that we can locate a common thread of Indigenous thinking/being as lying in an appreciation of relationships/connectedness across all manifestations of existence (cf. Adyanga & Romm, 2022; Cajete, 2000; Chilisa, 2017, 2020; Escobar, 2016; Harris & Wasilewski, 2004; Moreton-Robinson, 2013; Ritchie, 2015; Rosiek et al., 2020; Ryser et al., 2017; Smith, 1999;Wilson, 2008). In outlining core features of Indigeneity, Harris and Wasilewski point to the importance of relationality in the

> …profound sense that we human beings are related, not only to each other, but to all things, animals, plants, rocks—in fact, to the very stuff the stars are made of. This relationship is a kinship relationship. Everyone/everything is related to us as if they were our blood relatives. We, thus, live in a family that includes all creation, and everyone/everything in this extended family is valued and has a valued contribution to make. (2004, p. 492)

They also make the point that while Indigenous people who uphold and practice Indigenous values are not averse to entertaining new ideas as part of the process of generating "new energy" in the world, it is also important to hold onto "core values" when interacting with others – they feel that this is the only way for Indigenous people to survive (p. 490). One of these core values is indeed the importance of admitting a relational worldview (where we value relationships and their strengthening) and a way of knowing where ideas can be shared and generated in processes of dialogue. But they qualify that "globalization in its present form, however, is more about imposition, about forced participation, than an open invitation to participate [in genuine dialogue] (p. 500). This has implications for the relationship between Western-oriented styles of knowing and Indigenous-oriented accounts of what is involved in knowing together as a collective process, with reference also to respecting "all of creation" (or "all our relations" in Lakota terminology).

As Chilisa notes, the global context of knowledge formation thus far has still for the most part relegated Indigenous epistemologies to "junk status" (in academic as well as everyday discourses) through "narratives of 'backward', 'irrelevant' superstitious, and inferior knowledge which is 'less' than Western Science" (2017, p. 813). Or as Dey adds, the Euromodern systems of knowledge production have privileged the myth of a "non-situated ego" as the center of knowledge construction (an ego bereft of values and of sociocultural influences) as the route to true knowledge generation. This is opposed to what are seen as culturally embedded and oft- times superstitious claims in Indigenous contexts. Hence, Western philosophy and science have produced a myth about their seeking a "truthful universal knowledge" to which their scientific knowing processes have better access (more so than in other knowledge systems which therefore are interiorized). Dey cites Gordon (2019) as pointing, *inter alia*, to "invisibility linked to voice and invisibility linked to knowledge" (Dey, 2022, p. 3).

With this dominant vision of truth-making for the most part ruling the day, how can we define the contribution of IKS not only locally but also on the world arena as a needed (indeed crucial) contribution? How can a genuine dialogical encounter be established between different systems of knowing such that learning across them might be possible? Should we simply propose that each knowledge system has the right to exist as a system in its own right – or should we attempt to find a route to assessing knowledge systems which may contribute best to our quest to create a better society (as Ngara, 2017 indicates, so that we can contribute to creating a more beautiful world in terms of social and ecological justice concerns). Do we need – as part of a global project – to update our approaches to knowledge generation so that we (those concerned) can choose more responsibly which values we wish to live by? This would mean that we recognize that knowing is always linked to what is valued, as indeed is stressed by those propounding Indigenous knowing as admittedly linked to practical ways of living.

My objectives in this chapter are to:

- Unpack what it might mean to integrate Western-oriented processes of science with Indigenous ones without relegating the latter to an inferior status – on the contrary, recognizing the possibilities that IKS provide for helping us out of current crises of our times.
- Revisit the epistemic hegemony/dominance of Euro-American thought in educational systems that have functioned primarily to serve the needs of global capitalism and that have led to rendering irrelevant for the current era the value of IKS and its options for forwarding (and performing) a better world.
- Consider what Indigenous ways of knowing-and-being as built on relationality imply for new economic practices (new in relation to the dominant understanding of economics as built on a conception of a selfish "homo-economicus") – with implications for more inclusive and sustainable living.

With this in mind, I firstly turn to a discussion around Escobar's (2016) conception of the import of *Epistemologies of the South* (named as such by Santos, 2015). *The South* in this context is not intended to refer solely to a geographical region in the Southern hemisphere.

Rather, as Santos states,

> … it is a South that exists [also] in the geographic North (Europe and North America), in the form of excluded, silenced and marginalized populations, such as undocumented immigrants, the unemployed, ethnic or religious minorities, and victims of sexism, homophobia, racism and Islamophobia. (2016, p. 19)

Epistemologies of the South as a concept and framework enables us to appreciate what Escobar (2016, p. 13) calls "other knowledges, other worlds" – other than those purporting to offer knowledge of a single reality existing "out there" for discovery (and attendant control). I elaborate on this in the next section in terms of implications for advancing social and ecological justice.

BACKGROUND: A RELATIONAL EPISTEMOLOGICAL APPROACH AS A ROUTE FOR RENDERING SOCIAL AND ECOLOGICAL EXISTENCE MORE BEAUTIFUL AND INCLUSIVELY BENEFICIAL

Escobar indicates that what is revealed when using the framework of *Epistemologies of the South*, is that our knowing in relation to the worlds which we inhabit "do not require the divide between nature and culture in order to exist – … they exist [as they do] only through the way they are enacted by practices that do not rely on such a divide" (Escobar, 2016, p. 18). He argues that a problem arises when we, as humans try to separate ourselves from nature seen as an inert "thing" to be controlled, by converting "things" into nature and "nature" into resources (2016, p. 18). Escobar laments that alternative ways of thinking about and being-in-the-world then become seen as lacking (inferior) or are "rendered utterly illegible and invisible" (2016, p. 13). And so, "One World" thinking expresses its dominance across the globe. A focus on *Epistemologies of the South* brings forth, legitimates, and shows the value of "other knowledges, other worlds" (2016, p. 13). Hence, Escobar suggests that:

> The Epistemologies of the South framework provides workable tools for all those of us who no longer want to be complicit with the silencing of popular knowledges and experiences by Eurocentric knowledge, sometimes performed even in the name of allegedly critical and progressive theory. The Epistemologies of the South might also be useful to those who have been at the receiving end of those colonialist categories that have transmogrified their experiences [and] translated them into lacks. (2016, p. 13)

He clarifies (2016, p. 17) that *Epistemologies of the South* when regarded not as lacking, but indeed as offering a way of understanding our deep connection with creative life forces, suggest that the way to know the world is by "being alive" to the relational world (2016, p. 18). This requires a recognition (intuitively felt) that living beings of all kinds constitute each other's conditions of existence and need to be cherished as such (2016, p. 18).

However, the (Western) ways of knowing associated with the advent of industrialization and attendant colonization went hand in hand with a controlling stance in relation to other beings (human and more-than-human) – as also pointed out by, for example, Monnet and Arnsberger (2022). In terms of this approach, workers in capitalist modes of production became seen primarily as (exploitable) labor within and between countries (with racialized workers being super-exploitable); and nature became seen as what Moore aptly calls "cheap nature" (2019, p. 249) – all for the benefit of profit making. Escobar (2016) gives an example:

> Such is the case with industrial shrimp farming schemes and oil palm plantations for agro-fuels, which have proliferated in tropical regions in many parts of the world, often built at the expense of mangrove and humid forest lands, with the aim to transform them from "worthless swamp" to agro-industrial complexes (Ogden, 2010). Here, … [we see the] effacing of the life-enabling materiality of the entire domains of the inorganic and the non-human, and its treatment as "objects" to be had, destroyed, or extracted; and linking the forest worlds so transformed to "world markets" for profit. (2016, p. 19)

By operating in terms of the One World philosophy (treating people and "nature" as controllable in terms of profit potential) we "mess up" the world on social and environmental levels – far from bringing progress in terms of overall human and planetary wellbeing (Escobar, 2016, p. 14).

Escobar's concerns about the silencing of Indigenous approaches to knowing resonate with those of Cajete, who argues that the naming of IKS as *science* means that we are inclined to take it more seriously (2020, p. 1). His 2020 article on Indigenous science is based on his keynote address presented at an international conference held in Taiwan (2019) on climate change, Indigenous resilience and Local Knowledge Systems. The purpose of the presentation was to challenge the audience to "initiate new thinking about how ... Indigenous science .... may be integrated with appropriate forms of Western science in new and creative ways that serve to sustain and ensure survival [and wellbeing]" (2020, p. 1).

Cajete states that in contrast to the "high context" orientation of Indigenous science, Western science perceives from a "low-context" view, reducing context to a minimum with a focus on "material objectivity [the idea that we can understand without the involvement of our values], either-or logic [the idea that "things" are either one thing or another], and reproducibility [the idea that we must try to find universal patterns free of dependence on local contexts]" (2020, p. 2). It is notable that he suggests that Indigenous science can be integrated with *some forms of Western science – but not all forms*. I will examine this statement further below by unpacking what may be considered as appropriate forms of Western science, with the emphasis on helping to serve an inclusive social and ecological justice agenda.

Cajete makes the point, as do other Indigenous scholars to whom I have referred above, that what is distinct about Indigenous knowledge is that it is:

> ...guided by spirituality, ethical relationship, mutualism, reciprocity, respect, restraint, a focus on harmony, and acknowledgment of interdependence. This knowledge is integrated with reference to a particular people and "place" toward the goal of sustainability and perpetuation of culturally distinct ways of life through generations. (2020, p. 2)

This concurs with Ngulube and Onyancha's statement – as cited by Ngara (2017, p. 335) – that the knowledge (and knowing practices) of Indigenous communities "has roots in a particular community and is situated in a broader social context in which all life forms are the result of interaction between social and spiritual relations". This relational world is dynamic and likewise our knowing (as part of this process) too has to be constantly adapting to changing local dynamics (Ngara, 2017, p. 335). This implies that relational balance is not always manifest and has to be striven for, including on a social level where (more) balance can and should be sought. For instance, Chilisa argues that when examining Indigenous cultures (indeed any culture), we should not treat them as static. As far as researching Indigenous cultures go, we should include the:

> ...voices and knowledge systems of the subgroups within Indigenous essentialized cultures [insofar as we might reify them] that are potentially excluded within the already marginalized Indigenous cultures and research paradigms. (Chilisa, 2017, p. 824)

In other words, we must not contribute as researchers to ratifying marginalization as experienced in social life in terms of social categorizations such as gender, ethnicity, etc. With regard to relational principles when studying Indigenous cultures, we should facilitate keeping alive their dynamic and evolving character, ideally evolving to generate more balance (cf. Collins, 2000, p. 261; Omodan, 2024, p.124; Romm, 2017, p. 32). The principle of harmony implies that balance in relationships (human-to-human and human to more-than-human) is the guiding ethic, which directs us as knowers and actors to encourage more harmony/reciprocity in cultural repertoires for action while appreciating cultural diversity.

In the current era, Cajete suggests that education for sustainable development, which can include some integration of Indigenous and Western forms of science, must aim to "maintain cultural diversity, protect human health, create sustainable economic relationships, reconcile social issues non-violently and most essentially protect the environmental life support system" (2020, p. 3). He adds that education systems which are culturally responsive (in the sense of recognizing the worldviews offered through Indigenous science) "require the inclusion of Indigenous ways of knowing on an equal par with modern Western science" (p. 3).

Nevertheless, although he speaks of Western science and IKS being treated preferably on an equal par, he also recognizes the drastic consequences of a "science" (Western-oriented) that treats the natural world as "resources". He emphasizes that when practising Indigenous science, "plants, animals, and the natural world are not viewed as resources but as valued relatives that have the right to exist and be cared for responsibly" (2020, p. 3). Like other Indigenous authors, Cajete's focus is on highlighting, and thereby imploring us (his audience) to revitalize, an Indigenous ethic of "mutual reciprocal relationship and responsibility [between humans] towards one another and the natural world". He sees this as providing a "foundation for creating different kinds of educational, leadership, and social-economic activities that strengthen community while simultaneously mitigating the challenges of climate change for all" (2020, p. 3).

This orientation, however, clearly requires a move away from capital-driven uses of science, where research is all-too-often accompanied by damaging the sensitive balance of a myriad of forces by isolating variables that are likely to serve the interests of capital for profit. The massive amount of research dedicated to creating artificial fertilizers, pesticides, and GMO seeds for the mega-corporations involved in industrial agriculture is a case in point. The "scientific" research examines the effect of fertilizers without considering its potential damage to the structure of the soil and the organic and inorganic matter that makes it up, thus degrading the delicate balance (this is not part of the research process); it examines the effect of pesticides in killing certain pests, without considering the potential problem that pests might develop resistance and that what Shiva (2022) calls superpests then arise (this is also not included in the time frame – nor indeed is the damage to health in terms of air, soil, and water pollution caused by the pesticides); the research examines the effects of GMOs on avoiding certain plant diseases, without considering (and indeed riding roughshod over) the manner in which farmers have created ways of developing a diversity of seeds and sharing their knowledge about this, such that they can compare their resilience in different circumstances (Wright, 2014). The "expert" science is geared to removing this diversity and at the same time renders farmers at the mercy of the seeds supplied by the mega-corporations through what Shiva terms their "poison cartel" (2022). The science is presented to local farmers as expert knowledge grounded in superior methods of understanding, while disregarding their community experiences and continued experimentation done in communities in working together with their environment to render themselves resilient. (Of course, this resilience becomes all the more difficult in the face of the destructive consequences of climate change not of their making.) Speaking about the underlying values motivating "sustainability science", Popa, Guillermin and Dedeurwaerdere aver that

> ...without prior clarification of ... values (e.g., social justice, intergenerational equity, intrinsic value of nature, etc.) sustainability research risks confusion or ambiguity in its choice of methods, criteria of performance and normative orientation. (2015, p. 50)

From a relational perspective, the Western-oriented scientific way of isolating variables without taking seriously the ways of knowing of local people and their mode of living with the variety of Earth beings, can be viewed as pseudoscience (Shiva, 2012, 2022; Wright 2014; Stephens, Taket & Gagliano 2019). Far from it being superior to Indigenous science (and being supposedly an objective account of "what works") its creation of a world of capitalist industrial agriculture and other capitalist endeavors (through its performances in shaping the world) has much to be desired.

Wright, with her account of a study set in the Philippines (2014), offers an example of how alternative research which supports the practices of local organic farmers in a huge network called MASIPAG, can serve to perform a different world – a world in this case of organic agriculture. The research process involves a co-operation between the farmers in their communities and across the organization, where the farmers together with researchers from universities, have managed to undercut the illusion of the expert science of the so-called Green Revolution of industrial agriculture and hence to nurture (and perform) a growing organic agriculture movement. (See also *Equator Initiative*, 2022, which reports on the formation of Agro-Eco Philippines and its nature-based actions advancing sustainable development.)

In parallel vein, Shiva in her various writings and keynote speeches also reports on efforts in India to undercut the pseudoscience of experts of the Green Revolution, which she indicates is far from Green. She highlights (2022) that the Gates Foundation still supports this type of research with its links with the huge multinational companies benefitting from the chemical products, while the experiences of its impact as understood by those farmers who have become caught up in its aftermath are rendered invisible in the research reporting. The science is still presented as the resolution to food insecurity – despite all the experiences to the contrary.

In Africa, similar issues are arising as those involved in the mega business of industrial agriculture linked to the use of artificial fertilizers, pesticides, and GMOs try to present their research findings as objective science (cf. https://www.scientificamerican.com/article/bill-gates-should-stop-telling-africans-what-kind-of-agriculture-africans-need1/; and https://geneticliteracyproject.org/2022/09/19/power-hungry-meglomaniac-promoting-philanthrocapitalism-in-fundraising-pitches-anti-gmo-activist-groups-blast-bill-gatess-renewed-commitment-promoting-biotechnology-to-addres/).

But resistance is underway by local farmers supported by organic movements worldwide. One such organization is Participatory Guarantee Systems (PGS) – an organization that supports farmers in certifying their products as organic – with this being monitored by PGS (*Organic Without Boundaries*, 2020). PGS is also operative in South Africa, and has been working in various provinces to train farmers in organic farming methods and also to support the development of Indigenous seed banks while enabling farmers to share experiences with one another and with PGS (Wirawan et al., 2023). An example with which I am familiar through participating in a community-engaged project at the University of South Africa (UNISA), which also includes the University of Adelaide, is a case where PGS is involved with a community in Venda, expedited by a community organization called Dzomo la Mupo (DLM). The organization was founded by Indigenous leader Mphatheleni Makaulule in 2007 to protect Mupo – the voice or spirit of nature (http://www.thedzomolamupo.org/).

Currently the organization is also partnering, amongst its other partnerships such as Earthlife Africa and the Wild Law Institute of South Africa, with the University of Venda (UNIVEN). The aim of this partnership is to facilitate intergenerational learning, including the Elders and youth, women and men, university scientists and students, community farmers and community social entrepreneurs. For instance, Netshandama and Nevhudoli (2021) indicate that in one of the initiatives involving UNIVEN and Dzomo La Mupo (DLM), a mapping process of a "cultural eco-calendar" was undertaken:

> The mapping process … involved elders, knowledge holders [respected sages in the community] as well as the youth and school children as the latter were encouraged to participate through integrating what they have learnt in schools into the discussions. Elders shared their life experiences, their knowledge, while the younger generation implemented the knowledge shared in order to realize the needs of local communities. For example, in the study, knowledge was shared on the impact of climate change on agricultural practices, the potential of the establishment of mining on people's livelihoods, and the significance of sacred sites. (2021, p. 43)

Netshandama and Nevhudoli's reference in their last sentence above to the "potential of the establishment of mining" is made in the context of a planned Musina-Makhado Special Economic Zone (MMSEZ) which was stamped in December 2017 by then-Minister of Trade and Industry (Rob Davies). It would involve (if implemented) multinational mining conglomerates in a tight nexus with the government operating according to the modernist logic of "development", which has been shown elsewhere to be devastating in the mode of extraction of Africa's wealth. Earthlife Africa has been drawing public attention to the flaws in the project and indicated that people in the affected communities are "particularly opposed to this huge, anti-environmental metallurgical giant project because the affected section of the Vhembe Biosphere Reserve is home to diverse fauna and flora, which can only be found in the Boabab belt". The mining project would also destroy the community sacred sites (such as forests, wetlands, etc. which are home to the ancestors). Therefore, resistance has been mounting (https://earthlife.org.za/when-a-dream-turns-into-a-nightmare-why-communities-fight-to-protect-the-rich-heritage-of-musina-makhado/). Recently, the United Nations Development Programme (UNDP) which had formerly agreed to support the project, has acknowledged its likely destructive effect on the Indigenous communities and on the environment – and has pulled out of supporting it (cf. Carnie, 2024).

UNIVEN for its part, in partnership with DLM, has been raising awareness in the community, and meanwhile helping to nurture a decolonial consciousness which enables people to define together what futures are likely to be viable. Netshandama and Nevhudoli point out that the purpose of the various learning encounters that they set up was to "raise awareness of the importance of integrating intergenerational learning programs into mainstream education as a way to embrace intercultural, Indigenous, and current scientific knowledge systems" (2021, p. 43). Hence professional scientists from UNIVEN and their students were brought into contact with community members, whom DLM helped to recruit from Makaulule's involvements in the community. DLM has built up a respected position in the community in the Vhembe District due to countless activities undertaken to preserve the local ecosystems and the sacred forests – of which Makaulule is a custodian in terms of the local customs (cf. https://africa.cgtn.com/2019/03/27/faces-of-africa-the-protector-of-the-sacred-sites/).

The community in Venda are not alone in facing the onslaught of capitalist-led projects pushed in terms of narrow economic interests. As Cajete remarks: "Many small communities are suffering particular hardships, and Indigenous cultures, traditions and languages are facing major challenges to their existence" (2020, p. 3). Cajete expresses some hope in that certain Indigenous community leaders (along with scholars) have recognized that "Western notions of development and its paradigm of 'progress' with little regard for social, cultural, and ecological consequences is an extension of colonialism" (2020, p. 3). He points out that their understandings have been gaining momentum. He states that in many Indigenous leaders (in relation to their communities) are beginning to forge strategies such that:

> Indigenous science, traditional environmental knowledge, and communal ingenuity may be engaged toward the development of sustainable Indigenous education and community building at the community level. The use of traditional ecological knowledge in a balanced relationship with other forms of knowledge to address the challenges of climate change is an essential activity. (2020, p. 5)

He suggests that these "other" forms of knowledge include "contemporary forms of education, with science [Western and Indigenous] and other skills learned in a contemporary setting" (2020, p. 5). As far as education in a contemporary setting is concerned, he notes that we need to develop innovative models for it, and that "much of this work is already underway among some Indigenous educators" (2020, p. 7). I would add that besides such work being undertaken by certain Indigenous educators, progressive relational thinkers from the Global North (such as Dole, 2020; Gergen, 2009; Gill & Gergen, 2020; Savransky, 2021; & Schildermans, 2023) too have been exploring options for re-organizing education as a relational process – so as to cater for dialogues around better ways of being-in-the-world. More research is needed on how indeed this can be effected, including how these models of education resonate with the advocacy of Indigenous views and practices of education. For now, I offer four brief examples of practices taken from the Global South of how an awareness of multiple ways of knowing can be encouraged via our educational systems.

SOME EXAMPLES OF EDUCATION AIMED AT GENERATING APPRECIATION OF MULTIPLE WAYS OF KNOWING

Use of Indigenous Water Cleaning Methods Integrated with (Western-Derived) Measures of Testing Water Quality

Seehawer and Breidlid (2022) provide an example of a teacher in a South African science class (Grade 10) attempting to integrate IKS into his teaching, as required by the South African science curriculum (albeit that guidelines for doing this are not clear to many teachers – as also noted by Adyanga and Romm, 2016). The teacher in this case attempted this integration by inviting students to report upon and share Indigenous techniques that they had learned outside the classroom. One such exercise was asking them to consider Indigenous ly-derived options for purifying water, which they could learn more about, by asking their parents or others in the community about such practices. As a result, the students were able to present several local techniques, such as filtering the water through sand or limestone (p. 6). (Interestingly, Seehawer and Breidlid note that many of the children grew up with their grandmothers who might use Indigenous practices even more actively than their parents in faraway cities.)

The teacher then introduced Western-style laboratory experiments to consider which of the Indigenous practices as described and illustrated by the students produced the cleanest water (by measuring it in terms of certain criteria). Seehawer and Breidlid comment that in this way the teacher was enabling the students to recognize that both ways of knowing can complement one another. However, Seehawer and Breidlid also do comment that the students were advised that Indigenous water cleaning methods based on local materials such as sand and lime were both chemical- and cost-free – an advantage of such methods. Seehawer and Breidlid also point out that moving water purification from the specific community contexts to the classroom could imply that they become decontextualized and that Western scientific (laboratory-type) experiments do not take into account contextual or spiritual meanings that may be attached to such purification. They themselves do not try to conclude how differences and tensions between the ways of knowing can be addressed in the classroom or educational setting – but they suggest that this is a subject "that requires further exploration". One of their ideas is that any assessment of the repository of strategies for sustainable livelihoods must cater for the "holistic communal nature of IKS as well as allowing for possible interaction between knowledges [and ways of knowing]" (p. 7). We can conclude that at least the teacher was alerting the students to the notion that one way of assessing the relevance of the knowledge systems is if they are ecologically-friendly and also affordable (in this case, cost free).

Learning About Geometry From Pre-Colonial India: Continued Relevance

Dey (2020), in referring to the rope system of doing measurement and geometry in precolonial India suggests that "the rope trick" can be assessed in certain contexts as better than the Western approach insofar as it makes scientific and mathematical thinking more inclusive and also insofar as it is more ecologically-friendly science. As he states: "the rope trick reduces the cost of purchasing geometry boxes. So, scientific learning can be opened up for the poor as well" (slide 17). Furthermore, considered in terms of practicing "Ecology/Furtnomy-friendly science, the use of rope trick in Indian geometry reduces the cost of paper and also ensures reusability" (2020, slide 17). The usefulness of such a science can be considered in this light. All in all, we can suggest that students can be taught that the likely ecological impacts of techniques/technologies need to be taken into account and also that *the purpose of using technology is not to enrich the few – but to serve a more inclusive good.* Insofar as knowledge systems are combined (based on the deemed advantages of doing so in practice), this too must serve social and ecological justice ends.

A Note on Reconsidering the Binary Between (Past-Oriented) "Tradition" and (Enlightened) "Modernity"

Prempeh, considering how social justice education and communalism can be fostered in the context of Ghana, suggests that we must remember that IKS – which the Enlightenment associates with "the past" – is not a static system of knowledge production: we must be careful of referring to binaries of "past" and "modern", where the former is seen as a domain of tradition and the latter a domain of reason (2022, 1). He refers (p. 2) to the Akan proverb *Tete wo bi kyere*, meaning the past has lessons for going forward. IKS is a lived system of knowledge which helps us to go forward. An eclectic mixing of forms of knowledge production (drawn from different cultures) is possible – but we must be careful of relying on Western technocentric solutions to current crises as if they are more reasonable; we need to continue to draw lessons from the eco-centric type of science and technology associated with Indigenous onto-epistemologies.

Learning Across Generations Supported by the University of Venda (UNIVEN)

In the section above regarding a Relational Epistemological Approach, I referred to the article by Netshandama and Nevhudoli (2021) where they indicate how UNIVEN has been collaborating via a community engagement (CE) programme with Dzomo la Mupo (DLM). As background to their account of the UNIVEN-DLM collaboration, they refer to one case where Cortina and Earl (2020) showed how Elders and the younger generation could learn from one another in focus group and round table discussions, where they were encouraged to focus on exchanges of knowledge from their diverse experiences. Focus groups were set up by Cortina and Earl to encourage an appreciation not only of ecological biodiversity but also cultural biodiversity – which was a subject of discussion amongst the participants (Cortina & Earl, 2020, as cited in Netshandama & Nevhudoli, 2021, p. 42).

Netshandama and Nevhudoli highlight as important from Cortina and Earl's project that:

> …the concept of pluriversity and subversity are constructed not as a reified version of traditional customs, but rather as a continual process of critical dialogue and reflection that also renew and strengthen Indigenous political and moral philosophies. (2021, p. 42)

That is, they stress that *we should not reify tradition* as if it is a static set of knowledge content – we must recognize the dynamic character of "tradition" as it grows through engagement with the biodiversity of the Earth and with cultural biodiversity that enables options for ways of living to become augmented. They remark that their approach in the study of the UNIVEN-DLM partnership "allowed for the creation of space for dialogue between different generations" (2021, p. 49).

As just a taster of the results of their study (due to space limitations in this chapter), I refer to their first paragraph of their discussion hereof:

> An IKS Heritage Day symposium was hosted by University of Venda-Community Engagement in association with UNIVEN AIME, Marude Secondary School, and Dzomo La Mupo. The symposium was aimed at creating Intergenerational Learning Spaces. It was a fun-filled, lesson- or activity-packed environment which allowed learners opportunity to interact with elders in a university learning environment. The symposium was also aimed at educating learners about eco-mapping, nature conservation, preservation of respectful manners and Ubuntu, as well as Indigenous entrepreneurship development and career choices. (p. 49)

Here they point out that the study was used as an opportunity to express/revitalize Ubuntu and also to offer ideas regarding Indigenous entrepreneurship – that is, a form of entrepreneurship grounded in a relational axiology such as Ubuntu. In the next section, I take up the issue of how "business" can be (and has been) practiced accordingly.

CRITIQUE OF THE KNOWLEDGE ECONOMY BUILT ON PROFIT-ORIENTED KNOWLEDGE PRODUCTION

In discussing the concept of a knowledge economy, Hudson (1999) notes that "the creation of knowledge has been integral to the competitive dynamic of capitalist economies since they were first constituted as capitalist" (1999, p. 60). That is, capitalism rests on a capacity to "create new commodities and new ways of producing them". In tune with other authors as mentioned above, Hudson indicates that the "knowledge economy" is set within the "contradictory character of the class relations between capital and labor", where capital resists efforts to "challenge its quests for profits" (1999, p. 66). He furthermore points to the power and the wealth of "massive transnational corporations which dominate the global economy" (p. 66). Considering the trajectory of capitalist development, he avers that "the social relations of capitalism are at least as deeply marked by social inequality as they were then [at the inception of capitalism]" (p. 68). This inequality is buttressed by "discourses and ideologies that represent this [capitalist relations] as the natural course of socio-economic development" (p. 68).

As indicated above, the dominant narratives across the globe – buttressed and indeed performed through mainstream economic theory and practice – present the path towards "modernity" as being universal (or as Escobar, 2016, states, as the only possible world to inhabit). This pretence of universality becomes, in Hudson's terms, "a cloak behind which the harsher realities of capitalism may become hidden" (1999, p. 70). While the hegemonic presentations of universality appear to simply offer *descriptions* of "the world", the descriptions become *world forming* insofar as people accept the accounts. The *Epistemologists of the South* draw our attention to ways in which we can contribute to shaping a pluriverse rather than accepting a One World way of thinking. But let us consider more closely the historical trajectory of how this One World understanding of economics came to the fore.

The mainstream understanding of economic life is premised on the idea that people are essentially self-interested. In considering the individualist assumptions infusing conventional economic theory, Gintis questions its notion of "*homo economicus*" as a basis to understand economic activity. A person as *homo economicus* within conventional economic theory, is understood as "car[ing] about social interactions only insofar as they [their interactions] affect their final consumption [of goods] and wealth" (Gintis, 2009, p. 151). Murove similarly indicates how the notion that the economy runs on self-interested motivation came into being at a certain point in history, with the industrial revolution. He cites Polanyi's (1944) and Robertson's (2001) critical analyses of this individualist understanding of people's economic motivations, and argues that it is regrettable that certain "early modern economists attempted to divest economics from morality" and in this process the "idea of sympathy as social logic" was "hardened into a theory of self-interest" (Murove, 2005, p. 154).

Nevertheless, despite the prevalence of this theory and the (non-relational) practices which it legitimizes, diverse economic performances continue to be found in both the South and the North, such as: solidarity economies (Fonteneau et al., 2020; Sarmento, 2020); social enterprises as economic activities that take into account social and environmental justice concerns (Mair, Battilana & Cardenas, 2012; Ngatse-Ipangui & Dassah, 2019); and generative circular economies geared to regenerating the environment (Mullins et al., 2020; Lethole et al., 2022). In addition, at Harvard University (Center for Health and the Global Environment) Norris (2015) has been promulgating that business practices should be assessed in terms of what he calls a *handprint* as generating a positive contribution in ecological life as opposed to a *footprint* (a destructive impact). All such performances (and others not explored here) can be regarded as making up the tapestry of pluriversality.

Before I close this section, I wish to note that in offering a conception of pluriversality to counteract the universalistic assumptions of "the modern", Kothari et al. qualify that when they speak about "modernity", they do not mean to suggest that "everything modern is destructive or iniquitous, nor that all [Indigenous] tradition is positive". They argue that "modern elements such as human rights … are proving [to be] liberatory [in effect] for many people" (2019, p. xxii). I have suggested in this chapter with reference, for example, to the argument of Harris and Wasilewski (2004, p. 490), that indeed "new energy" can be created by people interacting across cultural heritages. As Gergen (2022) puts it, interacting across different cultural heritages is "similar to being asked to join a dance you have never heard of. Understanding people from other cultures is akin to finding a mutually agreeable way of dancing together" (p. 28). Gergen suggests that learning to dance together again points to our propensity for *relational being*. All in all, like the Indigenous authors to whom I have referred above, he urges us, in order to forge a way forward for a better future, to forego the individualist premises of treating people as separate entities and instead to focus on our being-in-relation (p. 35).

FUTURE RESEARCH DIRECTIONS

Thus far in this chapter I have made a case for why we should draw wisdom from Indigenous onto-epistemologies in order to disrupt the hegemonic power of One World Thinking. I have pointed to some options for generating more pluriversality in our thinking-and-being, so as to expand the repository of diverse approaches for performing diverse economies (not guided by "homo-economicus"), while experimenting with developing reciprocal relations with the more-than-human-world (to supplement Gibson-Graham and Dombrosky, 2020). The onto-epistemology of Indigenous approaches goes hand in hand with a suggestion that those situated in (Indigenized) academia should work alongside lay participant-researchers to explore likely social and environmental consequences of ways of doing "science" so that it is consciously tied to supporting mutually beneficial and inclusive ways of living.

Considering agricultural production specifically in terms of issues of sustainability, further research needs to be done on what the UN Environment Programme (UNEP 2020) calls the "true costs of [industrial] agriculture and food systems" which would include a "full assessment of zero budget natural [organic] farming". Currently, as I pointed out with reference to Vandana Shiva's naming of the poison cartel which supports the interests of big capital, this cartel is still avoiding/ignoring the experiences of many small farmers who have experienced decreases in their yields while becoming increasingly dependent on artificial fertilizers, pesticides and manufactured seeds, while the nutrition content of their diet is compromised and the water and air becomes polluted, leading to health implications (or what the UNEP calls health hazards). The UNEP calls this all the "negative externalities of conventional systems". More funding needs to be invested in exploring the advantages of natural farming by local scientists working in conjunction with communities who have developed (and continue to develop) ways of practising sustainable farming to protect the wellbeing of the planet. (In this regard, an example from Africa is a conference held in Zambia in 2012, where participants shared research confirming positive experiences of organic farming practices, including ethno-veterinary ones, and how the various practices included resilience in the face of climate extremes – Auerbach, Rundgren and Scialabba, 2013, p. iii.)

In relation to the recognition of the value of Indigenous science and Indigenous ways of knowing for sustainability science, Estrada et al. argue that there is a growing recognition among researchers and conservationists that Indigenous perspectives, knowledge systems and histories hold globally important conservation lessons (2022, p. 2). This is because they practice "low-impact and resilient" land use (p. 4). Additional research is needed to investigate – via an eco-friendly lens – current options for implementing low impact economic activity in both rural and urban areas.

Besides research on implementing low-impact activity, for example through green circular economies, research needs to be undertaken on how educational policies and processes can be established with participants and stakeholders to foster an awareness among (global) citizens of the potential for communitarian ecological democracy (so named by David, 2023) to protect the global commons. Possibilities for fostering of this orientation via our educational processes still need further research and examples of fruitful attempts in this direction.

As far as the shaping of a communitarian ethos is concerned, Barkin (2022) refers to networks of resilience which have been established by "many Peasants and Indigenous peoples in Mexico and in other parts of the Global South" (p. 1). He argues that "these peoples constitute a social and economic force that is altering the social and productive dynamics in many countries" (2022, p. 1). Continued research can add detail on how these communities are forging their networks worldwide and what strategies they are using to share understandings and provide political support for each other.

Finally, ways in which the media in public life can be harnessed towards these transformational ends, is also an area for needed additional research. (Ndlovu-Gatsheni provides some hints as to what this might imply for decolonizing journalism education in the context of South Africa – 2022, pp. 18-20).

CONCLUSION

In this chapter I have argued that while IKS is often relegated to inferior status on the world stage, the relational worldviews and ways of knowing that it advocates as a value-orientation offer important contributions to addressing both local and global current crises. Instead of treating Western-style scientific knowing as offering better "truths", I have shown how it can be argued to offer pseudoscience insofar as it is tied (as it so often is) to the knowledge economy introduced under industrial capitalism and associated colonialism. I have also shown that while IKS is associated with tradition, implying that it is static and reified, the tradition can be regarded as dynamic and evolving, as indeed is implied in the value-laden quest to renew and transform the world in terms of relational principles (as succinctly expressed by Ngara, 2017, p. 345). As Córdoba-Pachón likewise emphasizes (2024, p. 42), tradition is always being re-created and therefore can express creativity in new historical contexts. This means that we must desist from treating accounts of Indigenous ways of knowing and living as being a somewhat romantic portrayal of the past which is irrelevant for the future. The options for relational being-in-the-world as presented by Indigenous sages and scholars (as I interpret them) are *forward looking – seeking more balanced relationships* in various areas so as to strive for more beautiful and inclusively beneficial worlds. I have offered some implications for how knowledge exchanges and learning along these lines can be effected, for example, via re-directed educational processes in formal institutions in conjunction with community practices of sharing and discussing experiences. I have also indicated some pointers for conceiving and practising economic activities that encompass a quest for pluriversality and that can serve as examples of activating diverse social and productive dynamics in various societies, which undercut mainstream economic theorizing and its One World premises.

The chapter concludes that Indigenous onto-epistemologies and associated axiologies offer a hopeful route out of the crises that are now being increasingly highlighted not only by Indigenous sages and scholars, but by non-Indigenous scholars and activists' intent on pursuing/performing a more relationally-oriented involvement in world-formation.

DECLARATION OF INTEREST

There was no conflict of interest in my preparing this chapter.

REFERENCES

Addae, D., & McIntyre-Mills, J. J. (2022). Balancing individualism and collectivism: a conversation on Afrocentricity. In Kwapong, O. A. T. F., Addae, D., & Boateng, J. K. (Eds.), *Reimagining development education in Africa* (pp. 55–74). Springer. DOI: 10.1007/978-3-030-96001-8_4

Adyanga, F. A., & Romm, N. R. A. (2016). Researching Indigenous science knowledge integration in formal education: Interpreting some perspectives from the field. *International Journal of Educational Development*, 3(1), 1–14. DOI: 10.25159/2312-3540/2696

Adyanga, F. A., & Romm, N. R. A. (2022). Reflections upon our way of invoking an Indigenous paradigm to co-explore community mobilization against irresponsible practices of foreign-owned companies in Nwoya District, Uganda. *The Qualitative Report*, 27(7), 13. Advance online publication. DOI: 10.46743/2160-3715/2022.5147

Auerbach, R., Rundgren, G., & El-Hage Scialabba, N. (2013). (Eds). *African experiences in resilience and sustainability.* Food and Agricultural Organization (FAO) of the UN. https://www.fao.org/4/i3294e/i3294e.pdf

Barkin, D. (2022). Shaping a communitarian ethos in an era of ecological crisis. *Frontiers in Sustainability*, 3, 1–14. DOI: 10.3389/frsus.2022.944252

Cajete, G. A. (2000). *Native science: natural laws of interdependence*. Clear Light Books.

Cajete, G. A. (2020). Indigenous science, climate change, and Indigenous community building: A framework of foundational perspectives for Indigenous community resilience and revitalization. *Sustainability (Basel)*, 12(22), 9569. DOI: 10.3390/su12229569

Capra, F. (1975). *The Tao of physics*. Shambala.

Carnie, T. (2024). UNDP moves to scrap support for "risky" R165bn Limpopo heavy industry zone. *The Daily Maverick*, 25 February 2024. https://www.dailymaverick.co.za/article/2024-02-25-limpopo-heavy-industry-plan-on-shaky-ground-as-un-agency-rethinks-support/

Chilisa, B. (2017). Decolonizing transdisciplinary research approaches: An African perspective for enhancing knowledge integration in sustainability science. *Sustainability Science*, 12(5), 813–827. https://link.springer.com/article/10.1007/s11625-017-0461-1. DOI: 10.1007/s11625-017-0461-1

Chilisa, B. (2020). *Indigenous research methodologies* (2nd ed.). Sage.

Collins, P. H. (2000). *Black feminist thought* (2nd ed.). Routledge.

Córdoba-Pachón, J. R. (2024). *Ritual and systems thinking: managing an initial encounter*. Taylor & Francis.

Cortina, R., & Earl, A. K. (2020). Advancing professional development for teachers in intercultural education. *Education Sciences*, 10(12), 1–12. DOI: 10.3390/educsci10120360

David, I. N. (2023). *The potential of Namibian education policy to promote education for communitarian ecological democracy*. [Unpublished doctoral thesis]. University of the Free State.

Dey, S. (2020). *Western science v/s humanitarian science: decolonization and Indigenization.* Lectures presented at Yonphula Centenary College.

Dey, S. (2022). *Green academia: towards eco-friendly education systems*. Routledge. DOI: 10.4324/9781003301950

Dole, D. (2020). Collaborative, appreciative and experiential pedagogy in educational settings. In McNamee, S., Gergen, M. M., Camargo-Borges, C., & Rasera, E. F. (Eds.), *The Sage handbook of social constructionist practice* (pp. 352–362). Sage. DOI: 10.4135/9781529714326.n34

Equator Initiative. (2022). Nature-based actions advancing sustainable development. United Nations Development Programme. Retrieved on 20 October 2024 from https://www.equatorinitiative.org/2020/04/24/solution11417/

Escobar, A. (2016). Thinking-feeling with the Earth: Territorial struggles and the ontological dimension of the Epistemologies of the South. *Revista de Antropologia*, 11(1), 11–31.

Estrada, A.. (2020). Global importance of Indigenous Peoples, their lands, and knowledge systems for saving the world's primates from extinction. *Science Advances*, 8, 1–20. https://www.science.org/doi/10.1126/sciadv.abn2927 PMID: 35947670

Fonteneau, B., Neamtam, N., Wanyama, F., Pereira Morais, L., & de Poorter, M. (2010). *Social and solidarity economy: Building a common understanding*. ILO Job Creation and Enterprise Development Department Cooperative Programme (EMP/COOP). Retrieved on 20 October 2024 from https://www.ilo.org/publications/social-and-solidarity-economy-building-common-understanding

Gergen, K. J. (2009). *Relational being: beyond self and community*. Oxford University Press.

Gergen, K. J. (2022). The relational imperative. Taos Institute. Retrieved on 2 October 2024 from https://www.taosinstitute.net/product/the-relational-imperative-resources-for-a-world-on-edge

Gergen, K. J. (2023). *An invitation to social construction: co-creating the future* (4th ed.). Sage.

Gibson-Graham, J. K., & Dombroski, K. (2020). Introduction to the handbook of diverse economies: Inventory as ethical intervention. In Gibson-Graham, J. K., & Dombroski, K. (Eds.), *Handbook of diverse economies* (pp. 1–24). Edward Elgar. DOI: 10.4337/9781788119962.00008

Gill, S., & Gergen, K. J. (2020). Educational evaluation: a relational perspective. In McNamee, S., Gergen, M. M., Camargo-Borges, C., & Rasera, E. F. (Eds.), *The Sage handbook of social constructionist practice* (pp. 402–413). Sage. DOI: 10.4135/9781529714326.n39

Gintis, H. (2009). Beyond homo economicus. In by C. L. Spash (Ed.), *Ecological economics, Vol. 1* (pp. 150-166). Routledge.

Goduka, N. (2012). Re-discovering Indigenous knowledge—Ulwazi Lwemveli for strengthening sustainable livelihood opportunities within rural contexts in the Eastern Cape province. *Indilinga*, 11(1), 1–19.

Gordon, L. R. (2019). Four kinds of invisibility from Euromodernity. Ted-x Talks. Retrieved on 12 October 2024 from https://www.youtube.com/ watch?reload=9&v=bW_G3-DwtQw&t=189s

Harris, L. D., & Wasilewski, J. (2004). Indigeneity, an alternative worldview. *Systems Research and Behavioral Science*, 21(5), 489–503. DOI: 10.1002/sres.631

Hudson, R. (1999). The learning economy, the learning firm and the learning region. *European Urban and Regional Studies*, 6(1), 59–72. DOI: 10.1177/096977649900600105

Inoue, C. Y. A., Ribeiro, T. L., Goncalves, V. K., Basso, L., Moreira, P. F. (2023). Indigenous and traditional communities' ways of knowing and being in planetary justice. *Environmental Politics*, 1–20 (Online First): . 2023. 22934 37DOI: 10. 1080/ 09644 016

Kothari, A., Salleh, A., Escobar, A., Demaria, F., & Acosta, A. (2019). Introduction: finding pluriversal paths. In Kothari, A., Salleh, A., Escobar, A., Demaria, F., & Acosta, A. (Eds.), *Pluriverse: A post-development dictionary* (pp. xxi–xl). Tulika Books.

Lethole, P. V., McIntyre-Mills, J. J., Wirawan, R., & Makaulule, M. (2022). AA!! Venda women and social enterprise: stepwise process to regenerative and sustainable living. In McIntyre-Mills, J. J., & Corcoran-Nantes, Y. (Eds.), *Transformative education for regeneration and wellbeing* (pp. 353–372). Springer. DOI: 10.1007/978-981-19-3258-8_17

Mabunda, P. L., & McKay, V. I. (2021). Educational curriculum and multispecies relations. In McIntyre-Mills, J. J., & Corcoran-Nantes, Y. (Eds.), *From polarisation to multispecies relationships* (pp. 369–389). Springer. DOI: 10.1007/978-981-33-6884-2_16

Mair, J., Battilana, J., & Cardenas, J. (2012). Organizing for society: A typology of social entrepreneuring models. *Journal of Business Ethics*, 111(3), 353–373. https://link.springer.com/article/10.1007/s10551-012-1414-3. DOI: 10.1007/s10551-012-1414-3

Mbembe, A. (2016). Decolonizing the university: New directions. *Arts and Humanities in Higher Education*, 15(1), 29–45. DOI: 10.1177/1474022215618513

McIntyre-Mills, J. J., Lethole, P., Makaulule, M., Wirawan, R., Widianingsih, I., & Romm, N. R. A. (2023). Towards eco-systemic living: learning with Indigenous leaders in Africa and Indonesia through a community of practice: implications for climate change and pandemics. *Systems Research and Behavioral Science*, 40(5), 779–786. https://onlinelibrary.wiley.com/doi/full/10.1002/sres.2976. DOI: 10.1002/sres.2976

Molefe, M. (2014). Critical reflections on Gyekye's humanism. In Imafidon, E., & Bewaji, J. A. I. (Eds.), *Ontologized ethics: New essays in African meta-ethics* (pp. 121–132). Rowman and Littlefield.

Monnet, T. A. S., & Arnsperger, C. (2022). Modernity's temporality. *Text Matters*, 12, 7–31.

Moore, J. W. (2019). World-Ecology: A global conversation. *Sociologia Urbana e Rurale*, 120, 9–21. DOI: 10.3280/SUR2019-120002

Moreton-Robinson, A. (2013). Towards an Australian *Indigenous* women's standpoint theory: A methodological tool. *Australian Feminist Studies*, 28(78), 331–347. https://www.tandfonline.com/doi/abs/10.1080/08164649.2013.876664. DOI: 10.1080/08164649.2013.876664

Mullins, R., Dettmer, S., Eisenbardt, M., & Ziemba, E. (2020). Using knowledge exchange between prosumers and enterprises to implement circular economy activities in businesses. [OJAKM]. *Online Journal of Applied Knowledge Management*, 8(2), 71–90. DOI: 10.36965/OJAKM.2020.8(2)71-90

Murove, M. F. (2005). *The theory of self-interest in modern economic discourse*. [Unpublished doctoral thesis]. University of South Africa (Unisa).

Ndlovu-Gatsheni, S. J. (2022). On the meaning of decolonization. In Rodney-Gumede, Y., Chasi, C., Jaffer, Z., & Pono, M. (Eds.), *Decolonizing journalism education in South Africa: critical perspectives* (pp. 9–24). Unisa Press.

Netshandama, V. O., & Nevhudoli, N. D. (2021). Creating intergenerational learning spaces: A collaboration between UNIVEN community *engagement* programme and Dzomo la Mupo. *Critical Studies in Teaching and Learning*, 9(2), 39–63. DOI: 10.14426/cristal.v9i2.462

Ngara, R. (2017). Multiple voices, multiple paths. In Ngulube, P. (Ed.), *Handbook of research on theoretical perspectives on Indigenous knowledge systems in developing countries* (pp. 332–358). IGI Global. DOI: 10.4018/978-1-5225-0833-5.ch015

Ngatse-Ipangui, R., & Dassah, M. O. (2019). Impact of social entrepreneurs on community development in the Cape Town metropolitan municipality area, South Africa. *The Journal for Transdisciplinary Research in Southern Africa*, 15(1), 1–10. https://td-sa.net/index.php/td/article/view/474/973. DOI: 10.4102/td.v15i1.474

Ngulube, P., & Onyancha, B. O. (2011). What's in a name? Using informetric techniques to conceptualize the knowledge of traditional and Indigenous communities. *Indilinga*, 10(2), 129–152.

Norris, G. (2015). *Handprint-based NetPositive assessment*. T.H. Chan Center for Health and the Global Environment.

Ogden, L. (2012). *Swamplife: people, gators, and mangroves entangled in the everglades*. University of Minnesota Press.

Omodan, B. I. (2024). *Research paradigms and their methodological alignment in social sciences: A practical guide for researchers*. Taylor & Francis. DOI: 10.4324/9781003484066

Organic Without Boundaries. (2020). *How PGS changed the law on organic agriculture in the Philippines!* Retrieved on 20 October 2024 from https://www.organicwithoutboundaries.bio/2020/06/24/how-pgs-changed-the-law-on-organic-agriculture-in-the-philippines/

Polanyi, K. (1944). *The great transformation: the political and economic origins of our time*. Farrar & Rinehart.

Popa, F., Guillermin, M., & Dedeurwaerdere, T. (2015). A pragmatist approach to transdisciplinarity in sustainability research: From complex systems theory to reflexive science. *Futures*, 65, 45–56. DOI: 10.1016/j.futures.2014.02.002

Prempeh, C. (2022). Polishing the pearls of Indigenous knowledge for inclusive social education in Ghana. *Social Sciences & Humanities Open*, 5(1), 100248. DOI: 10.1016/j.ssaho.2022.100248

Ramoupi, N. L. L. (2014). African research and scholarship: 20 years of lost opportunities to transform higher education in South Africa. *Ufahamu. Journal of African Studies*, 38(1), 269–286.

Ritchie, J. (2015). Counter-colonial research methodologies drawing upon postcolonial critique and Indigenous onto-epistemologies. In Cannella, G. S., Pérez, M. S., & Pasque, P. A. (Eds.), *Critical qualitative inquiry* (pp. 77–90). West Coast Press.

Robertson, A. F. (2001). *Greed: gut feelings, growth, and history*. Blackwell.

Romm, N. R. A. (2015). Reviewing the transformative paradigm: A critical systemic and relational (Indigenous) lens. *Systemic Practice and Action Research*, 28(5), 411–427. https://link.springer.com/article/10.1007/s11213-015-9344-5. DOI: 10.1007/s11213-015-9344-5

Romm, N. R. A. (2017). Researching Indigenous ways of knowing-and-being. In Ngulube, P. (Ed.), *Handbook of research on theoretical perspectives on Indigenous knowledge systems in developing countries* (pp. 22–48). IGI Global. DOI: 10.4018/978-1-5225-0833-5.ch002

Romm, N. R. A. (2018). *Responsible research practice*. Springer https://link.springer.com/book/10.1007/978-3-319-74386-8

Romm, N. R. A. (2020). Eliciting children's/young people's (group) engagement with scenarios as participatory research practice for exploring and extending responses to climate change. *Participatory Educational Research*, 7(1), i–xiv. DOI: 10.17275/per.20.0.7.1

Romm, N. R. A. (2021). Responsibly and performatively researching multi-species relationships. In McIntyre-Mills, J. J., & Corcoran-Nantes, Y. (Eds.), *From polarisation to multispecies relationships* (pp. 223–260). Springer. DOI: 10.1007/978-981-33-6884-2_12

Romm, N. R. A. (2024). An Indigenous relational approach to systemic thinking and being: Focus on participatory onto-epistemology. *Systemic Practice and Action Research*, 37(6), 1–32. https://journals.sagepub.com/doi/abs/10.1177/1077800419830135. DOI: 10.1007/s11213-024-09672-4

Romm, N. R. A., & Lethole, P. V. (2021). Prospects for sustainable living with focus on interrelatedness, interdependence and mutuality: Some African perspectives. In McIntyre-Mills, J. J., & Corcoran-Nantes, Y. (Eds.), *From polarisation to multispecies relationships* (pp. 87–114). Springer. DOI: 10.1007/978-981-33-6884-2_6

Rosiek, J. L., Snyder, J., & Pratt, S. L. (2020). The new materialisms and Indigenous theories of non-human agency: Making the case for respectful anti-colonial engagement. *Qualitative Inquiry*, 26(3-4), 331–346. https://journals.sagepub.com/doi/abs/10.1177/1077800419830135. DOI: 10.1177/1077800419830135

Ryser, R. C., Gilio-Whitaker, D., & Bruce, H. G. (2017). Fourth World theory and methods of inquiry. In Ngulube, P. (Ed.), *Handbook of research on theoretical perspectives on Indigenous knowledge systems in developing countries* (pp. 50–84). IGI Global. DOI: 10.4018/978-1-5225-0833-5.ch003

Santos, B. S. (2015). *Epistemologies of the South: justice against epistemicide*. Routledge. DOI: 10.4324/9781315634876

Santos, B. S. (2016). Epistemologies of the South and the future. *From the European South*, 1, 17-29.

Sarmento, D. A. (2020). *Slowly infecting capitalism with solidarity economy: exploring Members' perceptions in a multiple case study in Urban São Paulo* [Unpublished MSc International Development Studies]. University of Amsterdam.

Savransky, M. (2021). *Around the day in eighty worlds: politics of the pluriverse*. Duke University Press.

Schildermans, H. (2023). *Experiments in decolonizing the university*. Bloomsbury.

Seehawer, M., & Breidlid, A. (2021). Dialogue between epistemologies as quality education: Integrating knowledges in Sub-Saharan African classrooms to foster sustainability learning and contextually relevant education. *Social Sciences & Humanities Open*, 4(1), 100200. DOI: 10.1016/j.ssaho.2021.100200

Shiva, V. (2012). *Making peace with the Earth*. Fernwood.

Shiva, V. (2022). *Dr. Vandana Shiva's keynote talk at the 2022 People's Food Summit*. Retrieved on 2 October 2024 from https://www.youtube.com/watch?v=cKM1QgOsb4I

Smith, L. T. (1999). *Decolonizing methodologies: research and Indigenous peoples*. Zed Books.

Stephens, A., Taket, A., & Gagliano, M. (2019). Ecological justice for nature in critical systems thinking. *Systems Research and Behavioral Science*, 36(1), 3–19. https://onlinelibrary.wiley.com/doi/abs/10.1002/sres.2532. DOI: 10.1002/sres.2532

UN Environment Programme (UNEP). (2020). What does the UN Environment Programme bring to the table in the zero-budget natural farming debate? UN story, 24 February 2020. Retrieved on 12 October 2024 from https://www.unep.org/news-and-stories/story/what-does-un-environment-programme-bring-table-zero-budget-natural-farming

United Nations (UN). (2009). *State of the world's Indigenous peoples*. Department of Economic and Social Affairs: Division for Social Policy and Development Secretariat of the Permanent Forum on Indigenous Issues.

Wilson, S. (2008). *Research is ceremony: indigenous research methods*. Fernwood.

Wirawan, R., McIntyre-Mills, J. J., Makaulule, M., Lethole, P. V., Pitsoane, E., Arko-Achemfuor, A., & Romm, N. R. A. (2023). Together we can grow: Resourcing the commons through pathways to wellbeing. *Systemic Practice and Action Research*, 36(5), 641–690. DOI: 10.1007/s11213-022-09613-z

Wright, S. (2014). Quantitative research performing other worlds: Lessons from sustainable agriculture in the Philippines. *The Australian Geographer*, 45(1), 1–18. https://www.tandfonline.com/doi/abs/10.1080/00049182.2014.869293. DOI: 10.1080/00049182.2014.869293

ADDITIONAL READING

Andreotti, V. (2023). Climate, complexity, and relational intelligence. Presentation for the Center for Teacher Education at the University of Kassel, Germany, October 5, 2023, https://www.youtube.com/watch?v=AWfHF3MiMT0

Chilisa, B. (2024). Philosophical underpinnings of mixed methods: decolonizing evaluation practice through decolonizing paradigms. In Y. Shan Y (Ed,), *Philosophical foundations of mixed methods research: Dialogues between researchers and philosophers* (pp. 54-82) Routledge.

Randazzo, E., & Richter, H. (2024). Political action in planetary times: Extinction activism, Anthropocene ontopolitics, Indigenous complexities. *Political Geography*, 112, 103107. DOI: 10.1016/j.polgeo.2024.103107

Stein, S., Andreotti, V., Suša, R., Ahenakew, C., & Čajkova, T. (2022). From "education for sustainable development" to "education for the end of the world as we know it.". *Educational Philosophy and Theory*, 54(3), 274–287. DOI: 10.1080/00131857.2020.1835646

KEY TERMS AND DEFINITIONS

Culture: A repository of meanings that can be drawn on by people as they define meaningful and valued ways of living.

Relational Onto-Epistemology: A recognition that reality consists of a web of relationships in which people – along with more-than-human others – are enmeshed and that our knowing processes contribute to shaping relations in-the-making.

Relational Axiology: An understanding that knowing and acting must be consciously tied to (re) generating reciprocal relations with all that we encounter as we together participate in enacting worlds.

Performative Stance: A recognition that we are always complicit through our knowing processes in the shaping of worlds and that we should reflect with others on our potential influence. An Indigenous onto-epistemology upholds such a stance.

Chapter 4
Indigenous and Transformative Paradigms:
Permeable Borders for a Just and Sustainable Future

Donna Marie Mertens
https://orcid.org/0000-0002-0251-8618
Independent Researcher, USA

ABSTRACT

The Indigenous paradigm provides a framework to guide methodological decisions through its axiological, ontological, and epistemological assumptions. This paradigm should be prioritized in research conducted in Indigenous communities because it encompasses beliefs related to spiritual connectedness and decolonization that are not found in other research paradigms. Its permeable borders are illustrated by an example of an Indigenous/transformative study from Africa that puts the Indigenous and transformative paradigms in conversation with each other to advance justice.

INTRODUCTION

The world is facing many challenges; some that seem to have been with us since early days and others that are emerging and worsening in the current time. Challenges based on inequities in access to health care, education, housing, employment, food, and land can be documented across the history of humanity due to colonization and discriminatory practices. These inequities are being exacerbated by the climate crisis that is increasing in intensity, as well as by an escalation of violence that drives people from their homelands. Researchers have a choice to make: Will they continue with "business as usual" or will they adopt changes in their philosophies and methodologies of research in order to be part of the solution? (Magro & van den Berg, 2021).

In this chapter, I make the argument that paying more attention to the previously excluded voices of Indigenous people provides greater potential for researchers to become part of the solution. The methodology explains the search strategy used in the literature review. A rationale for using the Indigenous paradigm is first provided at the philosophical level and then buttressed with an example of research that combined an Indigenous -transformative approach to increase justice in Indigenous communities. The

DOI: 10.4018/979-8-3693-8447-3.ch004

literature review provides a description of the Indigenous and transformative paradigmatic assumptions. A conversation is presented between Indigenous voices and advocates for justice who have developed a transformative approach to research to increase the potential contribution to a more just and sustainable future. A subsequent section includes a critical analysis of an exemplary study that illustrates how the combination of Indigenous and transformative assumptions can yield solutions to what are seemingly intractable problems.

METHODOLOGY

The methodology was a thematic literature review in which databases were used that include philosophical, theoretical, and methodological issues in the conduct of research in marginalized and Indigenous communities. Data bases included: Academic Search Complete, ERIC, ProQuest Central, PsycARTICLES, and Sage journals. Search terms included Indigenous, transformative, social justice, economic justice, environmental justice, paradigms, axiology, ontology, epistemology, and methodology. The author screened abstracts to determine if they addressed the core elements of Indigenous and/or transformative paradigms and their application in research.

In this chapter, literature is reviewed that focuses on Indigenous scholars from the United States (Lafrance & Nichols, 2010; Waapalaneexkweew [Bowman: Mohican/Lunaape, & Dodge-Francis, 2018]), Africa (Chilisa, 2020; 2024), Australia (Moreton-Robinson, 2016) and Aotearoa (New Zealand) (Chouinard & Cram, 2020; Cram & Mertens, 2015; Cram et al., 2018). They provide different perspectives based on their countries of origin. Thus, this chapter is based on scholarly writings of Indigenous scholars in North America, Africa, New Zealand, and Australia. An important limitation to be noted is that across the world, Indigenous peoples come from countries that are not represented here, such as India, Arabic countries, Persian countries, China, and Japan. Their voices are not represented in this chapter.

PHILOSOPHICAL ASSUMPTIONS OF THE INDIGENOUS AND TRANSFORMATIVE PARADIGMS

Academic imperialism has prevailed in the form of valuing the knowledge and theories that emanate from Euro-Western scholars and dismissing and/or denigrating the knowledge that comes from Indigenous scholars (Chilisa, 2020).

> For colonized, historically oppressed and marginalized groups, intellectual imperialism speaks to the tendency to exclude and dismiss as irrelevant knowledge embedded in the cultural experiences of the people…Most colonized societies were thought of as primitive, barbaric, and incapable of producing useful knowledge. *(Chilisa, 2020, p. 58)*

This contrasts with descriptions of Indigenous knowledge that characterize it "as old as the hills and the valleys, the mountains and the seas, and the deserts and the lakes that Indigenous people bind themselves to as their places of belonging" (Cram, Chilisa, & Mertens, 2013, p. 11).

Kolawole (2022) extends the value of Indigenous knowledge into the realm of addressing the climate crisis and its accompanying increase in inequities and destruction. Indigenous peoples created knowledge about the land, water, plants, animals, solar systems, and human nature to survive for centuries in balance with nature. Their experimentation with agriculture, architecture, music, textiles, and medicinal plants are recognized as having relevance today. Kolawole (2022) emphasizes the significance of Indigenous knowledge in:

> ...addressing global environmental issues such as climate change and problems associated with inorganic agriculture...Many efforts emphasizing the need to move away from chemical agriculture but revert to organic farming or conservation agriculture enshrined in IK [Indigenous knowledge] are already underway. *(pp. 135-136)*

Developments in bringing to visibility scholarship related to Indigenous knowledge about the philosophy and methodology of research bring with them the opportunity to understand how Indigenous researchers can contribute to a more just world, along with other advocates for justice. The remainder of this section describes the assumptions associated with the Indigenous paradigm and compares them with the transformative paradigm. Many Indigenous scholars have contributed to understanding this paradigm, such as Smith (2012), Battiste (2000), Wilson (2008), Chilisa (2012; 2020), and Cram, Tibbitts, and LaFrance (2018), among others. The African Evaluation Association, supported by the Bill and Melinda Gates Foundation, convened an international meeting with the purpose of furthering understandings of the meaning of a "Made in Africa" approach to research and evaluation (South to South Cooperation, 2018). The transformative paradigm was introduced in 1998 (Mertens, 1998); it is now viewed as one of major paradigms in the research literature (Mertens, 2023a; 2024a; Molina-Azorin & Fetters, 2019).

THE STRUCTURE OF PARADIGMS IN SOCIAL RESEARCH

Guba and Lincoln (1989; 2005) provided an important framework for researchers as a mechanism for them to understand the assumptions that guide methodological decisions. They introduced the social research community to the concept of paradigms and their associated philosophical assumptions. These assumptions include:

Axiology-the nature of ethics and values
Ontology-the nature of reality
Epistemology-the nature of knowledge and the relationship between the researcher and stakeholders
Methodology-the nature of systematic inquiry

This framework provides a powerful structure to understand the assumptions that are being used by the social research community, writ large. I use this framework to explore the assumptions of the Indigenous paradigm and the transformative paradigm.

INDIGENOUS PARADIGM: PHILOSOPHICAL ASSUMPTIONS

Indigenous scholars write about the uniqueness of assumptions that underlie their worldviews and their research methodologies (Waapalaneexkweew: Bowman: Mohican/Lunaape & Dodge-Francis, 2018; Chilisa, 2020 & 2024; Moreton-Robinson, 2016; Chouinard & Cram, 2020; Cram & Mertens, 2015; Cram et al. 2018). While there is variation across Indigenous communities, commonalities in their assumptions about research have emerged. There is an emphasis on spirituality and the importance of recognizing the interconnectedness of all people, the land, the sea, the animals, and the plants. These values are integrated into their philosophical assumptions.

Indigenous Axiological Assumption

The Indigenous axiological assumption includes the important values that direct an ethical stance. These values have been identified as including relationality, respect, reverence, responsibility, reciprocity, reflexivity, responsiveness, and decolonization (Adyanga & Romm, 2022; Chilisa, 2020 & 2024; Chilisa & Mertens, 2021; Chouinard & Cram, 2020; Cram, et al., 2018; Louis, 2007; Weber-Pillwax, 1999). The concept of relationality is understood to encompass the relationship between humans and the physical and spiritual worlds. This assumption underlies the emphasis in Indigenous research to be attentive to not only humans, but to all beings in the natural world when making decisions about how and what to research (Ramose 2020). This value has implications for conducting research that addresses the climate crisis as part of seeking increased environmental, economic, and social justice (Mertens, 2023b & 2024b).

The value of decolonization is unique to the Indigenous paradigm and involves the need to acknowledge and redress the negative consequences of colonization. Edwards, Barnes, McGregor and Brannelly (2020) provide guidance on the meaning of doing decolonizing research, especially in contexts in which Indigenous and non-Indigenous researchers work together. Researchers who commit to decolonization need to challenge their assumptions and accept that Indigenous people have alternative ways to understand the world that need to be valued. Chilisa (2020) writes that:

> Decolonization is a process of centering the concerns and worldviews of the colonized Other as that they understand themselves through their own assumptions and perspectives…Decolonization is thus a process of conducting research in such a way that the worldviews of those who have suffered a long history of oppression and marginalization are given space to communicate from their frames of reference. *(p. 11)*

Indigenous Ontological Assumption

The Indigenous ontological assumption aligns somewhat with the constructivist assumption that there are multiple realities, however, there are important differences. Indigenous communities (and researchers) hold that there is a spiritual reality that binds all of us together (interconnectedness) with each other and with the physical and spiritual worlds. Cram et al. (2018) provide this commentary about spirituality in evaluation that has parallel implications for researchers:

> Of all the components of IE (Indigenous Evaluation), spirituality is among the most distinguishing and challenging for evaluators. The presence of spirit, a topic that is usually ignored in evaluation or perhaps tentatively broached by those working in some religious context, is central to Indigenous peoples' worldview and thus to IE. Indigenous well-being inevitably involves a spiritual component that may or may not be recognized by Western religion (Kennedy et al., 2015; UN Permanent Forum on Indigenous Peoples, 2008). A prime example of spirituality is the use of protocols that cover all forms of tribal meetings and gatherings, including those conducted for the purpose of evaluation. These protocols are about keeping all those who are gathered safe in a spiritual sense. Research and evaluation protocols that spell out the need for respectful engagement with Indigenous peoples often try to put into non-spirit terms the importance of proper spirit-related protocols. However, for many

Indigenous peoples the roots of these protocols are in the recognition of the sacred spirit present in all things. (Cram & Mertens, 2016; Weber-Pillwax, 1999) *(Cram, et al. 2018, p, 11)*

The Indigenous ontological assumption means that the reality of Indigenous peoples needs to be given priority. Chilisa and Mertens (2021) bring the meaning of prioritizing Indigenous realities into greater clarity by describing the ontology of a Made in Africa Evaluation (MAE):

> The ontological assumption associated with an MAE approach holds that Africans are to play a greater role in solving their own problems; thus, questions on who prioritizes, initiates, and designs community programs and projects are essential. All areas of culture, living experiences, and Indigenous knowledge systems must be utilized to come up with a methodology through which the realities can be known. Reality is contextual and cultural bound. The evaluator should examine the history of the program, the location and its people, all connections and interconnections and interrogate how spirituality, relational power, political discourse, and historical temporal power (Cavino, 2013) shape the evaluand. Under this assumption, the main question addressed by the evaluation is: Whose priorities and aspirations are addressed by the evaluation? *(pp. 246-247)*

The Indigenous ontological assumption prompts researchers to ask: Whose realities are being given privilege? What are the consequences of accepting a Euro-Western version of reality over an Indigenous reality? How can researchers make visible the realities of the Indigenous communities so that they become the basis for methodological and programmatic decisions?

Indigenous Epistemological Assumption

Kolawole (2022) introduced the concept of Indigenous knowledge earlier in this chapter. This framing of knowledge aligns with the Indigenous axiological and ontological assumptions in that knowledge is viewed as relational and inclusive of spirituality, visions, and dreams, as well as scientific knowledge (Chilisa 2020; Chilisa & Mertens 2021; Waapalaneexkweew (Bowman: Mohican/Lunaape), & Dodge-Francis 2018). Centuries of colonization have left many Indigenous people without the necessary knowledge about their own cultures and languages. This suppression of their cultural knowledge, along with stealing their land, is related to the inequities that are currently experienced in Indigenous communities in health care, education, housing, education, and safe living conditions. The Indigenous epistemological assumption calls for valuing Indigenous knowledge and nurturing relationships that address power differences and historical inequities. "Knowledge is not viewed as simply an objective phenomenon that can be measured quantitatively, rather, knowledge is imbued with the sense of spiritual connection and is built through an understanding of history and cultural connections" (Mertens, 2023b, p. 17).

Edwards et al. (2020) describe Indigenous epistemologies as being diverse and evolving sets of knowledges and intellectual traditions. Thus, prompting researchers to critically assess their methodologies to be sure that they are embedded in the Indigenous context and will bring benefit to the community. This might include multiple ways to conceptualize the knowledge and spiritual traditions and encompass ethical practice. The epistemological assumption asserts that knowledge needs to be inclusive of the history of colonization and the injustices that have resulted from that historical legacy. Indigenous knowledge is needed to demonstrate the resilience of the communities and formulate solutions that consider social and environmental processes.

The Indigenous epistemological assumption that knowledge is relational means that researchers are responsible to their own relations, defined as all those who came before us, those who are with us now, and those who will come in the future. It also means that knowledge does not just come from human beings, but it is inherent in the cosmos (Wilson, 2008). It has implications for the relationship between researchers and the communities with whom they work. Relationships need to be attended to throughout the duration of the study in ways that consciously address power differences and provide strategies for inclusion of community members in ways that are culturally respectful and that value the knowledge that they bring.

Indigenous Methodological Assumption

The Indigenous methodological assumption reflects the methodological implications of the axiological, ontological, and epistemological assumptions in that it includes the use of a decolonizing methodology and makes use of Indigenous strategies for engaging with community and collecting data (Chilisa, 2020; 2024). The relational epistemological assumption leads to a methodological assumption that relationships be nurtured that allow for culturally respectful interactions that value Indigenous knowledge. Chilisa and Mertens (2021) recommend a "transformative participatory lens for mixing Indigenous qualitative and quantitative methods with Western quantitative and qualitative methods" (p. 246).

Indigenous Native American and Maori researchers have added to the literature around Indigenous methodologies by their discussion of Culturally Responsive Indigenous Evaluation (CRIE) (Waapalane-exkweew (Bowman, N., Mohican/Lunaape) & Dodge-Francis, 2018; Chouinard & Cram, 2019). CRIE methodologies emphasize the importance of gathering data about the historical and legal context in order to provide insights into the lasting legacy of colonization. Chilisa (2020) suggests several possible data collection methods that are reflective of the Indigenous methodological assumption: using "Proverbs and metaphors as conceptual frameworks, storytelling methods, sons and poems, talking circles, or Indigenous knowledge-based interview guides." (pp. 212)

PERMEABLE BORDERS: INDIGENOUS PARADIGM AND THE TRANSFORMATIVE PARADIGM

Mertens (2024a, b, & c; 2023a & b, 2022, 2021, 2018), and Mertens, Hall and Wilson (2025) explain the transformative paradigm as one that arose because members of marginalized communities expressed concern about the way research was being done "to" or "on" them. They described research as either ignoring them (marginalizing) or misrepresenting them in ways that were harmful to their communities (Harris, et al., 2020). The transformative paradigm's assumptions share common ground with the Indigenous paradigm, with notable differences as well. In this section, I describe the common ground, the differences, and implications for non-Indigenous researchers to learn from Indigenous researchers to improve their ability to contribute to increased social, economic, and environmental justice.

There is a point of tension when putting the Indigenous paradigm in conversation with the transformative paradigm. This arises from the critical importance of prioritizing the Indigenous worldview when research is conducted in Indigenous communities. That being said, the ethical principle found in the Indigenous paradigm, and made explicit in the "Made in Africa" approach, provides an avenue for prioritizing the Indigenous worldview while embracing other knowledge systems (Chilisa, 2020).

> Built on the principle of motho ke motho ka batho (a person is because of others, I am because we are), MAE has no boundaries between Africa and the rest of the world. This ethical principle incorporates the ideas of wholeness and relationality leading to the position that the evaluator has

an obligation to promote the transformation of all humans and the physical world (Ramose, 2020). Thus, the Indigenous paradigm manifest in the Made in Africa approach is suitable to integration of other paradigmatic perspectives. (Chilisa & Mertens, 2021, p. 246)

Therefore, I move forward with the argument that the common ground shared by Indigenous and transformative researchers provides motivation to support each other in the pursuit of methodologies that lead to increased social, economic, and environmental justice for all marginalized communities, including Indigenous communities. For example, both paradigmatic frameworks lead to methodological decisions that include building relationships with the full range of stakeholders in culturally responsive ways in order to address power inequities and set the stage for valuing the community-based knowledge (Cram & Mertens 2015; Mertens & Cram 2016). They also both include collection and analysis of contextual data that provides a historical framing with insights into legislation and existing social systems that impact the identified problems (Mertens 2020; 2023b; 2024b). Both emphasize the importance of conducting research in ways that provide benefits to the community (reciprocity) and a basis for action that increases justice.

The uniqueness of each paradigm provides food for thought on how philosophical assumptions and methodological decisions can be critically examined by putting the two frameworks into conversation with each other. The Indigenous axiological assumption adds the ideas that we are all interconnected in a spiritual way and that we have a responsibility to be good stewards of the environment, as well as to be responsible to our ancestors (those who came before), those who are here now, and those who will come in the future. This perspective raises issues that were not salient in the transformative paradigm (or in other Euro-Western paradigms). As academics, we can ask ourselves: What criteria are used for ethical review? How do they center Indigenous assumptions about ethics? How do the ethical review criteria address cultural issues, reciprocity, and contributing to social, economic, and environmental justice? What measures are taken to ensure that the research is not being complicit in sustaining an oppressive status quo that, for Indigenous people, is due to the legacy of colonization?

The Indigenous paradigm adds the uniqueness of including history related to sovereignty, decolonization, and repatriation as a way of understanding the realities faced by Indigenous people. It also includes the concept of spirituality as a dimension of reality. As academics, we can ask ourselves: What versions of reality are taught to students in their disciplines? How is the concept of challenging versions of reality based on Western/colonization included in the research curriculum? What strategies are taught to support students to make visible Indigenous versions of reality in their research? What risks do we take when we exclude the spiritual elements of people's lives and of nature in our research work?

The Indigenous epistemological assumption provides guidance on how to develop culturally responsive relationships. This includes attention to cultural rituals and protocols. For academics, this leads to such questions as: To what extent do I teach students strategies for engaging with Indigenous communities in their research? How do I model culturally responsive engagement with Indigenous communities? How do I teach students to give value to Indigenous knowledge in their research? Whose priorities are used to initiate a study, design a program, and determine data collection, analysis and use strategies? How can these be critically assessed to ensure that they reflect the priorities of the Indigenous community? Other questions that emanate from the Indigenous epistemological assumption include:

> How do I address power differences so that those who have been historically excluded are included in respectful ways? What strategies are used that clearly value knowledge from those with lived experience in the cultural context in which we work? What is the historical legacy that is present in the community in which we work? How does our methodology allow us to value community-based

knowledge and ensure that this is the basis for the development or revision of interventions? How does the data we collect embody the full range of knowledge of relevance in this context including spiritual knowledge? How are we being responsive to the community as well as the earth, the sky, the waters, the plants, the animals, and all of creation? *(Mertens, 2023b, p. 18)*

The Indigenous paradigm adds the unique element of decolonizing methodologies and ensuring that traditional Indigenous strategies are used to engage with communities. Academics can ask themselves: How do we teach research design in ways that value community strengths and builds relationships that prioritize Indigenous knowledge to inform the full lifespan of the research study, including establishing the purpose, questions, methodologies, and data collection, analysis, and dissemination? How do we teach the assumptions of the Indigenous and transformative paradigms as frameworks for research? How do I include building relationships and contextual analysis in the design of studies? What strategies do I teach so students can increase the impact and sustainability of their work? How do I guide students' search strategies and criteria for judging the quality of information? How does the study include reciprocity in the form of increased community capacity and increased social, economic, and environmental justice? We can also ask:

> Who leads the evaluation effort? What measures are we taking to build capacity to prepare community members to be leaders in evaluation? How do we design the study to value community strengths and build culturally responsive relationships? How do we include strategies for valuing community-based knowledge? To what extent are community members legitimately involved in deciding the purpose, questions, methodologies, data collection strategies, reporting, and dissemination of the study? How do we provide for reciprocity in the form of benefiting the community and contributing to a more just world? To what extent have we considered the historical and legal context and the continuing effects of oppression? How do we share power with members of communities who have traditionally been excluded from the evaluation funding, planning, and implementation? *(Mertens, 2023b, p. 18)*

The questions presented in this section are designed to stimulate thinking (and action) as to how Indigenous philosophical assumptions can contribute to a healthier and sustainable future. In the next section, an example is used to demonstrate the application of an Indigenous paradigm, accompanied by the transformative paradigm, in the design and conduct of the research. This example demonstrates the permeability of borders across the Indigenous and transformative paradigms.

RESEARCH PRIORITIZING INDIGENOUS ASSUMPTIONS: CONVERSATION WITH THE TRANSFORMATIVE PARADIGM

In this section, I critically analyze a research study that prioritized the Indigenous paradigm with connections with the transformative paradigm. It is from Africa (South Africa) and occurs in the agricultural and economic development sectors (Arko-Achemfuor, Romm & Serolong, 2019; McIntyre-Mills, Karel, Arko-Achemfuor, Romm, & Serolong, 2019).

Arko-Achemfuor et al., (2019) and McIntyre-Mills et al., (2019) reported on a study that took place in the Northwest Province of South Africa. The idea for the research came from Lesogo Serolong, an Indigenous woman from that region, who wanted to develop sustainable agriculture and entrepreneurship skills for farmers. To that end, she founded an impact investment company called Bokamoso Impact Investments to support the farmers. She deliberately started from an Indigenous positionality, believing that the farmers Indigenous knowledge and experience would be important to solving the economic problems they faced. The early phase of her study involved building relationships and conducting a contextual analysis. In accordance with Indigenous protocols in the area, she approached the village chief and asked permission to work with him and community members. She conducted a year-long contextual analysis to understand the full picture of the farmers' needs, challenges, and strengths. She analyzed quantitative data from extant databases that provided information about unemployment rates, literacy levels, climate conditions, and economic levels. Simultaneously, she conducted qualitative data collection through the review of government reports and observing community meetings and interviewing community members.

In keeping with the cyclical nature of transformative research, Serolong used the data gathered in the early stages of the study to refine the study focus. The data revealed several relevant facts that changed the nature of the project. First, she uncovered a version of reality held by government officials that the region had no economic viability and that government programs and policies provided very little in the way of practical remedies. Also, she found that the farmers had been promised life-changing interventions by the government, NGOs, and foundations, only to be disappointed. However, Serolong challenged this version of reality based on her experience with this community and the rich resource of Indigenous knowledge that was present there. She expressed this transformative version of reality thus:

> These villages and their surrounding areas have vast arable lands that are not being fully utilized due to lack of adequate public investments. This can be regarded as presenting a great opportunity for growth and economic development within these communities. Hence, instead of reproducing ways of speaking which point to the non-viability of these communities in terms of economic thriving, Bokamoso is intent on forwarding a "version of reality" based on seeing (and activating) untapped potential in recognition that, as Mertens (2017, p. 21) reminds us, there are consequences associated with accepting one version of reality over another. *(Arko-Achemfuor et al., 2019, p. 8)*

The nature of the study was also changed based on the contextual analysis of the literacy levels of the farmers. Her initial intentions were to jump right into teaching sustainable agriculture and entrepreneurship skills. However, the contextual data revealed that the farmers' literacy levels were too low to benefit from such training. When it became clear that she would have to start with literacy training, she expanded the project and built relationships with a wider group of experts. To achieve this goal, she expanded the relationships that were part of the project. She recruited Arko-Achemfuor and Karel who taught at the University of South Africa's (UNISA) Department of Adult Education and Youth Development. She also invited Romm, a researcher from South Africa who advocates for an Indigenous transformative approach to research, and McIntyre, a South African with an academic position with Flinders University in Australia, to join the team.

It was at this point that the implicit Indigenous and transformative framing for the study became explicit. They described how they aligned themselves with the Indigenous epistemological assumption:

> This way of working together is in keeping with certain Indigenous authors' promulgation of what is called an Indigenous research paradigm, which respects that 'knowing' in Indigenous settings is relational and takes place in communities as people interact with each other in relation to felt challenges. This epistemology (and attendant ethic) has been propounded by Indigenous scholars across the globe. (e.g., Arko-Achemfuor & Dzansi 2015) *(Arko-Achemfuor et al., 2019, p. 2)*

As the plans for literacy training developed, the influence of the transformative paradigm's consciousness about intersectionality was visible. Half the farmers in the training were male and half were female. When the farmers demonstrated competency on the literacy skills, they began training in horticulture. In keeping with the goals of sustainable economic development, the training included information on the production of vegetables on a commercial scale, along with training in marketing and sustainable farming. Data were continuously collected through attendance at the community meetings run by the chief where information was obtained about the farmers' desires for the future. This data was used to develop additional training opportunities about how to run a cooperative so they could sell their products to a wider range of consumers outside of their community. The interventions continued to evolve based on community input, economic indicators, and climate data about temperatures and water levels.

The Indigenous transformative approach supported transformative changes that were framed by Indigenous knowledge. The participants are earning a sustainable livelihood, and some are participating in cooperatives. They have requested additional training in the use of computers for business purposes and to learn English. Not everything went smoothly. For example, growing watermelons for market failed because they did not have a truck to carry the watermelons to market. However, such setbacks could be addressed with the chief and community because of the strong relationships and the strategies for engagement that they had developed. The benefits of the project are widespread; food insecurity is reduced; hundreds of people attend the community meetings; and migration to over-crowded cities is not seen as the only option for employment. They concluded:

> Practitioners become "academics" insofar as they co-research with others (including those situated in higher education institutions and also communities in the field) options for inputting constructively into the social and ecological fabric of life, where knowing is linked to action as part of the definition of "knowledge". We have also shown how this epistemological orientation is in keeping with transformative as well as Indigenous paradigmatic views of the knowing process, where knowing is tied to ethical action as those involved in the inquiry process become agents of justice-oriented interventions in the fabric of life. *(Arko-Achemfuor et al. 2019, p. 8)*

DISCUSSION: FUTURE RESEARCH DIRECTIONS

Indigenous epistemologies provide an avenue for creating a more sustainable and just future in Africa and other countries impacted by a legacy of colonization. When Indigenous epistemologies are placed in a paradigmatic framework, the full range of assumptions that guide methodological decisions becomes visible. Hence, the Indigenous paradigm is articulated in terms of four sets of assumptions: axiological, ontological, epistemological, and methodological. The axiological assumption shapes the three subsequent assumptions; just as the ontological assumption shapes the epistemological assumption. All three assumptions inform the methodological assumption. Clarity about these assumptions provides guidance in research designed to improve the future of Africa and other colonized countries.

The Indigenous paradigm is characterized as having "permeable borders", meaning that the Indigenous worldview dominates in Indigenous research, but Indigenous researchers can integrate, adopt, or adapt guidance from other paradigmatic frameworks. It also means that non-Indigenous researchers can improve their understandings of how to design research that contributes to a more just world by being in conversation with Indigenous researchers. The overlap of the Indigenous paradigms highlights the importance of being culturally responsive, building trusting relationships with the full range of stakeholders, consciously addressing power differentials, challenging oppressive versions of reality, attending to heterogeneity and intersectionality, and producing research that supports transformative action to increase social, economic, and environmental justice.

This valuing of Indigenous worldviews creates an opportunity for the academy to be a part of the solution by critically examining what is taught, by whom, and who is supported within its walls. The literature reviewed in this chapter documents the rich sources of Indigenous knowledge, not just about the cosmos, but also about research itself. The Indigenous paradigm promotes contextually relevant and relational research, rests on African self-determination, values spirituality and Indigenous knowledge, and includes the ethical principles of relationality, responsibility, reverence, reciprocity, respectful representation, reflexivity, decolonization, and interconnectedness.

The academy has a role to play in the pursuit of increased justice. They can do this through the way research is taught and conducted, through valuing and providing access to Indigenous knowledge, and through supporting Indigenous scholars and faculty. The benefits of partnerships with Indigenous community organizations were demonstrated in the example used from the Northwest Province of South Africa. The lead for the project was an Indigenous woman; the academy played a role in providing capacity building for the Indigenous farmers and also as guides to increase the usefulness of the research that was a part of the project. Lucero and colleagues (2018, 60) noted that community based participatory research "as a whole has adopted transformative ideas of endorsing community partners as knowledge co-creators and of applying research findings toward social justice, Indigenous methodologies add more focused attention to culturally driven epistemologies and community control over research, such as ensuring Tribal (or community) data ownership after research is completed".

When research is taught and conducted through an Indigenous lens, in conversation with the transformative paradigm, the design prioritizes forming culturally responsive relationships, and recognizing that this is an evolving process. The circle of people may need to enlarge as the direction of the research becomes clearer based on community input. Contextual analysis becomes a bigger part of the research process in order to understand the history, culture, and capacity of individuals participating in and affected by the research. Interventions may need to be modified as the study progresses. Outcomes need to be critically examined in ensure that these are the priorities of the community. Hence, challenges lie ahead in formulating strategies for research rooted in Indigenous worldviews so that designs can be flexible and responsive as new knowledge and insights are gained.

REFERENCES

Adyanga, F. A., & Romm, N. R. (2022). Reflections upon our way of invoking an Indigenous paradigm to co-explore community mobilization against irresponsible practices of foreign-owned companies in Nwoya District, Uganda. *The Qualitative Report*, 27(7), 1359–1389. DOI: 10.46743/2160-3715/2022.5147

Arko-Achemfuor, A., & Dzansi, D. Y. (2015). Business doing well by doing good in the community. *The Journal of Commerce*, 7(2), 53–68.

Arko-Achemfuor, A., Romm, N., & Serolong, L. (2019). Academic-practitioner collaboration with communities towards social and ecological transformation. *International Journal of Transformative Research*, 6(10), 1–9. DOI: 10.1515/ijtr-2019-0002

Battiste, M. (2000). *Reclaiming Indigenous voice and vision*. University of British Columbia Press.

Cavino, H. M. (2013). Across the colonial divide: Conversations about evaluation in Indigenous contexts. *The American Journal of Evaluation*, 34(3), 339–355. DOI: 10.1177/1098214013489338

Chilisa, B. (2015). A synthesis paper on the "Made in Africa" evaluation concept. African Evaluation Association. Retrieved February 4, 2020, from https://afrea.org/wp-content/uploads/2018/06/MAE2-Final-31st-august.pdf

Chilisa, B. (2020). *Indigenous research methodologies* (2nd ed.). Sage.

Chilisa, B. (2024). Philosophical underpinnings of mixed methods: decolonizing evaluation practice through decolonizing paradigms. In Shan, Y. (Ed.), *Philosophical foundations of mixed methods research* (pp. 54–82). Routledge.

Chilisa, B., Major, T. E., Gaotlhobogwe, M., & Mokgolodi, H. (2017). Decolonizing and indigenizing evaluation practice in Africa: Toward African relational evaluation approaches. *The Canadian Journal of Program Evaluation*, 30(3), 312–328.

Chilisa, B., & Mertens, D. M. (2021). Indigenous made in Africa evaluation frameworks: Addressing epistemic violence and contributing to social transformation. *The American Journal of Evaluation*, 42(2), 241–253. DOI: 10.1177/1098214020948601

Chilisa, B., & Tsheko, G. N. (2014). Mixed methods in Indigenous research: Building relationships for sustainable intervention outcomes. *Journal of Mixed Methods Research*, 8(3), 222–233. DOI: 10.1177/1558689814527878

Chouinard, J. A., & Cram, F. (2019). Culturally responsive approaches to evaluation. *Sage (Atlanta, Ga.)*.

Cram, F., Chilisa, B., & Mertens, D. M. (2013). The journey begins. In Mertens, D. M., Cram, F., & Chilisa, B. (Eds.), *Indigenous pathways into social research* (pp. 11–41). Left Coast Press.

Cram, F., & Mertens, D. M. (2015). Transformative and Indigenous frameworks for multimethod and mixed methods research. In Hesse Biber, S., & Johnson, R. B. (Eds.), *The Oxford handbook of multimethod and mixed methods research inquiry* (pp. 91–109). Oxford University Press.

Cram, F., Tibbitts, K. A., & LaFrance, J. (2018). A steppingstone in Indigenous evaluation. In F. Cram, A. Tibbetts & J. LaFrance (Eds.), *Indigenous Evaluation. New Directions in Evaluation*, 159, pp. 7-16.

Edwards, R., Barnes, H. M., McGregor, D., & Brannelly, T. (2020). Supporting Indigenous and non-Indigenous research partnerships. *The Qualitative Report*, 25(13), 6–15. DOI: 10.46743/2160-3715/2020.4754

Guba, E., & Lincoln, Y. S. (1989). Fourth generation evaluation. *Sage (Atlanta, Ga.).*

Guba, E., & Lincoln, Y. S. (2005). Paradigmatic controversies, contradictions, and emerging confluences. In Denzin, N. K., & Lincoln, Y. S. (Eds.), *Handbook of qualitative research* (3rd ed., pp. 191–216). Sage.

Harris, R., MacGlaughlin, H., Mertens, D. & Perez, J. (2020). Research Ethics for Sign Language Communities [ASL Translation]. *Deaf Studies Digital Journal*, 6.

Held, M. (2019). Decolonising research paradigms in the context of settler colonialism: An unsettling mutual collaborative effort. *International Journal of Qualitative Methods*, 18, 1–16. DOI: 10.1177/1609406918821574

Kennedy, V., Cram, F., Paipa, K., Pipi, K., Baker, M., Porima, L., Sauni, P., & Tuagalu, C. (2015). Beginning a conversation about spirituality in Māori and Pasifika evaluation. In Hood, S., Hopson, R., & Freirson, H. (Eds.), *Continuing the journey to reposition culture and cultural context in evaluation theory and practice* (pp. 151–178). Information Age Publishing.

Kolawole, D. (2022). Is local knowledge peripheral? The future of Indigenous knowledge in research and development. *Alternative*, 18(1), 132–140. DOI: 10.1177/11771801221088667

Lafrance, J., & Nichols, R. (2010). Reframing evaluation: Defining an Indigenous evaluation framework. *The Canadian Journal of Program Evaluation*, 23(2), 13–31. DOI: 10.3138/cjpe.23.003

Loud, M. L. (2021). Conclusion: Transforming evaluation practice for "business unusual". *The Canadian Journal of Program Evaluation*, 36(2), 187–190. DOI: 10.3138/cjpe.72730

Louis, R. P. (2007). Can you hear us now? Voices from the margin: Using Indigenous methodologies. *Geographical Research*, 45(2), 130–139. DOI: 10.1111/j.1745-5871.2007.00443.x

Lucero, J., Wallerstein, N., Duran, B., Alegria, M., Greene-Moton, E., Israel, B., Kastelic, S., Magarati, M., Oetzel, J., Pearson, C., Schulz, A., Villegas, M., & White Hat, E. R. (2018). Development of mixed methods investigation of process and outcomes of community-based participatory research. *Journal of Mixed Methods Research*, 12(1), 55–74. DOI: 10.1177/1558689816633309 PMID: 29230152

Magro, C., & van den Berg, R. (2021). Of portals and paradigms: evaluation, systems thinking and the pandemic. In R. van den Berg, C. Magro, & M.H. Adrien, (Eds.), *Transformational evaluation,* pp. 371-394. International Development Evaluation Association (IDEAS).

Mertens, D. M. (2017). Transformative research: Personal and societal. *International Journal for Transformative Research*, 4(1), 18–24. DOI: 10.1515/ijtr-2017-0001

Mertens, D. M. (2018). *Mixed methods design in evaluation. Sage (Atlanta, Ga.).*

Mertens, D. M. (2020). Transformation as a goal of mixed methods research in the Caribbean. *Caribbean Journal of Mixed Methods Research*, 1(1), 16–28.

Mertens, D. M. (2023a). *Mixed methods research.* Bloomsbury. DOI: 10.5040/9781350273191

Mertens, D. M. (2023b). The pursuit of social, economic, and environmental justice through evaluation: Learning from Indigenous scholars and the fifth branch of the evaluation theory tree. *Journal of Multidisciplinary Evaluation*, 19(44), 11–23. DOI: 10.56645/jmde.v19i44.749

Mertens, D. M. (2024a). *Research and evaluation in education and psychology* (6th ed.). Sage.

Mertens, D. M. (2024b). Mixed methods research for social betterment and a more just world. *Journal of Mixed Methods Research*, 18(3), 216–224. DOI: 10.1177/15586898241256834

Mertens, D. M. (2024c). The philosophical foundations of a transformative approach to mixed methods. In Shan, Y. (Ed.), *Philosophical foundations of mixed methods research* (pp. 30–53). Routledge.

Mertens, D. M., & Cram, F. (2016). Integration tensions and possibilities: Indigenous research and social transformation. *International Journal of Qualitative Research*, 9(2), 185–191. DOI: 10.1525/irqr.2016.9.2.185

Mertens, D. M., Hall, J., & Wilson, A. T. (2025). *Program evaluation theory and practice: a comprehensive guide* (3rd ed.). Guilford Press.

Molina-Azorin, J. F., & Fetters, M. D. (2019). Building a better world through mixed methods research. *Journal of Mixed Methods Research*, 13(3), 275–281. DOI: 10.1177/1558689819855864

Moreton-Robinson, A. (2016). Race and cultural entrapment: Critical Indigenous studies in the twenty-first century. In Moreton-Robinson, A. (Ed.), *Critical Indigenous studies: engagements in first world locations (Critical Issues in Indigenous Studies)*. University of Arizona Press. DOI: 10.2307/jj.423485.10

Ramose, M. B. (2020). *Motho ke motho ka batho*, as African perspective on popular sovereignty and democracy. In Jenco, L. K., Idris, M., & Thomas, M. C. (Eds.), *The Oxford handbook of comparative political theory*. Oxford Press.

South to South Cooperation. (2018). Addressing asymmetries in global evaluation: elevating the knowledge and influence of the global South. Final proposal. Phase 1: 2018–2021. Retrieved On 24 October 2024 from https://afrea.org

UN Permanent Forum on Indigenous Peoples. (2008). *Indigenous peoples and the indicators of well-being and development: Preliminary report.* United Nations.

United Nations. (2007). *United Nations declaration on the rights of indigenous peoples.* Author.

Waapalaneexkweew (Bowman, N., Mohican/Lunaape), & Dodge-Francis, C. (2018). Culturally responsive Indigenous evaluation and tribal governments: understanding the relationship. In F. Cram, K.A. Tibbetts, & J. LaFrance (Eds.), *Indigenous evaluation. new directions for evaluation*, 159, (pp. 17–31). Wiley Periodicals, Inc.

Weber-Pillwax, C. (1999). Indigenous research methodology: Exploratory discussion of an elusive subject. *The Journal of Educational Thought*, 33(1), 312–323.

Widianingsih, I., & Mertens, D. M. (2019). Transformative research and the sustainable development goals: Challenges and a vision from Bandung, West Java. *International Journal of Transformative Research*, 6(1), 27–35. DOI: 10.1515/ijtr-2019-0005

Wilson, S. (2008). *Research is ceremony: Indigenous research methods*. Fernwood.

Chapter 5
Indigenous Knowledge Epistemologies as Futuristic Imperatives

Joy Ndwandwe
Independent Researcher, South Africa

ABSTRACT

This chapter originates from the author's research thesis, which investigated Indigenous dignity from prehistoric epochs through the concept of time, space and knowledge, Moreover, Indigenous constitutive rules, that is, peace, ecology, inclusive development and Indigenous leadership, transforming Indigenous dignity through triangulating rethinking thinking, transdisciplinarity, and cognitive justice into Indigenous people's dignity. Meta-synthesis and genealogy analysis guided by these constitutive rules when applied to qualitative studies data from prehistoric this is prior to the historic advent of and contemporary epochs during systemic negation of Indigenous knowledge. Thus, leading discussions through restorative action by integrating UNSDGs and AU Agenda 2063 into transforming Indigenous people's dignity into Indigenous knowledge epistemologies. Towards futuristic imperatives for integrating within modernity in research and public policy for implementation in the Indigenous Knowledge Hub.

INTRODUCTION

This chapter originates from the author's research thesis on Aba-Ntu Governance and a People's Sovereignty Theory as a Futuristic Imperative: Towards Restorative Action in Regional Policy (SADC), Ndwandwe (2023). A research thesis emanating from the University of South Africa/South African Research Chair Initiative in Development Education [UNISA/SARChI] (Hoppers, 2008-2018) following goals: moving beyond post-colonial theorization to transformative interventions, committed to transforming the university from the ivory tower to civic space, making transdisciplinarity a core facet in leadership, robust exposition of the constitutive rules and norms that control current thinking and

DOI: 10.4018/979-8-3693-8447-3.ch005

practice; and finally represent an epistemology of hope-hope that probes the future and thereby illuminates the possibilities of the present.

This thesis postulated evidence on contextualizing Indigenous knowledge as the epistemology of hope from prehistoric (Ndwandwe, 2023), historical (Ndwandwe, 2023) and contemporary (Ndwandwe, 2023) epochs; hence, it is a transformative intervention moving beyond post-colonial theorization. Evidently, the chapter's emphasis will not focus on the slave trade, missionaries and colonialism but on Indigenous dignity embedded within mythologies, legends, values, adages, religious and cultural rights and heroes, including rights of passages. This transformative intervention's impetus is the concept of time, space and knowledge [TSK] (Petranker, 1997), advocating for non-dogmatic openness for Indigenous people's dignity to emerge from the following epochs. Prehistoric epoch, which is prior to the systemic negation of Indigenous knowledge (Ndwandwe, 2023); the historical epoch marking the advent of the systemic negation of Indigenous knowledge (Ndwandwe, 2023); and the contemporary epoch, when the systemic negation of Indigenous people's dignity was cultural lived experience (Ndwandwe, 2023). TSK as transformative intervention enables this chapter to transform Indigenous dignity into Indigenous people's dignity and thereafter contends these are epistemologies of hope.

On the UNISA/SARChI in Development Education, the goal of transforming the university from an ivory tower to a civic space creates appropriate context for Indigenous knowledge originating from the following epochs: Prehistoric at a time when the unwritten constitution was a cultural lived experience (Ndwandwe, 2023); historical when Indigenous people's dignity was a way of life (Ndwandwe, 2023); and contemporary epochs when European colonization and imperialism systemically negated Indigenous knowledge (Ndwandwe, 2023). In essence, this offers new inspiration (Petranker, 1997) for analyzing Indigenous knowledge rooted within Indigenous people's dignity that was and continues to be systemically negated in the post-colonial era. The transformation from ivory tower to civic space emerges when TSK (Petranker, 1997) takes the scientific spirit of inquiry into domains where science has refused to venture when analyzing mythologies, legends, values, adages, religious and cultural rights and heroes, including rights of passages. These transformative interventions enable Indigenous knowledge to emerge from the 'nothingness' rooted within the understanding of how knowledge is inseparable in time and space.

Consequently, transdisciplinarity is a core facet of leadership, which is explained by Nabudere (2011) as implying the rejection of all dogma, ideology, and closed systems of thought. Transdisciplinarity is critical during the transformative intervention beyond post-colonial theorization and university from the ivory tower into civic space. Evidently, transdisciplinarity and TSK postulate epistemology of hope rooted in Indigenous knowledge embedded within mythologies, legends, values, adages, religious, cultural rights and heroes, including rights of passages. These two theoretical frameworks reject all dogma, ideology, and closed systems of thought (Nabudere, 2011) and instead adopt non-dogmatic openness. Thus, it provides academic space for Indigenous people's dignity to emerge within Indigenous constitutive rules, that is, peace, ecology, inclusive development, and Indigenous leadership.

The constitutive rules explicated by Searle (2018) are the antecedent of existing and new forms of behaviour, which this chapter contends is embedded within Indigenous people's dignity. When emerging from the following Indigenous constitutive rules, peace, ecology, inclusive development, and Indigenous leadership enable the robust exposition of the constitutive rules and norms that control current thinking. This exposition manifests from systemic negation, as espoused by Tottie (1991) on natural language emerging as suggestions and denial of assertions. Evidently, this chapter rejects all dogma, ideology, and closed systems of thought through non-dogmatic openness, enabling Indigenous constitutive rules to emerge after centuries of systemic negation. This transformative intervention enables Indigenous knowledge rooted within Indigenous people's dignity and constitutive rules to be transformed into Indigenous knowledge epistemologies. These epistemologies are futuristic imperatives anchoring hope that probes into the future, thereby illuminating the possibilities in the present towards transforming the university from an ivory tower into a civic space.

In the literature review, this chapter's transformative intervention is done through triangulation of three geographic locations: North America, Australia, and Africa. This geographic triangulation creates the appropriate baseline for Indigenous dignity to be transformed into Indigenous people's dignity. Thereafter, a multi-epistemic and meta-synthesis lens was used, taking the oral and written contexts through research methodology, triangulating, rethinking thinking, transdisciplinarity, and cognitive justice. The triangulated Indigenous dignity results in rules, preserved as governance, sacredness and wholeness rules that leads into Indigenous people's dignity.

Meta-synthesis is the overall methodology of this chapter, transforming Indigenous dignity through the constitutive rules into the following Indigenous people's dignity: Indigenous democracy, transformation model, ethical leadership, ecological ethics and sacred covenant. When applying the systemic analysis of the genealogy analysis within prehistoric, historical and contemporary epoch's qualitative studies from United Nations Education, Scientific and Cultural Organization (UNESCO) General African History books. An example of the prehistoric epochs there emerged how Indigenous knowledge was embedded with remarkable statutes of organized self-governing inland and coastal settlements etc. (Ndwandwe, 2023). During the historical epochs, Indigenous knowledge was negated through mutation and differentiation in the economy, social organizations, culture and archaeology (Ndwandwe, 2023), to mention a few. In the contemporary epochs when Indigenous knowledge was negated through colonial and apartheid policies (Ndwandwe, 2023) such as predatory economy, money nexus, power and domination, and racial theory, to mention but a few. Evidently, this transformative intervention requires triangulating, meta-synthesis, and genealogy analysis supported by the core SARChI conceptual frameworks and methodologies, as well as rethinking thinking, transdisciplinarity, and cognitive justice.

LITERATURE REVIEW

This section of the literature review is based on the geographic triangulation of mythologies, legends, values, adages, religious, and cultural rights and heroes, including rights of passages within North America, Australia, and Africa. Literature review premised on the TKS (Petranker, 1997) concept of transparency evokes an openness of mind that values wonders and curiosity towards the off experience alongside or with the ordinary. This is critical when using a multi-epistemic and meta-synthesis lens, which requires openness of mind and wondrous and curious souls in elucidating Indigenous dignity from these three geographically triangulated areas. Indigenous dignity rooted in the context of no one has a monopoly over God is best articulated in rethinking thinking (Hoppers & Richards, 2011) instead of seeing Islam, Christianity, 'one love' Rastafarianism, and non-religious ethical humanism as four competing ideologies, seeing them as four among many complementary cultural resources.

This statement manifests in Indigenous dignity rooted within the concept of a Supreme Being evident in North America. Waters et al. (2004), virtually all tribes had a single Supreme Being. In Africa, Parrinder (1982) African people believe in a Supreme Being, the creator of all things, while in Australia, Smith (1970) Children that there is a Great Spirit, therefore affirming how Indigenous dignity is rooted in the Supreme Being. This concept is further supported by rethinking thinking (Hoppers & Richards,

2011). instead of quibbling over doctrine, we should vie with one another in good works, evidently anchoring Indigenous dignity.

This Indigenous dignity is elucidated through cosmology and astronomy, which is evident in the geographic triangulation in North America (Cajete, 2000) from the Navajo perspective on the creation of the universe, solar systems, and earth. In Australia (Berndt, 1952), although there are three principal Ancestral Beings, the African cosmology and astronomy explicated by Parrinder (1982) suggests that African religion can roughly be depicted as a triangle (cosmology). At the top, the head of all powers is God (Supreme Being). On the two sides of the triangle are the next greatest powers (ancestors)... and humanity. Evidently, in North America, Australia and Africa, the triangular matrix symbolizing interconnectedness and interdependency is critical in elucidating Indigenous dignity.

Indigenous Dignity Rules

In the introduction, this chapter stated that it is premised on four Indigenous constitutive rules, that is, peace, ecology, inclusive development and Indigenous leadership, which enabled the triangulation of Indigenous dignity rules into governance, sacredness and wholeness. These Indigenous dignity rules underwent geographical and methodology triangulation, which will be discussed in the methodology chapter. This led to Indigenous governance rules, which included the African constitution, vision and spirituality, chaos theory, and natural rights. Indigenous sacredness rules are sacred origins, sacred elements, nature's rules and sacred knowledge. Lastly, the Indigenous wholeness rules are cosmic debt, interconnectedness and interdependency, plurality and complementarity and metaphysics.

Indigenous Governance Rules

Firstly, on peace, the African Constitution (Williams, 1974) related to the one executive head transcendental powers and Indigenous democracy, the people are the first and final source of power. This is affirmed in Cajete (2000) Navajo philosophy of harmonious balance and Brendt (1952) explicating it as gender equality. In ecology, Knappert (1990) postulates that the earth, as a female deity creates harmony between the masculine and feminine transcendental powers, moreover, affirmed by Cordova (2007) and Berndt (1952) on the procreation of animals and humans. Therefore, Indigenous democracy was rooted in Indigenous knowledge anchoring feminine power, symbolizing Mother Earth as a living organism and the procreation of animals and humans, the manifestation of Indigenous governance and spirituality.

On inclusive development, Khoza (2011) avers Indigenous democracy is rooted in common consensus, consultative and communal decision-making, which is affirmed in Waters (2004), interdependent relational bonds between land and people and dualism (Maddock, 1972) in existence. On Indigenous leadership, Khoza (2011) contends that Ubuntu expresses the philosophy of ethical leadership and Cajete (2000) 'Mitakuye Oyasin' we are all related as a community, personhood and communal soul, further affirmed in Brendt (1952) gender inclusivity with masculine and feminine power. Manifestly, Indigenous democracy was rooted in Indigenous knowledge, anchoring Ubuntu in personhood and communal soul, which is the manifestation of Indigenous governance, spirituality, and diplomacy.

Secondly, on peace Vision and Spirituality (Cajete, 2000) the Sun and Moon symbolize masculine and feminine power affirmed through human morality (Parrinder, 1982) is behaviour is society and supported by Massola (1971) the manifestation of rainmaking ethics for dignifying the land and people. On ecology, Knappert (1990) African spirituality is embedded with nature, such as the baobab tree affirmed in Cajete's (2000) traditional Native healing structures and Massola (1971), the Aborigines used natural resources in sacred spaces with relationships between humans and nature. Therefore, Indigenous knowledge was rooted within vision and spirituality, evident in collective memory, the manifestation of Indigenous governance and spirituality.

On inclusive development, Ubuntu (Khoza, 2011) I am because we are, affirmed in Cajete (2000) the core experience of spirituality…philosophy of life, the foundation of community, and Massola (1971) diplomatic messengers. On Indigenous leadership, Parrinder (1982) King Sobhuza of Swazi (Swati)'s Indigenous spirituality and transcendental powers, Lankford (1987) on warriors of war and peace and Massola (1971) the significance of natural resources, manifesting sacred covenant with nature and creator. Evidently, Indigenous knowledge rooted within vision and spirituality is the manifestation of Indigenous governance, diplomacy and spirituality.

Thirdly, in Chaos theory (Knappert, 1990), Chakijane or Hlakanyane, the trickster, evokes chaos similar to Waters (2004). Coyotes, like Thales, manifest instability by forgetting their place in governance, while the Sun (Berndt, 1952) creates chaos through the procreation of humans and plants. On ecology, Parrinder (1982) King Mogodu of Luvedu prophesies on democracy destabilized by chaos forces, affirmed by Cajete (2000) unpredictability and randomness of the creative process and Massola (1971) in the dreams of medicine men evoking spirituality. Evidence of Indigenous knowledge emanating from chaos theory in land-based governance systems rooted within Indigenous governance and spirituality.

Inclusive development, Zulu (2014), the unflattering desert flower Mponeng, transcending material limitations, which was negated by the chaos of winner takes all (Cordova, 2007) and Massola (1971) medicine men's different approaches in destroying enemies. On Indigenous leadership, Knappert (1990) Nosenga's jealousy over Dzivanguru's wealth was affirmed by Cajete (2000). Indigenous thought in understanding chaos is also evident in Berndt (1952) in the mythology of the Djanggawul, the manifestation of chaos and balance. Thus, Indigenous knowledge emanating from vision and spirituality for managing instability from self-destructive behaviour, the manifestation of Indigenous governance, negated by monoculture, lack of diversity and unethical leadership.

Lastly, on peace and natural rights (Knappert, 1990), dignifying the people and land through growing rain dance, such as Incwala traditional ceremonies were affirmed in Waters (2004) land-based religions with their cosmologies and Berndt (1952) Yarrikalla myth of fertility, evoking the procreation of animals and humans. On ecology, Knappert (1990) on the Khoi of Namibia Sun Ram mythology, the Native American (Cajete, 2000) Tree of Life and the V formation hunting rituals (Massola, 1971) and expeditions. Thus, Indigenous knowledge rooted in natural rights is evident in the green philosophy of life, ecological ethics, and the sacred covenant with nature and creator, which are the manifestation of Indigenous spirituality and cultural technologies.

Inclusive development on natural rights, Zulu (2014) contends the fermentation process symbolises a catalyst for growth, negated by Europeans (Waters, 2004) economics and property dominion and their cultural infiltration (Massola, 1971) manifesting patriarchy marginalizing Indigenous natural rights. Finally, Indigenous leadership, Zulu (2014) avers on the negation causing a drift amongst buffalos, symbolized by Cajete (2000) as a divergence from living in ethical participation in nature, and affirmed in Djanggawul (Berndt, 1952) myth, negating ecological ethics and Indigenous spirituality. Thus, Indigenous knowledge was rooted in natural rights evident in the sacred covenant with nature and creator, the procreation of animals and humans and gender inclusivity, the manifestation of Indigenous governance and spirituality.

Indigenous Sacredness Rules

The first rule, sacred origins on peace, Zulu (2014) states how the discovery of desert knowledge is embodied in visions of transcending adversities; this is affirmed in Waters (2004) on the experiential learning from journeying and returning to centre, the origins, and the Djanggawul Myth (Berndt, 1952) on the procreation of humans and animals. On ecology, Knappert (1990) elucidates the Dzavunguru with magic manifesting into wealth; this is affirmed in Berndt (1952) in the concept of wongar totemic symbol and sacred tree of life (Cajete, 2000), a cosmological and structural symbol of life and nature centred living. Evidently, Indigenous knowledge was rooted in transcending material limitations, a sacred covenant with nature and creator and a green philosophy of life seeking life as the manifestation of Indigenous spirituality.

Inclusive development, Zulu (2014) relates how the sorghum seed emerges as the cosmic egg and mystical (Cajete, 2000) relationship between humans and animals, further affirmed in the sacred tree of life (Massola, 1971) in Aborigines. On Indigenous leadership, Knappert (1990) on the people of rain forests the natural community of place, ecological dynamics and natural relationships (Cajete, 2000) affirmed in Berndt (1952) when understanding climatic conditions the manifestation of ecological dynamics and natural relationships. Evidence of Indigenous knowledge is rooted in the transformation model with gender inclusivity, dynamic balance and harmony of the natural world, the sacred covenant with nature, and the creator, the manifestation of Indigenous spirituality.

Second, is the sacred elements for peace rules, elucidated by Parrinder (1982), the reeds or Umhlanga mythology and cosmology, supported by Cajete (2000) masculine and feminine power, the blessing and protection way, affirmed in Berndt (1952) embodied within totemic designs and emblems of rangga poles and body painting during rituals and ceremonies. The ecology (Knappert, 1990) evident during ancestral libations affirmed in Cajete (2000) on life seeking life and Massola (1971) when explicating vegetables are found everywhere. Thus, Indigenous knowledge was evident in the sacred covenant with nature and creator, the procreation of animals and humans, ecological ethics, the people's ecology and the green philosophy of life seeking life, the manifestation of Indigenous spirituality.

On inclusive development in Zulu (2014) postulates on the intellectual and technological prowess, supported in Native structures (Cajete, 2000) indicating available resources within landscape, affirmed in Massola (1971) on waterholes and underground reservoirs. In Indigenous leadership, sacred elements (Zulu, 2014) experiential learning embedded within children of the soil who embody spiritual gifts, affirmed within Earth (Cajete, 2000) elements of shelter, nature and resources, and totemic relation with the Sun (Berndt, 1952) including sexual organs. Thus, Indigenous knowledge was rooted in a sacred covenant with nature and creator, a green philosophy of life seeking life, a transformation model by transcending beyond material limitations, and the manifestation of Indigenous spirituality.

Nature's rules are third on the sacredness rules; Knappert (1990) brings into sacredness the concept of procreation through royal incest; this is affirmed in Cajete (2000) on procreation and cosmology connected to Earth, and Berndt (1952) on buggalili the concept of sacred convocation. On ecology, Knappert (1990), through the revered lion, edifies on transcendental and totemic powers; this is affirmed in Cordova (2007) on the Milky Way infused with all divinity and Berndt (1952) on the concept of Dreaming. Thus, Indigenous knowledge was rooted in transcendental and totemic powers, masculine and feminine power, sacred covenant with nature and creator, gender inclusivity and experiential learning, and the manifestation of Indigenous governance and spirituality.

Inclusive development, Knappert (1990) postulates the mythology of the scared tree, affirmed in Waters (2004) on Mother Earth nurturing all her children and in Massola (1971) human relationship with nature, the manifestation of dynamic balance and harmony with the natural world evoking Ubuntu. On Indigenous leadership, Zulu (2014) on the desert flower with lessons on transcending material limitations, Cajete (2000) on the concept of bio-philia and Berndt (1952) on tidal web and flow cycles, the manifestation of mastering environmental conditions. Evidently, Indigenous knowledge was rooted in ecological ethics; I am because we are all related to the transformation model and ecological ethics, the manifestation of Indigenous governance and spirituality.

Fourth is sacred knowledge of peace. Zulu (2014) states that the desert is fertile ground for producing new knowledge, thus requiring transcending material limitations while evoking ecological ethics affirmed by Cajete (2000) in the metaphor of the hunter, the good heart and Massola (1971) on sacred spaces and elements such as fire. On ecology, on the discovery of light, Knappert (1990) elucidating sacredness affirmed Cajete (2000) sacred ceremonies endowed with self-knowledge and Berndt (1952) on the importance of sex and food. Thus, Indigenous knowledge was rooted in the transformation model for transcending material limitations, a sacred covenant in nature and creator and sacred ceremonies restoring health and healing, the manifestation of Indigenous governance, spirituality and cultural technologies.

Inclusive development, Ndwandwe (2023) avers on risks embodied within rituals and ceremonies in sacred spaces for spirituality and mastery; this is affirmed by Cajete (2000) on pre-hunting rituals through visions and spiritual questing, and Berndt (1952) on the sacred aura of Djanggawul. Indigenous leadership, Ndwandwe (2009), young males participating in bull catching ritual, Cajete (2000) on the ecology of Native community affirmed in Berndt (1952) the earth symbolises human and animal drive projected into a non-human and inanimate environment. Thus, Indigenous knowledge was rooted in the transformation model, ecological ethics, ecology of community, communal, environmental ethics of Ubuntu, and the manifestation of Indigenous governance and spirituality.

Indigenous Wholeness Rules

The first wholeness rule, cosmic debt, in peace constitutive rules, on the soulful power of blood Knappert (1990) in seeking revenge, affirmed in Cajete (2000) explicated through individual and collective vision, and Massola (1971) the role of medicine men in balancing resilience and negativity. On ecology, when Zulu (2014) presents the analogy of predatory lions with a survival advantage, affirmed in Cajete (2000) on the ecological ethic of paying back rooted in mutual reciprocity and Massola (1971) on the concept of no man's land during peace and war. Therefore, Indigenous knowledge is rooted in human consciousness, self-organization, and creativity, enabling the understanding of chaos forces and maintaining cosmic and social order, as well as the manifestation of Indigenous governance, spirituality, and diplomacy.

Inclusive development, on the analogy of food and hunger, Knappert (1990) postulates human consciousness; this is affirmed in Cajete (2000) on the universal cycle of death, birth and rebirth and Massola (1971) on subsequent death after a man's death. On Indigenous leadership, Zulu (2014) teaches on lion and hyena hunting expeditions exemplifying balancing tensions, conflicts and wars, affirmed by Cajete (2000) on the self-effacement and humility of a hunter, and Massola (1971) on the dualism of weapons. Indigenous knowledge was rooted in Ubuntu ethics of humanness, human morality and values, mutual reciprocity, dynamic balance and harmony, and the manifestation of Indigenous governance.

The second rule is interconnectedness and interdependency on peace, explicated by Ndwandwe (2023) on the intergenerational regiments' legacy evoking harmonious human existence, is affirmed in Cajete (2000) on nature and human activities on earth guided by the sun and moon in knowledge creation, and Massola (1971) reveals evidence of nature and human relationships. On ecology, nature, and environment evoking gifts on earth explicated in Knappert (1990), this is affirmed by Cordova (2007) on earth as a living being and Berndt (1952) on symbols of fertility and live-giving qualities. Indigenous knowledge was rooted in self-organization and creativity in Indigenous social institutions, anchored through the environment, language, and identity, as well as the procreation of animals, nature, and humans, which is the manifestation of Indigenous governance and spirituality.

Inclusive development, Knappert (1990) relates the analogy of Tule as a chaos force transforming society by spreading water on earth, which is affirmed in Cajete (2000) on honouring and understanding reciprocal relationships with nature, and Massola (1971) on the geologist in unearthing sacred heritage and treasures on earth. On Indigenous leadership, Ubuntu (2011) on the moral, intellectual, emotional and social living of individual and group morality, affirmed by Cajete (2000), we are all related to a guiding principle in the spiritual ecology, and Berndt (1952) on reciprocal relationship with nature. Evidently, Indigenous knowledge was rooted in self-organization and creativity, individual and collective vision, and ecological ethics, and I am because we are all related, the manifestation of Indigenous governance and spirituality.

Third is plurality and complementarity rule, on peace the desert analogy of earthly constraints creating space of unrestrained desires, uninhibited thoughts and static basis by Zulu (2014). This is affirmed by Cordova (2007) in the concept of bridges uniting Indigenous communities and Massola (1971) the unity embodied by Indigenous messengers. On ecology, on the moon frog mythology in Knappert (1990), the frog envoy connecting moon and earth, affirmed by Cajete (2000) on an ecological person with long-lasting solutions rather than instant, and Berndt (1952) on the importance of the sexual act in conception and pregnancy. Therefore, Indigenous knowledge was rooted in environment, identity and language, sacred covenant with nature and creator, human consciousness, and procreation of animals and humans, the manifesting Indigenous spirituality and diplomacy.

Inclusive development, Zulu (2014) on the analogy of imphephu or incense connecting ancestral land with earth to access ancestral knowledge, is affirmed by Cordova (2007) the idea of a society in the context of individual and all contribution towards a greater whole, and Berndt (1952) on complementarity in myth and rituals. Indigenous leadership, Khoza (2011) on the concept of Ubuntu evoking elegance, harmony and balance in social order and thought, affirmed by Cajete (2007) on Indigenous communities' complementary relationship evoking masculine and feminine power, and Massola (1971) on the division of labour amongst sexes. Thus, Indigenous knowledge was rooted in the human consciousness of an ethical person, Ubuntu cosmology, philosophy and mythology, gender inclusivity, ecological ethics and the sacred covenant with nature and creator, the manifestation of Indigenous spirituality and governance.

Metaphysics is the last wholeness rule, on peace constitutive rule, Williams (1974) relates to the highly human African warfare, is affirmed in life and divinity of land (Cajete, 2000) with earth procreative powers, and Smith (1970) on the children of the Great Spirit guided by morals emanating from the soul. On ecology, Knappert (1990) traces the African rainmakers while Cajete (2000) on the natural process of nature and dynamic state of health and wholeness, and Berndt (1952) on the Djanggawul myth on the sacredness of sexual intercourse. Thus, Indigenous knowledge was rooted in human consciousness, Ubuntu ethics of humanness, rooted in life seeking life, divinity in qualities of land, procreation of animals and humans, feminine powers, and sacred covenant of nature and creator, the manifestation of Indigenous governance and spirituality.

Inclusive development, Zulu (2014) postulates on the desert flower's resilience in transcending material limitations while preserving memories. This is affirmed by Cajete (2000) on the theology of nature formed by spirit and nature and Berndt (1952) of the spiritual qualities embodied in Djanggawul myth in social conscience, thought and religious beliefs. On Indigenous leadership, Khoza (2011) on the transformative potential of collective knowledge sharing, while Cajete (2000) on plants as the primal connection to nature that humanity depends on, and Aboriginal (Maddock, 1971) on the plan of life as cosmology supposes nature and culture. Evidence of how Indigenous knowledge was rooted in transformation models and human consciousness preserved in Ubuntu epistemology as collective memory, the green philosophy of life seeking life and the sacred covenant with nature and creator, the manifestation of Indigenous spirituality and cultural technologies.

In conclusion, this literature review affirmed the Indigenous dignity rules of governance, sacredness and wholeness, while Indigenous people's dignity was emerging through geographic triangulation. Indigenous people's dignity, that is, Indigenous democracy for managing chaos forces to maintain cosmic and social order; transformation model with transcendental powers for balancing chaos through self-organization and creativity, including gender inclusivity; ethical leadership a combination of common consensus and human consciousness; ecological ethics of the people's ecology and sacred covenant of nature and creator. These were examined through prehistoric, historical and contemporary epochs through research methodology for results in the following sections.

METHODOLOGY

This chapter is of the view that conceptual framework and methodology are dependent upon each other. Moreover, this research adopted a meta-synthesis approach, as the boundaries of conceptual and methodology are narrow. Evidently this study introduced inclusiveness and wholeness as stated in the introduction this is through the concepts of time, space and knowledge, epistemology and Indigenous consciousness, evident in the above literature review section.

In essence, during my nurturing by Professor Catherine Odora Hoppers at UNISA/SARCHi in Development Education we were urged to "own" and "own up" to our works in the academy. As SARCHi students we were mentored to go beyond confronting social science methodology into problematique without accepting data analysis and theory formation. This journey of immersion was transformative, with regard to undergirding our studies with reflections; hence, beneath that is epistemology, which is part of Indigenous civilization.

In accordance with the SARChI objectives in the introduction, the meta-synthesis research methodology is supported by SARChI conceptual framework, rethinking thinking. Moreover, SARChI methodologies, rethinking thinking, cognitive justice, and transdisciplinarity enabled the philosophic and geographic triangulation of qualitative studies from North America, Africa and Australia. This research study further unearths genealogy research from the Southern African Region in prehistoric, historic and contemporary epochs published in the UNESCO General African History Volumes through the conceptual framework and methodologies.

Meta-synthesis, the research methodology, allows for systematic review and analysis of qualitative studies, which is supported by triangulation and reflectivity. The systemic review analysis was on the mythologies, legends, values, religious and cultural rights and heroes, and rites of passage triangulated in Indigenous communities of North America, Africa and Australia, according to themes, peace, ecology, inclusive development and Indigenous leadership. This approach is supported by Cajete (2000) when elucidating that ancient Indigenous people, through the coded form of mythology, conveyed their understanding. Hence, triangulation enhances the quality of Indigenous dignity. Research method and methodological triangulation were applied to UNISA/SARChI/DE methodologies, rethinking thinking, cognitive justice and transdisciplinarity on geographic data manifesting in Indigenous dignity rules, governance, sacredness and wholeness.

Moreover, the qualitative studies in UNESCO General History Volumes written by different scholars were systematically reviewed according to the Indigenous dignity constitutive rules, that is, peace, ecology, inclusive development and Indigenous leadership within the three epochs. Meta-synthesis was supported by theoretical and methodological triangulation research methods for transforming Indigenous dignity into Indigenous people's dignity. It is important to note that triangulation (Healy, 2013) enables the research to increase confidence and a comprehensive picture as a rigorous approach. The data and theoretical triangulation, including data source triangulation on the prehistoric, historic and contemporary epochs, were research methods used when transforming Indigenous dignity into Indigenous people's dignity.

The conceptual framework, rethinking thinking (Hoppers & Richards, 2011), contends that in knowledge production, it's important to figure out the point where thinking is knotted, and toxic elements are engrained in human thought. This is critical when researching Indigenous knowledge epistemologies requiring Indigenous consciousness and lens when knowledge production defaults to unblock and release knots, contributing towards clearing nasal passages (Hoppers & Richards, 2011) of the academy. Research methodology and cognitive justice (Visvanathan, 2015) enable different knowledge to exist in a dialogue, thus creating space within the academy to unblock and release knots while enabling epistemology to escape from the laboratory (Visvanathan, 2015) to clear nasal passages in knowledge production.

This, supported by the transdisciplinarity research methodology expounded by Nabudere (2011), is about rejecting all dogma, all ideology and closed systems of thought and creating a culture of questioning and accepting temporal response. Evidently, when researching Indigenous knowledge epistemologies that emanate from the cradle of humanity, this creates academic space for different knowledges to coexist, enabling the culture of questioning. Restorative action will lead the discussion section contextualized by Luutu (2015), which provides the possibility to understand state institutions within their own cultural and ideological bedrock, thus enabling Indigenous knowledge epistemologies to be futuristic.

RESULTS

The result of the philosophic, methodological and geographic triangulation within Indigenous communities of North America, Africa and Australia were the Indigenous dignity rules, governance, sacredness and wholeness leading the discussion in the literature review section. The Indigenous governance rules were the African constitution, vision and spirituality, chaos theory and natural rights. Indigenous sacredness rules included sacred origins, sacred elements, nature's rules, and sacred knowledge. Lastly, Indigenous wholeness rules were cosmic debt, interconnectedness and interdependency, plurality and complementarity, and metaphysics. These Indigenous dignity rules were geographically triangulated within the Indigenous constitutive rules of peace, ecology, inclusive development and Indigenous leadership. Secondly, the Indigenous dignity rules were systemically reviewed through data from prehistoric, historical and contemporary epochs into Abantu Indigenous people's dignity, which was affirmed or negated in each epoch.

In the prehistoric systemic review, which are epochs prior to the systemic negation of Indigenous knowledge, the unwritten African constitution was a cultural lived experience in inclusive and democratic societies. The historic epoch, where Indigenous knowledge experienced systemic negation, led to the contemporary epoch, where this negation was further exacerbated by European imperialism and colonial policies. Evidently, this chapter unearths through systemic review the following Indigenous people's dignity: Indigenous democracy, transformation model, ethical leadership, ecological ethics and the sacred covenant with nature and creator emanating from the Abantu migration.

Abantu Indigenous Democracy

Indigenous democracy was affirmed in managing chaos forces to maintain cosmic and social order through Indigenous governance and spirituality rooted in transcendental powers anchoring gender inclusivity, rainmaking ceremonies for population control and social behaviour with tributary systems. Subsequently, Abantu chaos forces were managed through inclusive economic, social and cultural development in gender-inclusive regimental ceremonies and rituals for mitigating disaster and environmental phenomena. The managing of Abantu chaos forces sustained cosmic and social order supported by Indigenous governance anchored through common consensus in decision-making within geographic landscape anchoring economic activities and Indigenous lifestyles.

In the prehistoric epoch, Indigenous democracy was negated by poverty consciousness, greed, and materialism emanating from stringent foreign trade controls, leading to uncontrollable slave trade and slave capture. Indigenous knowledge was further negated by subjugating Indigenous science accumulated during coastal migration while claiming that innovation and development originate from international commerce. Manifestly, Abantu's cosmic and social order was negated by patriarchy, power and dominance emanating from assimilation in Western religion and exploitation of Indigenous institutions for land invasion. Lastly, Abantu's cosmic and social order was negated through a divide-and-rule strategy stemming from unplanned towns, unfair trading and the supplanting of Indigenous knowledge.

In historical epochs, the negation of Indigenous democracy was evident when colonial chaos forces manipulated Indigenous leadership into signing treaties and concessions, leading to loss of sovereignty with the Indigenous elite captured by settlers as their cronies. Subsequently, Abantu chaos forces originating from the Mdlathule famine led to Mfecane tribal wars and conflicts, resulting in regional detribalization, which was infiltrated by the Great Trek raids, killings, invasions and captures. Later in this epoch, Abantu's cosmic and social order was negated by poverty consciousness arising from colonial forces, leading to instability in Indigenous leadership, transforming land possession and custodianship into ownership, resulting in land dispossession and landlessness.

Abantu were further negated through divide and rule strategy coming from manipulation of Mfecane conflicts with humanitarianism while creating imperial settlements anchoring monopoly over religion, politics and culture, including western education for protecting colonial interest. Lastly, Abantu were negated through patriarchy originating from colonial chaos forces attacking and killing Indigenous leadership, anchoring divide and rule strategy when some Abantu peacefully submitted to colonialism, calling for the abolition of Western education and domination.

In contemporary epochs, Indigenous democracy was negated by exclusive and undemocratic policies anchoring racial theory and infringing Indigenous rights and infrastructure development for mineral resource exploitation. Further negation by colonial chaos forces emerges in poverty consciousness proceeding the African landlessness and denial of land and mineral rights, as well as the emergence of the colonial money economy. Fundamentally, this fragmented Abantu cosmic and social order manifests divide and rule strategy, inducing Abantu inferiority and insecurity while systemically negating Indigenous knowledge embedded within Indigenous spirituality in healing, social, cultural and religious life.

Moreover, Abantu's cosmic and social order was negated by colonial forces through patriarchy in the universal theory of colonialism while transforming land into an economic asset. This was further exacerbated by colonial forces' poverty consciousness in capitalist institutions and the money economy anchoring economic inequality, leading to anti-colonialism and the demands for justice. Lastly, Abantu cosmic and social order was negated through divide and rule strategy emanating from the abolishing Indigenous leadership and appointing colonial stooges, creating new boundaries anchoring conflicts, including championing radical social and international ideology, which undermined Indigenous governance, spirituality and diplomacy.

Abantu Transformation Model

The transformation model was confirmed as self-organization and creativity and balance and order emanating from Abantu experiential learning while transcending material limitations during nonlinear migration, an iconic demographic revolution while mastering environmental changes and adaptation. This experiential learning enables Abantu balance and order through Indigenous agricultural productivity rituals and ceremonies such as Incwala and uMhlanga rain ceremonies rooted in unifying the divided endowed with ethos and moral values.

Intriguingly, Abantu's self-organization and creativity emerged through the environment, language, and identity within strategic geographic locations, which facilitated economic diversity by anchoring common heritage and collective memory. The transcending material limitation anchored active participation in trading with strategic alliances and partnerships affirming political ideology rooted in Indigenous knowledge anchoring metaphysics, led by regiments for peace manifesting into the establishment of Indigenous churches, the founders of African liberation struggles.

In the prehistoric epoch, the transformation model was infiltrated and later negated by poverty consciousness emanating from slave traders, invaders and settlers, manifesting in transforming Indigenous land possession and custodianship into ownership.

In the historical epoch, the Abantu transformation model was negated by the divide and rule strategy, originating from missionary teachings and treaties targeting land dispossession while creating buffer communities that were controlling the slave trade leading to the introduction of slavery. Successively, Abantu balance and order were negated by poverty consciousness emanating from land, humans and animal life capture, leading to the marginalization of Indigenous agriculture and pastoral productivity. Moreover, negation arising from patriarchy through supplanting and not supplementing Indigenous religions with Western religion became the coping mechanism for urban depression and loss of sovereignty.

Intriguingly, Abantu's self-organizing and creativity were negated during the Mfecane population explosion, land shortage, and climatic and ecological changes. This resulted in poverty consciousness through the Great Trek mercenary approach to land, mineral and labour exploitation while creating unfair internal trading practices. Lastly, balance and order were negated by Abantu power and dominance arising from Indigenous conquests and subjugation of existing chiefdoms through military and spiritual tactics.

In the contemporary epoch, the transformation model was negated by patriarchy, power and dominance emanating from political, social, religious and cultural rights evoking Pan-African liberation movements. Furthermore, Abantu's self-organization and creativity were negated through exclusive and undemocratic policies in Western religion, education, civilization, and urbanization, which is evident in urban planning and institutional housing designs and locations manifesting landlessness.

Evidently, this was exacerbated by excluding Africans in competitive subsidies and industrial and commercial banking sectors, creating unemployment and urban poverty. Leading to further negation through divide and rule strategy when recruiting foreign labour marking the advent of moral degeneration, internal displacement, xenophobia and discrimination. Lastly, the transformation model is negated through exclusive and undemocratic policies emanating from the deprivation of political and social liberties and the systemic exclusion of Africans from land and minerals.

Abantu Ethical Leadership

Abantu's ethical leadership was affirmed through common consensus and human consciousness emanating from Abantu, transcending material limitations while traversing environmental and climatic conditions and creating collective memory and knowledge. This Indigenous knowledge was preserved in the creation and environmental mythologies shared through gender-inclusive regimental institutions for eliminating conflicts and wars, including xenophobia. Hence, the collective memory and knowledge preserved is evident in Ubuntu, I am because we are all related, including all living things.

Evidently, Ubuntu manifestation in Abantu common consensus in decision-making and ethics under Indigenous leadership anchoring economic, social and political power for adaptation and integration. Thus, feminine power was evoked in strategic alliances and partnerships anchored through Indigenous production and cultural technology, which were critical in post-Mfecane wars and conflicts. The Indigenous knowledge for nation building and social cohesion is endowed with Indigenous metaphysics for integrating societies through environment, language, and identity of regimental systems and intermarriages as strategic appliances and partnerships.

In the prehistoric epoch, ethical leadership was affirmed; however, Abantu human consciousness was negated by patriarchal strategy through assimilation into international trade, commerce and social systems, and poverty consciousness by mining intruders circumventing controls and violating mining laws rooted in Indigenous knowledge.

In the historic epoch, ethical leadership was negated by the divide-and-rule strategy emanating from the removal of Indigenous leadership and the manipulation of Indigenous diplomacy to create protectorate states, leading to the loss of sovereignty, self-government and land rights. Thus, the manifestation of Abantu resisting landlessness through armed resistance against colonialism and Christianity. Successively, Abantu human consciousness was negated by settlers when they transformed land possession and custodianship into ownership, leading to the disruption of Indigenous wholeness. Lastly, the Abantu common consensus was negated through exclusive and undemocratic policies originating from missionary manipulation of religion, culture, and trade for human exploitation through labour and taxation while fragmenting mutual beneficiation embedded within Indigenous trade, social and economic participation.

In the contemporary epoch, ethical leadership was negated through exclusive and undemocratic policies of the colonial political structures, parliament, and government serving European interests in political systems, evoking liberation struggles. Noticeably, the Abantu common consensus was further negated by exclusive and undemocratic policies through the Land Appropriation Act, the money nexus of capitalism, discriminatory service delivery and education while destroying Indigenous lifestyles and food production. Manifestly negating Abantu human consciousness by divide and rule arising from land exploitation, restricting productivity and access to mineral resources while Abantu lived in the economic periphery on poverty wages.

In essence, patriarchy was effective in negating Abantu common consensus emanating from Indigenous governance while supplanting Indigenous knowledge instead of supplementing it with foreign religions. This negation led to the establishment of independent churches and the translation of the Bible into Indigenous languages, seeking to redress the divide-and-rule strategy evoked by tensions between Indigenous leadership and educated elites. The tension manifests from elites negating Indigenous governance and knowledge systems when championing Pan-Africanism, constitutionalism and colonial legislature to lead the liberation struggles.

Abantu Ecological Ethics

Ecological ethics are affirmed within people's ecology emanating from Abantu, anchoring all elements of earth, the epithet of Indigenous metaphysics in civilization, industrial and economic activity, anchoring gender inclusivity, including collective memory and knowledge in sacred spaces of geographic environment, language and identity. Furthermore, Abantu's dynamic balance and harmony in the natural world are evident during population movement, preserved within Indigenous productivity and cultural technology and governance, and rain dances are derived from Ubuntu's reciprocal relations in Indigenous knowledge.

Abantu people's ecology is evident in Indigenous cultural technology originating from environmental resources for economic activities and settlement construction, manifesting the environment as a strategic resource in nation-building. Also, Abantu dynamic balance and harmony in the natural world emanating from wealth consciousness preserved in Indigenous governance and cultural technology for agricultural and mining productivity, including pre-colonial coastal trading and international trading.

Additionally, Ubuntu, I am because we are all related stemming from pre-colonial reciprocal trade and exchange with links between coastal and interior evidence of regional trade. Moreover, the Abantu people's ecology was affirmed in socio-political institutions led by Indigenous leadership who were supported by rural communities in rebellion and struggle against colonialism. Lastly, Abantu dynamic balance and harmony in the natural world emanating from urban/rural linkages anchoring political and social development rooted in Indigenous metaphysics, celebrating common identity and collective memory.

In the prehistoric epoch, ecological ethics were affirmed; however, in the historic epoch, the Abantu people's ecology was negated by patriarchy emanating from wars and conflicts, enslavement and civilization agenda, and settler land ownership. This was further exacerbated by the divide-and-rule strategy through religious assimilation, leading to the supplanting of Indigenous knowledge. Marking the advent of Abantu resisting missionary, imperialist and colonial capture who were developing exclusive and undemocratic economic policies violating human and land rights, transcendental powers for healing the land through imperialism injustices, natural disasters and landlessness.

The Abantu people's ecology was negated by poverty consciousness anchoring land pressures while subjugating African trade interdependent on the ecology and culture. This was further worsened by the ecological effects arising from natural resource competition and climate changes, including coastal trade control. What is more, the Abantu dynamic balance and harmony in the natural world was negated by exclusive and undemocratic policies advancing economic exploitation with tax imposition together with land structural violence, emerging within urbanization and international cultural supremacy anchoring patriarchy.

In the contemporary epoch, the people's ecology was negated by patriarchy when subjugating and humiliating Indigenous knowledge through teachings separating religion and culture. This was aggravated by the divide-and-rule strategy during the introduction of Western medicines while systemically negating Indigenous healing rooted in wholeness. Furthermore, Abantu dynamic balance and harmony in the natural world was negated by poverty consciousness emanating from the exploitation of Indigenous science by the capitalist system while fragmenting feminine power and, most importantly, failure to integrate Indigenous knowledge.

In addition, the Abantu people's ecology was negated by the divide-and-rule strategy evident in the unequal distribution of social services, neglecting rural communities while subjugating Indigenous homestead designs and architecture. Likewise, Abantu anchoring all elements on earth was negated through exclusive and undemocratic policies of the capitalist world, economic dependency on primary goods, and racist political doctrine. Lastly, the Abantu people's ecology was negated by poverty consciousness emanating from economic investments for mineral extraction in industry, advancing imperialistic interest while marginalizing Indigenous commercial activity and creating enslavement.

Abantu Sacred Covenant with Nature and Creator

The sacred covenant with nature and creation was affirmed in the Abantu cosmology of the full moon, exemplified in sacred ceremonies and rituals, as well as rain dances, for dignifying the land and nation. This Indigenous knowledge evolved during the Abantu migration, which arose from mastering environmental and climatic conditions through rainmaking. Nonetheless, Abantu environment, language and identity were evident in sacred collective memory preserved as Indigenous metaphysics in geographic elements, arts, music and culture. In essence, Abantu life and divinity in the land are qualities anchoring gender inclusivity in matrilineal and economic power. This Indigenous wholeness is affirmed in ceremonies and rituals with the king's seclusion and initiation rites, anchoring the green philosophy of life-seeking.

Ubuntu, I am because we are all related, which was rooted in governance systems within social institutions and ceremonies evoking feminine power and the procreation of animals and humans. Abantu's green philosophy of life seeking life is affirmed within Indigenous metaphysics in homesteads while transcending material limitations in the natural environment with traceable migration routes, which is evidence of regional linkages. Furthermore, this green philosophy is evident in Indigenous leadership's preservation of land custodianship through tributary systems anchoring the procreation of animals and human life. Ubuntu evoked through resisting systemic negation of Indigenous knowledge emerging in anti-colonial ideology and protests.

Abantu's sacred covenant with nature and creator was affirmed in the prehistoric epoch; however, Abantu land and the divinity of land qualities were negated by poverty consciousness. This is evident in transforming land possession into ownership, the humiliation of Indigenous leadership and capturing political authority over the land. Furthermore, the Abantu green philosophy of life seeking life was negated by poverty consciousness of foreign land domination, marginalizing the Indigenous land possession as a communal asset for future generations.

In the historical epoch, the sacred covenant was negated through the divide-and-rule strategy of religious conversion and assimilation, evoking the inferiority complex while negating Ubuntu, leading to Indigenous spiritual resistance in anti-colonial movements. Moreover, I am because we are all related was negated by the divide and rule strategy when supplanting and systemically negating Indigenous knowledge by embracing Western religion without integration.

Evidently, this prevents Indigenous knowledge from being a source of social justice, wholeness, and improvement in quality of life. Likewise, Abantu's green philosophy of life-seeking life was negated through exclusive and undemocratic policies of land exploitation and predatory economic policy, manifesting in poverty and oppressive systems eroding dignity. Lastly, Ubuntu was negated by poverty consciousness emanating from the transformation of political and cultural life, manifesting in moral decay and structural violence that eroded community wealth and dignity.

In contemporary epochs, Abantu's green philosophy of life-seeking life was negated by poverty consciousness through the native land act, introducing minority population land ownership for agriculture and holdings. In essence, Ubuntu was negated by a divide and rule strategy when introducing western medicines to marginalize Indigenous medicines while colonists provide social health services in geographic areas of economic interest. Equally, Abantu, I am because we are all related, was negated through exclusive and undemocratic policies emanating from Western education and religion predominately for promoting labour exploitation, enabling African elite to be civil servants and adopt urban lifestyles leading to moral decay.

Additionally, Ubuntu was negated by exclusive and undemocratic policies in amalgamations, oligopolies, monopoly capitalism, offshore funds and investment reserves, including racial discrimination. Similarly, Abantu, I am because we are all related, was negated by poverty consciousness subjugating inter-Africa trade, extractive infrastructure, marginalization of Indigenous productivity and cultural technologies. Lastly, Ubuntu negated through divide and rule strategy emanating from a negation of Indigenous metaphysics, religion and witchcraft without understanding its psychological context.

DISCUSSION

In this chapter, the above results manifest Indigenous knowledge as Indigenous governance, spirituality, cultural technology and diplomacy manifesting from Indigenous people's dignity, Indigenous democracy, transformation model, ethical leadership, ecological ethics and sacred covenant. That sustained peace, ecology, inclusive development and Indigenous leadership during Abantu migration and settlement, which was systemically negated, particularly in historical and contemporary epochs. Evidently, the research methodology, restorative action, will enable these Indigenous knowledges to be transformed into epistemologies as futuristic imperatives. Restorative Action (Luutu, 2015) relates to the emergence of "natural democracy", which, in the context of this chapter, will transform Indigenous people's dignity and Indigenous knowledge into epistemologies as futuristic imperatives.

Restorative action is leading this discussion from the concept of natural democracy and, most importantly, as further explicated by Luutu (2015) as humility retraces and regenerates, understood from the feminine deity principle of self-renewal and self-creation that occurs through decomposition, translating into compost or renewal or regeneration. This chapter, as a transformative intervention, required humility to retrace and regenerate Indigenous knowledge epistemologies emanating from Indigenous people's dignity and Indigenous dignity. This retracing and regenerating of Indigenous knowledge occurred through geographic and methodological triangulation. Through meta-synthesis, it was revealed how Indigenous knowledge was systemically negated during the partly prehistoric, and most significantly in historical and contemporary epochs.

Evidently, this led to the robust explosion of the Indigenous people's dignity constitutive and Indigenous knowledge is being transformed through restorative action into Indigenous knowledge epistemologies such that they become the compost or renewal, or regeneration required by the academy to transcend beyond the ivory tower into civic space. This discussion will be in two aspects in the chapter. The first is on how Indigenous knowledge epistemologies can be strategic interventions in public policy, particularly the Africa We Want Agenda 2065. Moreover, Vision 2030 on United Nations Sustainable Development Goals (SDGs), Agenda 2030, social inclusion goals 1-7, economic growth 8-11, environmental responsibility 12-15, and governance 16-17. The second discussion will be regarding integrating Indigenous knowledge epistemologies into public policy, the case of the Abantu Eswatini Dignity Institute (AEDI), the Indigenous Knowledge Hub.

The Indigenous people's dignity and Indigenous knowledge are transformed into Indigenous knowledge epistemologies through restorative action as follows. Indigenous democracy in the context of restorative action (Luutu, 2015) allows understanding state institutions within their own cultural and ideological bedrock. This bedrock was systemically negated in the three epochs, evident in the African Aspirations for Agenda 2063 (AU, 2015) statement on peace, when Indigenous democracy is preserved within the African Constitution (Williams, 1974) on land custodianship as our gift from God. Hence, through Indigenous democracy, there are rules for managing chaos forces towards maintaining cosmic and social order, which is in line with UNSDGs.

Transformation model theory in the context of restorative action (Luutu, 2015) requires unlearning and returning to normative wisdom as an inclusive paradigm for Agenda 2063 on Pan-Africanism and African Renaissance. This model theory is preserved in the African Constitution (Williams, 1974) on the right to free land for economic use with Chiefs and Kings as custodians for fair distribution, evoking self-organization and creativity with balance and order, in line with UNSDGs. Ethical leadership in the context of restorative action (Luutu, 2015) speaks truth to all on the generative positive and negative impact of policies on the ground, reflected in the Agenda 2063 on peace and ecology. This ethical leadership is preserved in the African Constitution (Williams, 1974) on the community as a whole, anchoring common consensus on the manifestation of human consciousness, in line with UNSDGs.

Ecological ethics in the context of restorative action (Luutu, 2015) on how policymakers have no monopoly over knowledge as communities continue to exist without public policy, which is in line with the Agenda 2063 ideals of Pan Africanism and vision of African Renaissance. These ethics are preserved in the African Constitution (Williams, 1974) in African religion being evident in all institutions, symbolizing the people's ecology through dynamic harmony and balance with the natural world in sync with UNSDGs. Sacred covenant with nature and creator in the context of restorative action (Luutu, 2015) as communities continue to exist without public policy or state guidance, which aligns with Agenda 2063, inclusive and sustainable development. This sacred covenant is preserved in the African Constitution (Williams, 1974) on the trouble of one is the trouble of all, symbolizing I am because we are all related, Ubuntu, in sync with UNSDGs.

The AEDI, Indigenous Knowledge Hub, has embarked on resource mobilization to implement the integration of Indigenous knowledge epistemologies with modernity in the following areas. On peace, security and good governance, inclusive development, leadership development, gender and youth development, ecology, environment and climate change, by providing academic and research space for interdisciplinary, multidisciplinary, transdisciplinarity policymakers and researchers, advocacy, and technological innovation. The current plans for the hub are to integrate the five Indigenous people's dignity and Indigenous knowledge embedded within governance, spirituality, diplomacy, and cultural technologies.

FUTURE RESEARCH DIRECTIONS

In furtherance of the Indigenous knowledge embedded with Indigenous governance, spirituality, diplomacy and cultural technologies. Through the University of the State of African Diaspora, where I am the dean of the royal chamber and faculty on the historical legacy. I am conducting research in conversations with African kings, queens and traditional leaders on their roles in Indigenous governance, spirituality and diplomacy, including the artefacts.

CONCLUSION

In conclusion, the Indigenous Knowledge Hub anchor projects are collaborating with public policymakers and researchers on integrating Indigenous knowledge into existing land policy and the development of a national Indigenous knowledge policy. In the pipeline is a gender and youth community engagement partnership between AEDI and Taiwan Africa Vegetable Initiative in the kingdom on decolonizing, Africanizing and indigenizing citizen's nutrition and health through Indigenous foods in chieftaincies within the kingdom. Evidently, the AEDI project is a transformative intervention for moving beyond post-colonial theorization, transforming university into civic space with Indigenous knowledge epistemologies as futuristic imperatives.

REFERENCES

Angus, F. G. (1967). *South Australia. A.H. and A.W.* Reed Publishers.

AU (2015). African Union Agenda 2063: The Popular Version @African Union Commission

Berndt, R. M. (1952). *Djanggawul: An Aboriginal Religious Cult on North Eastern Arnhem Land.* Routledge and Kegan Paul Ltd.

Cajete, G. (2000). *Native Science: Natural Laws of Interdependence. Santa FE.* Clear Light Publishers.

Cordova, V. F. (2007). *How It Is: The Native American Philosophy of V.F. Cordova.* The University of Arizona Press.

Healy, R. (2013). *Understanding Triangulation. Evidence Based Nursing Journal Hoppers, C.O. and Richards, H. (2011). Rethinking thinking.* UNISA Press.

Khoza, R. J. (2011). *Attuned Leadership: African Humanism as Compass.* Penguins Books.

Knappert, J. (1990). *The Aquarian Guide to African Mythology.* Aquarian Press.

Lankford G.E. (1987). Native American Legends. August House/Little Rock.

Luutu, B. M. (2015). Restorative Action into Public Policy: An African Perspective. Human and Social Dynamics Research Seminar, HSRC.

Maddock, K. (1972). *The Australian Aborigines: A portrait of their society.* Allen Lane, Penguin Press.

Massola, A. (1971). The Aborigines of South Eastern Australia: As They Were. Australia.

Mountford, C. P. (Ed.). (1970). *The Dreamtime Australian Aboriginal Myths.* Angus & Robertson.

Nabudere, D. (2011). *Afrikology, philosophy and wholeness.* AISA.

Ndwandwe, J. D. (2009). *Akusiko Kwami Kwebantfu unearthing King Sobhuza II's Philosophy.* Mgan-gatho Publishers.

Parker, K. L. (Ed.). (1965). *Australian Legendary Tales.* Angus & Robertson.

Parrinder, G. (1982). *African Mythology.* Hamlyn.

Petranker, J. (Ed.). (1997). *Light of Knowledge, Essays on the Interplay of Knowledge, Time and Space.* Dharma Publishing.

Searle, J. (2018). *Constitutive Rules. Argumenta.* University of Sassari.

Smith, R. W. (1970). *Myths and Legends of the Australian Aboriginals.* George G. Atarrap & Co.

Tottie, G. (1991). *Negation in English Speech Writing: A Study in Variation.* Academic Press.

Vasvanathan, S. (2015). Cognitive Justice in Public Policy: A View from the South. HSRC Human and Social Dynamics Research Seminar, HSRC.

Waters, A. (Ed.). (2004). *American Indian Thought: Philosophical Essays.* Blackwell Publishing.

Williams, C. (1974). *The Destruction of Black Civilization: Great Issues of A Race from 4500 B.C. to 2000 A.D.* Third World Press.

Zulu, Z. (2014). *The Sacred Knowledge of the Desert.* Madisebo University Co.

Chapter 6
Global Higher Education:
The Sustenance and Hushed Annihilation of Indigenous Knowledges and Identities

Jacob Mapara
https://orcid.org/0000-0002-7386-3929
Chinhoyi University of Technology, Zimbabwe

Ethel Siamena
Chinhoyi University of Technology, Zimbabwe

Crispen Mazodze
https://orcid.org/0000-0002-2423-0420
Bindura University of Science Education, Zimbabwe

ABSTRACT

This chapter argues that there is a sustained onslaught on epistemologies and identities of the inhabitants of the Global South through the hegemony of Western ways of knowing through the promotion of globalization. It posits that peddling this myth sustains Western epistemological hegemonies that have perpetuated and sustained the extirpation of those of the Global South. Through these subtle weapons of mass destruction, the researchers observed that most formerly colonized Indigenous communities have lost their identities and cultures. They further contend that scholarships by the countries in the Global North, including the new geo-political arrivants China and Russia continuously pummel and destroy indigenous identities under the guise of building relationships. Through these observations, the chapter accentuates the observation that while people are not being decimated physically, this occurs subtly. The chapter concludes by emphasizing the importance of embracing and harnessing all epistemologies to build better more sustainable and resilient communities.

INTRODUCTION

Globalization is anti-diversity. In practice one notes that it is basically about the United States of America and largely its European allies wanting to dominate the world. It is an extractive culture that is driven and underpinned by capitalism. It rides on the myth of promoting development when in reality it

DOI: 10.4018/979-8-3693-8447-3.ch006

seeks to sustain the centre-periphery dichotomy in a manner that benefits the centre, which in this case is the Global North that has persistently imposed its will and standards on every other country on the globe. Globalization goes against some international treaties such as the United Nations Educational, Scientific and Cultural Organization (UNESCO) 2005 Convention on the Protection and Promotion of the Diversity of Cultural Expressions. This critical global treaty is ignored as the Global North imposes its will on the Global South, which are basically former colonies of most European countries through subtle means such as education and so-called cultural exchanges and language learning programmes. These exchanges also filter through other forms such as scholarship offers and training opportunities where the formerly colonized are invited to be further immersed in the ideologies of the former colonialists and their allies. Globalization is thus a new name for an old and enduring culture of imperialism. There is therefore a need for the formerly colonized to embark on epistemic disobedience and delink from the shackles of colonialism that still persist to this day, and one way of doing this is through developing curricula that speak to the needs and aspirations of nations and peoples of the Global South, both physically and conceptually. Through these acts, the formerly colonized can avoid or mitigate the impacts of epistemicides and identicides that have ravaged them and continue to subvert their humanity. Considering the foregoing, this chapter serves to argue a case for the opening up of the epistemological spaces through the bringing on board of Indigenous epistemologies that have been marginalized, or at best mutilated by the Western educational system. It submits that a deliberate policy shift in Indigenous communities' curriculum planners will significantly contribute to the emergence of a new generation that embraces globalization as a platform for all to be visible and not one where the West's ways of knowing become the only one.

But what are epistemicides and identicides?

EPISTEMICIDES AND IDENTICIDES

These are critical terms that need to be unpacked so that their significance is better understood and appreciated. This is essential because these two have persisted under globalization. After all, the so-called interconnectedness of the world has seen not only a flow of raw materials from the Global South reach epic proportions but has also experienced a mass exodus of professionals and other staff to the Global North. The Global South has witnessed the flow of a deluge of cultural goods like films and other finished goods from the other side. This has contributed to the epistemicides and identicides that characterize most communities in the formerly colonized spaces.

Epistemicide

This is a term that was coined by Boaventura de Sousa Santos (2014) to characterize the systematic obliteration of Third World knowledge by Western science (Bennett, 2015). It thus means the killing, silencing, annihilation, or devaluing of any knowledge system that is not Western. It occurs where there are epistemic injustices that are persistent, systematic, and collectively work as a planned subjugation of particular ways of knowing, especially those of the colonized or formerly colonized. Epistemicide has been in existence for at least five centuries, years in which people of color have been persistently subjected to European knowledge systems and where through brutal religious conversions and use of homes, Indigenous ways of knowing have not only been sidelined but have been exterminated, in some instances with their practitioners.

This scourge is captured in the following words:

> The energy that propels diatopical hermeneutics comes from a destabilizing image that I designate epistemicide, the murder of knowledge. Unequal exchanges among cultures have always implied the death of the knowledge of the subordinated culture, hence the death of the social groups that possessed it. In the most extreme cases, such as that of European expansion, epistemicide was one of the conditions of genocide. (de Sousa Santos, 2014, np)

These words speak about a European edifice that has ravaged not just other knowledge systems but has also exterminated willfully communities that were and still are perceived as lesser human. It is these communities that today are enticed through offers of scholarships and are daily bombarded by Western cultural goods such that they persistently remain unhinged from themselves as a people. Epistemicides have therefore been sustained through the education system that the former colonized inherited after independence (Simbi, 2022). Tied to epistemicides is the issue of memoricides which is the destruction of both physical, mental and cultural spaces of memory that are done through education.

Identicide

Identicide is a term that Sarah Jane Meharg coined in 1999. Meharg (1999) states that identicide involves the:

> … conscious intent to destroy the physical manifestations of a society in order to destroy cultural identity and to erase any signs of previous existence. When a community is affected by identicide, and significant cultural places and symbols are destroyed, there is a weakening of cultural identity because the cues of being have been altered or destroyed and disorientation and disassociation result. (Meharg, 1999, p. i)

These words are critical in that colonialism did exactly that through the destruction and sacrilege acts on Indigenous sacred spaces. The destruction of the ancient city of Benin by the British in 1897 is a good example of imperial acts of identicide. The writers of this chapter go on to further state that it is not just about the cultural destruction of spaces and symbols but also of a people in terms of connection to their children and their knowledge systems. This destruction has been done through Western education that was imposed by colonialism and the Christian Church.

Meharg (2011) makes clear the fact that identicide is a considered and calculated act that is routinely carried out as a tactic of armed conflict. It is however, more specifically a tactic of combat that intentionally aims to destroy the cultural elements of a people through a multiplicity of ways for the sole purpose of contributing to the eventual acculturation, elimination and/or complete obliteration of a particular group or collections of people. Identicide also includes the destruction of a people's contested signs and symbols, behaviours, principles and belief systems, heritages, places and performances. It is also the deliberate butchery of the link between people and place that eliminates the bond, which reinforces individual, community and national identity. Identicide manifests in different forms and all serve a sole purpose which is to negatively affect the connections between people and their places. Identicide can be a forerunner of mass extermination, but it does not inevitably end in genocide.

GLOBALIZATION: IMPERIALISM BY ANOTHER NAME

Globalisation is an ideology that supposedly favours the primacy of the free market, nominal state regulation of the economy and the supremacy of the individual as the rational economic player. Quijano (2000) opines that globalisation has cemented the dominance of Eurocentrism in the Global South. This dominance is not just in the form of goods, but more importantly, it has to be understood as the United States of America's efforts to get goods produced at a lesser cost in locations outside its borders, and also where the laws, especially relating to the environment are lax. It is also a form of imperialism that has been rechristened. There is nowhere where the subtle effects of globalization are realized other than in the former colonized states' education systems. Take for example, in higher education, it is Eurocentric perspectives that are often implied requirements for academic promotion, publishing and accessing funding. For funding, it has almost become mandatory that countries in the Global South have to partner with at least a university or research centre in the Global North. Some countries are blacklisted from applying for funding for political reasons because they are accused of having pursued interests that do not dovetail with those of the US and Europe. However, in cases where applications are successful, most African scholars are reduced to data hunters and gatherers for theoretical formulation in Europe and North America because non-Eurocentric theories are derisively dismissed as layman criticism or ideological polemics (Mazodze, Mapara and Tsvere, 2021; Ndlovu-Gatsheni, 2018). As a result, the higher education institutions in the Global South continue to be the sites on which Euro-American epistemologies are reproduced culturally, ideologically, psychologically and materially.

Most publications that come out of these joint studies are required to be published in Western-owned journals, or those affiliated to the Western ones. There is the academic imperial hook that has created the fallacy that high impact journals are only those hosted by the West. Academics in the Global South have not reflected on who owns the journals, nor have they investigated how the high impacts are achieved. Chawla (10 February 2023) reports a case where a paper was rejected because the author had cited only four of the papers published by the journal out of his 150 sources. This act is indicative of how some of these so-called high impact journals achieve those high citations. Although Elsevier (a Dutch publisher), who are the host of this journal, the *International Journal of Hydrogen Energy* (IJHE) have said, "We take coercive citation practices and peer-review manipulation very seriously and have retracted 131 papers from Elsevier journals over the last three years for peer-review manipulation", the question remains of how many other journals are not as strict as those under the Elsevier umbrella.

Besides the issue of citations and high impacts, scholars in the Global South have been made to accept and believe that high impacts are in citations. While this is true to some extent, a great academic should impact his/her community more than the so-called international academic fraternity (Barnett, 2024). To impact the global, and not the local is to miss the point. Academics should positively impact their societies through their acts, and also their writings. These impacts should lead to positive change in their communities. To continue to pursue the Western concept of high impact means to carry research for and on behalf of the West's institutions.

The above paragraph emphasizes the actuality that education plays a significant role in sustaining epistemicides and identicides. In line with this observation, Ngugi (1992) elucidates that Eurocentrism is maintained in the academic curriculum through various platforms which include textbooks, pedagogy and theories at the expense of Indigenous knowledges and subjectivities. These textbooks celebrate Western science, arts and culture as the universal achievements of humanity and the result is a persistent colonization of the mind and at the end of practice. It thus means that coloniality captures and annexes the minds and bodies of the subjugated people. This is because it forces them to shun and dump their cultural practices and epistemologies in favour of Western ones, an act that results in the concretisation of Western ideas in the former colonised people's minds. Ngugi (1997) argues that higher education in postcolonial Africa has continued to perpetuate colonial languages and identity formations by blindly following Eurocentric institutionalised systems without questioning their functions and purposes.

Ngugi's assertions are equally echoed by Olatunji (2010) who notes that language is central to decolonization. This is because history has shown that colonial processes are embedded in language. Language is equally significant because it is vital to the production of knowledge and also to ensure epistemic and cultural identity. Languages are also essential in that they serve as windows through which different worldviews are perceived and understood. Therefore, the deployment and use of local languages where there are robust language policies promotes and sustains the sociolinguistic culture and environment of the country (Alidou et al., 2006). These scholars emphasize that the negation of Indigenous African languages, culture and ways of knowing has resulted in the inferiorization of the Indigenous people who are largely found in the Global South.

In cultural spaces, for instance, Euro-American films present Africans as stooges, continuing with the thread that commenced with writers like Rider H. Haggard (*King Solomon's Mines*, 1885 and *She*, 1887) and Defoe (1719) in his novel *The Adventures of Robinson Crusoe*. It is thus in the same vein that Mazodze, Mapara, and Tsvere (2021) illustrate that universities in the Global South serve as the locus of forming and preserving systems of thought that have contributed to the extermination of Indigenous educational institutions. This same space has allowed for the disarticulation, abrogation and subalternization of Indigenous ways of knowing. It is also for the same reason that Mapara (2023a) argues that an African university does not exist, positing that what is there is a cheap imitation of the Euro-American version because the curriculum and schemes of association that are there peripherize the African institution and elevate that from the Global North. This reality is further amplified when it is noted that it is largely scholars from the Western academic institutions that are invited to come and speak as guest and keynote speakers. The Global South's willingness to be academic pariahs is also reflected in the fact that most scholars in the institutions of the Global South prefer to go to universities and colleges of the Global North for contact and sabbatical leave. This shows how much institutions in the Global South are inflicted by the house nigger complex.

The continued promotion of the Eurocentric epistemological canon in the Global South has also resulted in grand-scale epistemicides of Indigenous knowledges (Mignolo, 2009; de Sousa Santos, 2014). The academic and co-curriculum programmes in African universities continue to promote Eurocentric onto-epistemologies because of coloniality and globalisation. The texts that are used are not only those that are largely from the Global North but also curriculum decisions in the Global South are dictated by the need to meet the needs and expectations of those in the Global North. Thus, through the coloniality of power, there is the enduring of Eurocentric academic disciplines and a co-curriculum that has produced a petit bourgeoisie leadership in the Global South. That petit bourgeoisie leadership is incapable of transforming the political, economic and social institutions that were inherited from colonialism into its own image rooted in Indigenous ways of knowing and traditions (Ndlovu-Gatsheni, 2018). This is further buttressed by an academic curriculum that is anchored on Western models and more often than not, is also supported by visiting academics who are more of tourists, who know next to nothing but only that of their countries of birth. These academics are however accommodated because of some hoped-for links with a Western institution. Therefore, for higher education to produce graduates who fit into their societies there is a need for a major epistemological and ontological decolonisation of the

African mindsets. This calls for a deliberate and genuine epistemic disobedience and de-linking so that Indigenous ways of knowing can get space in the canon that is under the hegemony of that knowledge from the Global North (de Sousa Santos, 2014).

COLONIALITY OF POWER, KNOWLEDGE AND BEING IN THE GLOBAL SOUTH

As earlier indicated, globalisation is imperialism rechristened (Mapara & Mutasa, 2011). It is a vehicle through which coloniality of power, knowledge and being is perpetuated in the Global South. There is therefore the need to delink and de-westernize, which means embarking on serious and genuine epistemic disobedience, and disregarding the dictates of the hegemony of the Western academy. Concerning this study, the formerly colonized, especially Zimbabwe, coloniality of power also looks into the psycho-social dimensions of the curriculum where Eurocentrism has continued to exert its destructive influence on epistemic, ontological and axiological domains of education (Simbi, 2022). This is seen through mental colonisation or cognitive domination, which is expressed through racist dehumanisation, objectification and inferiorization of Indigenous people as portrayed in textbooks, films and co-curricular activities. Most of these productions are cultural goods which are given as donations to higher education institutions in the Global South. These donations also pour in as scholarships and partnership opportunities. However, more important, is the fact that Western hegemony is realized through the sustained marginalization of epistemologies of the South, which is realized through some exchange programmes. This development has resulted in Indigenous people shunning their blackness by using, for instance, skin-lightening creams to appear as Whites as exemplified by the identity crisis leading to the rise of stereotypes such as 'salads', 'nose brigades' and 'yellow bones'. It has also seen most women having so-called hair extensions. In Southern African countries, female students who lighten their skins are known as 'yellow bones' (Makobela, 2017; Mazodze, Mapara and Tsvere, 2021). Some studies have shown that in Black communities, a woman with a lighter complexion i.e., a yellow bone is privileged with a higher social status than one with a darker skin (Makobela, 2017). All these are results of the greatly under-appreciated cultural goods like films and books that students and staff in the Global South partake in.

Due to this idealisation, African female youths strive to achieve this 'yellow bone' status often with disastrous consequences because some skin lighteners damage their skin. Ndlovu-Gatsheni (2021) posits that the quest by African youths to ape whiteness indicates that Europeans succeeded in implanting negative images of blackness by modelling Europeanization into an aspiration for most African youths. This in itself is an indictment of African education, both at primary and secondary levels as well as at higher and tertiary.

It is in light of the above observation that Ndlovu-Gatsheni states:

> … the long term effect of this social engineering and epistemological process that was marked by epistemicides that invaded the core imaginary of the African psyche and culture to the extent that Africans today produce cultural Europeanization without direct tutelage of Europeans. (Ndlovu-Gatsheni, 2021, p. 194)

The above quotation indicates that the psycho-social and epistemological damage that was done is quite deep and the challenge that faces African theorists is how to inverse this mental impairment. At the cultural and epistemological levels, most Africans are still stuck in Eurocentric thought and they breathe coloniality daily. This is indicative of the capture and destruction of a people both physically and in spirit. It can therefore be noted that a person like a "yellow bone" mentioned above is nothing but a grotesque and physically impaired person because s/he does not represent the original but the disfigured and badly modified. To remedy this situation the researchers concur with Mignolo (2009) who advocates for epistemic delinking from the West to lead to the upliftment of Indigenous epistemologies as new sites of knowledge production in the Global South. This means an epistemic shift that creates space for other epistemologies and other principles of knowledge and understanding to be appreciated and embraced.

Coloniality of knowledge relates to the epistemic subjugation of Indigenous knowledges and cultures. It also means either a complete rejection of that knowledge or a grudging acceptance that is realized through the theft and repackaging as well as smuggling back of Indigenous epistemologies and practices in modified forms such as zero tillage in agriculture, and the so-called green religion. Further, the coloniality of knowledge denotes the enduring structures of racism in the construction and reconstruction of knowledge which favour the Western epistemic canon against Indigenous knowledges. This has killed the imagination and creativity of especially Africans to the extent that they have simply succumbed to the idea that their contribution to knowledge creation is to adopt Eurocentrism as the only valid knowledge production system. The death of creativity and imagination among most Africans is in fact borne out of the belief by most in a global village where they are bystanders and not contributors. They do not contribute because of inferiority and also because of a successfully planted consumer culture that makes them to believe that foreign, and especially Western is best. This behaviour is unAfrican because in Africa, one does not go to the male meeting place in the village with nothing. This has its roots in a curriculum that re-awakens the cultivated consumer to go back to her/his original senses, an awakening that is always critical. This goes back to the issue of identicides and epistemicides where values entrenched in consuming that which comes from one's sweat have been eroded. To this end, Ndlovu-Gatsheni (2021) amplifies this by stating that in addition to genocides Africa has suffered epistemicides, linguicides, culturecides and identicides in which Indigenous African ways of knowing, languages, cultures and identities were systematically destroyed. This is worsened by the former colonized, for instance, in Africa where they accept foreign labels like Anglophone, Francophone and Lusophone to place them in blocks that the West can control, and yet the Africans, because of these blocks cannot work together (Mapara, 2023b).

Ndlovu-Gatsheni (2018) further posits that the destruction of African ways of knowing and policies was a deliberate process of dissocialising and dismembering African people from their cultural and historical contexts. The author adds that:

> Western enlightenment regarded the non-west variously known as the ancient, the orient, the primitive world, the third world, the under-developed world, the developing world and now the global south … primarily as a place of parochial wisdom, of antiquarian traditions. (Ndlovu-Gatsheni, 2018, p. 18)

The above quote shows that colonialism came with the practice of binaries, a "the civilized and the barbaric" dichotomy that persists to this day, even in academia where it manifests in various shades and colours. Important to note is also the fact that Ndlovu-Gatsheni's (2018) assertion is that universities in sub-Saharan Africa are delivering curricula that denigrate their epistemologies and belief systems on behalf of the West. This is so because as has been highlighted earlier, there were no major curricular reforms after the gaining of independence. Even if universities wanted to implement curricular reforms that embed Indigenous knowledges, they faced the hegemonic influence of globalisation and Eurocentrism, epitomised through the control of knowledge production and validation. This was worsened by the deluge of scholarships that were (and still continue to be) offered to the newly "independent", as well as "advisory" services by the former colonizers, especially in all education and training sectors, including in the security services.

The European colonisers committed epistemicides by promoting a Western epistemic canon that invisibilized and inferiorized Indigenous ways of knowing. This is consistent with the "Western and Cartesian-Newtonian understanding of the academy, sciences, and humanities, Indigenous knowledges have been viewed as primitive, folkloric, anecdotal and unscientific" (Knopf, 2015, p. 179). It is for this reason that even Indigenous political, social, economic and legal structures were marginalized or re-engineered to fit the needs of the colonialists. In light of this, it means then that invisibilization implies that epistemological and theoretical priority is given to Eurocentric perceptions, scientists and academics that are touted as the best scholars, and their centres become fulcrums of excellence (de Sousa Santos, 2014). In fact, having succeeded in pushing the Indigenous ways of knowing to the periphery of the epistemic canon they proceeded to make Western ways of knowing the only authentic and scientifically verifiable ways of comprehending the universe. The presentation of Western epistemologies as the only ones that are genuine defy logic, because their assumption is wrong since there are various ways of looking at and appreciating the world and what it offers. Coloniality of knowledge reveals that Eurocentric knowledge production and neo-liberal practices have continued to silence other epistemologies by insisting on their universal validity (de Sousa Santos, 2014). A critical analysis of the coloniality of knowledge is therefore of fundamental importance in the pursuit of decolonising knowledge production and validation in the Global South.

Ndlovu-Gatsheni (2021) presents the problems of coloniality of knowledge as characterised by several traits. Among these are the realities that:

i) Eurocentric knowledges are privileged. It is because of this privilege that they are regarded as universalistic, neutral and objective. The fact that they are European or Western is not taken into consideration, yet this is the reality – that they are situational and accordingly socio-cultural and thus native to those climes. In addition, there is the reality that;
ii) Eurocentric racist hierarchies are established and maintained through a Eurocentric definition of knowledge and internalised through hegemonic discourses. This is given an impetus in higher education in the Global South where it is sustained as well through the curriculum (Ndlovu-Gatsheni, 2021, p. 33).

Ndlovu-Gatsheni's views speak about the reality that is captured in the Shona axiom, *Itsitsi dzei tsvimborume kubvisa mwana wemvana madzihwa?* (What sort of kindness is it that makes a bachelor wipe the nose of a single mother's child?). Western epistemology is implanted into the formerly colonized's curricula for a reason which is largely to maintain subtle control that becomes manifest through donations and related activities like participating in military trainings together. However, despite these tags that are kept on the former colonies, there is really no equality between the parties because even the so-called international law is meant to further entrench capitalist and imperialist interests, and they are the champions of this ideology. It is for this and other related reasons that Mazodze, Mapara and Tsvere (2021) posit that there is a locus of epistemic annunciation where knowledge is situated in the West at the expense of the subaltern ones that are largely situated in the Global South. All this occurs in an environment where there is the creation of a Eurocentric myth that posits that knowledge can be neutral and unlocated. This has led to the tragedy that Western epistemologies are presented as the only universal truth, and subaltern non-Western knowledge is regarded as particularistic.

Coloniality of power in the Global South is also maintained through the silencing of other theories such as postcolonialism, Indigenous knowledge systems and decolonisation (Grosfoguel, Margarita and Rodriguez, 2002; Ndlovu-Gatsheni, 2018.; de Sousa Santos, 2014). These theories are silenced by being regarded as outdated and thus are considered to be lacking validity (de Sousa Santos, 2014). Hence, the coloniality of knowledge explains that colonial domination does not only include political and economic hegemony, but it can also be epistemological through the privileging of Euro-American knowledge as universal (Mazodze, Mapara and Tsvere, 2021). Coloniality is the bedrock of modernity because it represents the rationalisation of concepts and forms of knowledge that support relationships of subordination (Gilroy, 1996; Hendrick and Young, 2017). Coloniality of knowledge also defines the structural aspects of domination in which oppression and exploitation are based on epistemological and racialist lines (Smith, 2019). Unfortunately, this is made possible by the inherent belief and acceptance by most in the Indigenous community of scholars who do not write or choose to be guided by the standards and guidelines that help sustain Western epistemologies, in the process subverting Indigenous ones.

It is in this context that decoloniality seeks to address the problem of coloniality of knowledge which manifests itself in the form of the mental invasion of the colonized people. As already noted, this has been through the education system, coupled with the Western version of Christianity that among other bad influences promotes what can best be described as the Jesus Complex where believers are made to forget about the value of family and community and put all trust and faith in Jesus Christ. Knowledge production is one of the domains where Western epistemic hegemony is still dominant hence the need to redress this through the promotion of Indigenous epistemologies. This, as noted in the first sentence of this paragraph can only be possible if Indigenous scholars and also members of the communities deliberately write and re-tell their own native stories. The same act of re-telling these stories should be extended to the cultural and creative industries because great stories and visuals have huge impacts on the minds of the viewers and listeners.

Coloniality of being equally involves the exclusion of ontology and practices of the subjugated people for the benefit of the coloniality of power (Mignolo, 2011). Coloniality has divided the world into zones of being and non-being and this frame provides useful analytical evidence to explore the various levels of our global geo-politics (Maldonado-Torres, 2017). What comes out is that coloniality has an over-arching bearing on the intra-psychic as well as structural levels of human development in the Global South. In support of this view, Smith (2019) quotes Cesaire thus, "I am talking of millions of men, who have been skilfully injected with fear, inferiority complexes, trepidation, servility, despair and abasement" (Smith, 2019, p. 11). This fear leads to an unwillingness to delink and embark on epistemic disobedience because of the inferiority complex that has been planted in Indigenous people. There is this fear of failing because of the severing of connections to poisoned curricula. The injected fear has killed the spirit of adventure, exploration and freedom. The people in the Global South fail to realize that the knowledge that they get from the Global North helps the North and not the South to grow economically. The South trains human resources that are then taken by the North in what is euphemistically called brain circulation, when in reality the South hardly gets human resources from the North, save maybe for prospectors and other fortune hunters. Zimbabwe is a good example of a country that has trained manpower that has hardly benefited her.

The above-referenced scenario also proves that coloniality and globalisation come as a dehumanising experience on the peoples of the Global South because their cultures and identities are obliterated leading to what Cesaire termed 'Thingification' of Indigenous people which is quite a dehumanising experience (Freire and Shor, 1987). Thingification is achieved through the disabling and dismemberment of psycho-social capabilities emanating from the negation of being, culture and personality (Ngugi, 1997). Ndlovu-Gatsheni (2018) in addition, hypothesizes that Eurocentric modernity initiated the colonization of being through its classification and hierarchization of human beings according to race whereby whites were categorised as superior to other races. This racist categorisation has always been used to deny Africans and other Indigenous people in the Global South the completeness of being to justify their treatment as inferior.

Ndlovu-Gatsheni (2018) adds that:

> African subjectivity that emerged from these processes of racialisation and inferiorisation of blackness is one that has diminished ontological density. It became a subject that was characterised by a catalogue of deficits and a series of lacks ... such as lack of souls, writing, civilisation, democracy, development, human rights and ethics. (p. 197)

The above words buttress the sad historical fact that Africans, like all former colonized peoples, were seen as less human and of even lesser significance when compared to the white man's dog as is depicted in the song, *Sendekera* (Frederikse, 1982). Based on these racist processes, Indigenous peoples were treated as the doomed societies of the earth and this scenario is graphically portrayed by Frantz Fanon (1965) in his book *The Wretched of the Earth.* In another treatise, *Black Skin, White Masks*, Frantz Fanon (2008) argues that the subjectivity of the colonised is constructed through white assimilation leading to the fragmentation of the colonised. This policy of assimilation was vigorously pursued by the French and the Portuguese and it in some way persists to this day. Following up on the issue of assimilation, Fanon depicts the colonised African as wearing a white mask which produces in him/her a deep sense of self-doubt, inferiority and ultimately ontological insecurity. This dehumanising experience faced by Africans which was presented as a necessary process of civilisation resulting in the denial of being to the Africans has persisted well after political decolonisation and is presented in this chapter as coloniality of being. The construct that inspired Fanon's work was to overcome this condition of inferiority and for Fanon, the embracing of blackness was the key to achieving ontological security. This same coloniality of being continues through private colleges that offer the Cambridge curriculum and not the Zimbabwean one.

Colonisation implanted into the minds of the Africans the belief that it came to lighten their darkness and to some extent the belief that if settlers were to leave, they would fall back to savagery and degradation (Ndlovu-Gatsheni, 2018). This type of coloniality is preserved in textbooks, media and other outlets of Euro-American ideas in institutions of higher learning in Africa (Ndlovu-Gatsheni, 2018). In support of this argument, Ngugi (1997) advocates for decoloniality to redress the condition in which Eurocentric ways of knowing and values have invaded the mental universe of the colonized peoples resulting in the removal of the hard disc of previous African memory reminiscence which had been replaced by the insertion of the software of European and American forms of knowledge. Perceiving herself/himself as the shadow of the white colonizer, the African has accepted and internalized a sense of inferiority. He strives to attain acceptance by the White person, and not by fellow Blacks. For instance, the former British colonies still want to remain within the orbit of Britain through the empire that has been rechristened the Commonwealth of Nations. Yet it still has the British monarch as its head.

Coloniality has demeaned Indigenous Africans onto-epistemologies and cultures in an attempt to silence them. Mignolo (2011) posits that through coloniality Eurocentrism has spread its Western practices and values. These have now been imposed on Africans and other peoples of the Global South as personal and cultural identities for them to adopt resulting in the subalternization of their cultures, languages, epistemologies and ontologies. For example, in West Africa, French is now accepted as an African language, while in South Africa and Namibia, the same is true of Afrikaans (Shizha, 2015). The

African people's social, political, economic, cultural and psychological lives have been modelled in ways that reflect Euro-American life styles and world views. African people have continued to be alienated by Euro-American ways of knowing and systems of subjectivity that oppress and exploit them in terms of political power in addition to areas such as epistemic and cultural values that are esteemed and protected.

FURTHER RESEARCH DIRECTIONS

Education is often touted as the key to success. However, no one has ever asked whose education and whose success. These are pertinent questions because as far as the education of the formerly colonized has been nothing other than miseducation. This miseducation needs to be countered by developing new tools of education and research becomes critical in this regard. Among other areas that need research, and development is the issue of language. It has to be noted that in terms of science and technology, no nation has ever developed by relying on the languages of foreigners, and worse still former colonizers. The examples are galore. Chemla (2019) notes this reality about the scientification and technologization of Chinese even though there was the myth of the European languages as the only ones "characterized by the fact that they offered appropriate tools that enabled scientists to define and enact classifications" (p. 112).

The reason why language becomes critical is because it is a carrier of worldviews, and these are critical in technology development because technologies that are devoid of a people's senses and sensibilities are not only disruptive but also destructive. It is through languages that people can access Indigenous epistemologies that for instance relate to agriculture, medicine, metallurgy and even knowledge about the universe, which is becoming more and more important, especially in the face of climate change. The death and exclusion of Indigenous languages from the development agenda has the effect of marginalizing the communities and the knowledge that is embedded within their tongues. The results are usually catastrophic. Events in Chimanimani for example, in March of 2019 when Cyclone Idai destroyed lives and livelihoods speak to such dangers (Siamena, Mapara & Chiweshe, 2024).

Besides issues relating to language, there is a need to further unravel the Indigenous practices related to health delivery. It is disheartening to note that Western medical practice is largely premised on the curative paradigm instead of the preventive. This is largely motivated by the capitalist desire for profits and not the health of communities. Equally significant and worth pursuing are further studies in the areas of restorative justice and restitution as informed by an Indigenous lens.

CONCLUSION

The analysis outlined above exposes the coloniality of power, knowledge and being as epistemic racism that goes beyond the suppression of African people's affective and imaginative potential to what Mungwini (2017)) describes as onto-epistemic terrorism. This is because Eurocentrism has created an enclosure on knowledge construction and reconstruction, and this larger has evicted other epistemologies from the realm of knowledge creation. Therefore, scholars in the Global South should be wary of the enduring epistemic racism infused in Eurocentric ideology that guides academic and cultural practices in their institutions. The writers argue that globalisation with its neo-liberal ideology is being used by the West to entrench Eurocentric epistemologies and ontologies while those of the Global South are being invisibilised. The place of epistemologies of the South in the global village has been peripherized and relegated to underneath the table where like the biblical poor Lazarus, the people of the South feed on the table of Eurocentrism. Decoloniality presents an alternative way of theorising to effectively decolonise power, knowledge and being. Decoloniality should therefore guide the Global South in its efforts to dismantle the coloniality of power matrices posed by globalisation that continue to exist after the fall of the empire.

Put differently, it is important that academics and thinkers in the Global South, both physically and conceptually engage with matters relating to knowledge justice. This is whereby all spheres are accepted and acknowledged as areas and places from which knowledge can be generated. This also means that the universalization of Western epistemologies should stop, and it should be acknowledged as one form of knowledge among others.

REFERENCES

Alidou, H., Boly, A., Brock-Utne, B., Diallo, Y. S., Heugh, K., & Wolff, H. E. (2006). Optimizing learning and education in Africa – the language factor: a stock-taking research on mother tongue and bilingual education in sub-Saharan Africa. Paper presented at the ADEA Biennial Meeting, Libreville, 27-31 March. Retreieved on 10 February 2023 from https://biennale.adeanet.org/2006/doc/document/B3_1_MTBLE_en.pdf

Barnett, A. (2024). Why I've removed journal titles from the papers on my CV. *Nature*. Advance online publication. DOI: 10.1038/d41586-024-02596-y PMID: 39122980

Bennett, K. (2015). Towards an epistemological monoculture: mechanisms of epistemicide in European research publication. In Plo Alastrué, R., & Pérez-Llantada, C. (Eds.), *English as a Scientific and Research Language: Debates and DiscoursesEnglish in Europe* (Vol. 2, pp. 9–36). De Gruyter Mouton. DOI: 10.1515/9781614516378-004

Chawla, D. S. (10 February 2023). Elsevier journal under fire for rejecting papers that didn't cite enough of its old papers. *Chemical & Engineering News*. Retreieved on 10 February 2023 from https://cen.acs.org/policy/publishing/Elsevier-journal-under-fire-for-rejecting-paper/101/web/2023/02

Chemla, K. (2019). Needham and the issue of chinese as a language for science: Taking a linguistic turn materially. *Isis*, 110(1), 109–115. DOI: 10.1086/702915

De Sousa Santos, B. D. (2014). *Epistemologies of the South: Against Epistemicide* (1st ed.). Routledge.

Frederikse, J. (1982). *None But Ourselves: Masses vs Media in the Making of Zimbabwe*. Ravan Press.

Friere, P. (2005). *Education for critical consciousness*. Seabury Press.

Gilroy, P. (1996). Route work: the black Atlantic and the politics of exile. In Chambers, I., & Curti, L. (Eds.), *The post-colonial Question: Common Skies, Divided Horizons* (pp. 17–30). Routledge.

Grosfoguel, R. (2003). Introduction: geopolitics of knowledge and coloniality of power- thinking Puerto Rico and Puerto Ricans from the colonial difference. In *Colonial subjects: Puerto Ricans in a global perspective*. University of California Press.

Hendrick, A., & Young, S. (2017). Decolonising the curriculum, decolonising ourselves: Experiences of teaching in and from the "third space". *Advances in Social Work and Welfare Education*, 19(2), 9–24.

Knopf, K. (2015). The turn toward the Indigenous: Knowledge systems and practices in the academy. *Amerikastudien. American Studies (Lawrence, Kan.)*, 60(2/3), 179–200.

Makobela, T. (2017). *Perceptions of black men in Katlehong about female "yellow bones": A Case Study*. Creative Commons.

Maldonado-Torres, N. (2007). On coloniality of being: Contributions to the development of a concept. *Cultural Studies*, 21(2-3), 240–270. DOI: 10.1080/09502380601162548

Maldonado-Torres, N. (2017). Frantz Fanon and the decolonial turn in psychology. Modern / colonial methods to the decolonial attitude. *South African Journal of Psychology. Suid-Afrikaanse Tydskrif vir Sielkunde*, 47(4), 432–441. DOI: 10.1177/0081246317737918

Mapara, J. (2023a). The african university and the urgent need for decoupling from the global north. In Ndofirepi, A., Vurayai, S., & Erima, G. (Eds.), *Unyoking African University Knowledge: A Pursuit of the Decolonial Agenda* (pp. 43–63). Brill. DOI: 10.1163/9789004548107_004

Mapara, J. (2023b). Tragic labels, catastrophic consequences: colonial treachery and the Cameroonian calamity. In Mavengano, E., & Mhute, I. (Eds.), *Sub-Saharan Political Cultures of Deceit in Language, Literature, and the Media* (Vol. II). Palgrave Macmillan., DOI: 10.1007/978-3-031-42883-8_13

Mapara, J., & Mutasa, D. E. (2011). Avoiding potential knowledge death: Protecting Indigenous knowledge in a globalising world. *Southern African Journal for Folklore Studies*, 21(2), 12–25.

Mazodze, C., Mapara, J., & Tsvere, M. (2021). Mainstreaming Indigenous African epistemologies into student development in higher education: A case of case. *East African Journal of Education and Social Sciences*, 2(3), 98–210. DOI: 10.46606/eajess2021v02i03.0118

Meharg, S. J. (1999). *Making it, Breaking it, and Making it Again: The importance of Identity in the Destruction and Reconstruction of War-torn Societies*. Master of Arts Thesis. Kingston, Ontario: Royal Military College of Canada.

Meharg, S. J. (2011). Restoration and reconstruction for environmental security. In Machlis, G., Hanson, T., Špirić, Z., & McKendry, J. (eds) *Warfare ecology*. NATO Science for Peace and Security Series C: Environmental Security (pp. 177-188). Springer. DOI: 10.1007/978-94-007-1214-0_12

Mignolo, W. D. (2009). Epistemic disobedience, independent thought and decolonial freedom. *Theory, Culture & Society*, 26(8), 159–181. DOI: 10.1177/0263276409349275

Mungwini, P. (2017). "African know thyself": Epistemic injustice and the quest for liberative knowledge. *International Journal of African Renaissance Studies*, 12(2), 5–18. DOI: 10.1080/18186874.2017.1392125

Naidu, T. (2024). Epistemic disobedience-undoing coloniality in global health research. *PLOS Global Public Health*, 4(4), e0003033. DOI: 10.1371/journal.pgph.0003033 PMID: 38648248

Ndlovu-Gatsheni, S. J. (2018). *Epistemic Freedom in Africa: Deprovincialisation and Decolonisation*. Routledge. DOI: 10.4324/9780429492204

Ndlovu-Gatsheni, S. J. (2021). Beyond coloniality of internationalism. *International Relations*, 6(1), 1–8.

Ngugi wa Thiong'o. (1986). *Decolonising the Mind: The Politics of Languages in African Literature*. Zimbabwe Publishing House.

Ngugi wa Thiong'o. (1992). *Moving the Centre: The Struggle for Cultural Freedoms*. James Currey & Heinemann.

Ngugi wa Thiong'o. (1997). *Detained: A Writer's Prison Diary*. Heinemann.

Olatunji, S. A. (2010). Thematic changes in postcolonial Africa literature: From colonialism to neocolonialism. *Sino-US English Teaching*, 7(10), 125–134.

Patin, B., Sebastian, M., Yeon, J., Bertolini, D., & Grimm, A. (2021). Interrupting epistemicide: A practical framework for naming, identifying, and ending epistemic injustice in the information professions. *Journal of the Association for Information Science and Technology*, 72(10), 1–13. DOI: 10.1002/asi.24479

Shizha, E. (2015). Reclaiming Indigenous cultures in Sub-Saharan African education. *Indigenous Education: Language. Culture and Identity*, 2(1), 301–317. DOI: 10.1007/978-94-017-9355-1_15

Siamena, E., Mapara, J., & Chiweshe, M. K. (2024). IDEWS and the building of resilient communities in the eastern highlands of Zimbabwe. *Journal of African History. Archaeology and Tourism*, 2(1), 75–97. DOI: 10.31920/2753-3204/2024/v2n1a4

Smith, L. H. (2019). Epistemic decoloniality as a pedagogical movement in Social Work: A turn of theory to anti-colonial theorists. *Critical Social Work*, 1(5), 1–20.

Walsh, C. (2018). The decolonial for: resurgences, shifts and movements. In Mignolo, W. & Walsh, C (Eds). *On Decoloniality: Concepts, Analytics, Praxis* (pp. 15-32). Duke University Press.

ADDITIONAL READINGS

Gilroy, P. (1993). *The Black Atlantic: Modernity and Double Consciousness*. Verso.

Kyriakakis, I. (2024). *Colonizing the Mind: The Witchcraft of the Global Class System*. The Colleagues' Publications.

Mignolo, W. D., & Walsh, C. E. (2018). *On Decoloniality: Concepts, Analytics, Praxis*. Duke University Press.

Ndlovu-Gatsheni, S. J. (2013). Coloniality of power in postcolonial africa: myths of decolonization. CODESRIA (Council for the Development of Social Science Research in Africa).

Ndlovu-Gatsheni, S. J. (2013). The entrapment of Africa within the global colonial matrices of power: Eurocentrism, coloniality and deimperialization in the twenty-first century. *Journal of Developing Societies*, 29(4), 331–353. DOI: 10.1177/0169796X13503195

Ngugi wa T. (2009). *Something Torn and New: An African Renaissance*. Basic Civitas Books.

Ngugi wa T. (2011). *Globaletics: theory and politics of knowing*. Columbia University Press.

Smith, L. T. (2021). *Decolonizing methodologies: research and Indigenous peoples*. Zed Books. DOI: 10.5040/9781350225282

Woons, M., & Weier, S. (Eds.). (2017). *Critical epistemologies of global politics*. E-International Relations Publishing.

KEY TERMS AND DEFINITIONS

Coloniality: Coloniality is related to colonialism but the two are not the same. While colonialism signifies a political and economic relation in which the independence of a nation or a people is made subservient to that of another power or nation, coloniality focuses on the long-standing patterns of power that materialized as a consequence of colonialism. Coloniality thus refers to relations sustained and transmitted through culture, labour, intersubjective relations, and knowledge production. These are activities that go well beyond the stringent confines of colonial governments. Consequently, coloniality lives on long after colonialism. It is something that is sustained and kept alive in books, in the benchmarks that are accepted and promoted as essential for academic performance. Coloniality also lives on in cultural patterns, in the so-called 'common sense', in the self-image of peoples, in ambitions of self, and so many other phases of our modern experience. It is also actively promoted by the former colonizers through scholarships and so-called aid and advisory services. "In a way, as modern subjects, we breathe coloniality all the time and everyday" (Maldonado-Torres, 2007, p. 243).

Decoloniality: It is linked to coloniality and how it manifests itself. It strives to comprehend the doggedness of coloniality, not just in the legacies of imperialism but in the precise constitution of the world. Decoloniality focuses on resisting the adoption and use of epistemologies produced by coloniality. It seeks liberation through the decentring and delinking from the colonial matrix of power and involves an approach of evaluation that contests the supposed universalism of coloniality and its associated systems. According to Catherine Walsh (2018, p. 17), 'decoloniality seeks to make visible, open up, and advance radically distinct perspectives and positionalities that displace Western rationality as the only framework and possibility of existence, analysis and thought'. It is thus driven by the argument that the West does not have a monopoly on knowledge and knowledge generation.

Delinking: To delink means to detach from the overall structure of Western knowledge so as to engage in an epistemic reconstruction. This means the reconstitution of ways of thinking, languages, and ways of life and being in a world where the rhetoric of modernity is disavowed and the logic of coloniality is implemented. The focus of this idea is therefore to disconnect and deliberately so from colonial biases of what constitutes knowledge, ways of knowledge production and how it is shared. There is also a need to delink from approaches to research, teaching, and learning. This means that it becomes essential to revisit the basic frameworks of research, principally the concept of objectivity and think through the possibility of taking on more reflexivity within the existing inquiry frameworks.

Epistemic Disobedience: This is part of the Decolonial movement and emphasizes the importance of the Global South researchers creating their own space for research. It places the burden of funding research on the institutions of the South so that they are not dictated to by funders from the Global North. It rejects what scholars in the Global South perceive as the repressive and exclusionary knowledge production machinery of the Global North. Epistemic disobedience also demands that scholars of the Global South set up their own journals and publishing houses that are not subjected to the gatekeeping practices of the Global North's scholars and publishers. When it comes to funding calls, scholars in the Global South should avoid those that oblige them to collaborate with Global North researchers to access large grants. Epistemic disobedience argues that such approaches "are blatantly inequitable, representing modern iterations of colonial knowledge and resource extraction and suppressing Global South researchers' agency" (Naidu, 2024, np).

Epistemicides: This term is derived from two terms, *episteme* which means knowledge, and *–cide*, a stem that means to kill as in suicide. Epistemicide, singly, or epistemicides thus means an act or acts of destroying knowledge, especially that of the othered. This term is used as part of the decolonial vocabulary that speaks on the negative results of colonialism on the knowledge of the formerly colonized. Their knowledge was not just spurned, but it was actively marginalized and destroyed and labelled as non-knowledge and the hallucinations of the backward and uncivilized. Patin et al. (2021, p. 2) underscore this tragedy when they state, "Epistemicide is the killing, silencing, annihilation, or devaluing of a way of knowing". It is a form of genocide where the knowledge owners are not necessarily exterminated physically, but in the Global South, people cease to exist when they lose their knowledge including languages.

Global North: The term Global North is not a geographical designation but a conceptual one that is used to refer to the societies of Europe, North America, Australia, Israel, and South Africa, amongst others that are viewed as economically developed. These are countries that in the recent past were tagged as First World. The Global North is powerful both militarily and economically such that it controls and owns most corporates that believe that the Global North owns and controls the world. One of the main tools that the Global North has used to dominate and exploit others, especially those in the Global South is education or knowledge production. The knowledge of the Global North is what has been universalized and other epistemologies have been marginalized, and their existence has not been acknowledged. The Global North seeks to perpetuate its hegemony on the world through the coloniality of power, and this power manifests itself through education and Euro-American modernity which they parade as the highest achievement of human civilization.

Global South: This term is used to refer to countries that are not necessarily in the Global South, although the majority of them are. These are countries that are viewed and accepted as economically backward, and they are found in countries in Africa, India, China, Brazil, and Mexico amongst others. The term however goes beyond the economic and also refers to countries that generally imbibe knowledge that is generated and produced in the Global North. Countries of the Global South are viewed as incapable of coming up with real knowledge, but of late, through theories like decoloniality, the Global South has started speaking back to the Global North and is demanding that there be epistemic justice.

Identicide: This term was coined by Sarah Jane Meharg in 1999. Identicide is the carefully considered, methodical and targeted annihilation of for example places, symbols, and objects, including ideas, values and aesthetica, that epitomize the identity of a people. The destruction is deliberately meant to erase the cultural chronicles and memories of those people to demoralize or absorb them into another cultural/political entity. There is also a huge possibility of getting rid of those people from a given area (Meharg, 1999). In Decolonial studies, the term can be extended to the area of epistemology where the formerly colonized have by and large been exterminated through not only the destruction of their symbols and languages but also by way of the marginalization and trashing of their epistemologies.

Memoricide: The word was coined by Miro Grmek in 1992. Memoricide is the erasure of the history of one people to write that of another people's over it. In the case of the formerly colonized, memoricide was committed and experienced when the colonialists erased the memories and histories of the oppressed. This has been through the renaming of places, for instance, and linguicides where Indigenous languages have been replaced by those of the former colonizers. In addition to that, memoricide has been through the deletion of Indigenous epistemologies and the imposition of Western knowledge and practices as supreme and unquestionable. This has seen the establishment of Western-style education institutions that have also overshadowed and even decimated Indigenous ways of knowledge impartation.

Peripherization: This word means to push to the margins or the periphery. It refers to acts of in-siginification. In the context of decolonization and decoloniality, the term means the marginalization of Indigenous epistemologies. They are labelled as non-knowledge in a world that is dominated by the knowledge of the Global North. Peripherization is part of the epistemic injustice and memoricides that have been visited upon the Global South by the Global North.

Chapter 7
Dismantling the Eurocentric Cognitive Empire and Re-Centering African Knowledges in Pursuit of Cognitive Justice in the 21st Century

Moloi Lehasa
University of South Africa, South Africa

ABSTRACT

This chapter provides a theoretical study through exploring the concept of cognitive empire alongside Afrocentricity, highlighting how the European construction of the modern world system has perpetuated a triumphalist narrative. Since the 15th century, Europe has systematically undermined African intellectual traditions, disrupting Indigenous cognitive processes and facilitating conquest. This historical trajectory has led to the dominance of European epistemologies while marginalising non-European contributions. In response, decolonial movements in Africa aim to reclaim oppressed humanity and challenge the moral decline associated with Eurocentric paradigms. The chapter argues that Indigenous knowledge systems are crucial for envisioning alternatives to Western-centric views and advocating for cognitive justice. Ultimately, it positions Afrocentricity as vital for reimagining knowledge production in the African context.

INTRODUCTION

This chapter aims to contextualize the politics surrounding knowledge production in Africa by examining the concepts of the cognitive empire and Afrocentricity as counterarguments to eurocentrism, which is posited as a hegemonic framework. The chapter advocates for the perspective that all forms of knowledge are inherently situated within the socio-cultural and geographical contexts from which they emerge. Given the historical backdrop of European colonization in Africa, knowledge itself remains colonized. Consequently, decolonial movements advocate for a critical re-examination of existing

DOI: 10.4018/979-8-3693-8447-3.ch007

knowledge systems to facilitate renewal and to transcend oppressive paradigms, particularly within the Global South and Africa specifically.

Knowledge serves as both foundational and essential for envisioning a future that diverges significantly from the present. Consequently, a community devoid of its own epistemologies lacks both a historical narrative and a future of its own design. Despite the critical role of knowledge in shaping the destinies of peoples, the ascendancy of Western-centric modernity has delegitimized alternative epistemologies and ways of knowing that exist outside the Western framework of perceiving, conceptualizing, and understanding the world. According to Moya (2011) the knowledge we produce is intimately influenced by how we conceptualize our shared social order and who we understand ourselves to be in that world. Thus, "our conceptual frameworks are inseparable from how we comprehend ourselves in terms of our gender, culture, race, sexuality, nationality, religion, age, and profession- even when we are not consciously aware of how these aspects of ourselves affect our interpretative horizons" (Moya, 2011, p. 81).

I have chosen to sub-divide my reflections into two sections: firstly, I introduce the concept of cognitive empire and its implications for knowledge production. Secondly, I discuss Afrocentricity, Knowledge decolonization and the significance of asserting African agency in pursuit of cognitive justice.

COGNITIVE EMPIRE AND THE POLITICS OF KNOWLEDGE PRODUCTION

According to Ndlovu-Gatsheni (2021), the cognitive empire is known by various names. Ngugi wa Thiong'o (1986) defined it as "the metaphysical empire," which, unlike the physical empire, had to "bend the body," compelling "a distorted consciousness of the relationship of their actual reality" (as cited in Ndlovu-Gatsheni, 2021, p. 45). Gildea (2019) elaborates on the cognitive empire as the "empire of the mind," capturing how epistemology frames ontology in a more profound historical manner. Ndlovu-Gatsheni (2021) highlights that at the centre of the cognitive empire is the imperial epistemology that facilitated the enslavement and colonization of the global South. This imperial epistemology was also underpinned by imperial racist science, which was extractive in its orientation. The term epistemological extractivism is clearly expressed by Collyer, Connell, Maia, and Morrel (2019) in their book *Knowledge, and global power: Making new sciences in the South*:

> The conquest of the world by Europeans and North American power, over 500 years of modern empire and globalization, not only produced material wealth for the imperial powers. It also produced a rich dividend of knowledge. The colonized world was a fabulous mine of information, and the colonizers began sending back information and specimens as early as they sent spices, silver and gold. (Collyer *et al.,* 2019, p. 9)

Much of this raw data became reserved in European libraries, art museums, research institutes, universities, and government agencies. Of interest is also how they expropriated African martyrs' body parts and kept them in their museums; for example, "it was only in June 2022 that the Belgium government decided to return the tooth of the anti-colonial leader of Congo, Patrice Lumumba as evidence of colonial atrocities" (Gonçalves, 2023, p. 1). As part of exposing the brutal character of the European cognitive empire, Asante (2014) argues that 'the deliberate inferiorization of Africa by Western scholars was part of the philosophical and cultural fabric of Europeanization. It was as if the definition of the West must include a discourse on Africa as a negation. According to George, Reddy and Searle-Chatterjee (1990, p. 1), the essence of the cognitive empire was about advancing ethnocentrism, defined as 'the tendency to view one's own ethnic group and its social standards as the basis for evaluative judgements concerning the practices of others – with the implication that one views one's own standards as superior.

Ndlovu-Gatsheni (2021) argues that the concept of the cognitive empire should be understood within the context of historical processes characterized by colonial and neocolonial economic dominance, which, in turn, has provided ideological justification for such dominance.

George et al (1990) argues that the persistence of Eurocentrism as a hegemonic paradigm has had the following negative effects on the non-European counterparts, and Africa in particular (p. 3):

(i) It has damaged non-European societies through the colonisation of their intellectuals
(ii) It has impoverished the academic disciplines themselves which remain unaware of alternative sources of knowledge outside mainstream development
(iii) It functions regardless of intention, to legitimate international systems of inequality

Ake (2011), in his book titled Social Science as Imperialism: The Theory of Political Development, clearly captures how the Western Social Sciences foist capitalistic values and capitalistic development on what was labelled as the Third World, now referred to as the Global South, to serve imperialist ends. As a result, the choice of subject matter, theorization, methodology, and epistemology in Social Sciences have nothing to do with the context of those they are colonizing rather they all reflect the constraints of the context which produced them and seek to impose their understanding of the world to advance Western mono-culturalism. Thus, what we have in most fields of study is Eurocentric indoctrination, which marginalises other people's ways of knowing and pervades stereotypes about them. The coloniality of knowledge in Africa mainly was achieved through the colonial education system. This education system contributed to the creation of false consciousness among the colonized. George et al (1990, p. 3) highlights that an important part of the strategy of domination was to convince the colonised that knowledge, whether in the sphere of culture, science or technology, could be acquired only through mediation of the colonial rulers. This belief remains strong in post-colonial African societies faced everyday with the demonstration of the marvels of western science, and technology used to foist inferiority amongst Africans that they have no innovative capacity, thus subjecting them to psychological dependence on Europeans. Asante (2007) argues that instead of seeking to produce independent thinkers, colonial education in the continent of Africa has always sought to produce subjects who would easily consent to their own colonization and subjugation. By extension. According to Badat (2023, p. 1) the Westernised universities in Africa produced African leaders who became administrators of the colonial system even in the absence of settler governments after the demise of juridical-administrative colonialism and apartheid in Africa. For this reason, the success achieved through colonial education has been to dislocate the victim's epistemic location to speak as though they were Europeans and not Africans.

In the realm of knowledge production, colonial epistemology transformed African knowledge producers into native informants, guiding Western thinkers on how to formulate theories that ultimately serve to further marginalize African communities.

Ngugi wa Thiongó (2009) uses the concept of cognitive empire to explicate on how colonialism operates in the sphere of the colonized mind. The cognitive empire specialized in the invasion of the mental universe of the colonized people, he elaborates on the operative logic of this empire in *Something Torn and New* in the following manner: "Get few natives, empty their hard disk of previous memory, and download into them software of European memory" (Ngugi wa Thiong'o, 2009, p. 21). Ndlovu-Gatsheni (2021) also attests that "the key consequences of all these processes have been epistemicides (which constitutes killing of existing endogenous knowledges), linguicides (killing of existing Indigenous languages and the imposition of colonial languages), culturecides (killing of Indigenous cultures and setting afoot cultural imperialism) as well as alienation (exiling of the natives from their languages, histories, cultures, and even from themselves). The ultimate agenda of the predatory cognitive empire is to fully control its targets into full submission and obedience, thus turning them into objects, which means. These slaves only take an instruction without question. It will take Africans to challenge their condition by calling for the decolonization of knowledge to reclaim their minds from dislocation and to completely dismantle the cognitive empire as they unchain from its captivity.

Ama Mazama, in her article titled *Afrocentricity and the Critical Question of African Agency*, highlights the following point about how Eurocentrism as a colonizing framework:

> There are at least four assumptions that European meta-paradigm has advanced which have played a major and negative role as far as African people are concerned: 1) all human beings evolve along the same line; 2) the European experience is universal; 3) Europeans are superior; and 4) "others" are defined by their experiences with Europeans. In other words, the European meta-paradigm rests, among other things, on the belief in the superiority and universality of the European experience. (Mazama, 2014, p. 1)

For this reason, European knowledges are informed by Eurocentric assumptions to the exclusion of others and have played a major role in making Africans secondary, even to themselves. Mazama (2001) attests in *The Afrocentric Paradigm: Contours and Definitions,* that "we thus find ourselves relegated to the periphery, the margin, of the European experience, we have become spectators of a show that defines us from without. We thus do not exist on our own terms rather on borrowed, European ones. We are therefore dislocated and having lost sight of ourselves during European decadence and madness, it has become difficult for us to orient our lives in a positive and constructive manner" (2001, p. 387).

Lander and Past, in their book titled Eurocentrism, Modern Knowledges, and the Natural Order of Global Capital (2002) highlight that the main assumptions of the perspective of Eurocentric knowledge can be summarized in the following terms:

1. Eurocentric knowledge is based on the construction of multiple and repeated divisions or oppositions. The most characteristic and significant of these—but not the only ones, to be sure—include the basic, hierarchical dualisms of reason and body, subject and object, culture and nature, masculine and feminine.
2. European regional or local history is understood as universal history. According to this perspective, Europe serves as the model or reference for every other history, representing the apex of humanity's progress from the "primitive" to the "modern",
3. Differences from *others* are converted into value differences, time-space distances, and hierarchies that define all non-European humans as inferior ("savage," "primitive," "backward," "underdeveloped"). The category of race as an instrument for classifying the different peoples of the world—on a scale from superior to inferior—plays a central role here.
4. Scientific knowledge and technological development advance in an upward linear direction toward ever higher levels of knowledge and greater ability to usefully transform the environment.

(Lander and Past, 2002, p. 20).

Without deconstructing the Eurocentric epistemic framework, Africa is unlikely to realize a future that deviates from its current trajectory or envision an alternative future beyond the prevailing conditions. The persistence of global colonial domination in knowledge production impedes this transformation. This is attributable to the fact that the contemporary system of knowledge production is not structured to facilitate change; rather, it is designed to maintain a state of equilibrium that upholds a hierarchical power structure. In this system, the Western world benefits from the established order, while non-Western regions remain marginalized. It is within this context that it becomes necessary to explore the role of Afrocentricity as a locative philosophical standpoint in the arena of knowledge inquiry to overcome the European created cognitive empire. Without relooking at the negative effects of colonial education system it will remain impossible to deepen decolonisation in Africa.

AFROCENTRICITY, KNOWLEDGE DECOLONIZATION AND ASSERTION OF AFRICAN AGENCY IN KNOWLEDGE PRODUCTION

The theory of Afrocentricity has become the dominant intellectual paradigm in the study of African phenomena in at least the past two decades. Although the concept had existed before and was used by thought leaders such as Kwame Nkrumah and WEB du Bois, it catapulted to the status of the philosophical paradigm in the 1970s through the brilliance of Molefi Kete Asante, along with a group of African American scholars referred to as Tempel circle of the Afrocentrists (Moloi, 2024, p. 27). Afrocentricity is a paradigm based on the idea that African people should re-assert a sense of agency to attain sanity against the ravaging effects of Eurocentrism as a hegemonic ideological framework (Asante, 2009, p. 1). In his book The Afrocentric Manifesto, Asante defines Afrocentricity as "a paradigmatic intellectual perspective that privileges African agency within the context of African history and culture trans- continentally and trans-generationally" (Asante 2007, p. 2). The primary role of Afrocentricity has been to seek to depose the dominant Eurocentric paradigm, which has been labelled as a leading cause of the de-centering and dislocation of Africans from their historical and cultural centre. Thus, Afrocentricity proposes the relocation and the re-centering of the African persons within their own historical, political, and cultural context as a remedy for their liberation from Eurocentric domination in pursuit for epistemic decolonization. More than anything else, Afrocentricity is a paradigm of agency and action, it seeks to remove Europe, its knowledge system, and its hegemonic practices from the centre of African reality and replace it with an African-based knowledge system informed by African history and culture as framed within the lived experiences of African people trans-continentally and trans-generationally.

It is important to note that according to Ndlovu-Gatsheni (2015) anti-colonial struggles of the 18th century mainly brought an end to direct white colonial administration but did not dismantle colonization in its indirect form, namely, neocolonialism. To that effect, Mazama (2003, p. 4) argues that Africans are still mentally subjugated. The reason for this sad mental situation is that we have been fighting against the evil colonization as an economic and political problem rather than a total conceptual distortion leading to confusion. For this reason, Mazama (2003, p. 4) contends that colonization must be analysed within the broader context of the European cultural ethos how they have deployed their knowledge base as a stratagem to conquer the African minds to generate economic exploitation and political suppression. Afrocentricity, therefore, challenges the idea that African people must ideologically be dependent on Eurocentrism in conceptualizing their reality, instead they must be grounded on their own knowledge informed by their history, culture, and must assert their agency to wield control. The main problem that must be challenged is Western solipsism in matters of thought whereby the Western voice and opinion is in monologue with itself without a critical interrogation that leads to self-introspection. Asante (2014, p. 3) re-emphasizes that "what remains the problem with Western thought is its inability to allow space for other knowledges, and therefore, it becomes self-absorbed in some notion of Europe as the categorical universal for the world. Such self-centredness has left the rest of the world searching for a theoretical corrective.

As part of the strategy to dismantle the Eurocentric epistemic empire, Asante's (2017, p. 26) Afrocentricity advances the following five principles which undergird it and its philosophical approach to knowledge inquiry:

- Afrocentricity asserts that Africans ought to have an intense interest in psychological or emotional location as determined by symbols, motifs, rituals, ceremonies, and signs. The African thinker should be seeking to determine by icon, myth, motifs, symbols, etc., where the text, person or event is located. This literally implies that no phenomena can be apprehended adequately without locating it first.
- Afrocentricity asserts that the scholar should possess a commitment to finding the subject-place of Africans in any inquiry be it political, social, economic, religious, and architectural.
- Afrocentricity insists that there must be a defence of African cultural elements as historically valid in the context of art, music, education, and literature. To defend African cultural elements means that we must study African cultures and be experts in our field of inquiry. Afrocentricity remains a form of cultural criticism that examines etymological use of words and terms to understand the researcher's location.
- Afrocentricity celebrates "centredness" and "agency" and activates commitment to lexical refinement and reconstruction that eliminates any misperceptions about Africans. According to Mazama (2001), "the concept of center occupies a critical place in the Afrocentric conceptual apparatus. It is fundamentally based on the belief that one's history, culture, and biology determine one's identity. That identity, in turn, determines our place in life, both material and spiritual" (Mazama 2002, p. 397).
- Afrocentricity remains a powerful imperative from historical and social sources to revise the collective text of the African people. Afrocentricity advocates re-education of Africans to overcome the negativity that has been conferred on Africans by European scholarship to liberate knowledge from Eurocentric hegemony.

In addition to the five principles explicated above, Mazama (2001, p. 399) has further articulated four critical cardinals which must inform Afrocentric research, which include cosmology, epistemology, axiology, and aesthetics of African people and must be centred in African experiences. Regarding cosmology, Asante (1990, pp, 8-9) asserts that the place of African culture in the myths, legends, literatures, and oratories of African people makes up, at the mythological level, the cosmological issue within the Afrocentric enterprise. The critical question given Afrocentrists, is what role does African culture play or contribute to the African's interface with the cosmos? To this effect, Mazama (2001) maintains that the essence of life is spiritual for African people. Consequently, "Afrocentric methods as well as Afrocentrically generated knowledge must reflect the primacy of the spiritual, the relationship between the physical and the spiritual, as well as the interconnectedness of all things" (Mazama 2001, p. 399). Therefore, the focus in Afrocentricity is holism. This communicates the idea that systems (physical, biological, chemical, social, economic, mental, and linguistic) and their properties should be viewed as a whole, not just as a collection of parts. Therefore, this implies that Afrocentricity rejects reductionism, which is mostly embedded within Eurocentrism, which has not only dismembered the African continent, but also their view of life.

The second cardinal is African epistemology. To question 'if there is such a thing called an 'African epistemology' is to question the very existence of the people called Africans, but it is to question their ability to produce any knowledge unique to their circumstances, in line with their life experiences. It is therefore impossible to imagine any future for Africans if we assume Africans do not exist. Asante (1996, p. 256) states that by the prefix "African" he is referring to a "composite African" not a specific discrete African orientation which would rather mean ethnic identification, i. e. Yuroba, Zulu, Nubia, etc. Afrocentricity, therefore, affirms the ability of Africans to think and be their own agents, using their own knowledge system informed by their history and lived experiences. The idea of Epistemology, according to Wong (2008), means the branch of philosophy that examines the nature of knowledge and the processes we use to acquire and value it. Alkebulan (2007, p. 414) also emphasise that epistemology refers to knowledge or truth in an Afrocentric inquiry, and he further states that music, language, rituals, and dance all embody what is considered truth. These modes convey the knowledge and lessons of African people and their culture. To this effect, Mazama (2001) affirms that an Afrocentric inquiry begins with self-knowledge but must also be conducted through the interaction between the examiner and the subject. The Afrocentric knowledge is validated through the combination of historical understanding and intuition; that is, knowing is both rational and suprarational. Furthermore, she contends, "one must remember always that knowing with one's heart is superior to all and priceless" (Mazama 2001, p. 399).

Senghor (1964) also supports Mazama (2001), when he states that African modes of cognition are more emotionally laden than being rational. He adds that European forms of thought or reasoning rely solely on rationalism; it is analytic and discursive whereas African reasoning is intuitive. African knowledge therefore emerges from cooperation of all human faculties and experience, and not only through reasoning. Akbar (1984) also emphasizes that within an Afrocentric mode of thought emotional reaction, as a means of knowing and as a balance for rationality, is legitimate. Therefore, the method through which the African arrive at trustworthy knowledge of reality is intuitive and personal experience (Ruch and Anyanwu, 1981).

The third critical aspect in Afrocentric inquiry as stated by Mazama (2001) is an axiological aspect. Axiology, which is also called the theory of value; it is the philosophical study of goodness, or value. Within an Afrocentric inquiry, axiology concerns what is acceptable and what is not in the context of the African worldview. Afrocentrists believe that the practice of research must hold African history and the lived experiences of the Africans in the highest esteem. What is in the best interest of the African people, in particular, their liberation from all forms Eurocentric hegemony should occupy the heart of the researchers. Hence, the focal point in Afrocentricity is the relocation and re-centering of Africans within their cultural context, to dislodge European ways of thinking, being, and feeling and restoring them to the modes that are germane to their cultural perspective. To this effect, Mazama (2001, pp, 387-388) argue that without re-centering us to the African cultural context we remain dislocated and having lost sight of ourselves in the mist of European decadence and madness, it becomes increasingly difficult for us to orient our lives in a positive and constructive manner.

Lastly, the important issue in Afrocentricity relates to aesthetics; according to Asante (1990, p. 11) aesthetics exist as the leading elements of the African response to art, plastic or performing. Welsh-Asante (1994) also asserts that an understanding of aesthetics facilitates the paradigm for artistic, literary, and philosophical criticism and scholarship. In addition, Welsh-Asante (1994) maintains that the cultural dynamics of a people creates a specific aesthetic complexion. For this reason, African aesthetics play a critical role in asserting the ability of African people to be innovative and contribute to the building of their own societies. African people should, therefore, not merely be seen as objects that must be fashioned into the image and artistic expressions of European cultural ethos. Rather they must affirm their right to be their own agents within the multicultural global context.

In line with the five principles articulated by Asante and the four cardinals of Afrocentric research articulated by Mazama (2001), the foundational and goal of Afrocentricity is to expose and resist white racial domination over Africans in all matters of African life and to restore Africans to their cultural centre as a legitimate ground on which to build. African knowledge must be at the centre of an African life and be treated with utmost respect as a way of validating the humanity of Africans.

FURTHER RESEARCH DIRECTIONS

Numerous Afrocentrically inspired philosophical frameworks are emerging within the field of African studies, aiming to transform the self-perception of African individuals. Among these is Afrofuturism, which encompasses a cultural aesthetic, a philosophy of science, and a historical perspective that examines the convergence of African culture with scientific and technological advancements. Gallagher (2006) posits that key figures in the development of Afrofuturism include science fiction authors such as Octavia Butler and Samuel R. Delany, as well as jazz musician Sun Ra, who crafted a mythical persona blending science fiction with Egyptian mysticism. This sense of "otherness" is central to Afrofuturism, enabling Africans to reconceptualize their lives and futures in novel ways and to challenge prevailing Eurocentric narratives about African existence. Consequently, this framework facilitates the reclamation of African consciousness and the advancement of a distinct cosmological perspective. This scholarly domain offers a valuable opportunity to reevaluate the politics of knowledge production from an African perspective.

CONCLUSION

In conclusion, this chapter underscores the profound impact of cognitive empire and Afrocentricity in understanding the historical and contemporary dynamics of knowledge production. It highlights how the European construction of the modern world system has fostered a triumphalist narrative through colonial knowledge, systematically undermining African intellectual traditions since the 15th century. This historical context has led to the dominance of European epistemologies while marginalizing non-European voices. In response, decolonial movements in Africa strive to reclaim oppressed identities and challenge the moral decline linked to Eurocentric paradigms. The chapter advocates for recognizing Indigenous knowledge systems as essential for envisioning alternatives to Western-centric perspectives and for promoting cognitive justice. By emphasizing the need for epistemic restitution and cognitive equity, it positions Afrocentricity as crucial for reimagining knowledge production within the African context.

REFERENCES

Ake, C. (2011). *Social science as imperialism: the theory of political development.* Ibadan University Press.

Asante, M. K. (2007). *An Afrocentric manifesto: toward an African renaissance.* Polity Press.

Asante, M. K. 2009. *Afrocentricity.* Retrieved on 20 October 2024 from http://www.asante.net/articles/1/afrocentricity

Asante, M. K. (2014). *Facing south to Africa.* Lexington Books.

Asante, M. K. (2017). Afrocentricity and knowledge: Adjusting the narrative lens. *The Thinker Magazine*, 74, 24–26.

Badat, S. 2023. Colonialism shaped modern universities in Africa – how they can become truly African. *The Conversation.* Retrieved on 22 October 2024 from https://theconversation.com/colonialism-shaped-modern-universities-in-africa-how-they-can-become-truly-african-214494 downloaded in 22/10/2024

Collyer, F., Connell, R., Maia, J., & Morrel, R. (2019). *Knowledge and global power: making new sciences in the South.* Wits University Press.

Gallagher, E. (2006). Afrofuturism. Tate. Retrieved on 22 October 2024 from https://www.tate.org.uk/art/art-terms/a/afrofuturism

George, J., Reddy, V., & Searle-Chatterjee, M. (1990). Eurocentrism in the social sciences. *Race & Class*, 31(4), 1–15. DOI: 10.1177/030639689003100401

Gonçalves, B. (2023, August 15). Belgium's return of Lumumba's tooth: A new moment for anti-colonial struggles? *EJIL Talk!* Retrieved on 22 October 2024 from https://www.ejiltalk.org/belgiums-return-of-lumumbas-tooth-a-new-moment-for-anti-colonial-struggles/

Gordon, J. (2002). Towards an African historiography. In Asante, M. K., & Mazama, A. (Eds.), *Egypt vs. Greece and the American academy: the debate over the birth of civilization* (pp. 25–36). African American Images.

Hegel, G. W. F. (1956). *The philosophy of history* (Sibree, J., Trans.).

Lander, E., & Past, M. (2002). Eurocentrism, modern knowledges, and the "natural" order of global capital. *Nepantla*, 3(2), 245–268.

Lerner, D. (1958). *The passing of traditional society: modernizing the Middle East.* Free Press.

Mazama, A. (2001). The Afrocentric paradigm: Contours and definitions. *Journal of Black Studies*, 31(4), 387–405. DOI: 10.1177/002193470103100401

Mazama, A. (2014). *Afrocentricity and the critical question of African agency.* Dyabukam. Retrieved on 22 October 2024 from https://www.dyabukam.com/index.php/en/knowledge/philosophy/item/136-afrocentricity

Moloi, L. (2024). *Developing Africa? New Horizons with Afrocentricity*. Anthem Press., DOI: 10.2307/jj.12638978

Moya, P. M. (2011). Who we are and from where we speak. *Transmodernity: Journal of Peripheral Cultural Production of the Luso-Hispanic World*, 1(2), 5–16. DOI: 10.5070/T412011809

Ndlovu, M. (2018). Coloniality of knowledge and the challenge of creating African futures. *Ufahamu. Journal of African Studies*, 40(2), 95–116.

Ndlovu-Gatsheni, S. J. (2021). The cognitive empire, politics of knowledge, and African intellectual productions: Reflections on struggles for epistemic freedom and resurgence of decolonisation in the twenty-first century. *Third World Quarterly*, 42(5), 882–901. DOI: 10.1080/01436597.2020.1775487

Ngugi wa Thiong'o. (1986). *Decolonizing the mind: the politics of language in African literature*. Portsmouth.

Ngugi wa Thiong'o. (2009). *Something torn and new: an African renaissance*. Basic Civitas Books.

Nkrumah, K. (1965). *Neo-colonialism: the highest stage of imperialism*. Heinemann.

Olusoga, D. (2015, September 8). The roots of European racism lie in the slave trade, colonialism—and Edward Long. *The Guardian*. Retrieved on 22 October 2024 from https://www.theguardian.com/commentisfree/2015/sep/08/european-racism-africa-slavery

Sundberg, J. 2009. *International Encyclopedia of Human Geography*. Retrieved on 22 October 2024 from https://www.sciencedirect.com/topics/social-sciences/eurocentrism on 22/10/2024

Welsh-Asante, K. (1994). *The African aesthetic: keeper of the traditions*. Praeger.

ADDITIONAL READINGS

Asante, M. K. (1991). The Afrocentric idea in education. *The Journal of Negro Education*, 60(2), 170–180. DOI: 10.2307/2295608

Asante, M. K. (2015). African pyramids of knowledge: Kemet, Afrocentricity and Africology. Retrieved 4 November 2024 from https://www.thefreelibrary.com/Asante%2c+Molefi+Kete.+African+Pyramids+of+Knowledge%3a+Kemet%2c...-a0498734481

Ndlovu-Gatsheni, S. (2018). *Epistemic freedom in Africa: deprovincialization and decolonization* (1st ed.). Routledge., DOI: 10.4324/9780429492204

Smith, A. X. (Ed.). (2023). *Afrocentricity in AfroFuturism: toward Afrocentric futurism*. University Press of Mississippi., DOI: 10.14325/mississippi/9781496847836.001.0001

KEY TERMS AND DEFINITIONS

Cognitive Empire: The cognitive empire is known by various names. Ngugi wa Thiong'o (1986) defined it as "the metaphysical empire," which, unlike the physical empire, had to "bend the body," compelling "a distorted consciousness of the relationship of their actual reality" (as cited in Ndlovu-Gatsheni, 2021, p. 45). Gildea (2019) elaborates on the cognitive empire as the "empire of the mind," capturing how epistemology frames ontology in a more profound historical manner. Ndlovu-Gatsheni (2021) highlights that at the center of the cognitive empire is the imperial epistemology that facilitated the enslavement and colonization of the global South.

Afrocentricity: Afrocentricity is a paradigm based on the idea that African people should re-assert a sense of agency to attain sanity against the ravaging effects of Eurocentrism as a hegemonic ideological framework (Asante, 2009, p. 1). Asante in his book *The Afrocentric Manifesto*, defines Afrocentricity as a paradigmatic intellectual perspective that privileges African agency within the context of African history and culture trans- continentally and trans-generationally" (Asante 2007, p. 2).

Cultural Bomb: "The biggest weapon wielded and actually daily unleashed by imperialism against the collective defiance is the 'cultural bomb'. The effect of the cultural bomb is to annihilate a people's belief in their names, in their languages, in their environment, in their heritage of struggle, in their unity, in their capacities and ultimately in themselves. It makes them regard their own past as 'a wasteland of non-achievement' and it makes them to distance themselves from that wasteland. It makes them to want to identify with that which is furthest removed from them; for instance, with other peoples; languages rather than their own. It makes them to identify with that which is decadent and reactionary, all those forces which would stop their own springs of life. It even plants serious doubts about the moral rightness of struggle. Possibilities of triumph or victory are seen as remote, ridiculous dreams. The intended results are despair, despondency and a collective death-wish" (Ngugi wa Thiong'o 1986, p. 26).

Eurocentrism: Eurocentrism refers to a worldview that centers European culture and history as the primary or most significant force in global affairs. This perspective often marginalizes or ignores the contributions and perspectives of non-European cultures, presenting European standards and values as the benchmark against which all other societies are measured. Historically, Eurocentrism has influenced how history is recorded and interpreted, often portraying European achievements as central to the development of world civilization while undervaluing or overlooking the importance of other cultures (Sundberg, 2009).

Neocolonialism: Refers to a form of indirect control or influence that powerful countries or entities exert over less powerful nations, often following the end of formal colonial rule. Unlike classical colonialism, which involved direct political and administrative control, neocolonialism operates through economic, political, and cultural means (Nkrumah, 1965).

Chapter 8
Indigenous Theories and Theorisation in African Scholarship in the Quest for Epistemic Freedom

Patrick Ngulube
https://orcid.org/0000-0002-7676-3931
University of South Africa, South Africa

ABSTRACT

This systematic literature review focuses on using Indigenous theories and theorising in African scholarship. African scholars have tended to use Western and other theories without critically evaluating them for their relevance and ability to give a full explanation of the dynamics of their Indigenous phenomena. A clear conceptualisation of these terms can assist African scholars in theorising and using theories that explain their knowledge claims appropriately and contextually. This systematic review aimed to track the use of Indigenous theories in African scholarship and to understand the extent to which Indigenous theories have been employed. A thematic synthesis of the studies showed limited use of Indigenous theories and theorisation in African scholarship. The chapter mapped and provided insights into the use of theory in African scholarship and provided strategies on how African scholars can theorise and use Indigenous theories in their research in the quest for epistemic freedom.

INTRODUCTION AND BACKGROUND

The continued dominance of Western theories and methodologies (i.e., the two key pillars of knowledge production) is one of the symbols of white supremacy and knowledge subjugation and marginalisation of Indigenous epistemologies. The situation has been referred to as "academic" and "intellectual" imperialism" (Chilisa, 2012, p. 55). The use of Indigenous theories in the academy is essential, and it also has the potential to give indigent African scholars epistemic freedom. Fanon (1964) urges the academic

DOI: 10.4018/979-8-3693-8447-3.ch008

community to reshape society in a just and equal way so that citizens' thinking is not constrained by Western paradigms.

The quest for epistemic freedom in the African university is essential. Indigenous people comprise most of the student and academic population in many African universities. The critique of the African university by scholars such as Chimakonam (2019), Mbembe (2016), Ndlovu-Gatsheni (2017, 2018), Nyamnjoh (2012), and the campaign for the decolonisation of universities by the students in South Africa, between 2015 and 2016, serve as a manifestation of the rejection of colonial symbols and the colonial knowledge narrative and all its forms of theorisation (Gumbo & Msila, 2017; Meda, 2020; Tran, 2021).

The unequal treatment of knowledge claims from different cultures regarding knowledge use and production processes manifests the phenomenon of epistemicide (Mitova, 2021, p. 24). Agitating against epistemicide entails "the right to think, theorise, interpret the world, … and write from where one is located and unencumbered by Eurocentrism" (Ndlovu-Gatsheni, 2018, p. 3). Theorisation is a powerful tool that is used by humans "to rationalise, to explain and to master the world" (Popper, 1980, p. 59). According to Indigenous authors such as Ngara (2017), the idea should not be to try to master (control) the world. The idea is to enter a reciprocal engagement with the world. Ngara (2017) states that:

The overriding focus of life is to be in harmony with the forces of life. Harmony implies living life – not fighting or controlling life. It means steering life while understanding that there are other forces that will, in part, determine the vehicle and the direction of travel (p. 344).

Such an assertion negates Popper's belief that the purpose of science is to be able to make predictions based on universal laws to be able to control outcomes (Popper, 1980). Poper was a popular philosopher to many (Godfrey-Smith, 2003). His influence gave rise to the universalisation of Western theories in many African academic spaces, which aligned with the objectivist paradigm that dominated the Western hegemonic ways of knowing. This prevented many African scholars from taking the "epistemic decolonial turn" (Grosfoguel, 2013) or "decolonial turn" (Maldonado-Torres & Cavooris, 2017), as universalism thinking stopped them from theorising from their domain when explaining a social phenomenon and, in turn, academia has failed to produce knowledge that sustained the future of Africa. This is except for scholars such as Chilisa (2012), who used Indigenous constructs and models to contribute to scholarship that sustains the future of Africa.

Imposing the universal theory mantra on African academia by colonial scholars was an indictment that deprived African scholarship of the essence of theorisation. That made Western ways of theorising become "deeply embedded" to the extent that they were "seen as 'natural' or appropriate norms rather than historically evolved social constructs" (Scheurich & Young, 1997, p.7). It is important to note that theories are "versions of the world that may change and are developed and refined through research" (Ngulube, 2018, p. 7). Jaggar and Rothenberg (1993) state that: "every human society develops theories designed to organise reality in ways that make it intelligible" (p. 75). The problem is that a specific way of making it intelligible can be imposed as if it is the only way. This can exclude dialogue around ways of understanding realities. For example, Western constructs of economic growth as a route to generating well-being can become imposed so that people may believe that they should consider conventional economic theory as evident - forgetting that economies can be, and have been, performed in alternative ways more in keeping with Indigenous philosophies of life.

The generation and use of Indigenous theories may result in a more adequately explained local phenomenon or issue. Knowledge production based on Western ways of knowing and theories is no longer sustainable. It is high time for African scholarship to move away from the parochial universalism of Western theories and embrace theories that resonate with their local contexts. In that regard, there is a need for mainstreaming the construction of Indigenous theories in academia to produce knowledge relevant to local contexts that can contribute to inclusive and sustainable futures. That fact is emphasised by Van de Ven, Meyer and Runtian (2018) when writing about the importance of engaged Indigenous scholarship in management and organisation studies. Van de Ven, Meyer and Runtian (2018) accentuate the need for research that incorporates local language, subjects, and constructs to build theories relevant to local social and cultural contexts.[1]

Including Indigenous theories may result in an adequately explained local phenomenon or issue. In that regard, there is a need for mainstreaming Indigenous theories in academia to produce knowledge relevant to local contexts that can contribute to sustainable futures. Making any theories universal implies the exclusion of others, and it is unscholarly (Ngulube, 2020). A theory cannot explain every social phenomenon, and its universal applicability is highly questionable. As Venter (1997, p. 62) aptly stated:

> People need to accept that there is no one unique truth, which is fixed and found, but rather a diversity of valid, and even conflicting versions of a world in the making.

Although these wise words do not specifically apply to theorising, they underscore the need to be accommodative of the contexts under which the world claims arise rather than imposing alien and misguided theorising assumptions on all contexts. The need to recognise and respect the context when making knowledge claims negates the universalism of theories and promotes pluriversality (Mignolo & Walsh, 2018;Smiet, 2022).

Many research epistemologies in the Global South are based on positivism and poststructuralisms resulting from the history and culture of the people who identified as a white race, in terms of the construct of race, which enabled them to call others non-white and to inferiorise them accordingly.[2] This "*reflects* and *reinforces* that social history and that social group, and this has negative results for the people of color in general and scholars of color in particular" (Scheurich, 1997, p.141). African academia should reflect on this before promoting and propagating Western theories as universal constructs that can explain African contexts.

The advancement of Africa and the development of Indigenous epistemologies will partly depend on the use of Indigenous theories in research and the movement of the relevant discourse beyond tokenism (Keane et al., 2023). Tokenism, that is, paying lip service to Indigenous causes and a lack of action when it comes to decolonising knowledge, were recently highlighted as prevalent in the field of science (Akinbosede, 2022; Laird, 2022; Snapp, 2022). Action is needed beyond just dishing out honorary degrees to famous Indigenous people, displaying images of a few black scientists, citing scholars from the Global South, inviting them to conferences and co-authoring with them (Akinbosede, 2022; Snapp, 2022). Indigenous scholars and knowledge should be engaged and centred in knowledge production to decolonise knowledge production in word and deed. There should be brave conversations about affirming Indigenous epistemologies and dislodging hegemonic Western ways of knowing in African scholarship.

Contrary to the views of Keane et al. (2023), no significant evidence suggests that African academia has developed and adopted Indigenous theories in their research. The use of Indigenous theory in research in Africa does not have a rich genealogy and archaeology. The knowledge of the Indigenous people continues to be marginalised and abused. In the foreword of a recently published textbook on research in South Africa, Sefotho (2022) points out that, in Africa, books on research methodologies can benefit from including Indigenous theories and epistemologies. Although not many books include Indigenous theories and epistemologies, Chilisa (2012, 2019), Mbah et al. (2022) and Omodan (2024) are important exceptions. For instance, Omodan (2024) discusses many different paradigms in-depth and gives much credence to the Indigenous and transformative paradigms, offering excellent examples.)

The predicament of overlooking Indigenous epistemologies does not seem to result from a lack of effort or motivation; the difficulty lies in how that can be achieved. Indeed, many Africanist scholars and decolonial thinkers acknowledge the need for the formerly colonised and subjugated Indigenous human societies to reclaim their right to theorise from their own local contexts. However, they rarely explain how to accomplish Africanisation and decoloniality realistically. For instance, Africanist academics like Ngugi wa Thiong'o, Kwasi Wiredu, and Catherine Odora Hoppers advocate for the Africanisation of academia without explaining how to put their admirable ideas into practice (Chimakonam, 2019).

Mignolo engages in decolonial thinking and decolonial doing and ends by advocating for the notion of pluriversality, which acknowledges the diversity and peaceful coexistence of various forms of knowledge (Mignolo & Walsh, 2018;Smiet, 2022). The delinking and de-Westernisation processes in Mignolo's theory of deconstruction are a continuation of the discourses of a need to affirm silenced forms of knowledge, such as the Indigenous ways of knowing and practice, without giving details of how it can be achieved in practical terms.

What sets this contribution apart from most existing literature is that it uses a systematic review approach to establish how African scholarship applies Indigenous epistemologies in their theorisation and use of theory in knowledge production and offers workable solutions to this vexing issue. A multi-layered framework of affirming Indigenous epistemologies in theorisation and the utilisation of theory in the academia of Africa is proposed. Such an approach holds the promise of support for cognitive justice and epistemic freedom. The strength of a systematic review lies in locating, selecting, and combining available empirical research on a certain subject, using reproducible, methodical and explicit procedures with the least bias possible (Chandler et al., 2022; Gough, Oliver & Thomas, 2017). The use of meta-analysis and systematic reviews to develop Indigenous epistemologies and understand their use is recommended in the literature (Levers, 2009; Li, Sekiguchi & Zhou, 2016; Wei, Liu & Chen, 2023). For instance, Mbah, Bailey and Shingruf (2023) used the same strategy to investigate the relational research methods used in Indigenous contexts. Despite their shortcomings, these literature reviews offer reliable, robust and reproducible evidence on a topic (Siddaway, Wood & Hedges, 2019).

The purpose of this study is to identify and narratively synthesise articles in which the Indigenous theories in African scholarship have been used to meet the objectives of this chapter using the systematic review approach. Based on a systematic review, the objectives of this chapter are as follows:

- Raise awareness of the importance of incorporating Indigenous theories into the planning and execution of research in African scholarship
- Affirm the suitability of Indigenous theories to African scholarship
- Accentuate that theory use in Africa should be based on the experiences of the Indigenous people
- Demonstrate that limited use of Indigenous theories in African scholarship may produce irrelevant and counterproductive knowledge

The following section begins with a comparison between Indigenous theories and decolonial epistemologies, followed by the need for theory in African scholarship; it then moves on to theorising in African scholarship, the conceptual framework, and the problem statement and methodology. The focus then turns to the analysis and interpretation of the findings before outlining the conclusions and the implications of the findings for research.

THE NEED FOR THEORY IN AFRICAN SCHOLARSHIP

Indigenous theory is needed for African scholarship to contribute to knowledge production and epistemic freedom. Indigenous theory provides novel explanations for international behaviour and a better understanding of social phenomena at the local or Indigenous level (Davison & Díaz Andrade, 2018). Indigenous theory also acknowledges, affirms and respects the diversity and complexity of Indigenous epistemologies, worldviews, cultures and traditions (Coates, Trudgett & Page, 2023; Davison & Díaz Andrade, 2018). The colonial and oppressive frameworks that marginalise Indigenous peoples can be contested through Indigenous theory, which can also help Indigenous peoples gain self-determination and empowerment (Celidwen, 2022; Wilson, 2001). Indigenous theories provide in-depth and incisive insights into a particular setting (Bruton, Zahra, Van de Ven & Hitt, 2021). Theory use and theorisation should aim to revise and modify, "potentially supplement and enrich or even supersede and replace the Western theories" (Li, Sekiguchi & Zhou, 2016, p. 583).

The Indigenous institutional theory, which merges Indigenous standpoint theory with institutional theory, has been used to explore Indigenous leadership in higher education in Australia (Coates, Trudgett & Page, 2023). This approach provides a nuanced understanding of how Indigenous leaders navigate and challenge institutional structures. Developing Indigenous management theories rooted in local contexts offers valuable insights into different regions' specific cultural, social, and economic conditions (Bruton, Zahra, Van de Ven & Hitt, 2021). These theories reveal how local cultural practices influence business operations and decision-making processes, leading to a more realistic and relevant understanding of management in non-Western settings. These examples highlight how Indigenous theories contribute deep, context-specific insights, enriching our knowledge of various environments and fostering more effective and culturally sensitive practices.

Understanding what constitutes Indigenous theory is essential to appreciate its place in African scholarship. On the one hand, a school of thought assumes that Indigenous theory is anchored on Indigenous epistemologies, worldviews, cultures and traditions. It is not imported from other contexts in its quest to explain Indigenous reality in a specific context or culture. Indigenous theory is purposely developed for those living in that context or culture (Davison & Díaz Andrade, 2018). This can be considered as the purist view. On the other hand, a growing corpus of scholarship supports a collaboration of theories by integrating Western and Indigenous theories, which is in line with the two-eyed seeing framework, which will be explained later (Reid, 2020).

To understand Indigenous theories and how they are developed, one must first understand what a theory is. Theory-driven research has been criticised and the nature of what theory is contested, but theory is a device that assists researchers (including those situated in academia and lay researchers who become research participants/co-researchers in research projects) to define their objectives and research questions, choose a methodological stance, provide a framework for data collection, analysis and interpretation, and help in explaining the findings (Collins & Stockton, 2018; Ngulube, Mathipa & Gumbo, 2015; Kawulich, 2009).

There are various definitions of theory, but for this chapter, theory is defined as "an integrated set of concepts, formed into propositions, that explains particular conditions or events in the world around us" (Schneider, 2006, p. 2). Each claim in the theory consists of concepts and specifications of the relationships between concepts that predict what will happen to instances of the phenomena in specific situations (Dubin, 1969; Dul & Kak, 2007). However, unlike in the positivist tradition, people can still contest the proferred explanations as these are always open to further dialogue. Indigenous epistemologies do not purport that explanations offered at any point in time are final (Collins, 2009; Dotson, 2015). For instance, Patricia-Hill Collins (2009) says other people can and should be invited to write into the stories generated through the research process. She calls them stories so that they are not considered finalised (Collins, 2009).

INDIGENOUS THEORIES COMPARED TO DECOLONIAL EPISTEMOLOGIES

The discussion on decoloniality and Africanisation and their role in academia is unnecessary to advance this chapter's objectives. Suffice it to say that Africanisation, decoloniality and indigenisation share one lineage: being confrontational to the hegemonic Western knowledge systems (Chikoko, 2021; Hungwe & Mkhize, 2022). Although decolonial theories are part of epistemic disobedience (Mignolo, 2009), they are not necessarily equivalent to Indigenous theories. The same applies to the critical race theory, which Chilisa (2012) deemed as a "potential decolonising" tool.

The understanding of decoloniality and decolonial theory in Africa differs from how it is conceived in the United Kingdom, for instance (Hall, Ansley, & Connolly, 2023). Similarly, the understanding of critical race theory differs from context to context, and its utility as an analytical framework should be revisited because of a digital technological revolution that has taken place (Lawrence & Hylton, 2022). This implies that it is important to consider the context when discussing critical race theory, decoloniality and decolonial theory. There is no room for universalisation: "decoloniality has to be diversely understood and diversely framed depending on the context you are working on" (Abdi, 2020). The implication is that there is no core decoloniality theory or critical race theory, just like Indigenous knowledge systems.

Decoloniality emerged out of the critique of all the processes of positivism and its narrow view of the world, colonisation, racialisation and hegemonic tendencies, and to "create and keep alive modes of knowing, being, and relating that these processes seek to eradicate" (Stein & Andreotti, 2016, p. 1). In other words, decoloniality aims to challenge and dismantle the lingering effects of colonialism in various aspects of society, including knowledge, power, and culture. On the other hand, Indigenous epistemologies did not emerge as a critique of Western hegemonic systems. Indigenous knowledge systems have existed from time immemorial. This implies that Western knowledge systems in the Global South are a recent phenomenon and "newer" (Martin, 2012). Decolonial thought and other progressive theories give Indigenous scholars tools towards deconstructing the current system of knowledge production and dissemination, but this does not make them necessarily Indigenous epistemologies.

In a nutshell, decoloniality, critical race theory,[3] and Indigenous epistemologies are interrelated and socially situated concepts and practices that question hegemonic discourses and power dynamics to establish a different world. Indigenous theorisation and theories have the potential of decentring and "deterritorialising" (Le Grange, 2016, p. 6) the colonial project, in line with decoloniality. In this regard, Indigenous epistemologies and decoloniality share the same concerns, even if their assumptions differ slightly.

THEORY BUILDING IN AFRICAN SCHOLARSHIP

According to Dubin (1969), research aims to develop theories. The process of theorising is a neglected area in many fields (Kawulich, 2009; Schneider, 2006; Ngulube, 2020). This has led to limited theory-building in various domains, including Indigenous knowledge. Theorising assists in understanding the "systematic reasons for a particular occurrence or non-occurrence" (Sutton & Staw, 1995, p. 378).

Theory building is the "process of knowing, perceiving and conceiving relationships, derived from systematic inquiry, which can transcend paradigms if a comprehensive and consistent 'worldview' is developed" (Kettley, 2010, p.10). Different Indigenous groups can theorise, organise, conceptualise, and validate their knowledge differently. As Indigenous people theorise, they should explain their reality, on the one hand, and theorise against what Chilisa (2012, p.301) refers to as "Western archival knowledge and its colonizing and 'othering' ideologies", on the other.

Theory building is not a monopoly of any academic or any privileged scholar. The work of Glaser and Strauss (1967) was crucial in developing the "liberating thesis that anyone can create their own theory" if it was based on empirical research (Alvesson & Skoldberg, 2000, p. 16). The four components of a theory are the phenomenon or object of the study, the concepts, the proposition or claims connecting the concepts, and the domain (Dubin, 1969; Schneider, 2006; Zaltman, Pinson, & Angelmar, 1973).

Theory building is based on theorising. Theory development should be regarded as a process rather than a product. Theorising is "making certain claims in the form of statements that reflexively apply specific rules of formation to constitute a discourse within a field" (Hassan, Lowry & Mathiassen, 2022, p. 8). Theory development may start "with guesses and speculations and ends with explanations and models" (Weick, 1995, p. 386). The process of theorising starts with developing a problem and reading diverse literature that defines theory (Schneider, 2006; Sutton & Staw, 1995), and drawing on the context of the phenomenon under study and the researcher's experiences and those of the Indigenous communities including the concepts and language they use (Ngulube, 2020; Dana-Sacco, 2010). The theorising process does not generally produce complete theories (Hassan, Lowry & Mathiassen, 2022; Weick, 1995), although it creates a theory to explain the problem.

Indigenous researchers prioritise collaborating with research participants, particularly those from marginalised groups within society (Omodan, 2025; Romm, 2018). This approach ensures that research is relevant, respectful, and beneficial to the community (Chilisa, Major & Khudu-Petersen, 2017; Dana-Sacco, 2010; Pidgeon & Riely, 2021). By involving the community in framing research problems, the research aligns more closely with their needs and perspectives, leading to theories that resonate with the community's unique context (Pidgeon & Riely, 2021). Such theories are likely to reflect the specific circumstances of each Indigenous community and promote culturally appropriate theorisation. This process is sensitive to the "cultural expectations of responsibility, relevance, and respect for Indigenous knowledge" (Pidgeon & Riely, 2021).

CROSS-INDIGENISATION GEOCENTRIC APPROACH

This approach acknowledges that scholars can use many knowledge systems and practices to produce interdisciplinary and integrated knowledge. Adopting the integrative approach entails using more than one theoretical perspective and combining theories "as an interpretive handyman or *bricoleur*" (Kaomea, p.

16; see also Levi-Strauss, 1966; Ngulube, Mathipa, & Gumbo, 2015). Scholars can adapt theories from Indigenous perspectives and Western epistemologies, like a *bricoleur* who uses whatever is available.

For instance, ways of knowing and practice such as Indigenous African knowledge systems [ubuntu, interconnectedness] (Msila & Gumbo, 2017), Afrikology [epistemology of knowledge production and dissemination] (Nabudere, 2011), and Afrocentricity [African-centeredness] (Asante, 2009; Mazama, 2001); First Peoples' theory [Indigenous wholistic theory] (Absolon, 2010; Marsden, 2005) and Fourth World theory (Ryser; Gilio-Whitaker & Bruce, 2017); and Indigenous Eastern epistemological system [Yin-Yang] (Li, 2016) can be integrated with Western knowledge, to develop a geocentric meta-framework. Chilisa (2012, p.51) also advocated for this integrative approach to knowledge production "with a balanced borrowing from Euro-Western paradigms and postcolonial Indigenous paradigms".

Joullié and Gould (2021, p.1) view this approach as a theoretical plurality. The authors describe theoretical pluralism as developing new frameworks, testing current ones,[4] or expanding the prevailing "orthodoxy" into other domains based on various ontological, epistemological, and methodological traditions. Theoretical plurality accommodates theory borrowing, as well as combining Western frameworks with Indigenous frameworks, to create hybrid or geocentric theories. Li et al. (2016) and Jukka, Blomqvist, Li and Gan (2017) advocate that future studies should focus on the "cross-indigenization geocentric approach" and on developing more geocentric theories. That stance recognises the interdependence of knowledge systems.

The geocentric, *yin-yang*, duality lens, described by Li (2008), which advocates for the cross-indigenisation geocentric approach, is consistent with the two-eyed seeing or *Etuaptmumk* framework that was developed by Mi'kmaq Elders, Albert and Murdena Marshall, from Unama'ki, in Canada (Wright et al., 2019). According to Elder Albert Marshall (cited by Reid, Eckert, Lane et al., 2021, p. 245), it means:

> …learning to see from one eye with the strengths of Indigenous knowledges and ways of knowing, and from the other eye with the strengths of Western knowledges and ways of knowing … and learning to use both these eyes together, for the benefit of all.[5]

Thus, the framework considers the more potent aspects of Indigenous and Western epistemologies. Through their mutual borrowing from one another, the combined effects of two elements can result in a more complete and balanced understanding of the phenomenon.

Insights on 'decolonial thinking' (Mignolo, 2011) can also assist in linking Western theories to Indigenous theorising and lead to the development of postcolonial Indigenous theories and critical Indigenous theories, for instance. These theories provide opportunities to expand epistemologies from the Global North to include the marginalised and silenced Indigenous ways of knowing from the Global South.

The stance expressed in this chapter warns against falling victim to the same colonial epistemology and pushing uncritically for the indigenisation of the theories, one of the unfounded criticisms levelled against Afrocentricity by Mbembe (2016). Instead, alternative theorising epistemologies that balance Western knowledge with Indigenous ways of knowing are suggested. This will further the role that theories can play in achieving social and intellectual freedom in the Global South.

UNIQUE THEORISING AND THEORY USE IN AN INDIGENOUS CONTEXT

The Indigenous theories developed through this approach are unique from the predominately Western ones. However, this framework promotes 'Indigenous compatibility' and unique theorising (Yang, 2000, p. 250). That implies that "the researcher's concepts, theory, methods, tools, and results adequately represent, reflect, or reveal the natural elements, structure, mechanism, or process of the studied phenomenon embedded in its context" (Yang, 2000, p. 250). Such unique theorising and theory use will "generate new pedagogical language and praxis that go beyond the normalised grammar of the colonial structures of African universities" (Zembylas, 2018, p. 1). However, the centrality of the involvement of the research participants in contributing to the story that is being constructed should not be underestimated so that the knowledge that is produced and theories that are formulated can lead to sustainable futures. Theorising, taking into consideration the Indigenous context, is also in line with Afrocentrism, which assumes that Africans and African scholars must define their reality and worldview when theorising. Accordingly, "Afrocentricity is a mode of thought and action in which the centrality of African interests, values, and perspectives predominate" (Asante, 2003, p. 2).

While the previous section favours the use of cross-indigenisation geocentric theories, scholars such as Li et al. (2016) acknowledge that relevant Indigenous theories should be developed before the formulation of geocentric theories can effectively take place.

Quantitative and mixed methods research approaches are suggested as strategies that can be utilised to test uniquely Indigenous theories. Adding Indigenous research methods into the mix of testing such theories might guarantee a complete break from the mainstream Western methodologies.[6] It should be noted that the mixed methods research approach can also provide useful tools for such an exercise, as recommended by Chilisa (2012, 2019), Chilisa and Tsheko (2014) and Chilisa (2024) insofar as provided it continues to expand beyond the Anglo-Saxon tradition and incorporate other ways of knowing (Creswell & Plano Clark, 2018; Martel, Shepherd & Goodyear-Smith, 2021; Ngulube, 2022).

CONCEPTUAL FRAMEWORK

A conceptual framework (CF) is needed to provide a tool to analyse the pertinent studies. A CF can assist in making the scope and focus of the systematic review explicit and improve the validity and dependability of the outcomes (Hallinger, 2013; Schreiber & Cramer, 2022). As it was difficult to find a theory that adequately addressed the current research problem, a CF was formulated from concepts from the extant literature, the experiences of the researchers and the context of the study - as suggested by Ngulube (2020), Salawu, Bolatitio and Masibo (2023) and Van der Waldt (2020).

The CF "drums" combine various concepts to bring about a shared understanding and diversity, as illustrated in Figure 1. The CF is based on some of the pillars of conceptualising Indigenous research, as Li et al. (2016) and the extant literature suggested. Specifically, it captures the Indigenous constructs and models, including their prevalence and transparent use, at one level, as shown in Figure 1. The primary concern is to determine the extent to which the theories that are used in African scholarship emanate from the culture of the context of the study and the "religion, cultural traditions, norms, language, metaphors, Indigenous knowledge systems, community stories, legends and folklores, social problems… or public policies", for instance (Chilisa, 2012, p.102). The other two levels illustrate the integrative cross-indigenisation geocentric approach, the innovative unique Indigenous theorising, and theory use.

Figure 1. Conceptual framework of the study

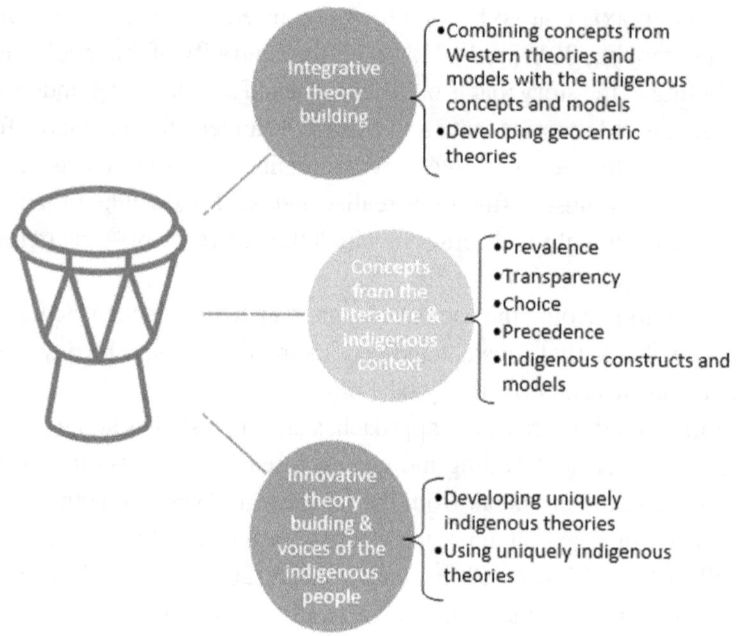

(Author)

STATEMENT OF THE PROBLEM

Relevant concepts and theories should inform research in an Indigenous environment to produce relevant and context-driven knowledge that can support sustainable development and reinforce epistemic freedom. However, little has been written on the extent of Indigenous theories and theorising in African scholarship. This chapter is consistent with the repeated calls for contextualising theories used in Indigenous research and the need to elevate Indigenous theories to the same status level currently afforded to other knowledge systems. The questions below were instrumental in achieving the objectives of this study.

- Does African scholarship use Indigenous constructs and models in theorisation?
- Is African scholarship transparent in its preference of Western theories over Indigenous theories and vice versa?
- Does African scholarship combine and supplement concepts from Western theories and models with the Indigenous concepts and models, to develop geocentric theories?
- To what extent does African scholarship develop uniquely Indigenous theories for explaining an Indigenous phenomenon?

METHODOLOGY

This systematic review concedes that Indigenous knowledge is interdisciplinary, with no core database available and various potentially relevant sources. However, the African Journals Online (AJOL), Scopus and Web of Science (WoS) databases and the academic search engine Google Scholar (GS) were used to identify the materials for the study. AJOL was heavily relied on because it hosts hundreds of research journals published in Africa, and it is the world's largest and most preeminent platform for African-published scholarly journals (AJOL, sa). Although GS cannot be solely relied on when conducting systematic reviews, it was used to supplement the data and to ensure effective and sufficient coverage (Bramer, Rethlefsen, Kleijnen & Franco, 2017; Justesen, Freyberg & Schultz, 2021). The details of the search results are given in Figure 2.

The "Preferred Reporting Items for Systematic Reviews and Meta-Analyses" (PRISMA) statement was followed when conducting this review (Page et al., 2021). This entailed four phases, as illustrated in Figure 2: identification, screening, eligibility, and inclusion.

The first phase (identification) started with the process of identifying eligible studies by searching for common labels of Indigenous knowledge, identified by Maluleka and Ngulube (2019), and Onyancha et al. (2018), namely, "Indigenous knowledge", "traditional knowledge" and "local knowledge", in conjunction with "theories", theoretical frameworks", "theoretical perspectives", "conceptual framework", "analytical framework" and "Africa" in the title, abstract and keywords of articles. The year of publication was used to filter the records. The choice of 2016 as the starting point was made since that year saw the onset of the #FeesMustFall movement, which called for an end to the dominance of the hegemonic Western knowledge systems in academia and knowledge production. No records were identified through citation searching, as the retrieved records were assessed for quality and relevance and found sufficient for analysis, following the recommendations of Chandler et al. (2022) on systematic reviews.

Figure 2. PRISMA flow diagram

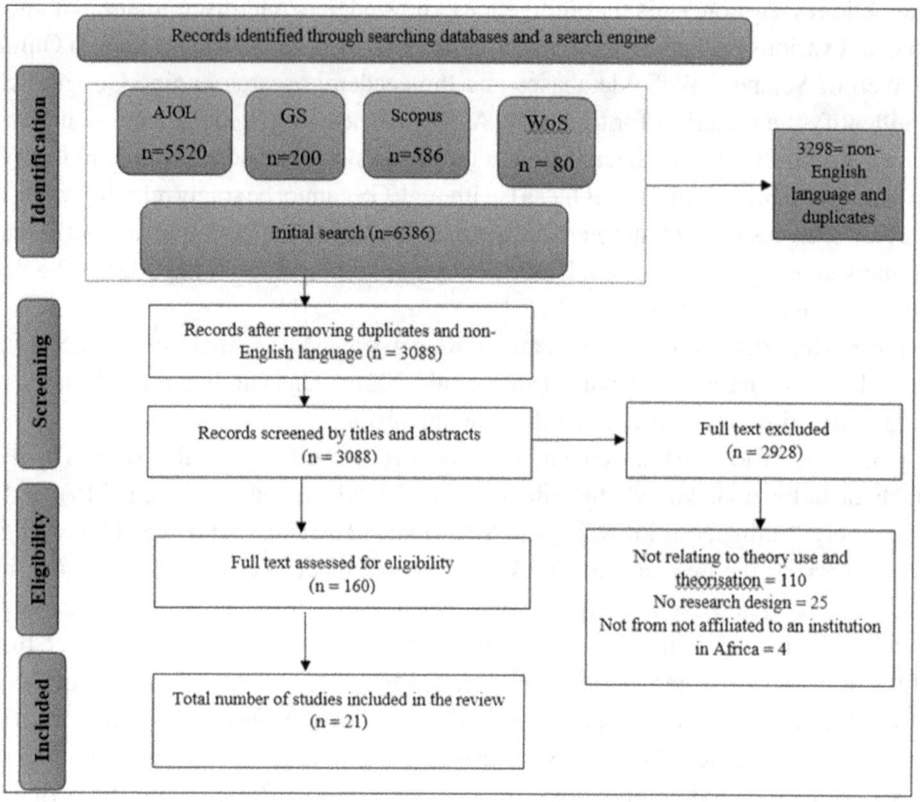

(Author)

Screening of the Records

Studies were selected based on the title, abstract, keywords, year, and authors of the identified records, which were exported to a citation manager, Endnote®. The citation manager identified and deleted duplicate versions of the same work. Reading the titles and abstracts was followed by reading the full text to determine the studies to be included in the analysis. Table 1 shows the inclusion and exclusion criteria.

Table 1. Inclusion and exclusion criteria

Criteria	Inclusion	Exclusion
Language and duplicates	Articles published in English	Articles published in a language other than English and duplicates
Publication type	Peer-reviewed articles based on empirical research	Editorials, letters, conference proceedings, meeting abstracts, conceptual papers, short communications, obituaries, dissertations, discussions, book reviews and systematic reviews (e.g., systematic, scoping, and rapid reviews) and reports
Publication date	Articles published between 2016 and 2023	Articles published before 2016 and after 2023
Specification of design	Mentions the design in one or all the following areas: title and abstract and methods section (quantitative (QUANT), qualitative (QUAL) and mixed methods research (MMR)	No design specified
Specification of theory, including models and constructs/concepts	Mentions theory or constructs or models in one or all the following areas: title and abstract and body of the article	No theory, constructs/concepts or models specified
Affiliation, as stated by the author(s)	From an institution in Africa	Not from an institution in Africa

Data Extraction

A data extraction table was constructed. The articles were subjected to content analysis to extract the information based on the conceptual framework. The Appendix details the elements that were the focus of data extraction.

Quality Assessment and Mitigating the Risk of Bias

The quality of the coding was evaluated separately by three postgraduate fellows. The outcomes were compared, and discrepancies were discussed until an agreement was reached. The inter-rater agreement was calculated using Cohen's kappa coefficient and was determined to be 0.7928. That implies that the level of agreement was substantial (McHugh, 2012; Viera & Garrett, 2005).

FINDINGS

The findings are arranged according to the objectives that informed the study. Although there was no specific objective on the profile of the articles, The Appendix gives details of the authors[7], the authors' country of affiliation, the journal's publisher, and the research design used in the article. As illustrated in Figure 3, many authors came from South African institutions. Qualitative designs were dominant. Although Unisa Press and Routledge were dominant, the articles were published by a diverse cross-section of publishing houses. The number of authors per article ranged from one to eight. However, single-authored articles (8) were predominant, followed by articles written by two authors.

Figure 3. Profile of the articles and research designs

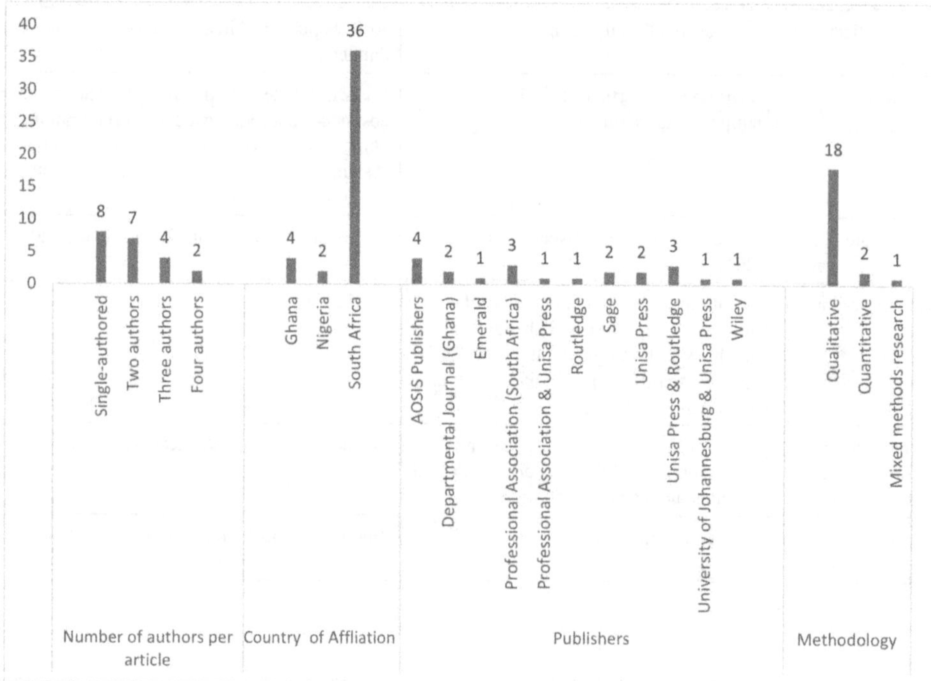

A total of 13 articles used Indigenous constructs and models, in contrast to 8 that exclusively used Western models and constructs. The Indigenous constructs made use of theories such as Afrocentricity (Makhanya & Zibane, 2020), eZiko (Mwanza, 2022) and Ubuntu (Lefike et al., 2023; Steenkamp & Rensburg, 2018). Constructs from decolonial and postcolonial theories were used in articles such as Mandikonza (2019), Mokotso (2022) and Seroto (2018; 2019).

One of the objectives sought to determine if African scholarship was transparent in the preference of Western theories over Indigenous theories and vice versa. Only four articles demonstrated why Indigenous theories were used to explain the phenomenon under study (Mahlatsi et al., 2021; Makhanya & Zibane, 2020; Seroto, 2019; Shokane & Masoga, 2018). A total of 17 articles uncritically used Western theories without showing how they were fit for purpose in an Indigenous context. Thirteen articles justified their theory choice without reflecting on the Indigenous context.

Theoretical transparency was established by the extent to which the scholars stated the theory's claims and the degree to which those claims aligned with the study's objectives or research questions. A total of thirteen (62%) articles noted the claims of the theories they used. Four of the thirteen did not frame the objectives and conclusions around the theories' claims (Du Toit, 2019; Mumpande & Barnes, 2019; Meda, 2020; Mandikonza, 2019). Eight (38%) out of twenty-one did not state the theories' claims (Du Toit, 2019; Mandikonza, 2019; Mbatha et al., 2023; Meda, 2020; Mumpande & Barnes, 2019; Mwanza, 2022; Seroto, 2018; Shokane & Masoga, 2018).

A total of seven (33%) articles authored by Ibrahim, Adjei and Boateng (2019), Lefike, Turpin and Matthee (2023), Mokotso (2022), Makhanya and Zibane (2020), Moropa (2018), Mwanza (2022) and Shokane and Masoga (2018) combined Western and Indigenous concepts to explain the phenomenon under study. None of the articles came up with new geocentric theories. Ibrahim, Adjei and Boateng (2019) used the realist theory of conflict and nicely combined it with the Indigenous wholistic theory but fell short of developing a new framework.

Only one (4%) article written by Mahlatsi, Pienaar, Nare and Mulaudzi (2021) developed a uniquely Indigenous theory after demonstrating that Western theories had several shortcomings, resulting in ineffective outcomes in Indigenous contexts.

DISCUSSIONS

The study's findings show that African scholarship relies heavily on Western theories. There was no theoretical transparency in using Western theories over Indigenous ones, as the scholars did not demonstrate why they opted to use Western theories to explain Indigenous contexts. Researchers must be transparent in their theory selections to make their findings reproducible and relevant (Ngulube & Mosha, 2023).

The prevalence of theory borrowing in African scholarship and the failure to develop uniquely Indigenous theories implies that developing Indigenous knowledge systems as a respectable discipline will take time to reach maturity. A cognitive field matures through the development of shared theories and practices. Theories serve as the foundation upon which disciplines build their knowledge (Hirschheim & Klein, 2003). By identifying essential factors and suggesting connections, theories shape the direction of research. The introduction of new theories can lead to paradigm shifts, which are vital for the advancement of a discipline. These shifts allow for the replacement of old paradigms with new ones that better explain the phenomena being studied (Kuhn, 1970). Therefore, the ongoing improvement and development of theories will remain a critical aspect of the evolution and maturation of disciplines over time.

A study by Gebru and Wario (2022) discovered that African scholars did not use Indigenous theories for fear of being rejected in a scholarly landscape that is dominated by Western theories. African scholars were using these theories to gain acceptability instead of basing their choice on the adequacy of the theories to explain the issues they were investigating. Such an approach is inimical to developing Indigenous theories and the assumption of epistemic freedom. Gebru and Wako (2022) also found that many scholars did not believe it was possible to formulate their Indigenous theories. However, the idea that theory development was the exclusive domain of particular academics was dismissed by Glaser and Strauss (1967), although they were rather positivist inclined in that they believed that somehow empirical data alone can be used to generate theory as they do not indicate that there are always a variety of interpretations as articulated by another constructivist grounded theorist such as Charmaz (2017) who highlighted the collaborative process of knowledge creation that occurs between the researcher and the participants.

Indeed, one article in this study formulated an Indigenous theory (Mahlatsi et al., 2021, which further confirmed that any researcher can develop a theory from empirical data and interpretive constructs as long there is a phenomenon to be explained. The researchers utilised the African Indigenous Health Research Framework (AIHRF), which emphasises the importance of grounding research in its authentic context to preserve indigenous meanings (Pienaar, 2017). This study was conducted as a negotiated partnership, allowing the Indigenous Khoisan community to determine the extent of their participation as research participants. The researchers employed orature—such as storytelling, metaphors, and praise poems—for data collection. Several themes emerged from the data, including spoken word, active communication, and the principles and attitudes surrounding communication. These themes underscored the significance of who communicates with whom, when, and how. The ultimate framework was created through collaborative efforts between researchers and the community, emphasising relational knowledge.

As illustrated in Figure 3, African scholarship seems to rely heavily on qualitative designs. Qualitative designs provide a fertile ground for theory building, along with mixed methods research used by Mbatha et al. (2023). That is partly the case because these designs accommodate Indigenous epistemologies.

The use of Indigenous constructs and models was as limited as the use of Indigenous theories. It is challenging to develop Indigenous theories in the absence of the relevant constructs. African scholarship should use Indigenous constructs as a basis of theory development. Logic dictates that African scholarship should be based on "concepts, ideas, theories, and philosophies that are centred on Africa, her people, and their culture" (Shai, 2023, p. 380) if it is going to support the growth of Indigenous epistemologies and Indigenous theories.

Apparently, African scholarship also failed to develop geocentric theories that integrated Western and Indigenous theories. Cross-indigenisation geocentric theories focus on integrating local (emic) and global (etic) perspectives into a shared global core (Li, 2012). These theories frequently include decolonial viewpoints that challenge the dominance of Western knowledge systems while also promoting the validation of Indigenous knowledge (Smith & Babich, 2023). That implies that theoretical pluralism was not evident in African scholarship. The creation of new frameworks, the evaluation of already existing ones, or the expansion of the orthodoxy currently in place into new fields are all examples of theoretical pluralism (Wight, 2019). Theoretical pluralism can take several forms, such as the development of new frameworks, the assessment of pre-existing ones, or the extension of the accepted theories into new domains. Multiple epistemologies and epistemological diversity may result in theoretical pluralism and integrative scholarship.

The findings revealed that some studies used postcolonial theories and decolonial theories. Such theories allow African scholarship to disengage from the colonial intellectual syndrome. Research informed by postcolonial theory enables African scholars to theorise a rejoinder and to conduct theory-driven studies that give a voice to the voiceless (Chilisa, 2012). However, postcolonial theory must be used cautiously, as it might prevent African scholars from moving beyond the colonial intellectual boundary (Grande & McCarty, 2018). The argument is that the postcolonial theory is rooted in a Western mindset rooted in individualism and secularisation (Grande & McCarty, 2018). For this reason, Chilisa (2012, p. 48) suggests the use "postcolonial Indigenous theory", instead of postcolonial theory without a qualifier. Theorising a rejoinder implies affirming, centring and understanding theory from previously "other(ed)" standpoints (Smith, 2012; Thambinathan & Kinsella, 2021). Chilisa and Mertens (2021) suggest that we give more weight to this kind of theorising in cases when theories and ways of theorising seem incompatible.

FUTURE RESEARCH DIRECTIONS

The limitation of this study is that it only used texts written in English and studied a certain period in the development of African scholarship. Comprehensive studies include a longer time span, which is necessary. The other limitation is the methodology used. While the methodology identified that Indigenous constructs and models appear to have been used, how they were developed, including the extent of engagement with research participants, was not revealed through the methodology. Other methodologies can shed more light on the use of indigenous theories in African scholarship, including how those involved in academia can and should engage with research participants to co-produce insights that will be meaningful and relevant for them.

It is necessary to conduct comparable investigations on theory development and use, employing a different methodology, such as mixed methods research, to get the views of sampled African scholars. For instance, an explanatory mixed methods research study can explain the patterns uncovered by this systematic review. The quantitative phase can include the sample of authors identified by this study, and then the qualitative sample may be drawn from the editors of the journals where these authors published their work. Alternatively, part of the sample of the identified authors can be used in the qualitative phase using the identical sampling strategy depending on the nature of the results to be explained. The interpretation of the results will also benefit from more in-depth attention to the articles identified as using Indigenous constructs and models so that the manner in which the theorising was built up can be qualitatively elucidated (and assessed in terms of criteria such as relational orientation and transformative potential).

CONCLUSION AND RECOMMENDATIONS

This study provides a comprehensive understanding of the use of theory and theorization in African scholarship. It offers crucial data on the application of Indigenous theories and highlights ways to enhance current and future implementation practices. Theory development does not appear to be a primary focus in African scholarship. Contrary to the claims made by Keane et al. (2023), the findings indicate that the use of Indigenous theories is limited, with a concerning trend of excessive borrowing from external theories. If this situation continues unchecked, a theory crisis is likely to occur. This chapter emphasizes the need for a paradigm shift in the application of theory and theorization within African scholarship. Such a shift could lead to the creation of relevant knowledge that supports sustainable futures and promotes epistemic freedom and social justice.

REFERENCES

Abdi, M. (2020). Decolonising higher education. In: *Decolonising education: Let's talk about it*. YouTube channel. Retrieved on 25 July 2023 from: https://www.youtube.com/watch?v=2Y0eBb0UuvQ&feature=youtu.be

Absolon, K. (2010). Indigenous wholistic theory: A knowledge set for practice. *First Peoples Child & Family Review*, 5(2), 74–87. DOI: 10.7202/1068933ar

Adjimah, H. P., Atiase, V., & Dzansi, D. Y. (2023). Do government incentives increase Indigenous innovation commercialisation? Empirical evidence from local Ghanaian firms. *International Journal of Entrepreneurial Behaviour & Research*. Advance online publication. DOI: 10.1108/IJEBR-02-2022-0157

African Journals Online (AJOL) (sa). Welcome to AJOL. Retrieved on 25 June 2023 from: https://www.ajol.info/index.php/ajol

Akinbosede, D. (2022). Acknowledge racist roots in science. *Nature*, 612(7938), 178–179. DOI: 10.1038/d41586-022-04123-3 PMID: 36424502

Alvesson, M., & Skoldberg, K. (2000). Reflexive methodology: New vistas for qualitative research. *Sage (Atlanta, Ga.)*.

Asante, M. K. (2003). *Afrocentricity: A theory of social change* (2nd ed.). African American Images.

Asante, M. K. (2009). *Afrocentricity*. SUNY Press.

Balogun, O. K., & Adeseke, A. E. (2022). Advancing Indigenous African values and ethos for film directing and production in Nollywood. [IJCRH]. *International Journal of Current Research in The Humanities*, 26(1), 87–97. DOI: 10.4314/ijcrh.v26i1.6

Bramer, W. M., Rethlefsen, M. L., Kleijnen, J., & Franco, O. H. (2017). Optimal database combinations for literature searches in systematic reviews: A prospective exploratory study. *Systematic Reviews*, 6(1), 245. DOI: 10.1186/s13643-017-0644-y PMID: 29208034

Bruton, G. D., Zahra, S. A., Van de Ven, A. H., & Hitt, M. A. (2021). Indigenous theory uses, abuses, and future. *Journal of Management Studies*, 1–17. DOI: 10.1111/joms.12755

Celidwen, Y. (2022). Why we need Indigenous wisdom: The rich perspective of Indigenous peoples has a crucial role to play in advancing climate solutions rooted in an understanding of our interdependence. In: *Insights Journey into the heart of contemplative science*. Retrieved June 2023 from: https://www.mindandlife.org/insight/why-we-need-Indigenous-wisdom/

Chandler, J., Cumpston, M., Thomas, J., Higgins, J. P. T., Deeks, J. J., & Clarke, M. J. (2022). Introduction. In: J. P. T. Higgins, J. Thomas, J. Chandler, M. Cumpston, T. Li, M.J. Page, & V. A. Welch (eds.), *J Cochrane Handbook for systematic reviews of interventions version 6.3* (updated February 2022). Cochrane, 2022. Retrieved 9 July 2023 from: www.training.cochrane.org/handbook

Charmaz, K. (2017). The power of constructivist grounded theory for critical inquiry. *Qualitative Inquiry*, 23(1), 34–45. DOI: 10.1177/1077800416657105

Chikoko, V. (2021). Re-visiting the decolonising of the South African higher education question: A systematic literature review. *South African Journal of Higher Education*, 35(1), 21–36. DOI: 10.20853/35-1-4420

Chilisa, B. (2012). Indigenous research methodologies. *Sage (Atlanta, Ga.)*.

Chilisa, B. (2019). *Indigenous research methodologies* (2nd ed.). Sage.

Chilisa, B. (2024). Philosophical underpinnings of mixed methods: Decolonizing evaluation practice through decolonizing paradigms. In: Y. Shan Y (Ed,), *Philosophical foundations of mixed methods research: Dialogues between researchers and philosophers* (pp. 54-82) Routledge.

Chilisa, B., Major, T. E., & Khudu-Petersen, K. (2017). Community engagement with a postcolonial, African-based relational paradigm. *Qualitative Research*, 17(3), 326–339. DOI: 10.1177/1468794117696176

Chilisa, B., & Mertens, D. M. (2021). Indigenous Made in Africa evaluation frameworks: Addressing epistemic violence and contributing to social transformation. *The American Journal of Evaluation*, 42(2), 241–253. DOI: 10.1177/1098214020948601

Chilisa, B., & Tsheko, G. N. (2014). Mixed methods in Indigenous research: Building relationships for sustainable intervention outcomes. *Journal of Mixed Methods Research*, 8(3), 222–233. DOI: 10.1177/1558689814527878

Chimakonam, J. O. (2019). Decolonisation, Africanisation and transformation: Why we need 'That' African contribution to world history and civilisation. In: *Ezumezu: A System of Logic for African Philosophy and Studies* (pp. 181–205). Springer. DOI: 10.1007/978-3-030-11075-8_11

Coates, S. K., Trudgett, M., & Page, S. (2023). Indigenous institutional theory: A new theoretical framework and methodological tool. *Australian Educational Researcher*, 50(3), 903–920. DOI: 10.1007/s13384-022-00533-4

Collins, C. S., & Stockton, C. M. (2018). The central role of theory in qualitative research. *International Journal of Qualitative Methods*, 17(1), 1609406918797475. Advance online publication. DOI: 10.1177/1609406918797475

Collins, P. H. (2009). *Black feminist thought: Knowledge, consciousness, and the politics of empowerment*. Routledge Classics.

Creswell, J. W., & Plano Clark, V. L. (2018). *Designing and conducting mixed methods research* (3rd ed.). Sage.

Dana-Sacco, G. (2010). The Indigenous researcher as individual and collective: Building a research practice ethic within the context of Indigenous languages. *American Indian Quarterly*, 34(1), 61–82. DOI: 10.1353/aiq.2010.a370587

Davison, R. M., & Díaz Andrade, A. (2018). Promoting Indigenous theory. *Information Systems Journal*, 28(5), 759–764. DOI: 10.1111/isj.12203

de Sousa Santos, B. (2014). *Epistemologies of the South: Justice against epistemicide*. Routledge., DOI: 10.4324/9781315634876

Dotson, K. (2015). Inheriting Patricia Hill Collins's *Black Feminist* epistemology. *Ethnic and Racial Studies*, 38(13), 2322–2328. DOI: 10.1080/01419870.2015.1058496

Du Toit, A. (2019). The potential of the South African consumer studies curriculum to foster indigenous knowledge. *Journal of Consumer Sciences*, 47, 27–39.

Dubin, R. (1969). *Theory building*. 2nd ed. Free Press, Collier-Macmillan

Dul, J., & Hak, T. (2007). *Case study methodology in business research*. Routledge., DOI: 10.4324/9780080552194

Fanon, F. (1964). *Toward the African revolution*. Grove Press.

Gebru, A., & Wako, W. (2022). Walking the theories we talk: Utilizing African social work theories in African research. *African Journal of Social Work*, 12(4), 189–198.

Glaser, B., & Strauss, A. (1967). *The discovery of grounded theory: Strategies for qualitative research*. Sociology Press.

Godfrey-Smith, P. (2003). *Theory and reality*. University of Chicago Press. DOI: 10.7208/chicago/9780226300610.001.0001

Gough, D., Oliver, S., & Thomas, J. (2017). Introduction to systematic reviews. In Gough, D., Oliver, S., & Thomas, J. (Eds.), *An introduction to systematic reviews* (pp. 1–19). Sage.

Grande, S., & McCarty, T. L. (2018). Indigenous elsewheres: Refusal and re-membering in education research, policy, and praxis. *International Journal of Qualitative Studies in Education: QSE*, 31(3), 165–167. DOI: 10.1080/09518398.2017.1401144

Grosfoguel, R. (2013). The structure of knowledge in Westernized universities: Epistemic racism/sexism and the four genocides/epistemicides of the long 16th century. *Human Architecture*, 9(1), 73–90.

Hall, R., Ansley, L., & Connolly, P. (2023). Decolonising or anti-racism? Exploring the limits of possibility in higher education. *Teaching in Higher Education*. Advance online publication. DOI: 10.1080/13562517.2023.2201676

Hallinger, P. (2013). A conceptual framework for systematic reviews of research in educational leadership and management. *Journal of Educational Administration*, 51(2), 126–149. DOI: 10.1108/09578231311304670

Hassan, N. R., Lowry, P. B., & Mathiassen, L. (2022). Useful products in information systems theorizing: A discursive formation perspective. *Journal of the Association for Information Systems*, 23(2), 418–446. DOI: 10.17705/1jais.00730

Hirschheim, R., & Klein, H. K. (2003). Crisis in the IS field? A critical reflection on the state of the discipline. *Journal of the Association for Information Systems*, 4(5), 237–293.

Hungwe, J. P., & Mkhize, T. M. (2022). The Africanisation of universities in Africa: Reclamation of humanity and rationality. *Critical Studies in Teaching and Learning*, 10(1). Advance online publication. DOI: 10.14426/cristal.v10i1.487

Ibrahim, M. G., Adjei, J. K., & Boateng, J. A. (2019). Relevance of indigenous conflict management mechanisms: Evidence from Bunkpurugu-Yunyoo and Central Gonja Districts of Northern Region, Ghana. *Ghana Journal of Development Studies*, 16(1), 22–45. DOI: 10.4314/gjds.v16i1.2

Jaggar, A. M., & Rothenberg, P. (1993). *Feminist frameworks: Alternative theoretical accounts of the relations between women and men*. McGraw-Hill.

Joullié, J.-E., & Gould, A. M. (2021). Theory, explanation, and understanding in management research. *BRQ Business Research Quarterly*, 0(0). Advance online publication. DOI: 10.1177/23409444211012414

Jukka, M., Blomqvist, K., Li, P. P., & Gan, C.-M. (2017). Trust-distrust balance: Trust ambivalence in Sino-Western B2B relationships. *Cross Cultural & Strategic Management*, 24(3), 482–507. DOI: 10.1108/CCSM-01-2016-0019

Justesen, T., Freyberg, J., & Schultz, A. N. Ø. (2021). Database selection and data gathering methods in systematic reviews of qualitative research regarding diabetes mellitus - an explorative study. *BMC Medical Research Methodology*, 21(1), 94. DOI: 10.1186/s12874-021-01281-2 PMID: 33941105

Kaomea, J. (2003). Reading erasures and making the familiar strange: Defamiliarizing methods for research in formerly colonized and historically oppressed communities. *Educational Researcher*, 32(2), 14–23. DOI: 10.3102/0013189X032002014

Kawulich, B. (2009). The role of theory in research. In Garner, M., Wagner, C., & Kawulich, B. (Eds.), *Teaching research methods in the social sciences* (pp. 37–47). Routledge.

Keane, M., Raciti, M., van der Westhuizen, G., Motala, S., Stanton, S., Gilbey, K., Manathunga, C., Qi, J., & Msimango, S. (2023). Indigenous knowledge systems in South Africa and Australia: Transforming doctoral education. *Curriculum Perspectives*, 43(S1), 83–93. Advance online publication. DOI: 10.1007/s41297-023-00183-1

Kettley, N. (2010). Theory building in educational research. *Continuum*.

Kuhn, T. S. (1970). *The structure of scientific revolutions* (2nd ed.). University of Chicago Press.

Laird, P. (2022). Fits and starts: Anti-racism efforts in the biosciences. *Nature*, 612, 177–178.

Lawrence, S., & Hylton, K. (2022). Critical race theory, methodology, and semiotics: The analytical utility of a "race" conscious approach for visual qualitative research. *Cultural Studies Critical Methodologies*, 22(3), 255–265. DOI: 10.1177/15327086221081829

Le Grange, L. (2016). Decolonising the university curriculum. *South African Journal of Higher Education*, 30(2), 1–12.

Lebakeng, L. T., Phalane, M. M., & Nase, D. (2006). Epistemicide, institutional cultures and the imperative for the Africanisation of universities in South Africa. *Alternation (Durban)*, 13(1), 70–87.

Lefike, M., Turpin, M., & Matthee, M. (2023). A systems framework to analyze the impact of corporate social investment projects with an information technology focus. *The Electronic Journal on Information Systems in Developing Countries*, 89(5), e12273. DOI: 10.1002/isd2.12273

Levers, L. L. (2009). A meta-theoretical framework for counselling people of African ancestry: An ecologically oriented response to the special section articles. *Journal of Psychology in Africa*, 19(4), 503–506. DOI: 10.1080/14330237.2009.10820322

Levi-Strauss, C. (1966). *The savage mind*. University of Chicago Press.

Li, P. P. (2008). Toward a geocentric framework of trust: An application to organizational trust. *Management and Organization Review*, 4(3), 413–439. DOI: 10.1111/j.1740-8784.2008.00120.x

Li, P. P. (2012). Toward an integrative framework of indigenous research: The geocentric implications of Yin-Yang balance. *Asia Pacific Journal of Management*, 29(4), 849–872. DOI: 10.1007/s10490-011-9250-z

Li, P. P. (2016). The global implications of the Indigenous epistemological system from the East: How to apply Yin-Yang balancing to paradox management. *Cross Cultural & Strategic Management*, 23(1), 42–77. DOI: 10.1108/CCSM-10-2015-0137

Li, P. P., Sekiguchi, T., & Zhou, K. (2016). The emerging research on Indigenous management in Asia. *Asia Pacific Journal of Management*, 33(3), 583–594. DOI: 10.1007/s10490-016-9484-x

Mahlatsi, K. S., Pienaar, A. J., Nare, N. E., & Mulaudzi, T. M. (2021). A conceptual framework for psychosocial health management grounded in the therapeutic merits of Indigenous Khoisan health dialogues. *Health SA = SA Gesondheid*, 26(0), a1626. DOI: 10.4102/hsag.v26i0.1626 PMID: 34522395

Makhanya, T., & Zibane, S. (2020). Students' voices on how indigenous languages are disfavoured in South African higher education. *Language Matters (Pretoria)*, 51(1), 22–37. DOI: 10.1080/10228195.2020.1711533

Maldonado-Torres, N., & Cavooris, R. (2017). The decolonial turn. In Poblete, J. (Ed.), *New approaches to Latin American Studies: Culture and power* (1st ed.). Routledge., DOI: 10.4324/9781315158365-8

Maluleka, J. R., & Ngulube, P. (2019). Indigenous knowledge in Africa: A bibliometric analysis of publishing patterns. *Publishing Research Quarterly*, 35(3), 445–462. DOI: 10.1007/s12109-019-09656-6

Mandikonza, C. (2019). Integrating indigenous knowledge practices as context and concepts for the learning of curriculum science: A methodological exploration. *Southern African Journal of Environmental Education*, 35(1), 1–16. DOI: 10.4314/sajee.v35i1.13

Marsden, D. (2005). *Indigenous wholistic theory for health: Enhancing traditional-based Indigenous health services in Vancouver. Unpublished Doctor of Philosophy*. University of British Columbia.

Martel, R., Shepherd, M., & Goodyear-Smith, F. (2021). He awa whiria – A "braided river": An Indigenous Māori approach to mixed methods research. *Journal of Mixed Methods Research*. Advance online publication. DOI: 10.1177/1558689820984028

Martin, D. H. (2012). Two-eyed seeing: A framework for understanding Indigenous and non-Indigenous approaches to Indigenous health research. *Canadian Journal of Nursing Research*, 44(2), 20–42. PMID: 22894005

Mazama, A. (2001). The Afrocentric paradigm: Contours and definitions. *Journal of Black Studies*, 31(4), 387–405. DOI: 10.1177/002193470103100401

Mbah, M. F., & Bailey, M. (2022). Decolonisation of research methodologies for sustainable development in Indigenous settings. In: M. F. Mbah, W. Leal Filho, S. Ajaps. (eds.), *Indigenous methodologies, research and practices for sustainable development*. World Sustainability Series. Springer. DOI: 10.1007/978-3-031-12326-9_2

Mbah, M. F., Bailey, M., & Shingruf, A. (2023). Considerations for relational research methods for use in Indigenous contexts: Implications for sustainable development. *International Journal of Social Research Methodology*, 27(4), 431–446. DOI: 10.1080/13645579.2023.2185345

Mbatha, N. T., Majola, Y. L., & Gumede, Z. S. (2023). Language maintenance: Factors supporting the use and maintenance of isiZulu in Soshanguve. *Literator*, 44(1), a1930. DOI: 10.4102/lit.v44i1.1930

Mbembe, A. J. (2016). Decolonizing the university: New directions. *Arts and Humanities in Higher Education*, 15(1), 29–45. DOI: 10.1177/1474022215618513

McHugh, M. (2012). Interrater reliability: The Kappa statistic. *Biochemia Medica*, 22, 276–282. DOI: 10.11613/BM.2012.031 PMID: 23092060

Meda, L. (2020). Decolonising the curriculum: Students' perspectives. *Africa Education Review*, 17(2), 88–103. DOI: 10.1080/18146627.2018.1519372

Mignolo, W. D. (2009). Epistemic disobedience, independent thought and decolonial freedom. *Theory, Culture & Society*, 26(7-8), 159–181. DOI: 10.1177/0263276409349275

Mignolo, W. D., & Walsh, C. E. (2018). *On decoloniality: Concepts, analytics, praxis*. Duke University Press.

Mitova, V. (2021). How to decolonise knowledge without too much relativism: Colonial insights into the South African experience. In Kumalo, S. H. (Ed.), *Decolonisation as democratisation* (pp. 24–47). Human Sciences Research Council. DOI: 10.1515/9780796926166-007

Mokotso, R. I. (2022). Considering the Basotho indigenous education and school system as resources for peace-building education in Lesotho. *Journal for Transdisciplinary Research in Southern Africa*, 18(1), 1–10. DOI: 10.4102/td.v18i1.1213

Moropa, K. (2018). Depiction of indigenous food in translated isiXhosa folk narratives and biographies. *South African Journal of African Languages*, 38(3), 247–254. DOI: 10.1080/02572117.2018.1518017

Morton, N., Hill, C., Meiring, D., & De Beer, L. T. (2019). Investigating the factor structure of the South African Personality Inventory-English version. *SA Journal of Industrial Psychology*, 45(1), 1–13. DOI: 10.4102/sajip.v45i0.1556

Msila, V., & Gumbo, M. T. (Eds.). (2017). *African voices on indigenisation of the curriculum: Insights from practice*. Reach Publishers.

Mumpande, I., & Barnes, L. (2019). Revitalisation of the Tonga language in Zimbabwe: The motivational factors. *Language Matters (Pretoria)*, 50(3), 46–69. DOI: 10.1080/10228195.2019.1691634

Mwanza, A. J. (2022). Can digital technology be used for addressing digital literacy in the dissemination of Indigenous Knowledge in rural areas? *South African Journal of Library and Information Science*, 88(1), 1–11.

Nabudere, D. W. (2011). *Afrikology, philosophy and wholeness: An epistemology*. Africa Institute of South Africa.

Ndlovu-Gatsheni, S. J. (2017). The emergence of trajectories of struggles for an "African university": The case of unfinished business of African epistemic decolonisation. *Kronos (Bellville, South Africa)*, 43(1), 51–77. DOI: 10.17159/2309-9585/2017/v43a4

Ndlovu-Gatsheni, S. J. (2018). *Epistemic freedom in Africa: Deprovincialization and decolonisation*. Routledge. DOI: 10.4324/9780429492204

Ngara, R. (2017). Multiple voices, multiple paths. In Ngulube, P. (Ed.), *Handbook of research on theoretical perspectives on Indigenous knowledge systems in developing countries* (pp. 332–358). IGI Global. DOI: 10.4018/978-1-5225-0833-5.ch015

Ngulube, P. (2018). Overcoming the difficulties associated with using conceptual and theoretical frameworks in heritage studies. In Ngulube, P. (Ed.), *Handbook of research on heritage management and preservation* (pp. 1–23). IGI Global. DOI: 10.4018/978-1-5225-3137-1.ch001

Ngulube, P. (2020). Theory and theorising in information science scholarship. In Ngulube, P. (Ed.), *Handbook of research on connecting research methods for information science research* (pp. 18–39). IGI Global. DOI: 10.4018/978-1-7998-1471-9.ch002

Ngulube, P. (2022). Using simple and complex mixed methods research designs to understand research in Information Science. In Ngulube, P. (Ed.), *Handbook of research on mixed methods research in information science* (pp. 20–44). IGI Global. DOI: 10.4018/978-1-7998-8844-4.ch002

Ngulube, P., Mathipa, E. R., & Gumbo, M. T. (2015). Theoretical and conceptual framework in the social sciences. In Mathipa, E. R., & Gumbo, M. T. (Eds.), *Addressing research challenges: making headway in developing researchers* (pp. 43–66). Mosala-MASEDI Publishers & Booksellers., DOI: 10.13140/RG.2.1.3210.7680

Ngulube, P., & Mosha, N. F. V. (2023). Transparency in the application of theoretical frameworks to the advancement of knowledge in selected library and information science journals: A systematic review. *African Journal of Library Archives and Information Science*, 33(2), 129–146. DOI: 10.4314/ajlais.v33i2.2

Nyamnjoh, F. (2012). Potted plots in greenhouses: A critical reflection on the resilience of colonial education in Africa. *Journal of Asian and African Studies*, 47(2), 129–154. DOI: 10.1177/0021909611417240

Omodan, B. I. (2024). *Research paradigms and their methodological alignment in social sciences: A practical guide for researchers*. Routledge., DOI: 10.4324/9781003484066

Omodan, B. I. (2025). Building reciprocal relationships through decolonial practices in academic research. *Cogent Social Sciences*, 11(1), 2443558. Advance online publication. DOI: 10.1080/23311886.2024.2443558

Onyancha, O. B., Ngulube, P., Maluleka, J. R., & Mokwatlo, K. I. (2018). Towards a uniform terminology for Indigenous knowledge concepts: Informetrics perspectives. *African Journal of Library Archives and Information Science*, 28(2), 155–167.

Page, M. J., McKenzie, J. E., & Bossuyt, P. M.. (2021). The PRISMA 2020 statement: An updated guideline for reporting systematic reviews. *BMJ (Clinical Research Ed.)*, 372(71). PMID: 33782057

Pidgeon, M., & Riely, T. (2021). Understanding the application and use of indigenous research methodologies in the social sciences by indigenous and non-indigenous scholars. *International Journal of Education Policy and Leadership*, 17(8). Advance online publication. DOI: 10.22230/ijepl.2021v17n8a1065

Pienaar, A. J. (2017). Learning and asserting an African Indigenous health research framework. In Ngulube, P. (Ed.), *Handbook of research on theoretical perspectives on Indigenous knowledge systems in developing countries* (pp. 85–99). IGI Global. DOI: 10.4018/978-1-5225-0833-5.ch004

Popper, K. (1980). *The logic of scientific discovery*. Abingdon-on-Thames: Routledge.

*References used in the systematic review.

Reid, A. J., Eckert, L. E., Lane, J.-F., Young, N., Hinch, S. G., Darimont, C. T., Cooke, S. J., Ban, N. C., & Marshall, A. (2021). "Two-eyed seeing": An Indigenous framework to transform fisheries research and management. *Fish and Fisheries*, 22(2), 243–261. DOI: 10.1111/faf.12516

Reid, B. (2020). Positionality and research: "two-eyed seeing" with a rural Ktaqmkuk Mi'kmaw Community. *International Journal of Qualitative Methods*, 19, 1609406920910841. Advance online publication. DOI: 10.1177/1609406920910841

Romm, N. R. A. (2018). *Responsible research practice: Revisiting transformative paradigm in social research*. Springer International Publishing. DOI: 10.1007/978-3-319-74386-8

Ryser, R. C., Gilio-Whitaker, D., & Bruce, H. G. (2017). Fourth World Theory and methods of inquiry. In Ngulube, P. (Ed.), *Handbook of research on theoretical perspectives on Indigenous knowledge systems in developing countries* (pp. 50–84). IGI Global., DOI: 10.4018/978-1-5225-0833-5.ch003

Sabati, S. (2019). Upholding "colonial unknowing" through the IRB: Reframing institutional research ethics. *Qualitative Inquiry*, 25(9–10), 1056–1064. DOI: 10.1177/1077800418787214

Salawu, R. O., Bolatitio, A. S., & Masibo, S. (2023). Theoretical and conceptual frameworks in research: Conceptual clarification. *European Chemical Bulletin*, 12(12), 2103–2117.

Sartori, G. (2009). Guidelines for concept analysis. In Collier, D., & Gerring, J. (Eds.), *Concepts and method in social science. The tradition of Giovanni Sartori* (pp. 97–150). Routledge.

Scheurich, J. J. (1997). *Research method in the post-modern*. Falmer Press.

Scheurich, J. J., & Young, D. M. (1997). Coloring methodologies: Are our research epistemologies racially biased? *Educational Researcher*, 26(4), 4–16. DOI: 10.3102/0013189X026004004

Schiller, U., & de Wet, G. (2018). Communication, indigenous culture and participatory decision making amongst foster adolescents. *Qualitative Social Work: Research and Practice*, 17(2), 236–251. DOI: 10.1177/1473325016662329

Schiller, U., & de Wet, G. (2019). Indigenous cultural and social work practices creating inequalities among foster care adolescents in decision-making processes. *International Social Work*, 62(1), 22–34. DOI: 10.1177/0020872817710548

Schneider, M. A. (2006). *The theory primer: A sociological guide*. Rowman and Littlefield.

Schreiber, F., & Cramer, C. (2022). Towards a conceptual systematic review: Proposing a methodological framework. *Educational Review*, 1–22. Advance online publication. DOI: 10.1080/00131911.2022.2116561

Sefotho, M. M. (2022). Foreword. In Adu, E. O., & Okeke, C. I. O. (Eds.), *Fundamentals of research in humanities, social sciences and science education: A practical step by step approach to a successful research journey* (p. v). Van Schaik.

Seroto, J. (2018). Dynamics of decoloniality in South Africa: A critique of the history of Swiss mission education for indigenous people. *Studia Historiae Ecclesiasticae*, 44(3), 1-14. *Seroto, J. (2019). The embodiment of the indigenous people by European travel writers at the Cape colony, Southern Africa. *Education as Change*, 23(1), 1–24.

Shai, K. B. (2023). "Pathetic scholarship" in Gukurahundi and related studies: A silver lining from social media cloud. *[APORTAL]*. *African Perspectives of Research in Teaching & Learning*, 7(2), 378–386.

Shokane, A. L., & Masoga, M. A. (2018). African indigenous knowledge and social work practice: Towards an Afro-sensed perspective. *Southern African Journal of Social Work and Social Development,* 30(1), 1-l18. https://upjournals.co.za/index.php/SWPR

Siddaway, A. P., Wood, A. M., & Hedges, L. V. (2019). How to do a systematic review: A best practice guide for conducting and reporting narrative reviews, meta-analyses, and meta-syntheses. *Annual Review of Psychology*, 70(1), 747–770. DOI: 10.1146/annurev-psych-010418-102803 PMID: 30089228

Smiet, K. (2022). Rethinking or delinking? Said and Mignolo on humanism and the question of the human. *Postcolonial Studies*, 25(1), 73–88. DOI: 10.1080/13688790.2022.2030595

Smith, C., & Babich, C. (2023). *Theories of individual and collective learning*. Campus Ontario. Retrieved 12 December 2024 from https://ecampusontario.pressbooks.pub/ticl/

Smith, L. T. (2012). *Decolonizing methodologies: Research and Indigenous peoples*. Zed Books.

Snapp, S. (2022). Change the reward structure. *Nature*, 612, 178.

Steenkamp, H., & Rensburg, R. (2018). Utilising an Ubuntu-centred communication management framework to analyse CSR messages on SNSS. *Communicatio*, 44(3), 17–40. DOI: 10.1080/02500167.2018.1541914

Stein, S., & de Andreotti, V. O. (2016). Decolonization and higher education: Introduction. In Peters, M. (Ed.), *Encyclopedia of educational philosophy and theory* (pp. 1–6). Springer. DOI: 10.1007/978-981-287-532-7_479-1

Sutton, R. I., & Staw, B. M. (1995). What theory is not. *Administrative Science Quarterly*, 40(3), 371–384. DOI: 10.2307/2393788

Thambinathan, V., & Kinsella, E. A. (2021). Decolonizing methodologies in qualitative research: Creating spaces for transformative praxis. *International Journal of Qualitative Methods*, 20, 16094069211014766. Advance online publication. DOI: 10.1177/16094069211014766

Tran, D. (2021). *Decolonizing university teaching and learning: An entry model for grappling with complexities*. Bloomsbury Academic. DOI: 10.5040/9781350160040

Van de Ven, A. H., Meyer, A. D., & Jing, R. T. (2018). Opportunities and challenges of engaged Indigenous scholarship. *Management and Organization Review*, 14(3), 449–462. DOI: 10.1017/mor.2018.28

Venter, E. (1997). A philosophy of education in a new South Africa. *South African Journal of Higher Education*, 11(1), 57–64.

Viera, A. J., & Garrett, J. M. (2005). Understanding interobserver agreement: The kappa statistic. *Family Medicine*, 37, 360–363. PMID: 15883903

Wei, X., Liu, R., & Chen, W. (2023). Meta theories of technological innovation based on the analysis of classic texts. *Heliyon*, 9(6), e16779. DOI: 10.1016/j.heliyon.2023.e16779 PMID: 37292334

Weick, K. E. (1995). What theory is not, theorizing is. *Administrative Science Quarterly*, 40(3), 385–390. DOI: 10.2307/2393789

Wight, C. (2019). Bringing the outside in: The limits of theoretical fragmentation and pluralism in IR theory. *Politics*, 39(1), 64–81. DOI: 10.1177/0263395718815784

Wilson, S. (2001). What is an Indigenous research methodology? *Canadian Journal of Native Education*, 25(2), 175–179.

Wright, A. L., Gabel, C., Ballantyne, M., Jack, S. M., & Wahoush, O. (2019). Using two-eyed seeing in research with Indigenous people: An integrative review. *International Journal of Qualitative Methods*, 18, 1609406919869695. Advance online publication. DOI: 10.1177/1609406919869695

Yang, K. S. (2000). Monocultural and cross-cultural Indigenous approaches: The royal road to the development of a balanced global psychology. *Asian Journal of Social Psychology*, 3(3), 241–263. DOI: 10.1111/1467-839X.00067

Zaltman, G., Pinson, C. R. A., & Angelmar, R. (1973). *Metatheory and consumer research*. Holt, Rinehart and Winston, Inc.

Zembylas, M. (2018). Decolonial possibilities in South African higher education: Reconfiguring humanising pedagogies as/with decolonising pedagogies. *South African Journal of Education*, 38(4), 1699. Advance online publication. DOI: 10.15700/saje.v38n4a1699

ADDITIONAL READINGS

Akin-Otiko, A. (2024). Hu Ìwà Sí i bí Ènìyàn lẹ́ran-Ara a (act towards the other as one with human flesh): A Yoruba theory of social interactions. *Social Sciences & Humanities Open*, 10, 100936. DOI: 10.1016/j.ssaho.2024.100936

Antoine, A., Mason, R., Mason, R., Palahicky, S., & Rodriguez de France, C. (2018). *Pulling together: A guide for Indigenization of post-secondary institutions*. BCcampus. Retrieved on 24 July 2023 from https://opentextbc.ca/indigenizationcurriculumdevelopers/

Karki, K. K. (2016). Walking the complexities between two worlds: A personal story of epistemological tensions in knowledge production. *Qualitative Social Work: Research and Practice*, 15(5–6), 628–639. DOI: 10.1177/1473325016652678

Moko-Painting (Ngāti Manu, T. P., Hamley (Ngāti Rangi, W., Hikuroa (Ngāti Maniapoto, T., Le Grice (Ngapuhi, T., McAllister (Te Aitanga a Māhaki, N. P., McLellan (Whakatōhea, N. T. R., Parkinson (Ngāti Hine, N. P., Renfrew (Te Rarawa, N., & Rewi (Ngāpuhi, N. H. (. (2023). (Re)emergence of Pūtaiao: Conceptualising Kaupapa Māori science. *Environment and Planning F*, 2(1-2), 11–37. DOI: 10.1177/26349825231164617

Peltier, C. (2018). An application of two-eyed seeing: Indigenous research methods with participatory action research. *International Journal of Qualitative Methods*, 17(1), 1609406918812346. Advance online publication. DOI: 10.1177/1609406918812346

Rosiek, J. L., Snyder, J., & Pratt, S. L. (2020). The new materialisms and Indigenous theories of non-human agency: Making the case for respectful anti-colonial engagement. *Qualitative Inquiry*, 26(3-4), 331–346. DOI: 10.1177/1077800419830135

Scott Thomas, D. (2022). Applying one dish, one spoon as an Indigenous research methodology. *Alternative*, 18(1), 84–93. DOI: 10.1177/11771801221087864

Suddaby, R. (2010). Editor's comments: Construct clarity in theories of management and organization. *Academy of Management Review*, 35(3), 346–357.

KEY TERMS AND DEFINITIONS

Cross-Indigenisation geocentric: Cross-indigenisation geocentric theories focus on integrating indigenous knowledge systems with global perspectives to foster a more inclusive and holistic understanding of various phenomena. By emphasising contextual understanding, this approach ensures that global theories are not applied in a one-size-fits-all manner but adapted to local realities.

Holism: Indigenous knowledge encompasses emotional, spiritual, cognitive, and physical dimensions, recognising that these aspects are interconnected and collectively contribute to human development and well-being.

Indigenous Epistemologies: Indigenous epistemologies represent the unique ways Indigenous peoples understand and interpret the world, rooted in their cultural traditions, experiences, and worldviews. These distinctive approaches to knowledge emphasise several key elements, such as *relationality, holism, the interconnectedness of the sacred and secular, and the significance of oral traditions and storytelling*. These concepts collectively embody a rich and comprehensive perspective underpinning Indigenous knowledge systems.

Oral Traditions and Storytelling: Knowledge is transmitted through generations via oral traditions and storytelling, which preserve historical and cultural understanding while reinforcing community values and social norms.

Relationality: refers to the interconnectedness of all beings and elements in the universe, encompassing relationships with the natural environment, the spiritual world, and the community.

Sacred and Secular Interconnection: it acknowledges the inseparability of spiritual and secular dimensions, with spirituality being an integral part of both learning and understanding.

ENDNOTES

1. According to Chilisa and Tsheko (2014), developing constructs regarding relational epistemology with the research participants is important. It is not just scholars situated in academia who do the theorising - but this must include the involvement of the communities.

2. Care must be taken when speaking of the white race, as races are social categorisations. The concept of race has no biological backing - the colour of skin (phenotype) does not reflect genotypical features. But with the invention of the term race and the attendant division of people into different races, people who were labelled as white were able to assign a superior status to themselves. Before the concept was invented, people did not see colour as an issue. The concept supported the Western colonisation of those called non-White.

3. Note that critical race theorists will never speak of the white race: they want to highlight how race is a social construct - so they speak of racialised groupings or racial formations - but not of the "white race" as if this is a thing independently of how race is used to structure social relations—this way of theorising undercuts popular uses of terms that create stereotypes.

4. Of course, there is contention around what "testing" involves. Positivists say it can be tested with reference to empirical occurrences but do not consider that experiences are always interpreted and can be interpreted in different ways. That is why Indigenous epistemologies call for dialogue to create collectively generated insights.

5. This is arguably also a tenet of an Indigenous paradigm - the knowing process must be used towards creating an inclusive benefit - and those most marginalised in the social and ecological fabric are critical participants in defining what is of benefit.

6. The concept of testing espoused here differs significantly from the ideas on testing as Popper (1980) advanced, for example. Testing does not lie only in the hands of professional researchers.

7. Although *The Cochrane Handbook for Systematic Reviews of Interventions* does not explicitly mandate including author names in the analysis, author names were included to enhance transparency and credibility in the systematic review (Chandler et al., 2022).

APPENDIX

Table 2. Data extraction and coding

Author(s)	Publisher	Affiliation	Research Design	Theory /model/ concepts	Justification of the choice	Transparency	Western models and constructs	Indigenous models and constructs	Combining concepts (Integrative)	Uniquely Indigenous (innovative)
Seroto (2018)	Church History Society of Southern Africa and Unisa Press	South Africa (SA)	Historiography (historical inquiry)	Model of coloniality Grosfoguel	No	No	Yes	No	No	No
Shokane & Masoga (2018)	Unisa Press	SA	Ethnographic design	Cross-cultural perspective	No	No	Yes	Yes	Yes	No
Makhanya & Zibane (2020)	Unisa Press	SA	Qualitative design	Anti-colonial discursive framework and Afrocentric theory	Yes	Yes	Yes	Yes	Yes	No
Seroto (2019)	University of Johannesburg and Unisa Press	SA	Content analysis	Decolonial theoretical framework Grosfoguel	Yes	Yes	No	Yes	No	No
Mahlatsi, Pienaar, Nare & Mulaudzi (2021)	AOSIS	SA	Indigenous African research design	Various concepts	Yes	Yes	No	Yes	No	Yes
Mwanza (2022)	Library & Information Association of South Africa	SA	Action research	Andragogy, situated learning and eZiko	No	No	Yes	Yes	Yes	No
Adjimah, Atiase & Dzansi (2023)	Emerald Publishing Limited	SA& Ghana	Survey	Below-the-radar theory of innovation	Yes	Yes	Yes	No	No	No
Schiller & de Wet (2018)	SAGE Publications	SA	Qualitative	Interactional Communication Theory	No	Yes	Yes	No	No	No
Mokotso (2022)	OASIS	SA	Indigenous auto-ethnography	Critical interculturality	Yes	Yes	Yes	Yes	Yes	No

continued on following page

Table 2. Continued

Author(s)	Publisher	Affiliation	Research Design	Theory /model/ concepts	Justification of the choice	Transparency	Western models and constructs	Indigenous models and constructs	Combining concepts (Integrative)	Uniquely Indigenous (innovative)
Schiller & de Wet (2019)	SAGE Publications	SA	Phenomenology	Developmental social welfare approach & systems theory	No	Yes	Yes	Yes	No	No
Lefike, Turpin & Matthee (2023)	Wiley	SA	Case study	Soft systems methodology Ubuntu philosophy & autopoiesis	Yes	Yes	Yes	Yes	Yes	No
Moropa (2018)	Routledge	SA	Content analysis	Descriptive translation studies (DTS) approach	Yes	Yes	Yes	Yes	Yes	No
Morton, Hill, Meiring & De Beer (2019)	AOSIS	SA	Survey	Six-factor SAPI framework	Yes	Yes	Yes	No	No	No
Meda (2020)	Unisa & Routledge	SA	Case study	Rawls' (1971) theory of justice	No	No	Yes	No	No	No
Steenkamp & Rensburg (2018)	Unisa & Routledge	SA	Phenomenology	Ubuntu-centred communication practices	Yes	Yes	Yes	Yes	No	No
Mumpande & Barnes (2019)	Unisa & Routledge	SA	Case study	Human needs theory	Yes	No	Yes	No	No	No
Du Toit (2019)	South African Association of Family Ecology and Consumer Sciences	SA	Content analysis	'Two sides of the same coin' framework	Yes	No	Yes	Yes	No	No
Mandikonza (2019)	Environmental Education Association of Southern Africa	SA	Qualitative	Socio-cultural theory	No	No	Yes	Yes	No	No

continued on following page

Table 2. Continued

Author(s)	Publisher	Affiliation	Research Design	Theory /model/ concepts	Justification of the choice	Transparency	Western models and constructs	Indigenous models and constructs	Combining concepts (Integrative)	Uniquely Indigenous (innovative)
Balogun & Adeseke (2022)	Department of Communication Studies, Faculty of Arts, College of Humanities and Legal Studies, University of Cape Coast, Ghana	Nigeria	Qualitative	Cultural norms and agenda-setting theories	No	No	Yes	No	No	No
Ibrahim, Adjei & Boateng (2019)	Faculty of Integrated Development Studies of SD Dombo University of Business and Integrated Development Studies, Ghana	Ghana	Case study	Realist theory of conflict & Indigenous wholistic theory	Yes	Yes	Yes	Yes	Yes	No
Mbatha Majola & Gumede (2023)	AOSIS	SA	Mixed methods approach (Convergent)	Haugen's theory of language ecology & Giles's model of ethnolinguistic vitality	No	Yes	Yes	No	No	N/A

Chapter 9
Deterritorialization and Reterritorialization of Indigenous Knowledge:
Reclamation of Misappropriated Indigenous Heritage Among Ethnocultural Communities, Southern Africa

Tlou Maggie Masenya
https://orcid.org/0000-0001-5400-0549
Durban University of Technology, South Africa

Collence T. Chisita
https://orcid.org/0000-0002-7375-8627
University of South Africa, South Africa

ABSTRACT

This chapter theorises cultural communities' alienation from their cultural roots due to the effect of deterritorialization, colonization, globalization and westernization, and this has also given rise to cultural change or acculturation. The chapter examined conversations concerning the reclamation of misappropriated heritage in cultural communities in Southern Africa. The study employed an Afronographic approach, which is an Afrocentric paradigm that places the interests of the African people at the centre of the analysis of the deterritorialization and reterritorialization of knowledge intending to reclaim ways of knowing and culture. Afrocentricity theory was employed to underpin this study because it is a novel research methodology that is deeply rooted in the African people's experiences, unlike Eurocentric theories that disorder Africans in the sideline of epistemology.

DOI: 10.4018/979-8-3693-8447-3.ch009

INTRODUCTION AND BACKGROUND

Western civilization, globalization and deterritorialization had negative impacts on knowledge and traditional cultures, and as a result, many of the past lessons in ethnocultural communities in Southern Africa are slowly fading away. There has been a cantankerous and vituperative effort to denigrate anything since the onslaught of colonialism in Africa, and this has dealt a serious blow to efforts to revitalise the African Knowledge system (Chisita, 2012). Western culture and the effect of globalization, cultural imperialism and cultural assimilation have precariously contaminated African traditional cultural values, norms, language, expectations and beliefs. Mgbeoji (2006) argued that modern history is replete with a myriad of forms of colonialism and conquest inspired by cultural chauvinism and economic exploitation, but none has been as violent and enduring in its radical transformation of the globe as European colonialism in the global South. On a similar note (Mgbeoji, 2011) contested that imperial Europe adopted a religious and geopolitical campaign that dichotomized world into a civilised Europe and a savage 'other' precipitating the unprecedented dispossession and genocide of colonised peoples and territories. This dispossession has morphed into the dispossession of the already dispossessed because of the neo-colonial onslaught on the people of the global South. Western culture began to creep into the African socio-cultural milieu with the contact of Europeans with Africa, a consequence of the Berlin conference (1884-1885) in the quest for imperial pilfering of African resources and later, consolidated by the unstoppable wave of globalization (Arowolo, 2010). The majority of ethnocultural communities in Southern Africa have assimilated the European or Western culture and have appropriated it so much that it now becomes almost part and parcel of their lives, and this has led to cultural change or acculturation and cultural homogenization among these communities. Pritchard (2009) further noted that ethnocultural communities lose their authenticity in favour of the dominant Western culture, particularly American norms and values, as cultural change or acculturation and homogenization, and global stratification is exacerbated. Suarez (2013) described cultural acculturation as a process in which members of one cultural group adopt the beliefs and behaviour of another group. Cultural homogenization refers to the idea that different cultures transform and become more similar to each other as globalization progresses (Hassi & Storti, 2012). Examples of cultural homogenization include the development of global culture, McDonaldization of cultures and global Americanization.

The colonial onslaught on ethnocultural communities in Southern Africa thus inflicted a devastating impact on their ways of being, living and knowing. Ethnocultural communities in Southern African countries such as Botswana and Namibia are often among the most marginalized communities and reterritorialized cultures in many societies and they are often excluded from participation in socio-economic life and rarely have access to political power while also encountering obstacles to manifesting their identity (United Nations High Commissioner for Refugees) (UNHCR) (2018). As also noted by UNHCR (2018), these obstacles are multiplied during forced displacement and amplified discrimination and that affected some members of communities more than others, particularly women, children, persons with disabilities and older persons

Hogan (2007) further noted that Indigenous knowledge or heritage have largely been marginalized, neglected and suppressed due to ignorance and arrogance, politics and the dominant ideology of a particular historical period, despite its importance to sustainable and equitable development. Southern African Indigenous knowledge has thus suffered neglect due to the impact of coloniality and post-coloniality which completely ignored the lived experiences of those in the global South. Morgan (2003) also observed that Indigenous knowledge was suppressed by the Western system which allowed the appropriation and exploitation of this knowledge for the benefit of the colonizers. The Southern African region was devastated by the arrival of the Portuguese, Dutch and British colonisers from 1400 to 1800, in terms of control of the means of production. In addition, the region lost its socio-cultural and epistemic identity. The study by Vantura and Rueter (2018) about health in Canada have found a connection to culture, including languages, which helps create resilience to health problems such as HIV, diabetes and suicide that disproportionately affect people. However, as reported by the UNESCO Atlas of the World's Languages in Danger, 230 languages went extinct between 1950 and 2010. Nowadays, a third of the world's languages have fewer than 1,000 speakers left and every two weeks a language dies with its last speaker while 50 to 90 per cent of them are predicted to disappear by the next century (Strochlic, 2018).

It has therefore been observed that many African countries implemented an exoglossic approach implying use of a non-Indigenous language in dealing with the language question by declaring the European language as their sole official language (Mohlahlo & Ditsele, 2022). Through xenoglossic language policies non-Indigenous languages were denigrated and marginalised. However, as noted by Mohlahlo and Ditsele (2022) South Africa opted for an endoglossic language policy against this tradition by adopting an endoglossic approach by officializing their own African languages. Even though non-Indigenous languages have been declared official the colonial languages still remain the main languages of upward mobility because of asymmetric power relations. Language policies are important in overcoming colonial damage because intangible heritage is encapsulated in language. The UNESCO Decade of Languages (2022-2032) emphasizes the urgent need to protect, support, revitalize and promote languages around the world and recognize their wider and deeper role in peacebuilding, good governance, the protection of the environment, the safeguarding of culture in all its forms and the sustainable development and reconciliation processes within society, based on the observations below which have contributed to the proclamation of the International Decade of Languages. However, the question that arises is how languages as carriers of knowledge can be leveraged to realise epistemic justice in line with the UNESCO Decade of Languages (2022-2032).

The culture of ethnocultural communities in Southern Africa is thus the essence of who they are, who they belong to, where they come from, and how they relate to one another. It is the basis of traditions, customs, protocols, values, spirituality, ceremonies, language, ways of knowing and being, and connections to the land and the life-sustaining resources of the land. Indigenous heritage or knowledge is intrinsic to Indigenous people's well-being and it includes ideas, experiences, worldviews, objects, forms of expressions, practices, spirituality, kinship ties and places valued by people. The Indigenous Heritage Circle (IHC) (2019) describes Indigenous or cultural heritage as ideas, objects, artistic expressions, practices, languages, knowledge and places that are valued because they are culturally necessary and meaningful, connected to shared memory, or linked to collective identity. Otite and Oginwo (2016) further described culture as the complex whole of human's acquisition of knowledge and ways of knowing, morals, beliefs, arts, customs and technology, which are shared and transmitted from generation to generation. However, culture is always in a state of constant flux, for example in many African countries have begun rethinking critical issues, for example, gender relations, child marriages, genital mutilation and use of marihuana as a remedy for health problems. Narratives in rituals, proverbs, songs, revered traditions, myths, and folktales invite participants and their communities to engage in identity discourses, deconstruct and reconstruct some of the negative ideologies that have caused psychological harm, embarrassment, or humiliation, and bring to the central positive stories that discourage shame and re-establish trust with people's cultures (Chilisa, Major, & Khudu-Petersen, 2017).

UNESCO (2009) further described culture as the whole complex of distinctive spiritual, material, intellectual and emotional features that characterise a society or social group, not limited to the arts and including modes of life, the fundamental rights of the human being, value systems, traditions and beliefs. Olive and Nevhudoli (2021) argued that current curriculum processes and systems lack intergenerational learning from elders and Indigenous knowledge holders (IKH), which is fundamentally flawed. Several findings from Moichela's study (2017: 162) suggest the role of elders should be considered when planning a curriculum, as they serve as valid bearers of remarkable renaissances such as the African Family Renaissance (a nexus of Indigenous Knowledge Systems) for posterity's sake. Various contexts can facilitate intergenerational learning, including those in which young people and the elderly engage in activities together in mutual respect (Olive & Nevhudoli, 2021). Cultural heritage is fundamental to all aspects of people's lives and it has spiritual, historical and social value by connecting people to their countries and thereby also to particular social relationships and custodial obligations. It also has wide-ranging scientific and environmental value, especially through informing environmental management and protection and essential to the overall well-being of communities and members (Heritage Chairs of Australia and New Zealand) (HCOANZ) (2020). Therefore, losing one's culture or being disassociated from the culture can contribute to a breakdown of health and well-being, especially if living within one's culture and being mostly involved in rituals connected to religious activities, traditions and values, which are the essence of identity, health and well-being. The dawn of independence from the 1960s to the 1990s heralded a novel era of hope amongst the cultural communities as they optimistically looked forward to reclaiming their lost socio-cultural and economic heritage. However, such hopes seem to have been dampened because of the neo-colonial ties with the former colonial powers thus dashing away any hopes for the reclamation of the intellectual space rooted in epistemologies. If ethnocultural communities in Southern Africa do not look back or gaze back to reclaim their past, they will not only perpetuate instability and poverty but may also be reduced to mental slaves and intellectual zombies in their continent, as stated by Quan-Baffour (2008). There is thus a need for reclamation and reterritorialization of African Indigenous knowledge that has been misappropriated and reterritorialized due to the effect of colonization and Westernization. Ethnocultural communities in Southern Africa should come out of their cultural logjam and take a 'backward gaze' into their cultural practices as well as values and norms. The United Nations Declaration on the Rights of Peoples (UNDRIP) (2007) further established a universal framework of minimum standards for the survival, dignity and well-being of the peoples of the world, and it elaborates on existing human rights standards and fundamental freedoms as they apply to the specific situation of peoples. The purpose of this chapter is thus to investigate how misappropriated Indigenous heritage can be reterritorialized and reclaimed among ethnocultural communities in Southern Africa

LITERATURE REVIEW

The following section x-rayed the literature review with specific reference to sources that explored the concept of deterritorialization and reterritorialization of African Indigenous heritage or knowledge.

Deterritorialization and Reterritorialization of African Indigenous Heritage

Ethnocultural communities in Southern Africa have been alienated from their cultural roots due to the effect of deterritorialization, colonization, globalization and Westernization. This demonization of knowledge systems has impacted negatively and eroded the African values, norms, purpose, growth, sustainability and improvement of cultural communities. Furthermore, Masoga (2005) opines that numerous communities in Africa were directly and indirectly affected by colonisation, which led to the loss of individual and community identity. In 1970, Mr. Seretse Khama argued that a nation without a past is a lost nation, and a people without a past is a people without a soul and further made a call for Africans to discover and write their personal history to show that our past is rich in wisdom and knowledge. Former South African president, Mr. Thabo Mbeki also noted in his speech, in 1998, that: 'We must embrace the culture of the globe while ensuring that we do not discard our own'. However, with Africa subjugated and dominated, the Western culture and European mode of civilisation began to thrive and outgrow African cultural heritage (Arowolo, 2010).

African cultural and traditional practices and ways of doing things were regarded as being primitive and archaic, and as a result, these practices became reterritorialized and were fully westernised. Deleuze and Guattari (1972) described deterritorialization as the process by which a social relation, called a *territory*, has its current organization and context altered, mutated or destroyed. As stated by Canclini (1990) deterritorialization speaks of the loss of the natural relation between culture and the social and geographic territories and describes a deep transformation of the link between our everyday cultural experiences and our configuration as preferably local beings. It is one of the central forces of the modern world as it brings labouring populations into the lower-class sectors and spaces of relatively wealthy societies, while sometimes creating exaggerated and intensified senses of criticism or attachment to politics in the home state (Appadurai,1990). Thus, from the colonial epoch to this modern day, knowledge systems and heritage have been reterritorialized, lost and misappropriated, and as noted by Mdhluli et al. (2021) if they remain unpreserved, they will go extinct. This underscores a need to reterritorialize the misappropriated and reterritorialized heritage among ethnocultural communities in Southern Africa.

Reterritorialization is regarded as the restructuring of a place or territory that has experienced deterritorialization. The idea of reterritorialization was also developed and proposed by Deleuze and Guattari (1972), and it is when people within a place start to produce an aspect of popular culture themselves, doing so in the context of their local culture and making it their own. Deleuze and Guattari (1972) observed that the understanding of the psyche was revolutionized by Sigmund Freud's concepts of libido and polymorphous perversity, and because the psyche was initially deterritorialised, they then conceptualized a new territory, the Oedipus complex, an understanding of tension in the psyche, that is in favour of repression by therefore reterritorializing it.

METHODOLOGY

This study is based on an Afronographic methodology, which is an Afrocentric paradigm for exploring African experiences with their interest at heart as they seek to answer epistemological questions (Rodgers, 2023). Asante (2007) describes Afronography as a method of recording and writing the African experience from an Afrocentric perspective and it provides the epistemological perspective to intellectually unpack African phenomena using African lenses. This chapter adopted an Afronographic approach to analyse the deterritorialization and reterritorialization with a focus on the reclamation of misappropriated heritage among ethnocultural communities in Southern Africa. Reviere (2001) argued that it was critical to note that an Afrocentric approach encourages the researcher to challenge the use of traditional Eurocentric research criteria of objectivity, reliability and validity. As stated by Reviere (2001), the researcher who opts for the Afrocentric research paradigm must scrutinize and place in the centre of the study all foibles and prejudices that would be hidden and hence stealthily influence the research endeavour. As a result of colonial Eurocentric views being incorporated into African research, these studies do not necessarily reflect Africans ways of thinking (Mkabela, 2005). Africans are noted for their great emphasis on collective ideals, notably a shared sense of duty. The "collective ethic" of African Indigenous tribes has often been ignored, misinterpreted, or rejected in research. The community ethic recognises that social harmony is vital for life, and that all activities must be taken within a communal context to maintain the harmony and balance of an interconnected and basically egalitarian society (Mkabela, 2005, p.185).

The Afrocentric method was developed and proposed by Asante (2007) as an appropriate method for conducting African cultural research, hence this method was deemed suitable for this study. The late Senegalese scholar Cheikh Anta Diop was the pillar of the Afrocentric paradigm, whose work simply argues that there are two cradles of civilization, the Southern cradle, represented by Africa, and the Northern cradle, represented by Europe, as cited by Davis (2023). Afrocentrism as a paradigm originates from the Afrocentricity conceptual framework, which argues that African culture and assumptions of human behaviour are pivotal to any scrutiny involving the study of African Americans (Asante, 2013; Oyebade, 1990 & Sherr, 2006). In Afrocentricity theory, Africans are expected to adapt their realities, their history and ultimately their destiny to fit their unique context. The African-centred paradigm emphasizes the use of healing approaches that are rooted in their belief systems (Dzingirayi, 2022). The study thus aimed to reconnoitre the deterritorialization and reterritorialization of Indigenous knowledge or heritage in an African context. Mabvurira and Makhubele (2018) are of the view that Afrocentric methodology shares the same characteristics as qualitative research methods. The Afronographic approach underpinned by the Afrocentric paradigm was chosen for this study because both resonate with the interpretivist because of the focus on comprehending and apprehending social phenomenon. The two methods assume that people use interpretive schemes which must be understood and that the character of the local context must be articulated (Mabvurira & Makhubele, 2018).

However, the Afrocentric methodology should be guided by key rules that are based on Asante (2007)'s Afrocentric principles of *Ma'a*t (the quest for justice, truth, and harmony), and *Nommo* "the productive word. These key rules of Afrocentricity include:

- *Ukweli (truth)-the study is grounded in Africanism with a focus on the deterritorialization and reterritolisation of knowledge. It aims to explore possibilities of reclaiming the cultural intellectual space of a stolen legacy*
- *Utulivu (justice)- the researchers will strive to exaggerate or foment divisions in communities but aim to realise a harmonious equilibrium by setting things right for the good of humanity*
- *Uhaki (harmony)- the study is for the people and provides a basis for further research.*
- *Ujamaa (community)- the researchers are inseparable from the communities and aim to contribute towards the unity of purpose. The research becomes transformative and is not a matter of articulating the local context but also contributing to rendering an inclusive existence (within and between different cultural centres)*

- *Kujitoa (commitment)- the researchers are committed to the upliftment of knowledge and its communities for an inclusive society*

Therefore, any inquiry concerning African experiences must satisfy the above-mentioned rules or canons to be truly legitimate, and employing all these new Afrocentric perspectives will necessarily provide more useful and honest research criteria.

FINDINGS AND DISCUSSION

The findings were presented in three themes: the impact of colonialism and deterritorialization on Indigenous peoples'culture, decolonial and intergenerational strategies to reclaim misappropriated Indigenous heritage and the role of cultural continuity in reclaiming cultural identity and misappropriated Indigenous heritage.

Impact of Colonialism and Deterritorialization on Indigenous Peoples' Culture

The impact of colonialism and deterritorialization are some of the primary ways that continue to influence people's cultures and identities, in the Global South and African continent, while they can also be marginalized through various forces (Das & Semaan, 2022). African countries were robbed of decades of opportunities such as opportunities for self-development, opportunities for self-government and opportunities for self-styled technological developmental pace (Arowolo, 2010). Colonialism is the process through which people's lives are upended by external forces that exert ways of living and perspectives on others (Loomba, 2015). Arowolo (2010) further described colonialism as an imposition of foreign rule over traditional political settings and foreign dominance and subjugation of African people in all spheres of their social, political, cultural, economic and religious civilisations. It refers to the policies and practices where foreign powers migrate to other lands and alter the social structures, and thus identities, of local populations (Das & Semaan, 2022).

The history of colonization is thus one defined by the enslavement, rape, and genocide of local peoples (Antles-Osortia, 2018). The impact of colonialism on knowledge systems includes environmental degradation, economic instability, ethnic rivalries and human rights violations issues that can outlast one group's colonial rule (Blakemore, 2019). It has brought about deep and far-reaching changes in the economic, political, educational and cultural spheres. Colonialists regarded knowledge as primitive, uncivilized, backward, superstitious and savage (Briggs & Sharp, 2004), and this knowledge was viewed as irrelevant to development and an obstacle rather than a force of change. On the other hand, decoloniality focuses on challenging how coloniality has shaped ontologies and epistemologies, while decolonization pays explicit attention to removing colonizing relationships, exploring and rejecting their influence, and shifting towards re-establishing independent nations (Das & Semaan, 2022).The idea idea is to ensure that relations of domination are subverted as people regain power by moving from what Fanon (2008) calls the zone of non-being to the zone of being . Fanon (2008) draws attention to colonial racism, where colonists often viewed native cultures as inferior to that of the colonial rulers, thus actively working to replace native cultures with that of the colonial power. The replacement of cultures has thus led to cultural assimilation denying ethnocultural communities and their later generations, particularly in Southern Africa, opportunities for understanding their own cultures and preventing them from developing an independent sense of identity, through the destruction and theft of cultural sites and artefacts. Smith (1999) further argues that we still live in a world that continues to portray whiteness as an indication of superiority. Cultural assimilation equates whiteness with superiority and pureness, while blackness is equated with inferiority and leads colonized people to see themselves as subhuman, and perpetually experience psychological trauma as it also affects their identities (Fanon, 2008). For example, cosmetic products advertised through digital media continue to perpetuate whiteness through the promotion of lighter skin that enforces and re-enforces beauty ideals shaped by whiteness (Hunter, 2007). Some of the members of ethnocultural communities in Southern Africa are also obsessed with lighter skin colour, showing how the legacy of European or Western colonialism endures.

The language of colonisers, which is English was also introduced during the British colonial era and it is still used as an official language in educational and administrative contexts despite colonial rule ending in 1947. This dominant language is increasingly used by many people as a preferred mode of communication, and in many African countries, English remains an official language, which continues to marginalize regional or Indigenous languages (Joseph, 2011). Indigenous people thus lose their sense of identity and belonging if there is no crucial connection to their linguistic and cultural history. Masaka (2018) further argued that the epistemicide instituted by agents of neo-colonialism has undersized the growth and success of the epistemological paradigm of the African indigenes. The total exclusion of knowledge from the school and university curricula confirms the deliberate and cantankerous initiatives as evidenced by the predominance of Western ways of knowing or finding solutions to problems because the dominant culture is presented as the only authentic producer of knowledge (Masaka, 2018).

Intergenerational Initiatives and Programmes as a Special Vehicle for Knowledge Transfer

As stated by Greengross (2008), intergenerational initiatives are essential to the development and maintenance of genuine communities inclusive of all their representative parts, and reciprocity in support and care is an essential part of all communities, whether local, regional, national or international. Libraries as social spaces are ideal for fostering a culture of intergenerational dialogue among various demographic groups as a way to transfer knowledge and diffused to create novel ways of interpreting reality among the younger generation while enabling them to participate in collective meaning-making, considering that Africa's population is constituted 70% of young people (United Nations, 2023). By pooling knowledge and skills and transmitting living memory and experience between generations, cognitive and cultural solidarity can be bolstered (Stričević & Ksibi, 2012). The aged and the young population demographic in Southern Africa are important in any study of deterritorialization and reterritorialization of heritage that was plundered by colonialism and replaced with the occidental symbols of supremacy during the colonial era. Among the heritage of Africa and other parts of the world are ideas, experiences, worldviews, objects, forms of expression, practices, knowledge, spirituality, kinship bonds, and places of value.

Chisita (2012) highlighted the need for community or public libraries to create physical and virtual spaces for intergenerational dialogue through such programmes as family reading time and storytelling as a way to enhance social cohesion. Storytelling can be an effective way to capture children's attention when sharing Indigenous knowledge about food sovereignty and foodways. As adults, it is our responsibility to ensure that we are sharing and discussing healthful stories with children (Bujold, 2022). The initiatives or programmes mentioned above are useful in sharing and preserving knowledge from the old to the young demographic groups who are trapped in a culturally homogenised society disguised as globalisation. Intergenerational dialogue is critical for sustaining heritage among the indigenes in Africa and the world. Heritage should be viewed as a basic human right indispensable to one's security, the wisdom of history, and distinctiveness and in the context of deterritorialization peoples and other minority populations are at a loss when their heritage is vilified, pillaged, and disfigured or wrecked (Nicholas, 2022).

Knowledge Piracy as a Setback to the Reclamation and Preservation of Indigenous Heritage

Biopiracy refers to means by which companies or researchers (usually from the Global North) steal the genetic resources and traditional knowledge of countries and peoples and local communities (usually from the global South) without their consent, and, typically, patent this information to enable knowledge to be fenced and monetised for profit (Robinson, 2010). It should be noted that in Africa the African Regional Intellectual Property Organisation (ARIPO) has urged member states to protect and stop the biopiracy of traditional or Indigenous knowledge (Swakopmund Protocol on the Protection of Traditional Knowledge and Expressions of Folklore, 2010). Shivani, and Mishra, (2022) argued that the critical concern of the third world and fourth-world countries was the biopiracy of their intellectual resources and hence the need to curb the problem by formulating holistic means to safeguard knowledge. The dire need to protect such knowledge was felt in the year 1978 after which WIPO and UNESCO came together with a joint initiative for the protection of traditional knowledge which led to the adoption of WIPO-UNESCO Model Provisions for National Laws on the Protection of Expressions of Folklore against Illicit Exploitation and Other Prejudicial Actions in 1982 (Dutfield, 2005). There are numerous classic cases of biopiracy in Southern Africa, for example, the Hoodia *gordonii*, *Aspalathus linearis* (rooibos), *Sceletium tortuosum* (kougoed) and *Pelargonium sidoides* each with histories of biopiracy charges (Wynberg, 2023).

Ageh and Lall, (2019) highlighted that the government of the Republic of Zimbabwe amended its constitution as a means to guarantee the recognition and preservation of traditional knowledge as a right to cultural heritage. The state must take measures to preserve, protect and promote knowledge systems, including knowledge of the medicinal and other properties of animal and plant life possessed by local communities and people (2013 Amended Constitution of Zimbabwe Chapter 2, Section 33). The above-mentioned clause confirms Zimbabwe's enforceable constitutional recognition of the traditional knowledge system (Ageh & Lall, 2019). Furthermore, Ageh and Lall (2019) cited the Republic of South Africa's Intellectual Property Laws Amendment Bill of 2008 since its preamble highlights the importance of the traditional knowledge system by explicitly stating that the wealth of traditional knowledge held by the people of South Africa is recognised, preserved, protected and promoted and made accessible to the public. *Hoodia gordoni* case which effectively concluded in 2011 following negative clinical trial results is a classic example of biopiracy. Other examples included the study on Aspalathus *linearis* (rooibos) and *Pelargonium sidoides.*Table 1 outline key features characterizing the biopiracy charges of selected plant species

Table 1. Overview of key features characterizing the biopiracy charges of selected plant species (Wynberg, 2023).

Pant Species	Biopiracy Charge	Origin	Commercial Use	Traditional use& knowledge	Benefit-sharing agreements (BSA) & governance	Governance
Hoodia gordonii	TK of San used in CSIR patent application without consent. NGO Biowatch alerted SASC who brought the challenge.	Southern Africa	Food Supplements to suppress appetite	Used traditionally by San to quench thirst (and hunger).	BSA between the state-based Council for Scientific and Industrial Research (CSIR) and the Society for Arts, Social Sciences and Culture (SASC).	Trust is set up to distribute benefits. Now discontinued.
Aspalathus linearis (rooibos)	100-year-old rooibos tea Industry based on Traditional Knowledge of San and Khoi without sharing benefits. Challenge brought by SASC and NKC. Nestle was accused of biopiracy due to the unauthorized use of rooibos in the R&D program and the lodging of 5 associated patents. Challenge brought by NGOs Natural Justice and the Berne Declaration	Western Cape, South Africa	Food and Beverage, Supplements, wide application for antioxidant and other properties	Used traditionally for various purposes. Many harvesting practices are traditionally based.	BSA between the Rooibos Council, SASC and NKC. BSA between Nestl´e South Africa, SASC and the National Kho and San Council (NKC)	The Andries Steenkamp Trust administers benefits on behalf of the SASC. A "Khoikhoi Peoples Rooibos ABS Trust" administers benefits on behalf of the NKC.
Pelargonium sidoides	Biopiracy charge against Schwabe led by African Centre for Biosafety, Berne Declaration and traditional healers.	South Africa and Lesotho	Medicinal remedy to treat bronchitis	Widely used as a traditional medicine	BSA exists between German pharmaceutical company Schwabe, South African company Parceval Pharmaceuticals and two traditional councils. Funds are transferred to the chiefs and traditional leadership through the King Sandile Development Trust and Imingcangathelo Community Development Trust.	Unclear how funds are distributed to the community

It should thus be noted that turning a blind eye to biopiracy is tantamount to aiding and abetting the continuous dispossession of the already dispossessed Indigenous African communities, hence the need to protect them and save humanity. Wynberg (2023) did an excellent job of educating the present generations concerning the pilfering of African Indigenous knowledge and the current study builds on that energy to generate more positive energy in restoring the intellectual property rights of Indigenous people in Southern Africa. Using biopiracy, Wynberg (2023) pointed out that for over 2000 years, the global North has plundered the global for resources. In addition to building colonial empires and facilitating financial prosperity and scientific advancement in Europe and North America, unauthorized plant collections have served to build colonial empires. Wynberg's (2023) study provides examples of biopirated charges relating to Indigenous plants, for example, *Hoodia gordoni and* Aspalathus *linearis (rooibos)* among others. The author highlighted the need for fair benefit sharing agreements to ensure that Indigenous people benefit from the proceeds of sui generis knowledge.

Decolonial and Intergenerational Strategies to Reclaim Misappropriated Heritage

It is important to emphasise that urgent and more decisive steps need to be taken to reclaim, reterritorialize and reverse this evanescent trend of cultural emptiness, without which Africa may experience seasons of cultural extinction and drought of African values. However, one of the major challenges confronting Sothern African ethnocultural communities is the formation of intergenerational and decolonial strategies to reclaim their knowledge or cultural heritage.

Africanization and Decolonization of the Curriculum in Higher Education

Murirwa and Higgs (2023) and Mohamedbhai (2013) concur that in Africa there exists an affluence of local information encapsulated in cultural and ecological diversity, and indigenes have used this knowledge to solve a variety of developmental and environmental problems for centuries. African principles, values, wisdom and knowledge are deeply embedded and rooted in African scholarship. However, the arrival of missionaries and colonial political powers in Africa from the mid-1800s onwards and their attitudes towards African cultural institutions impacted negatively on the viability of the African system of education, which was condemned as barbaric, heathen and an impediment to the consolidation of Christianity and Western culture on the continent (Nduka, 1980; Ball, 1983; Comaroff & Comaroff, 1986). As noted by Higgs (2012) the liberation of Africa and its peoples from centuries of racially discriminatory colonial rule and domination have far-reaching implications for educational thought and practice. Briggs (2005) observed that the integration of knowledge into the education system that is dominated by the Western worldview occupies a lesser position than the other. Colonial education has replaced native languages with English, native spiritual practices with Christianity, and native family and community practices with Westernized values and beliefs, through the practices of settler colonialism and assimilation (Lomawaima & McCarty, 2006). Higgs (2012) further posits that the marginalization of African values in African education has resulted in the general westernisation of education theory and practice in Africa. The goal of the settler colonial educational system where the settler narrative is at the forefront of policy, curriculum, instruction, assessment and discipline was to silence ways of knowing and being. The settler is described as the perspective that privileges the abilities, values, and knowledge base of the white settler as well as denies the historical racial subjugation and systematic oppression enacted towards people (Arowolo, 2010).

Yosso (2002) describes colonial education as a deficit-laden traditional curriculum that does not recognize scholars or people as holders and creators of knowledge and does not appreciate the collective orientation to knowledge-creation. This colonial education has turned African nations and school graduates into not only aliens to their own culture and society but also unemployable and can hardly contribute to the socio-economic development of their communities and the continent at large (Mart & Toker, 2010). Morgan (2003) thus emphasized that those who were once colonized should initiate decolonization, a process that allows the revaluing and recovery of the lost. There is thus an urgent need to transform and decolonize education that is responsive to students' unique cultural needs or reflects on the type of education that would be more relevant for the continent and its people. Policymakers, planners, politicians and educators need to transform education by incorporating relevant African cultural values and practices such as patriotism, humaneness, respect for fellow humans, cooperation and the use of languages side by side with the dominant foreign language. Educators in higher education should reclaim some important African values and gaze backwards into their education system expounded by Sankofaism philosophy and blend its valuable aspects with the adopted formal Western education. The concept of 'Sankofaism' represents the re-emergence of benefiting African knowledge systems and cultural practices infused with Western ideas of morality (Benagr & Ofosu, 2016).

The United Nations Educational, Scientific, and Cultural Organization (UNESCO) (2016) has also called for educational reforms that reflect, and respect local contexts, traditions, and values based on the local needs and relevance to the learners. Despite the rhetoric of reform, cultural education in Southern Africa remains marginalized in preference of Western ideals and models that continue to dominate. In 2005 Sousa Santos (2005) further argued that: "In the name of modern science, many alternative knowledges and sciences have been destroyed, and the social groups that used these systems to support their autonomous paths of development have been humiliated. In short, in the name of science, epitemicide has been committed, and the imperial powers have resorted to it to disarm any resistance of the conquered peoples and social groups." (p. xviii). Du Toit (2000) thus argues that if colonialism remains untackled the legacies of intellectual colonisation and racialisation are a threat to academic freedom. This underscores the need to decolonise and Africanise curriculum in higher education. Yende (2020) further mentioned that higher education should interrogate decolonisation as an essential transformative and reconstructive tool to promote and strengthen African scholarship from within. Pete, Schneider and O'Reilly (2013) described decolonization as the process of deconstructing colonial ideologies of the superiority and privilege of Western thought and approaches, and it involves dismantling structures that perpetuate the status quo and addressing unbalanced power dynamics. As observed by Higgs (2012) different scholars, writers and advocates have had an opportunity and a right to redefine higher education and its curriculum within the African context in the attainment of independence, since the struggle for decolonisation in higher education in South Africa. However, colonial education has blinded ethno-cultural communities in Southern Africa to rediscovering their scholarship rooted in a different way of knowing, which is more relationally oriented. The increase in the interconnectedness of countries of the world, which facilitates the exchange of culture, education, and ideas, among others, is also a threat to the success of decolonization and decoloniality.

Revitalization of African Indigenous Languages

Language is the foundation and vehicle of a culture, and it holds the stories, songs, dances, protocols, family histories and connections. Language proficiency is declining in African communities as they are compelled to embrace Western culture and civilisation and when a language dies, the link to the cultural and historical past also disappears. Many practices throughout history put pressure on ethnocultural communities to abandon the use of their languages. For example, when the federal government set up boarding schools in the 19th century to assimilate Native American children into American culture, one of the objectives was to get them to turn away from the use of their native languages. This also included the forced cultural assimilation that resulted from the Indian Civilization Act of 1819. This act established Indian boarding schools to teach subjects such as math and science while suppressing the use of languages and cultures. The development of boarding schools has also led to cultural assimilation and their effect was devastating for communities and their languages. Some of the communities have lost their languages due to assimilation efforts. Additionally, the forced relocation of the communities from their places of origin has also caused these communities to become fragmented due to some families remaining behind or being exempt from relocation. These factors also increased the stress on ethnocultural communities to simply survive.

Re-Appropriation and Reclamation of Indigeneity and Cultural Belonging in the Post-Colonial Dispensation

Communities and people across Southern Africa endure the highest rates of landlessness, political alienation, cultural appropriation, impoverishment and marginalisation in the world due to the effect of deterritorialization, colonialization and globalization. The growth of neoliberal capitalism throughout the region has thus increased rather than reduced inequalities, rendering assertions of belonging or indigeneity ever more claims for the control of land, resources and nature. With the guidance of the Afrocentric paradigm and its principles of *utulivu (justice),* the study used ethnic names such as Khoisan that dignify people as equal and dignified human beings and hence avoid derogatory or pejorative terms like Bushmen. Most of the Khoisan have lost or been denied or deprived of access to their lands and natural resources. For example, Khoisan were being pressured to move away from their lands in other parts of northern Namibia by more powerful actors who are determined to privatise these lands, and they contest their voluntary evictions from Etosha National Park that were enacted for commercial reasons.

San and Khoisan in Botswana were also cut off from most government services and removed from Bakgalagadi of the Central Kalahari Game Reserve (CKGR). Their experiences also encompass the exclusion from formal recognition of the unique histories of people, equal rights before the law, citizenship rights, rights to education, representation etc. People have felt they lack a sense of belonging and a sense of rootedness in place. Sense of belonging is of central significance to the people of Southern Africa including the or San who articulate their indigeneity in many ways. longing to belong that has become one of the most valued ideological commodities in the era of late capitalism. In 2003, anthropologist Adam Kuper noted that for people to be able to receive rights and land in today's world, they, have to claim an essentialised, 'authentic' identity and culture associated with spiritualism rather than materialism, which they show that they are in tune with nature. However, the refusal of Namibia, Botswana and South Africa to recognise officially the unique status of the region's original inhabitants, via their policy position that all of the people in their countries are has led some of the descendants of colonists, the white populations of southern Africa to consider themselves and to mobilise discourses of indigeneity within their claims to belonging (Koot, Hitchcock & Gressier, 2019) As also noted by Koot, Hitchcock & Gressier (2019) such claims are staked through notions of birth, citizenship, political dominance or environmental and economic stewardship of the land. Against all these challenges and forces, communities are thus engaged in active struggles for land and natural resources and are affirming their rights by reclaiming their cultural identity and belonging.

The governments of South Africa, Botswana and Namibia are also attempting to assist communities, mostly San people who are marginalised and disadvantaged in reclaiming their belonging and identity, through various state-sponsored programmes, such as the Community-Based Natural Resource Management (CBNRM) initiative. CBNRM encompasses natural resources such as water, forests, fisheries, range land and wildlife on which people depend for their livelihood. The project also outlines the rules or regulations that must be understood and agreed to by communities and their members, and recognized and respected beyond the community, governing how, when or in what quantity the natural resources can be used. CBNRM thus sets out to empower communities in Botswana, Mozambique, Namibia, Zambia, and Zimbabwe by providing rights over land and natural resources, building skills and capacity, establishing community decision-making bodies, promoting community advocacy, and enabling women to take up leadership positions. Mundy

Analysis and Reframing of African Indigenous Knowledge Systems Policy and Protocols

As noted by Simpson (2004) communities can disentangle themselves from the oppressive control of colonizing state government through policy and decolonized strategies. African laws and protocols were promulgated to govern African communities, through the apartheid regime in South Africa. Several departments such as the Department of Science and Technology (DST) (2004) approached the cabinet in pursuit of formulation of a knowledge policy such as the Protection, Promotion, Development and Management of Knowledge Act 6 of 2019, to protect the Indigenous knowledge from unauthorised use and misappropriation, to regulate the equitable distribution of the benefits of the use of knowledge and to provide for the documentation knowledge. Several legislative initiatives were also taken in the education sphere to ensure that the education system was aligned with the constitutional goal of promoting multilingualism. Among others are: the National Education Policy Act 27 of 1996, the Language in Education Policy (LiEP) of 1997 and the Norms and Standards regarding Language Policy (1997), which were formulated in terms of the South African Schools Act of 1996. The review and analysis of these policies and protocols helped in understanding whether knowledge policy enacted by policymakers in government protects and supports knowledge practices.

The government of South Africa has also countered the bio-piracy of knowledge and resources by passing laws which protect knowledge systems (Masango, 2010). However, some knowledge systems laws, acts and policies were also replaced with alien laws that outlawed cultural practices, replaced traditional leadership and governance with elected leaders and municipal-style governance, moved communities onto reserves and took away their independence. All these factors have an impact on cultural continuity. The loss of other laws and systems of governance means some ethnocultural communities may not have that knowledge to draw upon as many of these communities move towards a return to self-government. Even though policies and laws have been promulgated to give recognition to the rights of people, much still needs to be done as there is a lack of practical recognition and implementation. As a result, policymakers should begin to spot the undesirable influence of Western civilization and culture on ethnocultural communities in Southern Africa, so that they may begin to re-evaluate the policies that contribute to Africa's cultural famine or negation of the values of cultural revitalization.

The Role of Cultural Continuity in Reclaiming Cultural Identity and Misappropriated Heritage

Indigenous peoples are those who self-identify as members of one or more communities, have historical continuity with pre-settler societies, territories, languages, cultural practices and political systems, and resolve to maintain and reproduce their ancestral environments and systems as distinctive peoples and communities (UNPFII, 2006). However, MacNeil (2008) noted that the colonial policies which have led to an erosion of language, and cultural traditions that lead to cultural continuity have created a loss of cohesion and identity in communities and their families. MacNeil (2008). Colonial policies related to residential schools the erosion of language and cultural traditions that lead to cultural continuity that have also led to a loss of cohesion and identity in communities, and this has had an impact on the family. Recognizing the devastating effects of colonization on their ancestors, for example, youth in countries such as Canada, in a special committee of the Royal Commission on Aboriginal People, identified a sense of loss of culture and identity and perceived this as an important factor in healing (Thura, 2000). Morris (2007) also identifies the devastating effects of cultural loss on youth and there is consensus among these young people that they are suffering from a loss of culture and a loss of family.

As noted by Morris (2007) there is a disconnect from the cultural values and traditions espoused in communities due to colonization and cultural assimilation. Cultural identity has a major influence on our confidence and self-esteem. Cultural disconnection and lack of cultural continuity create a loss of confidence at the individual level in understanding how to live life and make decisions (MacNeil, 2008). Brown (2003) describes cultural continuity as the ability to preserve the historical values and ways of life of a culture and carry them forward with that culture into the future, and it is closely linked to the concept of cultural identity that has a major influence on our confidence and self-esteem. Cultural continuity entails the preservation of a community or groups cultural values whereby key aspects of culture are passed on from one generation to another. However, colonial societies have lost the power and sense of cultural continuity and not only certain aspects of the culture, and it became practically impossible to recover and revitalize lost or misappropriated heritage. Providing youth with access to the arts which are important in their culture may be regarded as one of the strategies to promote cultural continuity. The cultural continuity of ways of life can be sustained through the adaptation of the ways of communication, for example, folk media that encompasses oral traditions (sound including, songs of happiness and sadness, dance, drama, puns and everything that constitute the art of conversation. Mundy and Compton (1991) highlighted the importance of endogenous means of communication as opposed to exogenous communication that falls short of conveying ideas to audiences among the Indigenous people. In an endogenous communication network, information is exchanged naturally among its members rather than imposed from outside (Cason, Sheremeta, & Zhang, 2017). Lo Liyong (1991) notes that oral traditions always have been revered and utilised as the media for the preservation of history and renewing of cultural traditions, and genealogical traditions of communities. Generally, Western scholarship has tended to look down upon oral traditions on the basis that they lack validity but there has been a paradigm shift as emphasis now shifts towards the integration of oral traditions in the educational curriculum.

RECLAIMING STOLEN LEGACY- INDIGENOUS HERITAGE

There is a growing awareness and raging debate among developing countries to reclaim their stolen legacy back to its original habitat (Ankoma, 2020). The repatriation of stolen cultural artefacts comes against the background of the United Nations (UN) 1995 Convention on Stolen or Illegally Exported Cultural Objects (UNIDROIT Convention) includes the terms ' or other communities' its recital and it recognises that cultural property was taken during the colonial era and therefore could include the cultural heritage of African peoples. The Convention embraces property which may have been removed over 100 years ago. The subsequent paragraphs aim to amplify initiatives undertaken by governments in Zimbabwe and South Africa to reclaim their stolen legacy. The case studies highlighted below capture classic examples in the drive to reclaim stolen Indigenous heritage.

Case Studies From Zimbabwe

Machiridza and Thondhlana (2022) highlighted that several soapstone carvings with spiritual and cultural significance were looted from the ancient city of Great Zimbabwe, and they were displaced in different parts of the world during the early 19th century by the colonial agents. Since the attainment of independence in 1980, most of these soapstone birds gradually found their way back into the country. It was interesting to note that the 7th Soapstone bird had been broken during the plunder with one part remaining in the country and another being displaced to Germany through efforts by the governments of Germany, Belgium and Zimbabwe the soapstone bird was finally reunited and repatriated back home in 2000 (Machiridza & Thondhlana, 2022). The "Mythical" Mukwati walking stick presents a classic example of plunder and the unflinching desire to return stolen legacies to the people of Zimbabwe. Mataga (2014) noted that Mukwati, a prominent spirit Medium of the Mwari cult who organised anti-colonial and imperial resistance against the British colonisers in 1896 -1897 had his magical stick confiscated and displaced to Europe. It was through the efforts of his descendants with the cooperation of the family of the looters that the magical stick was reconnected with its owners through the National Museums and Monuments of Zimbabwe (NMMZ).

Case Studies From South Africa

Snowball, Collins and Nwauche (2022) cited the high-profile repatriation case in South Africa of the body of Sarah Baartman, a Khoisan woman who was taken to France in the 1800s and displayed as an anthropological curiosity, dubbed 'The Hottentot Venus' even though such a title goes against the Afrocentric paradigm of *utulivu(justice)* that promotes dignity for all African people. According to the information available, the remains of the late great Sarah Baartman were eventually returned from France to South Africa in 2002 and buried there in 2014. Even though Sarah Barrtman's case does not relate to artefact repatriation, but to the return of human remains, it provides an excellent *locus standis* to analyze the return of cultural artefacts to Africa.

The Steve Biko Case in South Africa

Snowball, Collins and Nwauche (2022) cited the case of Steve Biko's letter signed by Biko in 1973 that was addressed to the Magistrate of East London, requesting leave to travel to visit his wife came up for an online auction in Britain. Hahndiek and Isaacs (2018) the South African Heritage Resources Agency (SAHRA) contacted the auction house since their view was that the letter, in addition to being state property, constituted part of the 'national estate'. Snowball, Collins and Nwauche (2022) noted that the auction house would not reveal the name of the seller, and SAHRA failed to prevent the final auction of the letter to an unknown buyer. Hahndiek and Isaacs (2018) bemoaned the susceptibility of cultural objects and highlighted the need to strengthen collaboration between South African Heritage Resources Agency (SAHRA), South African Museums, the National Archives, State Institutions, Auction houses as well as Customs and Excise officials. The other cultural artefacts that are yet to be repatriated to South Africa include, a small (18 cm high) wooden carving of a woman carrying a child on her back in the traditional Zulu manner. It included decorative beadwork and resin and was pillaged by Sir Theophilus Shepstone, whose secretary was Rider Haggard, in 1939 and was later donated to the British Museum (Snowball, Collins & Nwauche, 2022).

FURTHER RESEARCH DIRECTIONS

Indigenous heritage and traditional cultures need to be recovered and revitalized so that it can be re-territorialized and re-appropriated in creative and relevant ways for our 21st century spaces. This chapter determined the impact of colonialism and deterritorialization on Indigenous peoples' culture. Indigenous researchers, Indigenous knowledge holders, ethnocultural communities, academics, government and traditional leaders need to work together in dismantling the imperialism and colonial manifestations by engaging in anti-imperialism and anti-colonial strategies for the recovery and reterritorialization of the destroyed and misappropriated Indigenous knowledge. Reterritorialization has been regarded as decolonial strategy to recover and reclaim misappropriate and reterritorialized Indigenous heritage among ethnocultural communities in the global South and South Africa is not an exception. Future research should thus focus more on developing indigenization, anti-colonial and anti-imperialism strategies in the reterritorialization, reclamation and re-appropriation of Indigenous heritage. The Indigenous heritage research should also follow a community-based participatory and action research, which adopts fundamental principles of Sankofaism and Afrocentrism paradigms.

CONCLUSION AND RECOMMENDATIONS

The study has confirmed that the deterritorialization and reterritorialization of heritage in selected countries in Southern Africa is a reality that the present generation must deal with, hence the adaption of the Afrocentric paradigm (Asante, 1987) and its five canons to unpack the mystery concerning stolen legacies. The principles of the Afrocentric paradigm namely: ukweli (truth), utulivu (justice, uhaki (harmony), ujamaa (community), and kujitoa (commitment) have been given adequate and meticulous attention. In reality, Africa lost its legacy as a result of colonial deterritorialization and reterritorialization. It is up to the present generation to reclaim the African cultural heritage and preserve it for posterity with modern technology and social capital embedded in the extended African family. Memory institutions have a critical role in ensuring that the collective Indigenous knowledge and memory of Indigenous people in South Africa are preserved and revitalized. This study recommends that:

- There is a need for more scientific research focusing on the reclamation of pillaged heritage among ethnocultural communities. Such studies should be premised on research methodologies that are deeply rooted in African epistemology and ethos, such as Afrocentricity or Afrikology.
- Memory institutions, namely: galleries, libraries, archives and museums (GLAM) in Africa should rethink and recalibrate their strategies concerning the preservation of Indigenous knowledge including sui *generis works* in synch with the Afrocentric paradigms five cannons concerning truth, justice, harmony, community and commitment as highlighted in previous sections of the study.
- Memory institutions should descend from their ivory towers of elitism and anachronism by rethinking effective ways of reasserting epistemic justice for marginalised communities who are part of the magic formula for the realisation of the United Nations' Sustainable development goals.
- Exogenous channels of communicating with communities using the Indigenous languages and means of communication should not be viewed as an option but rather epistemological priori.
- The reclamation of physical, digital and intellectual space should be viewed as the basis for the recovery of people's cultural heritage.
- More studies should be undertaken concerning people and cultural continuity and creative ways of addressing this in all aspects of socio, economic and political activity.

REFERENCES

Ageh, P. A., & Lall, N. (2019). Biopiracy of plant resources and sustainable traditional knowledge system in Africa. *Global Journal of Comparative Law*, 8(2), 162–181. DOI: 10.1163/2211906X-00802003

Amenyedzi, S. B. (2023). The Afrocentric-Youth Paradigm. *Journal of Youth and Theology*, 1, 1–12.

Ankoma, A. (2020). "Repatriation of African Art." *Arts24*, 9 25 December 2020). Retrieved 10 October 2025 from News24.com/amp/life/arts-and-entertainment/arts/its-time-for-African-countries-to-take-advantage-of-the-moment-and-coordinate-a-decisive-position-on-the-custodianship-of-African-art.

Appadurai, A. (1990). Disjuncture and difference in the global cultural economy. *Theory, Culture & Society*, 7(2-3), 295–310. DOI: 10.1177/026327690007002017

Arowolo, D. (2010). The effects of Western civilisation and culture on Africa. *Afro-Asian Journal of Social Sciences*, 1(1), 1–13.

Asante, M. K. (1991). The Afrocentric idea in education. *The Journal of Negro Education*, 60(2), 170–180. DOI: 10.2307/2295608

Asante, M. K. (2005). Afronography. *Encyclopaedia Of Black Studies*, 76-77.

Asante, M. K. (2007). An Afrocentric manifesto: Toward an African renaissance. *Polity*.

Asante, M. K. (2013). *The African American people: a global history*. Routledge. DOI: 10.4324/9780203145807

Atiles-Osoria, J. (2018). Colonial state crimes and the CARICOM mobilization for reparation and justice. *State Crime*, 7(2), 349–368. DOI: 10.13169/statecrime.7.2.0349

Benagr, S., & Ofosu, T. B. (2016). Ghanaian screendance perspectives: The nuance of 'Sankofaism'as emerging aesthetics and rejection of orthodoxy. *The International Journal of Screendance*, 6, 78–93. DOI: 10.18061/ijsd.v6i0.4913

Berkes, F. (2004). Rethinking community-based conservation. *Conservation Biology*, 18(3), 621–630. DOI: 10.1111/j.1523-1739.2004.00077.x

Brown, J., Knol, D., Prevost-Derbecker, S., & Andrushko, K. (2007). Housing for Aboriginal youth in the inner city of Winnipeg. *First Peoples Child & Family Review*, 3(2), 56–64. DOI: 10.7202/1069464ar

Bujold, R. (2022). *Sharing intergenerational food stories to engage children in indigenous food sovereignty*. Masters diss., Dalhousie University, Halifax, Nova Scotia.

Burchardt, H. J., & Dietz, K. (2014). Neo- extractivism: A new challenge for development theory from Latin America. *Third World Quarterly*, 35(3), 468–486. DOI: 10.1080/01436597.2014.893488

Canclini, G. N. (1990). *Culturas híbridas: estrategias para entrar y salir de la modernidad*. Consejo Nacional para la Cultura y las Artes/Grijalbo.

Cason, T. N., Sheremeta, R. M., & Zhang, J. (2017). Asymmetric and endogenous within-group communication in competitive coordination games. *Experimental Economics*, 20(4), 946–972. DOI: 10.1007/s10683-017-9519-2

Chisita, C. T. (2011). Role of libraries in promoting the dissemination and documentation of Indigenous agricultural information: Case Study of Zimbabwe. In Paper of World Library And Information Congress: 77th IFLA General Conference And Assembly. Retrieved 10 October 2025 from http://conference. ifla.org/past-wlic/2011/78-chisita-en. pdf, diakses.

Chisita, C. T., & Abdullahi, I. (2010, August). The challenges and opportunities of preparing LIS students for orally based communities. In World Library and Information Congress: 76th IFLA General Conference and Assembly (pp. 10-15). Retrieved 10 October 2025 from https://www.ifla.org/past-wlic/2010/152-chisita-en.pdf

Daniel, R. A., Wilhelm, T. A., Case-Scott, H., Goldman, G., & Hinzman, L. (2022). What is "Indigenous Knowledge" and why does it matter? integrating ancestral wisdom and approaches into federal decision making.". The White House. December, 2. Retrieved 10 October 2025 from https://www.whitehouse.gov/ostp/news-updates/2022/12/02/what-is-indigenous-knowledge-and-why-does-it-matter-integrating-ancestral-wisdom-and-approaches-into-federal-decision-making/

Das, D., & Semaan, B. (2022, April). Collaborative identity decolonization as reclaiming narrative agency: Identity work of Bengali communities on Quora. In *Proceedings of the 2022 CHI Conference on Human Factors in Computing Systems* (pp. 1-23). DOI: 10.1145/3491102.3517600

Davis, L. (2023). *A case study of gun violence in Philadelphia: an Afrocentric analysis of cultural factors*. PhD. Temple University.

de Sousa Santos, B. (Ed.). (2005). *Democratizing democracy: beyond the liberal democratic canon* (Vol. 1). Verso.

Deleuze, G., & Guattari, F. (2009). *Anti-Oedipus: capitalism and schizophrenia*. Penguin.

Du Toit, A. (2000). From autonomy to accountability: Academic freedom under threat in South Africa? *Social Dynamics*, 26(1), 76–133. DOI: 10.1080/02533950008458687

Dutfield, G. (2005, May). Harnessing traditional knowledge and genetic resources for local development and trade. In Draft Paper Presented at the International Seminar on Intellectual Property and Development. In *Draft Paper Presented at the International Seminar on Intellectual Property and Development*. 2005. Retrieved on 14 June 2023 from https://www.wipo.int/edocs/mdocs/mdocs/en/isipd_05/isipd_05_www_103975.pdf

Dzingirayi, P. (2022). The position of Afrocentric Paradigm in the contemporary globalised professionalised counselling services in Zimbabwe. *The Zimbabwe Journal of Health Sciences (ZJHS)*, 193.

Fanon, F. (2008). *Black skin, white masks*. Grove Press. Global action plan the international decade of languages (IDIL2022-2032) Retrieved on 14 June 2023 from https://en.unesco.org/sites/default/files/global_action_plan_int_decade__languages_en.pdf

Hahndiek, C., & Regina, I. (2018). Leadership and role SAHRA can play to facilitate connectivity with and between museums. *The Heritage Portal*, Retrieved on 14 June 2023 from https://www.theheritageportal.co.za/article/leadership-and-role-sahra-canplayfacilitate-connectivity-and-between-museums

Heritage Chairs of Australia and New Zealand. (HCOANZ). (2020). Dhawura Ngilan: a vision for Aboriginal and Torres Strait Islander heritage in Australia and the best practice standards in cultural heritage management and legislation, Australian Government Department of Agriculture, Water and the Environment, Canberra.

Higgs, P. (2012). African philosophy and the decolonisation of education in Africa: Some critical reflections. *Educational Philosophy and Theory*, 44(sup2), 37–55. DOI: 10.1111/j.1469-5812.2011.00794.x

Hunter, M. (2007). The persistent problem of colourism: Skin tone, status, and inequality. *Sociology Compass*, 1(1), 237–254. DOI: 10.1111/j.1751-9020.2007.00006.x

Indigenous peoples, voices (United Nations Permanent Forum on Issues). Retrieved on 19 June 2023 from https://www.un.org/esa/socdev/unpfii/documents/unpfiibrochure_en07.pdf

(2012). Intergenerational Solidarity in libraries: La Solidarité Intergénérationnelle. InStričević, I., & Ksibi, A. (Eds.), *Les Bibliothèques*. De Gruyter Saur.

Joseph, M. (2011). India faces a linguistic truth: English spoken here." New York Times (2011). Retrieved on 14 June 2023 from https://www.nytimes.com/2011/02/17/world/asia/17iht-letter17.html

Koot, S., Robert, H., & Catie, G. (2019). Belonging, indigeneity, land and nature in Southern Africa under neoliberal capitalism: An overview. *Journal of Southern African Studies*, 45(2), 341–355. DOI: 10.1080/03057070.2019.1610243

Kugara, S. L., & Sekgothe, M. (2022). Challenges presented by digitisation of VhaVenda oral tradition: An African Indigenous knowledge systems perspective. *Hervormde Teologiese Studies*, 78(3), 1–8. DOI: 10.4102/hts.v78i3.7428

Lo Liyong, T. (1991). *Culture is Rutan*. Longman.

Loomba, A. (2015) Psychoanalysis and colonial subjects. *Colonialism/Postcolonialism*, 139–153.

Machiridza, L. H., & Thomas, P. T. (2022). Lost and found property: Evaluating efforts to resurrect and resuscitate 'institutionalised garbage in post-colonial Zimbabwean museums. *Curator (New York, N.Y.)*, 65(2), 451–470. DOI: 10.1111/cura.12462

MacNeil, M. S. (2008). An epidemiologic study of Aboriginal adolescent risk in Canada: The meaning of suicide. *Journal of Child and Adolescent Psychiatric Nursing*, 21(1), 3–12. DOI: 10.1111/j.1744-6171.2008.00117.x PMID: 18269407

Mart, C. T., & Alpaslan, T. (2010). How did British colonial education in Africa become a reason for decolonization? In *Second International Symposium on sustainable development.*

Masaka, D. (2018). The prospects of ending epistemicide in Africa: Some thoughts. *Journal of Black Studies*, 49(3), 284–301. DOI: 10.1177/0021934717753731

Masoga, M. (2005). South African research in Indigenous knowledge systems and challenges of change. *Indilinga*, 4(1), 15–30.

Mataga, J. (2014). Practices of pastness, post wars of the dead, and the power of heritage: museums, monuments and sites in colonial and post-colonial Zimbabwe, 1890. Retrieved on 14 June 2023 from https://open.uct.ac.za/bitstream/handle/11427/12843/thesis_hum_2014_mataga_j.pdf?sequence=1&isAllowed=y

Mgbeoji, I. (2008). The civilised self and the barbaric other: imperial delusions of order and the challenges of human security. In *International Law and the Third World.* Routledge-Cavendish.

Mgbeoji, I. (2011). *Collective insecurity: the liberian crisis, unilateralism, and global order*. UBC Press.

Mkabela, Q. (2005). Using the Afrocentric method in researching Indigenous African culture. *The Qualitative Report*, 10(1), 178–189.

Mohamedbhai, G. (2013, March). Indigenous knowledge must be harvested for development. University World News. Issue 262. In British Council's Going Global 2013 Conference in Dubai (pp. 4–6).

Mohlahlo, A. L., & Ditsele, T. (2022). Exploring multilingualism at the national department levels in South Africa post the Use of Official Languages Act of 2012. *Literator-Journal of Literary Criticism. Comparative Linguistics and Literary Studies*, 43(1), 1819.

Morris, K. (2007). Re-examining issues behind the loss of family and cultural and the impact on Aboriginal youth suicide rates. *First Peoples Child & Family Review*, 3(1), 133–142. DOI: 10.7202/1069532ar

Mpanza, C. D. (2023). The emergence of semilingualism in South African schools–Are the Stipulations of the language in education policy to blame? *African Journal of Development Studies*, 13(1), 179.

Mundy, P., & Compton, L. (1991). Indigenous communication and Indigenous knowledge. *Development Communication Report*, 74(3), 1–3.

Murwira, S., & Higgs, P. (2023). Afrocentric epistemic systems and higher education curriculum. The case of a university in Zimbabwe. *South African Journal of Higher Education*, 37(2), 283–297. DOI: 10.20853/37-2-5334

Nicholas, G. (2022). Protecting Indigenous heritage objects, places, and values: Challenges, responses, and responsibilities. *International Journal of Heritage Studies*, 28(3), 400–422. DOI: 10.1080/13527258.2021.2009539

Oyebade, B. (1990). African studies and the Afrocentric paradigm: A critique. *Journal of Black Studies*, 21(2), 233–238. DOI: 10.1177/002193479002100208

Pete, S., Schneider, B., & O'Reilly, K. (2013). Decolonizing our practice: Indigenizing our teaching. *First Nations Perspectives*, 5(1), 99–115.

Quan-Baffour, K. P. (2008). The wisdom of our forefathers: Sankofaism and its education lessons for today. *Journal of Educational Studies*, 7(2).

Republic of South Africa Intellectual Property Laws Amendment Bill of. 2008. (2017), Retrieved on 14 June 2023 from https://www.publishsa.co.za/file/1446644266ord-ip-laws-amendment-act-2013.pdf

Reviere, R. (2001). Toward an Afrocentric research methodology. *Journal of Black Studies*, 31(6), 709–728. DOI: 10.1177/002193470103100601

Robinson, D. (2010). *Confronting biopiracy: challenges, cases and international debates*. Routledge.

Rogers, N. N. (2023). An Afrocentric examination of Afrocentric schools: status, agency, and liberation. Temple University. DOI: 10.4324/9781849774710

Sherr, M. E. (2006). The Afrocentric paradigm: A pragmatic discourse about social work practice with African Americans. *Journal of Human Behavior in the Social Environment*, 13(3), 1–17. DOI: 10.1300/J137v13n03_01

Shivani, S. & Mishra, S. (2022). Traditional knowledge: much more than what meets the eyes. *International Journal of Health Sciences*, 6(S5)), 1570–1581.

Smith, L. T. (2021). *Decolonizing methodologies: research and Indigenous peoples*. Bloomsbury Publishing. DOI: 10.5040/9781350225282

Snowball, J., Collins, A., & Nwauche, E. (2022). Ethics, values and legality in the restoration of cultural artefacts: The case of South Africa. *International Journal of Cultural Policy*, 28(5), 513–530. DOI: 10.1080/10286632.2021.1995377

Snowball, J., Collins, A., & Nwauche, E. (2022). Examining restitution and repatriation options for cultural artefacts: An empirical enquiry in South Africa. *International Journal of Cultural Policy*, 28(5), 531–545. DOI: 10.1080/10286632.2021.1995378

Strochlic, N. (2018). The race to save the world's disappearing languages. *National Geographic*, •••, 16.

Swakopmund Protocol on the Protection of Traditional Knowledge and Expressions of Folklore Within the Framework of the African Regional Intellectual Property Organization (ARIPO). (2010). Adopted by the Diplomatic Conference of ARIPO at Swakopmund (Namibia) on August 9, 2010. Retrieved on 14 June 2023 from https://wipolex-res.wipo.int/edocs/lexdocs/treaties/en/ap010/trt_ap010.pdf

The Amended Constitution of the Republic of Zimbabwe. (2013). Retrieved on 14 June 2023 from https://www.constituteproject.org/constitution/Zimbabwe_2013.pdf

United Nations Permanent Forum on Issues Report on the fifth session (15-26 May 2006) Economic and Social Council Official Records Supplement No. 23. Retrieved on 14 June 2023 from https://documents-dds-ny.un.org/doc/UNDOC/GEN/N06/384/35/PDF/N0638435.pdf?OpenElement

Ventura, J., & Rueter, A. (2018). *Healthcare workers use language to build trust and break down barriers*. CBC.

Wynberg, R. (2023). Biopiracy: Crying wolf or a lever for equity and conservation? *Research Policy*, 52(2), 104674. DOI: 10.1016/j.respol.2022.104674

Yende, N. E. (2020). The quest for decolonisation in higher education in the context of globalisation and internationalisation. *African Journal of Development Studies*, 10(3), 277. DOI: 10.31920/2634-3649/2020/10n3a14

Young People's Potential, the Key to Africa's Sustainable Development. Retrieved on 14 June 2023 from https://www.un.org/ohrlls/news/young-people%E2%80%99s-potential-key-africa%E2%80%99s-sustainable-development

KEY TERMS AND DEFINITIONS

Colonialism: A control by one power over a dependent area or people, and it occurs when one nation subjugates another, conquering its population and exploiting it, often while forcing its own language and cultural values upon its people.

Cultural Acculturation: It is a process in which members of one cultural group adopt the beliefs and behaviour of another group.

Cultural Continuity: It is the ability to preserve the historical traditions of a culture and carry them forward with that culture into the future, and it is closely linked to the concept of cultural identity that has a major influence on our confidence and self-esteem

Decolonization: It is a continuous process of anti-colonial struggle that honours Indigenous approaches to knowing the world, recognizing Indigenous land, Indigenous peoples and Indigenous sovereignty.

Ethnocultural Community: It is a group of ethnic people that has one or more shared characteristics such as language, religion, geographical region of origin or national identity.

Indigenous Knowledge: It is the traditional, cultural and community knowledge produced and owned by local people in their specific communities and passed on from one generation to the next generation, through practice and oral channels.

Indigenous or Cultural Heritage: It is described as ideas, objects, artistic expressions, practices, languages, knowledge/ways of knowing, and places that are valued because they are culturally necessary and meaningful, connected to shared memory, or linked to collective identity.

Sankofaism: The re-emergence of advancing African knowledge systems and cultural practices suffused with Western ideas of morality

Chapter 10
Integrating CARE and FAIR Principles in the Digital Preservation of Indigenous Knowledge to Achieve the Sustainable Development Goals

Josiline Chigwada
https://orcid.org/0000-0003-0964-3582
University of South Africa, South Africa

Patrick Ngulube
https://orcid.org/0000-0002-7676-3931
University of South Africa, South Africa

ABSTRACT

Preserving and disseminating indigenous knowledge in the digital age presents unique ethical challenges. There is a need to use frameworks such as the CARE and FAIR principles in the digital preservation of Indigenous knowledge. A literature review was conducted to unpack how the principles can be integrated into the digital preservation of Indigenous knowledge. The databases searched were Scopus, Web of Science, and Google Scholar. The findings show that integrating both principles require community-led governance models, culturally appropriate metadata systems, and collaborative partnerships between indigenous communities, researchers, and technologists. These approaches ensure that indigenous groups retain authority over sharing knowledge while benefiting from digital advancements. This, emphasising that there is a need for a paradigm shift that respects Indigenous epistemologies and the development of technology that prioritises Indigenous communities by all stakeholders.

DOI: 10.4018/979-8-3693-8447-3.ch010

INTRODUCTION

In the digital age, the preservation of Indigenous knowledge (IK) is both a technical and ethical endeavour (Aiseng, 2023; Makananise, 2024). Indigenous knowledge encompasses Indigenous communities' unique cultural practices, languages, ecological insights, and traditions (Tippeconnic & Faircloth, 2010; Wagner & de Clippele, 2023). IK plays a crucial role in achieving the United Nations Sustainable Development Goals (SDGs) (Chigwada & Ngulube, 2023; Ngulube, 2023) as it offers locally rooted, time-tested practices that can complement modern scientific and technological approaches, as shown in Table 3. IK is a cornerstone of cultural heritage and a critical resource for achieving sustainable development goals, particularly education, environmental conservation, and community well-being. With rapid digital advancements, Indigenous knowledge systems (IKS) face increasing risks of misrepresentation, exploitation, and erasure if not carefully managed (Sioui, 2020). To ensure that these systems are preserved ethically and effectively, two sets of guiding principles have emerged, that is the CARE (Collective Benefit, Authority to Control, Responsibility, and Ethics), and FAIR (Findable, Accessible, Interoperable, Reusable) principles. CARE reflects the values and rights of Indigenous communities (Global Indigenous Data Governance, 2024), while FAIR aims to improve the usability of data (Wilkinson et al., 2016) but may inadvertently challenge Indigenous sovereignty over their knowledge. This chapter shows how CARE and FAIR principles can be integrated into the digital preservation of Indigenous knowledge systems. This is caused by the tensions between these principles, and there is a need to ensure that the rights and benefits of Indigenous groups are respected while promoting accessibility responsibly and ethically.

Indigenous communities around the world are increasingly concerned about the digitization of their cultural heritage and knowledge (Francis & Liew, 2010; Oyelude, 2023). The adoption of open access models and the FAIR principles, while making data and knowledge more accessible, can often overlook the issues of Indigenous knowledge preservation (Oguamanam, 2020). This could lead to ethical challenges, including the unauthorized use of culturally sensitive information, and may undermine the collective rights of Indigenous peoples to control their knowledge. Conversely, the CARE principles advocate for Indigenous communities' authority over their knowledge but may create friction with FAIR's ideals of openness and reusability (Carroll et al., 2020). Therefore, the problem lies in finding a balance between openness and control, accessibility and sovereignty, ensuring that Indigenous communities can benefit from technological advancements without compromising their rights and cultural integrity. This study discusses how CARE and FAIR principles can be harmonised to achieve ethical, inclusive, and impactful IK preservation. The case studies that would be analysed help in providing actionable recommendations that would guide stakeholders in developing sustainable digital preservation practices that align with the SDGs while safeguarding the rights and interests of Indigenous communities. It is such a background that the chapter sought to achieve the following objectives:

1. Analyse the role of IK in achieving SDGs.
2. Examine the role of digital technologies in IK preservation.
3. Identify the constraints of preserving Indigenous knowledge in the digital era.
4. Analyse case studies where FAIR and CARE principles have been successfully integrated with IK preservation.

HISTORY OF FAIR PRINCIPLES

The FAIR data principles were established to enhance the management and sharing of scientific data, fostering better data reuse across research disciplines (Wilkinson et al., 2016). These principles emerged in response to the growing need for robust and transparent data stewardship to promote global open science and research collaboration. Caballo-Garcia and Boté-Vericad (2022) gave a summary of the activities that led to the development of the FAIR data principles as follows:

a) In 2014 in Leiden, Holland, a workshop called "Jointly Designing a Data Fairpot" brought together academic and private stakeholders from research infrastructure and policy, publishing, semantic web, and life science research. They aimed to support existing communities trying to find solutions to problems in scientific data retrieval and use (Data Fairport, 2014).
b) The result of the above workshop was a set of agreed-upon guiding principles and practices to enhance discovery, access, integration, and recycling of data. At the end of the meeting, there was a draft of fundamental principles (Wilkinson et al., 2016).
c) The draft was improved by a working group composed of the FORCE 11 community, which was a cooperation and discussion platform for new ways to transform scholarly communication through technology (Martone, 2015).
d) In 2016, the "FAIR Guiding Principles for Scientific Data Management and Stewardship" was published by the journal Scientific Data as shown in Table 1 (Wilkinson et al., 2016).

Table 1. FAIR data principles (Wilkinson et al., 2016)

To be Findable	**F1.** (Meta)data are assigned a globally unique and persistent identifier **F2.** Data are described with rich metadata (defined by R1 below) **F3.** Metadata clearly and explicitly includes the identifier of the data they describe **F4.** (Meta)data are registered or indexed in a searchable resource
To be Accessible	**A1**. (Meta)data are retrievable by their identifier using a standardised communications protocol **A1.1** The protocol is open, free, and universally implementable **A1.2** The protocol allows for an authentication and authorisation procedure, where necessary **A2**. Metadata is accessible, even when the data are no longer available
To be Interoperable	**I1**. (Meta)data use a formal, accessible, shared, and broadly applicable language for knowledge representation. **I2**. (Meta)data use vocabularies that follow FAIR principles **I3**. (Meta)data include qualified references to other (meta)data
To be Reusable	**R1**. (Meta)data are richly described with a plurality of accurate and relevant attributes **R1.1**. (Meta)data are released with a clear and accessible data usage license **R1.2**. (Meta)data are associated with detailed provenance **R1.3**. (Meta)data meet domain-relevant community standards

Findable means data should be easily discoverable by humans and machines, and this requires clear metadata, a unique and persistent identifier, e.g., a Digital Object Identifier (DOI), and data repositories with searchable capabilities (Go FAIR, 2024). Once data is found, it must be accessible, not necessarily free or unrestricted, but there should be clear conditions under which the data can be accessed, whether openly or under certain restrictions, with standardised protocols for access (Wilkinson et al., 2016). Data must be in a format that can be integrated with other datasets and tools using standard formats and ontologies so that data can be easily interested, compared, and combined with other datasets from different sources. FAIR data principles aim to make data reusable for future research, and the data must be well-described, with clear licensing terms and detailed provenance, to allow others to assess the quality and reliability of the data for their use (Wilkinson et al., 2016). By making data more findable, accessible, interoperable, and reusable, the FAIR principles contribute to a more open, transparent, and efficient research environment.

HISTORY OF CARE PRINCIPLES

The CARE principles were introduced by the Global Indigenous Data Alliance (GIDA) in 2019 as a complementary framework to the FAIR principles, which focus on making data findable, accessible, interoperable, and reusable (Carroll et al., 2020; Carroll et al., 2021). They are also known as CARE principles of Indigenous data governance, which were developed as a framework to ensure that Indigenous peoples have control over the governance, collection, ownership, and use of their data (Robinson et al., 2021). Members of the International Indigenous Data Sovereignty Interest Group of the Research Data Alliance "acknowledged the tension between protecting Indigenous rights and interests in data and supporting open data" (Carroll et al., 2020, p4). While the FAIR principles were designed to ensure the usability and accessibility of data, they do not explicitly address the needs, interests, and rights of Indigenous peoples, who have historically been marginalised in global data practices. The CARE principles emerged in response to growing concerns that global data initiatives, including the open data movement, often overlook the unique rights and cultural values of Indigenous communities (Carroll et al., 2020). Indigenous data governance recognises the need for a distinct framework that prioritises Indigenous values and knowledge systems, ensuring that data management practices respect Indigenous sovereignty. Carroll et al. (2020) summarised the key documents that contributed to the development of the CARE principles as follows:

a) Principles of Māori Data Sovereignty (Te Mana Raraunga, 2018).
b) Indigenous Data Sovereignty Communique (Maiam nayri Wingara, 2018).
c) United States Indigenous Data Sovereignty Network Indigenous Data Governance Principles (United States Indigenous Data Sovereignty Network, 2020).
d) OCAP®: Ownership. Control, Access, and Possession (First Nation's Information Governance Centre, 2018).
e) Principles of Ethical Métis Research (Metis Centre, 2011).
f) National Inuit Strategy on Research (Inuit Tapiriit Kanatami, 2018).
g) Open Data Charter Principles (Open Data Charter, 2015).
h) FAIR Guiding Principles for scientific data management and stewardship (Wilkinson et al., 2016)
i) STREAM properties for industrial and commoditized data (Demchento et al., 2018).

Factors Leading to CARE Principles

While initiatives like the FAIR principles emphasise openness and accessibility, they often fail to recognise the unique cultural sensitivities surrounding Indigenous data (Carroll et al., 2021). The CARE principles were created to address this gap, ensuring that data practices are culturally appropriate and equitable for Indigenous communities (Carroll et al., 2020). Indigenous communities faced historical injustices for centuries, where lands, resources, and knowledge were exploited (Carroll et al., 2019). Colonial practices often led to the extraction of Indigenous data without consent, leading to the misuse or misrepresentation of Indigenous knowledge and cultural practices (Raine et al., 2019; Williamson et al., 2023). This legacy of data extraction contributed to mistrust between Indigenous peoples and external researchers or institutions. As a result, the concept of data sovereignty emerged as a means for Indigenous people to assert their rights over their data. Data sovereignty emphasises that Indigenous communities should have the authority to make decisions about how their data is collected, accessed, and used, aligning with their self-determination and governance structures (Carroll et al., 2020; Garba et al., 2023; Hudson et al., 2023; Oguamanam, 2020)

Table 2. The CARE principles of Indigenous data governance (Carroll et al., 2020)

Collective Benefit	For inclusive development and innovation
	For improved governance and citizen engagement
	For equitable outcomes
Authority to control	Recognising rights and interests
	Data for governance
	Governance of data
Responsibility	For positive relationships
	For expanding capability and capacity
	For Indigenous values and world views
Ethics	For minimising harm and maximising benefit
	For justice
	For future use

Collective benefit states that the use of Indigenous data should provide benefits to Indigenous peoples (Carroll et al., 2020; Griffiths et al., 2021). This principle emphasises that data should contribute to the social, economic, and environmental well-being of Indigenous communities, recognising the collective nature of Indigenous knowledge systems (Carroll et al., 2021; Robinson et al., 2021). Data initiatives should enhance the self-determination of Indigenous peoples, supporting their development and governance goals. Authority to control points out that Indigenous people have the inherent right to govern and control their data (Carroll et al., 2020; Griffiths et al., 2021). This principle asserts that Indigenous communities should have full decision-making authority over how their data is collected, accessed, and used. It respects the legal rights of Indigenous peoples to manage their data following their governance structures, laws, and priorities. Responsibility stipulates that those handling Indigenous data must act responsibly, ensuring that the data is used in ways that respect Indigenous rights, interests, and values (Carroll et al., 2020; Hudson et al., 2023). This principle emphasises the ethical obligations of researchers, governments, and organisations to engage with Indigenous communities in culturally appropriate ways, building respectful and trustworthy relationships. Finally, Indigenous data governance should align with ethical practices that respect the values, protocols, and cultural heritage of Indigenous communities (Carroll et al., 2020; Oyelude, 2023). This principle highlights the importance of obtaining free, prior, and informed consent when working with Indigenous data, ensuring that data practices do not perpetuate harm or exploitation, as shown in Table 2.

RESEARCH METHODOLOGY

The research adopted a comprehensive literature review to unpack how FAIR and CARE principles can be integrated into Indigenous knowledge preservation to achieve sustainable development goals. Case studies of institutions that are preserving Indigenous knowledge in the digital form were identified during the literature search. The databases that were searched were Scopus, and Web of Science, which provide information sources that went through a rigorous peer review process before publishing, and Google Scholar to give access to information sources that are not available on the other two databases.

Figure 1. PRISMA flow chart

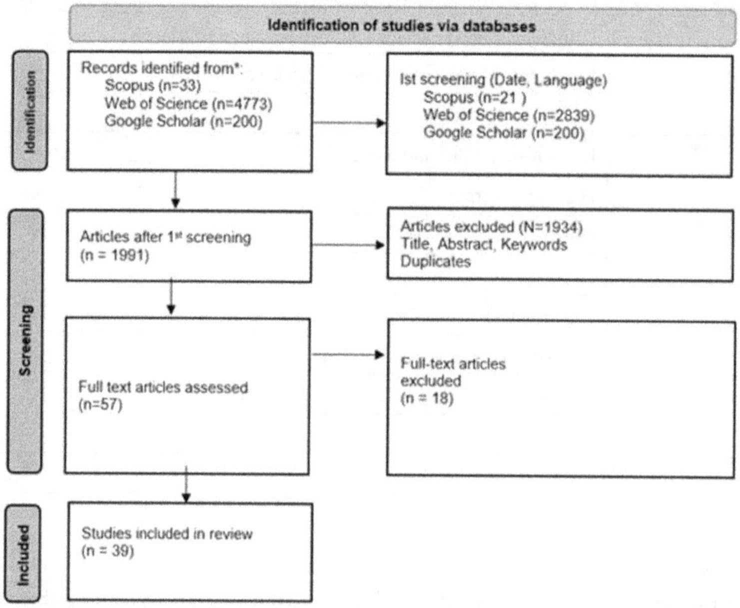

The search terms that were used include "Indigenous knowledge AND digital technologies", "IK digital preservation", "FAIR data principles", "CARE data governance principles", and "FAIR and CARE principles". The inclusion criteria included information sources in the English language to avoid the language barriers published between 2014 and 2024. These dates were chosen since the FAIR data principles were introduced in 2014, while the CARE data governance principles were introduced in 2018. There was no limit on the geographical coverage. Duplicates were removed using the Zotero reference management software. The total number of sources that were retrieved was 5006, and 1991 were left after the initial screening. Of the sources that were considered for full-text screening were 57, and 39 were included in the review, as shown in Figure 1. A data extraction template was used whereby two independent reviewers performed data extraction to deal with the risk of bias from the 39 sources included in the review. Web content analysis was applied to the websites of institutions that archive Indigenous knowledge to assess if they have integrated the two principles. Content analysis was utilised to analyse the data presented according to the chapter objectives.

FINDINGS

The findings showed that Indigenous knowledge represents times of cultural evolution, ecological stewardship, and social systems developed by Indigenous communities (Jessen et al., 2020). As the United Nations notes, Indigenous knowledge is vital for global biodiversity, climate resilience, and sustainability (UNDP, 2024). However, its preservation is threatened by globalization, colonization, and modernization (Mdhluli et al., 2021). It was also noted that IK is important in achieving SDGs, as shown in Table 3.

Table 3. IK and sustainable development goals

SDG	Use of IK	Sources
2 Zero hunger	Indigenous knowledge supports sustainable agriculture through traditional farming methods, crop diversification, and the use of local plant varieties, which can lead to food security. Communities often know how to adapt crops to local climates, pests, and environmental changes, enhancing resilience in food production.	Antonelli, 2023; Buthelezi & Hughes., 2014; Imoro et al., 2021; Kom et al., 2022; Materechera, 2021; Mohan et al., 2021; Mzimela & Moyo, 2024; Vijayan et al., 2022; Vilakazi et al., 2022; Zuza et al., 2024
3 Good health and wellbeing	IK is often used in traditional medicine, offering solutions based on natural remedies and holistic approaches to health. Indigenous communities rely on plants, herbs, and knowledge of the environment to treat diseases, contributing to accessible healthcare in rural areas.	Bodeker & Kariippanon, 2020; Dahlberg & Trygger, 2009; Hossain & Ballardini, 2021; Kennedy et al., 2022; Mahomoodally, 2013; Masango & Nyasse, 2015; Ndhlovu et al., 2023; Redverse et al., 2023; Sibandze & Dludlu, 2022
4 Education	Indigenous communities have their education systems that they use to teach their learners in line with what they believe in. Schools are now incorporating Indigenous ways of teaching and learning into the curriculum.	Ann Rahman & Cochrane, 2023; da Silva et al., 2023; Kennedy et al., 2022; Mawere et al., 2022; Seehawer, 2018; Tom et al., 2019; Yip & Chakma, 2023; Ziphora, 2017
6 Clean water and sanitation	Many Indigenous communities have long understood water management and conservation techniques, such as rainwater harvesting, water purification using plants, and sustainable irrigation practices. Their water conservation methods help manage scarce water resources, especially in arid regions.	Basu & Dasgupta, 2023; Behailu et al., 2016; Masika et al., 2022; Zvobgo et al., 2022
11 Sustainable cities and communities	Indigenous knowledge contributes to building resilient communities through sustainable land use, disaster risk reduction, and housing practices. Their expertise in sustainable construction using local materials and techniques can offer low-impact housing solutions.	Bansal et al., 2024; Imoro et al., 2021; Kohsaka & Rogel, 2019; Mazzocchi, 2020; Tom et al., 2019
12 Responsible consumption and production	Indigenous communities often live by principles of sustainability, practising responsible consumption and ensuring that resources are used in a way that does not harm the environment or deplete future supplies. Their practices promote recycling, reducing waste, and maintaining ecological balance.	Bansal et al., 2024; Mazzocchi, 2020; Tom et al., 2019
13 Climate action	Indigenous peoples have a deep understanding of environmental changes and ecological cycles. Their knowledge of ecosystems, biodiversity, and natural resource management is vital for adapting to and mitigating the impacts of climate change. This knowledge can inform climate resilience strategies.	Imoro et al., 2021; Kom et al., 2022; Redvers et al., 2023; Vilakazi et al., 2022; Zuza et al., 2024; Zvobgo et al., 2022

continued on following page

Table 3. Continued

SDG	Use of IK	Sources
15 Life on land	Indigenous knowledge helps conserve biodiversity by protecting ecosystems and sustainably managing natural resources. Indigenous peoples are custodians of many of the world's ecosystems, such as forests, mountains, and wetlands, which are vital for maintaining biodiversity and mitigating deforestation.	Greenhood & Lindsay, 2019; Jessen et al., 2022; Mazzocchi, 2020; Nitah, 2021; Tom et al., 2019
16 Peace, justice, and strong institutions	Indigenous knowledge systems are built around principles of governance that emphasise community participation, conflict resolution, and social equity. They foster strong, inclusive institutions that encourage peaceful coexistence.	Ayeni & Aborisade, 2022; Boadu, 2022; Maunganidze & Halsall. 2016; Tom et al., 2019
17 Partnerships for the goals	Indigenous knowledge can facilitate partnerships between local communities, governments, and international bodies to address global challenges. Involving Indigenous peoples in decision-making processes enhances collaboration for sustainable development.	Bansal et al., 2024; Haywood et al., 2018; Kohsaka & Rogel, 2019; Tom et al., 2019; Wheeler & Root-Bernstein, 2020

ROLE OF DIGITAL TECHNOLOGIES IN IK PRESERVATION

Indigenous knowledge (IK) refers to the wisdom, traditions, and cultural practices passed down through generations within Indigenous communities (Ajani et al., 2024). It encompasses various domains, such as agriculture, medicine, environmental management, and social practices, that have been adapted to local ecosystems. However, as globalization and technological advancements spread, many Indigenous cultures and their knowledge systems are under threat (Fernández-Llamazares et al., 2021; Mdhluli et al., 2021; Obiero et al., 2023). Digital technologies offer promising tools to document, preserve, and disseminate IK, ensuring that it remains accessible to future generations while also contributing to global knowledge systems, as shown in Table 4.

Table 4. Role of digital technologies in IK preservation

Role of digital technology	Sources
Digital archiving and documentation	Ajani et al., 2024; Balogun, 2023; Balogun & Kalusopa, 2022; Chigwada & Ngulube, 2024; Greyling & Zulu, 2009; Jethro, 2024; Kugara & Mokgoatšana, 2022; Mdhluli et al., 2021; Mukurtu, 2024; Oyelude, 2023.
Enhancing access and dissemination	Ajani et al., 2024; Ariyani et al., 2022; Babagana & Kemboi, 2024; Chigwada & Ngulube, 2024; Karasimos, 2022; Olaare, 2024; Owiny et al., 2014; Richards et al., 2024; Shiri et al., 2022; Warner, 2021

Digital Archiving and Documentation

One of the most apparent roles of digital technologies in Indigenous knowledge preservation is in archiving and documentation (Mdhluli et al., 2021; Oyelude, 2023). Digital archives, databases, and repositories serve as central locations where Indigenous practices, stories, songs, and skills can be captured in digital formats, ranging from audio and video recordings to digital manuscripts (Ajani et al.,

2024; Balogun & Kalusopa, 2022; Chigwada & Ngulube, 2024; Jethro, 2024). These archives not only ensure that the knowledge remains accessible to future generations but also help to prevent distortion and misinterpretation, which often occur with oral traditions. Institutions such as libraries, museums, and universities are developing digitization projects to collect Indigenous artefacts, historical records, and oral narratives (Balogun, 2023; Greyling & Zulu, 2009; Kugara & Mokgoatšana, 2022). The digitization of Indigenous knowledge is more than just a technical process since it requires an understanding of how to present the knowledge without stripping it of its cultural significance. For example, projects like Mukurtu, a digital platform tailored for Indigenous communities, offer culturally sensitive tools for archiving Indigenous knowledge in ways that give control back to the communities themselves, including setting parameters around who can access specific parts of the archive (Mukurtu, 2024).

Enhancing Access and Dissemination

Beyond documentation, digital technologies play a significant role in making Indigenous knowledge accessible to a broader audience, including younger generations who might not engage with traditional oral transmission methods (Ajani et al., 2024). Social media platforms, mobile applications, websites, and digital learning platforms have become vital in disseminating Indigenous knowledge (Chigwada & Ngulube, 2024; Owiny et al., 2014). These platforms allow communities to share their heritage with the world and their diaspora, maintaining their relevance even in far-flung locations. One example of this is the use of mobile apps for language preservation (Ariyani et al., 2022; Olaare, 2024; Richards et al., 2024). Indigenous languages are disappearing at an alarming rate, and technology has offered innovative solutions for their preservation. Mobile applications and platforms such as Duolingo or Google's Endangered Languages Project provide Indigenous communities with tools to learn and share their languages (Karasimos, 2022; Warner, 2021). By incorporating digital storytelling, podcasts, and blogs, digital tools empower younger generations to participate in knowledge transmission and create interactive, multimedia-based repositories (Shiri et al., 2022). Moreover, digitization increases the visibility of Indigenous knowledge on global platforms, allowing for wider recognition and respect for these cultures. It also opens the door for collaboration between Indigenous communities, scholars, and activists working toward preserving Indigenous heritage (Babagana & Kemboi, 2024). However, these technologies also present ethical challenges, particularly regarding intellectual property, cultural sensitivity, and the risk of exploitation.

CHALLENGES OF IK PRESERVATION IN THE DIGITAL ERA

The preservation of Indigenous knowledge (IK) in the digital era is troubled with several challenges despite the promise of technology to safeguard and disseminate cultural heritage (Evens & Hauttekeete, 2011). While digital technologies offer tremendous opportunities for preserving Indigenous knowledge, they also come with concerns around intellectual property (IP) and cultural integrity. Indigenous knowledge is often shared communally, with traditions and practices deeply embedded within the cultural fabric of communities (Bruchac, 2014; Mdhluli et al., 2021). As such, digitizing and sharing this knowledge raises important questions about who owns it and who has the right to control its use. These challenges are multifaceted and range from technical issues, such as access and sustainability, to cultural and ethical concerns surrounding ownership and control.

The Digital Divide and Technological Access

One of the most significant challenges is the digital divide, which refers to the gap between those who have access to digital technologies and those who do not (Lythreatis et al., 2022). Indigenous communities, particularly those in remote or rural areas, often lack access to the necessary infrastructure, including high-speed internet, reliable electricity, and modern digital devices. Without basic access to technology, Indigenous groups are at a disadvantage when it comes to participating in the digitization and preservation of their knowledge (Chisa & Hoskins, 2014). This lack of access can prevent communities from fully engaging with and benefiting from digital tools meant to preserve their cultural heritage. Furthermore, disparities in digital literacy compound the problem, as many communities may not have the skills required to use digital tools effectively.

Cultural Integrity and Representation

The digitization of Indigenous knowledge raises concerns about how such knowledge is represented in digital formats. Indigenous knowledge is often deeply intertwined with cultural, spiritual, and environmental contexts, and stripping it from those contexts can lead to misrepresentation or distortion (Magni, 2017). For example, when oral traditions, sacred rituals, or medicinal knowledge are recorded digitally, they are often translated into formats such as text, video, or audio that may not fully capture their cultural significance. This can lead to the commodification of knowledge that is sacred or sensitive (Wagner & de Clippele, 2023). Furthermore, some aspects of Indigenous knowledge are meant to be shared only within specific cultural or spiritual contexts, and digitizing them may expose them to unintended audiences, which can violate cultural protocols.

Intellectual Property and Ownership

One of the most quarrelsome issues surrounding the digital preservation of Indigenous knowledge is the question of intellectual property (IP) rights. Indigenous knowledge is typically shared communally, and the Western framework of individual IP rights does not always align with these communal traditions (Cross, 2010). The risk of misappropriation or exploitation is high when Indigenous knowledge is digitized and shared on open-access platforms. For example, pharmaceutical companies might use traditional medicinal knowledge without compensating the communities from which that knowledge originated. Protecting the rights of Indigenous communities in the digital realm requires robust legal frameworks that account for communal ownership and the unique nature of Indigenous knowledge. Many countries lack clear legal and policy frameworks to address the complexities of Indigenous knowledge preservation in the digital realm. Existing intellectual property laws are often ill-suited to protect communal knowledge or cultural heritage, and Indigenous communities may struggle to assert their rights within these frameworks (Chisa & Hoskins, 2016). Moreover, international conventions and agreements on Indigenous rights, such as the United Nations Declaration on the Rights of Indigenous Peoples (UNDRIP), do not always translate into national policies that protect Indigenous knowledge in the digital space (United Nations, 2007). This leaves many Indigenous communities vulnerable to exploitation and loss of control over their cultural heritage.

Consent and Control Over Knowledge

The issue of consent is critical in Indigenous knowledge preservation (Australian Government, 2024; Mittermeier, 2024; Tagle, 2009). Digital technologies often make knowledge more widely accessible, but not all Indigenous knowledge is meant to be shared openly. Some knowledge is sacred or restricted to specific members of the community and making it publicly accessible can be a violation of cultural norms (WIPO, 2019). Ensuring that Indigenous communities retain control over how their knowledge is shared and who can access it is a significant challenge. In many cases, external researchers or institutions may undertake digitization efforts without fully consulting the community or obtaining their informed consent. This can lead to a loss of control over cultural heritage, as communities may find that their knowledge is being used in ways that they did not authorize.

Sustainability of Digital Projects

The long-term sustainability of digital preservation projects is another challenge (Allison-Cassin et al., 2023; Ngulube, 2023). Digital technologies are constantly evolving, and the formats used to store and present Indigenous knowledge today may become obsolete in the future. Without careful planning and ongoing maintenance, digital archives can become inaccessible due to outdated formats or discontinued platforms. For instance, files saved in now-outdated formats may not be readable on future devices, and platforms created for specific digitization projects may shut down due to a lack of funding. Ensuring the sustainability of Indigenous knowledge in digital formats requires not only initial investment but also ongoing resources for updating and maintaining the digital infrastructure (Balogun & Kalusopa, 2021; Masenya, 2023).

Resource Limitations

Indigenous knowledge preservation in the digital era often requires substantial resources, including funding, technological infrastructure, and skilled personnel (Ajani et al., 2024; Boamah & Liew, 2017). Many Indigenous communities and organizations lack the financial means to invest in large-scale digitization projects, and external funding is often needed to support such efforts (Chowdhury et al., 2023). Moreover, even when funding is available, there may be limitations in terms of technical expertise. Digitization requires specialized skills, such as knowledge of data management, digital archiving, and content creation, which may not be readily available in Indigenous communities (Perry, 2014). This can lead to reliance on external experts who may not fully understand the cultural issues of the knowledge they are preserving.

Balancing Commercialisation With Preservation

While digitization offers the potential for economic opportunities, such as the commercialization of Indigenous art or tourism, there is a delicate balance between economic development and cultural preservation (Craven, 2020; Hossain & Ballardini, 2021). Some digital platforms may encourage the commodification of Indigenous knowledge, turning cultural practices into commercial products. This commodification risks reducing the cultural and spiritual significance of Indigenous knowledge, turning it into a commodity for profit rather than a living tradition (Quang et al., 2022; Tuyen, 2023). Striking a balance between leveraging digital platforms for economic development and ensuring the respectful preservation of Indigenous heritage is a complex challenge.

Ethical Implications

The ethical implications of digitizing Indigenous knowledge are profound (Francis & Liew, 2010; Manžuch, 2017). Decisions about what to digitize, how to categorize it, and how to make it accessible must be made with sensitivity to cultural values and norms. Ethical dilemmas often arise when external institutions, such as universities, libraries, or museums, take the lead in digitization efforts without fully consulting the Indigenous communities involved. There is also the question of how to balance public access to Indigenous knowledge with the need to protect sensitive or sacred information (Chigwada & Ngulube, 2024; Hossain & Ballardini, 2021; Oyelude, 2023). In some cases, well-intentioned efforts to preserve knowledge can lead to its commercialization or misappropriation, undermining the very communities they seek to help. Digital platforms need to adopt ethical frameworks that recognize the communal nature of Indigenous knowledge and allow Indigenous groups to have control over how their knowledge is accessed and used. Open access repositories must be mindful of how Indigenous knowledge is categorized and distributed, ensuring that it is not misappropriated or commercialized without consent. Tools like Creative Commons licenses or protocols like the Traditional Knowledge (TK) Labels provide frameworks to protect the rights of Indigenous communities in the digital realm (Khan, 2018). These systems give communities the ability to label their knowledge with culturally appropriate terms and restrictions, thereby controlling access and usage.

CASE STUDIES THAT INTEGRATED CARE AND FAIR IN IK PRESERVATION

Three projects were considered for this study: Mukurtu in Australia, Southern African Intangible Cultural Heritage (SAICH) in Zimbabwe, and Ulwazi programme in South Africa. The projects represent the integration of IK in both developing and developed countries.

Mukurtu

Mukurtu is an Indigenous knowledge content management system (CMS) developed to address the specific needs of Indigenous communities in managing, sharing, and preserving cultural heritage and traditional knowledge (Mukurtu, 2024). Originating in the Warumungu community of Australia, Mukurtu CMS empowers Indigenous peoples to control how their cultural materials are stored, accessed, and shared, ensuring respect for traditional protocols and community preferences. A significant aspect of Mukurtu's design is its alignment with the CARE and FAIR principles of data governance, which provide a framework for the ethical management of Indigenous data and digital cultural heritage (Christen et al., 2017). Mukurtu, meaning "dilly bag" in Warumungu language, refers to the traditional woven bag used by Indigenous communities to hold sacred items, shared only with authorized people (Christen et al., 2017; Montenegro, 2023). The idea of Mukurtu CMS arose from this cultural metaphor, symbolizing a secure and respectful repository for sacred knowledge. The CMS was developed because of collaborations between Indigenous communities, technology developers, and cultural heritage experts.

Indigenous communities around the world often face significant challenges when it comes to controlling the use of their cultural heritage, especially when these materials are stored in digital formats on external platforms that do not align with their cultural protocols (Nicholas, 2021). Mainstream content management systems tend to prioritize open access and global sharing, often overlooking the need for restricted access to sacred or culturally sensitive materials. Mukurtu addresses these gaps by offering a culturally appropriate platform that allows Indigenous peoples to define, monitor, and enforce access to their knowledge based on traditional values and governance practices. It is designed to integrate Indigenous protocols into the digital world, ensuring that cultural knowledge remains in the control of the community (Machulak, 2020). It enables users to establish tiered access to digital materials, with permissions based on factors such as age, gender, status, or other culturally defined roles. Communities also can change these permissions as their protocols evolve over time (Christen et al., 2017).

Mukurtu and CARE Principles

Mukurtu CMS integrates the CARE principles into its design to enable communities to determine how their knowledge is shared, ensuring that it is used to support collective community goals. For example, by allowing Indigenous communities to set access protocols, Mukurtu ensures that the knowledge is shared only with individuals or institutions that have the community's trust and that will use the data to benefit the collective (Mukurtu, 2024). One of the core features of Mukurtu is its ability to embed community-defined protocols for accessing, managing, and sharing cultural materials. Indigenous communities have full authority to control who sees, uses, or shares their knowledge (Machulak, 2020). They can define rules for different users, manage multiple levels of access, and easily change permissions as necessary. This aligns with the CARE principle that Indigenous peoples should maintain control over their data. Mukurtu promotes responsible stewardship by embedding cultural protocols into the system itself. The platform provides clear guidelines and responsibilities for users, emphasizing the importance of respecting Indigenous knowledge and upholding ethical practices when accessing or sharing it (Hayward et al., 2021). In this way, Mukurtu ensures that responsibility for data management is shared between the community and users. Ethical considerations are at the core of Mukurtu's design by ensuring that cultural materials are handled in a manner consistent with their values, avoiding exploitation or misuse (Christen et al., 2017). By prioritizing the ethical use of data, Mukurtu ensures that Indigenous knowledge is respected and protected.

Mukurtu and FAIR Principles

The FAIR principles, while initially developed for scientific data, have been adapted to align with the needs of Indigenous communities through Mukurtu's design. These principles focus on making data Findable, Accessible, Interoperable, and Reusable, with respect to Indigenous sovereignty and governance frameworks. Mukurtu ensures that Indigenous knowledge is findable by incorporating traditional knowledge labels and metadata that are culturally specific (National Humanities Alliance, 2024). For example, items within the CMS are tagged according to community-defined terms, allowing materials to be located based on Indigenous naming conventions. This approach respects Indigenous knowledge systems while making the data more discoverable within the context of the community's protocols. Additionally, Mukurtu enables Indigenous communities to manage how their cultural items are represented and discovered online (Christen et al., 2017). By allowing for controlled access to materials, communities ensure that only trusted individuals or institutions can find and engage with their cultural heritage (Machulak, 2020). In this way, Mukurtu balances findability with the need to respect cultural protocols. It also provides access to Indigenous cultural materials in ways that are consistent with community-defined rules. Access is restricted based on cultural protocols, ensuring that only those who have been given permission can view certain materials. This supports the CARE principle of Authority to Control, while still aligning with the FAIR goal of making data accessible to authorized users. Moreover, the platform is built with an emphasis on accessibility for community members, regardless of their technical background (Christen, 2020). Mukurtu's user-friendly interface ensures that Indigenous peoples can engage with the system and manage their knowledge without needing advanced technical skills, democratizing access within the community itself.

Mukurtu is designed to be interoperable with other data systems and platforms while maintaining community control over cultural materials (Shanahan, 2019). For example, Mukurtu can work with systems like Omeka and other content management platforms to enable the sharing of Indigenous knowledge with broader audiences, where appropriate. This interoperability is key to enabling data integration across systems while respecting Indigenous protocols for access and use. Reusability in the context of Mukurtu means that cultural materials can be shared, where appropriate, in ways that align with community goals (Christen et al., 2017). Mukurtu ensures that materials are reusable only under conditions set by the community. For example, specific cultural materials may be made available for academic research, but only if the community agrees that the research will contribute positively to Indigenous knowledge and cultural preservation.

Mukurtu's approach to integrating CARE and FAIR principles offers a unique solution for Indigenous data governance. It ensures that Indigenous peoples retain full control over their knowledge while enabling findability, accessibility, interoperability, and reusability within the context of community-defined protocols. This dual framework ensures that digital cultural materials are managed ethically and effectively, balancing the needs of Indigenous communities with the broader goals of digital preservation and access. Mukurtu's system also emphasizes ongoing collaboration with Indigenous communities to ensure that the platform evolves alongside their needs. By prioritizing community control, Mukurtu not only safeguards cultural heritage but also empowers Indigenous communities to engage with digital technologies on their terms, ensuring long-term sustainability and ethical use of Indigenous knowledge in the digital age (Institute of Museum and Library Services, 2024). Mukurtu CMS represents a pioneering approach to Indigenous data governance, aligning with both the CARE and FAIR principles. It provides Indigenous communities with a powerful tool to manage and share their knowledge while ensuring that their cultural protocols are respected (Chigwada & Ngulube, 2023). Therefore, Mukurtu promotes a model of ethical, culturally grounded digital stewardship that prioritizes Indigenous rights and sovereignty over their cultural materials.

SAICH Project

The SAICH project at Chinhoyi University of Technology (CUT) in Zimbabwe is an initiative to preserve and promote African Indigenous knowledge and cultural heritage using digital technologies within seven African countries (SAICH, 2024). The seven participating countries in this project are Botswana, Eswatini, Lesotho, Malawi, Namibia, Zambia, and Zimbabwe. This project recognises the importance of Indigenous knowledge systems (IKS) in the development and identity of African communities and seeks to safeguard this knowledge for future generations. By incorporating digital tools and platforms, the SAICH project enables communities to manage and control their cultural elements, aligning with the CARE and FAIR principles of Indigenous data governance. The SAICH project was initiated to address the increasing need to preserve African Indigenous knowledge systems, which face numerous challenges due to modernization, urbanization, and the effects of globalization (Masenya, 2024). In Africa, Indigenous knowledge has traditionally been transmitted orally from one generation to another, making it vulnerable to loss in the absence of a formalized system of preservation. Furthermore, the encroachment of foreign cultural systems has led to the marginalization of local knowledge, highlighting the need for structured preservation efforts. The project's mission is to build a vibrant sub-regional platform for the protection, preservation, and promotion of intangible cultural heritage that will grow into a regional UNESCO centre (SAICH, 2024). The project involves collaboration with Indigenous communities, cultural institutions, and local experts, ensuring that Indigenous people are central to the process of collecting, curating, and managing their cultural data.

SAICH Project and CARE Principles

To ensure the sustainability of this digital repository, the SAICH project integrates community-driven protocols for the preservation, sharing, and use of Indigenous knowledge. The project uses digital technologies not only to archive cultural materials but also to make them accessible to future generations and researchers in a way that respects the values and customs of Indigenous communities. Importantly, the project integrates both the CARE and FAIR principles into its framework, ensuring ethical and sustainable data governance practices. In the context of the SAICH project, these principles ensure that Indigenous communities retain control over their cultural assets and benefit from the data's use. The SAICH project prioritizes the benefit of Indigenous communities in its approach to safeguarding cultural heritage (Chimhundu, 2019). Collective benefit emphasizes that the digitization and sharing of Indigenous knowledge should provide tangible and intangible benefits to the communities involved. SAICH fosters community development by ensuring that cultural assets are not commodified or exploited without the approval and benefit of the people to whom the knowledge belongs. In addition, the repository facilitates knowledge-sharing within and between communities, fostering inter-generational knowledge transmission and empowering Indigenous peoples to strengthen their cultural identity. Community members can access the repository to learn about their heritage, ensuring that Indigenous knowledge is used for the collective advancement of the people (SAICH, 2024).

The SAICH project gives Indigenous communities the authority to control how their cultural materials are digitized, accessed, and used. Instead of relying on external researchers or institutions, Indigenous groups are central to decision-making about their cultural heritage (SAICH, 2024). This principle ensures that communities retain the right to determine who can view or use their cultural assets and under what conditions. The project enables community leaders and cultural custodians to define protocols for access to sensitive materials (Chimhundu, 2019a). For example, knowledge that is sacred or restricted to certain members of the community may be kept private, with access limited to authorized individuals. This control prevents the inappropriate use or exploitation of Indigenous knowledge, ensuring that it remains in the hands of those who created and maintained it. SAICH promotes the responsible stewardship of Indigenous cultural data by embedding clear responsibilities for data management into its framework. Communities, researchers, and custodians have distinct roles and responsibilities in ensuring that Indigenous knowledge is protected and used following ethical principles (SAICH, 2024). The project emphasizes the need for cultural sensitivity when dealing with Indigenous knowledge, particularly in cases where the knowledge is sacred or tied to specific cultural practices.

By ensuring that community-defined guidelines are followed, the SAICH project upholds the principle of responsibility and safeguards the integrity of the knowledge being preserved. Ethical considerations are a cornerstone of the SAICH project's approach to data governance. The project ensures that Indigenous cultural materials are handled in ways that respect the values, beliefs, and customs of the communities involved (SAICH, 2024). This includes preventing the exploitation or commercialization of knowledge without consent and ensuring that knowledge is not misappropriated by external entities. The SAICH project also ensures that ethical research practices are followed when working with Indigenous communities. Researchers and partners involved in the project are required to adhere to community protocols and to obtain proper permissions before using or disseminating Indigenous knowledge. In this way, the project protects the cultural integrity of the materials in its repository.

SAICH Project and FAIR Principles

The FAIR principles aim to enhance the accessibility and utility of data, making it easier to find, access, and reuse. Within the context of the SAICH project, these principles are applied to Indigenous knowledge while ensuring that community-defined protocols are respected. The SAICH project ensures that Indigenous cultural assets are findable through its comprehensive database (SAICH, 2024). Metadata is created for each item in the repository, allowing it to be indexed and found easily by community members, researchers, or authorized users. The project incorporates culturally relevant metadata and traditional knowledge labels to ensure that the items are categorized and searchable according to Indigenous naming conventions and knowledge systems. At the same time, the project respects the need for controlled access. While items are findable within the system, they are only accessible to those who have been granted permission by the community (SAICH, 2024). While the SAICH project makes Indigenous knowledge accessible to authorized users, it also ensures that access is restricted based on cultural protocols. Communities can determine which materials are open to the public and which are restricted, ensuring that knowledge remains accessible only to those with the proper cultural standing or authority. Accessibility is also ensured using a user-friendly platform, enabling community members to engage with the repository without needing advanced technical skills. This ensures that Indigenous communities themselves can access and benefit from their cultural assets.

The SAICH project is designed to be interoperable with other digital systems and platforms, allowing for the integration of Indigenous knowledge with other databases or research networks (SAICH, 2024). This interoperability ensures that cultural materials can be shared, where appropriate, with wider audiences, including academic researchers or other cultural institutions, while still maintaining control over how the knowledge is used. The SAICH project ensures that cultural materials are reusable in ways that align with community goals. Indigenous knowledge can be shared with researchers or educators, provided that it is used ethically and following community-defined protocols. This ensures that the knowledge is not only preserved but also utilized in ways that benefit the community, such as through educational programs or cultural revitalization efforts. Therefore, the SAICH project represents a significant effort to preserve and protect African Indigenous knowledge systems. By integrating the CARE and FAIR principles of Indigenous data governance, the project ensures that cultural materials are managed in a way that respects the rights of Indigenous peoples while promoting access and usability for future generations. Through its focus on ethical stewardship, community control, and the responsible use of data, the SAICH project serves as a model for Indigenous knowledge preservation in Africa and beyond.

Ulwazi Programme

The Ulwazi programme, an Indigenous knowledge management initiative based in Durban Municipality, South Africa, represents a pioneering effort to preserve and promote Indigenous knowledge (IK) through digital technologies (McNulty, 2013). Launched by the eThekwini Municipality's libraries, the project seeks to document, archive, and disseminate Indigenous knowledge in a way that strengthens community identity and supports the preservation of African cultural heritage. The Ulwazi programme aligns with the CARE and FAIR principles of Indigenous data governance, ensuring that the documentation and sharing of Indigenous knowledge respect community rights while promoting accessibility and usability. The Ulwazi programme, meaning "knowledge" in Zulu, was established to create a digital repository of local Indigenous knowledge (McNulty, 2014). It operates through a participatory model where community members are directly involved in the collection and sharing of information. The project focuses on recording oral histories, cultural practices, traditional medicine, Indigenous languages, folklore, and other forms of local knowledge that are at risk of being lost due to globalization. The project's primary objective is to ensure the preservation of this knowledge for future generations while making it accessible to a broader audience, including local communities, students, researchers, and the public. The platform is hosted by the eThekwini Municipality and operates as part of the municipal library system, giving it a strong institutional foundation (McNulty, 2014). Additionally, it empowers communities to take ownership of their knowledge by contributing content to the platform, thus decentralizing the knowledge production process. The project also aims to bridge the digital divide by making technology accessible to local communities. Through the platform, people are encouraged to record and upload videos, audio files, and written content that documents their traditions and experiences, ensuring that the knowledge is stored in a variety of formats for posterity.

Ulwazi Programme and CARE Principles

The Ulwazi programme integrates the CARE principles into its structure and processes by emphasising that the digitization and dissemination of Indigenous knowledge should benefit the communities that own it (Ulwazi programme, 2024). One of the primary aims of the project is to foster community development and cultural preservation by ensuring that the knowledge is not only preserved but also actively used to promote community well-being. The project promotes collective benefits by facilitating access to the digital repository, allowing communities to learn from each other, and encouraging cultural exchange between different Indigenous groups. It also contributes to education and economic empowerment, as knowledge about traditional practices, such as farming techniques or local medicinal plants, can be used to enhance livelihoods (Greyling & McNulty, 2012). Moreover, the project enables communities to reclaim and revitalize their cultural identity, particularly younger generations who may not have been exposed to certain aspects of their heritage. A central feature of the Ulwazi programme is its participatory model, which ensures that communities retain control over how their knowledge is collected, managed, and shared. The project allows contributors to establish guidelines on how the knowledge can be accessed and used, respecting traditional governance structures.

Communities have the authority to determine which knowledge should be shared publicly and which should remain restricted or only available to certain members. This principle is crucial, particularly in cases where cultural knowledge is sacred or considered sensitive. By allowing communities to define access protocols, the Ulwazi programme respects their right to control how their cultural heritage is used, ensuring that the knowledge remains in the hands of its rightful custodians. The Ulwazi programme embeds responsibility into the way Indigenous knowledge is managed, ensuring that it is handled with care and respect. The project encourages contributors to follow community-approved protocols when sharing their knowledge (McNulty, 2014). It also provides training for individuals involved in the collection of knowledge to ensure that ethical standards are maintained during the documentation process. Responsibility is shared between the community members, the project administrators, and the researchers or individuals accessing the knowledge. Contributors are responsible for ensuring the accuracy and appropriateness of the information they provide, while project managers are responsible for upholding ethical guidelines and safeguarding the integrity of the knowledge in the repository (McNulty, 2013). Ethical data governance is at the heart of the Ulwazi programme since the project ensures that the collection, storage, and sharing of Indigenous knowledge adhere to cultural norms and values. Contributors are encouraged to respect the ethical boundaries established by their communities, ensuring that sensitive information is not disclosed inappropriately. Additionally, the project prevents the commercialization or exploitation of Indigenous knowledge without the consent of the communities involved. Researchers or external users must obtain permission before using the knowledge for academic or commercial purposes, and the project's administrators ensure that any use of the knowledge aligns with community values and benefits its custodians.

Ulwazi Programme and FAIR Principles

The FAIR principles are widely adopted in data management to ensure that data is organized in ways that enhance its utility and reach. The Ulwazi programme incorporates these principles while ensuring that the sovereignty of Indigenous communities over their knowledge is respected. The Ulwazi programme makes Indigenous knowledge findable by creating a searchable digital repository. The platform uses metadata and tagging systems that make it easier to locate specific information. The project also ensures that knowledge is findable within community-specific contexts. By allowing contributors to tag content with culturally relevant labels, the platform aligns with the CARE principle of collective benefit, as it ensures that the knowledge is accessible to community members who can understand and use it. Accessibility is a key feature of the Ulwazi programme, ensuring that knowledge is available to a wide audience, including community members, researchers, and the public. However, access is controlled by community-defined protocols, ensuring that only authorized users can view certain materials. This prevents the misuse or exploitation of sensitive knowledge (McNulty, 2014).

The project also promotes digital inclusion by providing access to technology for rural or underserved communities. Public libraries play a significant role in this, acting as access points where people can engage with the digital platform, thus bridging the digital divide. This focus on accessibility aligns with the FAIR principle while ensuring that communities remain in control of how their knowledge is shared. The Ulwazi programme is designed to be interoperable with other digital platforms and content management systems. The repository is built using open-source software, allowing it to be integrated with other databases or systems that researchers or institutions may use. This interoperability ensures that Indigenous knowledge can be shared across different platforms, provided that the community grants permission. It enables wider dissemination of knowledge while adhering to the principle of responsibility, as the project ensures that data shared with external platforms still follows ethical guidelines. The Ulwazi programme ensures that Indigenous knowledge is reusable under specific conditions set by the community. Knowledge shared on the platform can be reused for educational, research, or cultural revitalization purposes, provided that the community benefits from the use and gives consent. This ensures that knowledge is not misappropriated or commercialized without proper acknowledgement or compensation. By ensuring that Indigenous knowledge is reusable in ethical and responsible ways, the Ulwazi programme aligns with the Indigenous data governance principles.

From these case studies, it can be noted that balancing openness with Indigenous control is challenging but achievable. Collaborative governance models and culturally appropriate metadata systems are crucial for ensuring Indigenous communities retain control over their knowledge while benefitting from digital preservation.

CARE AND FAIR PRINCIPLES IN IK PRESERVATION

The CARE principles are rooted in the rights and interests of Indigenous communities (Global Indigenous Data Alliance, 2024). Collective benefit ensures that Indigenous communities reap the rewards of digital preservation projects. Authority to Control emphasizes that Indigenous peoples should decide how their knowledge is used. Responsibility highlights the duties of researchers and institutions to protect this knowledge, while Ethics demands respect for the cultural significance of the information being preserved. FAIR principles aim to enhance the usability of data (Go FAIR, 2024). Data should be Findable, Accessible, Interoperable, and Reusable, enabling it to be used by researchers and institutions worldwide. While these principles are effective in scientific and data management contexts, they may clash with the CARE principles, especially when data pertains to sensitive cultural knowledge (Carroll et al., 2021). FAIR principles promote openness and usability and often disregard issues related to intellectual property and cultural sensitivity in Indigenous knowledge preservation. CARE principles, while protecting Indigenous rights, may limit data accessibility. It is, therefore, important to integrate these two principles when dealing with the digital preservation of Indigenous knowledge. From the review findings, it is suggested that the following can be done.

Findability With Collective Benefit

Integrating FAIR and CARE principles into Indigenous knowledge preservation can create a framework that respects both scientific standards for data management and the rights and values of Indigenous communities. Findability can be aligned with collective benefit by ensuring that metadata and Indigenous knowledge data are findable by using standardised, machine-readable formats and assigning persistent identifiers. The knowledge must be findable in a way that benefits the Indigenous communities and ensures that access to the data directly serves the community's interests such as promoting cultural heritage and supporting economic opportunities for the Indigenous knowledge holders and community. It is important to design systems where Indigenous communities define how their knowledge is represented and discovered, for example, create metadata that includes Indigenous terms and language, ensuring the community has control over how knowledge is contextualised and who finds it. The use of Indigenous-led repositories or platforms that give Indigenous groups authority to determine the reach and benefits of sharing their knowledge would be beneficial to the Indigenous community.

Accessibility With Authority to Control

Accessibility can be paired with authority to control by ensuring that Indigenous knowledge is accessible to those who need it using standardised access protocols. The Indigenous peoples must retain the right to control who accessed their knowledge and under what conditions. This can be done by implementing tiered access where different levels of permission are needed and given, and access is granted on community consent. Technologies like blockchain can be used to create decentralised access protocols that ensure that no one can bypass the permissions set by Indigenous authorities. It is also important to develop culturally sensitive access rules ensuring that knowledge considered sacred or private remains under community control.

Interoperability With Responsibility

Interoperability can be used together with responsibility where data should be interoperable using standardised formats and vocabularies that facilitate reuse and integration with other systems. In addition, the responsibility to preserve, share, and use Indigenous knowledge lies with both data stewards and users, ensuring that it is done in ways that respect the cultural context and values of the originating communities. Involving Indigenous communities in the creation of interoperable data formats that include culturally appropriate protocols would be useful in achieving this. This may involve creating ontologies that represent Indigenous worldviews or working with Indigenous data sovereignty organisations. When integrating Indigenous knowledge with other datasets, ensure that the context is not lost, and that the knowledge is not misrepresented or decontextualised.

Reusability With Ethics

Indigenous knowledge data should be reusable, with clear terms for how it can be applied in other contexts, under appropriate licenses. Ethical considerations must govern the reuse of Indigenous knowledge, ensuring it is not misappropriated or commercialised without the consent of the originating community. Using reusability with ethics can be done by collaborating with Indigenous communities to develop culturally appropriate terms of use and guidelines that prioritise ethics in the reuse of Indigenous knowledge. There is a need to apply for licenses that align with ethical guidelines specific to Indigenous knowledge, such as traditional knowledge or Indigenous data sovereignty licenses. These ensure that reusers are bound by community-specific rules regarding how the knowledge can be used.

FUTURE RESEARCH DIRECTIONS

Investigating the impact of digital preservation on Indigenous language revitalization is crucial in understanding how modern technologies can aid in the survival and transmission of endangered languages. Additionally, analysing the role of blockchain technology in securing Indigenous knowledge offers insights into how decentralized systems can protect sensitive cultural information from unauthorized access and misuse. Furthermore, exploring the intersection of intellectual property law and Indigenous knowledge in the digital age highlights the challenges of safeguarding traditional knowledge while navigating global legal frameworks. To ensure these efforts are beneficial, developing new tools for measuring the ethical impact of digital preservation projects on Indigenous communities is essential to assess whether these initiatives truly support Indigenous autonomy and cultural integrity.

CONCLUSION AND RECOMMENDATIONS

It can be concluded that IK is important in achieving the SDGs since it is used in many ways as shown in this study. As a result, the preservation of Indigenous knowledge in the digital era requires an approach that respects the rights of Indigenous communities while also embracing the benefits of data accessibility and reusability. Integrating CARE and FAIR principles provides a pathway for achieving this balance, but it requires collaborative efforts between Indigenous communities, researchers, and technologists. The way forward involves creating governance structures and digital tools that respect Indigenous sovereignty, cultural values, and intellectual property while leveraging the advantages of digital technologies for knowledge preservation. Digital technologies offer a range of tools to capture, store, and share Indigenous knowledge in ways that respect its cultural contexts and ownership. Therefore, people working with Indigenous people should "Be FAIR and CARE" to ensure that Indigenous rights and interests of Indigenous communities are respected when preserving Indigenous knowledge in the digital era. The following are the recommendations made by the authors when integrating the CARE and FAIR principles in IK preservation.

The openness encouraged by FAIR might conflict with the closed, community-specific control encouraged by CARE. Therefore, it is important to incorporate controlled openness, where Indigenous communities determine the extent of openness based on their values and goals. This can be done through creative licensing or agreements that combine aspects of open science with Indigenous data sovereignty. The following recommendations can be used to integrate FAIR with CARE in Indigenous knowledge preservation in the digital era. These call for the use of digital technologies that support both FAIR and CARE principles which can be in the form of Indigenous knowledge management systems that incorporate Indigenous governance models while ensuring that data remains FAIR. By weaving together, the technical standards of FAIR with the ethical, community-driven focus of CARE, Indigenous knowledge preservation would respect both the scientific community's needs and the rights and values of Indigenous people.

1. **Develop Community-Led Governance Models:** Indigenous communities should be empowered to lead the governance of their digital knowledge systems, ensuring that CARE principles are prioritized.
2. **Adopt Culturally Appropriate Metadata Systems:** Metadata systems should reflect Indigenous epistemologies, allowing for the representation of cultural nuances.
3. **Foster Collaborative Partnerships:** Institutions, technologists, and researchers should build long-term partnerships with Indigenous communities to ensure that digital preservation projects align with community goals.

4. **Create Legal Frameworks to Protect Indigenous Knowledge:** Policymakers should establish legal protections that prevent the unauthorized use or exploitation of Indigenous knowledge in digital formats.
5. **Develop Indigenous-Centric Digital Technologies:** The development of digital tools should be guided by the needs and values of Indigenous communities, ensuring that technology serves their interests.
6. **Capacity Building:** Provide training to both data stewards and Indigenous communities on how to implement and balance FAIR and CARE.
7. **Community Engagement:** The stakeholders should support Indigenous communities in developing their digital infrastructure and protocols for managing and preserving their knowledge.

REFERENCES

Aiseng, K. (2023). Challenges and opportunities of preserving African Indigenous knowledge using digital technologies: the case of Bogwera. In Masenya, T. (Ed.), *Digital preservation and documentation of global Indigenous knowledge systems* (pp. 123–141). IGI Global., DOI: 10.4018/978-1-6684-7024-4.ch007

Ajani, Y., Oladokun, B., Olarongbe, S., Amaechi, M., Rabiu, N., & Bashorun, M. (2024). revitalizing Indigenous knowledge systems via digital media technologies for sustainability of Indigenous languages. *Preservation. Digital Technology & Culture*, 53(1), 35–44. DOI: 10.1515/pdtc-2023-0051

Allison-Cassin, S., Bateman, R., Bhardwaj, R. K., & Chisita, C. T. (2023). Introduction. *The Serials Librarian*, 84(5–8), 62–65. DOI: 10.1080/0361526X.2024.2313400

Ann Rahman, E., & Cochrane, T. (2023). Pedagogy and Indigenous knowing and learning. *Oxford Review of Education*, 49(4), 429–445. DOI: 10.1080/03054985.2023.2228640

Antonelli, A. (2023). Indigenous knowledge is key to sustainable food systems. *Nature*, 613(7943), 239–242. DOI: 10.1038/d41586-023-00021-4 PMID: 36627421

Ariyani, F., Putrawan, G. E., Riyanda, A. R., Idris, A., Misliani, L., & Perdana, R. (2022). Technology and minority language: An Android-based dictionary development for the Lampung language maintenance in Indonesia. *Tapuya : Latin American Science, Technology and Society*, 5(1), 2015088. Advance online publication. DOI: 10.1080/25729861.2021.2015088

Australian Government. (2024). Respectful use of Indigenous knowledge. Retrieved on 10 October 2024 from https://www.ipaustralia.gov.au/understanding-ip/Indigenous-knowledge/how-to-engage-Indigenous-knowledge

Ayeni, A. O., & Aborisade, A. G. (2022). African Indigenous knowledge systems and the world. In Oloruntoba, S. O., & Falola, T. (Eds.), *The Palgrave handbook of Africa and the changing global order*. Palgrave Macmillan., DOI: 10.1007/978-3-030-77481-3_8

Babagana, U. I., & Kemboi, G. (2024). Digital preservation of Indigenous Knowledge in Ramat Library (University of Maiduguri, Nigeria). *iPRES*. Retrieved on 10 October 2024 from https://ipres2024.pubpub.org/pub/o5shuj0u

Balogun, T. (2023). Digitization of Indigenous knowledge systems in Africa: The case of South Africa's National Recorded System (NRS). *Records Management Journal*, 33(1), 88–102. DOI: 10.1108/RMJ-10-2022-0029

Balogun, T., & Kalusopa, T. (2022). Web archiving of Indigenous knowledge systems in South Africa. *Information Development*, 38(4), 658–671. DOI: 10.1177/02666669211005522

Bansal, S., Sarker, T., Yadav, A., Garg, I., Gupta, M., & Sarvaiya, H. (2024). Indigenous communities and sustainable development: A review and research agenda. *Global Business and Organizational Excellence*, 43(4), 65–87. DOI: 10.1002/joe.22237

Basu, M., & Dasgupta, R. (2023). Exploring Indigenous water knowledge, values, and practices: insights and examples. In Basu, M., & DasGupta, R. (Eds.), *Indigenous and Local Water Knowledge, Values and Practices*. Springer., DOI: 10.1007/978-981-19-9406-7_1

Behailu, B. M., Pietilä, P. E., & Katko, T. S. (2016). Indigenous practices of water management for sustainable services: Case of Borana and Konso, Ethiopia. *SAGE Open*, 6(4), 2158244016682292. Advance online publication. DOI: 10.1177/2158244016682292

Boadu, E. S. (2022). Indigenous knowledge and community development in Africa: a framework to explore the approaches, philosophies and practices in Ghana. In: Mbah, M.F., Leal Filho, W., Ajaps, S. (eds) *Indigenous methodologies, research and practices for sustainable development*. World Sustainability Series. Springer. https://doi.org/DOI: 10.1007/978-3-031-12326-9_9

Boamah, E., & Liew, C. L. (2017). Conceptualising the digitisation and preservation of Indigenous knowledge: the importance of attitudes. In: Choemprayong, S., Crestani, F., Cunningham, S. (eds) *Digital libraries: data, information, and knowledge for digital lives*. ICADL 2017. Lecture Notes in Computer Science, vol 10647. Springer. https://doi.org/DOI: 10.1007/978-3-319-70232-2_6

Bodeker, G., & Kariippanon, K. (2020, February 28). Traditional medicine and Indigenous health in Indigenous Hands. *Oxford Research Encyclopedia of Global Public Health*. Retrieved on 10 October 2024 from https://oxfordre.com/publichealth/view/10.1093/acrefore/9780190632366.001.0001/acrefore-9780190632366-e-155

Bruchac, M. M. (2014). Indigenous knowledge and traditional Knowledge. In Smith, C. (Ed.), *Encyclopedia of global archaeology*. Springer., DOI: 10.1007/978-1-4419-0465-2_10

Buthelezi, N. N., & Hughes, J. C. (2014). Indigenous knowledge systems and agricultural rural development in South Africa: Past and present perspectives. *Indilinga*, 13(2), 231–250.

Carballo-Garcia, A., & Boté-Vericad, J. J. (2022). Fair data: History and present context. *Central European Journal of Educational Research*, 4(2), 45–53. DOI: 10.37441/cejer/2022/4/2/11379

Carroll, S. R., Garba, I., Figueroa-Rodríguez, O. L., Holbrook, J., Lovett, R., Materechera, S., Parsons, M., Raseroka, K., Rodriguez-Lonebear, D., Rowe, R., Sara, R., Walker, J. D., Anderson, J., & Hudson, M. (2020). The CARE principles for Indigenous data governance. *Data Science Journal*, 19(43), 1–12. DOI: 10.5334/dsj-2020-043

Carroll, S. R., Herczog, E., Hudson, M., Russell, K., & Stall, S. (2021). Operationalizing the CARE and FAIR principles for Indigenous data futures. *Scientific Data*, 8(1), 108. DOI: 10.1038/s41597-021-00892-0 PMID: 33863927

Carroll, S. R., Rodriguez-Lonebear, D., & Martinez, A. (2019). Indigenous Data Governance: Strategies from United States Native Nations. *Data Science Journal*, 18(31), 1–15. DOI: 10.5334/dsj-2019-031 PMID: 34764990

Chigwada, J., & Ngulube, P. (2023). Stakeholders in the acquisition, preservation, and dissemination of Indigenous knowledge projects. *Information Development*, 0(0), 02666669231192851. Advance online publication. DOI: 10.1177/02666669231192851

Chigwada, J., & Ngulube, P. (2024). Librarians' role in the preservation and dissemination of Indigenous knowledge. *IFLA Journal*, 50(2), 242–256. DOI: 10.1177/03400352231217270

Chimhundu, H. (2019). UNESCO/ Flanders project on strengthening Sub-regional cooperation and national capacities in seven Southern African countries for implementing the 2002 convention for the safeguarding of intangible cultural heritage. Retrieved on 10 October 2024 from www.saich.org/home/downloads/SAICH-Platform-Narrative-Report-final-draft-amended-291019.pdf

Chimhundu, H. (2019a). SAICH news 2019: on the tracks of living heritage in Southern Africa. Retrieved on 10 October 2024 from https://ich.unesco.org/doc/src/SAICH_NEWS_2019.pdf

Chisa, K., & Hoskins, R. (2014). The effects of information and communication technologies on Indigenous communities in South Africa: A library and information science perspective. *Mousaion*, 32(3), 49–68. DOI: 10.25159/0027-2639/1676

Chisa, K., & Hoskins, R. (2016). African customary law and the protection of Indigenous cultural heritage: Challenges and issues in the digitization of Indigenous knowledge in South Africa. *Indilinga*, 15(1), 1–15.

Chowdhury, G., McLeod, J., Lihoma, P., Teferra, S., & Wato, R. (2023). Promoting access to Indigenous information in Africa: Challenges and requirements. *Information Development*, 39(3), 611–623. DOI: 10.1177/02666669211048488

Christen, K. (2020). Mukurtu: a digital platform that does more than manage content. CAS in the media, November 13. 2020. Retrieved on 11 October 2024 from https://cas.wsu.edu/news/2020/11/13/mukurtu-a-digital-platform-that-does-more-than-manage-content/

Christen, K., Merrill, A., & Wynne, M. (2017). A community of relations: Mukurtu hubs and spokes. *D-Lib Magazine : the Magazine of the Digital Library Forum*, 23(5/6). Advance online publication. DOI: 10.1045/may2017-christen

Craven, J. M. (2020). Indigenous approaches to economic development and sustainability. *International Critical Thought*, 10(1), 126–137. DOI: 10.1080/21598282.2020.1742529

Cross, J. T. (2010). Property rights and traditional knowledge. [PELJ]. *Potchefstroom Electronic Law Journal*, 13(4), 12–47.

da Silva, C., Pereira, F., & Amorim, J. P. (2023). The integration of Indigenous knowledge in school: A systematic review. *Compare: A Journal of Comparative Education*, 54(7), 1210–1228. DOI: 10.1080/03057925.2023.2184200

Dahlberg, A. C., & Trygger, S. B. (2009). Indigenous medicine and primary health care: The importance of lay knowledge and use of medicinal plants in rural South Africa. *Human Ecology: an Interdisciplinary Journal*, 37(1), 79–94. DOI: 10.1007/s10745-009-9217-6

Data Fairport. (2014). Data FAIRport. Retrieved on 11 October 2024 from https://www.datafairport.org/index.html

Demchenko, Y., Los, W., & de Laat, C. (2018). Data as economic goods: Definitions, properties, challenges, enabling technologies for future data markets. *ITU Journal : ICT Discoveries*, 2(23). Retrieved October 18, 2024, from https://www.itu.int/en/journal/002/Documents/ITU2018-12.pdf

Evens, T., & Hauttekeete, L. (2011). Challenges of digital preservation for cultural heritage institutions. *Journal of Librarianship and Information Science*, 43(3), 157–165. DOI: 10.1177/0961000611410585

Fernández-Llamazares, Á., Lepofsky, D., Lertzman, K., Armstrong, C. G., Brondizio, E. S., Gavin, M. C., Lyver, P. O., Nicholas, G. P., Pascua, P., Reo, N. J., Reyes-García, V., Turner, N. J., Yletyinen, J., Anderson, E. N., Balée, W., Cariño, J., David-Chavez, D. M., Dunn, C. P., Garnett, S. C., & Vaughan, M. B. (2021). Scientists' Warning to humanity on threats to Indigenous and local knowledge systems. *Journal of Ethnobiology*, 41(2), 144–169. DOI: 10.2993/0278-0771-41.2.144

First Nation's Information Governance Centre. 2018. The First Nations principles of OCAP®. Retrieved on 12 October 2024 from https://fnigc.ca/ocap

Francis, K. D., & Liew, C. L. (2010). Digitised Indigenous knowledge in cultural heritage organisations in Australia and New Zealand: An examination of policy and protocols. *Proceedings of the American Society for Information Science and Technology*, 46(1), 1–21. DOI: 10.1002/meet.2009.145046025

Garba, I., Sterling, R., Plevel, R., Carson, W., Cordova-Marks, F. M., Cummins, J., Curley, C., David-Chavez, D., Fernandez, A., Hiraldo, D., Hiratsuka, V., Hudson, M., Jäger, M. B., Jennings, L. L., Martinez, A., Yracheta, J., Garrison, N. A., & Carroll, S. R. (2023). Indigenous Peoples and research: Self-determination in research governance. *Frontiers in Research Metrics and Analytics*, 8, 1272318. DOI: 10.3389/frma.2023.1272318 PMID: 38033627

Global Indigenous Data Alliance. (2024). CARE Principles for Indigenous data governance. Retrieved on 18 October 2024 from https://www.gida-global.org/care

Go, F. A. I. R. (2024). FAIR principles. Retrieved on 11 October 2024 from https://www.go-fair.org/fair-principles/

Greenwood, M., & Lindsay, N. M. (2019). A commentary on land, health, and Indigenous knowledge(s). *Global Health Promotion*, 26(3, suppl), 82–86. DOI: 10.1177/1757975919831262 PMID: 30964411

Greyling, E., & McNulty, N. (2012). How to build an Indigenous digital library through community participation: the case of the Ulwazi programme, Information for Sustainable Development in a Digital Environment. *XXth Standing Conference of Eastern, Central and South Africa Library and Information Associations (SCECSAL).*

Greyling, E., & Zulu, S. (2009). Content development in an Indigenous digital library: a case study in community participation. *IFLA Congress*, Milan, Italy 23-27 August 2009. https://www.ifla.org/annual-conference/ifla75/index.htm

Griffiths, K. E., Blain, J., Vajdic, C. M., & Jorm, L. (2021). Indigenous and tribal peoples data governance in health research: A systematic review. *International Journal of Environmental Research and Public Health*, 18(19), 10318. DOI: 10.3390/ijerph181910318 PMID: 34639617

Hayward, A., Sjoblom, E., Sinclair, S., & Cidro, J. (2021). A new era of Indigenous research: Community-based Indigenous research ethics protocols in Canada. *Journal of Empirical Research on Human Research Ethics; JERHRE*, 16(4), 403–417. DOI: 10.1177/15562646211023705 PMID: 34106784

Haywood, L. K., Funke, N., Audouin, M., Musvoto, C., & Nahman, A. (2018). The sustainable development goals in South Africa: Investigating the need for multi-stakeholder partnerships. *Development Southern Africa*, 36(5), 555–569. DOI: 10.1080/0376835X.2018.1461611

Hossain, K., & Ballardini, R. M. (2021). Protecting Indigenous traditional knowledge through a holistic principle-based approach. *Nordic Journal of Human Rights*, 39(1), 51–72. DOI: 10.1080/18918131.2021.1947449

Hudson, M., Carroll, S. R., Anderson, J., Blackwater, D., Cordova-Marks, F. M., Cummins, J., David-Chavez, D., Fernandez, A., Garba, I., Hiraldo, D., Jäger, M. B., Jennings, L. L., Martinez, A., Sterling, R., Walker, J. D., & Rowe, R. K. (2023). Indigenous peoples' rights in data: A contribution toward Indigenous research sovereignty. *Frontiers in Research Metrics and Analytics*, 8, 1173805. DOI: 10.3389/frma.2023.1173805 PMID: 37215248

Imoro, Z. A., Imoro, A. Z., Duwiejuah, A. B., & Abukari, A. (2021). Harnessing Indigenous technologies for sustainable management of land, water, and food resources amidst climate change. *Frontiers in Sustainable Food Systems*, 5, 691603. DOI: 10.3389/fsufs.2021.691603

Institute of Museum and Library Services. (2024). Mukurtu software preserves Indigenous digital heritage through technologies of today. Retrieved on 11 October 2024 from https://www.imls.gov/grant-spotlights/mukurtu-software-preserves-Indigenous-digital-heritage-through-technologies-today

Inuit Tapiriit Kanatami. (2018). National Inuit Strategy on Research. Retrieved on 11 October 2024 from https://www.itk.ca/wp-content/uploads/2018/03/National-Inuit-Strategy-on-Research.pdf

Jessen, T. D., Ban, N. C., Claxton, N. X., & Darimont, C. T. (2020). Contribution of Indigenous knowledge to ecological and evolutionary understanding. *Frontiers in Ecology and the Environment*, 20(2), 93–101. DOI: 10.1002/fee.2435

Jethro, D. (2024). History uploaded: Digital archives after thirty years of democracy. *South African Historical Journal*, •••, 1–24. Advance online publication. DOI: 10.1080/02582473.2024.2351828

Karasimos, A. (2022). The battle of language learning apps: A cross-platform overview. *Research Papers in Language Teaching and Learning*, 12(1), 150–166.

Kennedy, A., Sehgal, A., Szabo, J., McGowan, K., Lindstrom, G., Roach, P., Crowshoe, L., & Barnabe, C. (2022). Indigenous strengths-based approaches to healthcare and health professions education – recognising the value of Elders' teachings. *Health Education Journal*, 81(4), 423–438. DOI: 10.1177/00178969221088921 PMID: 35531386

Khan, M. (2018). Traditional knowledge and the commons: the open movement, listening, and learning. Retrieved on 11 October 2024 from https://creativecommons.org/2018/09/18/traditional-knowledge-and-the-commons-the-open-movement-listening-and-learning/

Kohsaka, R., & Rogel, M. (2019). Traditional and local knowledge for sustainable development: empowering the Indigenous and local communities of the world. In: Leal Filho, W., Azul, A., Brandli, L., Özuyar, P., Wall, T. (eds) *Partnerships for the Goals. Encyclopedia of the UN Sustainable Development Goals.* Springer, https://doi.org/DOI: 10.1007/978-3-319-71067-9_17-1

Kom, Z., Nethengwe, N. S., Mpandeli, S., & Chikoore, H. (2022). Indigenous knowledge indicators employed by farmers for adaptation to climate change in rural South Africa. *Journal of Environmental Planning and Management*, 66(13), 2778–2793. DOI: 10.1080/09640568.2022.2086854

Kugara, S. L., & Mokgoatšana, S. (2022). Challenges presented by digitisation of VhaVenda oral tradition: An African Indigenous knowledge systems perspective. *Hervormde Teologiese Studies*, 78(3), 1–8. DOI: 10.4102/hts.v78i3.7428

Lythreatis, S., Singh, S. K., & El-Kassar, A. (2022). The digital divide: A review and future research agenda. *Technological Forecasting and Social Change*, 175(121259), 121359. Advance online publication. DOI: 10.1016/j.techfore.2021.121359

Machulak, E. (2020). Mukurtu: A digital platform that does more than manage content. *Humanities (Washington)*, 41(4). Retrieved October 11, 2024, from https://www.neh.gov/article/mukurtu-digital-platform-does-more-manage-content

Magni, G. (2017). Indigenous knowledge and implications for the sustainable development agenda. *European Journal of Education Research. European Journal of Education*, 54(4), 437–447. DOI: 10.1111/ejed.12238

Mahomoodally, M. F. (2013). Traditional medicines in Africa: An appraisal of ten potent African medicinal plants. *Evidence-Based Complementary and Alternative Medicine : eCAM*, 617459, 1–14. DOI: 10.1155/2013/617459 PMID: 24367388

Maiam nayri Wingara. (2018). Indigenous Data Sovereignty Communique. Indigenous Data Sovereignty Summit. 20th June 2018, Canberra: Australian Capital Territory. Retrieved on 11 October 2024 from https://static1.squarespace.com/static/5b3043afb40b9d20411f3512/t/5b6c0f9a0e2e725e9cabf4a6/1533808545167/Communique%2B%2BIndigenous%2BData%2BSovereignty%2BSummit.pdf

Makananise, F. O. (2024). Digital preservation of Indigenous languages and participatory epistemic knowledge systems of the Global South: insights from South African experiences. In Makananise, F. O., & Madima, S. E. (Eds.), *Digital media and the preservation of Indigenous Languages in Africa: toward a digitalized and sustainable society* (pp. 21–48). Lexington Books.

Manžuch, Z. (2017). Ethical Issues in digitization of cultural heritage. *Journal of Contemporary Archival Studies*, 4(4). https://elischolar.library.yale.edu/jcas/vol4/iss2/4

Martone, M. E. (2015). FORCE11: Building the future for research communications and e-scholarship. *Bioscience*, 65(7), 635. DOI: 10.1093/biosci/biv095

Masango, C. A., & Nyasse, B. (2015). Documenting Indigenous knowledge about Africa's traditional medicine: A myth or a reality? *Indilinga*, 14(1), 52–66.

Masenya, T. M. (2023). Revitalization and digital preservation of Indigenous knowledge systems for sustainable development of Indigenous communities in South Africa. *The Serials Librarian*, 84(5–8), 86–102. DOI: 10.1080/0361526X.2023.2277962

Masika, D. M., Anyona, G. O., & Kowenje, E. A. (2022). Indigenous knowledge and practices for sustainable water resources management: a case of Luo and Banyala in Kenya. In Kowenje, C. (Eds.), *From Traditional to Modern African Water Management*. Springer., DOI: 10.1007/978-3-031-09663-1_13

Materechera, S. A. (2021). Relevance of Indigenous knowledge systems in sustaining agriculture under smallholder farming in sub-Saharan Africa. In: Leal Filho, W., Pretorius, R., de Sousa, L.O. (eds) *Sustainable development in Africa*. World Sustainability Series. Springer. https://doi.org/DOI: 10.1007/978-3-030-74693-3_3

Maunganidze, L., & Halsall, J. (2016). A moral compass that slipped: Indigenous knowledge systems and rural development in Zimbabwe. *Cogent Social Sciences*, 2(1), 1266749. Advance online publication. DOI: 10.1080/23311886.2016.1266749

Mawere, J., Kugara, S. L., & Tshamano, W. (2022). Curriculum transformation in South Africa: An Indigenous knowledge systems perspective. *Journal of African Education*, 3(3), 11–27. https://hdl.handle.net/10520/ejc-aa_jae_v3_n3_a2. DOI: 10.31920/2633-2930/2022/v3n3a1

Mazzocchi, F. (2020). A deeper meaning of sustainability: Insights from Indigenous knowledge. *The Anthropocene Review*, 7(1), 77–93. DOI: 10.1177/2053019619898888

McNulty, N. (2013). The Ulwazi Programme: bridging the gap with digital skills and local content. Retrieved on 11 October 2024 from https://www.niallmcnulty.com/2013/06/the-ulwazi-programme-bridging-the-gap-with-digital-skills-and-local-content

McNulty, N. (2014). The Ulwazi Programme: a case study in community-focused Indigenous knowledge management. In Osei-Bryson, K. M., Mansingh, G., & Rao, L. (Eds.), *Knowledge management for development. Integrated Series in Information Systems* (Vol. 35). Springer., DOI: 10.1007/978-1-4899-7392-4_13

Mdhluli, T. D., Mokgoatšana, S., Kugara, S. L., & Vuma, L. (2021). Knowledge management: Preserving, managing and sharing Indigenous knowledge through digital library. *Hervormde Teologiese Studies*, 77(2), a6795. DOI: 10.4102/hts.v77i2.6795

Metis Centre. (2011). Principles of ethical métis research. National Aboriginal Health Organization. Retrieved on 11 October 2024 from https://ruor.uottawa.ca/bitstream/10393/30555/1/2011_04_ethics.pdf

Mittermeier, C. (2024). Free, prior and informed consent in context: respecting Indigenous peoples' rights is essential to supporting our mission of helping societies to responsibly and sustainably care for nature. Retrieved on 11 October 2024 from https://www.conservation.org/projects/free-prior-and-informed-consent-in-context

Mohan, A., Aswathi, K. R., & Chitra, K. P. (2021). Indigenous farming systems and global sustainability. In: Leal Filho, W., Marisa Azul, A., Brandli, L., Lange Salvia, A., Gökçin Özuyar, P., Wall, T. (eds) *Reduced inequalities. Encyclopedia of the UN Sustainable Development Goals*. Springer. https://doi.org/DOI: 10.1007/978-3-319-95882-8_129

Montenegro, M. (2023). Bridging histories through intercultural archiving. *Public History Weekly*, 11(2). Advance online publication. DOI: 10.1515/phw-2023-21185

Mukurtu. (2024). Welcome to Mukurtu CMS: a safe keeping place. Retrieved on 11 October 2024 from https://mukurtu.org/

Mzimela, J. H., & Moyo, I. (2024). Indigenous knowledge system and agricultural drought adaptation in the uMkhanyakude District Municipality, KwaZulu-Natal. *Journal of Asian and African Studies*, 0(0), 00219096241275391. Advance online publication. DOI: 10.1177/00219096241275391

National Humanities Alliance. (2024). Mukurtu CSM: an Indigenous archive and publishing tool. https://humanitiesforall.org/projects/mukurtu-an-Indigenous-archive-and-publishing-tool

Ndhlovu, P. T., Asong, J. A., Omotayo, A. O., Otang-Mbeng, W., & Aremu, A. O. (2023). Ethnobotanical survey of medicinal plants used by Indigenous knowledge holders to manage healthcare needs in children. *PLoS One*, 18(3), e0282113. DOI: 10.1371/journal.pone.0282113 PMID: 36972257

Ngulube, P. (2023). From Orality to Digital Assets: Managing Indigenous Knowledge in Africa in the Wake of the Open Science Movement. *The Serials Librarian*, 84(5–8), 71–85. DOI: 10.1080/0361526X.2023.2296880

Nicholas, G. (2021). Protecting Indigenous heritage objects, places, and values: Challenges, responses, and responsibilities. *International Journal of Heritage Studies*, 28(3), 400–422. DOI: 10.1080/13527258.2021.2009539

Nitah, S. (2021). Indigenous peoples proven to sustain biodiversity and address climate change: Now it's time to recognize and support this leadership. *One Earth*, 4(7), 907–909. DOI: 10.1016/j.oneear.2021.06.015

Obiero, K. O., Klemet-N'Guessan, S., Migeni, A. Z., & Achieng, A. O. (2023). Bridging Indigenous and non-Indigenous knowledge systems and practices for sustainable management of aquatic resources from East to West Africa. *Journal of Great Lakes Research*, 49(1), S128–S137. DOI: 10.1016/j.jglr.2022.12.001

Oguamanam, C. (2020). Indigenous Peoples, Data Sovereignty, and Self-Determination: Current Realities and Imperatives. *The African Journal of Information and Communication*, 26(26), 1–20. DOI: 10.23962/10539/30360

Olaare, S. (2024). The role of technology in language preservation. *European Journal of Linguistics*, 3(2), 44–56. DOI: 10.47941/ejl.2046

Open Data Charter. (2015). Open Data Charter Principles. Retrieved on 11 October 2024 from https://opendatacharter.net/principles/

Owiny, S. A., Mehta, K., & Maretzki, A. N. (2014). The use of social media technologies to create, preserve, and disseminate Indigenous knowledge and skills to communities in East Africa. *International Journal of Communication*, 8, 234–247.

Oyelude, A. A. (2023). Indigenous knowledge preservation as a sign of respect for culture: Concerns of libraries, archives and museums. *Insights: the UKSG Journal*, 36(1), 1–21. DOI: 10.1629/uksg.628

Perry, S. R. (2014). Digitization and digital preservation: A review of the literature. *School of Information Student Research Journal*, 4(1). Advance online publication. DOI: 10.31979/2575-2499.040104

Quang, T. D., Noseworthy, W. B., & Paulson, D. (2022). Rising tensions: Heritage-tourism development and the commodification of "Authentic" culture among the Cham community of Vietnam. *Cogent Social Sciences*, 8(1), 2116161. Advance online publication. DOI: 10.1080/23311886.2022.2116161

Raine, S. C., Kukutai, T., Walter, M., Figueroa-Rodrigues, O. L., & Walker, J.. (2019). Indigenous data sovereignty. In Davies, T., Walker, S., Rubinstein, M., & Perini, F. (Eds.), *State of open data: histories and horizons* (pp. 300–319). African Minds., DOI: 10.5281/zenodo.2677801

Redvers, N., Aubrey, P., Celidwen, Y., & Hill, K. (2023). Indigenous peoples: Traditional knowledges, climate change, and health. *PLOS Global Public Health*, 3(10), e0002474. DOI: 10.1371/journal.pgph.0002474 PMID: 37831713

Richards, M., Jones, C., Lardy, J., Godden, A., Godden, W., Bock, S., & Lardy, H. (2024). Warrma Mangarrayi: Co-designing an app for learning Mangarrayi, an Indigenous language of Northern Australia. *International Journal of Human-Computer Interaction*, 0(0), 1–18. DOI: 10.1080/10447318.2024.2381924

Robinson, C. J., Kong, T., Coates, R., Watson, I., Stokes, C., Pert, P., McConnell, A., & Chen, C. (2021). Caring for Indigenous data to evaluate the benefits of Indigenous environmental programs. *Environmental Management*, 68(2), 160–169. DOI: 10.1007/s00267-021-01485-8 PMID: 34046755

SAICH. (2024). Southern African Intangible Cultural Heritage Cooperation. Retrieved on 11 October 2024 from www.saich.org

Seehawer, M. (2018). South African science teachers' strategies for integrating Indigenous and Western knowledges in their classes: Practical lessons in decolonisation. *Educational Research for Social Change*, 7(special issue), 91–110. DOI: 10.17159/2221-4070/2018/v7i0a7

Shanahan, K. (2019). Mukurtu CMS: differential access for the ethical stewardship in cultural and digital heritage. Retrieved on 11 October 2024 from https://sustainableheritagenetwork.org/system/files/atoms/file/NAMSI-Mukurtu%20History%20and%20Intro.pdf

Shiri, A., Howard, D., & Farnel, S. (2021). Indigenous digital storytelling: Digital interfaces supporting cultural heritage preservation and access. *The International Information & Library Review*, 54(2), 93–114. DOI: 10.1080/10572317.2021.1946748

Sibandze, G. F., & Dludlu, M. N. (2022). Interfacing Indigenous knowledge with scientific knowledge for improved health outcomes: lessons from Eswatini. In Pullanikkatil, D., & Hughes, K. (Eds.), *Socio-ecological systems and decoloniality*. Springer., DOI: 10.1007/978-3-031-15097-5_3

Sioui, M. (2020). *Indigenous water and drought management in a changing world*. Elsevier.

Tagle, Y. R. (2009). The Protection of Indigenous knowledge related to biodiversity: the role of databases. Retrieved on 11 October 2024 from https://www.sylff.org/wp-content/uploads/2009/03/sylff_p131-146.pdf

Te Mana Raraunga. (2018). Principles of Māori Data Sovereignty Brief #1. Retrieved on 18 October 2024 from https://static1.squarespace.com/static/5d3799de845604000199cd24/t/5d6f940cc5442b00013e4dab/1567593484780/TMR%2BMa%CC%84ori%2BData%2BSovereignty%2BPrinciples%2BOct%2B2018.pdf

Tippeconnic, J. W., & Faircloth, S. C. (2010). *The education of Indigenous students*. International Encyclopedia of Education., DOI: 10.1016/B978-0-08-044894-7.00123-8

Tom, M. N., Sumida, H. E., & McCarty, T. L. (2019). Indigenous knowledges as vital contributions to sustainability. *International Review of Education*, 65(1), 1–18. DOI: 10.1007/s11159-019-09770-9

Tuyen, Q. D. (2023). Staging culture, selling authenticity: the commodification of the Cham community's traditions. In *Heritage conservation and tourism development at Cham sacred sites in Vietnam*. Springer., DOI: 10.1007/978-981-99-3350-1_8

Ulwazi programme. (2024). Indigenous knowledge important for innovation. Retrieved on 15 October 2024 from https://www.ulwaziprogramme.org

UNDP. (2024). Indigenous knowledge is crucial in the fight against climate change – here is why. Retrieved on 18 October 2024 from https://climatepromise.undp.org/news-and-stories/Indigenous-knowledge-crucial-fight-against-climate-change-heres-why#:~:text=Many%20Indigenous%20traditional%20practices%20offer,the%20rest%20of%20the%20world

United Nations. (2007). United Nations Declaration on the Rights of Indigenous Peoples. Retrieved on 16 October 2024 from https://www.un.org/development/desa/Indigenouspeoples/declaration-on-%20the-rights-of-Indigenous-peoples.html

United States Indigenous Data Sovereignty Network. (2020). Indigenous data governance. Retrieved on 17 October 2024 from https://www.policylink.org/sites/default/files/AdrianDominguez.pdf

Vijayan, D., Ludwig, D., Rybak, C., Kaechele, H., Hoffmann, H., Schonfeldt, H. C., Mbwana, H. A., Rivero, C. V., & Lohr, K. (2022). Indigenous knowledge in food system transformations. *Communications Earth & Environment*, 3(1), 213. DOI: 10.1038/s43247-022-00543-1

Vilakazi, B. S., Zengeni, R., & Mafongoya, P. (2022). Integrating Indigenous knowledge systems (IKS) with science-based strategies into climate variability mitigation and adaptation of farming communities in Semi-Arid Southern Africa. *Indilinga*, 21(1), 119–140.

Wagner, A., & de Clippele, M. S. (2023). Safeguarding cultural heritage in the digital era – a critical challenge. *International Journal for the Semiotics of Law*, 36(5), 1915–1923. DOI: 10.1007/s11196-023-10040-z

Warner, A. (2021). Google launches app for language preservation. Retrieved on 18 October 2024 from https://multilingual.com/google-launches-app-for-language-preservation/

Wheeler, H. C., & Root-Bernstein, M. (2020). Informing decision-making with Indigenous and local knowledge and science. *Journal of Applied Ecology*, 57(9), 1634–1643. DOI: 10.1111/1365-2664.13734

Wilkinson, M. D., Dumontier, M., Aalbersberg, I. J., Appleton, G., Axton, M., Baak, A., Blomberg, N., Boiten, J. W., da Silva Santos, L. B., Bourne, P. E., Bouwman, J., Brookes, A. J., Clark, T., Crosas, M., Dillo, I., Dumon, O., Edmunds, S., Evelo, C. T., Finkers, R., & Mons, B. (2016). The FAIR Guiding Principles for scientific data management and stewardship. *Scientific Data*, 3(1), 160018. DOI: 10.1038/sdata.2016.18 PMID: 26978244

Williamson, B., Provost, S., & Price, C. (2023). Operationalising Indigenous data sovereignty in environmental research and governance. *Environment and Planning. F, Philosophy, Theory, Models, Methods and Practice*, 2(1-2), 281–304. DOI: 10.1177/26349825221125496

WIPO. (2019). The Protection of traditional knowledge: draft articles facilitators' rev. Retrieved on 18 October 2024 from https://www.wipo.int/edocs/mdocs/tk/en/wipo_grtkf_ic_40/facilitators_text_on_tk.pdf

Yip, S. Y., & Chakma, U. (2024). The teaching of Indigenous knowledge and perspectives in initial teacher education: A scoping review of empirical studies. *Journal of Further and Higher Education*, 48(3), 287–300. DOI: 10.1080/0309877X.2024.2327029

Ziphora, M. K. (2017). *Integration of Indigenous knowledge systems in the curriculum for basic education: possible experiences of Canada*. University of South Africa., http://hdl.handle.net/10500/25096

Zuza, E. J., Lambert, K., Macmillan, T., Chiyemura, F., Araya, Y., Bowskill, V., Oluseye, A., Ng'endo Kanui, M., Keding, G., Butler, G., Breman, E., King, I. O., Matumba, L., Heddenhausen, J., Nordin, S., & Bhagwat, S. (2024). Crop species diversity: A key strategy for sustainable food system transformation and climate resilience. *Food and Energy Security*, 13(3), e558. DOI: 10.1002/fes3.558

Zvobgo, L., Johnston, P., Williams, P. A., Trisos, C. H., & Simpson, N. P. (2022). The role of Indigenous and local knowledge in water sector adaptation to climate change in Africa: A structured assessment. *Sustainability Science*, 17(5), 2077–2092. DOI: 10.1007/s11625-022-01118-x

ADDITIONAL READINGS

Moore, A. (2022). Data with principles: how to be FAIR, and why you should CARE. Retrieved on 18 October 2024 from https://www.lib.sfu.ca/help/publish/scholarly-publishing/radical-access/fair-and-care-data

Sterner, B., & Elliott, S. (2022). The FAIR and CARE data principles influence who counts as a participant in biodiversity science by governing the fitness-for-use of data. [Preprint]. Retrieved on 11 October 2024 from https://philsci-archive.pitt.edu/21039/1/FAIRandCAREData-April6-2022.pdf

Wang, M., & Savard, D. (2023). The FAIR principles and research data management. In: K. Thompson; E. Hill, E. Carlisle-Johnston, D. Dennie & É. Fortin, (eds.), *Research data management in the Canadian context: a guide for practitioners and learners* (pp. 31-33). Retrieved on 13 October 2024 from https://ecampusontario.pressbooks.pub/canadardm/

KEY TERMS AND DEFINITIONS

Accessible: Once data is found, it must be accessible under clear conditions, whether openly or under certain restrictions, with standardised protocols for access.

Authority to Control: Indigenous peoples have the inherent right to govern and control their data.

CARE Principles: A set of guidelines that prioritise Collective Benefit, Authority to Control, Responsibility, and Ethics in the management of Indigenous knowledge.

Collective Benefit: The use of Indigenous data should provide benefits to Indigenous peoples.

Culturally Appropriate Metadata: Metadata systems that reflect the cultural values and epistemologies of Indigenous communities.

Data Sovereignty: refers to the concept that data is subject to the laws and governance structures of the country or jurisdiction where it is collected, stored, or processed.

Data Governance: refers to the frameworks and practices that ensure Indigenous peoples have control over the collection, management, access, and use of data related to their cultural heritage, knowledge systems, and communities.

Digital Preservation: The process of maintaining and archiving digital information over time to ensure its accessibility and usability.

Ethics: Indigenous data governance should align with ethical practices that respect the values, protocols, and cultural heritage of Indigenous communities.

FAIR Principles: Guidelines designed to make data Findable, Accessible, Interoperable, and Reusable.

Findable: Data should be easily discoverable by humans and machines. This requires clear metadata, a unique and persistent identifier (e.g., DOIs), and data repositories with searchable capabilities.

Indigenous Data Sovereignty: the rights of Indigenous peoples to control and govern data about their communities, lands, resources, and knowledge.

Indigenous Knowledge (IK): Traditional knowledge developed by Indigenous communities over generations, encompassing cultural, ecological, and social practices.

Indigenous Peoples' Data: comprise of information and knowledge about the environment, lands, skies, resources, and non-humans with which they have relations; information about Indigenous persons such as administrative, census, health, social, commercial, and corporate and, information and knowledge about Indigenous Peoples as collectives, including traditional and cultural information, oral histories, ancestral and clan knowledge, cultural sites, stories, and belongings.

Indigenous Data Sovereignty: draws on the United Nations Declaration on the Rights of Indigenous Peoples (UNDRIP), which reaffirms the rights of Indigenous Peoples to control data about their peoples, lands, and resources.

Interoperable: Data must be in a format that can be integrated with other datasets and tools.

Responsibility: Those handling Indigenous data must act responsibly, ensuring that the data is used in ways that respect Indigenous rights, interests, and values.

Reusable: FAIR data aim to be reusable for future research.

Chapter 11
Promoting the Documentation and Preservation of African Indigenous Epistemologies for Posterity

Stanley Nil Murwira
Belvedere Technical Teachers College, Zimbabwe

ABSTRACT

Indigenous African epistemologies have contributed to the preservation and sustainable use of African heritage and resources for centuries. Indigenous African knowledge systems (IK) were transmitted orally by Indigenous knowledge-practising elders of the community who also served as IK custodians of the knowledge system. However, the marginalisation and displacement of African Indigenous epistemologies by Western knowledge systems have threatened their existence. This chapter argues that promoting the documentation and preservation of African Indigenous knowledge is crucial, not only for posterity but to protect it from exploitation, distortion and extinction. This chapter adopts the Afrocentric theoretical framework on African Indigenous epistemologies and calls for more research and documentation of African Indigenous knowledge systems to be conducted to promote the preservation and sustainable use of local heritage and resources. In addition, the rights to African Indigenous knowledge systems need to be protected by law, and patents should be given to local products.

INTRODUCTION

Indigenous African knowledge systems have existed in Africa since the beginning of human habitation on the continent. The local epistemologies have been the bedrock of human activities across different societies in Africa and have been transmitted orally from generation to generation within the African communities by members of specific communities in an interactive and active manner (Mkabela, 2005). It is important to note that most of the knowledge which has been generated over the centuries has remained in the private domain, as there was little need or demand for its documentation and preservation. African Indigenous knowledge is a trans-generational knowledge system deployed

DOI: 10.4018/979-8-3693-8447-3.ch011

for centuries by the Indigenous people of Africa. It is embedded within the boundaries of the people's culture and implemented with dexterity, wisdom, conservation and preservation, and sustainability and environmental harmony as its hallmarks (Ayeni & Aborisande, 2022). The dominance of Eurocentric or Western epistemologies, which have dominated the African landscape in the last century, originated with the commencement of colonialism in Africa.

African societies possess a rich heritage comprising their traditions, culture and practices. This heritage is embedded in both tangible and non-tangible aspects. However, not all information pertaining to these elements is documented or preserved for the benefit of future generations. The heritage encompasses both man-made and natural environments that support communities in their existential circumstances. Societies across the world depend on their Indigenous epistemology to address their day-to-day socioeconomic challenges, address numerous environmental problems and adapt to change (Mawere, 2015). Africa is not an exception in this regard as members of its communities continue to utilise local knowledge in the areas of medicine and agriculture for survival. Indigenous knowledge of the environment encompasses the intricate knowledge of the fauna, flora, social and physical environment, which is utilised for sustainable living in the community and an all-inclusive worldview that parallels the scientific discipline of ecology (Appiah-Opoku, 2005).

African Indigenous knowledge is situated within local communities and the inhabitants of particular localities are the custodians of these knowledge systems. The ideal scenario would be for this knowledge system to be preserved either in written hard copies or digital form. This preservation can provide future generations the means to access and benefit from research findings in areas such as medicine, agriculture, architecture and technology. The wealth of Indigenous knowledge that still subsists among the elders and other knowledge holders in local societies validates the vibrant intellectualism to which African researchers and intellectuals should turn (Kaya, 2013). Some African scholars express skepticism regarding the viability and relevance of the African Indigenous knowledge systems in addressing societal problems. African Studies scholars also disagree on the current state of African Indigenous knowledge and the role it plays in the lives of contemporary Africans. The central issue in this debate is not about the necessity or validity of Indigenous knowledge but rather whether Africans should continue to utilise Indigenous knowledge and the strategic role such knowledge can play in resolving Africa's problems (Heto & Mino, 2022). It can be argued that AIKs still have a role to play in the lives of the people on the African continent, as some rely on them for daily survival in terms of food, shelter and medicines. The purpose of this chapter is to explore methodologies for documenting, affirming or validating, and disseminating Indigenous epistemologies to African students for sustainability. The chapter focuses on Indigenous knowledge and African heritage in context and is organized into different segments or sub-headings detailing the transmission of Indigenous knowledge, world views of AIKs, the theoretical perspectives on AIKs, preservation and documentation of AIKS. Additionally, the chapter assesses the challenges in documenting AIKS and the rights to Indigenous knowledge.

INDIGENOUS KNOWLEDGE AND AFRICAN HERITAGE IN CONTEXT

Various scholars have proposed diverse definitions, interpretations, and understandings of Indigenous knowledge. Scholars conceptualise Indigenous knowledge (IK) from multiple viewpoints, such as traditional knowledge, local knowledge, cultural knowledge, farmer knowledge, elderly wisdom, traditional wisdom, native knowledge, and traditional science (Owolabi, Ovwasa, Ajayi & Odewale,

2022). Generally, Indigenous knowledge is associated with people of a particular locality and permeates various facets of human existence. Ndlovu (2023) posits that African Indigenous Knowledge (AIK), in particular, refers to an archive of experiences, cultural beliefs, values and norms of the people of African descent. In essence, African Indigenous knowledge possesses utility value and does not solely rely on generating knowledge from foreign ideas (Yip & Chakma, 2024). Indigenous knowledge (IK) also refers to the traditional knowledge systems, practices, and beliefs passed down from generation to generation within Indigenous communities. IK serves as a crucial source of information that informs how Indigenous communities interact with their environment, and it plays a significant role in promoting sustainable development and cultural diversity. In the African context, Indigenous knowledge originates from people's existential circumstances and local environments, supporting human existence sustainably. In this regard, Indigenous knowledge may be defined as the collective body of strategies, practices, procedures, tools, intellectual resources, explanations, beliefs, and values accumulated over time in a specific locality without the interference and impositions of exterior forces (Emeagwali, 2014). The Indigenous knowledge system is holistic as it encompasses various aspects of societal life extending beyond the material sphere to interconnects with spiritual and non-material realms of existence. In Africa, Indigenous epistemology constitutes a vital resource used by Africans to engage with their environment in nearly all human endeavours for subsistence (Olaopa & Ayadele, 2021). Indigenous epistemologies are central to the existence and sustenance of societies worldwide. Whether simple or complex, communities around the world possess unique knowledge systems specific to their contexts. Before the advent of neoliberal scientific knowledge and systems, Indigenous knowledge sustained communities in and around Africa (Khumalo & Baloyi, 2017). While international institutions and their agencies (working in alliance with local governments) acknowledge Indigenous knowledge systems, they tend to diminish the applicability of these knowledge systems to real-world situations. One approach to integrating local knowledge into development initiatives is to adopt a bottom-up strategy rather than a top-down approach and identify ideas within the body of Indigenous knowledge systems that should be incorporated into developmental agendas (Pophiwa & Saidi, 2000). Indigenous institutions of knowledge production, conservation and sharing, such as initiation schools, Indigenous games, agricultural systems, dances and songs, storytelling, proverbs, etc., continue to serve as pillars of Indigenous methods of knowledge acquisition (Kaya, 2013). Consequently, the sources and custodians of Indigenous knowledge reside within the community itself and are not external. Thus, AIKs can constitute an important component in the development agenda of local human activities.

Indigenous knowledge systems define African reality for the reason that Indigenous knowledge is a crucial resource Africans utilise to engage with each other and the environment for the mutual good of all, as well as their survival. Indigenous knowledge systems are deep-rooted in real-life experiences, which, in turn, define their worldview, relations and practices, generating a system far removed from the Euro-Western celebrated orthodox scientific systems (Asakitikpi, 2018). Higgs, Higgs and Venter (2003) point out that societies use Indigenous knowledge at the local level by societies as the foundation for decision-making pertaining to food security, human and animal health, education, natural resources management and other important aspects. The Indigenous knowledge system offers a totality of experiences to members of specific societies as it transcends a diversity of aspects which are prerequisites for the survival of society. Indigenous knowledge is also a key component of the social capital of the poor and it constitutes the main asset in their efforts to gain control of their own lives (Higgs, Higgs & Venter, 2003). According to Heto and Mino (2020), many individuals are increasingly turning to African Indigenous knowledge for inspiration on addressing Africa's intractable problems because they blame some of Africa's problems on the continued dominance of Western and Islamic knowledge systems. Indigenous knowledge is paramount to local communities, and this makes it imperative to avail this knowledge to a much broader spectrum of society.

African Indigenous knowledge has not been highly publicised and disseminated compared to Western knowledge. With Indigenous knowledge primarily shrouded in secrecy, the hope of it being disseminated to everybody and exploited for the public good is almost impossible Mdhluli et.al., 2021). The lack of publicity has affected Indigenous peoples' way of preserving, managing and sharing knowledge and it is argued that this has contributed to their knowledge being threatened with extinction. This could be affecting its uptake by people in Africa, who generally tend to make independent decisions when it comes to certain aspects of AIK, such as traditional medicine. Indigenous knowledge systems define African existence for the reason that Indigenous knowledge is a key resource Africans use to engage with each other and the environment for the common good of all, as well as their existence (Pophiwa & Saidi, 2022). The argument being advanced is that the knowledge should be preserved and disseminated into the public domain. AIKs should be documented and made easily accessible to people so that they can utilise the knowledge system to solve particular social and economic challenges emanating from their existential circumstances. The importance of Indigenous knowledge cannot be dismissed, particularly now that the world is facing many challenges that mainstream science cannot address (Khumalo & Baloyi, 2017). It implies that the African people also have their own African Indigenous Knowledge Systems (AIKs) that need to be acknowledged and recognised as they define how they understand their locality and challenges. In Africa, most AIK is conveyed in folktales and songs, passed on orally by the elders to the younger generation (Ndlovu, 2023). The challenge is that some of the valuable knowledge can be distorted or lost in the process of passing it orally from generation to generation.

Africa boasts a vast and varied heritage, encompassing both tangible and intangible elements. This rich cultural and natural legacy serves as the source of AIKs, which local communities draw upon to address their everyday challenges. Chirimuuta, Gughlanga and Bhukuvani (2012) point out that Zimbabwe, together with all other African nations, is endowed with Indigenous knowledge which permeates all facets of life, such as agriculture, climate issues, medicine, education and technology. African heritage encompasses languages, art, dance, music, religion, folklore, rituals and diverse cultural activities. Thus, African heritage transcends both tangible and non-tangible aspects of society. Intangible heritage is part and parcel of the realms of Indigenous knowledge, given that these values can only be expressed through local ways of knowing, such as traditions, customs and practices (Mawere, 2015). These may be referred to as the latent aspects of African heritage. The worthiness and importance of the AIKS are mostly acknowledged and appreciated by members of particular societies who utilise it for survival. Intangible heritage is a word used to describe aesthetic, spiritual, symbolic, and other social values that people in a given geographical location may associate with a specific place (Mawere, 2015). Intangible heritage can be Indigenous knowledge, which covers aspects of knowledge that are in the metaphysical realm, and different communities have different ways of exercising issues about spirituality and reverence to ancestral spirits. Such issues of spirituality require an insider's perspective to appreciate and understand its importance in the lives of local people.

It is crucial for local students and communities to be exposed to heritage education to help them appreciate their local heritage. Heritage education, a crucial component of the citizenship curriculum, is an educational approach that enables individuals to explore and understand heritage resources. This process allows students to gain insights into their cultural roots, tracing how past heritage has evolved throughout history. By connecting the past to the present, heritage education bridges temporal gaps in cultural understanding. Hossain and Ballardin (2021) observe that, for centuries, Indigenous peoples around the world have developed their own knowledge systems which relate to, but are not limited to, environmental management and nature conservation, and have accordingly adapted over time to various prevailing knowledge systems. The local heritage is a rich source of Indigenous knowledge which communities can rely on. Indigenous knowledge is dynamic and evolves alongside nature to address specific social changes and challenges in society.

Transmission of Indigenous Knowledge

African Indigenous knowledge emanates from local societies and as such, cannot be treated as a separate entity from society. African Indigenous knowledge has sustained the survival of societies for several centuries (Jain & Jibril, 2016). It informs and guides life in societies in terms of acceptable norms, values and beliefs. This means that Indigenous knowledge is culture-specific, vibrant and originates from age-old beliefs, traditions, experiences and observations. Traditional knowledge is at the core of Indigenous identity, culture, languages, heritage and livelihoods, and its transmission from one generation to the next must be protected, preserved and encouraged (United Nations, 2019). It is non-formal, orally transmitted and usually not documented. Indigenous knowledge falls under tacit knowledge, which is subjective knowledge that is problematic to put into words and can also be conveyed through action (Rasoreka, 2008). Knowledge is more of a lifestyle since one cannot separate society from knowledge. Hence, the two are indispensable. It is not codified, and it is not rules-based. Hence, it is challenging to transfer. No single person can claim sole proprietorship of the AIKS as all society members exploit the knowledge with access to it. Traditional African information and Indigenous knowledge are recorded through memory and are embedded in folklore, songs, stories, dances, proverbs, rituals, the local language, myths, beliefs, games, cultural values, agricultural knowledge, knowledge of local flora and fauna (Rasoreka, 2008). This forms a fundamental difference between Western and African epistemologies, as Western epistemologies are written down or documented, and people tend to view written documents as authentic and verifiable. This brings to the fore the need to record African Indigenous epistemologies for verifiability and for the benefit of future generations.

As alluded to, African Indigenous knowledge, historically, was not documented but transmitted orally, and this prompts the debate on its authenticity. The argument is on whether the oral transmission of Indigenous knowledge can be done without some valuable facts being distorted or lost. The most significant gathering exercise for African Indigenous knowledge is tradition, namely, the collective testimonies and recollections of the past, inherited from earlier generations and transmitted in various forms of oral testimony (Emeagwali, 2003). In this regard, the information is stored in the memory, and its authenticity depends on the ability of the individual to recall the facts which were passed on to him or her by Indigenous knowledge-practicing elders. According to Hikwa (2012), it has been argued and presumed that every developing society passes through a phase when oral tradition was the only storage, transmission and dissemination of information. Based on the analysis by Hikwa (2012), one must take note that the issue of information storage and dissemination has evolved due to technological changes in society. The world is not static but evolving, hence there is a necessity to preserve Indigenous knowledge in the form of documents stored in various forms, from hard copies to soft copies on different platforms. Ukwuoma (2015) notes that, certain aspects of Indigenous knowledge are oral, and are transmitted from generation to generation through word of mouth and rituals. African Indigenous knowledge is transmitted through a varied array of media that comprise religion, taboos, customs, riddles and children's games, proverbs and folk stories (Ndhlovu & Masuku, 2004). The dissemination of the African Indigenous knowledge is so diversified and addresses target audiences of varying ages from children to adults, depending on the nature of knowledge disseminated. It is imperative for the knowledge to be preserved in ways in which different age groups in society can access it, for example certain stories can be transmitted in cartoon form for young children.

The leading threat to African Indigenous knowledge pertains to methods of preservation which depends on or are sustained by elder members of the community (Mdhluli et al., 2021). Indigenous peoples' methods of preserving, managing and sharing knowledge are argued to have contributed to their knowledge being threatened with extinction because of its lagging behind time and lack of compatibility with present-day ways of disseminating knowledge. Because of this, Indigenous people's assets and ways of transmitting knowledge have become outdated and fall short of intergenerational preservation of knowledge (Mdhluli et al., 2021). Currently, in this contemporary society, the internet, digital satellite television, and social media technologies have changed the ways people communicate and access information and knowledge at anytime from anywhere they want. Social media applications provide great opportunities for librarians to create, manage, preserve, and share Indigenous knowledge (Jain & Jibril, 2016). African scholars and researchers should take advantage of the diversity in media applications to store the AIKSs for posterity.

DIFFERENT WORLDVIEWS ON AFRICAN INDIGENOUS KNOWLEDGE SYSTEMS

There are different world views as far as African Indigenous epistemologies are concerned. The most prominent or pronounced world views in terms of AIKSs are the Eurocentric or Western views versus the Afrocentric views on African Indigenous episteme. The Eurocentric or Western perspective tends to denigrate African Indigenous knowledge and treat it as not a worthwhile knowledge system. On the other hand, the Afrocentric perspective or worldview views the AIKSs as a worthwhile knowledge system which can be utilised to address Africa's socio-economic challenges.

Eurocentric or Western Views of AIKSs

The Western world has never accorded African Indigenous epistemologies and heritage the same importance accorded to their knowledge systems. The Western cosmology views African knowledge production as primitive, barbaric and descriptive (Afolabi, 2020). The marginalisation of Indigenous knowledge can also be accredited to prejudices and stereotypes brought about by colonialism (Khumalo & Baloyi, 2017). Colonial mentality still hovers all over the African continent as all knowledge and items which emanate from Europe and North America are treated with reverence and superiority by most Africans. The decolonization of the African Academy remains one of the prevalent challenges not only in terms of the curriculum, teaching strategies, and textbooks but also in terms of the democratization of epistemology and the regeneration, evaluation, and adaptation of old epistemologies to suit new post-colonial realities. Indigenous knowledge provides a beacon of light within the tunnel of Eurocentric dogma, misinformation, and untruths (Emeagwali, 2014). AIKSs were not embraced in schools, they were included in history textbooks, and they were not allowed into public domains. Rather, the public domain was reserved exclusively for knowledge, heritages, cultures, institutions, norms, and idiosyncrasies of Western society (Hoppers, 2021). The onus is upon African scholars to research, expose, preserve and advocate for the use of Indigenous knowledge in institutions of learning. Reliance on the modern Western knowledge system ultimately means that African countries, scholars, scientists and researchers wait for originations and knowledge generated in foreign localities (Khumalo & Baloyi, 2017). During colonialism, the gathering, creation, and dissemination of Indigenous knowledge by anthropologists and missionaries served as information for purposes of colonisation and, more generally, the promotion of colonizing agendas (Murove, 2018). The marginalisation and disempowerment can be attributed to various factors, including the growth and spread of Christianity and the adoption of colonial Western science and culture (Chiwanza, Musingafi & Mupa, 2013). In Africa, through formal colonialism and informal coloniality, Western knowledge system dominance has resulted in the debasement and near extinction of African knowledge systems (Afolabi, 2020). African researchers must be active in harvesting Indigenous knowledge in order to avoid its extinction and package it in ways which make it easily accessible to a wider audience.

Most communities in Africa still use Indigenous knowledge in their daily life. However, this is usually done in the private domain as this usually goes against foreign religious beliefs which now dominate the length and breadth of the African continent. However, by virtue of being utilised informally, Indigenous knowledge remains underutilized and side-lined. African scientists seem to care less about local Indigenous knowledge remedies and investigating or analysing African herbs for curing local diseases, among other issues (Khumalo & Baloyi, 2017). Western knowledge has been prioritised over traditional means of knowledge acquisition. Western relics, forms and values remain products of continuous and sustained domination of Africa and its intellect, especially in knowledge production, and are commonly justified and sustained by the petty African intellectual bourgeoisie (Afolabi, 2020). There is a need for mental emancipation and a wake-up call among the African policymakers so that they realize that not everything Western is good, and not everything African is inferior and evil. Therefore, the Western domination of knowledge and marginalization of African systems of knowledge continues to be an academic challenge that calls for a comprehensive evaluation, rigorous planning and watchful implementation of policies that ensure the acknowledgement and provisions of space for the local in the existing political, economic, cultural, and pedagogical domains (Osman, 2014). Efforts to ignore IKSs and paint it as primitive have been the foundation of some writings by some authors, thus leaving this unique knowledge to disappear.

Afrocentric Perspective

Afrocentricity is a paradigm that has at its core the understanding of the African identity as rooted, centred and located in the African culture in all facets: spiritual, social, political and economic (Owusu-Ansah & Mji, 2013). The Afrocentric perspective affords one the opportunity to view African knowledge epistemologies from the African cosmic view and avoiding western prejudices and biases. What is baffling is that Western or Eurocentric epistemologies still form the core of curriculum in most universities in Africa and AIKSs are at the periphery of most studies. The issue of Indigenous epistemologies needs to be approached from an African perspective or world view. To non-Africans, certain cultural or religious practices may be viewed as diabolic, too simplistic to inform societal life or not worthwhile. Indeed, the paradox of the universities in Africa is that they are situated in Africa, teach and do research in Africa but not much of what they teach, and research reflect the African context (Osman, 2014). Indigenous knowledge systems, including African epistemologies, have to date, not been allowed into public domains. Only the western episteme, which includes Eurocentric knowledge, heritages, cultures, institutions, norms and idiosyncrasies have been considered to be important in public institutions of higher learning (Higgs, Higgs & Venter, 2003). Therefore, African scholars should not shy away from Indigenous knowledge but should rather be at the forefront of campaigning for its preservation, dissemination and its use in the public domain.

African Indigenous Epistemologies and Globalisation

Globalisation as a phenomenon has brought both positive and negative aspects to African societies. African societies are now exposed to a Western episteme of life in comparison to their local knowledge. The process of globalisation is affecting all spheres of development and has eroded and undermined African Indigenous knowledge systems, including traditional modes of life that have been in existence over many centuries but have been able to survive even under the most relentless forms of suppression (Sifuna, 2008). Due to globalisation, there is an abundance of information emanating mostly from Western countries promoting a narrative which is contrary to African epistemologies. Global epistemology has been dominated by Eurocentrism and Western knowledge production paradigms and platforms (Afolabi, 2020). In Africa, the process of knowledge production has been muddled, supplanted, and ultimately subservient to the orthodox Western education forms and structures established by colonial authorities (Afolabi, 2020). The problem in Africa is that knowing, or knowledge is usually judged or assessed in terms of European or Western standards.

According to Chiwanza, Musangafi and Mupa (2013), the West considered Africa as a dark continent and hence despised its traditions, customs, belief systems and Indigenous knowledge systems as diabolic, barbaric and retrogressive. Indigenous knowledge is not accorded the same status as Western knowledge, which is perceived to be rational and scientific. The phenomenon of globalisation encompasses several entities and permeates various aspects such cultural heritage, economic, political, social and education. The repudiation of African Indigenous knowledge systems was the paradigm that define the curricula by missionaries and colonialists and apartheid authorities. Often dismissed as fetish, retrogressive, superstitious, and of no scientific validity, Africa's Indigenous knowledge systems were mostly ridiculed by the colonial administrations. Many Africans bought into the racial superiority rhetoric promoted through colonial education and became convinced that Western knowledge was the only knowledge that all humanity should strive to master. There was little critical consciousness of the implications of the imposition of an alien episteme on local communities possessing their own sophisticated Indigenous knowledge systems (Ezeanya-Esioba, Oguamanam & Ndungutse, 2021). A close analysis of the globalisation phenomenon reveals that society is constantly changing and gravitating towards Western lifestyles and knowledge due to the influx of a lot of sources of information and media from that sphere of the world. In this regard, the Afrocentric epistemologies are affected as well, as not much is shown and promoted on digital satellite is of benefit to Africa and worth knowing. This calls for effective creation of programmes which record and preserve the African knowledge systems without exhibiting any prejudices. Search engines, E-books, online courses and other virtual education provide people with the ability to learn about subjects that used to require a certain level of income or social positioning (Baxter, 2018). In this globalised world, knowledge is abundant and there is opportunity not only to receive information but to enter into academic spaces and contribute. Globalisation provides space for ideas to transcend borders and social strata (Baxter, 2018). African scholars need to exploit opportunities available due to globalisation to promote Indigenous epistemologies showing that they can contribute to the world of ideas and can co-exist with other world knowledge systems. Western globalisation of ideas and knowledge are more prevalent in the humanities and social sciences, particularly in the social construction of the individual and social realities of Africans. This is because Western ideas are culturally incongruent with African social realities (Afolabi, 2020). Africans should use the global media systems to preserve and promote AIKSs, and to make it available at the click of the button.

PRESERVATION OF AFRICAN INDIGENOUS EPISTEMOLOGIES

The importance of preserving Indigenous epistemologies is crucial in maintaining Africa grounded in its history and benefiting from abundant heritage and knowledge. Preservation of Indigenous knowledge is critical, because it ensures the perpetuation of the community and its knowledge (Owiny, Mehta & Maretzki;2014). Indigenous knowledge systems (IKSs) are mostly transmitted informally in societies (Maohi, 2012). It is imperative to formalise Indigenous knowledge systems and allow the preservation and dissemination of the knowledge to be governed by government policies and pronouncements. In post-colonial Africa, the procedure of establishing new institutions of higher learning and research centers has not stopped but of concern is the fact that the new institutions are dominated by the Western knowledge systems even though the library and literature of Indigenous African knowledge has started to flourish and draw attention and participation of some African academics (Osman, 2014). Ngara (2007) is of the view that certain contributions created by Africa and her inhabitants to history and civilization are conspicuously missing from textbooks for formal education, and not many people are aware of its existence. AIKSs should be documented and included in the prescribed books used in formal learning institutions. The fundamental objective of preserving Indigenous knowledge in the public library is to generate awareness among individuals and host communities and increase its utilization (Owolabi, Ovwasa, Ajayi & Odewale, 2022). Information professionals should ensure the longevity of the documented IK by devising preservation strategies (Ngulube, 2002). There is a need to use preservation methods which ensure that the information is stored and readily accessible by future generations.

Use of Technology in Indigenous Knowledge Preservation

Information storage and preservation has evolved throughout the history of mankind, from the papyrus scrolls, then written books, recordings and currently the digital mode. Libraries are using social software technology and Indigenous knowledge is preserved through establishing a community web portal using web 2.0 technology. A memory database is embedded in the portal as a wiki allowing collaborative writing and sharing of content (Jain & Jibril, 2016). The use of digital technology will ensure that knowledge is stored in its original form and is not prone to distortion or extinction. In this modern era, social media such as Instagram, YouTube, Facebook, Google docs, and Twitter can be used to produce, record, access, and share information or skills within the global community, thus they can reach a wider audience. The use of the wide spectrum of social media do not require expert skills or training but just basic literacy. However, some social media technologies require reading and writing abilities, so they are accessible only to literate individuals and are, therefore, limited to urban and educated rural populations (Owiny Mehta & Marekzki, 2014). Platforms such as YouTube allows users to upload, share, and view videos. It is possible for one to record worthwhile knowledge on human activities and upload the information on social platforms for the greater benefit of members of society. Furthermore, it was revealed that millions of records and information objects have been digitised at various institutions, but these items largely remain inaccessible because of a lack of ICT infrastructure, financial and human resources and policies (Owiny Mehta & Maretzki, 2014). One must be cognizant of the realities in most African communities where internet connectivity and access to electricity are still a pipe dream, leading to inaccessibility to information. Such inaccessibility to information assets has implications for education, healthcare, agriculture, industrial innovation and development, environmental sustainability and disaster response in Africa (Chowdhury et al., 2023).

Several externally and internally funded initiatives have taken place for building institutional repositories of digital content and digitising content held in memory institutions such as national libraries and national archives in different African countries over the past few years. However, lack of appropriate ICT infrastructure, metadata standards, policies, human resources, and user education make such content currently inaccessible to the public (Chowdhury et al., 2023). The situation can be referred to as a "tragedy for Africa" in the sense that, in this digital and technologically advanced age, we still have many people who have no access to vital ICT gadgets or internet connectivity and, thus, deprived of vital information which they use in solving their social and economic challenges.

Libraries and Knowledge Preservation

Although people can now easily access information via the Internet from wherever there are, libraries still play an important role in information preservation and dissemination in society. Overall, libraries act as mediators between the digital world and Indigenous communities, fostering equitable access to knowledge while respecting cultural values and ensuring that Indigenous knowledge remains a living, relevant and cherished resource for present and future generations (Chigwada & Ngulube, 2024). In recognising the importance of Indigenous knowledge, public libraries are particularly playing important roles in management and preservation of Indigenous knowledge all over the world including Botswana public libraries (Jain and Jibril, 2016). The lack of skilled librarians, especially those with the knowledge and willingness to incorporate oral culture in collections and services to rural communities, creates knowledge gaps and further limits the reach of library services in rural Africa (Owiny, Mehta & Maretzki, 2014). As in other countries, Botswana's Indigenous knowledge is derived from the cultural life of the local people and their experiences that have been passed from generation to generation (Jain & Jabril, 2016). The knowledge which is being passed orally should be transformed to the written word to preserve it in its original state.

Digital Preservation of Knowledge

Digitalisation is now a topical issue in the context of the globalized world. Global phenomena such as the introduction of the internet, information exchange, digitalisation, virtualisation and social media have made the digital transformation strategy inevitable for universities to stimulate student experience, particularly the delivery of education, meaning, the increasingly digital transformation process has become a dependable platform to design, develop and retain competitive advantages (Mohamed, Tlemsani & Mathews, 2021). Digitalization can be utilised in the documenting and preservation of knowledge. Acquiring, preserving and disseminating Indigenous knowledge in the digital era comes with various challenges for libraries. Addressing these challenges requires a collaborative and community-oriented approach, with a strong emphasis on building trust, respecting cultural protocols, and empowering Indigenous communities to play an active role in the preservation and dissemination of their knowledge (Chigwada & Ngulube, 2024).

Documentation of African Indigenous Knowledge

The main task facing Africa is the creation of sustainable strategies to preserve and disseminate Indigenous knowledge (Owiny, Mehta & Maretzki, 2014). Academics in institutions of higher learning and specialists in information technology should work together to document, record and preserve African Indigenous knowledge. Universities in Africa must critically analyze the epistemology they disseminate to students in a bid to adopt knowledge plausible to societal development (Murwira, 2020). The universities should strive to shrug off the Eurocentric tag and pursue research and teachings based on Indigenous African knowledge systems. African intellectuals should help Africa narrow the gap created over four hundred years of domination and marginalisation of Africa's Indigenous knowledge systems. This can be achieved by rejecting the utilisation of the dominant Western worldview of knowing and knowledge production as the only way of knowing (Kaya & Seleti, 2013). Proper documentation of Indigenous knowledge produces a written cultural heritage that can be passed on from generation to generation. The Indigenous knowledge documentation should show great promise for a better and more sustainable future (Chiwanza, Musingafi & Mupa, 2013). The onus of any African Indigenous knowledge centre is to extensively document before it is too late, a variety of accumulated experiences from African regions far and near, obscure or obvious (Emeagweli, 2003). Countries in Africa should have distinct centres and programmes dedicated to preserving AIKSs.

The documentation and preservation of AIKSs should not be a mere event but a process which ensures that the valuable is preserved and, at the same time, utilised in the development agenda of society. In shaping the future development of Africa, it must be acknowledged that for the continent to meet its developmental challenges, particularly the eradication of poverty, it is important to integrate Indigenous knowledge in the development process (Mohamedbhai, 2013). Throughout history, African Indigenous knowledge systems were often branded and dismissed as archaic and 'primitive' because of the slow pace Indigenous epistemologies conform to change (Mdhluli et.al., 2021). The digital revolution is now topical in our societies and the world over due to the emergence of new technologies which facilitate the speedy dissemination of information. In this context, the digitisation of African Indigenous knowledge has become increasingly necessary to prevent the erosion of authentic African heritage (Akinwale, 2017). Consequently, there is a need for academics and librarians to be encouraged to use these modern methods of preserving information. Such digitisation can promote effective management of natural resources in Africa and globally. The concept of digitisation is connected with an interest in the management of cultural heritage in the technological environment (Akinwale, 2017). It is imperative for the preservation and management of African Indigenous knowledge to conform to the digital revolution to attract usage by students who are information technology competent and compliant. Information can be stored using methods such as cloud storage, Google Drive, hard disk drives, memory cards, solid-state drives, and flash drives. Multi-storage devices may be utilised so as to provide a backup of information and avoid loss of information.

The preservation of the African epistemologies can potentially be done successfully. People's IKSs globally need to be recognised and not inferiorised and censored because it is about the unique ways in which groups of people understand concepts and interpret meanings related to their experiences and values (Ndlovu, 2023). Societies around the world are unique in terms of their heritage and culture. Hence, it is important to study and gain insight into those societies before condemning them. Preservation of IKSs is essential to avoid the loss and extinction of certain aspects of the knowledge system due to an influx of foreign teachings and doctrines which condemn everything African as inferior and not worth to be recognised. This implies that preservation helps in long time storage of valuable information on African epistemology and heritage. There are various methods in which Indigenous knowledge can be preserved for posterity in public libraries, which include restoration, medial reprography, preventative preservation, and pre-archival intervention. According to Ngulube (2002), it is evident that if IK is not recorded and preserved, it will be lost and remain inaccessible to other knowledge systems as well as to development workers. Preservation of the AIK will ensure that it will be available to researchers and students of diverse interests around the world.

Significance of Recording and Documenting African Indigenous Knowledge

AIKSs are quite diversified, and they are utilised by African societies to sustain a variety of human activities they partake in. For example, there is much Indigenous knowledge which can be valuable in the field of health sciences if preserved and verified by relevant practitioners in the field. Through Indigenous knowledge, local people are creatively involved in traditional medical practices such as the use of various herbs to cure different ailments, heat therapy, quarantine, incisions, bone setting and surgery to provide for their medical needs (Olaopa & Ayodele, 2021). Oral traditions, songs, chants and storytelling can be audio-recorded to capture the nuances of spoken languages and preserve the richness of oral traditions (Chigwada & Ngulube, 2024). Video recordings can be used to capture cultural rituals, ceremonies, traditional performances and traditional knowledge transmission between generations Traditional medicine should be accorded a place alongside Western medicine. The current pursuit of Indigenous knowledge systems development in Africa is aimed to not only counter this trend in neo-colonial terms but to produce knowledge that is relevant to the cultural contexts in which it is generated and disseminated (Murove, 2018). Using local medicines in Africa may ameliorate the shortage of vital life serving medicines in communities or hospitals. Local inhabitants of different localities should be exposed to their local heritage and how they benefit from the fauna and flora in their environment.

Academics should advocate and emphasise the documentation of AIKSs and their introduction to students in the education system. The inclusion of Indigenous African knowledge systems in the education system empowers students to be aware of and acknowledge the existence of diverse ways of knowing and not to depend on Western forms of knowledge as the only legitimate benchmark of knowing (Kaya & Seleti, 2013). In reality, students in African learning institutions learn more Western inspired knowledge than information emanating from their local environment. The disappointing fact is that the students will not use most of the Western information in life after school. The knowledge becomes an end in itself. There are deliberate efforts to eliminate contributions made through African Indigenous knowledge perspectives from documented sources because most documented sources originating from the West are not written with the framework of Africa in mind (Ngara, 2007). African scholars must be encouraged and motivated to write and document information from their perspective, that is an insider's perspective. Since Indigenous knowledge is usually verbal and not documented, and that it is society-centered and usually not easily measurable, it has been perceived by many as not worthwhile and not commensurate with ways of contemporary science (Emeagwali, 2003). Indigenous knowledge may assist recognise cost-effective and sustainable mechanisms for poverty mitigation that are locally manageable and meaningful (Chiwanza, Mupa & Musingafi, 2013). Promoting the utilization of African Indigenous knowledge and innovation (AIK&I), therefore, has great potential for solving the development challenges exemplified by unemployment, balance of payment problem, climate change, environmental degradation, poor resource management, hunger, disease, among other problems facing most countries in Africa (Olaopa and Ayodele,2021). The application of AIKSs in the different spheres of life entails the use of the local knowledge system in addressing different societal challenges. In this regard, it is important for this knowledge system to be recorded and documented for use by different members of society.

As Africans, it is imperative to believe in African Indigenous knowledge systems as part of the solution to problems bedeviling the African continent such as hunger, poverty and diseases. This can be done through the identification of innovative pathways to sustainable human development that enhance local communities and their environments. Indigenous knowledge is part of a global knowledge system (Ukwuoma, 2016) and proper documentation of Indigenous knowledge can produce a written cultural heritage that can be passed on from generation to generation. Local knowledge preservation may create an improved and more sustainable future (Mdhluli, Mokgoatsana, Kugara & Vuma, 2011). Documentation of the Indigenous knowledge can allow people to share the knowledge from an informed point of view, rather than speculate and share ignorance on different aspects of AIKSs. Indigenous knowledge systems discourse recognises that knowledge intellectualisation, and knowledge-production, should not be divorced from contextual, nor cultural inclinations or experiences (Murove, 2018). Preservation of IK is vital to various communities because this will make the unborn generations know about the people's history, culture, and norm. If there is no proper preservation of Indigenous knowledge, many societies will lose a significant value of their ways of life and cultural heritage (Owolabi, Ovwasa, Ajayi & Odewale, 2022). There is a call for preservation of IKSs as this helps to preserve the history and cultural heritage of African people and at the same time inform their development trajectory into the future. Thus, the importance of IK and its value for example, towards keeping the environment for securing herbal and medicinal plants for poor communities is critical for rural development (Maunganidze, 2016). This points to a very key paradigm in African traditional research which, despite not having laboratories and modern scientific tools and machinery, was able to generate, test and apply knowledge. The African continent was able to deal with a range of common disasters such as snake bites, diseases and sickness, hunger and able to construct houses, palaces, bridges prior to the advent of scientific knowledge (Khumalo & Baloyi, 2017). There was knowledge and civilisation that flourished on the content before the dawn of colonialism and Africa was not a dark continent as typically said by westerners. Before the advent of modern hospitals, Africa dealt with infertility, barrenness, and other misfortunes. Researchers, therefore, believe that investing more in researching and developing IKSs will largely lead to discoveries and solutions that can alleviate problems faced by Africans and other races and people (Khumalo & Baloyi, 2017). More research, recording and documenting of African Indigenous knowledge should be done in order to promote its practical use in both urban and rural populations.

Challenges of Documenting and Preserving Indigenous Knowledge

A myriad of internal and external challenges militates against the harvesting and preservation of Indigenous knowledge on the African continent. Of major concern is the issue of funding in researching, documenting and preserving AIKS. Investment in Indigenous knowledge research is a basis for development and knowledge production in Africa. Without investments, adequate financial support, research, experiments, tests and equipment needed, such activities are impossible (Khumalo & Baloyi, 2017). A variety of methods are used in preserving Indigenous knowledge, namely, documentation, digitization, video recording, and providing Internet access. These methods require meticulous skills, equipment and financial resources for success to be achieved. Documenting is the most used method of Indigenous knowledge preservation (Jain & Jibril, 2016).

As discussed earlier, Indigenous knowledge is always passed by word of mouth from one generation to another (Mohomedbhai, 2013). Many of the bearers or custodians of Indigenous knowledge are from the older generation and now find it difficult to communicate their beliefs and practices to the scientifically educated young generation, once the older generation passes away, the knowledge disappears with them. There may be impediments such as language barriers and interpretations during the documentation of oral evidence from the older members of society. Consequently, because of this marginalisation, IK was not documented resulting in it not being viewed or recognized (Mdhluli et. Al, 2021). Another major challenge faced in the quest to co-opt IK into developmental projects is the marginalisation of Indigenous knowledge by African scholars, governments, scientists, entrepreneurs and the learned (Khumalo & Baloyi, 2017). It is imperative therefore that, without delay, IKSs in Africa be protected, documented, studied, modified and then widely disseminated to promote development (Mohomedbhai, 2013). The protection and documentation of AIK requires the consented effort of the whole continent in recognising the knowledge system as a viable alternative to existing Western knowledge systems.

There is a need for a more rigorous search for information on African Indigenous knowledge so as to authenticate and document it. Emeagwali (2014) points out that African manuscripts, in museums in other parts of Europe such as France, Germany, and Belgium, continue to be repositories of African primary sources of knowledge. While systematic documentation captures and preserves orally transmitted knowledge for present and future generations, it exposes Indigenous communities to the risk of losing their intellectual property rights through piracy and commercial exploitation (Chiwanza, Musingafi & Mupa, 2013). According to Sananayake (2006), Indigenous knowledge is orally-transmitted, or transmitted through imitation and demonstration. Writing down African Indigenous knowledge can change some of its fundamental properties however, on the other it can create permanent easy to reference knowledge. African Indigenous epistemology is rooted in a particular place and set of experiences, and generated by people living in those places. The result of this is that transferring the Indigenous knowledge to other places runs the risk of dis-locating it. African Indigenous epistemologies face many challenges, including the lack of legal frameworks at the national and international levels to support the library efforts (Khumalo & Baloyi, 2017). National governments must come up with policies which will make it possible to regulate the harvesting and preservation of IKSs by libraries and other related institutions. Like all other revolutionary changes, digital transformation involves intense adjustment and re-adjustment (Hashim, Lemsan & Mathews 2021). In essence, there is a need to adopt and use digital technology in the preservation of African Indigenous technology for posterity.

Indigenous African Epistemologies and the Issue of Rights

The main characteristics of Indigenous knowledge systems are that it is oral, tacit and invisible. It is sustained through and protected by linguistic diversity and community ownership (Rasoreka, 2008). However, in this globalised environment, individuals are supposed to have patents and copyrights to the information and items they produce. African Indigenous epistemologies are ubiquitous on the African continent, utilised by all community members and the ownership of the knowledge is difficult to determine (Baum, 2001). Another challenge bedeviling the integration of IK into African development goals is the lack of a comprehensive IK legal framework (Khumalo & Baloyi, 2017). The knowledge is exploited by both locals and foreigners without any acknowledgements or compensation. This is quite prevalent in areas such as the arts, culture, medicine, songs and collection of artefacts. The protection of traditional knowledge also faces a challenge when it comes to patenting intellectual property rights. Most AIKSs are characteristically in the public domain, and no particular individual can claim original rights to knowledge. Foreign researchers can appropriate traditional knowledge and apply for a patent, claiming to have invented a new product since traditional knowledge has attributes of communal ownership and sometimes has spiritual significance (Chiwanza, Musingafi & Mupa, 2013). Many Indigenous communities are concerned about the appropriation of their knowledge by researchers within and outside Africa, without permission or respect for customary law and with little benefit to them (Mohamedbhai, 2013). Recording, digitising and giving patent rights to inventions and discoveries can go a long way in protecting Indigenous epistemologies. However, real thorns come from the fact that IKSs are oftentimes appropriated, adapted, and patented by scientists and industry with little or no compensation to the custodians of this knowledge, without their prior informed consent, and without any consideration for benefit sharing (Hoppers, 2021). Protecting the appropriation of AIKSs by people who are not indigenes of particular localities must be regulated. A case in point is the variety of African artifacts abound in museums across Europe which are not benefiting communities in Africa. In spite of the fact that sharing is the main means of disseminating Indigenous knowledge, there is IK that is unique to certain individuals, although they use that knowledge for the benefit of the whole community. Intellectual property rights should be upheld so that Indigenous communities can benefit from the commercial use of their traditional knowledge (Ngulube, 2002). A comprehensive IK legal framework can be one of the pillars for research and application of IK in the African continent (Khumalo & Baloyi, 2017). Intellectual property rights are a means for obtaining ownership of specific assets that are elusive. It, for the most part, includes the security of some type of innovation made by the human mind (Mdhluli et al., 2021). Preservation and giving patent rights to IK is an acceptable way for protecting the IK system from piracy and other forms of abuse. This indicates that the preservation of IK allows the communities to access the intellectual property rights of their artwork (Owolabi, Ovwasa, Ajayi & Odewale, 2022). It is important for African countries to network and come up with conventions that guide and protect African heritage and knowledge.

FURTHER RESEARCH DIRECTIONS

This chapter mainly focused on preserving African Indigenous knowledge; however, the chapter is not exhaustive. There is a need for further research on how to authenticate information gathered orally. More research needs to be done on appropriate methods of gathering information on tangible and non-tangible heritage and preserving it for posterity. Issues of innovation and technology in AIKS preservation are also areas of interest recommended for further research. By researching the areas recommended, future researchers can build upon the foundations laid in this chapter and advance our understanding of promoting the documentation and preservation of African Indigenous epistemologies for posterity.

CONCLUSION

African Indigenous knowledge is abounding in local communities and permeates all facets of human life in the communities. It guides human activities in agriculture, religious and cultural ceremonies, resource conservation, knowledge transmission systems, building, health, and a variety of other activities. The main challenge with AIKS is that it is mainly transmitted verbally, and elders of society are mostly active in practicing the knowledge system. The main concern is the need to come up with strategies to record, document, and preserve AIKSs so that they do not get distorted or extinct. Different forms, such as digital texts, audio recordings, video recordings, photographs, digital artefacts and objects, interactive multimedia, digital maps, digital archives and repositories, and digital exhibitions and online collections, can be used in the documentation process of Indigenous knowledge systems (Dare, 2022) A critical analysis of the chapter shows that, although some steps have been taken to preserve Indigenous knowledge within the African communities, much more still needs to be done in that regard. It is recommended that AIKSs be preserved through recording and documenting using technologically available methods and devices for the benefit of future generations. To reach a wider and younger audience, there is a need to exploit the opportunities presented on the internet and social media. There is also a need to promote the issue of rights and patents in the preservation of AIKs for posterity so as to prevent distortion and exploitation by foreigners.

REFERENCES

Afolabi, O. S. (2020). Globalisation, decoloniality and the question of knowledge production *HEA/RESA,* 18(1), 93-109.

Akinwale, A. A. (2017). Digitalization of Indigenous knowledge for natural resource management in Africa. *AERN SUMMIT 2012 Conference Paper.* Retrieved on 29 October 2024 from https://api-ir.unilag.edu.ng/server/api/core/bitstreams/f0e2186a-f930-471e-9d01-90056f1d17d9/content

Appiah-Opoku, S. (2005). *The need for environmental impact assessment: the case of Ghana.* The Edwin Mellen Press.

Asakitikpi, A. (2018). Western versus African medical knowledge systems: A comparative epistemological. *Analysis Synthesis Philosophica*, 33(1), 91–104. DOI: 10.21464/sp33106

Ayeni, A. O., & Aborisade, A. G. (2022). African Indigenous knowledge systems and the world. In Oloruntoba, S. O., & Falola, T. (Eds.), *The Palgrave handbook of Africa and the changing global order.* Palgrave Macmillan. DOI: 10.1007/978-3-030-77481-3_8

Baxter, B. (2018). The impact of globalization on communication & education. HIGHVIBE Network. Open Access. Retrieved on 29 October 2024 from https//medium.com.highgate.

Chigwada, J., & Ngulube, P. (2024). Librarians' role in the presentation and dissemination of Indigenous knowledge. *IFLA Journal*, 50(2), 242–256. DOI: 10.1177/03400352231217270

Chirimuuta, C., Gudhlanga, E., & Bhukuvani, C. (2012). Indigenous knowledge systems: A panacea in education for development? The place and role of Indigenous knowledge systems in curriculum innovations in Zimbabwe. *International Journal of Social Science Tomorrow*, 4(5), 131–138.

Chiwanza, K., Musingafi, M. C. C., & Mupa, P. (2013). Challenges in preserving Indigenous knowledge systems: Learning from the past experiences. *Information and Knowledge Management*, 2(23), 19–25.

Chowdhury, G., McLeod, J., Lihoma, P., Teferra, S., & Wato, R. (2023). Promoting access to Indigenous information in Africa: Challenges and requirements. *Information Development*, 39(3), 611–623. DOI: 10.1177/02666669211048488

Dare, K. O. (2022). Is local knowledge peripheral? The future of Indigenous knowledge in research and development. *Alternative*, 18(1), 132–140. DOI: 10.1177/11771801221088667

Emeagwali, G. (2003). African Indigenous knowledge systems (AIK): implications for the curriculum. In Falola, T. (Ed.), *Ghana in Africa and the world: essays in honour of Adu Boahen*. Africa World Press.

Emeagwali, G., & Dei, G. J. S. (eds.). (2014). *African Indigenous knowledge and the disciplines*. Sense Publishers.

Ezeanya-Esioba, C., Oguamanam, C., & Ndungutse, V. (2021). Marginalisation of Indigenous knowledge in Africa: the case of Rwandan traditional medicinal treatment for livestock. Working paper 24. Open AIR (African Innovation Research) Network. Retrieved on 29 October 2024 from https://www.oguamanam.com/publications/2021-2-18/marginalisation-of-indigenous-knowledge-in-african-education-the-case-of-rwandan-traditional-medicinal-treatments-for-livestock

Heto, P. P. K., & Mino, T. (2023). (Dis)continuity of African Indigenous knowledge. *Alternative*, 19(1), 71–79. DOI: 10.1177/11771801221138304

Higgs, P., Higgs, L. G., & Venter, E. (2003). Indigenous African knowledge systems and. innovation in higher education in South Africa. *SAJHE/SATHO,* 17(2), 40-45.

Hikwa, L. (2012). *Indigenous languages as a preserver of Indigenous knowledge in Zimbabwe*. NuSpace Institutional Repository. NUST.

Hoppers, C. O. (2021). Research on Indigenous knowledge systems: The search for cognitive justice. *International Journal of Lifelong Education*, 40(4), 1–18.

Hossain, K., & Ballardina, R. M. (2021). Protecting Indigenous traditional knowledge through a holistic principle-based approach. *Nordic Journal of Human Rights*, 39(1), 57–57. DOI: 10.1080/18918131.2021.1947449

Jain, P., & Jibril, L. (2016). *Expanding library services for Indigenous community posterity: a case of selected public libraries in Botswana*. IFLA.

Kaya, O. H. (2013). Integration of African Indigenous knowledge systems into higher education in South Africa: Prospects and challenges. *Alternation (Durban)*, 20(1), 135–153.

Kaya, O. H., & Seleti, Y. N. (2013). African Indigenous knowledge systems and relevance of higher education in South Africa. *The International Educational Journal: Comparative Perspectives*, 12(1), 30–44.

Khumalo, B. N., & Baloyi, C. (2017). African Indigenous knowledge: an underutilised and neglected resource for development. *Library Philosophy and Practice (e-journal)*. 1663. https://digitalcommons.unl.edu/libphilprac/1663

Maunganidze, L. (2016). A moral compass that slipped: Indigenous knowledge systems and rural development in Zimbabwe. *Cogent Social Sciences*, 2(1), 1–12. DOI: 10.1080/23311886.2016.1266749

Mawere, M. (2015). Indigenous knowledge and public education in Sub-Saharan Africa. *Africa Spectrum*, 50(2), 57–71. DOI: 10.1177/000203971505000203

Mdhluli, D. T., Mokgaotsana, S., Kugara, S. L., & Vuma, L. (2021). Knowledge management: Preserving, managing and sharing Indigenous knowledge through digital library. *Hervormde Teologiese Studies*, 77(2), a6795. DOI: 10.4102/hts.v77i2.6795

Mkabela, Q. (2005). Using the Afrocentric method in researching Indigenous African culture. *The Qualitative Report*, 10(1), 178–179.

Mkabela, Q. (2005). Using the Afrocentric method in researching Indigenous African culture. *The Qualitative Report*, 10(1), 178–179.

Mohamed, A. M. H., Tlemsani, I., & Mathews, R. (2021). Higher education strategy in digital transformation. *Education and Information Technologies*, 27(3), 3171–3195. DOI: 10.1007/s10639-021-10739-1 PMID: 34539217

Mohamedbhai, G. (2013). Indigenous knowledge must be harvested for development. *University World News*, 262. Retrieved on 30 October 2024 from https://www.universityworldnews.com/

Murove, F. M. (2018). Indigenous knowledge discourse and inclusivity: An Afrocentric quest for recognition in a globalised world. *Journal for the Study of Religion*, 31(1), 159–176. DOI: 10.17159/2413-3027/2018/v31n1a9

Murwira, S. (2020). Indigenous African knowledge systems and teaching and learning at the Catholic University of Zimbabwe: critical investigation.(Doctoral Dissertation, UNISA, Pretoria). Retrieved on 1 November 2024 from http://hdl.handle.net/10500/26722

Ndhlovu, F., & Masuku, J. (2004). *Mainstreaming African indigenous knowledge systems in Higher and Tertiary Education: The Case of Zimbabwe*. UNISA Press.

Ndlovu, S. (2023). Indigenous knowledge systems in Africa: Exclusion of students with disabilities in South African higher education. *Curriculum Perspectives*, 43(1), 127–140. DOI: 10.1007/s41297-023-00202-1

Ngara, C. (2007). African ways of knowing and pedagogy revisited. *Journal of Contemporary Issues in Education*, 2(2), 7–20.

Ngulube, P. (2002). Managing and preserving indigenous knowledge in the knowledge management era: Challenges and opportunities. *Journal of information. Development*, 28(2), 107–116.

Oguamanam, C. (2023). A critical examination of the African legal framework for Indigenous knowledge. *Journal of African Law*, 16(1), 34–57. DOI: 10.1017/S0021855323000049

Olaopa, O., & Ayodele, O. A. (2021). Building on the strengths of African Indigenous knowledge and innovation (AIK& I) for sustainable development in Africa. *African Journal of Science, Technology, Innovation and Development*, 14(5), 1–14. DOI: 10.1080/20421338.2021.1950111

Osman, A. (2014). Indigenous knowledge in Africa: Challenges and opportunities. *OAlib*, 2(3), 1–9.

Owiny, S. A., Mehta, K., & Maretzki, A. N. (2014). The use of social media technologies to create, preserve, and disseminate Indigenous knowledge and skills to communities in East. *The Pennsylvania State University International Journal of Communication*, 8(1), 234–247.

Owolabi, A. K., Ovwasa, D. E., Ajayi, T. B., & Odewale, M. O. (2022). Preservation and use of Indigenous knowledge practices in public libraries in Nigeria. *Public Library Quarterly*, 41(50), 485–502. DOI: 10.1080/01616846.2021.1938897

Owusu-Ansah, E. F., & Mji, G. (2013). African Indigenous knowledge and research. *African Journal of Disability*, 2(1), 1–5. DOI: 10.4102/ajod.v2i1.30 PMID: 28729984

Pophiwa, N. & Saidi, U. (2002). Approaches of embedding Indigenous knowledge system in made in Africa Evaluation. *African evaluation Journal,* 10(1).137-154. http://dx.doi.org/DOI: 10.4102/aej.v10i1.62

Rasoreka, K. (2008). Information transformation Africa: Indigenous knowledge securing space in the knowledge society. *The International Information & Library Review*, 40(4), 243–250. DOI: 10.1080/10572317.2008.10762789

Sifuna, N. D. (2008). From colonisation to globalisation" *TRANS. Internet-Zeitschrift fur Kulturwissenschaften*, 17/2008. Retrieved on 10 November 2024 from www://http.inst.at/trans/17Nr/9-3sifuna.htm

Steup, M., & Neta, R. (2020). Epistemology. In Zalta, E. N. (Ed.), *The Stanford Encyclopedia of Philosophy*. Stanford University.

Ukwuoma, U. C. (2016). Educational development of Africa: Changing perspectives on the role of Indigenous knowledge. *Research Pedagogy*, 6(1), 123–145. DOI: 10.17810/2015.28

United Nations. (2019). *Indigenous people's traditional knowledge must be preserved, valued globally*. UN.

Yip, S. Y., & Chakma, U. (2024). The teaching of Indigenous knowledge and perspectives in initial teacher education: A scoping review of empirical studies. *Journal of Further and Higher Education*, 48(3), 287–300. DOI: 10.1080/0309877X.2024.2327029

ADDITIONAL READINGS

Adesina, A. O., & Ocholla, D. N. (2019). The SECI model in knowledge management practices: Past present and future. *Mousaion*, 37(3), 1–34. DOI: 10.25159/2663-659X/6557

Ajayi, S. A. (2015). Indigenous knowledge and sustainable development in Africa. *Journal of Indigenous Studies*, 5(1), 1–5.

Buthelezi, C. S., Ocholla, D., & Dhlamini, P. (2024). Strategies of documenting and dissemination of indigenous knowledge at a southern African university. *South African Journal of Information Management*, 26(1). Advance online publication. DOI: 10.4102/sajim.v26i1.1648

Dondolo, L. (2005). Intangible heritage: The production of indigenous knowledge in various aspects of social life. *Indilinga*, 4(6), 110–126.

Mawere, M. (2017). *Indigenous knowledge and the management of natural resources in Africa*. Langaa Research and Publishing.

Odora-Hoppers, C. A. (2002). *Indigenous knowledge and the integration of knowledge systems*. New Africa.

KEY TERMS AND DEFINITIONS

African Ingenious Knowledge Systems: Refers to the knowledge, practices, beliefs and innovations developed and accumulated by African communities over centuries.

Afrocentric: A worldview or perspective which is premised on African culture, history and experiences, prioritizing African perspectives values and knowledge systems.

Epistemologies: Refers to the study and investigation of the different forms of knowledge, it is concerned with nature, sources scope and methods of acquiring and validating knowledge

Heritage: It refers to the cultural, social and natural legacy inherited from past generation, it encompasses the aspects of shred human experiences that we inherit, preserve and pass to future generations.

Indigenous Knowledge: It is the form of unique, traditional and cultural knowledge systems developed and passed down by Indigenous people through generations.

Marginalised Knowledge: Knowledge that is excluded, ignored, dismissed or overlooked by dominant knowledge systems. It is undervalued knowledge that is considered less important or relevant than other knowledge and is often suppressed.

Posterity: Refers to the preservation and transmission of information, ideas and information to future generations. P. It also entails documentation or recording knowledge in a lasting format, such as writing, printing or digital storage.

Western: Refers to the philosophical, scientific and cultural knowledge traditions which originate in Europe and North America.

Chapter 12
Driving the Indigenous Knowledge System Quest Through Music and Dance:
Promoting Zimbabwe's Heritage Through Indigenous Songs and Dance

Khatija BiBi Khan
https://orcid.org/0000-0003-2408-6657
University of South Africa, South Africa

ABSTRACT

This study aims to establish how music and dance can be used to drive IKS promotion to preserve national heritage. The study was undertaken to answer questions about the type of music and the views of both the musicians and listeners. Southern Africa has a rich history of indigenous knowledge systems (IKS) grounded in a strong oral tradition. IKS of Southern Africa can be traced from the arts, agriculture, traditional medicine, music, education, and so on. As the social-cultural values are predominantly expressed in the linguistic and musical interactions of the cultural practitioners and custodians, language, dance, and music may be seen as repositories and organic inventory systems for the Indigenous knowledge legacy. The musicians tell the stories of their lives: their identity, experiences, encounters, attachments, emotions, and sentiments. They portray the interconnectedness of their cosmology and spirituality. Making one question the understanding of indigenous identity.

INTRODUCTION

This chapter intends to reveal how music and dance, inseparable as they are, carry a lot of influence in the quest to promote IKS within the Southern African drive towards national heritage restoration and preservation. Southern Africa, specifically Zimbabwe, is a pluralistic society with diverse cultures that sometimes share the communal and social environment. Each culture has different values, musical styles, and dances that are quite significant within the culture. Most modern organisations have made knowledge management their credo, whereas knowledge related to local communities has been generally disregarded. However, there has been a rising understanding of the significance of Indigenous

DOI: 10.4018/979-8-3693-8447-3.ch012

knowledge (IK), particularly among rural communities, in promoting cultural identity and managing natural resources. Concern about the accelerating degradation of the Earth's ecosystems has intensified simultaneously, adding to concerns about food supply and world famine. Indigenous Knowledge (IK) is a major source of food and medicine for a sizable fraction of the world's population, particularly in underdeveloped nations.

This chapter looks at the background of Music and dance as well as its value within African society. This chapter is structured to provide a comprehensive exploration of how music and dance contribute to the promotion and preservation of Indigenous Knowledge Systems (IKS) in Southern Africa, with a specific focus on Zimbabwe. It begins with an introduction that sets the context for the importance of IKS and the historical suppression of Indigenous cultural practices under colonial regimes. The chapter then gives the background and historical overview of Indigenous music and dance, discussing their resilience and revival post-colonization. It further details the methodology employed in the study, which includes participatory methods to engage with musicians and communities actively preserving these traditions. The results section highlights specific examples of traditional dances and their significance, followed by a discussion on the implications of these findings for cultural preservation. Finally, the chapter concludes with a synthesis of the study's insights and implications for future policy and research in the field of Indigenous cultural preservation.

BACKGROUND AND OVERVIEW

Due to forced cultural assimilation (including music prohibitions) that went along with the larger oppressions of colonial occupation and imperial exploitation of Indigenous lands, Indigenous cultural practices like music and dance faced significant obstacles under colonisation (Harrison, 2020). Through censorship and legislation like the Witchcraft Suppression Act of 1899, the colonial government and Christian missionaries systematically attempted to eradicate Indigenous cultural legacy. African music and dance were seen as evil and prohibited from being worshipped and used in the educational system in Southern Africa during the colonial era (Nompula, 2011). As a result, it was taken out of the school curriculum. European music was introduced through churches and mission schools, and it quickly took over the "black" schools' music instruction programme. Western music was considered better, while African music was considered inferior. This is called ethnocentrism by sociologists and ethnomusicologists (ibid). Ethnocentrism is prevalent when individuals believe written culture or music is superior to oral tradition. According to Anderson and Campbell (1989), American instructors in America have often encouraged their pupils to think that there is only one musical system in the world: the European art system.

Growing "Africanness" awareness has been seen in the post-colonial period. African Indigenous music and dance are just as viable as Western music, as both ethnomusicologists and music educators in South Africa have realised (Nompula, 2011). Indigenous African children's songs, in particular, are reportedly vanishing and risk being forgotten. In order to speak out against Eurocentrism, the Southern African Music Educators Society (SAMES) was established in 1985 (Nompula, 2011).

Both ethnomusicologists and music educators have expressed concern about the lack of Indigenous songs in African school curricula (Gibson & Petersen, 1994; Oehrle, 1994, 1997). According to Gibson and Petersen, the present justification for music in Africa is founded on Western education and aesthetics that have been followed for the previous three hundred years. By necessity, the African music curriculum must depart from this Eurocentric paradigm and include values appropriate to the post-racial, post-sexist, and democratic world. This situation is quite urgent. According to the National Curriculum Statement (NCS), 2005, the Department of Education in South Africa included arts and culture as one of the subject areas for the foundation, intermediate, and senior phases after 1994.

In the post-1994 South African government, re-asserting people's cultural heritage is a national imperative (Ngubane, 2003). The first significant step in efforts to recognise, acknowledge, develop, promote, and safeguard Indigenous Knowledge Systems in South Africa was established when the Cabinet in the country approved the Indigenous Knowledge Systems (IKS) Policy in November 2004. Uncovering IKS's complexities and difficulties took longer than expected. Still, it gave those involved extraordinary experiences from which they emerged united in their awe of the depth and scope of South Africa's priceless Indigenous knowledge resources. Additionally, it improved the atmosphere of cooperation among all parties engaged in its creation, including individuals with expertise as well as representatives of government agencies and scientific councils. In its preamble, the IKS Policy of South Africa states.

> The Government of the Republic of South Africa registers its commitment to the recognition, promotion, development, protection, and affirmation of IKS. This Policy is the product of extensive consultation, scholarly reflection, debate, and participation from a range of stakeholders. The involvement of practitioners and holders of Indigenous Knowledge (IK) has been of critical importance. (2004)

For cultural reasons, there is a definite need for IKS recognition and protection in South Africa. Two ubiquitous circumstances often hinder such awareness. Repressing human rights based on race and culture gave rise to the first condition unique to South Africa's political history. The transformation of the knowledge institutions founded in that age and those that came before it to reflect an African intellectual personality is ongoing. The second circumstance results from quickening globalisation and impacts all IKS globally.

UNDERSTANDING INDIGENOUS KNOWLEDGE SYSTEMS

Since its introduction in Africa and other non-western countries, the Western worldview of "knowledge" has, according to Lander (2000) and Chavunduka (1995), lacked a grasp of the holistic character and approach of non-western methods of knowing and knowledge creation. Ngara and Mangizvo (2013) define Indigenous knowledge as a body of knowledge that has persisted over generations among Indigenous people belonging to a specific cultural group. According to Mapara (2009), Indigenous knowledge is information unique to a particular ethnic group and has been passed down through the centuries via rituals, dance, music, and other forms of culture (Ngara, 2007; Owusu-Ansah & Mji, 2013; Mapara, 2009). According to Nkondo (2012), the Western conception of Indigenous African knowledge is that it is just a repeating of practices without any theory to support the inaccurate portrayal of Western intellectual and cultural arrogance. According to African researchers, a traditional healer who can treat a specific sickness with a particular herb has knowledge of the plant species and their traits. In the words of Ekpo (2018), defenders of Africanism accept that modernity is indeed good for Africa, but Africa can and should modernise without losing its African soul, cultural uniqueness, and roots. In any case, the argument continues, it is not even possible for Africa to become modern and developed without first re-gathering or, in Ngugi's words, remembering its colonially dismembered and disqualified old roots. A typical saying that captures Africanism's primacy of cultural self-repossession as a precondition for modernisation/development goes like this: "If you cut your chain, you free yourself; if you cut your roots, you die."

The fundamental problem is that educational structures inherited from colonialism were based on cultural values different from those in most African Indigenous societies. The continued social, economic, and technological ties between African countries and their former colonising powers perpetuate this lack of relevance. Education in Africa is one of the main tools to foster this bond rather than reduce it. Reforms in African education are still conceived and implemented within the framework of this relationship by marginalising the integration of African cultural values and Indigenous languages into the education system at all levels (Smith, 2002; Walter, 2002).

The United Nations Educational, Scientific and Cultural Organisation (UNESCO), through the Convention for the Safeguarding of the Intangible Cultural Heritage, adopted in 2003, speaks of the need to protect heritage from vanishing and culture against change in light of the influence of modernisation, globalisation, urbanisation, and environmental degradation on cultural sustainability.

INDIGENOUS MUSIC AND DANCE

According to Nketia (1974) and Kamanga, (2008), the act of generating music is often communal in African civilisation. Group performances are prioritised over solo ones. In African communities, organised dance and musical events play a significant role in promoting communal life. They provide a venue for inter-communal interaction and collaboration. Nketia (1974: 21) states, "the degree of social cohesion in such communities is typically extreme." According to him, music-making is a social activity promoted in African civilisation because it fosters a sense of community and humanity known as "*ubuntu*," an African humanistic concept that emphasises relationships between individuals. Participating in these communal musical practices cultivates society and fosters the ubuntu sense of community.

Most African cultures do not distinguish music and dance as two different and distinct art forms. They even share the same title in several popular African languages due to their close ties. When "song and dance" are performed during rituals, festivals, and community festivities, they are referred to as *ingoma* in Zulu (Nompula, 2011). Likewise, "song and dance" is referred to as integrated cultural acts or processes in the Sotho languages *mahobelo, khiba*, and *pina*. Due to the popular misconception that Indigenous performing arts depend on the language of their practitioners to survive, dance and music traditions as parts of the living legacy have been overlooked. The tonal contour of words is used to denote meaning, purpose, and context in Indigenous African dance and music. The words "music" and "dance," which Westerners could refer to as organised sound (music) and movement (dancing), are redundant abstractions to the majority of Indigenous people in Southern Africa since they are continuations of linguistic history. As a result, the metalanguage of dance and music is linked to social function and cultural settings in most Indigenous groups, including the Shona, Ndebele, Zulu, Xhosa, Venda, Bapedi, Batswana, and Basotho. Therefore, using a socially and culturally integrated metalanguage may be helpful to describe Indigenous choreomusical processes and behaviours connected to participation and social engagement in performance.

AFRICAN MUSIC AND DANCE

Dance

Movement is an essential component of music production in the majority of African communities (Lundquist, 1989). Without dance, Indigenous music-making is incomplete. The dominant music and dancing at ceremonies signify the importance and purpose of the event. The relationship between the living and the dead is confirmed. The creation of music is incomplete without dance or movement. Dance music is typically incredibly rhythmic and rapid. Syncopated rhythms, polyrhythmic patterns, and underlying rhythms characterise it. A particular ethnic dance is associated with specific rhythms and melodic patterns. Drumming and clapping are frequently used to enhance and advance dance rhythms. Children learn new rhythmic patterns by playing the drums and dancing, storing them in their song schemata and expanding their repertoire of rhythmic patterns in the process.

Language as the Indigenous Driver of Knowledge Through Music and Dance

Since songs, ululations, and other vocal expressions are performed in the practitioners' native languages, language plays a vital role in performing and transmitting Indigenous musical legacy. Dance is choreographically and verbally using terminology from the Indigenous lexicon as a part of musical history. Identifying the community as Venda in most Indigenous groups in Southern Africa, for instance, the Venda people in South Africa and Zimbabwe, refers to the Venda people's language and culture. Tshivenda is the term used to describe the entire Venda culture, which consists of interwoven music, dance, language, symbols, rituals, beliefs, and stories that together make up the performed systems for creating meaning and substance of the Venda people's way of life, according to Gwerevende (2020).

According to Thanasoulas (2001), culture, the collection of socially acquired practises and ideas that shape our lives, gives language its existence. Vocabulary is essential to Indigenous peoples' cultural history, present, and future and serves as a map to social reality. Music and dance are also parts of their living legacy. According to Stern (2009), culture is viewed more effectively. It responds to three sets of needs: the individual's fundamental needs, the instrumental needs of society, and the symbolic and integrating needs of both the individual and the community.

In the culture carriers' language, music and dance are transmitted as components of Indigenous cultural ecosystems. Oral culture relates to what is said and sung, while auditory culture refers to what is heard, according to Patterson (2015, 4). Indigenous dance and music traditions rely on oral, acoustic, and performative modes of transmission. These techniques are virtually always present simultaneously in Indigenous societies and are crucial for efficiently conveying choreo-musical tradition. Therefore, it is difficult to ensure the preservation of Indigenous musical ecosystems in Southern Africa without considering the function of regional languages as a method of cultural expression and as a medium for its performance.

Language, dance, and music traditions are cultural heritage elements that exhibit cultural variety. Native languages aid Indigenous groups in encoding, communicating, and maintaining the knowledge of their cultural ecologies, including various performing arts, while adjusting to socioeconomic changes. These artistic forms share a particular reality by being moulded and adapted to the socio-ecological context (Maffi, 2005, p. 605). As a result, while discussing cultural variety, we must acknowledge that it is shaped not only by religious, political, environmental, and social variables but also by language ecology.

The loss of a language may indirectly impact the natural environment; it may be further claimed because Indigenous knowledge of cultural ecosystems is implicit in the languages of their people (Maffi, 2005, pp. 601-603). According to Maffi's additional justification, language is a storehouse for human cultural memory since it conveys ideas that cannot be articulated in any other 'code system' (Maffi, 2005). As a living culture, dance or music is preserved via memory.

The mother tongue is a vital conduit for Indigenous knowledge, customs, and values. It is frequently employed in rites and ceremonies as well as in the practice and transmission of live cultural legacy, particularly in oral cultures. Indigenous people who practise certain traditions frequently utilise highly specialist lexicons, concepts, phrases, and expressions in their mother tongue, revealing an innate connection between language and the Indigenous cultural heritage.

A THEORETICAL FRAMEWORK

Since the 1950s, the term "cultural ecology" has been used in anthropology. It alludes to the research on how people adjust to their social and physical circumstances. Ecosystems and music sustainability are modern notions introduced by the word ecology in ethnomusicology. According to Titon (2009) and Schippers and Bendrups (2015), the sustainability of music refers to the circumstances in which musical knowledge, sounds, practises, styles, and expressions, as well as the cultures most closely associated with them, endure across time. At the same time, music ecosystems create the environments necessary for music to flourish in society. From an Indigenous perspective, the performance of musical arts as cultural heritage for community sustainability involves the interaction of living and non-living beings (living-dead or ancestral spirits), as well as musical and non-musical elements (including language) (Gwerevende, 2022).

Although academics have been interested in the sustainability of Indigenous music ecosystems since the beginning of music anthropology and ethnomusicology over the past century, ethnomusicologist Jeff Todd Titon was the first to put forth a thorough ecological perspective on music and sustainability (Titon, 2009). He advanced the comparison that musical traditions functioned like ecosystems and broadened the prevailing paradigm from the ecology of the ecosystem in the 20th century. Although his work may not have been the first to integrate ecology with music study, it did lead to a rise in the popularity of the notion that the area of music may be seen from an ecological perspective. According to Titon (2009), a musical ecology entails people and groups interacting around a particular genre or type of music. The language of the population to which it belongs contains the majority of the inputs, processes, and outputs of the music ecosystem. Therefore, the promotion and development of Indigenous languages that "created" the values, experiences, and principles need to be considered as crucial measures in safeguarding the living cultural heritage to preserve the meaning of choreo-musical practice and ensure the sustainability of the cultural ecosystem. Indigenous music traditions range from traditional practises to fusions of Indigenous musical knowledge, practises, and ideas with virtually every type of popular music. They also constantly develop new Indigenous music traditions, reflecting the diversity of cultures and languages. Numerous of these musical cultures are represented in regional tongues and presented in conjunction with other forms of art, such as dance, theatre, and poetry, in particular socio-cultural settings. Any music ecosystem includes "both physical and cultural factors of the musical environment" (Titon, 2009, p. 120), such as "ideas about music, sound and sound-producing instruments, recording studios, media, venues, musical education and transmission, and the economics of music - indeed music as cultural production and a cultural domain, which relate to the health of musical individuals, populations, and communities." This ecological approach to music focuses on the details, patterns, and interactions within the more extensive system to demonstrate how ecosystem components interact. African music sustainability should concentrate on preserving Indigenous performing arts (dance, music, drama), subject to severe ecological and social challenges brought on by colonialism, imperialism, globalisation, and climate change. These challenges span the breadth and depth of Indigenous music ecosystems (Gwerevende & Mthombeni, 2023).

METHODOLOGY

Participatory methodologies and methods were emphasised, including local people's knowledge and opinions in designing and managing research activities in their contexts, even if we recognise that there is no one way to IKS research. These include interviewing, visualising, and group dynamics, among other things. When feasible, these methods forsake writing in favour of oral communication tools, including images, symbols, tangible objects, and group discussions.

This study is based on data gathered from interviews, discussions, participatory observation and a review of existing literature. Open-ended oral interviews were conducted, including formal and informal meetings between researchers and informants. The study was conducted in two phases. The first phase involved visiting and interviewing musicians who were still performing Indigenous music with their audience and observing the dance. The second phase included a library search to determine what others have written on the same issue. Both primary and secondary sources are referred to in the discussions. The ages of musicians interviewed ranged from 25-76 years. The present study employed a 'participatory' approach, which allowed the researcher to interact with musicians from the targeted cultures to establish the research working relationship. Participants' cultural activities were observed, and Indigenous musicians were interviewed so as to obtain first-hand information. During the research visits, the researcher was both a participant and an observer. This allowed the participants to share their experiences and expertise in Indigenous music with the researcher, enabling the researcher to obtain the broadest possible range of information. These participants were selected because of their knowledge and informative qualities.

RESULTS PRESENTATION AND ANALYSIS

Traditional dances have persisted to this day despite the assault of colonialism, Christianity, and westernisation, albeit with adaptations. One dance that has endured the test of time and Western influence is Mbende/Jerusarema in Zimbabwe. In many cultures of the world, dance occupies an essential role in the lives of the people. In Zimbabwe, traditional dances and performances such as Jerusarema are still important living traditions practised in many contexts and still revered by the local communities.

Description of the Dance

Mataga gives a good description of this dance-music performance:

The dance is characterised by sensual, acrobatic waist-shaking and hip movements by women in unison with men. Both dancers end with energetic thrusts of the pelvis directed towards each other, thereby creating exhilaration amongst the audience. Such movements made the dance unpopular among the missionaries, who interpreted them as sexually explicit and suggestive. The music is performed by one master drummer, well supported by clappers, rattles, and costumes. (Mataga, 2003, p. 97)

Overview of Jerusarema Dance and Music

The need to know the background of the musical dance festival was the first call during the interviews. All interviewees agreed that this dance is as old as their culture can remember. The traditional dance is performed for entertainment, for ritual purposes, at festivals and for commemoration and celebration. The widely accepted explanation of the origins of *Jerusarema* is that it was portrayed as a war dance and diversionary tactic by the Shona during military encounters, as supported by the research of Welsh-Asante, (2000a, 2000b). Currently, *Jerusarema* is performed during all festival occasions, such as weddings, celebrations, recreational competitions, funerals and political gatherings.

It was further established that despite threats from the colonial administration, missionaries, and Westernisation, the dance has persisted and is still an essential part of Shona society's social, political, and cultural fabric. One of the most notable cultural practises still carried out by the Zezuru people, who reside in Zimbabwe's Mashonaland East province's Murehwa area, is *Jerusarema/Mbende*.

Due to its significance, interviewees believed that is the reason the dance is no longer only performed by the villages of Murehwa; other Shona and non-Shona organisations, urban dance clubs, and traditional performance groups have all appropriated it for usage by tourists, political rallies, and other social occasions. It is now being performed at notable cultural events and tourist centres because it carries the identity of the Shona and functions to restore and preserve Indigenous knowledge of the Zezuru people and the nation at large. Various modifications have resulted from its development as a cultural manifestation.

The Meaning of Jerusarema Musical Dance

The study sought to establish the shared meaning of this musical festival among the Shona people of the Zezuru dialect in Zimbabwe. It serves as both a form of cultural expression and a symbol of the struggle for survival and, ultimately, for freedom through a turbulent history. The uniqueness of *Jerusarema* and its dramatic impact and meaning have made it liable for various interpretations, which continue to the present day. The corroborated responses reveal that the festival has been upgraded to the national level, thereby assuming too many variations and meanings away from the original. The dance's meaning and aim are frequently misconstrued; it is criticised for being overly seductive and sensual, and because of its enormous popularity in the community, it was considered a danger to the Christian Church's efforts to grow its membership. In the big metropolis, the dance has been made commercial, and new, greatly deformed forms have appeared.

The post-colonial administration has altered and modified the dance to produce an "acceptable" style for official political events. The *Mbende* traditional dance, now more commonly known by its Christian name, *Jerusarema*, is distinctive for the people of Murehwa and the entire nation as it is featured as a prominent performance at all national functions. To make the dance more palatable to Christian missionaries, it is commonly thought that the traditional name of *Jerusarema* was *Mbende*, with the former being a biblical derivation from the holy city of Jerusalem in Israel. Both the rural and urban areas still follow the Jerusarema tradition. There are several *Jerusarema* dance clubs in Murehwa and elsewhere, and the dance is still performed during celebrations, funerals, political rallies, and weddings. Urban dance clubs also frequently perform *Jerusarema, which is* one of their signature performances.

There are several accounts of the dance's pre-colonial history or the time before European arrival in the 1890s and its significance during that time. Some claim that it stood for family, sexuality, and fertility. Others connect it to conflict, particularly the military incursions carried out by Ndebele warriors on some areas of Mashonaland during the Mfecane wars in southern Africa in the 1830s (Welsh-Asante, 2000a). This story says that the dancing was a distraction technique. Before a fight, the enemy's attention was diverted by the seductive and energising dancing moves. It is seldom easy to determine the initial form and intent of such cultural performances, which frequently exhibit spatiotemporal alterations due to cultural practices' fluid and dynamic character.

However, Mbende's intriguing moniker tells us a lot about its ups and downs throughout the ages. Because they believed that *Jerusarema* was licentious, lusty, immoral, and provocative, the white missionaries worked with the Native Commissioners to get it outlawed. They believed that doing so would make it more difficult to convert the natives to Christianity. Because of its popularity, it also infuriated the early European settlers, who had the difficult challenge of finding African labour for the farms and mines. The renowned Zimbabwean author Chenjerai Hove (1996) recounts a local reflection on how *Jerusarema* was outlawed as follows in his award-winning novel Ancestors:

> Even the guys who had wished to avoid bothering the white man now had to forsake their preparations to travel to Johannesburg to work and raise taxes for the white man when "*Jerusarema*" arrived. When the white man came to seize them and force them to work on the roads and in the mines, they resisted and fought back. Some hid in the hills and mountains during the day and emerged at night to dance the "*Jerusarema*" with the villagers' ladies and men. The white man then added, "This dance is another problem," as he sat down one day. It must be stopped. These men, who refuse to labour in our mines or on our roads, gain too much pride from the dance. It must end immediately. Never again to be danced. Never again will the music that introduces this dance be played. Anyone who sings the tunes deserves death. Anyone who does the dance deserves death. So '*Jerusarema*' was put to death. These days, we only hear about it in far-off places. (Hove, 1996, pp. 128–29)

Influence of Jerusarema on the Identity and Culture of People

The performers and audience of this Musical dance festival were asked about their feelings as the event participants. They all identified with the musical festival and professed allegiance for many years. Because it has the capacity to immediately penetrate the human spirit, Indigenous music, regardless of its folk, art, classical, or entertainment genre, directly affects the emotional side of man and fosters a sense of patriotism. Songs honouring heroes, the nation, martyrs, and patriots pique huge consumer interest, boost pride, and make people feel more sensitive to the need to identify with their heritage. Such musical compositions are based on an arrogant melody and have a moving narrative. "The expression of identity through music would be an extraordinary atmosphere of sensitivity, specified with special means of artistic expression" (Hysi, 2005, p. 249).

National Meaning of the Jerusarema Musical Dance

The Zimbabwe Broadcasting Corporation (ZBC) chose Mbende drumming as the entrance and distinctive music for all radio and television news bulletins at the time of Zimbabwe's independence in 1980 because of its popularity (Vambe, 1999). Such was its power that when ZBC changed the *Jerusarema* melody in 2000, a public outcry ensued, and the national broadcaster was compelled by popular pressure to return the tune. The dance served as a unifying force within the community and a crucial link between the public and the freedom fighters during the 1970s conflict for independence from the British colonisers. The fighters, who mostly used guerilla tactics, would dress up and go to dance events where they could acquire intelligence and material assistance from the local populations. Following independence, this politicisation of the dance persisted, with appearances at political rallies and official events. Therefore, its current appeal stems not just from its social and cultural applications but also from its accepted political role.

Perusal of historical and documentary evidence revealed that Mbende is one of the 43 cultural expressions from all over the world that were named masterpieces of humanity's oral and intangible heritage in 2005. This recognition of the dance as a vital living cultural heritage that is fragile and perishable but necessary for the community's cultural identity and one that represents an exceptional example of the intangible cultural heritage of humanity, attesting to the world's cultural diversity and richness, places Mbende among the other 43 cultural expressions on this list. Mbende was acknowledged as one of the top still-practising cultural practices in the world. The *Makishi* masquerade from Zambia, the *Vimbuza* healing dance from Malawi, and *Gule Wamkulu* (Zambia, Malawi, and Mozambique) are among the other prominent performances from the Southern African region that are mentioned. Although Douglas Vambe, a renowned master drummer, was entered for the 2006 UNESCO Cultural Music Heritage Competition, where his drumming competed with other traditional musical performances worldwide, the dance was nominated for the UNESCO list in 2005.

The Musical Instruments Associated With Jerusarema/Mbende

It was refreshing to observe that it is not only the dance and music that carry the meaning and preserve Indigenous knowledge about this musical festival. The instruments and regalia of the performers are very much a part of the Indigenous knowledge system and its preservation.

The musical instruments are the main components of *Mbende* material culture. A *Mbende* musical instrument takes on a variety of meanings in its context, both as an object and as an instrument. The aesthetics of *Mbende* musical instruments are pleasant to the ear as well as the sight. What the instruments signify to the individuals who produce and play them became clear via an analysis of the music, performance circumstances, and the instrument itself. Studying the artefacts reveals a wealth of information about the performance's continuity and evolution, as well as nostalgia, complicated abilities, creative expression, and human/nature links.

The dance outfits represent the second important group of *Mbende* physical assets. These were formerly made of animal skins, but due to the influence of Western clothing, they have been altered, and modern dancers now mostly wear textile costumes combined with animal skin. The clothes' structure and materials have an aesthetic appeal and improve the dancing moves. The dancing moves are complemented by the *mbikiza/madhumbu*, which is formed of thin, flexible animal skin stripes and has a strong sexual draw for viewers. The women's stretchy animal hide skirts accentuated their hip thrusting and waist-shaking moves. These were created using the flexible, cosy, and simple-to-work leather of wild animals, including cheetahs, leopards, and monkeys. Although the quality of these garments is far lower than that of wild animals' skins, occasionally, the skins of other domestic animals, such as cattle, goats, and sheep, are utilised.

DISCUSSION

Traditional dance still plays a significant role in many African countries as a means of cultural expression. Some traditional dances have endured and continue to demonstrate tenacity in the face of pressure. Mbende maintains a prominent place not just in the region where it first emerged but has developed into a significant national expression with many traits deserving of promotion and preservation. An important intangible legacy is comprised of the distinctive dance motions, the dance's cultural importance in society at large, the items it employs, and the skills and craftsmanship required to manufacture them. There is an urgent need to preserve the Mbende cultural practices that are closer to the dance's original form since outside influences are badly impacting the dance itself. Promoting Mbende events will restore the dance's standing in society and guarantee its survival. By giving the required resources and knowledge, the dance's nomination as a UNESCO masterpiece would undoubtedly improve the preservation of this significant component of Zimbabwe's intangible heritage.

FUTURE RESEARCH DIRECTIONS

Future studies emanating from this current study might look at comparative studies across regions. An examination could be undertaken to compare how various Southern African nations employ music and dance as a means to advance and safeguard their native knowledge systems. This could encompass nations with comparable historical backgrounds but distinct present strategies for safeguarding cultural heritage. Another future study can also explore the utilisation of current media tools, such as social media, streaming services, and online archives, to promote and safeguard Indigenous music and dance genres. This study aims to investigate the efficacy of these platforms in engaging younger generations and diaspora populations.

CONCLUSIONS

African intellectuals should help Africa close the gap created by over four hundred years of domination and marginalisation of African people's knowledge systems by rejecting the utilisation of the dominant Western worldview of knowing and knowledge production as the only way of knowing. Ngugi *wa* Thiong'o (1986) expressed this problem in his key essay, "Decolonizing the Mind." African traditional knowledge should not just be considered an "alternative" form of knowledge but also one among many. The implication for higher education is that it is no longer acceptable to do research in Africa with local communities and individuals as if their opinions and life experiences are unimportant. Research should not be viewed as a purely academic endeavour, given the extraversion of African knowledge systems, but rather as an activity that takes place within a specific set of ideological, political, and social contexts.

This is because a large portion of the research that is now done on African communities is still influenced by Eurocentric preconceptions (Hountondji, 2002).

The use of IKS and the promotion of post-Africanism are answers to the obvious numerous failings of postcolonial modernity in Africa. In other words, who would care to look back if post-colonial culture had succeeded in bringing about a modernised and compassionate Africa if the African route to development had produced economic prosperity, greater living standards, peace, and political stability?

REFERENCES

Adam, A., Othman, N., & Halim, A. (2021). Indigenous knowledge transfer among Dusuns and Bajaus in Kota blue, Sabah, Malaysia: Approaches and challenges. *International Journal of Academic Research in Business & Social Sciences*, 11(11). Advance online publication. DOI: 10.6007/IJARBSS/v11-i11/11656

Bruchac, M. (2014). Indigenous knowledge and traditional knowledge., 3814-3824. DOI: 10.6007/IJARBSS/v11-i11/11656

Campbell, P. (1991). *Lessons From the World: A Cross-cultural Guide to Music Teaching and Learning*. Schirmer Books.

Carter, A., Sisco, S., & Fowler, R. (2023). Since we are, therefore, I am ubuntu and the experiences of black women leadership coaches. *Consulting Psychology Journal*, 75(1), 51–67. DOI: 10.1037/cpb0000227

Chavunduka, M. (1995). *The missing links. keynote address to the workshop on the study and promotion of Indigenous knowledge systems and sustainable natural resources management in Southern Africa.* Midmar [April.]. *KwaZulu-Nata*, 1, 24.

Chigangaidze, R., Matanga, A., & Katsuro, T. (2021). Ubuntu philosophy as a humanistic–existential framework for the fight against the covid-19 pandemic. *Journal of Humanistic Psychology*, 62(3), 319–333. DOI: 10.1177/00221678211044554

Chivaura, V. G. (2006). African Indigenous worldviews and ancient wisdom: a conceptual framework for development in Southern Africa. In Kunnie, J. E., & Goduka, N. I. (Eds.), *Indigenous Peoples' Wisdom and Power: Affirming our Knowledge Through Narratives*. Ashgate Publishing Aldershot.

Chowdhury, J., Wahab, H., Saad, M., Roy, P., Hamidi, M., & Ahmad, M. (2021). *Ubuntu philosophy*., 361-381. DOI: 10.4018/978-1-7998-7480-5.ch022

Čutura, M. (2020). Consumer ethnocentrism and social identity: theoretical backgrounds and empirical studies overview. Market-Tržište, 32(SI), 131-146. https://doi.org/DOI: 10.22598/mt/2020.32.spec-issue.131

Ekpo, D. (2018). Africa without Africanism: post-Africanism vs. Indigenous knowledge systems, culture/art. In *R. de Lange, I. Stevens, R. Kruger and M. Sirayi, The Arts and Indigenous Knowledge Systems in a Modernized Africa*. Cambridge Scholars Publishing.

Ela, J., Mudgway, M., Harré, N., & Hamill, J. (2022). Planetary health, Indigenous knowledge and healthcare: a scoping review. DOI: 10.1101/2022.11.28.22282853

Gade, C. (2012). What is Ubuntu? different interpretations among South Africans of African descent. *South African Journal of Philosophy*, 31(3), 484–503. DOI: 10.1080/02580136.2012.10751789

Gašević, D., Vranješ, M., & Tomašević, D. (2019). The influence of consumer lifestyle in the Republic of Serbia on consumer ethnocentrism. *Skola Biznisa*, (2), 78–94. DOI: 10.5937/skolbiz2-22136

Gipson, P., & Petersen, A. (1994). Music Education in South Africa: In *Proceedings of the 21st World Conference of the International Society for Music Education*, 193-287.

Grange, L. (2020). COVID-19 pandemic and the prospects of education in South Africa. *Prospects*, 51(1-3), 425–436. DOI: 10.1007/s11125-020-09514-w PMID: 33100404

Green-Thompson, L., McInerney, P., & Woollard, B. (2017). The social accountability of doctors: A relationship based framework for understanding emergent community concepts of caring. *BMC Health Services Research*, 17(1), 269. Advance online publication. DOI: 10.1186/s12913-017-2239-7 PMID: 28403860

Gwerevende, S. (2020). "Venda Cultural Expressions: An inclusive analysis of tshigombela as a performative enactment of ubuntu." Unpublished master's thesis, Choreomundus.

Gwerevende, S. (2022). "Music Ecosystems. Why Do They Matter? Centre for Music Ecosystems." Accessed January 23, 2023. https://www.centerformusicecosystems.com/music-ecosystems-builders/solomon-gwerevende

Gwerevende, S., & Mthombeni, Z. M. (2023). Safeguarding intangible cultural heritage: Exploring the synergies in the transmission of Indigenous languages, dance and music practices in Southern Africa. *International Journal of Heritage Studies*, 29(5), 398–412. DOI: 10.1080/13527258.2023.2193902

Harrison, K. (2020). Indigenous Music Sustainability During Climate Change. *Current Opinion in Environmental Sustainability*, 43, 28–34. DOI: 10.1016/j.cosust.2020.01.003

Hountondji, P.J. (1995). *Second Bashorum MKA Abiola Distinguished Lecture Address.* Reported in Khosa, Meshack M. 1996.

Hove Chenjerai. (1996). *Ancestors*. College Press Publishers.

Hutchinson, P., McIlduff, C. D., Legare, M., Keewatin, M., Hagel, M., Chapados, M., & Acharibasam, J. B. (2023). Indigenous knowledge mobilization: Reflection on context, content, and relationship. *Alternative*, 19(4), 902–913. DOI: 10.1177/11771801231198082

Hysi, F. (2005). *Estetike ne tri pikepamje*. Uegen.

Jurdi, H., & Houjeir, R. (2020). Cultural norms and the marketplace: Favor economies in the arab world. *Journal of Consumer Marketing*, 37(2), 226–235. DOI: 10.1108/JCM-05-2018-2681

Kamanga, R. (2008). Protection of Intellectual Property: Musical, Literary and Artistic work in Malawi. Unpublished article presented at the PASMAE Conference in Lilongwe, Malawi.

Lander, D. (2000). Eurocentricism and colonialism in Africa. *Nepantla*, 1(2), 510–532.

Lebaka, M. E. K. (2019). Understanding Indigenous Knowledge as Indigenous Practice, Skill and Know-How: The Case of Communal Music-Making in Bapedi Culture as a Social Practice. *European Journal of Social Science Education and Research* May - August 2019 6(2), 25. ISSN 2411-9563 (Print) ISSN 2312-8429 (Online)

Lundquist, B. (1989). Sub-Sahara Africa in Multicultural Perspectives in Music Education. In Anderson, W., & Campbell, S. (eds) Reston, VA:*Music Educators National Conference*, pp. 145-178.

Machida, S. (2020). Ethnocentrism and support for nuclear armament in Japan. *Asian Journal of Comparative Politics*, 7(1), 95–114. DOI: 10.1177/2057891120939314

Madonsela, B., Semenya, K., & Shale, K. (2024). A review of Indigenous knowledge systems and their application in sustainable solid waste management. *WORLD (Oakland, Calif.)*, 5(2), 219–239. DOI: 10.3390/world5020012

Maffi, L. (2005). Linguistic, Cultural, and Biological Diversity. *Annual Review of Anthropology*, 29(1), 599–617. DOI: 10.1146/annurev.anthro.34.081804.120437

Mapara, J. (2009). Indigenous knowledge systems in Zimbabwe: Juxtaposing postcolonial theory. *The Journal of Pan African Studies*, 3, 139–155.

Marinković, V. (2017). Effects of the Republic of Serbia's citizens' patriotism and animosity towards the European Union on consumer ethnocentrism. *Ekonomski Horizonti*, 19(1), 3–15. DOI: 10.5937/ekonhor1701003M

Masoga, M., & Shokane, A. (2019). Viewpoint: Indigenous knowledge systems and environmental social work education: towards environmental sustainability. *Southern African Journal of Environmental Education*, 35(1). Advance online publication. DOI: 10.4314/sajee.v35i1.14

McCleary, D., & Williams, R. (2009). Sociopolitical and personality correlates of militarism in democratic societies. *Peace and Conflict*, 15(2), 161–187. DOI: 10.1080/10781910902837248

McGregor, D. (2014). Lessons for collaboration involving traditional knowledge and environmental governance in Ontario, Canada. *Alternative*, 10(4), 340–353. DOI: 10.1177/117718011401000403

McGregor, D. (2021). Indigenous knowledge systems in environmental governance in Canada. *Kula Knowledge Creation Dissemination and Preservation Studies*, 5(1). Advance online publication. DOI: 10.18357/kula.148

Michailova, S., Piekkari, R., Storgaard, M., & Tienari, J. (2017). Rethinking ethnocentrism in international business research. *Global Strategy Journal*, 7(4), 335–353. DOI: 10.1002/gsj.1159

Mji, G. (2013). African Indigenous knowledge and research. *African Journal of Disability*, 2(1), 1–5. PMID: 28729984

Mohan, D. (2024). Wisdom of the ancestors: Harnessing Indigenous knowledge for contemporary challenges. *International Journal of Applied Research*, 10(3), 112–119. DOI: 10.22271/allresearch.2024.v10.i3b.11591

Mokhachane, M., Green-Thompson, L., George, A., Wyatt, T., & Kuper, A. (2023). Medical students' views on what professionalism means: An ubuntu perspective. *Advances in Health Sciences Education : Theory and Practice*, 29(3), 841–857. DOI: 10.1007/s10459-023-10280-5 PMID: 37710029

Molose, T., Goldman, G., & Thomas, P. (2018). Towards a collective-values framework of ubuntu: Implications for workplace commitment. *Entrepreneurial Business and Economics Review*, 6(3), 193–206. DOI: 10.15678/EBER.2018.060312

Mugovhani, G. N. (2013). African Renaissance, Indigenous African Music, and globalisation: collusion or collision? Mugovhani, N.G. (2008) a. Research Findings of the University of Venda Indigenous Music and Oral History Project 2003-2008. *Paper presented at the NIKSO EXPO* in Pretoria. Transvaal Museum.

Mugovhani, N. G. (2010). The role of Indigenous African choral music in the search for identity. With special reference to Mzilikazi Khumalo's music. *Muziki Journal for Music Research in Africa* vol 17, issue1 2010.

Ngara, R. (2012). Mukwerera, a Shangwe rainmaking ceremony: Its music, dance and symbols, *MA thesis. Eastern Cape*: University of Fort Hare.

Ngubane, B. S. (2003). Roll-out of Indigenous Music/Oral History Collection, Promotion and Preservation Project in Three Universities: Government *White Paper on Arts, Culture, Science and Technology*. Pretoria: Department of Arts, Science and Technology.

Nkondo, M. (2012). *Indigenous African knowledge systems in a polyepistemic world: The capabilities approach and the translatability of knowledge systems.* Paper presented at the Southern African Regional Colloquium on Indigenous African knowledge systems: methodologies and epistemologies for research, teaching, learning and community engagement in higher education. Howard College Campus: University of KwaZulu-Natal.

Nompula, Y. (2010). *Finding Our Voice: A Critique of Eurocentrism. Saarbrucken*. Lambert Academic Publishing.

Oehrle, E. (1994). The First South African National Music Educators' Conference in Context of World Music Education, in the Proceedings of the First National Music Educators' Conference. Durban, South Africa: SAMES.

Owusu-Ansah, F. E., & Mji, G. (2013). African Indigenous knowledge and research. *African Journal of Disability*, 2(1), 1–5. DOI: 10.4102/ajod.v2i1.30 PMID: 28729984

Patterson, E. E. (2015). "Oral Transmission: A Marriage of Music, Language, and Tradition." *The Research and Scholarship Symposium* 12. https://digitalcommons.cedarville.edu/research_scholarship_symposium/2015/podium_presentations/12

Pllana, D., & Aliu, H. (2018). The impact of consumer ethnocentrism on purchase intentions: The case of Kosovo. *International Journal of Research and Studies Publishing*, 8(10).

Porten, S., Loë, R., & McGregor, D. (2016). Incorporating Indigenous knowledge systems into collaborative governance for water: Challenges and opportunities. *Journal of Canadian Studies. Revue d'Etudes Canadiennes*, 50(1), 214–243. DOI: 10.3138/jcs.2016.50.1.214

Schippers, H., & Bendrups, D. (2015). Ethnomusicology, Ecology and the Sustainability of Music Cultures. *The World of Music*, 4(1), 9–19.

Shankarmahesh, M. (2006). Consumer ethnocentrism: An integrative review of its antecedents and consequences. *International Marketing Review*, 23(2), 146–172. DOI: 10.1108/02651330610660065

Smith, A. (2002). Power and hierarchy of knowledge. *Geoforum*, 40(1), 230–248.

Smylie, J., Martin, C., Kaplan-Myrth, N., Steele, L., Tait, C., & Hogg, W. (2004). Knowledge translation and Indigenous knowledge. *International Journal of Circumpolar Health*, 63(sup2), 139-143.

Stokes, M. (1994). *Ethnicity, Identity and Music - The Musical Construction of Place.* Bridgend: WBC Bookbinders.

Swanson, D. (2008). Ubuntu: An African contribution to (re)search for/with a 'humble togetherness'. *Journal of Contemporary Issues in Education*, 2(2). Advance online publication. DOI: 10.20355/C5PP4X

Titon, J. T. (2009). "Music and Sustainability: An Ecological Viewpoint." *The World of Music* 51 (1): 119–137. https://www.jstor.org/stable/41699866

Tshivhase, M. (2018). Love as the foundation of Ubuntu. *Synthesis Philosophica*, 33(1), 197–208. DOI: 10.21464/sp33112

UNESCO. 1953.1968. The Use of the Vernacular Languages in Education Monographs on Fundamental Education Vlll. UNESCO, Paris.

UNESCO. (2008). Education for Sustainable Development. Accessed May 12, 2023, from http://portal .unesco.org/education/

Vambe, M. (1999). 'The Dynamics of the 'Mbende' Dance', *MOTO Magazine*, September 1999, Issue, 2000, pp. 26-27

Vijayan, D., Ludwig, D., Rybak, C., Kaechele, H., Hoffmann, H., Schönfeldt, H. C., Mbwana, H. A., Rivero, C. V., & Löhr, K. (2022). Indigenous knowledge in food system transformations. *Communications Earth & Environment*, 3(1), 213. Advance online publication. DOI: 10.1038/s43247-022-00543-1

Wa Thiong'o, N. (1986). *Decolonising the Mind*. James Currey.

Walter, D. (2002). Colonial and post-colonial discourses in Social Sciences: Agricultural critique of colonialism. *Latin American Research Review*, 28(3), 120–134.

Welsh-Asante, K. (2000a). 'The 'Jerusarema' Dance of Zimbabwe', *Journal of Black Studies*, 15(4), African and African American dance, Music, and Theatre, June, 1985. pp.381–403.

Welsh-Asante, K. (2000b). *Zimbabwe Dance: Rhythmic Forces, Ancestral Voices: An Aesthetic Analysis*. Africa World Press.

Whyte, K., Brewer, J.II, & Johnson, J. (2015). Weaving Indigenous science, protocols and sustainability science. *Sustainability Science*, 11(1), 25–32. DOI: 10.1007/s11625-015-0296-6

Worthington, E.Jr, & Cowden, R. (2017). The psychology of forgiveness and its importance in South Africa. *South African Journal of Psychology. Suid-Afrikaanse Tydskrif vir Sielkunde*, 47(3), 292–304. DOI: 10.1177/0081246316685074

Zimbabwe Dance https://en.wikipedia.org/wiki/Zimbabwe_Dance Consulted (10/01/2007). Candidature file for the nomination of '*Mbende/Jerusarema* dance as a Masterpiece of Oral

KEY TERMS AND DEFINITIONS

Indigenous Knowledge Systems (IKS): A body of knowledge embedded in cultural traditions of regional, Indigenous, or local communities. IKS includes the arts, agriculture, traditional medicine, education, and more, which have been passed down through generations. Indigenous knowledge systems (IKS) encompass the cumulative body of knowledge, practices, and beliefs that Indigenous peoples have developed over generations through their interactions with the environment and their cultural contexts. These systems are characterised by their holistic and relational nature, integrating ecological, spiritual, and social dimensions that reflect the unique experiences and worldviews of Indigenous communities (Mohan, 2024; Bruchac, 2014; McGregor, 2021). IKS are not merely a collection of facts but are deeply embedded in the cultural identity and heritage of Indigenous peoples, serving as a framework for understanding their relationships with the land, resources, and each other (Masoga & Shokane, 2019; Adam et al., 2021). This knowledge is often transmitted orally and through practical engagement, emphasising community participation and experiential learning, which contrasts with the more individualistic and formalised approaches of Western knowledge systems (McGregor, 2014; Hutchinson, 2023). Furthermore, IKS play a crucial role in sustainable practices, particularly in areas such as environmental management and biodiversity conservation, where Indigenous peoples have historically maintained a balance with their ecosystems (Ela et al., 2022; Smylie et al., 2004; Madonsela, 2024). The recognition and integration of Indigenous knowledge into contemporary governance and policy-making are essential for fostering sustainability and respecting the rights of Indigenous communities, as these systems provide valuable insights into adaptive management and resilience in the face of environmental challenges (Whyte et al., 2015; Vijayan, 2022; Porten et al., 2016).

Ethnocentrism: The belief in the inherent superiority of one's own ethnic group or culture, often at the expense of other cultures. This term is used to describe the devaluation of African music and dance during colonial times in favour of European cultural practices. Ethnocentrism is a sociocultural phenomenon characterised by the belief that one's own ethnic or cultural group is superior to others, leading individuals to evaluate other groups based on their own cultural norms and values. This perspective often manifests in a dichotomous worldview, categorising people into "us" versus "them," which can foster in-group favouritism and out-group prejudice (Machida, 2020; McCleary & Williams, 2009). Ethnocentrism influences various aspects of social behaviour, including consumer preferences, where individuals may show a marked preference for domestic products over foreign ones, reflecting a sense of loyalty and identity tied to their cultural background (Čutura, 2020; Gašević et al., 2019; Pllana & Aliu, 2018). The concept has been explored across multiple disciplines, including psychology, sociology, and marketing, with implications for understanding intergroup relations, social identity, and consumer behaviour (Michailova et al., 2017; Shankarmahesh, 2006). Ultimately, ethnocentrism not only shapes individual attitudes and behaviours but also plays a significant role in broader societal dynamics, impacting everything from political ideologies to economic interactions (Jurdi & Houjeir, 2020; Marinković, 2017).

Ubuntu: An African humanistic philosophy that emphasises community, sharing, and generosity. It is mentioned in the context of music and dance, fostering a sense of community and humanity. Ubuntu is a profound philosophical concept rooted in African culture, particularly among the Zulu and Xhosa peoples, encapsulating the essence of human interconnectedness and communal relationships. The term "Ubuntu" is often translated as "I am because we are," reflecting the belief that an individual's humanity is intrinsically linked to the humanity of others, thus emphasising the importance of community, compassion, and mutual respect in human interactions (Carter et al., 2023; Chigangaidze et al., 2021; Chowdhury et al., 2021). This philosophy advocates for a relational understanding of personhood, where one's identity and moral standing are shaped by social connections and collective well-being (Tshivhase, 2018; Swanson, 2008). Ubuntu serves as a guiding principle in various domains, including ethics, education, and social justice, promoting values such as solidarity, dignity, and shared responsibility (Gade, 2012; Green-Thompson et al., 2017; Molose et al., 2018). In contemporary discourse, Ubuntu is increasingly recognised as a framework for addressing social challenges, fostering inclusivity, and enhancing community engagement, thereby reinforcing its relevance in both local and global contexts (Grange, 2020; Worthington & Cowden, 2017; Mokhachane, 2023).

Choreomusical: Relating to the integrated nature of music and dance within many African cultures, where these forms of expression are not distinct but interconnected parts of cultural and social practices.

Cultural Ecology: A field of study that examines how human societies adapt to their environments and how cultural practices and interactions affect ecological systems.

Music Ecosystems: A concept that refers to the environments necessary for music to flourish in society, including all physical and cultural factors that influence musical practice, transmission, and sustainability.

Chapter 13 Knowledge System and the Mopane Worm Economy Amongst the Kalanga of Mangwe in Southwestern Zimbabwe

Innocent Nyathi
Solusi Adventist University, Zimbabwe

ABSTRACT

The chapter examines the integration of Indigenous Knowledge Systems (IKS) in the context of the mopane worm (Imbrasia belina) economy amongst the Kalanga ethnic group in Mangwe District, southwestern Zimbabwe. The mopane worm, a vital resource in southwestern Zimbabwe, plays a crucial role in subsistence and a source of income for the Kalanga. In the context of this book, IKS, passed down from generation to generation, governs the harvesting, processing, and consumption of mopane worms. This chapter examines the role of Kalanga IKS in sustainably managing the mopane ecosystem, emphasising Kalanga ecological practices, harvesting patterns, preparation, and preservation techniques. It also analyses the socio-economic impact of the Mopane economy, looking at how IKS facilitates local trade and markets, provides livelihoods, and enhances food security in a semi-arid environment. By researching and documenting these, the study accentuates and illuminates our understanding of the importance of IKS in promoting sustainable development while preserving cultural heritage.

INTRODUCTION

The Kalanga are a Bantu ethnic group found in the semi-arid regions of southwestern Zimbabwe. The Kalanga are located where agricultural productivity is often constrained by unreliable rainfall and poor soil quality. Considering the climate in the area, Indigenous communities have developed resourceful and innovative strategies for survival. One such strategy, deeply embedded in local culture and ecology, is the harvesting and consumption of mopane worms (*Gonimbrasia/imbrasia belina*), the larvae of the

DOI: 10.4018/979-8-3693-8447-3.ch013

mopane moth (see Figure 1).. These worms, which feed on the leaves of the mopane tree (*Colophospermum mopane*), have long served as a vital source of nutrition and income for the people of this region. The mopane worm economy is not only an example of ecological adaptation but also a demonstration of how traditional knowledge systems (IKS) have enabled communities to navigate the challenges posed by both environmental unpredictability and economic marginalization and, therefore, demonstrates environmental awareness (Duri et al., 2007). As a high-protein, low-cost food source, mopane worms offer a solution to food insecurity, especially during times of drought or crop failure, which are increasingly common in the context of climate change. In addition to subsistence uses, mopane worms have become a key commodity in the regional economy, traded in local markets and sometimes even exported, thus providing a crucial income stream for rural households.

Figure 1. Mopane worms (imbrasia belina).

Picture by the author

This chapter explores the mopane worm economy as a survival strategy, examining how it intersects with broader socio-economic and environmental dynamics in southwestern Zimbabwe. By analysing the historical, cultural and ecological dimensions of this economy, the chapter sheds light on the resilience and resourcefulness of Indigenous communities. It also highlights the challenges and opportunities facing these communities as they navigate the dual pressures of globalization and climate change (Chifamba, 2020). Through this exploration, the chapter aims to contribute to a deeper understanding of how traditional resource-based economies serve as adaptive survival strategies in the face of contemporary challenges contextualised in the broader studies of livelihoods (Dovie, 2024).

BRIEF HISTORY OF THE KALANGA AND THEIR INDIGENOUS KNOWLEDGE SYSTEMS (IKS)

The Kalanga People: Origins and Historical Context

The Kalanga people are part of the larger Bantu linguistic group and are primarily found in the southwestern regions of Zimbabwe, particularly in Bulilima and Mangwe Districts, as well as parts of Botswana and South Africa. Their origins are closely linked to the migration and settlement patterns of Bantu-speaking peoples in southern Africa, with historical records suggesting that the Kalanga, like many other Bantu groups, moved southward from the Great Lakes region in the great Bantu migrations recorded in the 6-9th centuries A.D (Dube, 2020).

Over time, the Kalanga people settled in the semi-arid regions of southwestern Zimbabwe, where they adapted to the harsh environmental conditions of the savannah and woodland ecosystems (Dube, 2020). Their settlement area, straddling the border between Zimbabwe and Botswana, is characterised by rocky outcrops, dry plains, and the abundance of natural resources such as the mopane tree (Colophospermum mopane), which would become central to their economy (see Figure 2).

Figure 2. The rocky outcrops that punctuate Kalanga country in Mangwe. These have attracted illegal gold panning that not only despoils the environment but also impact negatively on mopane worm harvesting.

Picture by the author.

The rocky outcrops have spanned other destructive economic activities, such as artisanal gold panning, which have put traditional economies at risk. Again, due to climate change, the area inhabited by the Kalanga has experienced rising temperatures and even more frequent droughts, thus putting traditionally well-grounded economies such as mopane harvesting in a perilous position.

Historically, the Kalanga were known for their advanced agricultural techniques. These included terracing fields in order to conserve scarce water. There are known for successfully adapting drought resistance food crops such as millet and finger millet, and several sorghum varieties that survive in their climes. They also developed strong trade links with groups in southern Africa and beyond. Some of these trade links are still open, for example those with the Batswana. They exhibit a strong sense of communal existence, and very strong familial and kinship patterns buttressed in a patriarchal society. Kalanga women, however are highly respected in the society as there are sees as the custodians of their IKS. It is common to find that amongst the Kalanga, traditional healers and *sangomas*(diviners) are mostly women. Through personal observation, most storytelling and folklore telling is done by women in the Kalanga society, thus demonstrating their pivotal role in the development of Kalanga IKS and as a repository of Kalanga culture.

The Kalanga formed part of the Lemba and Karanga peoples, who were part of the broader Shona group in Zimbabwe. The Kalanga developed distinct identities through their interactions with neighbouring ethnic groups, such as the Ndebele, as well as with the British colonizers during the colonial era (Nyathi, 2022). Despite external pressures and conflicts, the Kalanga maintained their cultural and linguistic uniqueness, preserving traditions that remain central to their identity today (Dube, 2020).

METHODOLOGY

The study explores the role of Kalanga IKS in the mopane economy. Precisely, the Kalanga have used IKS to manage and utilise the mopane worm, which has been an important part of the Kalanga economic and social life for centuries. The study seeks to show how communities can use IKS to cope with traditional and contemporary economic realities.

This is an ethnographic qualitative study, observation, and semi-structured interviews that inform the data tools and data analysis. The interviews used open-ended questions to allow the participants to share their unsolicited experiences, which critically enriched the research. Critical ethnography is a relatively recent form of ethnographic research where the researcher is both an observer and an active participant in the research setting, especially when focusing on Indigenous methodologies (Denzin et al., 2008). The main positive of using this method is that it allows the researcher to fully immerse themselves in the research work and go through the experiences and thought processes of subjects. However, the main challenge of this approach is that the researcher may become sentimental and teleological trappings may inhibit rigour in the research. To counter that, it is important to base the findings of the research on observable and empirical data rather than on inferred assumptions, and contextualising findings is also important.

The study focused on the Kalanga of Mangwe in southwestern Zimbabwe. The study zeroes in on communities where the mopane worm is harvested. The population of interest include local peasant farmers, traditional leaders and women. On a broader scale, the buyers and processors of the worm, as well as other stakeholders, were consulted.

Interviews were conducted with community members. These included women and children, community elders and traditional leadership. The major issues discussed included the traditional methods of harvesting, preparation and preservation. In addition, the economic significance of the worm was discussed, including how it traded, as well as the innate social and ecological knowledge associated with its collection.

These focused on different community subgroups such as youth and women self-help groups, commonly known as *'societies'* in the region. Local entrepreneurs were also consulted, especially those who delve into the mopane business when there is an outbreak of the worm. These discussions captured the diverse views on how IKS has been mobilised in the mopane worm economy. Also, these discussions shed light on the challenges and opportunities in the mopane economy in light of changing ecological and economic realities.

The researcher spent time in the community observing mopane worm harvesting, processing and trade. This enabled the researcher to gain first-hand experience of the economy, especially the role of IKS and how these traditional practices have evolved over time to accommodate the changing ecological and economic realities.

Documents produced by other researchers, government agencies and non-governmental organisations were consulted and reviewed to provide this study's contextual and theoretical grounding. A few articles from newspapers helped capture the mood and sentiment, especially during an outbreak of the worm.

The data was broken down into themes and analysed, thus using NVivo and other qualitative data analysis software. Themes were arrived at by identifying patterns and categories with the data exhibiting similar traits. Key themes identified related to sustainability and economic outcomes.

Participants were fully informed about the research, its aims and its purpose. The data collection methods were fully explained to the participants, and their privacy rights were emphasised. Written and verbal consent was obtained from the participants.

The data collected will be kept confidential, and all personal identifiers or aspects that might expose the participants will be removed.

The researcher observed that the Kalanga are culturally sensitive people, and in that regard, great care was taken to respect the local cultures, traditions, and customs. This ensured that the study did not negatively portray the community.

CONTEXTUALISING KALANGA IKS AND THE MOPANE WORM ECONOMY

Clearly, there is a symbiotic relationship between Kalanga IKS and the mopane worm economy that stretches for generations (Gondo, 2018). This can be discerned in several ways. First, this relationship is seen in the sustainability of the mopane economy. For generations, the Kalanga have mastered the cyclical nature of the mopane ecosystem. In this regard, they have been able to dictate the patterns for the harvesting of the mopane worm. Certain times are chosen for the harvesting of the worm. This ensures that the worm is harvested at the most opportune time to avoid overharvesting that would endanger future generations of worms, thus showing awareness of environmental conservation (Mawere, 2013). From observation, the start of the harvesting of the worm is universally 'agreed' in the community, however, with seemingly no formal pronouncement from traditional leadership or otherwise. Notwithstanding, traditional authority is usually alert to the process and oversees it (The Herald, 13 April 2017). The worm is harvested selectively, and the Kalanga do not harvest everything from a particular tree. In this way, the balance between harvesting and conserving resources is balanced, thus ensuring sustainability and showing the centrality of IKS in resource exploitation.

Second, the harvesting of the worm is not merely an economic activity. There are cultural practices associated with it which have been observed for generations. The collection and preparation of the worm invokes the communal way of life of the Kalanga. When the outbreak is large, communities form harvesting parties which go out into the woods to camp and harvest, sometimes for weeks. In these camps, communal bonding is encouraged through shared resources and responsibilities. For instance, while some gather the worms in the forest, some are left guarding the harvested worms in the camp. In the evening, the members gather around fires to tell stories and share experiences as they disembowel the worms. In such settings, IKS is shared and passed from elders to the younger members of the party. The harvesting of the worm is associated with seasonal rituals and celebrations. These activities encourage respect for nature and sustainability. These practises help strengthen community relations and promote the preservation of Kalanga culture.

Third, Kalanga IKS acts as a bulwark against the dangers brought by the globalisation of markets. The globalisation of markets has the unintended consequence of promoting the overharvesting of the worm. The Kalanga use their IKS to discourage practises that lead to overharvesting; for example, they discourage the wasteful practice of sand frying the worms in favour of boiling them. Boiling is a tedious process, and this naturally leads to a shorter harvest. In the whole process, there are taboos associated with harvesting certain worms or the destruction of trees, as such actions are believed to disrupt the balance of nature and are frowned upon. To buttress the above, the mopane worm economy has taken on multifaceted significance for the Indigenous Kalanga and other ethnic communities in Mangwe District. It plays a central role not only in addressing nutritional needs but also in sustaining cultural practices tied to resource management and social exchange. The collection and processing of mopane worms are informed by IKS and passed down through generations, which govern the mopane economy.

THEORETICAL UNDERPINNINGS

Several theoretical ideas underpin our understanding of the Kalanga IKS and are designed to show how the Kalanga acquire, process and pass down knowledge, often rooted in their culture, environment and social ethos. The main theory used in this study is the Ecological Systems theory (EST) developed by Urie Bronfenbrenner in 1979. The Ecological systems theory is important in helping us appreciate the importance of traditional ecological knowledge systems in understanding human-environment relations. This theory extends to our understanding of the use of traditional knowledge on sustainability and conservation practices. Contextually, this theory helps us understand ecological knowledge in the context of global environmental challenges and Indigenous knowledges.

RESULTS AND DISCUSSION

An analysis of the mopane worm economy as a survival mechanism in Mangwe

The mopane worm is an important insect in Mangwe district. It has, for generations, acted as an important leeway for survival in an area that experiences intermittent droughts. Broadly speaking, the worm provides important nutritional, economic and ecological benefits to the Kalanga. Below is a discussion of some of the benefits.

For communities existing in semi-arid regions, IKS serves as a window to mitigate the marginal status. The mopane economy provides not only a healthy food source but also plays a huge role in supplementing the incomes of the families involved. Incomes and food are two resources in short supply in rural areas; hence, harvesting and dealing in the mopane worm serves as an important economic activity in rural areas where there are limited prospects for formal employment (Vedeld et al., 2007; Dube, 2024; Kozanayi & Prost, 2002). Since the 1980s, the harvesting of mopane worms has become an important cushion to those returning from South Africa or Botswana due to old age or, as Stack observed, frustrations over scarce job opportunities in those countries and the attractive income generated by the harvest of the worm (Stack et al., 2018). In the face of climate change, droughts and marginalisation, households manage diversified livelihood portfolios to ensure the availability of alternative resources in times of need as a buffer against the uncertainty of natural resource availability or as a coping strategy (Kamanga et al.,

2009). However, it is important to emphasise that, as an activity that has been passed from generation to generation, the economy is now generally viewed as an important repository of Kalanga culture.

It is plausible to argue, on empirical evidence provided by researchers such as Hobane *et al*, (1994; 2004) and Dube and Chitarika (2018; 2024), that the mopane worm business has positively enhanced livelihoods and food security in Mangwe. Dovie et al (2006) argue that the worm acts as source of income and food security in the area. These practices also enhance the economic status of women and children, arguably leading to a measure of financial and economic independence amongst the women and children in Mangwe. Additionally, the economic activity is generally easy to engage in, with little technical know-how expected of the participants, thus making it universally accessible to members of the Kalanga community. These non-timber forest products help promote informal markets in the rural areas which assist in mitigating poverty and unemployment (Dube, 2018). Over the years, rural food security has been an important discussion for government and policymakers; thus, such economic activities help promote food security in communities experiencing recurring droughts (Ndlovu et al., 2019). In that schema, the mopane worm acts as an important resource for income diversification. If properly done, the mopane worm economy can help promote sustainable development in areas experiencing similar ecological conditions as Mangwe (Ndlovu et al, 2024). Crucially, the mopane worm is renewable and thus sustainable, in addition to being marketable. As emphasised, the importance of the mopane economy to the Kalanga of Mangwe can be tied together with their resilience in the face of food insecurity and poverty(Bayegunhi et al, 2016).In explaining the above, the importance of the mopane worm to the socio-economic history of Mangwe offers an insight to how locals can harness local knowledges and resources to circumvent poverty and marginalisation and how the exploitation of local resources can be magnified to mirror the bigger picture of resource competition and the struggle to preserve Indigenous knowledge in rural areas.

Cultural Identity and the Mopane Worm Economy in Mangwe

The cultural significance of the mopane worm is closely interwoven in the IKS of the Kalanga of Mangwe. Traditional ecosystems knowledge is applied in the use of resources such the mopane worm. Mopane worms are seen as a sacred connection with nature amongst the Kalanga of Mangwe. There are beliefs that the worm is protected by the guardians of the land. In some instances, one observes that before going out to harvest the worm, rituals are performed for bountiful harvest of the worm, such as burning incense or reciting traditional clan names as one collects the worm. This is meant to connect the harvester with the guardians of the land and to show appreciation of nature, which in the long term promotes environmental awareness. Over-harvesters are usually frowned upon as being disrespectful to the land.

Shackleton (2000) posits that tradition inevitable accompanies the exploitation of resources such as the mopane worm. For example, one observes the gendered social dynamics of the harvesting of the worm in Mangwe. This reflects age-old practises where +women and children were assigned tasks deemed less heavy in a patriarchal society. Women involvement in the mopane economy provides them with economic leverage in a patriarchal society (Brockington, 2002; Madzudzo, 2015).

Amongst the Kalanga of Mangwe, women play an important role in the preservation of traditions and IKS (personal communication: Assah Khuphe, 18 April 2018). The evolution of gender roles in the harvesting and utilisation of the mopane worm attests to the reassignment of these roles in the area and society in general, and how IKS plays a role in that process. Historically, women's roles in the productive sector have been undervalued, particularly in the informal sector and subsistence (Reeves and Baden, 2000). This has led to the under researching or even ignorance of their contributions to rural livelihoods through activities such as harvesting mopane worms. In Mangwe, as is the case with most third world societies, women's labour and productivity is seen as 'elastic', providing fall back in times of crises (personal communication: William Borman Ncube, 22 April 2017). Traditionally, the mopane worm economy had been one of the significantly female (and child) dominated industries (Reeves and Baden, 2000). Often, women and young girls were at the forefront in the harvesting, preparation, and marketing of the worm. The skills of harvesting the worm are exchanged in the bush where the activity takes place. Like most forms of IKS it is learnt through observation. In that direction the economy has empowered the females in an area where economic power is mostly male dominated, thus showing how harnessing IKS can emancipate some sections of society.

Traditional knowledge systems guide in the collection, and preparation of the mopane worms. It can also be observed in the management of the worm (Shackleton & Shackleton, 2007). The management is embedded in local ethos and traditions that are primarily aimed at controlling the harvesting of the worm for sustainability (Dovie, et al, 2015). As alluded to earlier, women play a crucial role in the mopane economy, and it is their Indigenous knowledge that shapes their participation, processing and marketing (Madzudzo, 2015). IKS, especially those related to natural resources, can be very useful in promoting sustainable use of resources (Fairhead & Leach,1996). There is a need to balance ecological health with economic activities (Campbell & Shackleton, 2001). Additionally, the collection of the worm is usually done communally. This enables people to easily share the rituals of collecting the worm. For example, the community generally agrees on the timing of the harvesting. It is not a random activity, but it is done when the worms are at their peak, fully grown and about to hibernate. There are beliefs that certain birds provide the signal for the beginning of harvest (personal communication: Assah Khuphe, Makubu Village). There is also emphasis on the proper handling of the worm, especially when one is removing the guts of the worm. Generally, a harvester identifies a spot where they sit down and remove the guts in a single place to avoid polluting the environs. The mopane tree itself is not destroyed in the process as the tree is considered sacred and a source of identity to the community; hence, some areas are colloquially referred to as *Emaphaneni* (those of Mopane land).

Again, to cement the cultural significance of the worm, the techniques for the harvesting and preparation of the worm are passed down from elders to children, both boys and girls. The transfer of knowledge goes beyond the practical aspects but is, in essence, also about the passing of cultural values and norms. In this way, positive values such as respect for the environment and sharing are passed down to the next generation. In the mopane worm collection camps, stories of the community are told in the evening gatherings and this help preserve local traditions and cultural traits.

In some instances, the mopane worm is treated as sacred food. It is so because of its peculiarity to certain regions. For example, the first or last harvests are usually given to certain members of the family, or they may be prepared in a special way where rituals are performed in order to ensure the return of the worm in the next harvest. The worm is thus given a special place in the folklore of the Kalanga of Mangwe, where it is not uncommon to hear of stories where people survived on the worm in drier seasons or even droughts (Interview: Borman Ncube, Makubu Village). The worm, thus, is a potent reminder of the relationship between people and nature.

In extension, there is a strong belief that the harvest of the worm should show a clear understanding of local ecology as a way of balancing the activity with nature. The regenerative potential of the mopane forest is guarded by jealousy, and in some cases, some areas are left untapped to ensure that the worm completes its cycle.

CHALLENGES OF THE MOPANE WORM ECONOMY IN MANGWE: ENDANGERED IKS

The changing pattern of land use is having an impact on the mopane worm economy in Mangwe. To start with, the mopane worm depends on the mopane tree for its survival as if it feeds on its leaves. The destruction of the mopane tree's foliage may lead to the reduction of the mopane worm. The major problem arises from shifting rainfall patterns as a result of climate change (Mukwakwami & Moyo, 2019). With depleted foliage and little or no rainfall, the quantity and quality of the mopane worm will be compromised. If that occurs frequently, the entire industry suffers. Furthermore, deforestation for agriculture or fuel in the form of firewood is reducing the mopane forest. This, in turn, decreases the natural habitat of the worm, thus reducing its population. Without sustainable practices on the use of the mopane tree, anchored in age-old Kalanga IKS, the mopane worm industry will be in jeopardy (Makoni et al., 2020). In addition to the above, there are challenges associated with the commercialisation of the mopane worm. Local and international demand for the worm has fuelled the unapparelled harvesting of the worm (The Herald, 22 October 2016). New markets for the worm, such as in Asia and the greater Southern African Development Community (SADC), have led to the rise in the commercialisation of the worm that has put traditional harvesting norms at odds with immediate economic gains (Madzuro et al., 2019). Ecologically, overharvesting reduces the populations of the worm in Zimbabwe in general and in Mangwe in particular (Mugabe et al., 2017). Overharvesting during peak periods has been shown to greatly reduce the populations of the next outbreak of the worm (Mugabe & Gondo, 2017; personal communication: NamaVuka, Makubu village). The collection of worms without due regard to replenishment cycles is a dangerous practice that must be discouraged. In Mangwe, locals talk of a prolonged period without the worm which led to worries it had become extinct (The Herald, 20 December 2016). It is imperative, therefore, that sustainable harvesting practises, rooted both in traditional Indigenous knowledge and modern conservation techniques, are employed for the continuation of the mopane economy (Mujuru et al., 2019). Many communities in Mangwe still use traditional knowledge in the harvesting of the worm.

Despite the importance of the mopane worm to local economies, there are no clear or formal regulations to govern the practice outside the ambit of traditional custom. This regulatory vacuum creates unsustainable harvesting practises, including 'invasions' by outsiders with little regard for the environment and, indeed, local IKS as they come only for the short period of harvest. In the same vein, there appears to be little governmental support for the sector, as is seen in agriculture, for example (Ndlovu et al., 2018). There is no evidence of support in terms of extension services and research, at least at a local scale. However, traditional chiefs have sought to engage the government on the dangers of over-harvesting.

Apart from the ecological and human-induced challenges, logistical issues also impact the mopane economy. In Mangwe, for example, the marketing of the worm is hindered by the road infrastructure. Poor road networks restrict access to more lucrative markets in towns and cities (Chikova, 2016). Limited demand in the local communities, especially in times of harvest, reduces the price of the worm, thereby impacting heavily on local harvesters. In the same vein, when harvests are reduced, prices may rise sharply, thus hurting access by local consumers (Makoni & Nyahunzwi, 2015). For households that rely on the mopane worm, this can create economic instability.

On the other hand, the newly opened markets are not always easy to reach. The Asian market, for example China is difficult to access. Logistical difficulties, high transportation costs and red tape make it difficult to access these markets and reduce profitability, thus making communities in areas such as Mangwe fail to profit from their harvest (Gumbo, et al, 2022). Also, cultural and market awareness factors continue to limit the demand of the mopane worm in Asia.

In addition, the communities do not have proper infrastructure for preparation and storage, especially during peak harvest, which is usually in the rainy season. The lack of infrastructure also limits value addition in the production chain (Makombe et al, 2019). For example, value addition would result in canned or dried worms which could increase market access and improve income for local communities (Chikozho, 2020). Even though harvesters of the worm use traditional preservation methods such as salting and drying, the vagaries of nature more often than not lead to a sizeable amount of the worm going bad, thus reducing the quality and quantity of harvest. Additionally, the harvesting of the worm using bare hands is both risky and slow. Traditionally, this is the most common method, and many harvesters concurred that they find it easier that way.

SOCIAL AND GENDERED CHALLENGES

Like most rural-based economic practises, the mopane worm economy in Mangwe has not escaped the social and gendered challenges prevalent in most rural societies. Mainly, it is women and children who partake in this economy (Chikwanha, 2016). This placed a further burden on these groups of people, and like most patriarchal societies, these two groups have limited wherewithal to make financial decisions. In the end, they tend to benefit little from the proceeds of their labour. In addition, more preference is now given to formal employment and education, thus deriving the sector of important labour supplies. Cultural shifts have also resulted in the gradual decline of the importance of the mopane worm as a preferred food choice, especially amongst the younger generations. The young generation sees the mopane economy as backwards, labour-intensive, and not as rewarding as other 'modern' occupations such as gold mining (Maphosa, 2020).

Sometimes, mopane harvesting brings about disruptive behaviour, especially to school-going children. In the strictest senses, the activity could be construed as promoting child labour, especially when the profits accrued from the harvest do not reach the pockets of the children but are appropriated by the parents. However, as already alluded to, with little government intervention and the patriarchal nature of the society, children are seen as labour for the family and have little say on what they are expected to do to support family subsistence. The safety of harvesters is not guaranteed especially in the harvesting camps. There have been reports of snake or scorpion bites, as well as encounters with dangerous wild animals (personal communication: Borman Ncube, villager, Makubu, April 2024). However, it is important to stress the culturally embedded connections in the mopane economy that necessitate the participation of children, which guarantees the perpetuity of the exercise.

FURTHER RESEARCH DIRECTIONS

IKS plays an important role in the sustainable management of natural resources as shown by the Kalanga of Mangwe. Future research on this subject can provide crucial knowledge on the resilience and sustainability of livelihoods while highlighting the importance of cultural heritage. In that regard, there is a need to investigate further how these communities adapt the mopane worm economy to climate change and seasonal variations. This approach will help initiate community-based conservation policies to ensure the continued existence of the economy. Additionally, there is a need further to explore the cultural capital of the mopane economy. This will further enrich our understanding of how to intertwine the cultural capital of the mopane worm with the economic needs of the community. Additionally, the leveraging of the mopane worm economy in the economy chain remains scantily studied. There is a need to investigate the methods through which mopane worm-producing communities can increase income opportunities through value addition. Fundamentally, there is a need to research how the mopane worm economy can be integrated into regional and national policy networks. This would result in better support for the communities. Lastly, and equally important, is the need to look at how the mopane worm economy can be integrated with broader attempts at ecological restoration considering the climate 'crisis', using IKS.

CONCLUSION: KALANGA IKS AND THE MOPANE WORM ECONOMY

The chapter underscores the deeply seated relationship between culture, ecology and livelihoods in Indigenous communities in the southwestern part of Zimbabwe. The Kalanga people's understanding of the mopane worm (ambrosia Belina) demonstrates an intricate understanding of their ecosystems, resource management and sustainable patterns developed over generations. This shows the economy aligns with local cultural practices and rituals. From the foregoing, the mopane worm is not just a food source, but a viable economic activity with the potential to enhance the livelihood of its practitioners. Kalanga Indigenous knowledge systems (IKS) ensures that the mopane worm is harvested in a way that does not endanger its sustainability. With changing times, this knowledge has ensured that the Kalanga are able to integrate their mopane economy into regional and even global markets. Though this has raised concerns about sustainability, Kalanga IKS has been critical in ensuring that this activity continues. In conclusion, the mopane worm economy illustrates a successful integration of IKS and modern economic needs. The synergy of the two, IKS and economic needs provides a good case study of the use of IKS for economic development.

REFERENCES

Chifamba, N.. (2020). Climate Change and the Mopane Worm: Implications for rural livelihoods in Zimbabwe. *Zimbabwe Journal of Environmental Studies*, 45(2), 135–148.

Chikwanha, O. (2016). Gender dynamics in the Mopane Worm Economy: A case study of Mangwe District. *Gender and Development Studies*, 34(3), 205–218.

Denzin, . (2008). *Handbook of Critical and Indigenous Methodologies*. SAGE Publications. DOI: 10.4135/9781483385686

Dovie, D. B. K., Shackleton, C. M., & Witkowski, E. T. F. (2006). The role of mopane worm (imbrasia belina) in the livelihoods of the poor in Southern Africa. *Economic Botany*, 60(3), 301–310.

Dube, S. S., & Chitarika, M. (2024). Dynamics and socio-environmental Impacts of the Mopane Worm Harvesting in Rural Communities in Zimbabwe: Prospects for climate- smart approaches. *Frontiers in Sustainability*, 1(5), 1–10. DOI: 10.3389/frsus.2024.1414058

Dube T (2020), *The Kalanga in Historical Perspectives*, online publication.

Duri, F., & Mapara, J. (2007). Environmental Awareness and Management Strategies in Pre-colonial Zimbabwe. *Zimbabwe Journal of Geographical Research*, 1(2), 98–111.

Fairhead, J., & Leach, M. (1996). *Misreading the African Landscape: Society and Ecology in a forest-savanna mosaic*. Cambridge University Press. DOI: 10.1017/CBO9781139164023

Flavier, J. M.. (1995). The regional program for the promotion of Indigenous knowledge in Asia. In Warren, D. M., Slikkerveer, L. J., & Brokensha, D. (Eds.), *The cultural dimension of development: Indigenous knowledge systems* (pp. 479–487). Intermediate Technology Publications. DOI: 10.3362/9781780444734.043

Gondo R et al (2018), *Integration of Indigenous Knowledge Systems in Natural Resources Management in Hurungwe District, Zimbabwe*, International Journal of African Renaissance Studies - Multi- Inter- and Transdisciplinary 13(4): pp 1-10.

Hobane, P. A. (1994), *The Effects of the Commercialisation of the Mopane Worms: Natural Resources Management Project,* USAID Project no. 690 0251; by the same author (1995), '*Amacimbi: the gathering, processing, consumption and trade of edible caterpillars in Bulilimamangwe District*', Centre for Applied Social Studies Occasional Paper 67, Harare, Zimbabwe.

Kamanga, , Vedeld, P., & Sjaastad, E. (2009). Forest Incomes and Rural Livelihoods in Chiadzulu District, Malawi. *Ecological Economics*, 68(3), 122–131. DOI: 10.1016/j.ecolecon.2008.08.018

Kandiyoti, M.. (2021). The Ecological Impact of Overharvesting Mopane Worms: An analysis of Resource Depletion in Mangwe. *Environmental Sustainability Review*, 12(1), 46–71.

Kelman, I., Mercer, J., & Gaillard, J. (2012). Indigenous Knowledge and Disaster Risk Reduction. *Geography (Sheffield, England)*, 97(1), 12–21. DOI: 10.1080/00167487.2012.12094332

Kozanayi, W., & Frost, P. (2002). *Marketing of Mopane Worm in Southern Zimbabwe,', Institute of Environmental studies*. University of Zimbabwe.

Kugara., ... (2022), *The role of Indigenous Knowledge Systems (IKS) in Climate Change,* Handbook of Research on Protecting and Managing Global Indigenous Knowledge Systems, IGI-Global

Machingura, F. (2018). Market Volatility in the Mopane Worm Trade. *Journal of African Economic Studies*, 56(4), 134–147.

Magadza, S. N. (2006). Empowerment. *Mobilisation and Initiation of a Community Driven Project: Women and the Marula*, 7(7), 102–112.

Makombe, G.. (2019). *Economic and Infrastructure Challenges of the Mopane Worm Economy in Rural Zimbabwe,* Development. *Economic Review (Kansas City, Mo.)*, 25(2), 98–110.

Mapara, J. (2009). Indigenous Knowledge Systems in Zimbabwe: Juxtaposing the postcolonial theory. *The Journal of Pan African Studies*, 3(1), 23–19.

Maphosa, C. (2020). The Decline of Traditional Knowledge in Zimbabwe: The Case of Mopane Worm Harvesting. *Journal of Rural Sociology*, 22(3), 41–56.

Mapira, J., & Mazambara, P. (2013). Indigenous knowledge systems and their implications for sustainable development in Zimbabwe. *Journal of Sustainable Development in Africa*, 15(5).

Mary, K. (2018). Use of community cultural practices and beliefs in the conservation of Lake Baringo Ecosystem in Kenya. *International Journal Natural Resources Ecological Management*, 3(32), 14–34.

Matsika, C. (2012). *Traditional African education: Its significance to current education practises with special reference to Zimbabwe*. Mambo Press.

Mawere, M. (2010). *Indigenous Knowledge Systems' (IKS) Potential for establishing a moral, virtuous society: Lessons from selected IKS in Zimbabwe and Mozambique,* Journal of Sustainable Development in Africa (Volume 12, No.7) Zimbabwe'. *Zimbabwe Journal of Geographical Research*, 1(2), 98–111.

Mawere, M. (2013). A Critical Review of Environmental Conservation in Zimbabwe. *Africa Spectrum*, 2(48), 85–97. DOI: 10.1177/000203971304800205

Mawere, M., & Awuah-Nyamekye, S. (2015). *Between Rhetoric and Reality: The State and Use of Indigenous Knowledge in Post-Colonial Africa*. Langa RPCID. DOI: 10.2307/j.ctvh9vwc4

Mtetwa E et al (2013), *Poverty and Rural Development; Tapping from the Grassroots,* in International Journal of Humanities and Social Science Invention, vol. 2. Issue 5, pp 112-126.

Mukwakwami, S., & Moyo, M. (2019). Land Use and its effects on Mopane Worm Populations in Zimbabwe's Southern Regions. *Journal of Environmental Management*, 49(1), 32–45.

Munsaka E & Dube E (2018), *The contribution of Indigenous knowledge to disaster risk reduction activities in Zimbabwe: A big call to practitioners*, Jamba, 26;10(1):493. doi: DOI: 10.410.

Ndlovu, I., Nunu, W. N., Mudonhi, N., Dube, O., & Maviza, A. (2019). *Land use-land cover changes mopane worm harvest in Mangwe District in Plumtree*, Zimbabwe. *Environmental Systems Research*, 8(1), 98–112. DOI: 10.1186/s40068-019-0141-5

Ndlovu, J., ... (2024), *Mopane Worms(Imbrasia Belina) Value Chain: Experiences of Rural Women on the Impact of Climate Change and Environmental Sustainability in Mangwe District*, published online: http://hdl.handle.net/

Ndlovu, Z.. (2018). Government Support for Mopane Worm Harvesters in Zimbabwe: An overview. *Journal of African Policy and Governance*, 31(4), 185–199.

Noyoo N (2007), *Indigenous Knowledge Systems and their Relevance for Sustainable Development: A Case of Southern Africa,* in K Boon & L Hens, Indigenous Knowledge Systems and Sustainable Development: Relevance for Africa, Kamla-Raj Enterprises

Nyathi 1(2022), *Negotiating marginalisation: A socio-economic history of the Kalanga of Mangwe, Zimbabwe, 1940 to 2015,* unpublished PhD thesis, Rhodes University.

Pandey, D. N. (2002). (in press). Cultural resources for conservation science. *Conservation Biology*.

Reeves, H., & Baden, S. (2000), *Report No. 5,* prepared for the Department of International Development (DFID), United Kingdom.

Scoones I (2009), '*Livelihood Perspectives and Rural Development,*' in the Journal of Peasant Studies, 1(36), pp 145-158.

Shackleton, C. M., & Shackleton, S. E. (2007). The contribution of mopane worms to the rural economies of southern Africa.'. *African Journal of Ecology*, 45(1), 62–71.

Shizha, E. (2017). Indigenous knowledges and knowledge codification in the knowledge economy. In Ngulube, P. (Ed.), *Handbook of Research on Theoretical Perspectives on Indigenous Knowledge Systems in Developing Countries*. DOI: 10.4018/978-1-5225-0833-5.ch012

Sibanda, A. (2013). *Striving for Broad Based Economic empowerment in Zimbabwe: Localisation of Policies and Programmes to Matabeleland and Midlands*, a working paper submitted to Ruzivo Trust (Knowledge for Growth, Prosperity, and Life. Thrupp L.A. (1998). "*Legitimizing Local Knowledge: From Displacement to Empowerment for Third World People. Agriculture and Human Values*, (Summer Issue), 13–24.

The Herald, '*Bulilima hard hit by amacimbi extinction*', 20 December 2016. ZBC news online, '*Mopane worm expected to rack in US$500, 000,*' 7 April 2017.

The Herald, '*Kezi chief warns amacimbi harvesters.*' 13 April 2017.

Vedeld, P., Angelsen, A., Bojo, A., Sjastaad, E., & Berg, G. K. (2007). Forest Environmental Incomes and the Rural Poor. *Forest Policy and Economics*, 9(7), 156–172. DOI: 10.1016/j.forpol.2006.05.008

Warren, D. M. (1991). *Using Indigenous Knowledge for Agricultural Development*, World Bank Discussion Paper 127. Washington, D.C. World Bank, (1997). "*Knowledge and Skills for the Information Age, The First Meeting of the Mediterranean Development Forum*"; Mediterranean Development Forum.

Chapter 14
Harnessing Indigenous Entrepreneurship in the 21st Century for Sustainable Development

Bonny Ngakane
https://orcid.org/0000-0001-6832-910X
Independent Researcher, Botswana

ABSTRACT

This chapter explores the harnessing of indigenous entrepreneurship (IE) in Botswana for sustainable development in the 21st century. It examines two case studies of Indigenous entrepreneurs in Botswana, highlighting the use of traditional and modern knowledge. One of these Indigenous entrepreneurs sells donkey milk products, and another deals with fashion apparels. Data were collected through in-depth interviews with participants from two indigenous ventures in Botswana. Data were also gathered from published secondary sources with information on Indigenous entrepreneurship in Botswana and other countries like Canada, Nigeria, South Africa and Peru. The analysis revealed a lack of dedicated structures and policies for IE, with government support being generic. Indigenous Knowledge Systems (IKS) are identified as valuable capital for start-ups, aiding adaptation and innovation. Indigenous entrepreneurs primarily use IKS for product development and digital technologies for marketing. The chapter recommends enacting and implementing specific policies and structures to govern IE. It suggests the Ngakane model as a guide for policymakers to sustain IE through modern technology integration. The model also addresses challenges faced by Indigenous entrepreneurs, aiming to foster a sustainable and innovative Indigenous entrepreneurial ecosystem.

INTRODUCTION AND BACKGROUND

Indigenous entrepreneurship (IE) has been a currency of sustainable socio-economic development in African communities for centuries. Colbourne (2017) argues that Indigenous Entrepreneurship is the creation, management and development of new ventures by Indigenous people for their own benefit. Contemporary Indigenous entrepreneurs alter traditional patterns of behaviour and utilise resources in

DOI: 10.4018/979-8-3693-8447-3.ch014

the pursuit of self-determination and economic independence by self-employment (Colbourne, 2017). In Botswana, Indigenous entrepreneurship can be traced to the pre-colonial period when barter tradind was practised before the arrival of European traders and explorers. By then trade happened at a small scale among neighbours, friends, relatives and neighbouring communities. The principles of trade were deeply embedded in the cultural ethos of Ubuntu/Botho which underscores the interconnectedness of individuals within a community. Jones-Dube (1984) reveals that during the pre-colonial period, Batswana communities were largely self-sufficient, through pastoral farming, subsistence Agriculture and trading in craft products. British colonisation changed Indigenous entrepreneurship practices in the country that have sustained communities for centuries. British colonisers started to introduce European goods to Batswana natives in the early to the middle of the nineteenth century. Jones-Dube (1984) states that in the new form of trade, Western goods such as mirrors were exchanged for local products like ivory, gold and ostrich feathers. Schapera 1971(cited in Jones-Dube, 1984) indicates that Batswana also traded in livestock, meat, corn, and other food products. The trading system was not fair as colonial settlers exchanged products of insignificant value like mirrors and sugar for local products of high value like gold, livestock or grain.

As the influx of British colonial settlers and European traders continued the Batswana trading system diminished as barter trading was replaced by the trading system that involved money. People were forced off their land to give way to imperialist colonial settlers. Without access to Agricultural land where they could produce for self-sustenance, they became disempowered and were forced to work on white farms. Some migrated to seek work in South Africa as they were restricted from economic activities by colonial laws and regulations that were introduced (Jones-Dube, 1984). Some oppressive laws that were imposed on local people included the hut tax which forced men to migrate to South Africa to look for jobs in mines to generate money to pay taxes for running the European administrative tasks (Mogal-akwe, 2006). The introduction of these oppressive taxes destroyed subsistence farming and indigenous entrepreneurship that the Batswana community relied on for a living. Best (1970) reveals that after the Second World War, the colonial administration denied indigenous Batswana trading licenses and cited capitalisation, overly aggressive European competition, and lack of experience as justification for not issuing local trading licenses. Colonial laws and legislation restricted Indigenous entrepreneurship by limiting access to credit, land, markets, and education for Batswana natives (Jones-Dube, 1984). This marginalised and suppressed indigenous entrepreneurship throughout the colonial era.

Botswana gained independence in 1966 (Parson, 1984) and that was a great opportunity for the country to claim its identity back. Though colonisation through its harsh laws had negatively affected indigenous entrepreneurship, in the post-independence era the Batswana revived the spirit of entrepreneurship. The study conducted by Jones-Dube (984) found that the country promoted small and medium enterprises by enacting supportive policies such as Financial Assistance Policy (FAP), the Local Preference Scheme (LPS) and the Selebi-Phikwe Regional Development Programme (SPRDP). Botswana also established institutions such as Botswana Development Corporation (BDC) in 1970, and the Botswana Entrepreneurial Development Unit (BEDU) in 1974 (Jones-Dube, 1984). The latter asserts that LPS was introduced in 1978 to promote local purchases from indigenous manufacturers by the government, local authorities, and parastatals. This was followed by the introduction of FAP in 1982 which facilitated economic diversification through promoting local entrepreneurship which created employment and also trained citizens to upskill them (Jones-Dube, 1984).

In the 21st century, the government continued to enact policies that promoted local businesses. These policies include the Citizen Entrepreneurial Development Agency (CEDA) which was formulated in 2001 and Local Enterprise Authority (LEA) founded in 2003. Ngakane (2024) states that though the government can be commended for enacting policies that promote local businesses, no framework and policy addresses the unique needs of Indigenous entrepreneurship. Available policies focus on generic entrepreneurship and give no specific attention to indigenous entrepreneurship. Ngakane (2024) argues that Indigenous entrepreneurship is unique and different from general entrepreneurship as it uses traditional knowledge and most traditional raw materials that fall under protected or endangered species. Colbourne (2017) asserts that though Indigenous entrepreneurship encompasses a broad range of economic activities and business forms, generally, it is distinguished from the mainstream entrepreneurship by its emphasis on community, self-reliance and traditional culture and values. Hindle and Moroz (2010) view indigenous entrepreneurship as an activity focused on new venture creation or the pursuit of economic opportunity or both, for the purpose of culturally viable and community acceptable wealth creation. Ngakane (2024) realises that that Indigenous entrepreneurship has the potential of stimulating sustainable economic growth and unemployment reduction in communities.

Despite the potential that indigenous entrepreneurship has in creating employment and sustainable economic development, it still faces challenges in Botswana and other African countries. Ngakane (2024) states that indigenous entrepreneurs in Botswana face challenges such as poor pricing techniques, Lack of Copyrights and Intellectual Property Rights, lack of commercial space, geographic isolation, financial and tax problems. Quesnel (2019) argues that most indigenous entrepreneurs do not have collateral security for loans from financial institutions. Alo, Achem, Mohammed, and Abdulquadir, (2012); Oluwalana, Adekunle, and Okojie (2012) reveal that indigenous entrepreneurs in Nigeria who make black soap are not adequately supported with capital for their business ventures and this has adversely affected their operations. In Nigeria elderly women who are engaged in black soap production were not given the relevant support as they were considered old and less relevant to their societal growth (Akanle & Adewusu, 2023). The chapter discusses the harnessing of Indigenous entrepreneurship in the 21st Century to foster sustainable development with a focus on Botswana. The chapter was informed by the book written by Bonny Ngakane (2024) entitled *Praxis of Indigenous Entrepreneurship in Botswana: A shift towards a knowledge-based economy*. This chapter explores the praxis of Indigenous entrepreneurship in Botswana and examines how it can be harnessed to foster sustainable development in the 21st century. The main purpose of the chapter is to advocate for policy and financial support of indigenous entrepreneurship for economic development and sustenance. The chapter further advocates for the training of indigenous entrepreneurs and equipping them with skills and knowledge of using technology to innovate their product offering and access to markets. It begins with an introduction and background, setting the stage for the discussion. The chapter then conceptualises indigenous entrepreneurship and sustainable development through the perspectives of various scholars. It presents findings from selected case studies of indigenous entrepreneurs in Botswana. It further discusses the integration of indigenous entrepreneurship with modern innovation, drawing on literature from distinct scholars and benchmarking successful practices from other countries. The challenges faced by indigenous entrepreneurs are also discussed. In the end, the chapter develops and discusses a contextual framework for integrating Indigenous entrepreneurship with modern innovations in Botswana and concludes with a summary of findings and recommendations for policymakers, government, and other relevant stakeholders. The chapter is grounded on the sustainable development theory, because indigenous entrepreneurship is seen as a socio-economic capital that can be harnessed to attain sustainable development in communities.

INDIGENOUS ENTREPRENEURSHIP

Dana (2007), Dana (2015) and Ngakane (2024) assert that indigenous entrepreneurship is self-employment based on indigenous knowledge. Similarly, April and Itenge (2020) view indigenous entrepreneurship as involving any type of entrepreneurial activity or self-employment based on Indigenous knowledge by indigenous people. April and Itenge (2020) further note that through indigenous entrepreneurship communities use the resources available in their environment to achieve self-sufficiency and self-determination. Self-employment supports the United Nations (UN) Agenda 2030 goal 8 which advocates for decent work and economic growth through promoting inclusive and sustainable economic growth, full and productive employment and decent work for all.

Ngakane (2024) spells out that indigenous entrepreneurship has the potential to create employment and accelerate economic growth. The African Agenda 2063 reveals that Africa can develop through the utilisation of the ideas of Pan-Africanism and the vision of Africa's Renaissance. The African Renaissance calls for development based on African indigenous knowledge systems (Madlela, 2017). Indigenous entrepreneurship presents a good opportunity for Africans to venture into business and prosper using their knowledge which they already possess in abundance in their communities. The African Agenda 2063 advocates for African development that is driven by people, relying on the potential of African people. Ngakane's (2024) study reveals that though Indigenous entrepreneurship creates employment for local people and promotes socio-economic development, it still faces challenges. These challenges include laws that restrict Indigenous entrepreneurs from using traditional raw materials that are protected by law, lack of funding and policies that are tailored towards the interests and context-specific needs of Indigenous entrepreneurs.

SUSTAINABLE DEVELOPMENT

Bukovova and Foltinova (2020) assert that sustainable development means meeting the needs of present generations without jeopardizing the needs of future generations. Kapsdorferova, Kadlecikova and Svitova (2015) posit that sustainable development has been the leading motive behind mankind's political, economic and environmental activities for many years. It is a comprehensive approach that synergistically integrates socio-economic and environmental aspects, linking the present with the future (Kapsdorferova et al., 2015). Bukovova and Foltinova (2020) argue that the concept of sustainable development strives to balance socio-economic and environmental goals. Indigenous knowledge as a community resource can be used to stimulate and support sustainable development in communities.

The African Agenda 2063 Shared Strategic Framework for Inclusive Growth and Sustainable Development notes that the use of African resources would result in development that would benefit all Africans. Ngakane (2024) says that the resource that is already available in African communities is Indigenous knowledge that Africans can use to start their own Indigenous business ventures and income-generating projects to promote sustainable development in communities. The utilisation of indigenous knowledge to start up indigenous income-generating projects eradicates poverty and hunger. The 2030 Agenda for Sustainable Development's goal 1 calls for the end and eradication of poverty, and goal 2 calls for the end of hunger, and achievement of food security and improved nutrition and the promotion of sustainable Agriculture. Most local people already possess skills in Agriculture and with adequate support, if they embark on indigenous Agricultural entrepreneurial projects they will bring sustainable development to their communities. Development and self-reliance are some of the principles of Botswana's Vision 2036. The principle of development is linked with socio-economic advancement aimed at eradicating extreme poverty and inequality in the country. The principle of self-reliance underscores the importance of individual efforts leading to self-empowerment, which in turn contributes to the growth of the nation (Botswana Vision 2036 Presidential Task Team, 2016). Indigenous entrepreneurship can be used as a vehicle that facilitates sustainable development through self-reliance in communities.

SUSTAINABLE DEVELOPMENT THEORY

Zahedi (2019) asserts that the theory of sustainable development was put forward by the United Nations at a global level, and a considerable number of development experts accepted the theory and endeavored to develop and disseminate it. In the World Environment commission hosted by Norway in 1987 the world widely considered sustainable development. From then the word sustainable development has been the most repeated in several books and articles (Crowther, Seifi, & Moyeen, 2018). In 1992 at the Rio de Janeiro assembly which was considered as the largest global assembly in the 20th century, the term was reemphasised (Sedmak, 2013). Zahed (2019) argues that the integrated and balanced model of sustainable development theory is presented in a holistic perspective in subject, space and time pillars.

The subject pillar has political dimension, economic dimension, scientific and technological dimension as well as environmental and biological dimensions. Zahed (2019) says that political dimension is associated with the formulation of developmental policies from different points of views and participation of all individuals, groups, and organisations of society in the economic and political spheres of the country. Chang et al. (2016) and Koo (2017) argue that to promote sustainable development the government can support business investments with subsidies, public loans and favourable regulatory frameworks such as tariffs and tax incentives. Ngakane (2024) calls for the government of Botswana to enact tailor made policies and regulations that promote indigenous entrepreneurship in a sustainable way in the country. Zahedi (2019) views economic dimension in sustainable development as bringing about optimal and efficient use of resources and an equitable distribution of benefits. The expansion of infrastructure should be prioritised especially the infrastructure which are built on a long-term use and sustainable approach. Terra dos Santos, Frimaio, Giannetti, Agostinho, Liu, and Almeida (2023) argue that the transition to a sustainable future should balance economic performance, social inclusion, and environmental resilience, for the benefit of current and future generations. In the context of Botswana for vision 2036 to be achieved there should be inclusion of all citizens in economic activities including indigenous entrepreneurs. Ngakane (2024) states that indigenous entrepreneurs use natural resources including endangered ones that are protected species. In this scenario Ngakane (2024) advises the government to enact laws and policies that regulate the use of protected species by indigenous entrepreneurs. Doing so would balance economic development and environmental protection and preservation.

Lee and Mathews (2013) assert that technology and innovation play a critical role in expediting transition to a sustainable mode of development. Drejeris and Ozeliene (2019) view technologies focused on sustainability principles such as conservation of resources, reuse and recycling of products, improvement of energy efficiency, reduction of negative impacts on the environment and pollution as generally considered to be sustainable. Zahedi (2019) argues that the needs of the society are met through recognition, absorption, mastery of advanced technology and localisation. In this dimension, factors such as the degree of technological advancement, the level of automation, the characteristics and composition of the workforce, the degree of technical complexity and the digital divide in terms of geography between cities and villages, class, age, and gender should be taken into consideration. Botswana's vision 2036 calls for a shift towards the knowledge based economy. To achieve this there should be robust investment in science, innovation and technological infrastructure especially in rural areas where most indigenous entrepreneurs are based. Ngakane (2024) reveals that most rural areas in Botswana do not have strong technological infrastructure and most indigenous entrepreneurs are not affluent in the use of technology. There is need therefore for the government to develop technological infrastructure in rural communities and capacitate indigenous entrepreneurs with skills and knowledge of utilising technology to innovate their business processes.

Pertaining to environmental and biological dimensions Sharma and Singh (2020) assert that the United Nations's (UN) 2030 agenda for sustainable development is about economic, social and environmental justice and equity. The environmental dimension deals with the relationship between human and nature and is associated with the protection and enhancement of various physical and biological resources and ecosystems. Nature has its own complexities and regulations, and human beings are required to comply with the laws and requirements embodied in it to benefit from its resources Zahedi, (2019). In the context of Botswana since most indigenous entrepreneurs use raw materials from their immediate environments, it is necessary for the government to come up with laws that regulate the use of natural resources including endangered and protected species.

Pertaining to space pillar and time pillar Zahedi (2019) argues that the policies of an integrated and balanced sustainable development need to be implemented at local level national level, regional level and global. Ngakane (2024) realises that if the government could empower indigenous entrepreneurs with innovation and technological skills, they can be able to make strong inroads in the local, national, regional and global markets. Since sustainable development is not supposed to jeopardize the future, the development of the present generation should support future generations too. Therefore, the time pillar of sustainable development can be divided into past performance, current situation and prediction of the future (Zahedi, 2019). Ngakane (2024) views indigenous entrepreneurship as a business sector that sustained communities during the pre-colonial period, and still sustains communities at the present, and if well-funded and supported would sustain future generations.

METHODOLOGY

The study adopted a qualitative approach in which in-depth information was collected from the participants in their natural settings. The participants were purposively selected based on their practices in indigenous entrepreneurship. These were Donkey Milk Industry and Collections by Bk Proctor based in Gaborone, Botswana. The in-depth interviews with semi-structured questions were used during the interview with the director of the Donkey Milk industry and the Manager of Collections by Bk Proctor. Rutledge and Hogg (2020) assert that in-depth interviews are one of the qualitative techniques that seeks to gather detailed information on participants' experiences, and perspectives on a particular topic, process or issue. They are conducted on one which allows the researcher to observe emotions and non-verbal language (Rutledge and Hogg, 2020). The latter further addresses that in-depth interviews use questions that encourage detailed narratives. However, the weakness of the interview is that it uses a small sample size (Rutledge and Hogg, 2020). Observations were conducted on the products and services offered by indigenous entrepreneurs, as well as within their stores. Photographs of these products and services were taken to enhance and validate the data collected from interviews, thereby ensuring the study's trustworthiness.

Published secondary sources were used to further understand the praxis of indigenous entrepreneurship. This approach ensured a more robust analysis of the findings of the study. According to Bolderston (2008), the qualitative approach is relevant for organising and analysing data derived from secondary data. The latter is organised in themes in a study using qualitative. Secondary sources and case studies served as a base for the development of a contextual framework that is both empirically informed and theoretically sound, ultimately contributing to a more effective and sustainable framework for integrating Indigenous entrepreneurship with modern technologies. Published reviews are important because of their strength to gather and disseminate evidence from distinct sources and disciplines to inform a professional practice on a certain topic (Bolderston, 2008). Research ethics were observed as participation was entirely voluntary. Data were collected after participants signed their informed consent forms, agreeing to use their personal and company names. Creswell (2013) asserts that in adherence to research ethics, informed consent should be made available for participants and voluntary participation should be recognised.

INDIGENOUS ENTREPRENEURSHIP PRACTICES IN BOTSWANA IN THE 21ST CENTURY

Case studies of the Donkey Milk Industry and Collections by Bk Proctor in Botswana extracted from Ngakane (2024) are used to demonstrate the praxis of indigenous entrepreneurship in Botswana. This gives an overview of how indigenous businesses are operated and sustained in the context of Botswana. Botswana is a middle-income country where economic growth has historically depended on the diamond industry. Recently, the performance of diamond sector has declined, creating a gap in economic expansion and diversification. Botswana is also faced with the high rate of unemployment among the youth. Indigenous entrepreneurs are critical in reducing the unemployment currently faced in Botswana through job creations which can stimulate economic growth while contributing to the country's gross domestic product. Although this sector has received less attention, it is essential to foster broader economic growth locally and internationally. Given Botswana's vision 2036 of being an export-led economy, leveraging indigenous entrepreneurship can provide a significant advantage. Additionally, Botswana offers a conducive social environment for indigenous ventures. The widespread practice of IKS in the country further supports the context for indigenous entrepreneurship.

The selected indigenous businesses are based in Gaborone which is the capital city of Botswana and have expanded to other towns in Botswana. Even though the majority of Indigenous entrepreneurs in Botswana are based in rural villages, there is a gradual participation in indigenous business in towns and cities of Botswana. Both businesses started in the 21st century using indigenous knowledge inherited from their ancestors. In the past indigenous knowledge was transferred from one generation to another through oral transmission by elderly people. This was done to preserve and foster cultural practices. Therefore, entrepreneurial practices like the Donkey milk industry are based on indigenous knowledge which was passed down by parents while Collections by Bk Proctor heavily relies on both indigenous and modern technologies. Most importantly these businesses innovate their traditional products and services to meet the modern-day needs of the 21st century customer. This section unfolds as follows:

A CASE STUDY OF DONKEY MILK INDUSTRY

This is a family business founded by Mr Visagie who is the director of the company. It operates in two regions of Botswana, Werda which is in the Kgalagadi region and Gaborone which is in the South East region. Werda serves as a farming centre for producing donkeys and milk while Gaborone is a base for manufacturing final products and services. The business uses Secret Kalahari as the Brand slogan. The director emphasised that the meaning behind the secret Kalahari is derived from the secret of the Kalahari people who used to live for over 100 years. This secret was donkey milk which was used in families to keep them healthy and live longer. Furthermore, the testimony of the healing properties of donkey milk was demonstrated by the Directors' family who used it and passed it from one generation to another. Therefore, this was a source of inspiration drawn from elders who led healthy lifestyles which contributed to long life. The founder of this company is an ex-miner who started his company through market research and customer needs. The company was founded in 2015 after a market gap in donkey milk was identified. In the past donkey milk was consumed within family boundaries and this led to Mr Visagie establishing his company to mitigate this issue. Additionally, Mr Visagie reiterated that he was once diagnosed with heart enlargement and during this time, he consistently consumed donkey milk for six months. After the check-up review, Mr Visagie said that the results indicated that the heart was normal and the medical doctor advised him to continue using donkey milk.

As of today in 2024 the business is grown and operates in Gaborone and has agents all over Botswana. The business also operates online, with its products and services available on Amazon website (https://www.amazon.com/Kalahari-Secret-Moisturizing-Nourishing-All-Natural/product-reviews/B08P2CRTCR). Additionally, the business has expanded its market to South Africa, where its products can be accessed through online purchasing on the Makro Company website (https://www.makro.co.za/health-beauty/hair-care/shampoo-conditioners-masks/shampoo/secret-kalahari-donkey-milk-nourishing-shampoo-bar—100g/p/0283bf5f-363b-4d45-ab5c-28a2f175e79d).

The Director asserted that:

The donkey milk was tested and analysed by the University of Botswana based in Gaborone. The results proved that donkey milk is a healthy source of food which contains numerous nutrients and which can minimise stroke and heart diseases. This statement indicate that company follows health and safety measures of their products before they can be consumed by buyers.

Furthermore, the director claimed that the donkey milk is comparable to the human mother's milk. However, the company places a strong emphasis on the scientific examination of its indigenous products to ensure that they meet local and international safety standards before they are sold to the market. This is a good practice which can increase business trust and is likely to heighten business loyalty and maximise sales and profits. According to the Director the donkey milk was sold in unbranded bottles and later in 2016, they diversified their products into cosmetics such as bath soaps, body lotions, hand lotion, face wash and cream. He added that the cosmetic products are made from a blend of donkey milk and organic indigenous materials.

Moreover, their market and product research skills led to the introduction of supplementary herbal pills for non-communicable diseases in 2020. The latter is known as the period of COVID-19 which became an obstacle in many entrepreneurial businesses, however, this was not the case with donkey milk industry. Most of the company's herbal pills are branded with Setswana names to reflect the essence of the native language and the product's origin. African language is an important element and foundation of indigenous knowledge. Additionally, using the Setswana language to identify business products promotes culture and identity. Some of these herbal supplements names are; *Pelo* which is used to stabilise high blood pressure and reduce risks of heart attack; *Sukiri* regulates and balances blood sugar levels; *Botsogo jwa borre* enhances man's sexual desire through balancing hormones and increasing libido. These supplements also protect against prostate cancer and reduce high cholesterol; *Emang Basadi* is used for the treatment of fibroids, hormonal imbalances and period pains experienced by women (Ngakane, 2024). The company has a diverse target market including hotels which are supplied with soaps and lotions, hospitals and pharmacies supplied with supplementary herbal pills and donkey milk, and retail shops. In addition to this, they also target young and old individuals who use their products for their benefit. Some of their products images such as donkey milk, facial cream and soap are presented below:

Figure 1. Source: Secret Kalahari Facebook page, January 2024)

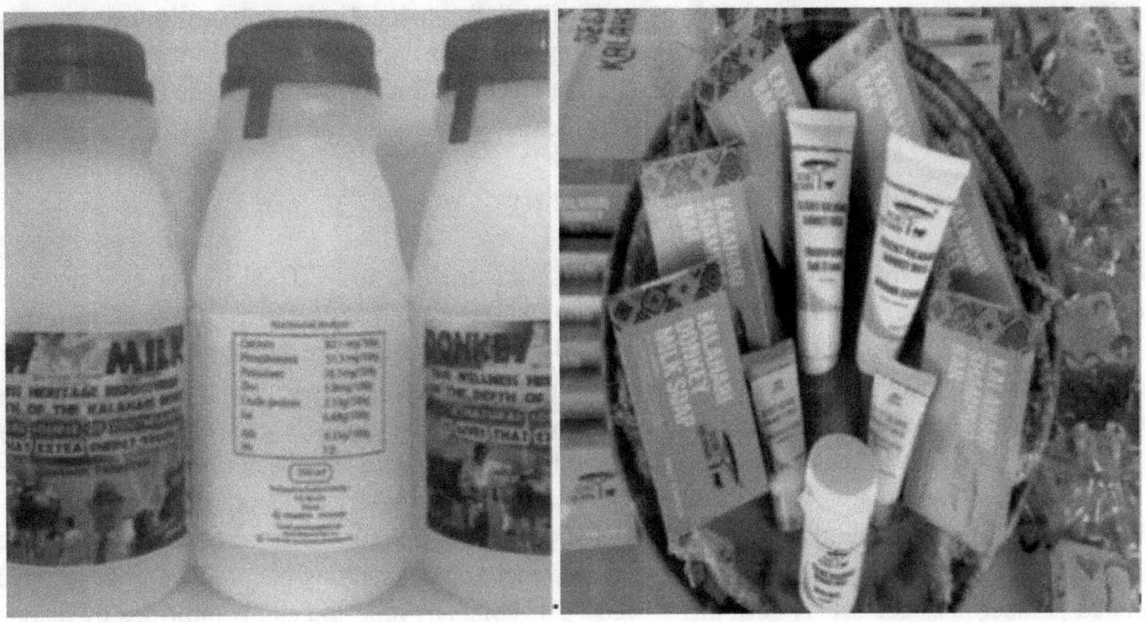

Challenges

The director reiterated that their businesses are exposed to several challenges such as a lack of understanding of indigenous knowledge by the government. Indigenous entrepreneurship is not taken as a sector with unique characteristics and practices. There are no regulated structures put in place to govern indigenous knowledge systems. Mr Visagie asserted that this gap makes their businesses suffer as they are not accommodated in governance. Furthermore, The Director contemplated that authorities require scientific knowledge to verify indigenous knowledge products while overlooking that indigenous knowledge practices are based on a spiritual belief system. Such circumstances generated difficulties in registering Donkey Milk supplementary herbal pills with the Botswana Medicines Regulatory Authority (BOMRA). The latter categorises these supplements under medicinal products despite them being supplementary health nutrition. Such confusion stems from the medicinal properties derived from the indigenous plants.

Other Challenges are associated with the lack of funds for upscaling and expansion and the community stigma over donkey milk products. This is because in the past donkeys have been over-exploited for farming activities and transportation which led people to underrate donkey milk which carries more nutritional properties. In addition to this, the lack of convenient land for breeding donkeys is scarce and this becomes a challenge because donkey milk is the main ingredient of the company's products. Since the milk is produced from Werda which is approximately 363km distance to Gaborone, transporting milk becomes a challenge. Donkey milk is perishable and needs proper transportation with adequate resources to keep the milk at the required temperatures. Another main challenge was the lack of intellectual and patent rights to protect their business ideas.

Even though the government has shortcomings as indicated in the above challenges, it supports the business in other dimensions such as invitations to local and international business expos and government approves products for public consumption. The donkey milk industry works closely with other structures such as Botswana Digital & Innovation Hub (BDIH), Botswana Investment and Trade Centre (BITC), Local Enterprise Authority (LEA), and USAID Trade Hub, on various initiatives.

Sustainable Development in the 21st Century

Donkey milk has strategic goals and practices to sustain their business. The Director indicated that company shares will be divided among his children so that they can also fully participate in the business. This is how the indigenous knowledge of donkey milk will be transferred to them as it has been a norm from the Directors' parents who passed such knowledge to them. Also the legacy of the Donkey milk industry will continue from one generation to another. Eventually, this will lead to the sustainability of the business in the 21st century and beyond. In addition, the business has been certified by the African Growth and Opportunity Act (AGOA) to trade in the Southern African Development Community (SADC), the United Kingdom, and Europe. To secure its financial growth, the company intends to be listed on the stock exchange. The director emphasised more on diversifying the product range to avoid limited product lines and satisfy customer's needs. The company relies on and is informed by market research before developing new products. Such research practices are relevant for business growth and development as well as sustaining it.

The Director asserted that they use digital technologies in marketing their products. They use branding strategy to make their products appeal to a customer in the 21st century and this is done through attractive and enhanced products packaging and colours. They use digital technologies such as social media to distribute and market their products. Their online presence is found on their Facebook page called Secret Kalahari and a website. They also use QR codes for their products as a modern tool in the 21st century. The QR code provides detailed information about their products. In addition to this Donkey milk industry collaborated with Brand Botswana which is an ambassador for promoting locally made products to the locals and international market. They co-branded with Donkey milk industry under their slogan Botswana 'our pride your destination'. This slogan promotes Botswana's identity, values, and unique attributes, fostering a sense of community and national pride while attracting global interest. Doing so is also part of branding which fosters modern technology in traditional businesses.

A CASE STUDY OF COLLECTIONS BY BK PROCTOR

Collections by Bk Proctor is a Motswana-owned business founded by Bokang Proctor who was transitioning from the music industry into being an entrepreneur in the fashion industry. They infuse elements of tradition and culture in their modern product design to resemble Batswana's cultural heritage. The business specialises in designing footwear, fashion apparel and accessories. The company name is derived from the Director's name with Bk being derived from his name Bokang. Batswana culture has a practice of using abbreviations to shorten one's name such as Bk is short for the full name Bokang. Proctor is the Director's surname while collections refers to the grouping of the products according to their similarities and name given to them. Collection by Bk Proctor was formed in 2018 and during this time it was offering footwear collections. The core principles of the business in which it was founded were research, innovation, creativity, fashion and unique designs in which every individual can express themselves. Such principles led to the business diversification into accessories and fashion apparel.

The company prides itself on its unique collections, which are designed to tell a story and foster uniqueness in the fashion industry. One of their policies is not to repeat a collection of clothing apparel, ensuring that each collection is unique to a particular period. For example, in 2022 they developed a leteisi collection which was composed of leteisi sneakers. Leteisi is commonly used to make dresses, skirts and head covers for women in particular married women and elderly people. However, Collections by Bk Proctor innovated the sneakers through an application of the leteisi on them to enhance their appearance. This Leteisi on the sneakers resembled an application of cultural heritage into modern designs. This approach promotes Collections by Bk Proctor as a native brand to the international markets while Leteisi serves as a symbol of Botswana's culture.

Another collection is the Sarona Collection which comprises traditional colours associated with the indigenous colors of Batswana. The Sarona designs are blended with modern designs to cater for customers in the 21st century. The recent collection is the Botswana collection in 2024 which consists of product designs and colors drawn from the national flag of Botswana. These products carry Blue, Black and White colours found in the national flag of Botswana. These colours are applied to the products as a sign of Batswana's national identity and origin. This is seen as a sign of using national assets such as the Flag as commodities for business gains while also respecting the community through applying their Indigenous knowledge in their products. Below are the leteisi sneakers and the jacket with Botswana flag colours (blue, black and white). Also below is the Botswana traditional dance group which assist Collections by Bk Proctor with marketing activities.

Figure 2. Source: Collections by Bk Proctor Facebook Page, October 2024

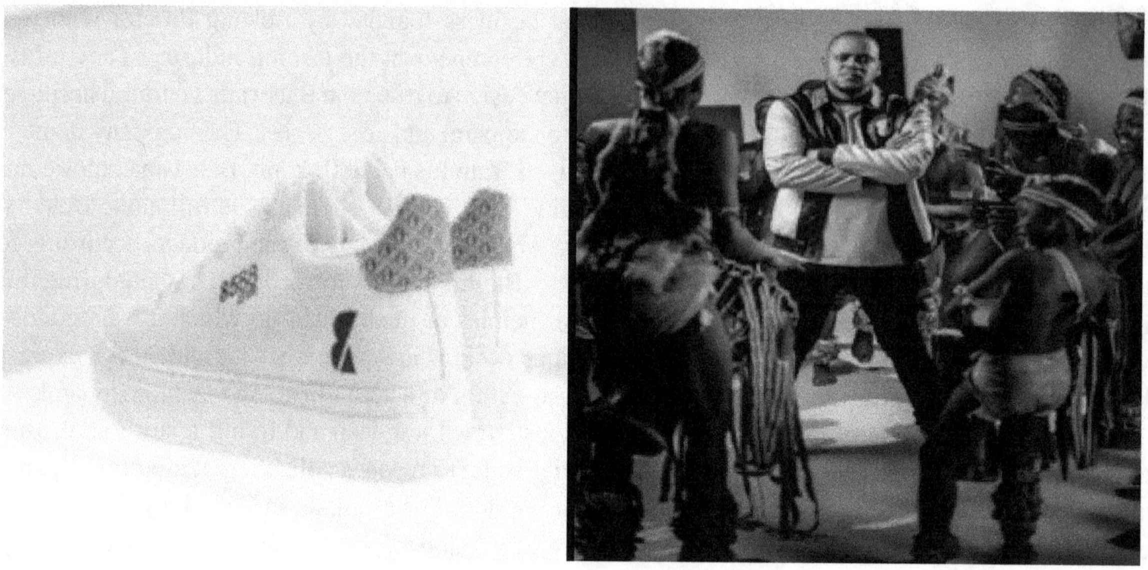

One of the managers in Collections by Bk Proctor was interviewed and asked to define indigenous entrepreneurship. He said:

> *'The concept of Indigenous entrepreneurship is multifaceted, encompassing the ability to create something from nothing with a specific intent, and the capacity to adapt existing products such as Indigenous products to serve a market niche that others may overlook'.*

This definition highlights key areas such as creativity and innovation drawn from the participant's words of the ability *to create something from nothing*. Most often indigenous entrepreneurs leverage the use of indigenous knowledge(IK) to create new products. The words *'with a specific intent'* indicate that the Indigenous entrepreneurs will have a purpose or goal to achieve such as cultural preservation, self-sufficiency and economic sustainable development. The phrase *'the capacity to adapt existing products' and 'to serve a market niche that others may* overlook' implies the ability to identify and adapt indigenous business niches that mainstream businesses can ignore. In a holistic view, this definition resonates well with indigenous entrepreneurship in modern society where IKS is used for economic gains while technology is used to foster creativity and innovation in business practices.

Challenges

Challenges occur in most entrepreneurial business practices and this does not exclude Collections by Bk Proctor. The latter face challenges associated with the lack of convenient access to raw materials such as the textile used in product design. This textile is produced outside Botswana and necessitates the import of such materials. In addition to this, the organisation does not have automation machines for manufacturing products in large quantities. In this regard, they design the prototype locally and send it to international manufacturers for mass production. Proctor relies more on research to design products and this attracts more costs. Marketing and advertising the business products attract more funds. In this regard, the company is faced with financial challenges which are incurred in research and marketing. On another note, some customers develop a negative attitude and mindset toward their products in terms of quality and pricing. However, the company is addressing this behaviour through ensuring consistency, reviews and feedback from clients. Even though this is the case, the government of Botswana is supportive of their business venture. CEDA assisted the venture with financial support and LEA has been helpful in promoting the business products and services.

Sustainable Development in the 21st Century

The company is guided by its vision which is underpinned by unique policies and procedures. They have a long term plan in which they want to change the landscape of their business operations and developments. In this plan, they have factored in the establishment of a warehouse in which their products will be manufactured. This warehouse will also be extended to assist local fashion designers. Currently, they want to grow their business and become the key leader in the fashion industry in Botswana and the international market. They also host an annual fashion Gala event to showcase and promote the business brand and interact with their customers under the slogan 'Express yourself'. Furthermore, the company collaborate with school going children to nurture talent and creativity at the grassroots level. The institution that they partnered with, aims to foster skills and knowledge of young designers and models at a tender age. Such collaborations are driving forces for sustainable development because nurturing talent at a young age is crucial for children's self-discovery and creativity. To expand into such education collaboration in schools, Collections by Bk Proctor can also bring in elements of cultural heritage to balance between traditional knowledge and skills of contemporary fashion designs.

The company invest heavily in digital technologies to mark their presence online through advertising their products on Facebook, Instagram, Twitter, WhatsApp, and their website. Their video adverts are made in a way that they present cultural traditions while selling the products to the people. They use other indigenous entrepreneurs such as popular celebrities to market their products. They also use local and international influencers to showcase their products. In the background of their video adverts appear traditional dancers, wearing their traditional clothes and playing traditional songs and dancing. This serves as a strength for showcasing culture and business to the local and international consumers.

Based on the above two case studies, the viability and sustainability of the business are guaranteed through business research, vision, and sustainable development plans. Diversification through innovation and creativity of their business products also plays a significant role in sustaining business practices. Cultural heritage preservation is an important element recognised by these businesses, as they integrate culture and traditions at all levels, including design, advertising, and sales. This approach helps avoid product resistance from local customers and presents unique products designed in Botswana to the international market. Digital inclusion in the 21st century makes the products relevant to contemporary consumers both locally and internationally, contributing to business growth and sustainability.

ECO-FRIENDLY INDIGENOUS BUSINESS PRACTICES FOR SUSTAINABLE DEVELOPMENT

Donkey milk businesses practice eco-friendly methods through emphasising the health benefits of donkey milk, a traditional practice that has existed for centuries without harming the environment. Herbs used in herbal supplement pills are harvested carefully, adhering to plant conservation management guidelines to ensure environmental sustainability. Collections by Bk Proctor also embody eco-friendly business practices by integrating indigenous and contemporary materials in their clothing designs. Although their clothing is manufactured internationally, such as in China, the raw materials used are readily available and environmentally friendly. These practices align well with Sustainable Development Goal 12, which encourages green business initiatives. By embracing the sustainable production processes outlined in Sustainable Development Goal 12, these companies contribute to a form of business development that ensures decent livelihoods, promotes economic development, and safeguards the environment (International Trade Centre, 2019).

INTEGRATING INDIGENOUS ENTREPRENEURSHIP AND MODERN INNOVATIONS FOR SUSTAINABLE DEVELOPMENT IN THE 21ST CENTURY

This section focuses on the nexus of indigenous entrepreneurship practices and contemporary innovations, highlighting their potential for sustainable development. Indigenous entrepreneurship is rooted in the application of indigenous knowledge (IK) to create or produce traditional products and services. Indigenous knowledge, as defined by Grenier (1998), is the unique cultural knowledge inherent to a specific society. This knowledge is traditionally transmitted across generations, a practice upheld by African elders who serve as the custodians of IKS. Consequently, Onwuegbuzie (2010) argues that elderly people as knowers and holders of traditional knowledge pose a threat to sustaining it over time. Equally African approaches of transferring IK such as oral communication are not compatible with long-term sustainability. Most of the knowledge has been lost from our ancestors who died with it. Similarly, Onwuegbuzie (2010); purports that IK is represented in diverse generations who carry critical experience and if it is not recorded and documented it can be lost and not recovered (Tabuti 2004; Tabuti, & Van Damme, 2012 cited in Madlela, 2017). However, in this scenario, Indigenous entrepreneurship can be incorporated into a formal school curriculum to teach children how to use IK in entrepreneurship. In this manner, it can improve documentation, reduce unemployment, promote creativity and innovation at the grassroots level and sustain IKS for future use.

Modern innovations in the context refer to the fourth industrial revolution technologies that can be used to enhance and innovate the products and services generated by indigenous entrepreneurs. Such technologies range from automation and digital technology such as manufacturing machines, computers, digital marketing technologies and Artificial Intelligence software and applications. The fourth industrial revolution is marked by digital transformation in particular to social media platforms which have changed business landscape for entreprenurs.Examples of the common social media used for marketing and reaching target market are Facebook, WhatsApp, You tube and instragram.Indigenous entrepreneurs can leverage on these platforms to reach internal and external markets without any huddles of traditional marketing which is costly and causes harm to the environment from promotional items such as magazines and flyers.

In the 21st century, indigenous knowledge faces a significant risk of extinction due to the pervasive influences of modernisation and globalisation. These forces exert considerable pressure on many societies, particularly in developing countries, to adopt Western cultural norms and practices. Consequently, contemporary generations find themselves in multicultural environments where Western culture predominates under the guise of modernity. This modernity poses a substantial threat to the sustainability of indigenous entrepreneurship, underscoring the necessity of integrating traditional practices with modern technologies to meet the needs of an ever-evolving and multi-cultural society. Furthermore, such integration will help in what I term the three 'R' Reviving, Restoring, and Reserving traditional knowledge and cultural heritage, which have been increasingly lost over the years due to the lack of formal practice and documentation. This is confirmed by Subba Rao (2006:) who states that ''IK is being produced and reproduced, discovered and lost'' whereas Nieto and Satamaria (2010) emphasise the integration of IK with modern knowledge for significant benefits.

The integration of IE and modern technologies can be catalysed through collaborations between indigenous entrepreneurs and modern technology developers. This is more likely to produce contextualised technologies that can fit into Indigenous business practices (James, 1988) since the high-tech innovation ends up being irrelevant to Indigenous entrepreneurs in the context of developing countries (Onwuegbuzie, 2010). In this regard, Onwuegbuzie (2010) acknowledges that Western corporations have started adjusting to fit in IK in developing products for developing countries. On the contrary, Africans should be self-reliant through the use of IK and modern knowledge to develop technologies that are context-specific in manufacturing and designing their business products. Over-reliant on developed countries for developing technologies that are irrelevant to Africans is likely to inhibit sustainable development solutions and economic transformation. Africans should benchmark with developed countries and assemble their technologies that best fit their cultural practices and resemble eco-friendly features considering the protection of the ecosystem which is currently facing challenges due to land exploitation and climate change. Taking India as an example, institutions that support indigenous entrepreneurship innovations and creativity were established increasing professional support and networking and mass production (Mashelkar, 2001; Gupta, Sinha, Koradia, Patel, Parmar, Rohit, Patel, Patel, Chand, James, Chandan, Patel, Prakash, & Vivekanandan, 2003). These institutions helped as the spearhead in integrating knowledge and innovations among indigenous communities (Gupta, 1995). The later indicates that these transformations improved the Indian economy and hence more likely to sustain other economies that incorporate such practices.

Challenges

Gupta et al. (2003) argue that Indigenous entrepreneurs operate on a micro-level scale due to a lack of funds and business education necessary for scaling and innovating their products. Ngakane (2024) also found that a lack of financial support is a significant challenge faced by indigenous entrepreneurs in Botswana. Consequently, there are no specific financial institutions, such as banks and initiatives, dedicated to assisting indigenous entrepreneurs with funding in Botswana. Nonetheless, Botswana has financed business start-ups through initiatives such as CEDA, Young Farmers Fund, Youth Development Fund, and Women's Economic Empowerment Programme. Although these initiatives represent an improvement in financial assistance, Indigenous entrepreneurs still lack adequate support because many are rural dwellers and illiterate, making it difficult for them to meet the requirements of such programmes, like the business plans required by CEDA. Additionally, most of these programmes focus on the youth, overlooking the majority of elders who participate in Indigenous entrepreneurship. These issues of inadequate financial support are likely to compromise the success and sustainability of indigenous entrepreneurship. In contrast, Indigenous entrepreneurs in Peru and Canada have access to funding and supportive trading agreements (Reyes Jalaskoski, 2023; Weir, 2007).

Africa faces a high rate of youth unemployment, implying that white-collar jobs have reached saturation. Despite the existence of indigenous knowledge, which can be utilised as a capital for start-up businesses, it is often overlooked, possibly due to a lack of innovation that renders it irrelevant Studies emphasize the role of indigenous entrepreneurship in tackling significant societal challenges such as crises, unemployment, and poverty, in particular in developing countries (Mafimisebi, 2016; Mafimisebi & Thorne, 2017; Amankwah-Amoah & Hinson, 2019; Onwuegbuzie & Mafimisebi, 2021). Warren (1992) reiterates that IKS which is built on generations of experience and unrecorded insights, present valuable opportunities for commercialization.

Ngakane (2024) found that Indigenous entrepreneurship and any form of IKS are not incorporated into the school curriculum from the grassroots level to tertiary education in Botswana. The formal school curriculum in Botswana is dominated by Western education, which has phased out traditional education. As of 2024, there are no institutions in Botswana specifically dedicated to training indigenous entrepreneurs. In contrast, countries like Samoa have the National University of Samoa, and Canada has Aboriginal institutions built on the cultural heritage and traditions of Indigenous communities. According to Weir (2007) cited in Ngakane (2024), the Aboriginal colleges and universities in Canada comply with the career interests of Aboriginal communities through offering programmes such as Aboriginal entrepreneurial, business, and economic development. In Botswana LEA and Botswana Innovation Hub (BIH) provide capacity-building programmes, mentoring, business incubation, and support for innovation and creativity. However, these services are generalised for the broader entrepreneurial community. In contrast, Canada has specific institutions and programmes dedicated to indigenous entrepreneurs. Therefore, it is crucial for LEA and BIH to develop targeted programmes that address the unique needs of indigenous entrepreneurs.

The National Entrepreneurship Policy (NEP) of Botswana (2019) emphasises the need for entrepreneurship education in schools. However, five years later, nothing has been done, particularly at the grassroots education level. Despite this, some tertiary institutions in Botswana have implemented entrepreneurship programmes. This is seen as a positive development, however, modern and Indigenous entrepreneurship can be integrated and embedded in school curricula to appreciate the entrepreneurial landscape of Indigenous people while strengthening and sustaining it through modern knowledge. Conversely, a curriculum should follow a bottom-up approach, where the foundation of the Indigenous entrepreneurship programme is applied from primary and secondary schools, advancing to tertiary education.

Srinivas and Sutz (2008) emphasise the need for developing specific policies in developing countries to achieve economic transformation. Similarly, Ngakane (2024) and Onwuegbuzie (2010) highlight the necessity for policies dedicated to indigenous entrepreneurs in Africa. Peru and Canada have successfully developed policies and programmes that specifically address the needs of indigenous entrepreneurs (Global Affairs Canada, 2022). Weir (2007), as cited in Ngakane (2024), asserts that the Business Development Bank of Canada (BDBC), through its Aboriginal banking programmes and services, sponsors an Internet-based national Aboriginal Youth Business Plan Competition. This competition includes online modules, mentoring, and business plan templates for Aboriginal youth in Grades 10-12. Such programmes enable indigenous entrepreneurs to incorporate e-commerce and business skills into IKS. However, in the context of Botswana, the National Entrepreneurship Policy (NEP) 2019 is generic and does not specifically address indigenous entrepreneurship. The policy focuses on general entrepreneurship and

Small, Medium, and Micro Enterprises (SMMEs) policies. Despite being designed for all entrepreneurs in the country, it has been shown that indigenous entrepreneurs do not benefit from the NEP.

Ngakane (2024), in her study comprising case studies of indigenous entrepreneurs in Botswana, indicated that there are no policies or structural frameworks dedicated to them, and they are uncertain of their place in the business sector. Additionally, they reported that some sectors, such as mainstream medical institutions, confuse their indigenous herbal products with scientific medicinal products. Within these narratives, it is evident that Indigenous entrepreneurship is unique from other scientific or mainstream entrepreneurship and thus requires its regulations and policies. Implementing such measures can contribute to the success and sustainable development of indigenous entrepreneurship, which is more likely to enhance economic growth in Botswana and Africa at large.

The vision of the NEP (2019) is "to place Botswana among the front-ranking entrepreneurial and innovative nations by 2036." This vision is crucial and relevant in the context of integrating Indigenous entrepreneurship with modern technologies to foster creativity and innovation among Batswana entrepreneurs. If Indigenous entrepreneurs are not innovative, it is unlikely that the goals of Vision 2036 will be achieved. Another important element is Indigenous entrepreneurship trading, as drawn from Canada. Canada supports Aboriginal businesses through the Canada-United States-Mexico Agreement (CUSMA) signed in 2018, which includes a chapter promoting Aboriginal participation in international trade. Although Botswana has made significant strides by becoming a member of the African Continental Free Trade Area (ACFTA) introduced by the African Union, it is unclear how this membership benefits Batswana, particularly indigenous entrepreneurs. The mandate of ACFTA is to facilitate free trade and eliminate non-tariff barriers among African countries to ensure the smooth flow of intercontinental trade. Despite the African Union's efforts to boost the African economy and be inclusive of all traders in Africa, some scholars have identified shortcomings such as over-dependence on exports, weak infrastructure, the continent's developing status, and high-security risks (Wapmuk and Ali, 2022).

Modern technologies encompassing digital and automated systems of the 21st century have drastically redefined the landscape of entrepreneurship. Selelo (2023) posits that Indigenous entrepreneurship practices, particularly in rural areas, have become increasingly dysfunctional due to the inadequate integration of technology. These Indigenous ventures gradually lose their relevance and, ultimately their viability. Selelo (2023) further argues that the root cause of this technological gap in rural village businesses is persistent technological illiteracy. To highlight the magnitude of this issue, Selelo (2023) hypothesises that some Indigenous entrepreneurs remain entrenched in practices echoing the first or second Industrial Revolutions. Globalization is not constant, and as such, indigenous entrepreneurs are likely to remain behind. To address these inconsistencies, a dual approach of technology adoption and adaptation alongside traditional business practices is necessary. Selelo (2023) supports this by stating that Indigenous entrepreneurship should be integrated with technology to create a hybrid structure.

Most importantly, Teffo (2019) accentuates that IKS can be used as a tool to bridge the digital divide and improve socio-economic conditions in rural areas. However, this is only relevant if IKS is applied within the field of indigenous entrepreneurship. In the context of Botswana, the National Vision 2036 highlights the need to achieve sustainable development goals through human development and transitioning Botswana from a middle-income to a high-income country through an export-led economy. This vision cannot be realised if digital illiteracy and a lack of innovation persist among Batswana indigenous entrepreneurs. Ngakane (2024) asserts that indigenous entrepreneurs have the potential to strengthen the export-led economy due to their unique products and services, and indigenous knowledge, such as traditional herbs, music, dance, and food and indigenous arts and crafts which cannot be found elsewhere.

Issues such as poor infrastructure, including network coverage and transport, are significant concerns for proper trading in rural areas. The Smart Botswana Connectivity programme plays a critical role in providing internet access and connectivity in both urban and rural areas. This programme ensures that internet access is available in community facilities such as healthcare centres, Kgotla, schools, malls, youth centres, government ministries and departments, libraries, and public areas. This connectivity is particularly relevant for indigenous entrepreneurs in rural areas, who can leverage free digital connectivity to promote and increase visibility for their products and services, attract target markets and increase access, acquire and address customer feedback, and respond to their needs. They can also use digital tools for the procurement of their products and services. Some indigenous entrepreneurs supply urban indigenous entrepreneurs with raw materials or products they produce. This distribution or supply chain can be efficiently managed using digital tools such as email, WhatsApp attachments, and Facebook for receiving and sending orders. Online banking tools, such as payment transfers, can bridge the gap in physical transactions, reducing transport and other unnecessary costs.

Internet technologies, such as Artificial Intelligence (AI), can stimulate innovative ideas and improve business management. AI can also serve as a translation tool from English to indigenous languages, aiding indigenous entrepreneurs in understanding new entrepreneurship developments, such as programs and policies. YouTube can assist them in connecting with other indigenous entrepreneurs worldwide who have demonstrated innovation in their services, serving as a learning tool. They can also share videos of their innovative indigenous products and receive feedback.

Indigenous entrepreneurs can form business groups to collaborate in terms of skills and knowledge and participate in initiatives such as crowdfunding for their business ventures. Although Indigenous entrepreneurs do not operate in isolation from non-indigenous entrepreneurs, digital technologies can facilitate networking and collaboration in various business operations and survival strategies. However, technology education is essential for indigenous people to learn how to operate and leverage online opportunities to enhance their business operations and expand their markets. Even though case studies of the Donkey Milk Industry and Collections by BK Proctor have incorporated digital technologies into their businesses, they still lack automation machines to design some of their products, which would reduce outsourcing costs and over-dependence from international entrepreneurs.

Of recent AI tools are being recognised and adopted and this necessitates training of indigenous entrepreneurs on the application of AI in their entrepreneurial practices. As of 2024, no studies have been conducted on the adoption of AI in indigenous entrepreneurship. Teoh and Goh (2023) articulate numerous benefits of implementing AI in business management, including enhancing decision-making in business strategies, marketing, and finance, and improving customer satisfaction. Additionally, AI can enhance accounting systems and analysis, business data generation, market research analysis, and reporting in indigenous ventures. Integrating IKS and entrepreneurship within the realm of AI poses challenges such as the risk of poaching and theft of business ideas. However, Botswana has developed a policy on intellectual property and patent rights. The Botswana Intellectual Property Policy(BIPP), operating under the Company Intellectual Property Authority(CIPA), is designed to protect the intellectual properties of indigenous entrepreneurs, including their indigenous knowledge, from being exploited and stolen by foreign investors, and to safeguard cultural products and ideas. CIPA protection extends to cultural products, ideas, and the intellectual properties of indigenous entrepreneurs, ensuring that they are secure from theft and exploitation by foreign investors, agents, and colonisers. The protection of cultural knowledge and products not only promotes local innovation and creativity but also gives

Botswana a competitive edge in local and international markets. This, in turn, can lead to economic growth and development in Botswana.

The BIPP was developed to sustain and promote the inclusivity of intellectual property. This policy can enhance the smooth integration of indigenous entrepreneurship with modern technologies without fear of losing their indigenous knowledge, ideas and products to Westerners. The participant in the Donkey Milk Industry case study highlighted the need for intellectual property rights. This can help prevent history from repeating itself, as in the past, colonisers have over-exploited Indigenous knowledge and products due to a lack of policies that govern and protect Indigenous people and their knowledge. For such policies to be even more relevant and effective they should be applied in line with the context of indigenous entrepreneurs.

Lessons Learned from Successful Countries in Indigenous Entrepreneurship

Australia, Canada, and New Zealand have successfully empowered indigenous entrepreneurs by recognizing indigenous communities in their constitutions and advancing their rights (Gover, 2011; Reyes Jalaskoski, 2023). Canada, for example, has implemented the Procurement Strategy for Indigenous Business (PSIB), which increases indigenous participation in government contracts, fostering job creation and economic growth (Global Affairs Canada, 2022). The Aboriginal Entrepreneurship Programme (AEP) further supports indigenous entrepreneurs through business advisory services, training, and market development (Global Affairs Canada ,2022). These initiatives, along with support from financial institutions, have significantly contributed to the success of indigenous businesses. Botswana should benchmark with countries like Australia, Canada, and New Zealand by introducing and implementing indigenous entrepreneurship programmes, procurement strategies, and financing mechanisms for indigenous businesses. Additionally, incorporating provisions for indigenous entrepreneurship into the Constitution of Botswana, as seen in these countries, would further support the development and empowerment of indigenous entrepreneurs.

In South Africa, government interventions such as the Traders Upliftment Project (ITUP) and policies for Small, Medium, and Micro-Size Enterprises (SMMEs) provide essential support for indigenous entrepreneurs (Zulu, 2015; Mukwarami & Tengeh, 2017). These programmes offer financing, infrastructure investment, and skill development, which are crucial for integrating indigenous entrepreneurship with modern innovation. Such support ensures sustainable development and economic participation of indigenous communities, offering valuable lessons for Botswana. Additionally, Ndivhuho (2021) asserts that policies such as Black Economic Empowerment (BEE) in South Africa and Indigenisation and Economic Empowerment (IEE) in Zimbabwe aim to promote indigenous participation in economic activities. However, the equity transfer approach of BEE has primarily benefited a few politically connected elites rather than significantly redistributing wealth to marginalized groups ((Chowa and Mukuvare 2013). Nevertheless, the BEE and IEE policies offer valuable lessons for Botswana, which currently lacks policies dedicated to indigenous entrepreneurs.

CONTEXTUAL FRAMEWORK FOR INTEGRATING INDIGENOUS ENTREPRENEURSHIP AND MODERN INNOVATION

In the absence of dedicated formal structures and frameworks that govern indigenous entrepreneurship in Botswana, there is a need to develop one. Therefore, based on Botswana's case studies above and benchmarks from other countries that have successfully implemented Indigenous entrepreneurship, such as Canada, India,Peru, South Africa and Zimbabwe as well as the literature reviewed, a framework for integrating Indigenous entrepreneurship with modern technologies for sustainable development was developed. The Kitso Boswa model in Ngakane (2024) also served as a reference tool in developing the model in Figure 1 for integrating indigenous entrepreneurship with modern technologies for sustainable development. This model has ten components that can serve as guidelines for policymakers, academics, non-governmental organisations, financers, indigenous entrepreneurs and enterprise authorities. These components are outlined as follows:

- Research and benchmarking on IE and modern innovation
- Enacting Indigenous Entrepreneurship Policy
- Curriculum Transformation
- Tailor-made funding opportunities
- Contextualisation of Automation and Digital Systems
- IE Continuing Professional Development Programme
- IE for Sustainable Development
- Evaluation and Reporting

Figure 3. Ngakane model for integrating Indigenous entrepreneurship and modern innovation

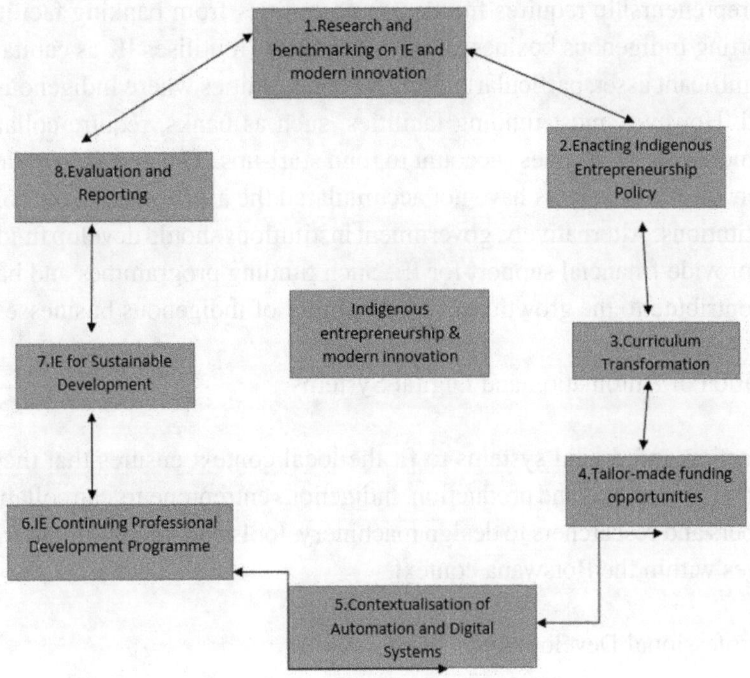

1. Research and benchmarking on IE and innovation

Researching integrating IE with modern technologies is necessary for valuable insights in the practices of IE and innovation. Through research, benchmarking with other countries who have implemented IE and modern innovation successfully can be benchmarked and best practices can be adopted and adapted to best fix the context of Botswana.

2. Indigenous Entrepreneurship(IE) Policy

Establishing policies that are dedicated to Indigenous entrepreneurship and benchmarking them against modern innovations is necessary for creating a supportive environment.

3. Curriculum Transformation

Including IE and innovation in the educational curriculum ensures that future generations are equipped with the necessary skills and knowledge for the Fourth Industrial Revolution and the 21st century. Doing so can also foster cultural preservation while generating income from traditional entrepreneurial practices. IE education will assist in documenting Batswana's cultural heritage, which is at risk of erosion. Incorporating IE in the curriculum can also help contextualise learning, generating home-grown solutions for entrepreneurship problems.

4. Tailor-made Funding Opportunities

Indigenous Entrepreneurship requires funding opportunities from banking facilities and initiatives dedicated to supporting Indigenous businesses. IE is unique as it utilises IK as capital for start-up businesses. This is a significant asset, particularly for rural communities where Indigenous entrepreneurship is widely practiced. However, most funding facilities, such as banks, require collateral security or a certain amount of money in the business account to fund start-ups. This requirement poses a setback for IE, as many Indigenous entrepreneurs have not accumulated the assets needed for collateral security in money-lending institutions. Alternatively, government institutions should develop funding initiatives that are tailor-made to provide financial support for IE. Such funding programmes and banking institutions can significantly contribute to the growth and sustainability of indigenous businesses.

5. Contextualisation of Automation and Digital Systems

Adapting automation and digital systems to fit the local context ensures that they are relevant and effective for indigenous operations and production. Indigenous entrepreneurs can collaborate with modern technology developers and researchers to design machinery, tools and systems that address their specific needs and challenges within the Botswana context.

6. Continuing Professional Development

Tailor-made ongoing training and development opportunities for indigenous entrepreneurs can help them stay updated with the latest technological advancements. Language of teaching and learning as well as learning materials should be applied according to the context of the indigenous entrepreneurs. The IE professional development programme should also consist of indigenous mentorship in particular to young indigenous entrepreneurs who can acquire indigenous knowledge and practices from elders. Indigenous entrepreneurs should be trained on intellectual property rights to safeguard their traditional knowledge and innovation from theft and over-exploitation by the mainstream knowledge. Training on available internet technologies, ICT, Cyber security and AI is crucial. Also, professional development in financial and records management, business management, banking and customer care is essential. Assessment and feedback should be provided during training to ensure that learning takes place.

7. IE for Sustainable Development

Sustainable indigenous business practices are essential for sustainable development. IKS has sustained communities for centuries, demonstrating their potential for sustainable development in intergenerational transmissions. Cultural preservation through indigenous entrepreneurship can facilitate the economic transfer of IK, reduce unemployment, and restore oral transmission while reviving IK practices. Integrating IE with modern technologies is a safe practice that values old traditions and modern advancements without compromising traditional culture while benefiting from modern technologies. These practices can economically empower Indigenous entrepreneurs as custodians of IKS and promote the socio-economic inclusion necessary for sustainable development. Creative and effective strategies should be implemented to promote sustainable development by merging indigenous knowledge and modern knowledge in entrepreneurial activities.

8. Evaluation and Report

Formative and summative evaluation throughout the process and reporting can help track progress and identify areas for improvement.

CONCLUSIONS

This chapter highlights the importance of integrating indigenous entrepreneurship with modern innovations to achieve sustainable development in Botswana. By leveraging technology for product design, packaging, and digital marketing, IKS can be preserved and made relevant for the 21st century. Incorporating indigenous entrepreneurship into mainstream curricula fosters creativity and counters perceptions of indigenous products as outdated. Achieving the UN Agenda 2030 and African Agenda 2063 goals requires this integration to promote inclusive economic growth and socio-economic development. Ngakane model framework is essential to guide stakeholders in implementing these practices to overcome current challenges and support indigenous entrepreneurs. Future research should focus on refining this Ngakane model Framework and exploring additional ways to integrate technology with indigenous knowledge system practices, ensuring sustainable development and cultural preservation.

RECOMMENDATIONS

- It is recommended that the challenges faced by IE should be addressed using the Ngakane model. This model, developed based on participants' views and benchmarks from successful countries in indigenous entrepreneurship, provides a comprehensive framework for solutions.
- Policymakers and the government of Botswana should adopt and adapt Ngakane model as a framework for designing tailored- made policies that address the specific needs of indigenous entrepreneurs.
- It is also recommended that tailor-made training programmes and funding initiatives dedicated for indigenous entrepreneurs should be developed.
- The chapter further recommend the integration of IE with modern innovations including automation and digital technology to foster inclusion on the forth industrial revolution. Doing so, can enhance product design and packaging and preserve cultural heritage while making IKS relevant to the intergeneration of modern-day society.
- The modern curriculum should incorporate IE into mainstream entrepreneurship programmes. This integration acknowledges both the unique features of IE and modern entrepreneurship practices, fostering business skills, creativity and innovation that align with the needs of 21st-century consumers.
- The Ngakane model should serve as a guideline for policymakers and other relevant stakeholders in Botswana to implement IE for sustainable development. This framework can also be adapted and contextualised by other African countries with similar contexts to Botswana.
- By adopting these recommendations, Botswana and other African countries can leverage on IE to achieve sustainable development goals, promote socio-economic growth, and address issues such as poverty and hunger, as outlined in the UN Agenda 2030 and the African Agenda 2063.

FUTURE RESEARCH DIRECTIONS

Given that this study employed a qualitative approach, future research could benefit from adopting quantitative methodologies to quantify the participation of Indigenous entrepreneurs in both urban and rural settings. Such studies could provide statistical insights into the prevalence and distribution of Indigenous entrepreneurship across different regions. Additionally, further research could explore the diverse practices of indigenous entrepreneurship within various sectors, including health, tourism, and the creative arts. This would offer a comprehensive understanding of how indigenous entrepreneurial activities are integrated and manifested in different industries. Moreover, there is a significant opportunity to investigate the role of Artificial Intelligence (AI) in enhancing indigenous entrepreneurship, particularly within the context of the Fourth Industrial Revolution and Globalisation. Studies in this area could examine how AI technologies can be leveraged to support and scale indigenous business ventures, thereby contributing to sustainable development.

REFERENCES

African Union. (2015). *African Agenda 2063 – The Africa we want, Framework document*. Addis Ababa.

April, W. I., & Itenge, D. I. (2020). Fostering indigenous entrepreneurship amongst San people: An exploratory case of Tsumkwe. *International Journal of Business and Globalisation*, 24(4), 496–512.

Best, A. C. (1970). General trading in Botswana, 1890–1968. *Economic Geography*, 46(4), 598–611.

Bolderston, A. (2008). Writing an effective literature review. *Journal of Medical Imaging and Radiation Sciences*, 39(2), 86–92.

Bukovova, S., & Foltinova, L. (2020). Sustainable development and its impact on the quality of human capital. In *SHS Web of Conferences* (Vol. 83, p. 01005). EDP Sciences.

Chang, Y., Fang, Z., & Li, Y. (2016). Renewable energy policies in promoting financing and investment among the East Asia Summit countries: Quantitative assessment and policy implications. *Energy Policy*, 95, 427–436.

Chowa, T., & Mukuvare, M. (2013). An analysis of Zimbabwe's indigenisation and economic empowerment programme (IEEP) as an economic development approach. *Journal of Economics*, 1(2), 2–18.

Cresswell, J. W. (2013). *Qualitative inquiry and research design: Choosing among five approaches* (3rd ed.). Sage Publications.

Crowther, D., Seifi, S., & Moyeen, A. (2018). The goals of sustainable development.

Collections by Bk Proctor. (2024, October). Figure 2. [Facebook page]. Retrieved from https://www.facebook.com/collectionsbybkproctor

Dana, L. (2007). A comparison of indigenous and non-indigenous enterprise in the Canadian sub-Arctic. *International Journal of Business Performance Management*, 9(3), 278–286.

Dana, L. (2015). Indigenous entrepreneurship: An emerging field of research. *International Journal of Business and Globalisation*, 14(2), 158–169.

Drejeris, R., & Oželienė, D. (2019). New approach to the technological aspect of corporate sustainable development. *Business: Theory and Practice*, 20, 363–371.

Global Affairs Canada. (2022). Indigenous business and federal procurement. Available from: https://www.sac-isc.gc.ca/eng/1100100032802/1610723869356. Accessed on 29 September 2024.

Global Affairs Canada. (2019). Trade and Indigenous Peoples Summary. Available from: https://www.iscsac.gc.ca/eng/1582037564226/1610797399865. Accessed on 15 November 2023.

Gover, K. (2011). Indigenous rights and governance in Canada, Australia, and New Zealand. Retrieved by Oxford Bibliographies.

Government of Botswana. (2016). [– *Achieving prosperity for all*. Lentswe La Lesedi] [PTY] [LTD.]. *Vision (Basel)*, •••, 2036.

Grenier, L. (1998). *Working with indigenous knowledge: A guide for researchers*. International Development Research Centre.

Gupta, A. (1995). People's knowledge for survival: Grassroots innovations for sustainable natural resource management. Paper presented at the IFADs International Conference on Hunger and Poverty.

Gupta, A. K., Sinha, R., Koradia, D., Patel, R., Parmar, M., Rohit, P., Patel, H., Patel, K., Chand, V. S., James, T. J., Chandan, A., Patel, M., Prakash, T. N., & Vivekanandan, P. (2003). Mobilizing grassroots' technological innovations and traditional knowledge, values and institutions: Articulating social and ethical capital. *Futures*, 35(9), 975–987.

Hindle, K., & Lansdowne, M. (2005). Brave spirits on new paths: Toward a globally relevant paradigm of indigenous entrepreneurship research. *Journal of Small Business and Entrepreneurship*, 18(2), 131–141.

Jones-Dube, E. (1984). *Indigenous and non-indigenous entrepreneurs in Botswana: Historical, cultural and educational factors in their emergence*. University of Massachusetts Amherst.

Kapsdorferova, Z., Kadlecikova, M., & Svitova, E. (2015). Utilization of quality managerial systems in business entities in the Slovak Republic. *International Journal of Qualitative Research*, 9(2), 197.

Koo, B. (2017). Examining the impacts of feed-in-tariff and the clean development mechanism on Korea's renewable energy projects through comparative investment analysis. *Energy Policy*, 104, 144–154.

Lee, K., & Mathews, J. (2013). Science, technology and innovation for sustainable development. CDP Background Paper Series (16).

Madlela, B. (2017). Incorporating indigenous knowledge systems into the formal natural science curriculum: A case study of schools in Pongola, KwaZulu Natal Province. University of South Africa. Available at https://hdl.handle.net/10500/28656 [Accessed on 28 September 2024].

Mashelkar, R. (2001). *The Indian innovation system. Development Outreach*. World Bank.

Mogalakwe, M. (2006). From pre-colony to post-colony: Continuities and discontinuities in political power relations and governance in Botswana. *Journal of African Elections*, 5(2), 5–20.

Mukwarami, J., & Tengeh, R. K. (2017). Sustaining native entrepreneurship in South African townships: The start-up agenda.

National Entrepreneurship Policy. (2019). *National Entrepreneurship Policy of Botswana*. Ministry of Investment, Trade and Industry.

Nieto, M. J., & Santamaría, L. (2010). Technological collaboration: Bridging the innovation gap between small and large firms. *Journal of Small Business Management*, 48(1), 44–69.

Ngakane, B. (2024). *Praxis of indigenous entrepreneurship in Botswana: A shift towards knowledge based economy*. Poeticblood Publishers.

Onwuegbuzie, H. N. (2010). Sustainable development strategy using indigenous knowledge and entrepreneurship. Available at *SSRN* 1841787.

Parson, J. (1984). *Liberal democracy and the labour reserve in southern Africa*.

Reyes Jalaskoski, R. (2023). Integrating indigenous entrepreneurs into the global economy: Identifying the main obstacles and opportunities of indigenous-led ventures in Peru.

Rutledge, P. B., & Hogg, J. L. C. (2020). In-depth interviews. *The International Encyclopedia of Media Psychology*, 1-7. Available at https://www.researchgate.net/publication/345737833_In-Depth_Interviews [accessed October 17 2024].

Secret Kalahari. (2024, January). Figure 1. [Facebook page]. Retrieved from https://www.facebook.com/secretkalahari/

Sedmak, C. (2013). Sustainability: Ethical perspectives. In *Sustainable entrepreneurship: Business success through sustainability* (pp. 51–65). Springer Berlin Heidelberg.

Selelo, M. E. (2023). Indigenous Entrepreneurship and the Fourth Industrial Revolution in South Africa. *African Journal of Development Studies*, 2023(si1), 129.

Sharma, P., & Singh, K. (2020). Sustainable development: Dimensions, intersections and knowledge platform. In *Sustainability* (pp. 43–68). Fundamentals and Applications.

Srinivas, S., & Sutz, J. (2008). Developing countries and innovation: Searching for a new analytical approach. *Technology in Society*, 30(2), 129–140.

Subba Rao, S. (2006). Indigenous knowledge organization: An Indian scenario. *International Journal of Information Management*, 26(3), 224–233.

Teffo, L. (2019). Harnessing IKS and technologies in the quest for sustainable rural economic development. *Indilinga*, 18(2), 245–256.

Terra dos Santos, L. C., Frimaio, A., Giannetti, B. F., Agostinho, F., Liu, G., & Almeida, C. M. (2023). Integrating environmental, social, and economic dimensions to monitor sustainability in the G20 countries. *Sustainability*, 15(8), 6502.

Tshikovhi, N. (2021). A theory of indigenous entrepreneurship in the modern African state: The case of South Africa and Zimbabwe. International Journal of African Renaissance Studies - Multi-, Inter- and Transdisciplinarity, 16(2), 21-60.

UNDP. (2015). *Sustainable development goals*. United Nations.

Wapmuk, S., & Ali, J. M. (2022). The African Continental Free Trade Area (AFCFTA) and regional economic integration: Prospects and challenges. *Zamfara Journal of Politics and Development*, 3(1), 15–15.

Weir, W. I. (2007). *First Nation small business and entrepreneurship in Canada*. National Centre for First Nations Governance.

Zahedi, S. (2019). Sustainable development theory: A critical perspective and an integrative model. *Journal of Economics and Sustainable Development*, 10(21), 43–52.

Zulu, L. (2015). Minister of Small Business Development, Small Business Development Budget Vote 2015/16. South African Government. Available at http://www.gov.za/speeches/minister-lindiwe-zulu-small-business-development-deptbudget-vote-201516-20-may-2015-0000

ADDITIONAL READING

Colbourne, R. (2017). Indigenous entrepreneurship and hybrid ventures. In *Hybrid ventures*. Emerald Publishing Limited.

Digital Connectivity Standards for Villages in Botswana. Available at https://smartbots.gov.bw/sites/default/files/202110/Digital_Connectivity_Standards_V4_2021.pdf. Accessed on 10 October 2024.

Dode, R.O. & Egugbo, C.C. (2019). Modernization theory and African development in the 21st century. *KIU Journal of Social Sciences, 5*(1): 85-94. Du Plooy, G. 2017. *Communication Research Methods*. Landowne: Juta.

Onwuegbuzie, H. N., & Mafimisebi, O. P. (2021). Global relevance of scaling African indigenous entrepreneurship. *Technological Forecasting and Social Change*, 166, 1–16.

Onwuegbuzie, H. N., Mafimisebi, O. P., Okupe, A., & Orighoyegha, E. (2022). Indigenous Knowledge and Africapitalism: An Unexploited Source for Sustainable Development. *Journal of Entrepreneurship and Innovation in Emerging Economies*, 8(2), 244–257.

Sustainable Development: Sustainable development is a holistic approach that aims to meet the needs of the present without compromising the future generations to meet their needs, through balancing socio-economic and environmental goals and integrating these aspects in sysnergy (Bukovova & Foltinova, 2020; Kapsdorferova, Kadlecikova & Svitova, 2015).

KEY TERMS AND DEFINITIONS

Indigenous Entrepreneurship: Indigenous entrepreneurship is the creation, management, and development of new ventures by indigenous people, based on indigenous knowledge, to achieve self-sufficiency, self-determination, and community benefit (Dana, 2015; Ngakane, 2024; April & Itenge, 2020; Hindle & Lansdowne, 2005).

Chapter 15
Locating the Doma Indigenous Knowledge Systems in Changing Environments:
An Exposition of the Doma Way of Life in the Zambezi Valley, Zimbabwe

Kwashirai Zvokuomba
https://orcid.org/0000-0002-8152-9621
University of Johannesburg, South Africa

ABSTRACT

The chapter examines how the Doma people in the Zambezi Valley maintain their traditional knowledge systems in a changing environment and the vulnerabilities associated with it. They retain a reclusive semi-nomadic culture and are less known to the outside world. An ethnographic design within the qualitative research approach was utilised. The analysis was guided by Antonio Gramsci's cultural hegemony theory and established that although the Doma people maintain part of their Indigenous traditions as reflected in their daily lives and copying strategies, field base evidence shows mounting pressure that erodes traditions and creates vulnerabilities amongst them, hence, the primary purpose and thrust of the study is to establish how the Doma knowledge system is being retained as well as the challenges impressed upon it.

INTRODUCTION

The study reflects on the surging importance of Indigenous Knowledge Systems (IKS) in development discourse, which is a paradigm shift from the colonial theory, which for a long time portrayed African traditions as 'backward,' 'savage', and 'uncivilised' (Chanza & De Wit, 2013). Any acknowledgement of the importance of IKS has been made in wildlife conservation, medicine and health care, agriculture and food security, and conflict and disaster management. This study situates the Doma culture and way of life in a broader context of a fast-changing environment characterised by the use of technologies, internationalised standards and best practices, food security and information dissemination and many other developments in the rural enclave. However, despite the ability to maintain their cultures and IKS,

DOI: 10.4018/979-8-3693-8447-3.ch015

the Doma is undergoing a mixture of experiences which have seen them being pushed to the 'margins of the margins' (Rutherford, 2017) with the revolving wheel of time as well as the global 'village' transformations. Climate change and the disasters induced by the same changes have serious implications for the natural environment, which hosts Indigenous knowledge systems in many parts of the world. Global, national and local developments, which affect, for example, the Amazon Forest, mining activities, dam constructions, urbanisation projects and the wide globalisation drive directly and indirectly impact traditional systems and conservative culture. Both classic and contemporary scholarship point to the importance of Indigenous knowledge systems and the need for communities to apply them in the wake of these identified transformations and catastrophes. Even at the international level, a myriad of policies and laws do not only acknowledge but show concerted efforts by global leadership on the critical importance of Indigenous knowledge systems in development. Bol and van Niekerk (2024) identify the Sendai Framework for Disaster Risk Reduction and the Paris Agreement on Climate Change. Thus, Therefore the study is not part of the romanticisation of IKS issues but explores its place and space in development discourse by zeroing in on the trends, trajectories and experiences of the Doma people located on the frontier communities in the northern part of Zimbabwe along the mid-Zambezi Basin. The study raises questions and answers by fronting empirical evidence on Doma's resilience and vulnerabilities within the context of the highlighted changes at both the local and international levels. Against this background, the study sought to explore and debate, unpacking issues through an exploratory and ethnographic engagement of the Doma culture, their Indigenous knowledge systems and how they adapt to changes in the Zambezi Valley.

CONTEXTUALISATION AND RATIONALE

The rationale of the study originated from the fact that few scholars have investigated the cultural, political and economic life of the Doma people. This is partly because the Doma is geographically located in an isolated area of the Zambezi Valley, which is difficult to navigate due to poor communication infrastructure, particularly on roads and telephones (Lan, 1985). This question of underdevelopment, highlighted by Lan (1985), made a progressive appearance in researchers' work, including Hasler (1996), Hulme and Murphree (2001) and recent work by Matema (2019) and Zvokuomba (2023). This limited literature scholarship concurs that the semi-arid region of the Zambezi Valley is characterised by harsh climatic conditions, very high temperatures and regions that are tsetse infested. Hence, these conditions discourage researchers. This is also noted by Sachez (2018) in an analysis of the Zambezi Basin water-food security nexus. Hasler (1996, p. 56) acknowledges that little has been written about the Doma despite the richness of their knowledge systems. In his book *Guns and Spirit Mediums*, Lan (1985) says, "They (the Doma people) are a small population, and in keeping with their elusive and recalcitrant style they appear only occasionally in this book" (p.17). Due to the nature of their lifestyle, the Doma people have been shrouded in mystery, racist mythology by colonially oriented writers and, in recent years, misconceptions even by local news reporters. For example, the 18 July 2012 *Herald Newspaper* had a headline titled, 'Doma people stuck in Stone Age' (*Herald Newspaper*, 18 July 2012). Chiwome, Mguni and Furusa (2000) concur with recent scholarship, Matema (2019) and Zvokuomba (2023), that myths about the Doma exist because little is written about them by researchers and policymakers. Hence, their stories have been told by travellers, hunters, and tourists who can transverse the bad terrain around the Zambezi Valley for non-scholarly purposes. As already alluded to above, the media has given a limited account of the Doma culture's situation. Where newspapers have done so, it has been intentional to sell news by depicting them as 'primitive'. Few studies about the Doma by Cutshall (1990), Nhira (1989), Hachipola (1998), Lan (1985), Hasler (1996), and Chiwome et al. (2000) did not focus on the Doma traditional knowledge system, which is central in this discussion. The post-2000 era is equally not endowed with studies about the Doma, except for Kanengoni (2017), Zvokuomba (2023) and Matema (2019). Undoubtedly, academic research on the Doma ought to be deliberately commissioned to narrow the knowledge gap about this special group. Some recent studies (Ndungu et al., 2023; Bola et al., 2014; Matema & Anderson, 2015; Ncube et al., 2017) on Communal Areas Management Programme for Indigenous Resources (CAMPFIRE) within the broad wildlife management mention the Doma people in the community-based wildlife management without devoting a broad and focused deliberation.

Therefore, this chapter about the Doma traditional knowledge system is meant to be part of building new knowledge on the Indigenous knowledge systems in Zimbabwe, with special attention to minority and isolated groups. It is equally part of building literature about the Doma people and adds to the already existing few. Thus, the sustained argument in this write-up is that the production of knowledge and documentation of Indigenous knowledge systems in pre-colonial and colonial periods has been skewed as anthropologists gave Eurocentric perspectives and worldviews. Contemporary literature ought to generate a de-colonised perspective based on new evidence. Slikkerveer (1999) says, "… the Indigenous systems are still very imperfectly recorded and understood, and much more research is required" (p. 83). Boesen, Maganga, and Odgaard (1999) also assert that "…it seems that despite the growing literature on Indigenous technical knowledge, the study of it still has a long way to go" (p. 83). Muwanga-Zake's (2014, 2017) argument also goes beyond recording Indigenous knowledge systems but introducing them into the local curriculum and generating a counter-Western value system in present and future generations. Therefore, the Doma study has been an 'ought to do' project in terms of generating new knowledge on culture and development. The lack of a clear policy framework in Zimbabwe on Indigenous knowledge systems, their role, reconstruction and preservation, despite an acknowledgement by policymakers of its importance in development discourse, has also made the Doma study justifiable. As mentioned earlier, the case of the Doma vulnerabilities has to be known beyond the Zambezi Valley and build a sense of agency and advocacy to galvanise scholars and policymakers, similar to the case of San or Bushmen which was roused initially by classical writer Kohler (1959) in the work titled 'The Study of the Gobabis district' as well as a more recent script by Ferguson (2013) titled the '*Declaration of Dependency*'. These academic writings brought the situation of the San on the world map; hence, it is within the same breath that the Doma case may be advocated.

The Doma fit into the definition of Indigenous people of the Zambezi Valley, as postulated by Kuhnlein et al. (2009) that Indigenous people "…are those who retain the knowledge of their land and food resources rooted in historical continuity within their region of residence" (p. 3). The definition is buttressed by The Inter-governmental Science Policy Platform on Biodiversity and Ecosystem Services (IPBES 2021), as cited by Muwanga-Zake and Kibukamusoke (2024) that "…it is the knowledge, practice and innovations embedded in the relationship of Indigenous peoples and local communities to nature" (p. 60). Posey (1999) qualifies the Doma as Indigenous people by saying, "communities…people… which, having been a historical continuity with pre-invasions and pre-colonial societies that have developed on their territories, consider themselves distinct from other sectors of societies now prevailing in those territories or parts of them" (p. 3). The Doma's oral history makes them the Zambezi Valley's Indigenous people. Cutshall (1990), however, adds a dimension to the debate and history by arguing that there are speculations about the origins of the Doma, who are believed to be the same group of people as the Chikunda but separated ways at some point unknown by historians and anthropologists. Isaacman (2000) provides another speculation that the Doma could have come from Chicoa or Songo, an area around Kabora Bassa gorge in Mozambique, sometime in history and moved to their present location well before the Chikunda moved to the Kanyemba area. This narrative fits in with many historians today who state that the Doma left Kabora Bassa gorge running away from the Portuguese slave raiders who had arrived and were capturing local populations, making them workers in their plantations. The third speculation raised by Beach (1980) is that the Doma are the descendants of the Shona sub-ethnic group, the Korekore, who could have decided to live in isolation around the Chewore Mountain to avoid raids from the Chikunda, the Kololo, the Lozi and the Ndebele armies. A study by Nhira (1989) states that the Doma lineage was linked to a grand ancestor called Nyamapfeka. Nhira, (1989) and Cutshall (1990) concur that the Doma might have been a low-status group in this era of raids and counter-raids, who depended on the rouged terrain to live an isolated and independent existence. This argument fits into the contemporary cultural practice of the Doma. Hence, the origins of the Doma people remain shrouded in mystery, with no clear explanations. The issue is even more complex as intermarriages have occurred with other groups, such that complexities about ethnic identity have become so rife, an argument supported by Hasler (1996) and Kanengoni (2017).

While the case of the Doma people is exceptional in Zimbabwe and Southern Africa, evidence from other regions of the world shows that through policy dialogues, research and publishing, the minority and Indigenous groups' situation attracts government, non-governmental and private players' attention. Perz and Smith (2019) identify the Yaminahua people of the Brazilian Amazon forest whose lifestyle has similarities with the Doma and said, "Their main settlement activities consist of gathering, hunting, fishing, the sale of crafts and Brazilian nuts and eventually the sale of labour to Brazil…" (p. 95). Athayde, Silva-Lugo, Schmink and Heckenberger (2017) equally studied another group in the same region, the Kaiabi group and pointed out that their IKS is well documented using digital technologies, films and photos. The Amazon Forest is transboundary as it covers the Brazilian part, Peru and Bolivia, and the Indigenous people's well-documented cultures (Fabiano et al., 2021). A more compelling case is that of the Maori in New Zealand and Australian aborigines. Keane, Raciti, van der Westhuizen, Motala, Stanton, Gilbey, et al. (2023) are Indigenous Australians and academics who not only write about the Aborigine's story but do advocate and participate in the Aboriginal Indigenous rights movement. Keane et al. (2023) regard them as First Nation people in Austrian identity politics. Among them, Kathryn Gilbey is a Director at Batchelor Institute while Aunty Sue Stanton is a cultural leader in the same academic institute. Due to this level of engagement and involvement in the Indigenous rights movement and national leadership, the Aborigines and the Torres Strait Islanders, as the main Indigenous groups, formed a political party which represented their interests in parliament (Keane et al. ibid). While the Doma people are far from this form of blending Indigenous knowledge systems in contemporary socio-economic and participating in the Indigenous rights movement, researching, documenting, and education anchored on their natural endowments becomes the foundation for preserving their IKS for transmission to future generations. This generic background to the study helps give context to the investigations done about the Doma and creates a link with the fieldwork analysis provided below.

METHODOLOGY AND THEORETICAL DELINEATION

The collection and generation of data were centred on ethnography in which the researcher carried out interviews and observations and involved research participants in a participatory fashion linked to Chambers's (1991) classic approach of 'putting the last first'. Denzin and Lincoln (2017) and Lopez-Deflory, Perron and Miro-Bonet (2022) note that participatory research provides a 'voice' to research participants, especially in the context of vulnerable groups. The researcher shared short but extended periods with the Doma people and participated in their daily activities, including hunting excursions and foraging in an ethnographic approach. The Doma research participants and community leaders led the research outings, sometimes conversing in their language, and assisted in carrying the research kit and using it to take photographs during the fieldwork. Thus, the research process was located in the qualitative principle of interacting with research participants in their natural setting, in line with Neuman's (2007) conceptualisation of naturalism as that dimension of qualitative research that involves observing events and social phenomena in the natural order. The three community leaders with smart cell phones also used their digital devices to augment the capturing of issues, whilst another used the researcher's camera. The Doma thus used media digital devices in the study as the researcher guided them. The participatory transact walks across the research site allowed the researcher and the team of Doma interlocutors to discuss with non-Doma people in the area: the Chikunda and the Shona-Korekore who live in the same Wards. A sense of responsibility, ownership, and representation of the whole Doma community was built during the data generation period. Hence, the study enhanced their confidence and a sense of agency about the importance of their IKS.

For the purpose of the study, a Ward is defined in terms of the Rural District Councils Act Chapter 29: 13 of 2002 as a subdivision of a district with clearly defined boundaries, represented by an elected member of the local council. The study focused on Wards 1, 2 and 11, the geographical spaces where the Doma people were domiciled. According to the ZimStat (2022) report, Ward 1 had 1 893 males and 1 880 females in terms of population. Ward 2 had 27716 males and 2915 females, while Ward 11 had the lowest population in the district with 827 males and 817 females. These population figures are for both non-Doma and Doma people since other ethnic groups are also resident in the areas. The Doma population remain estimated as some of their members do not participate in public engagements such as the national census. Besides spending time with the Doma, the study allowed for interactions with private safari operators who do their business in the Dande Forest and employ some as hunting trackers and anti-poaching volunteers. In all these field excursions, researcher observation was based on field note-taking supplemented by digital machine recordings; Data analysis became an ongoing process during the fieldwork, which followed the thematic analytical procedures. Common ideas that repeatedly develop into patterns were identified, reviewed, defined and named. The ideas and sub-themes eventually constituted the sub-headings in the data presentation and analysis sections below, which depict the lived experiences and storied data that allowed for the gleaning of deep insights. A good example is the sub-theme of the forest as a 'supermarket', which was the building block for the main theme of nutrition and food security.

Theoretically, the study was guided by Antonio Gramsci's philosophical ideas about society and led to the analysis of the experiences of the Doma people responding to hegemonic forms of cultures around their geographies and changes in the natural ecosystems. Being a neo-Marxist, Gramsci (1971) believed that the dominant ideology of a society is reflected in the belief systems of a society or community, and the beliefs are acceptable by the rest through consent via social institutions such as schools, churches, courts and the media. Cultural hegemony represents conditions in which a social class or group exert its influence and control over the wider society (El Aidi &Yechouti, 2017). The dominant group, through its mainstream views, norms, and values, seeks to indoctrinate the lower group by consent and "…consenting must somehow be truly convinced that the interests of the dominant group are those of society at large…" (Gramsci, 1971, p. 42). This is further simplified by Flynn (2021), who states that cultural hegemony is about dominant norms and values of life and the socio-cultural and economic conditions of society that are eventually viewed as natural and inevitable by the rest. In the context of this study, dominant views came from the mainstream ethnic groups, which pushed for the adoption of non-Doma language and systems of life by the Doma. Due to the vulnerabilities of the Doma, partly a result of climate change and modernisation, the Doma become choiceless and succumb to the hegemonic influence peddled through education and other socialisation institutions and eventually find themselves in more compromised lifestyles that did not represent their IKS. On the other hand, Gramsci's theory of cultural hegemony in this context also implied that hegemonic control came from the Doma leaders over the rest of the Doma communities around Dande Forest in which the ordinary people were not forced but consented to abide and retain the Doma traditional ways of life. The Doma's capacity to adapt to the changing environment and vulnerabilities thus come from hegemonic ideas from within and outside the group. Therefore, as depicted in the narratives below, Gramsci's theory of cultural hegemony is best attuned to explain the Doma people's vulnerabilities and resilience.

THE RESEARCH SITE

The descriptive location of the research site is identified in general terms as the Zambezi Valley. However, this study adopts the specific descriptions in a government report entitled 'The Statutory Instrument number 14 of 2023' on Ward boundaries. On the ground, the study established that there were three Wards in which the Doma population is domiciled, and this is on the fringes of the forest, as already alluded to in the above section. The main population was found in Chapoto-Kanyemba, which is Ward 1 on the northern side, while the other populations were in Angwa Ward 2 and Masoka community, which is Ward 11. These three wards are home to the Doma people and are part of the 17 wards that make up the Mbire District.

Figure 1. Map of Mbire district showing Dande Forest

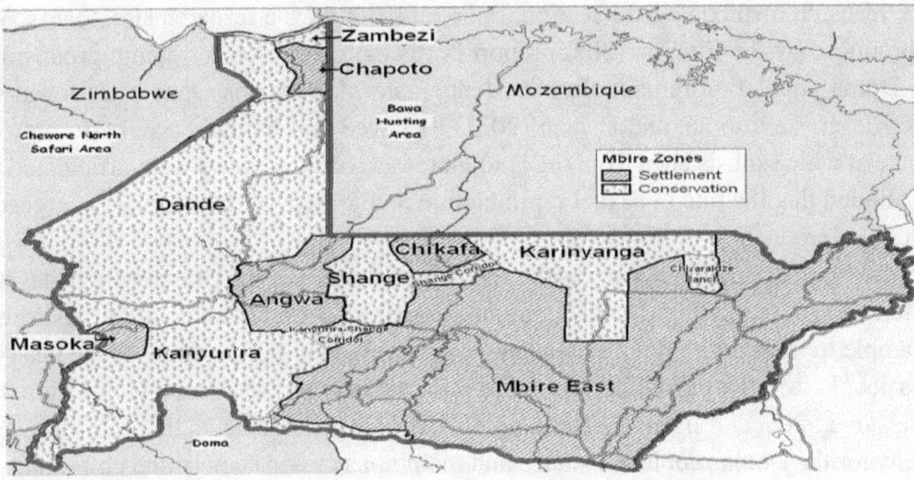

Source: Zvokuomba (2023)

Mbire district is on the northern tip of Zimbabwe, bordering Zambia and Mozambique, with the Zambezi River being the shared natural transboundary line. While Griffin et al. (2022) argue that these areas are endowed with a natural heritage which is the backbone of the Doma people's survival, Ncube, Kunguma, Nyahwo and Manombe (2017) state that "of these Wards namely Chapoto Ward 1, Angwa Ward 2 and Masoka Ward 11 were identified as the most food insecure Wards in food vulnerability assessments" (p. 2). The Doma people's relationship with the natural resources as a means of survival attracted the attention of the study gaze. Much of the data that came from this research site is presented and analysed below.

GENDERED SOCIAL ORGANISATION AND WARPED POPULATION

The population of the Doma has always been difficult to be accurately known because of the difficulty in carrying out a census. The chief of the area had bemoaned that even enumerators during the 2022 census could not reach them all and said:

> "The population of the Doma people cannot be accurate because a good number do not turn up for the exercise. They remained in the forest even during the previous head counting. Only a small number were available."

In the late 1980s, Nhira (1989) estimated them to be less than 500. A baseline survey was conducted to ascertain the food security situation and requirements of the Doma by Christian Care (Christian Care, 2010), working on a UN World Food Programme project, which pegged the population at 1500. However, a key informant in the Mbire Rural District Council alluded that the figure could not be accurate because of the people's reclusive attitude and lifestyle but estimated them to be around 2500 in the three Wards. A combination of the thick Dande Forest and bad roads for traversing the spaces does not only affect the population counting exercise but many other social service provision activities by government agencies and non-governmental organisations. Therefore, the study argues that the Doma population, which in most cases seems to be an estimation, may be outgrowing the capacity of the forest and its ability to be the sole source of food in line with Thomas Malthus' (1766-1834) ideas. As already alluded in the research methodological section, the population of the research site, that is, Wards 1, 2 and 11 was produced by the ZimStat (2022) report but is inclusive of other ethnic groups in the area. The specific Doma population remain elusive. Hence, neo-Malthusians Ophuls & Boyan (1992), as well as Ali, Ozdeser, Cavusoglu, and Usman (2021), believed that human capacity to reproduce may outpace the rate at which subsistence from the land increases. Similar views were articulated by Mellos (1988), who alluded that the rate of sexual reproduction and growth of the population exceeds the rate of sustainable food production.

Related to this population dimension was the social organisation of the Doma, which the study established to be an endogamous system. An endogamous marriage is a set of social relations where society encourages people to marry within the same clan or group (Green, 2021). One village head confessed that the Doma people socialise their children to develop a preference towards marrying within their ethnic group and the same totem and lifestyle. The type of marriage arrangements that the study established make it friendly for the Doma people to manage and maintain a closed society and community in which little interaction with the outside world is experienced, hence, the capacity to retain their Indigenous knowledge and cultural practices. One of the local informants who was engaged during fieldwork had this to say:

> "Yes, the Doma manage to maintain their cultural practices due to strict intermarriage rules but more so, the non-Doma do discourage their children from marrying our children. There is a big gulf between them and us."

The study further established that, unlike the neighbours, that is, the Chikunda and the Shona-Korekore, who follow a patrilineal social, cultural kinship practice, the Doma have limited social mobility in their marriage arrangements, which prohibits intermarriages and diversity. An extended discussion with a Doma senior man revealed that a man due for marriage is encouraged to make a marriage proposal to his sister's daughter, obtain a wife from his mother's brother's daughter or any other relative rather than marry from outside. In this arrangement, individual males would not be made to pay bride price or dowry but work for it within the in-laws' sphere of influence, for example, bringing a big hunt for the in-laws' consumption. Therefore, the marriage arrangements of the Doma are centred on the son-in-law providing bride service to his relatives in the form of constructing huts, hunting and foraging. Although hunting has become more secretive because of its illicit nature within the wildlife conservationist approach, the son-in-law fulfils his social obligations to his in-laws since being a successful hunter attracts great respect. The study also established that new couples reside near other affine, thus developing a cluster of homesteads that have a lot in common and share resources. In this context, the role of women is not restricted to homesteading; they get involved in foraging, picking wild fruits, insects, and small animals, and fishing. On an ordinary day, men would go on a hunt by themselves, but in other circumstances, both men and women would go out and hunt for many days and weeks, and this explains why some of their homesteads are semi-permanent.

The Doma social organisation is strengthened by common communal residence in village systems similar to other ethnic groups and falls under the jurisdiction of traditional leadership. While this may suggest a similarity with other contemporary village systems, observations during transect walks established that the Doma do not construct permanent houses or huts. Their huts are arranged in cluster formation or sometimes haphazardly, and when the time comes for them to migrate to other places in the forest, they abandon the huts and only return to their original geographical space after completing their hunting and foraging expedition. Such excursions could take weeks or months, and at times, they may not return to the exact location again. Therefore, the Doma places of residence are determined by what the situation dictates, especially in line with the food security determinants.

DOMA FOOD SECURITY SYSTEMS AND THE ADVENT OF CLIMATE CHANGE

In questioning and seeking to proffer answers about the Doma way of life, culture and traditional knowledge systems, the study zeroed in on Doma farming practices and foraging practices, which constitute nutritional and food systems. The Doma farming practices and crop production have always been key components of the survival of the Doma society in conjunction with hunting and gathering. Researcher observations during the transact walks revealed that the Doma do not practice any animal rearing of any sort. Unlike the San of Botswana, which keeps few goats, the Doma area of domicile is hostile to animal rearing. One of them said:

"We do not keep domesticated animals because it is a cultural practice which originated from living in a wildlife-infested area. One can only own an animal or animals when they are kept outside the Doma habitat."

The problem of tsetse flies that cause sleeping sickness in people and death in domesticated animals, as well as the abundance of wild predators like wild dogs, lions, cheaters, and hyenas, prohibit the Doma from venturing into animal rearing. The study established that the Doma engage in farming but on significantly small scales where they plant along the river banks. However, because of the population growth and competition for scarce resources in the forest as well as the restrictions on hunting by government and private safari operators, some of the Doma are slowly turning to mainstream villagisation. Their limited cultivation is done on a *dimba,* a piece of land situated along the banks or riverbed.

Figure 2. Picture of a dimba along Angwa River

Source: Field work camera

Through observations and specific interviews of key informants, the study established that the Doma people do not only depend on hunting and gathering but also on tillage of small portions, especially along the Mwazamutanda and Angwa Rivers, which are major sources of livelihood. The farming activities along the river bed and banks take place both in the wet and dry seasons and follow the principles of shifting cultivation. Rapoko, sorghum, beans, maize, legumes, and cassava are the most common cultivated plants. Thus, the study concurs with Bola et al. (2014) that the *dimba* tillage is a long traditional food security practice of the Doma although it is against conventional conservation farming practices.

However, because of climate change, the Zambezi Valley, like other dry regions of the world, is experiencing low rainfall, drying rivers and other sources of fresh water, which creates vulnerability for the Doma people. Mugambiwa, Rankoana and Tirivangasi (2023) as well as Chanza and Gundu-Jakarasi (2020) concur that IKS is an appropriate approach to managing climate change. This speaks to the Doma's situation, especially regarding food and their relationship with other forms of wildlife. By virtue of being Indigenous to the Zambezi Valley, they live in their natural ecosystem and notice the environment, including changes to it every day and make appropriate changes, adjustments and conclusions based on the observations. Nhemachena (2007) states that they even notice the behaviour of animals, birds and other species and use them to interpret weather changes. Hence, these natural ecosystems work as early warning systems for potential disasters. The Doma deploy similar Indigenous knowledge systems to survive in an arid area.

A small percentage of the Doma people are now forced to consider mainstream farming practices outside the *dimba* that have an assurance of seasonal food security but lack the required capacity. Observations made showed that there is a shift towards the take up of conventional farming techniques by a few who appreciate the new forms of life in line with Gramsci's (1971) ideas on cultural hegemony. However, the sustainability of such farming technologies is in conflict with wild animals destroying their plants. Therefore, the Doma find themselves wedged between the possibility of relinquishing their IKS for a different lifestyle or retaining their cultural systems wholly. Thus, the study concurs with Cortes-Ramirez (2015) that in such circumstances, there would be an experience of some cultural imperialism from the dominant groups who indirectly push towards the adoption of a mainstream lifestyle.

The study established that foraging and food processing are central to the day-to-day activities of the Doma life. Foraging includes the collection of fruits, and berries and hunting entails catching wild animals for household food consumption as well as honey collection and fishing. The Doma have been judged as backward and uncivilised, especially by outsiders, but living with them during fieldwork revealed that they have a wealth of experiences, expertise, knowledge and skills about wildlife. They know fruit trees by their names, those that are edible and poisonous. Knowledge of how to transverse the forest is their expertise. The forest is not a homogeneous barrier to their life. It provides a variety of resources and opportunities. Even mere walking in the forest, either for a hunt or for a collection of fruits and berries, is culturally defined and through socialisation, that knowledge is passed from one generation to another. Other traditional foods the Doma collect in the forests and are part of their nutritional support system include the *Banda panda* pods of the *mupasa* plant which is abundant in the forest. The plant produces a fruit which has seeds inside that are collected, dried in the sun and then cooked in the same method as beans and is used as a relish. Of great interest to the study was the *garatongo* tuber whose botanic name is *dioscorea asteriescus* which is not eaten under normal circumstances. It is only taken as food in times of food scarcity because the tuber is toxic and if not prepared to specifications, it can cause death to the consumer. Due to its toxic nature, the tuber is soaked in water for the night, and the water is thrown away in a pit so that no other person, particularly children or animals, may get in contact with it. The other way of preparing it is to boil it in water for not less than five times, throwing away the water until the toxic substance is finished. To test whether it is free from poison or not, it is left in the open and when flies can pitch on it, it will be considered ready for consumption. The pieces of the tuber are then crushed and turned into a white powder which is cooked in the same way as meal-meal for *sadza.* Based on the information about food from the forest, one Doma community leader described the forest as their 'supermarket' where they 'purchase' and bring home daily food provisions. However, whilst the forest may be a provider of food, it remains a forest, a jungle endowed with wildlife threats to human life. A village head narrated how a father and his sons lost their lives after a day's hunting and gathering which produced nothing.

It was a bad year and there was a severe food shortage as usual, a father and his male children who returned from a hunting expedition hungry got home and ate the boiled tube thinking that it was well-boiled and drained of toxic stuff. In the absence of those who had boiled it halfway, the team ate and they lost their lives.

The *bepe* is another edible but poisonous tuber, of which the poison can be leached from the tuber through boiling. The same way of preparing other tubers is done to the *guruvi* tuber which is consumed in the same way. *Musawu* tree produces *masawu* fruits which turn orange in colour when they get ripened. The local informants indicated that the *masawu* are part of the Doma people's staple food but are equally consumed by the Chikunda and the Korekore. Generally, the *musawu* trees are abundant in the Zambezi Valley and other low-lying areas of Zimbabwe. The fruits are collected and consumed during the walks in the forest but are also kept under storage and consumed in times of need. The Doma sometimes crush the fruit and its seed and the powder is used to prepare thick porridge consumed during breakfast, it can also be used to prepare *maheu* drink and sometimes an alcoholic beverage. *Mupama* is one other staple food of the Doma people. This is a tuber but not as highly poisonous as the *guruvi* and *bepe*. The poison can be removed from the tuber by boiling it in water not more than two times. When the boiled tuber is dried, it is then crushed into fine powder that is used to prepare *sadza* like substance which is the most common food in the Doma households. The white substance can also be used to prepare 'bread' of which the cooking process involves using big leaves.

The Doma elders and community leaders talked to revealed that the Doma people do not use money for the purchase of their daily provisions such as matches for lighting and salt for cooking. They use traditional technologies by collecting salty soil from specific sites in the forest and processing it to produce salt. They use the filtration process based on sun drying. They use the salt for cooking and preparing 'biltong' or dried meat. The preparation of dried meat is central to the Doma food security since they engage in hunting activities. They also dry meat using fire on a raised pole platform. When the meat is dry, it is then safely kept in traditional containers and consumed throughout the year. The fire which is used for cooking, drying of meat and warming the bodies in winter, is started by rubbing pieces of wood against each other continuously until a fire is started. However, many of them now use pieces of metal to spark a fire by tricking them together. The researcher participated in the fire-making process and it proved to be a difficult activity for the unfamiliar. How young boys and girls could easily make fire was an indicator of the ongoing social reproduction of culture through socialisation. However, the major question that remains lingering concerns the sustainability of the Doma culture in social, economic, political and ecological environments that are changing. Thus, from a Gramscian theoretical position, the dominant Shona ethnic group subtly distribute and spread their ideas and value systems about education, wildlife-human relations, economics, agriculture and other facets of life to the Doma.

Weaver (2023) concurs with the classical work of Boggs (1976) states that ideas of the main groups and the most powerful permeate society and shape societal values, attitudes, beliefs and morality.

The above social, cultural, economic, and environmental dynamics should be viewed in the context of government policy towards wildlife management. Through the Department of Parks and Wildlife Management, the government arm responsible for the management of wildlife, the Mbire Rural District Council and the private safari operators, the Doma community is slowly being forced to abandon their way of life characterised by foraging and gathering. These national and private wildlife management institutions are not only practising wildlife management from a national perspective but also as part of the global conservation machinery guided by international best practices. Consequently, the idea of romanticising wildlife, which has been growing within the context of the sprouting tourism industry, has exclusionary effects on local Indigenous traditions and practices. Interactions with the Doma traditional connoisseur showed that they perceive Western-oriented wildlife management as part of the wider white culture biased against and bent to displace them from their traditional and customary source of livelihood. This perception is based on the fact that for generations, their source of livelihood has been the forest, which they regard as their supermarket. So why not now?

Informal conversations with the wildlife rangers and non-Doma people augmented the study and revealed that the Doma are now regarded as poachers within their traditional habitat every time they are caught up between hunting and gathering lifestyle and the conservationist wildlife management approaches in the Dande Safari area. One elder amongst the Doma had this to say,

> "This is the lifestyle we were born in it is not easy to leave it and adapt to a new life system. The park rangers, police and private safaris are sometimes on us and this is taking us into severe desperation in terms of food for daily consumption."

Those with the responsibility of protecting wildlife, both from government and the private safari operators, have the mandate to protect wildlife, which partly contradicts the Doma way of life. Whilst some of their members have been encrypted into the local community rangers as community volunteers in the anti-poaching war, it does not remove the fact that many of them still live on hunting and gathering, and it creates a dilemma and irony about the Doma benefiting directly from the Community Area Management Programme for Indigenous Recourses (CAMPFIRE). However, the study established that the CAMPFIRE programme does not provide an alternative to the Doma but a contradiction. The study is in concurrence with Mamimine (2000) who argues that the CAMPFIRE project failed to take into account the basic needs and interests of the Doma in community development initiatives. While some Doma leaders acknowledge that the CAMPFIRE project is being adjusted to take into consideration Doma needs, for a long time since its inception in the late 1980s, its implementation failed to consider the fact that the forest has always historically and traditionally been part of the Doma physical, emotional and spiritual world. Comparatively, similar observations were made by Heater (1990) about the Glwi, Gllana, and Kung San groups of Central Kalahari Game Reserve in Botswana.

DOMA PEOPLE AND THE OUTSIDE WORLD

Interactions with various interlocutors and non-Domas in the Zambezi Valley, that is, the Chikunda and the Shona-Korekore, revealed that they still view the Doma as primitive and backward. This stereotypical description of the Doma group has had far-reaching consequences as the Doma seem to have been excluded and marginalised, including in the wildlife management projects funded by donors. As a result, the Doma have kept to themselves in the forest and continue not to be part of the mainstream development process. It would not have been easy to join the Doma community during the research process if not for the 'gatekeepers' who persuaded them to share their life with the ethnographer. They argue that previous studies by outsiders have never produced positive results for them. As such, the Doma community leaders treat researchers as part of the spy machinery on their way of life aimed at making the law enforcement agents know more about their hunting expeditions. Outsiders have always remained less knowledgeable about the Doma world and experiences. For example, footprint marks and animal waste found in the bush may be regarded as meaningless by an 'outsider' while to them, these would be key sources of information on who or what animal went where with whom and at what time of the day. The sand on the riverbed and dust on the footpath bear the marks of almost all activities of the day and are key sources of information and communication. Observations during hunting expeditions revealed that footprints tell whether others have already crossed the river or not and whether animals have gone to the river for water drinking or not, which is important for hunting. The human and animal tracks may mean nothing to the outsiders as well as to the uninitiated, but mean the difference between life and death to the Doma people because their life is dependent on them.

SPIRITUALITY AND RELIGIOUS LIFE

The Doma, like many other ethnic groups of Bantu origins, believe in and worship God. An engagement with Doma elders revealed that the Doma believed that all natural resources in the forest belong to God and, therefore should be harvested responsibly. Their worldview on spirituality fits in Mugambiwa's (2017) assertion about the role of spiritual leaders, that they do not only provide leadership but appropriate behaviour in the forest. Fabiano et al. (2021) provide the example of the Amazonian cosmovision in which territorial wetland spirits provide stewardship over the ecosystem of the Urarina people of the Chambira Basin on the Peru side. The leadership makes people aware of the 'dos and don'ts', and in some other cases, individuals may be harshly reprimanded. In that context, people's primary responsibility is to take care of nature and utilise it for their own good. Religious beliefs and spiritual life contribute to the maintenance of their tradition and customary practices. The Doma communicate with the spiritual world through the spirit mediums and perceive the universe as populated by powerful forces that have an interest in their lives. Interviews with key members of the community showed the prevalence of the belief that powerful forces monitor human behaviour and send punishments to those who violate moral rules, but if humans approach spirits appropriately, they may use their powers to confer benefits. The Doma religion is a way of dealing with uncertainties they cannot control in the hunter-gatherer lifestyle. It is also a way of providing meaning through explaining what cannot be explained by humans, for example, foretelling the future, death, and mysteries of life. Importantly, religion creates solidarity amongst the Doma; hence, they can maintain a closed way of life. The Doma believe that when elders pass away, they continue to influence the lives of the living through the spirit mediums. Through the genealogy that links them to Nyamapfeka, the spirit of the dead manifests itself through spirit mediums and eventually maintains a strong interest in the lives of the descendants and maintains order by sending sickness and misfortunes when rules of the Doma are broken. They believe that the spirit mediums give hunters advice on how to hunt and what to hunt, which is part of the Indigenous and traditional wildlife conservation system. When certain animals behave in certain ways, the Doma spirit mediums make conclusions based on such animal behaviour.

However, three major issues negatively affect the Doma spirituality: heavy influence of Christianity, formal education and climate change. Those who accept to be part of the mainstream village lifestyle find themselves exposed to the new Christian belief system and this is affecting the young generation with the potential of diffusion Doma Indigenous knowledge systems transition from one generation to another. A case of a woman who was very sick and was taken to a church for prayers and was prayed for to become completely healed attracted many people amongst the Doma to Christianity. The case presented itself as a challenge to the traditional ethno-medical and healthcare system much anchored on spiritualism. The few Doma children who pursue education at Mariga school, Angwa school and Masoka school become looped in Western values that despise the Indigenous knowledge system as inferior. Observations in the Masoka area showed that those who were selected to be village heads within the Doma communities had acquired primary education and were able to read and write. It is the same people amongst the Doma who now attempt to build permanent huts and partly leave foraging. Though the educational level is limited to primary schooling, it creates exposure to dominant values and ethos as expounded by Gramsci (1971). Smith (2002), in further explaining the cultural hegemonic theory, argued that the application of Westernised education models in African systems exposes communities to Western economic models that create vulnerabilities and poverty. When Indigenous communities fail to fit into the Westernised economic systems, they are rendered useless due to the elitism of Western systems. In attempting to adapt to the views of the dominant groups, norms and values of elites, a system of power and hierarchy is created in which the Indigenous people are located at the lowest ebb (Kaya, 2013).

UPGRADE OF KANYEMBA BUSINESS CENTRE INTO A TOWN AND DOMA ECONOMIC VULNERABILITIES

Evidence from key informants, the non-Doma people in the research site, revealed that the central government made a deliberate move to upgrade Kanyemba centre into a border town, but it is to be established how the Doma became part of that development trajectory. Strategically located at the border with Zambia and Mozambique along the Zambezi River, the development of Kanyemba into a big business and tourism hub in the Zambezi Valley will not only affect the lifestyle of the Doma in future but also the present. The Herald newspaper of the 2nd of August 2021 reported that the conversion of Kanyemba centre into a leisure town with leisure facilities such as mega golf courses and hotels already started to affect the Doma people. Although few, some members of the Doma community already provide their labour in construction services while others work as casual labourers as a way of breaking off from the semi-nomadic lifestyle. However, evidence shows that the Doma struggle to adapt to the 'new' lifestyle due to their lack of required skills and basic education. The Doma did not benefit from formal education for many years. The nearest primary school in Ward 1 is Kanyemba Primary School, which is in the range of 15 to 20km away. A major river, Mwazamutanda, divides the Doma community from the Kanyemba business centre where the school is located and during the rainy season, they may not cross it. However, the development of Kanyemba into a town saw a school being built in the Doma community to reduce the distance for pupils. By the time of field visits, Mariga School had two blocks which housed pupils from the Doma community heralding the beginning and training of a new generation of Doma people embracing mainstream ethos. While the case of Mariga School was a sign of embracing a different way of life, it also speaks volumes about the future of the Doma local knowledge systems which may in future be abandoned if the education system does not include the IKS in the curriculum.

The study draws similarities with Gordon's (2017) '*Taming Bushmen Farm Labour*', a study that was carried out in Namibia in which integrating the bushmen included employing them as cheap labour with rights however straddled upon. Deploying the classic Gramscian theoretical framework to explain the Doma circumstances, the exploitation of the Doma labourers is a form of hegemonic influence over the other, an expression of cultural manipulation. The Doma is being pushed to uphold the beliefs of the dominant groups, norms and values so that the whole group hold the worldview of the dominant (Urbinati, 1998). While the Bushmen in Gordon (2017) worked in a commercial farming environment in the Grootfontein and Gobabis districts, the Doma are employed casually in villages by the Shona-Korekore and the Chikunda people, the major cultural groups in the research site who require cheap labour. When they perceive that the abuse becomes too much, revert to the semi foraging and hunting systems. One Chikunda villager said:

"It is difficult to work with them because they do not understand issues, they are too used to the wild."

However, a contradictory sentiment was shared by a Doma male who worked as a labourer:

"The food they pay us with only lasts for a day, I cannot continue work on an empty stomach, hence, the need to go back to the forest."

Thus, in the final analysis, the upgrade of the Kanyemba centre into a town becomes an indirect and direct factor in the transformation of the Doma people. Hence, the question of the sustainability of their Indigenous knowledge systems remains questionable in future.

FURTHER RESEARCH DIRECTIONS

Studying the Doma people's Indigenous knowledge systems could not be the beginning and the end. Many scholars have written about them from various vantage positions; therefore, the study is a build-up of an already existing knowledge production system and bank. More so, this single study could not be exhaustive not only about the Doma knowledge systems but many other facets of Indigenous knowledge systems in Zimbabwe which require special attention. A lot of aspects deserve further research, for example, the Doma ethno-medicine and how they responded to pandemics such as COVID-19 and HIV and AIDS. Another area for further research is the Doma children's education since it has a lot of bearing on the transmission of local knowledge to the next generations. This is a very critical aspect which was highlighted by Matema (2019) in his PhD Thesis. Education has the potential to change the socio-cultural configurations of the Doma people. Hence, more needs to be explored within mind their rights as Indigenous people.

A highlight on the policy and legal issues also provides directions for the future. At the international level, there is evidence that the rights of Indigenous people are recognised. Most notable was the United Nations Declaration on the Rights of Indigenous Peoples, which was made in 2007 article 31 states that the Indigenous people have rights to maintain, control, protect and develop their intellectual property over their cultural heritage, traditions and cultural expressions (UN, 2009). In 2015, the UN General Assembly adopted the 2030 Agenda for Sustainable Development framed as 'Transforming Our World: The 2030 Agenda for Sustainable Development, anchored on the principle of 'not leaving no one behind' and further proclaimed on 9 August of each year as a day of Indigenous people (UN ibid). Even the ILO Convention number 169 affirms that Indigenous people have rights to education, health, employment and better standards of life (UN Human Rights, 2013). However, on spending time with the Doma people and reflecting on their situation and worldview, there appears to be a disconnection between these international legal and policy positions and the reality on the ground. Policy and legal scholars would argue that international policies only provide a framework for states to draw from the global perspective and adopt as well as develop local mechanisms for the promotion and protection of Indigenous knowledge systems. Zimbabwe is a signatory to international conventions and derives its legal positions from them. The Constitution of Zimbabwe of 2013 recognises the place of Indigenous people but does have a different definition from other scholars and international law. As cited by Chibememe Dhliwayo, Gandiwa, Mutisi, Muboko, and Kuipa (2014), an Indigenous community is "a community of persons that has inhabited Zimbabwe continuously since before the year 1890 and whose members share the same language and dialect or same cultural values, traditions or customs" (p.17). Further, Section 16(1) of the Constitution provides for the state institutions and agencies to promote and respect cultural values, practices, and heritage, and Section 33 says, "The state must take measures to preserve, protect and promote Indigenous knowledge systems….". Together with the Parks and Wildlife Act Chapter 20: 14 of 1996, the Wildlife-Based Land Reform Policy of 2006 and the Zimbabwe Policy on Wildlife of 1992, they set a policy and legal framework for protecting the Indigenous people's interests. However, it is another thing to find these issues being put into practice; thus, more needs to be done on the ground. Therefore, the study recommends that the state commission a variety of studies on Indigenous groups and some focused on the Doma not only for building, strengthening and filling the literature gap but provide a basis for evidence-based policy framing. The various studies would feed into policy positions on Doma nutrition and food security, IKS and climate change in Mbire district, health and education delivery systems and the economic implications of new developments in the area. The policy positions feed into the development of specific pieces of legislation and particularist implementation mechanisms relevant to local environments. In putting down all these systems, the government of Zimbabwe would lead the private corporations, non-governmental organisations and other stakeholders in promoting, protecting and upholding Indigenous people's rights.

CONCLUSION

The case of the Doma is of great interest as it reflects on the capacity of traditional society located in a postmodern era, however, holding to its traditional knowledge systems. The study has also shown how and why it is becoming more difficult to survive in a hunting and gathering fashion in contemporality. The study concludes that the Doma people are not backward and neither are they uncivilised. It is because the mainstream society chose not to see the world from their vantage point and Doma ontology. However, the survival and sustainable development of their way of life are now dependent on national and global trends in wildlife management as well as the political will to preserve Indigenous cultures and the livelihoods of minority groups living in fringes of forest reserves. The study also shows the vulnerability of minority ethnic groups, the level of marginalisation by mainstream society, and complexities around the issue of the rights of minority ethnic groups, particularly in maintaining their cultures.

The survival of the Doma culture also depends on its leaders' capacity to express its hegemonic influence over the wider Doma community. Spirit mediums still hold a lot of power and influence in the lives of the Doma as they determine what shall be done by the whole community, how it shall be done, where and when. The role of the spirit mediums is highly valued in the Doma world such that they believe that the spirits of the dead control the lives of the living. The Doma also follow strictly a taboo system that makes them observe wildlife conservation that has been acknowledged by the CAMPFIRE programme. Forests have been maintained because of the sacred sites and religious rites that are held there. The Doma has managed to live a self-reliant lifestyle in those areas that are regarded by 'outsiders' as barriers to development. They manipulate the environment such that it provides the required inputs; for example, wild plants, during the farming season, are not regarded as weeds, and therefore, their crops sometimes may not be destroyed by wild animals. Food production is done in a way that takes advantage of the abundant availability of tubers, fruits, honey, wild animals and small insects. As a result, the Doma have managed to live a life which is outside the money economy. Their day-to-day activities do not require the use of any form of money since they acquire their clothes through barter trade, produce salt and make fire through Indigenous knowledge systems. However, life in the marginal areas of the Zambezi Valley is not spared from natural hazards like droughts which are a common phenomenon in low-lying places. The paper concludes that the Doma while they face a myriad of challenges, they deserve to be supported by policies and legislations that promote and protect Indigenous knowledge systems.

ACKNOWLEDGEMENTS

Special acknowledgement goes to the research participants and key informants from Ward 1, 2 and 11 who played an important role in knowledge generation.

DECLARATION

There were no conflicts of interest nor was the study funded.

REFERENCES

Aliyu, U. S., Ozdeser, H., Çavuşoğlu, B., & Usman, M. A. M. (2021). Food security sustainability: A synthesis of the current concepts and empirical approaches for meeting SDGs. *Sustainability (Basel)*, 13(21), 11728. DOI: 10.3390/su132111728

Athayde, S., Silva-Lugo, J., Schmink, M., & Heckenberger, M. (2017). The same but different: Indigenous knowledge system retention, erosion and innovation in the Brazilian Amazon. *Human Ecology: an Interdisciplinary Journal*, 45(1), 533–544. DOI: 10.1007/s10745-017-9919-0

Beach, D. N. (1980). *The Shona in Zimbabwe 900-1850*. Mambo Press.

Boggs, C. (1976). *Gramsci's Marxism*. Pluto Press.

Bol, G., & van Niekerk, D. (2024). Application of African Indigenous knowledge systems and practices for climate change and disaster risk management for policy formulation. *International Journal of Disaster Risk Reduction*, 111, 104670. DOI: 10.1016/j.ijdrr.2024.104670

Bola, G., Mabiza, C., Goldin, J., Kujinga, K., Nhapi, I., Makurira, H., & Mashauri, D. (2014). Coping with droughts and floods: A case study of Kanyemba. Mbire district, Zimbabwe. *Physics and Chemistry of the Earth Parts A/B/C*, 67–69(1-2), 180–186. DOI: 10.1016/j.pce.2013.09.019

Chambers, R. (1991). *Rural development: Putting the last first*. John Wiley.

Chanza, N., & De Wit, A. (2013). Epistemological and methodological frameworks for Indigenous knowledge in climate science. *Indilinga*, 12(2), 203–216.

Chanza, N., & Gundu-Jakarasi, V. (2020). Deciphering the climate change conundrums in Zimbabwe: an exposition. In Tiefenbacher, J. (Ed.), *Global warming and climate change*. Intech Open. DOI: 10.5772/intechopen.84443

Chibememe, G., Dhliwayo, M., Gandiwa, E., Mtisi, S., Muboko, N., & Kuipa, O. (2014). *Review of national laws and policies that support or undermine Indigenous peoples and local communities*. Natural Justice and Ford Foundation.

Chiwome, E. M., Mguni, Z., & Furusa, M. (2000). *Indigenous knowledge and technology in Africa and diaspora communities. Multidisciplinary approaches*. National Council for Black Studies.

Christian Care. (2010). *Vulnerable group feeding end of programme report.* Christian Care.

Cortes-Ramirez, E. (2015). Cultural hegemony today: From cultural studies to critical pedagogy. *Postcolonial Directions in Education*, 4(2), 116–139.

Cutshall, C. R. (1990). *Kanyemba/Chapoto Ward: a socio-economic baseline survey of community households.* University of Zimbabwe Press.

Denzin, N., & Lincoln, Y. (2017). *The SAGE handbook of qualitative research.* Sage.

El Aidi, A., & Yechouti, Y. (2017). Antonio Gramsci's theory of cultural hegemony in Edward Saidi's orientation. *International Multidisciplinary Research Journal*, 6(5), 1–23.

Fabiano, E., Schulz, C., & Branas, M. (2021). Wetland spirits and Indigenous knowledge: Implications for the conservation of wetlands in Peruvian Amazon. *Current Research in Environmental Sustainability*, 3(1), 100107. DOI: 10.1016/j.crsust.2021.100107

Ferguson, J. (2013). Declaration of dependence: Labour, personhood and welfare in Southern Africa. *Journal of the Royal Anthropological Institute*, 19(2), 223–242. DOI: 10.1111/1467-9655.12023

Fernaandez-Llamazares, A., & Virtanen, P. (2020). Game masters and Amazonian Indigenous views on sustainability. *Current Opinion in Environmental Sustainability*, 43(1), 21–27. DOI: 10.1016/j.cosust.2020.01.004

Flynn, S. (2019). Revisiting hegemony: A Gramscian analysis for contemporary social work. *Irish Journal of Sociology*, 29(1), 77–96. DOI: 10.1177/0791603519884201

Gordon, R. (2017). Taming' Bushman farm labour: A villainous era in neo-liberal feudal Namibia? *Anthropology Southern Africa*, 40(1), 261–275. DOI: 10.1080/23323256.2017.1395708

Government of Zimbabwe. (2002). *Rural District Councils Act Chapter 29: 13.* Government Printers.

Government of Zimbabwe. (2023). *Statutory Instrument 14 of 2023. Proclamation of Ward Boundary Description.* Government Printers.

Gramsci, A. (1971). *Selections from the prison notebook* (Hoare, Q., & Nowell-Smith, G., Eds. & Trans.). Lawrence & Wishart.

Green, E. (2021). Explaining inter-ethnic marriages in Sub-Saharan Africa. *Journal of International Development*, 33(4), 627–654. DOI: 10.1002/jid.3535

Greene, S. (1999). Family concerns: Gender and ethnicity in pre-colonial West Africa. *International Review of Social History*, 44(1), 15–31. DOI: 10.1017/S0020859000115172

Griffin, C. T., Wynd, B. M., Munyikwa, D., Broderick, T. J., Zondo, M., Tolan, S., Langer, M. C., Nesbitt, S. J., & Taruvinga, H. R. (2022). Africa's oldest dinosaurs reveal early suppression of dinosaur distribution. *Nature*, 609(7926), 313–319. DOI: 10.1038/s41586-022-05133-x PMID: 36045297

Griswold, W. (1992). *Cultures and societies in a changing world.* Pine and Forge Press.

Hachipola, S. (1998). *A survey in the minority languages of Zimbabwe.* University of Zimbabwe Press.

Hasler, R. (1996). *Agriculture, foraging and wildlife resource use in Africa*. Kigan Paul Publishers.

Heater, D. (1990). *Citizenship: the civic ideals in world history, politics and education*. Longman.

Herald Newspaper, 18 July (2012). Zimbabwe Newspapers

Herald Newspaper, 2 August (2021). Zimbabwe Newspapers.

Hulme, D., & Murphree, M. (2001). *African wildlife and livelihoods: the promise and performance of community conservation*. James Currey.

Isaacman, A. (2000). *Chikunda transfrontiersmen and transnational migration in pre-colonial South-Central Africa, 1850-1900*. University of Minnesota Press.

Kanengoni, T. (2017). Doma community faces challenges as nature reserve designation restricts hunting. *Global Press Journal*. Retrieved on 5 November 2024 from https://globalpressjournal.com/africa/zimbabwe/doma-community-faces-challenges-nature-reserve-designation-restricts-hunting-fishing/

Kaya, H. (2013). African Indigenous knowledge systems and relevance of higher education in South Africa. *The International Educational Journal: Comparative Perspectives*, 12(1), 30–44.

Keane, M., Raciti, M., van der Westhuizen, G., Motala, S., Stanton, S., Gilbey, K., Manathunga, C., Qi, J., & Msimango, S. (2023). Indigenous knowledge systems in South Africa and Australia: Transforming doctoral education. *Curriculum Perspectives*, 43(1), 584–593. DOI: 10.1007/s41297-023-00183-1

Kohler, O. (1959). *The study of the Gobabis district. Union of South Africa*. Department of Bantu Administration.

Kuhnlein, V., Erasmus, B., & Sprigelski, D. (2009). *Indigenous people's food systems*. FAO.

Lan, D. (1985). *Guns and rain: guerrillas and spirit mediums in Zimbabwe*. Zimbabwe Publishing House.

Lopez-Deflory, C., Perron, A., & Miro-Bonet, M. (2022). A methodological and practical guide to study peripheral voices in qualitative research. *International Journal of Qualitative Methods*, 21(10), 1–20. DOI: 10.1177/16094069221100639

Mamimine, P. (2000). How far the destination? Decentralisation and devolution in governance of commons under CAMPFIRE. In E. M. Chiwome., Mguni, Z. & Furusa, M. (Eds). *Indigenous knowledge and technology in Africa and diaspora communities. Multidisciplinary approaches*. California, National Council for Black Studies.

Matema, C. (2019). *Zimbabwe's CAMPFIRE public investments: implications on education, adaption and preferences*. PhD Thesis University of Cape Town.

Matema, S., & Anderson, S. (2015). Why are lions killing us? Human-wildlife conflict and social discontent in Mbire district, northern Zimbabwe. *The Journal of Modern African Studies*, 53(1), 93–120. DOI: 10.1017/S0022278X14000664

Mellos, K. (1988). *Neo-Malthusian theory: critical essay*. Palgrave MacMillan., DOI: 10.1007/978-1-349-19598-5_2

Mombeshora, S., & Le-Bel, S. (2009). Parks-people conflicts: The case of Gonarezhou National Park and the Chitsa community in south-eastern Zimbabwe. *Biodiversity and Conservation*, 18(1), 2601–2623. DOI: 10.1007/s10531-009-9676-5

Mugambiwa, S. (2017). *Knowledge of climate change and the use of Indigenous knowledge system to adapt to climate hazards in Mutoko rural district of Mashonaland East Province, Zimbabwe,* MA Dissertation, University of Limpopo.

Mugambiwa, S. (2020). *Climate governance through IKS for sustainable development in Mutoko district of Mashonaland East Province, Zimbabwe,* PhD Thesis, University of KwaZulu Natal.

Mugambiwa, S., Rankoana, S., & Tirivangasi, H. (2023). Climate governance beyond the government: Indigenous knowledge systems in rural Zimbabwe's climate change adaption. *International Journal of Development and Sustainability*, 12(6), 238–249.

Muwanga-Zake, J., & Kibukamusoke, M. (2024). Sustainability of Africa through technological innovations and Indigenous knowledge systems: A discussion of key factors and way forward. *African Journal of Social Work*, 14(2), 59–69. DOI: 10.4314/ajsw.v14i2.2

Ncube, A., Kunguma, O., Nyahwo, M., & Manombe, S. (2017). Nutritional vulnerability: An assessment of 2010 feeding food programme in Mbire district, Zimbabwe and its impact on pregnant women. *Jàambá*, 9(1). Advance online publication. DOI: 10.4102/jamba.v9i1.406 PMID: 29955343

Ndungu, H., Zaba, P., & Ndlovu, E. (2023). Enhancing resilience to recurrent droughts and floods in semi-nomadic communities. a case of CMDRR amongst the Doma community in Mbire district, Zimbabwe. *International Journal of Research and Innovation in Social Sciences*, V11(1), 1532–1554. DOI: 10.47772/IJRISS.2023.70822

Neuman, W. L. (2007). *Basics of social research: qualitative and quantitative approaches.* Pearson Education Inc.

Nhemachena, C. (2007). Assessment of the economic impacts of climate change approach. Policy Research Paper. *World Bank Development Research Group.*

Nhira, C. (1989). A socio-economic appraisal study of the Chapoto ward-Guruve district. *University of Zimbabwe Working Paper Number 5.*

Ophuls, W., & Boyan, A. (1992). *Ecology and the politics of scarcity revisited.* W. H. Freeman.

Perez, C., & Smith, C. (2019). Indigenous Knowledge System and conservation of settled territories in the Bolivian Amazon. *Sustainability (Basel)*, 11(1), 6099. Advance online publication. DOI: 10.3390/su11216099

Posey, D. (1999). *Cultural and spiritual values of biodiversity.* UN Environmental Programme.

Rutherford, B. (2017). *Farm labour struggles in Zimbabwe: the grounds of politics.* Indiana University Press.

Sanchez, G. (2018). *The Zambezi River Basin: water resources management. Energy-food-water nexus approach,* MSc Thesis, Stockholm University.

Slikkerveer, L. (1999). Changing values and attitudes of social and natural scientists towards Indigenous peoples and their knowledge systems. In D. Warren., L. Slikkerveer. & S. Titilola. (Eds.), *Indigenous Knowledge Systems: Implications for agriculture and international* development (pp. 121-137). Studies in Technology and Social Change. Iowa State University Press.

Smith, A. (2002). Power and hierarchy of knowledge. *Geoforum*, 40(1), 230–248.

UN Development Group. (2009). *Guidelines on Indigenous people's issues*. UNDG Publications.

UN Human Rights, (2013). *Indigenous People's and the UNHR system.* UN.

Urbinati, N. (1998). From the periphery to modernity: Antonio Gramsci's theory of subordinating and hegemony. *Political Theory*, 26(3), 370–391. DOI: 10.1177/0090591798026003005

Warren, D. M., Slikkerveer, L., & Brokensha, D. (1999). *The cultural dimensions of development: Indigenous knowledge systems*. ITD.

Weaver, K. (2023). Bridging Indigenous and Western knowledge systems in Amazonian Indigenous communities: A systematic realistic review. *Development Studies Research*, 10(1). Advance online publication. DOI: 10.1080/21665095.2023.2203842

ZimStat. (2022). *Population and Housing Census Report. Population distribution by Ward* (Vol. 2). Government Printers.

Zvokuomba, K., Kabonga, I., & Chingozha, P. (2023). The state of the human rights of ethno-cultural minority groups in Zimbabwe: reflections on the Doma people of the Zambezi Valley. *Sverige, Raoul Wallenberg Institute,* 73-86.

ADDITIONAL READING

Ajani, Y., Oladokun, B., Olarongbe, S., Amaechi, M., Rabiu, N., & Bashorun, M. (2024). Revitalizing Indigenous knowledge systems via digital media technologies for sustainability of Indigenous languages. *Preservation. Digital Technology & Culture*, 53(1), 35–44. DOI: 10.1515/pdtc-2023-0051

Bird, S., Shobbrook, K., Flick, D., & Carroll, S., A. A., and A. Lim. (2019). Technology for Indigenous language revitalization: The state of the field. *Language*, 95(S1), e1–e29.

Hinton, L. (2019). Language revitalisation and new technologies: Culture of electronic mediation and the refiguring of communities. In J. Jaffe., D. Morita., Vogel, N. & G. Spolsky (Eds.), *Language ideologies and globalisation of standard languages* (pp. 244-264). Routledge.

Mogajane, N. M. (2022). Digital technology impact in the access, preservation and curation of indigenous artefacts at Mafikeng Museum. PhD Thesis. Northwest University.

KEY TERMS AND DEFINITIONS

Indigenous People: People located and domiciled in a particular geographical space, and they retain the knowledge of the land and the knowledge is rooted in historical continuity within that region or space.

Indigenous Knowledge: Awareness, familiarity, skills acquired through experience and practice by the local people, and the skills are embedded in the local communities and is gained through historical continuity.

Foraging: A socio-cultural practice by local people of collecting wild resources such as fruits, tubers and other aspects of the natural ecosystem for food, shelter making as well as other products for sustaining life.

Doma People: A minority group domiciled in the Zimbabwean side of the Zambezi valley. The Doma people are indigenous to the Zambezi Valley geographical space and live a semi-hunter-gatherer lifestyle in the fringes of the forests.

Zambezi Valley: A geographical space formed naturally along the Zambezi River, which flows from northern Zambia and parts of Angola, passing through as a border space for Zambia, Zimbabwe, and Mozambique before the river feeds into the Indian Ocean. It is in that lowveld area of northern Zimbabwe where the Doma people reside.

Chapter 16

Harnessing Language, Adult Education, and Indigenous Knowledge for Sustainable Urban Agriculture in Lagos

Amidu Owolabi Ayeni
https://orcid.org/0000-0001-7022-1667
University of Lagos, Nigeria

Adeola Funmilola Afonja
https://orcid.org/0000-0003-1573-2282
University of Lagos, Nigeria

John Olubunmi Faloju
https://orcid.org/0000-0003-0346-5703
University of Lagos, Nigeria

ABSTRACT

Language, adult education, and Indigenous knowledge are key pillars in shaping how urban communities engage with agriculture. On this note, this chapter aims to examine the importance of language, adult education, and Indigenous language among adult farmers for sustainable urban agriculture in Lagos. A structured interview was conducted with 114 adult urban farmers of 50 years and above from the six selected urban farms' clusters across Lagos urban areas. The findings of this research reveal that language, adult education, and Indigenous knowledge are sine qua non for sustainable urban agriculture that will alter antiquated methods in agriculture in Lagos and promote effective transmission and usage of improved agro-technologies by farmers and farm labourers. This chapter holds that for sustainable urban agriculture in Lagos to be achieved, language, adult education, and Indigenous knowledge should be integrated and employed in the usage of relevant science and technological innovations.

DOI: 10.4018/979-8-3693-8447-3.ch016

INTRODUCTION

Language is an inextricable part of human society. The human language implies the grammar, norms as well as other rules that permits people in a particular society to communicate in a manner that others (within and outside) the society will understand (Muin *et al.*, 2021). It is indeed an exclusively human property and it has been described as an effective and veritable vehicle of communication in any community, particularly in a multilingual linguistic community. It has also been described as the lubricating oil of the society and the chord that binds the society together (Faloju & Sobola, 2021). The *Merriam-Webster* dictionary (2009) also defines language as "a systematic means of communicating ideas or feelings by the use of conventionalized signs, sounds, gestures, or marks having understood meanings". Since language is a systematic means of communicating ideas, knowledge or feelings, therefore, it can be described as a means of achieving social, cultural, political and economic transformations in any society. Lawal (2015) asserts that linguistic communication is necessary for a human community because it helps humans possess social organisation. On the other hand, adult Education, in the view of Lakshmi Reddy (2000), can be defined as education that starts at a particular level or stage of education. Lindeman (1961) cited in Aggarwal (1989) succinctly avers that adult education starts when vocational education ends. Faure *et al.*, (1972) conceptualize adult education as the normal culmination of the education process. However, the United Nations Educational, Scientific and Cultural Organization (UNESCO) General Assembly (GA) in Nairobi (UNESCO, 1976, p.2) defines adult education as "the entire body of organized educational processes, whatever the content, level and method, whether formal or otherwise, including various roles, functions and levels".

Moreso, the disposition of people towards education across the world indicates its importance. Thus, it is not surprising that education is considered a fundamental human right by (United Nations Educational, Scientific and Cultural Organization (UNESCO, 2008). Education, and in particular, adult education, is a tool for individual and national socioeconomic growth and development as well as an instrument for social change by including traditional painting techniques, soil management practices and water conservation methods by Indigenous communities into its curriculum. Agriculture is one of the sources of income for many people in the rural and urban areas of Nigeria. However, development in the agricultural sector has not progressed at a fast pace since the discovery of oil in the 1970s, although today, it is considered one of the major contributors to the Nigerian Gross Domestic Product. It accounted for 22.36% of Nigeria's GDP in the first quarter of 2022 (Adeniran, 2022). According to a report by the Nigerian Bureau of Statistics, this figure is higher than the contribution in the first quarter of 2021 but lower than the fourth quarter of 2021, which stood at 22.35% and 26.84%, respectively (Adeniran, 2022).

Indigenous knowledge is referred to as unique knowledge confined to a particular culture or society (Senanayake, 2006; Tharakan, 2017; Tume *et al.,* 2019; Ogar *et al.,* 2020; Ramirez-Santos *et al.,* 2023). Indigenous knowledge has been applied from time immemorial in rural and urban communities to find a solution to the diverse problems in Agriculture, Medicine, Architecture, and Engineering as well as in other fields of human endeavours, and tremendous success has been recorded in all these fields. These successes include but are not limited to Agroforestry Research Project in Kenya; Neem Bio-Pesticides in Niger; Soil Regeneration in Rwanda; Agro-Forestry Projects in West Africa; Water-Harvesting in the Sahel, and Indigenous Societies and Conservation Areas (Lalonde 1991; Ayeni & Aborisade, 2022). However, it has been observed that most Indigenous knowledge systems vanish or are less applied. This arises as a result of the proliferation in the usage of Western technologies, particularly in urban agriculture sustainability and the loss of technology skills and expertise in Agriculture (Tharakan, 2017; Ogar *et al.,* 2020). Therefore, in Africa we need to pioneer new analytical systems for understanding our local and Indigenous communities, while also addressing the challenges we are likely to face in the process (Dei, 2012), and more so, develop epistemic freedom to challenge the myth that Europe is the world's teacher and Africa its pupil (Ngugi wa, 1986). Moreso, this is because, the people utilizing Indigenous knowledge possess a more profound knowledge and understanding of their situation and what works for them. More so, the concerns on language, adult literacy, and Indigenous knowledge instruction present significant challenges to sustainable urban food sufficiency and security. For instance, the region's linguistic diversity creates barriers to effective communication and education regarding urban agricultural practices (Tollefson & Tsui, 2014; Buarqoub, 2019). Moreover, low levels of adult literacy limit individuals' ability to engage in sustainable farming techniques and access relevant resources (Anthony & Omang, 2018; Charatsari *et al.,* 2022; Magesa *et al.,* 2023). Integrating Indigenous knowledge systems into agricultural education is crucial for leveraging traditional farming practices and enhancing food production resilience (Mabhaudhi *et al.,* 2019; Dramani & Godwell, 2023). The crux of the above argument is that linguistic diversity, low adult literacy levels, and the marginalization of Indigenous knowledge hinder the effectiveness of urban agriculture initiatives. The linguistic barriers among adults, particularly those 50 years and above impede the communication of sustainable practices, while low literacy rates restrict access to essential agricultural information. The undervaluation of Indigenous knowledge prevents the integration of proven, context-specific techniques into modern practices. Addressing these challenges requires promoting linguistic literacy, expanding adult literacy programmes, and integrating Indigenous instruction into formal and informal agricultural education initiatives.

Therefore, this chapter raises the following questions. What strategies can overcome linguistic barriers to communicate sustainable urban agriculture benefits to Lagos adult farmers? How do adult education programs impact the enhancement of agricultural literacy among urban farmers in Lagos? How can Indigenous knowledge be integrated to promote urban agriculture sustainability in Lagos? In addressing these questions, this study aims to examine the importance of Language, Adult Education, and Indigenous Language among adult farmers for Sustainable Urban Agriculture in Lagos. The objectives of this study are to (i) identify strategies for overcoming linguistic barriers and effectively communicate the advantages of sustainable urban agriculture among urban farmers in Lagos, (ii) assess adult education programmes and their impact on urban agricultural literacy enhancement, and (iii) explore the Integration of Indigenous knowledge for urban agriculture sustainability in Lagos.

THEORETICAL FRAMEWORK

In the context of sustainable urban agriculture, a multidimensional approach is necessary to address complex challenges such as food security, environmental sustainability, and economic resilience (Davis & Olayide, 2018; Badewa & Dinbabo, 2023; Kapucu, et al., 2024). Language, adult education, and Indigenous knowledge are pivotal elements that can shape and enhance the practices and policies of urban agriculture. This theoretical framework explores the interrelationship between these three critical factors, emphasizing how they contribute to creating a robust, sustainable urban agricultural system (Davis & Olayide, 2018; Fadhlur et al., 2023). While Language serves as the medium for knowledge transmission, Adult Education equips individuals with the skills and adaptability needed for sustainable practices (Anthony & Omang, 2018; Ozurumba et al., 2022). On the other hand, Indigenous knowledge, rooted in centuries of ecological wisdom, provides context-specific strategies for sustainable farming (Maweu, 2011; Kom, et al., 2022; Ebenezer, 2024). The multidimensional approach can be viewed as follows:

Language as a Tool for Knowledge Transmission and Empowerment

Language serves as a fundamental medium for communication, shaping how knowledge is transmitted, shared, and retained across generations (Tektigul et al., 2023; Ayeni and Oladeji, 2023). In urban agriculture, language plays a dual role: it is both a conduit for transmitting Indigenous agricultural knowledge and a tool for fostering effective communication within and between communities (Abdullah & Hassan, 2015; Badewa & Dinbabo, 2023). Urban farmers often come from diverse cultural and linguistic backgrounds, and the ability to use multiple languages can bridge gaps between communities, policymakers, and educators (Wadumestrige-Dona et al., 2021). The role of language extends beyond mere communication; it reflects cultural identities and social structures, influencing how urban farmers perceive and approach agricultural practices (Reeves et al., 2024). Local languages, in particular, serve as carriers of Indigenous knowledge, preserving the ecological wisdom and sustainable practices that have been handed down through generations.

Such knowledge often includes insights into crop selection, pest management, soil fertility, and water conservation—elements critical to the success of urban agriculture in diverse environments. Moreover, language can either enable or limit access to formal education and technical information related to urban farming. For adult learners, language proficiency is a significant determinant of their ability to engage with formal educational programs or acquire new skills. Consequently, multilingual educational approaches are essential in urban agricultural settings, ensuring that non-native speakers or those with limited formal education can benefit from advanced agricultural techniques (Reeves et al., 2024).

Adult Education as a Driver of Capacity Building and Innovation

Adult education provides a structured mechanism for capacity building, enabling urban farmers to acquire new skills, adapt to technological innovations, and respond to evolving market and environmental conditions (Anthony & Omang, 2018; Ozurumba et al., 2022). It involves teaching practical skills, but more importantly, it empowers individuals to critically engage with the socio-economic and ecological dimensions of urban agriculture.

In this framework, adult education plays a transformative role, equipping urban farmers with the knowledge needed to optimize agricultural productivity while minimizing environmental impacts. It fosters adaptive capacity, allowing farmers to implement sustainable practices such as organic farming, waste recycling, and water management. In addition, it promotes a critical understanding of food systems, encouraging participants to engage in discussions about food sovereignty, climate change, and urban resilience (von Kotze, 2022; Deskin & Harvey, 2023). Adult education programs tailored to urban agriculture should integrate Indigenous knowledge with modern agricultural science (Anthony & Omang, 2018; Ozurumba et al., 2022). This integration ensures that traditional ecological wisdom is not overlooked in the rush to adopt high-tech solutions. Instead, farmers learn to combine Indigenous practices, such as permaculture or agroforestry, with innovative techniques like hydroponics or vertical farming, optimizing both land use and crop yield in urban settings (Schneider et al., 2021).

Indigenous Knowledge as a Foundation for Sustainable Practices

Indigenous knowledge represents a wealth of sustainable practices, honed through centuries of interaction with local environments (Ayeni & Aborisade, 2022). These practices are not only relevant to rural areas but are increasingly being recognized as vital to urban agriculture. Indigenous knowledge offers context-specific insights into biodiversity conservation, natural resource management, and community resilience, which are essential to creating sustainable urban food systems (Maweu, 2011: Kom, et al., 2022; Ayeni & Aborisade, 2022; Ebenezer, 2024).

This therefore posits that Indigenous knowledge should not be viewed as static or outdated; rather, it should be integrated with contemporary scientific knowledge and technological innovations to enhance sustainable agricultural practices. Indigenous practices often emphasize the cyclical nature of ecosystems, the importance of biodiversity, and the need to maintain ecological balance. Such principles align with the goals of urban agriculture, which seeks to produce food in ways that are ecologically sustainable, economically viable, and socially inclusive (Ayeni et al., 2014; Ayeni & Olorunfemi, 2014; Bennedetti et al., 2023; Kanosvamhira, 2024)

For instance, traditional crop rotation methods or the use of natural pesticides can be adapted to urban environments, reducing the dependency on chemical inputs and enhancing soil health. Indigenous knowledge also emphasizes communal participation and resource-sharing, principles that can foster social cohesion and collective action in urban farming communities.

Synthesis: A Holistic Approach to Urban Agriculture

The interplay of factors stated above forms the foundation for a resilient and inclusive urban agricultural system, capable of addressing modern environmental and socio-economic demands. By integrating language, adult education, and Indigenous knowledge, this framework provides a holistic approach to sustainable urban agriculture. Language facilitates communication and the preservation of Indigenous knowledge, while adult education builds the capacity for innovation and adaptation. Indigenous knowledge serves as a foundation for sustainable practices, offering ecological wisdom that complements scientific advancements. Finally, the interplay recognizes the importance of preserving and integrating Indigenous knowledge, making it accessible through adult education, and facilitating its transmission through inclusive linguistic practices. Together, these elements can create resilient, sustainable urban agricultural systems that contribute to food security, ecological balance, and social equity.

METHODOLOGY

This study adopts the content analytical and qualitative research known as "the human as instrument approach" in harnessing the importance of language, adult education and Indigenous knowledge for sustainable urban agriculture. The "human as instrument approach" is a qualitative research methodology focusing on the researcher as the primary instrument for data gathering and analysis. This method is especially beneficial in the social sciences and humanities, where the researcher(s) is/are heavily involved in the data collection process, and the data is frequently subjective and context-dependent. The approach is based on understanding humans' richly textured experiences and reflections about those experiences (Lincoln & Guba, 1985; Jackson *et al.,* 2007; Ayeni & Aborisade, 2022). On this premise, relevant data/

information were gleaned from existing literature on Indigenous knowledge, adult education, and language as they relate to sustainable urban agriculture. The data used in this research were collected, reviewed, and arranged into specific samples of usage representing the cultural, socio-linguistic, environmental and political context to enhance the necessary reflection and understanding of language, adult education and Indigenous knowledge in urban agriculture.

In this chapter, a desktop review methodology on language, adult education, and Indigenous knowledge related to sustainable urban agriculture was adopted. This provided a theoretical framework and contextual background for the study. Data on keywords as well as other related and useful information on subject matters gleaned from literature that are sourced from online databases as well as available existing hardcopies journals, textbooks etc. Explicit online databases consulted for this study include but not limited to google, Universities' Repositories. These research search engines provided useful access to relevant peer-reviewed journals, encyclopedias, books that housed information on Indigenous knowledge, adult education, and language. From a total of 145 accessed literature, only 99 documents were used in this study while the remaining 46 documents were discarded due to duplication of already used in the study and/or irrelevant to the study. In addition, this methodology employed in selecting the accessed literature was meticulous and purpose-driven, ensuring that the final set of documents included in the study not only meets academic standards but also significantly contributes to the discourse on language, adult education and Indigenous knowledge for sustainable urban agriculture

For the case study relevant data, a combination of qualitative methods was utilised. These methods include in-depth interviews, participant observation, and document analysis. To ensure fairness representativeness and minimize bias for this study, people involved in farming were selected and enumerated using random sampling techniques between September and October, 2023. The process began by obtaining a comprehensive list of urban farming locations within the target area. From this list, a random sample of farming sites was selected using a random number generator, ensuring that each farming site had an equal chance of being included in the study. Within each selected farming site, respondents were randomly chosen from the pool of active urban farmers through systematic random sampling. This method ensured that every farmer within the selected sites had an equal opportunity to participate, thereby enhancing the generalizability of the study findings to the broader urban farming population. From these, 114 urban farmers, six urban farms' clusters (locations) were identified from 5 Local Government Areas (LGAs).

The target population for the research comprises adult farmers aged 50 years and above in Lagos. This demographic is chosen because these individuals often possess a wealth of Indigenous knowledge crucial for sustainable urban agriculture. With years of experience, they hold traditional farming practices and linguistic nuances invaluable for preserving agricultural methods adapted to the urban environment. Choosing this age group also aligns with the focus on adult education, as these farmers are typically more receptive to lifelong learning initiatives. Based on personal observation and interaction with adult farmers, they were willing to integrate modern techniques with traditional practices, fostering a holistic approach to urban agriculture. Also, this target population often plays a mentoring role within their clusters, therefore, making them ideal candidates for disseminating knowledge to younger generations.

Subsequently, structured interview was conducted with the 114 selected urban farmers based on the proportion of urban farmers' population in each cluster. The reason behind the interview was to incorporates and directly apply experiences of farmers who have benefited from applying language, adult education and Indigenous knowledge in the contexts of urban agriculture, process effectiveness and productivity. Conducting research survey by using primary data from farmers themselves will aid the identification of genuine causes of problems facing urban agriculture and how to intervene rather than solely relying on literature to alleviate the problems. On these premises, the following formed the case study questions asked during the interview: name of the farming cluster, demographic characteristics, duration of residence and years of practicing urban farming, impacts of language on farming activities, importance of Indigenous knowledge etc. Surveys instrument (questionnaire) were administered to complement the interviews, providing quantitative data to support qualitative insights. The surveys included questions similar to those in the interviews, allowing for cross-verification of responses. Meetings were held with groups of farmers from the selected clusters. The discussions provided a platform for farmers to share experiences, challenges, and strategies collaboratively. It also helped to uncover common themes and divergent views. In addition, authors' personal observations were also focused on farming practices, the use of language and Indigenous knowledge, and the overall environment of urban farming in each of the clusters. This observation helps at the farm clusters help to contextualize and validate the information gathered from interviews, surveys, and meetings.

It should be noted that for proper understanding of the questions and responses, the respondents who were not proficient in English language or pidgin were guided by local language interpreter. This approach aimed to facilitate meaningful engagement with participants and generate rich insights into the complexities of urban farming within diverse linguistic and cultural contexts.

Tables, graphs, and charts provide a visual and numerical snapshot of data, making complex information accessible and comprehensible. Tables systematically organize data into rows and columns, allowing for detailed comparisons and precise value analysis. Graphs, such as bar or line graphs, illustrate trends, patterns, and relationships over time or across categories, highlighting significant variations and outliers while pie charts offer a concise, visual representation of data distribution and proportions, facilitating quick, intuitive understanding of the dataset. In this study, these tools enhance data interpretation, revealing insights that might be overlooked in textual descriptions alone.

RESULTS

Language and its Role in Achieving Sustainable Agriculture

The usage of African languages is not considered essentially important in teaching farmers both young and old agricultural techniques and approaches by western countries (Ugboh, 2004). The comprehension of the usage of agricultural approaches and technology in both the rural and the urban communities in Nigeria to enhance sustainable agriculture in rural and urban areas is premised on language usage by farmers and farm labourers. There is indeed a need for the establishment of effective interactive communication between farmers, agro-industries stakeholders and representatives, government officials as well as farm labourers to attain sustainable urban agriculture. However, the importance of language in enhancing sustainable agriculture in urban Lagos has been relegated despite the position Lagos occupies in Nigeria. Lagos is a cosmopolitan city with a wide range of linguistic diversities. Therefore, there is a need to harness language, adult education, and Indigenous knowledge in achieving sustainable urban agriculture in Lagos Nigeria.

Adult Education

Adult education is key to the development of any society. The consequence of ignorance is oftentimes beyond devastating to human society. The extreme importance of adult education in the world cannot be overstressed and the influence in the global world is vast. UNESCO (2008) asserts that some of the multiple areas of everyday relevance of adult education include learning and development of skills, promotion of healthy living, facilitation of education for the people, employment opportunities, and continuous growth and development (UNESCO, 2008). Despite the realization that education is invaluable, there are still many people around the world (especially in poorer countries) who are without any form of education and a sizable percentage of them are adults, this can only be taken care of only by adult education. Nigeria as the most populous country in sub-Saharan Africa is not immune to the debacle of adult illiteracy. Extant national statistics revealed that illiteracy amongst adults increased from 24 million in 1991 to 41,261,920 in 2015 (Knoema.com/atlas).

The grand implication of illiteracy on any nation is stagnation and underdevelopment; this is because education (or literacy) is a critical recipe for human development without which capacity building will be minimal; therefore, adult illiteracy is a threat to economic growth. The need to reverse the trend of growing adult illiteracy has become paramount, especially as it has the potential to mitigate some of the daily basic challenges that are pulling the country down. Galadima (2012) considers adult education to be a form of education set up to cater for the adult members of a society. Kapur (2019) further describes it as an area of education that is concerned with providing educational opportunities for adults who were not opportune to have formal education while in the school-age bracket or who need to become educated about something new. Furthermore, "adult education is an ongoing learning process" which simply means that adult education is a continuous process, however, it is different from child education (Ekuri, *et al.,* 2006). The disparity between adult and child education suggests that adult education needs some level of re-engineering to make it attractive and ensure usage by members of society. This is necessary because it shows potential adult students the benefits of partaking in such educational programmes (Galadima, 2012). A major catch with adult education is that it is seen as a viable tool for inducing social change as it helps adults in acquiring education which is instrumental in developing competent skills that are useful in the alleviation of the numerous challenges affecting their general well-being (Kapur, 2019). Hence, Kapur (2019) opines that adult education is "a vehicle for acquiring knowledge and skills" which could be employed to overcome poverty, unemployment, degradation, exploitation and other problems confronting society. This view is further affirmed by Schleicher (2016) through a survey conducted by the Organization for Economic Cooperation and Development (OECD) on adult skills. The findings of the survey show that people's knowledge and how they use it have a huge effect on their life chances.

The factors that enhance and foster learning in adults are situated in their local needs, circumstances and other elements of culture (Diep *et al.,* 2019; Loeng, 2020). This is in tandem with the opinion of Galadima (2012) that when programmes are based on needs, adults are encouraged to learn especially if it tallies with their goals and aspirations. The thrust of the foregoing is that adults learn best in contexts that will improve their situation which further supports the claims that adult education is an instrument for social change. Adult education equips adults with the knowledge base for solving challenges that they may face them in urban agriculture by combining traditional knowledge with modern science.

AN OVERVIEW OF INDIGENOUS KNOWLEDGE

The United Nations defined the word Indigenous as "groups of people whose social, cultural and economic conditions distinguish them from other sections of the national communities, and whose status is regulated wholly or partially by their customs or traditions or by special laws or regulations" (The World Bank, 1998, p. 4). This connotes that Indigenous people are people who share socio-economic and cultural qualities which is largely or partly regulated by special laws or by their culture which separates them from others. While some countries prefer to use the term Indigenous, some other variants of the term are commonly used in other countries; notable ones are aboriginals, ethnic groups, tribes, and first people (The World Bank, 1998). At other times, they are given names that are geographic like nomads, peasants, hill people and hunters. Indigenous people across the world have special and peculiar ways of carrying out their daily activities. According to Eyong (2007), such pursuits include Indigenous healing practices, agricultural systems, flood risk management, disease vector control, food habitats, Indigenous conservation techniques (such as cultivation methods, food taboos, hunting habits, sacred forests etc.) and several others. These activities have been successfully implemented for many centuries using what is popularly known as Indigenous knowledge.

According to Senanayake (2006, p. 87), "Indigenous knowledge is the social capital of the poor and the main assets the poor in the society use to invest in their struggle for survival to produce food". Warren (1991) views Indigenous knowledge as the knowledge peculiar to the local people which is deployed to enhance the living conditions in a particular environment. Johnson (1992) describes Indigenous knowledge as a body of knowledge developed by a generation of people who live close to nature. It can also be seen as the "knowledge and knowledge systems belonging to a given culture (Ellen and Harris, 1996). Similarly, Melchias (2001) opines that the Indigenous system is the technical know-how of a people which they have executed repeatedly through error and across different generations. Apart from these, it can be regarded as an unconventional aggregation of knowledge that merges some aspects of theory with beliefs, practices and technologies generated to ease the management of farms without the involvement of a modern, formal or scientific establishment (Gilbert *et al.,* 1980; Chambers *et al.,* 1989). In the view of UNESCO (2002) cited in Simatimbe *et al.,* (2018) Indigenous knowledge is a potential source of creativity and enrichment embodied in several social and cultural identities, each of which expresses its uniqueness (Simatimbe *et al.,* 2018; UNESCO, 2002). The usage of Indigenous knowledge depicts its richness in agricultural practices. Simatimbe *et al.,* (2018) averse that from ancient times, Indigenous people have employed several Indigenous techniques for farming and these techniques are passed down from one generation to another orally. According to Abiola *et al.,* (2011) cited in Simatimbe *et al.,* (2018), contemporary studies have shown in detail how people apply their local knowledge to improve agriculture and achieve sustainable agriculture through informal experiments and deeper comprehension of their environments which they apply in the various agricultural practices such as Indigenous soil preparation and planting materials, Indigenous methods of controlling pests and diseases, Indigenous methods of maintaining soil fertility, Indigenous methods of controlling weeds as well as Indigenous methods of harvesting and storage.

One of the major drawbacks of Indigenous knowledge is its lack of documentation, yet, it has managed to survive extinction because of its firm rootedness in cultural elements such as traditional institutions, relationships, rituals and generational communal practices (Kolawole, 2001). However, while some of this knowledge is general and learned through individual experience and day-to-day activities, others are classified and entrusted to certain categories of Indigenous people (Bruchac, 2014). This category of

Indigenous people is referred to as "the keepers" (Lalonde, 1991) or "the gatekeepers" (Bruchac, 2014). The "keepers or gatekeepers" are charged with the responsibility of preserving specialized and sophisticated knowledge such as those that concern farming, hunting, gathering, fishing, artisanship, healing, cattle-rearing and several other skills that are required to enhance the efficient management of local resources and ensure human survival. Therefore, the anxiety about Indigenous knowledge is more on the undocumented nature of the specialized knowledge coupled with other factors like the demise of elderly custodians (gatekeepers) without worthy younger gatekeepers that are successors (Chikaire *et al.,* 2012). The danger now lies in the endless changes in the socio-economic and political environments caused by the continuous contact of Indigenous people with other local and international knowledge systems thereby impacting "the would-be younger generations of custodians" to abandon the traditional assets of Indigenous knowledge in search of greener pastures. The implication of this is that this knowledge system may eventually be lost and the chances of deploying it as a remedial alternative for combating the many modern-day activities that are currently driving the environment into chaos may become lost. For instance, the fast growing and continuous adoption of western technology has currently made Africa environment more burdened with irrelevant knowledge that disempowers rather than empowers individuals and communities (Ndlovu-Gatshen, 2015).

The Concept of Sustainable Urban Agriculture

Harris and Fuller (2014) describe agriculture as a term that is used to denote the several ways in which crop plants and domestic animals are used to uphold the human population through the provision of food and other utilities. According to *Oxford English Dictionary* (1971), agriculture is defined as the science and art of cultivating the soil, including the allied pursuits of gathering in the crops and rearing livestock, tillage, husbandry, farming (in the widest sense). Therefore, agriculture is a human activity that has a very broad spectrum given that it merges many other activities which includes cultivation, horticulture, domestication, and agriculture plus different forms of livestock management (Harris & Fuller, 2014). An agricultural system with the capacity to perpetually provide food and other resources for the growing global population is fundamental for the continued existence of mankind and its numerous preoccupation (Velten *et al.,* 2015). However, agriculture as human activity has continued to suffer from several problems that are impeding its ability to meet the needs of the human population. Some of the problems that are causing agricultural failures include climate change, soil erosion and land degradation, high rate of biodiversity loss, pollution (salinization and depletion of water resources), compaction, poverty and rural emigration (Velten *et al.,* 2015; Tume *et al.,* 2019). The persistence of these problems has led to the emergence of several approaches targeted at proffering workable solutions to tackle these threats. One of the prominent ideas which emanated from these efforts and more precisely from the publication of the Brundtland Report of 1987 was that of sustainable agriculture. MacRae *et al.,* (1989) stated that sustainable agriculture is a system that uses natural processes in the conservation of all resources to ensure waste minimization and environmental impact, check problems and improve agroecosystem resilience, self-regulation, evolution and the provision of nourishment at a sustained rate. It is an agricultural system targeted at producing food abundantly without causing any damage to the earth's resources or causing environmental pollution (Pearson *et al.,* 2015).

The U.S. Farm Bill (1990) defined it as an integrated system of plant and animal production practices having a site-specific application that will, over the long term: (a) satisfy human food and fibre needs; (b) enhance environmental quality; (c) make efficient use of non-renewable resources and on-farm resources and integrate appropriate natural biological cycles and controls; (d) sustain the economic viability of farm operations; and (e) enhance the quality of life for farmers and society as a whole. Therefore, sustainable agriculture can simply be described as agriculture-based on nature's principles. Based on the foregoing, sustainable urban agriculture is simply the implementation of sustainable agriculture in urban areas. However, further clarity on what sustainable urban agriculture entails can be obtained via explanations of urban agriculture. According to Mougeot (2000), urban agriculture refers to an industry located within (intra-urban) or on the outskirts (peri-urban) of a town or urban centre, a city or metropolis, which grows or raises, processes and distributes a diversity of food and non-food products. This industry primarily utilizes human and material resources, products and services found in and around that urban area, and in turn supplying human and material resources, products and services largely to that urban area. The New York State Energy Research and Development Authority (NYSERDA, 2013) states that urban agriculture is a tool for generating economic value from existing vacant or underused space. Most of the vacant sites used for urban agriculture are sites that are usually unsuitable for other purposes, but they can be manipulated for urban agricultural practices like greenhouse cropping, intensive animal industries, and horticultural practices (NYSERDA, 2013). In light of this, therefore, the principal concern of urban agriculture is to support the quest for local food security by augmenting fresh and healthy food supplies to parts of the cities with food availability and accessibility challenges (Martellozzo *et al.*, 2014; Mack *et al*, 2017). Thus, the urban agricultural strategy of making spaces available for food cultivation and also providing a market within cities will aid the growth of the local food economies. Urban agriculture can be operated at different scales. Pearson *et al.*, (2015) identified three such scales to include micro-scale (green roofs, walls, courtyards, backyards, and street verges); mesoscale (community gardens, individual collective gardens, urban parks etc.) and macro-scale which include commercial-scale farms, nurseries, greenhouses, floriculture and vegetables. The idea of sustainable urban agriculture, therefore, is centred on keeping the existing laboratories (sites) of urban agriculture sustainably productive without attendant negative outcomes like pollution.

The Lagos Experience

Lagos State, the commercial and economic centre of Nigeria considered as the third largest economy in Africa today (Premium Times, 2023), is still plagued with the problem of ensuring adequate food security. Nigeria as a country with a forecasted higher level of poverty and hunger than they are currently not on track to meet the United Nations Sustainable Goals by 2030 (DataPhyte 2022; Vanguard (2022). In order to eradicate poverty and hunger it is imperative to achieve sustainable urban agriculture in Nigeria and particularly in Lagos State. However, it is important to note that achieving sustainable urban agriculture in Lagos is increasingly becoming a knotty task due to the following factors: the population influx being experienced in Lagos Nigeria, high cost of accessing lands, the peculiar socio-linguistic situation of Lagos that depicts Lagos as a multilingual and multiethnic society, the non-application of non-formal type of adult education, and Indigenous knowledge to agriculture have impeded agriculture in developing at an exponential rate and these have adversely affected sustainable agriculture in Lagos. According to UN-Habitat and international development agencies, the estimated population of Lagos State was approximately 24.6 million people in 2015 (Lagos State Government, 2019) and this makes up 27.4% of the national urban population with a steady daily increase due to rural-urban migration (Lagos State, 2022). The covid-19 pandemic, the economic recession, the climate problem, and soaring prices of fertilizers have aggravated the problem thereby making sustainable agriculture in Lagos more difficult to achieve. Feeding this large population indeed poses a challenge not only to the state but to Nigeria at large. Staple food is being imported by Lagosians from neighboring states and countries to ensure food security. The poor road network and infrastructural deficit existing in Nigeria have exacerbated the problem of overcoming the problem of food security. It noteworthy to state that urban agriculture is being successfully practiced all over the world and it has brought about definite gains not only on the economy of such counties or states that practice it but it has also provided food security and increased the nutritional value of children (Futi, 2019). Therefore, Lagos State has to look inward and put measures in place to achieve sustainable urban agriculture. Indeed, ensuring food sustainability will help to provide more food for the people, serve as a source of foreign exchange, and also provide an opportunity to boost the economy. It is however imperative to state that food sustainability is not a feat that can be achieved easily in a rapidly growing economy and/or Mega City like Lagos. It requires years of dedication, hard work and the use of Indigenous knowledge in agricultural practices and development, the application of adult education training to farmers in Lagos State as well as solving the problem of language barriers that exist among the farmers and the farm herds. When food sustainability is ensured, the agricultural sector of the state will experience a boost and hunger levels will be significantly decreased and this will help Nigeria in meeting the UN SDG goals. To achieve this and also meet up with the feeding needs and demands of its people both now, and in the future, with the increasing population, Lagos State and Nigeria as whole need to invest heavily in agriculture and food sustainability and also employ the use of Adult Education and Indigenous knowledge in achieving sustainable agriculture.

Case Study: The Lagos Urban Adult Farmers

Resulting from the above, the information generated from 114 farmers revealed insightful findings regarding urban farming in Lagos. Out of the 114 sampled farmers, 31.6% are located in the Barracks urban farming cluster, marking it as the highest represented group, while only 1.8% are in the Tejuoso urban farming cluster, indicating the lowest representation (Table 1). This data sheds light on the distribution and concentration of urban farming activities across different clusters, offering valuable insights for urban agricultural planning and resource allocation.

Table 1. Selected urban farming clusters in Lagos

S/n	Urban farming clusters	LGAs	No of farmers (%)
1	Tejuoso	Lagos Mainland	2 (1.8)
2	Oko-Oba	Agege	12 (10.5)
3	Barracks	Ojo	36 (31.6)
4	Idi-Araba	Surulere	6 (5.3)
5	Alapere	Ketu	28 (24.6)
6	LASU/Iyana – Iba	Ojo	30 (26.3)
	Total		114 (100)

Delving into the demographic characteristics of the respondents, it was observed that all participants were adults, with the majority falling within the age range of 50 to 65 years. An analysis of the duration of residence and years of practicing urban farming revealed that a significant proportion (28.1%) of the sampled farmers have been residing and practicing urban farming for 11 to 15 years. This suggested a considerable level of experience and familiarity with the urban agricultural landscape. Conversely, 8.8% indicated that they had embarked on urban farming within the past five years, highlighting the influx of newcomers into this sector (Fig. 1).

Figure 1. Duration of residence and years of practicing urban farming

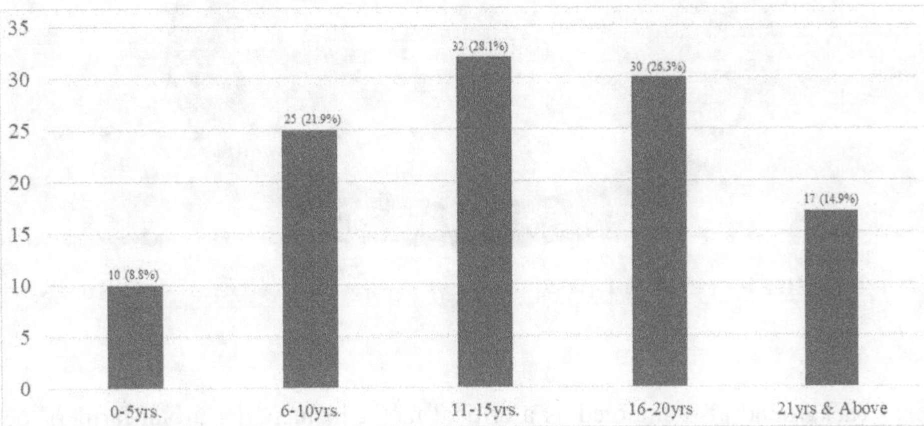

Language emerged as a pertinent issue affecting urban farming businesses, with approximately 68% of the sampled farmers expressing concerns regarding its impact, particularly in the context of english language proficiency. Many adult farmers cited difficulties in communication with agro-chemical representatives, potentially hindering their ability to interact directly when there is new opportunity to acquire, lease, and enter into contract due to inability to read and write. Finding further revealed that a majority (68%) of respondents attributed the significant impact of english to its status as the language of the elite and the official language in urban areas of Nigeria (Fig. 2). This observation is particularly noteworthy considering that a substantial portion of urban farmers are Hausa individuals who migrate from northern Nigeria to Lagos in pursuit of livelihood opportunities. Nonetheless, others (which notably refers to as pigins) accounted for 2%. It should be noted that pigins is very popular among Lagosian (i.e. Lagos populace) but majority of adults' farmers particularly the hausa that dominated urban agriculture still struggle to speak it for proper farming business vis-à-vis to alter antiquated methods in Agriculture in Lagos, promote effective transmission, and usage of improved agro-technologies by farmers and farm labourers. It was also observed that Pidgin English is ineffective among older urban farmers in Lagos due to their stronger attachment to native languages like Yoruba, Igbo, and Hausa etc., which they have spoken for most of their lives. Many lack proficiencies in Pidgin, viewing it as informal compared to their native tongues. Furthermore, generational differences and limited exposure to Pidgin through informal education or community interactions further reduce its usage and effectiveness within this demographic.

Figure 2. Impacts of language and literacy

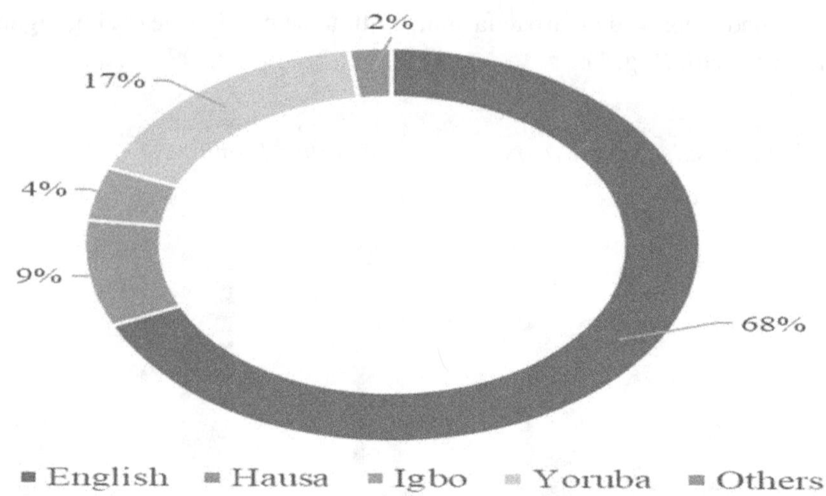

Educational background also emerged as a critical factor influencing urban farmers' perspectives and aspirations. Among the sampled respondents, only 20.2% reported having formal education which mainly consisting of primary leaving certificates while a staggering 79.8% claimed to have 'no formal education', underscoring the prevalence of educational disparities within the urban farming community. Furthermore, findings revealed a strong willingness (86.8%) among those without formal education are ready to embrace educational opportunities if afforded, recognizing its potential to enhance their urban farming practices and sales, technologies and agro-chemical prescriptions and usage among others. This highlights the importance of education as a catalyst for empowerment and socio-economic advancement within the urban farming sector.

Out of the 114 sampled respondents, only 19 (16.7%) claimed they valued Indigenous knowledge and practicing it to some extent while 63 (55.3%) argued that they value Indigenous knowledge but are not practicing due to the environment and available farm sizes for farming (Fig. 3). The remaining 28.1% claimed that Indigenous knowledge is not an option to them so far, their interest is to meet quick market demand and what would help achieve this e.g. fertilizer is within their reach.

Figure 3. Importance of Indigenous knowledge (IK)

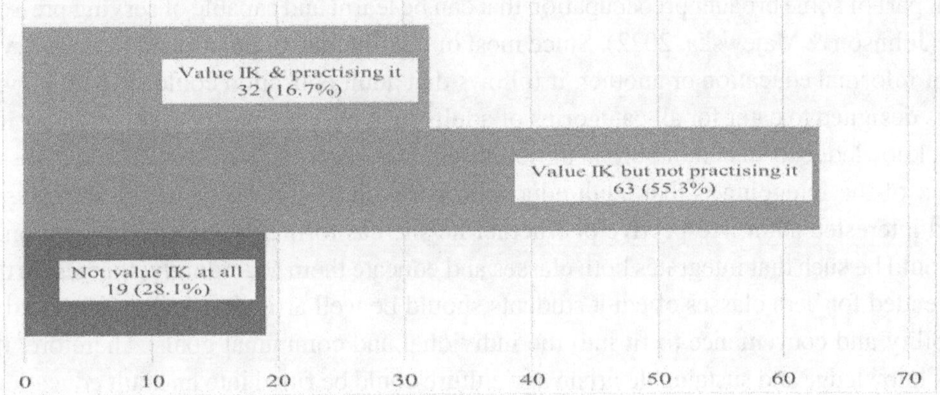

ADULT EDUCATION, INDIGENOUS KNOWLEDGE AND SUSTAINABLE URBAN AGRICULTURE

The United Nations Fund for Population Activities (UNFPA, 2007) prediction on African and Asian cities is gradually becoming evident as cities (urban areas) on these continents are experiencing a huge demographic surge in recent times. Statistically, over four billion people currently live in urban areas globally (Ritchie & Roser, 2018); and according to the population division of the United Nations Department of Economic and Social Affairs (UNDESA, 2018), the world's urban population is expected to experience further increase which is to be dominated by a few countries with India, China and Nigeria predicted to account for 35% of the projected growth between 2018 and 2050.

This projected increase is already playing out in Nigeria where the urban population has grown significantly in the past five decades by approximately 35.0% (from 18.2% to 52.0%) (Knoema, 2021). This implies that the rural areas which used to accommodate about 81.8% of the populace now account for 48% of the country's population. The driving factors for the continuous escalation in the urban population are the traditional push and pull factors which exert immense propulsive force on the rural dwellers (especially the youth) who are also attracted by the magnetic lure of the urban factors (Nweke, 2019). The consequences are conspicuous in both the rural and urban spaces; thus, while the rural areas have experienced abrupt economic decline which has led to persistent poverty and food insecurity, the urban areas are experiencing incessant socio-economic pressure induced by the continuous need for basic amenities like housing, water, employment, healthcare and other such necessities (Mini, 2001; Nweke, 2019). The needs of the communities (such as food and others captured in the preceding) have continued to increase across Nigerian urban areas and the government's response has been minimal (Galadima, 2012). Education is a viable medium through which some of these communal needs could be met. With a cumulative urban literacy rate of 62.02%, the adult population (people who are above 15 years) in Nigeria's urban areas that can benefit from adult education is very high. Furthermore, the 37.98% adult illiterate population can still be integrated into an adult education programme. This is because two classes of education have been identified, formal and informal education (Ogwo & Oranu, 2006). The informal education category described as all forms of education administered to youths and adults outside the formal school system which may also include any systemic educational activity operating separately or an essential part of some broader preoccupation that can be learnt and capable of serving pre-set learning objectives (Johnson & Majewska, 2022). Since most of the illiterate population in the country often has one form of informal education or another, it follows that adult education could be with its varieties of programs is designed to cater for all categories of adults making it an indispensable tool for integrating Indigenous knowledge to sustaining urban agriculture.

The crux of the foregoing is that adult education particularly in Lagos should be inclusive for all willing and interested adult irrespective of whether he/she has formal or informal education. The curriculum should be such that integrates both classes and educate them accordingly; themes on the subject matters intended for both classes of adult students should be well articulated, simplified and delivered with flexibility and convenience to fit into the individual and communal goals. Therefore, themes on Indigenous knowledge and sustainable urban agriculture could be fitted into an adult education curriculum and delivered to interested adult societal members. This means adult education could be adopted as the vehicle for conveying the interesting aspect of these themes to adult students. These themes should however be simplistic, easy to understand, practicable and designed in a way that would stimulate the continuous interest of the adult communal members who can help to drive the change needed (increase in food production) in the community. For instance, sustainable urban agriculture has some basic factors that represent stumbling blocks to its success, some of the critical ones as given by Jacobi *et al.,* (2000) are:

- **Natural Conditions:** This was categorized into hazards and pests and diseases.
- **Institutional Conditions:** This is made up of policy/legislation as well as access to markets
- **Physical Infrastructures and Services:** Extension services, as well as inputs like capital features here.
- **Socio-Cultural Conditions:** This is about public acceptance, as well as the danger of property theft and the level of organization.

Pests and diseases exist in practically every location where agriculture is being practiced hence urban agriculture cannot be an exception. Extension services are common for rural agriculture, urban agriculture will gain a lot from the availability of such services as professional extension officers could even be used as adult educators to teach adult learners the best approaches to urban agriculture. Jacobi *et al.,* (2000) highlighted the problem of water as an infrastructural challenge that needs to be tackled before any attempt to venture into urban agriculture. This is because water is needed for irrigation in countries where the average rainfall is low, and the dry season is extensive. There is no running away from the capital and other inputs like good seedlings, and planting sites (which for urban agriculture could be vacant land, rooftops, community gardens, nurseries, urban parks etc.); none of these will come cheap. Of the socio-economic conditions, the theft of property is one of the major limitations, hence the need for a strong institutional presence to forestall the loss of leased properties is key.

Some of the challenges listed above can be tackled via Indigenous knowledge as numerous sources in the academic literature gave notable examples of areas where the knowledge of the Indigenous people has been deployed with success. These include:

- Agronomy
- Soil management

- Water conservation and management
- Disease vector control
- Indigenous conservation techniques
- Prediction of seasons
- Farming practices: crop rotation, mixed/inter-cropping, polyculture, permaculture, etc.
- Agroforestry (Chambers *et al.,* 1989; Richards 1985; Kerr & Sanghi, 1992; Critchley *et al.,* 1995; Heim, 2020; Ayeni & Aborisade, 2022).

FUTURE RESEARCH DIRECTIONS

Environmental security, local knowledge and the implications for sustainable development in Nigeria can be a viable area for further research.

CONCLUSION

There is no gainsaying that there exist too many problems confronting mankind at this age and time but the available solutions are scanty and almost negligible. It is imperative to give a fair chance to systems that have a historical antecedence of success at combating some of these challenges; the knowledge system of the Indigenous people falls into this bracket. The fact remains that this system has been successful in the time past, therefore, there should be an extensive attempt at deploying it to achieve sustainable urban agriculture in Lagos. This is in line with Ayeni and Aborisade (2022) view that knowledge is not the exclusive preserve of any race, tribe or ethnic nationality rather it is an offshoot of capacity, intelligence, hard work and endurance. Therefore, devising new approaches for using Indigenous knowledge either via collaborations, intense research or other such efforts may just bring better results. Furthermore, the need for food security cannot be over-emphasized anywhere in the whole world, however, its continuous availability cannot be guaranteed and most especially in Nigeria. This is because the Nigerian urban areas currently stand at 17,196.23 Square Kilometres in land mass (World Bank, 2010) and accommodate 52% of the nation's population (Knoema, 2021). Yet the bulk of the food being consumed in the cities is still being expected from the rural areas that are beset with myriads of challenges, primary among which are the insecurity issues that have flooded the core agricultural states in the country in the last decade and is still very much raging. Besides that, mankind can no longer depend on the techniques of agriculture that have created so many unwanted legacies, therefore, the time is right for a change of tack and strategy as this is the only way to prevent the looming food insecurity. The application of language, adult education and Indigenous knowledge to sustainable urban agriculture within the urban environment can aid food productivity, particularly in Lagos.

ACKNOWLEDGMENT

The authors thank the University of Lagos for providing a conducive environment for this research work.

REFERENCES

Abdullah, H. J., & Hassan, T. K. (2015). The use of Indigenous knowledge in agriculture and its role in sustainable development. *International Journal of Scientific Research*, 4(10), 1312–1317.

Adeniran, A. (2022). A report on Nigerian Gross Domestic Product. Q1 NBS. Retrieved on 10 June 204 from https://nigerianstat.gov.ng/elibrary/read/1241175

Aggarwal, J. C. (1989). *Adult and social education in India: global perspective*. Doaba House.

Andreas, S. (2016). Better skills, better jobs, better lives. *Global Comparative Education: Journal of the WCCES*, 1(1), 144–149.

Anthony, G. B. & Omang, T. N. (2018). Adult literacy education: a pivot for environmental sustainability. *LWATI: A Journal of Contemporary Research*, 15(4), 172-193.

Ayeni, A., Soneye, A., & Badru, F. (2014). Adaptation to water stress in Nigeria derived Savanna area: The Indigenous knowledge and socio-cultural nexus of management and humanitarian services. [JMPP]. *Journal of Management Policy and Practice*, 15(3), 78–87.

Ayeni, A. O., & Aborisade, A. G. (2022). African Indigenous knowledge systems and the world. In *The Palgrave handbook of Africa and the changing global order* (pp. 155–173). Springer International Publishing., DOI: 10.1007/978-3-030-77481-3_8

Ayeni, A. O., & Oladeji, K. O. (2023). Pond water as a cultural element: Understanding Aringíyà's household activities and festivities. [JECA]. *Journal of English and Communication in Africa*, 6(3&4), 20–35.

Badewa, A. S., & Dinbabo, M. F. (2023). Multisectoral intervention on food security in complex emergencies: A discourse on regional resilience praxis in Northeast Nigeria. *GeoJournal*, 88(2), 1231–1250. DOI: 10.1007/s10708-022-10679-4 PMID: 35729954

Bennedetti, L. V., de Almeida Sinisgalli, P. A., Ferreira, M. L., & Lemes de Oliveira, F. (2023). Challenges to promote sustainability in urban agriculture models: A review. *International Journal of Environmental Research and Public Health*, 20(3), 2110. DOI: 10.3390/ijerph20032110 PMID: 36767476

Bruchac, M. M. (2020). Indigenous knowledge and traditional knowledge. In *Encyclopedia of Global Archaeology*. https://doi.org/DOI: 10.1007/978-3-030-30018-0_10

Buarqoub, I. A. S. (2019). Language barriers to effective communication. *Utopía y Praxis Latinoamericana*, 24(6), 64–77.

Chambers, R. ., Pacey, A., & Thrupp, L. A. (1989). *Farmers first*. Intermediate Technology Publications Ltd (ITP).

Charatsari, C., Fragkoulis, I., Anagnou, E., & Lioutas, E. D. (2022). Can adult education boost sustainability transitions? Some evidence from farmers and teachers. *Sustainability (Basel)*, 14(16), 9859. DOI: 10.3390/su14169859

Chikaire, J., Osuagwu, C. O., Ihenacho, R. A., Oguegbuchulam, M. N., Ejiogu-Okereke, N., & Obi, K. U. (2012). Indigenous knowledge system: The need for reform and the way forward. *Journal of Agricultural Science*, 1(8), 201–209.

Critchley, W. R. S., Reij, C., & Willcocks, T. J. (1994). Indigenous soil and water conservation: A review of the state of knowledge and prospects for building on traditions. *Land Degradation & Development*, 5(4), 293–314. Advance online publication. DOI: 10.1002/ldr.3400050406

DataPhyte. (2022). Can SDGs be achieved in Nigeria before 2030? According to data, not likely. Retrieved on 1 April 2024 from https://www.dataphyte.com/latest-reports/sdgs/can-sdgs-be-achieved-in-nigeria-before-2030-according-to-data-not-likely/

Davis, K. F., & Olayide, O. E. (2018). Multidimensional framework for achieving sustainable and resilient food systems in Nigeria. In Leal Filho, W. (Ed.), *Handbook of climate change resilience*. Springer., DOI: 10.1007/978-3-319-71025-9_115-1

Dei, G. S. (2012). Indigenous anti-colonial knowledge as 'heritage knowledge' for promoting Black/ African education in diasporic contexts. *Decolonization*, 1(1), 102–119.

Deskin, Z. Y., & Harvey, B. (2023). Critical food systems education in university student-run food initiatives: Learning opportunities for food systems transformation. *Frontiers in Sustainable Food Systems*, 7, 1230787. DOI: 10.3389/fsufs.2023.1230787

Diep, A. N., Zhu, C., Cocquyt, C., De Greef, M., Vo, M. H., & Vanwing, T. (2019). Adult learners' needs in online and blended learning. *Australian Journal of Adult Learning*, 59(2), 223–253.

Dramani, J. M. F., & Godwell, N. (2023). Farmers' choice for Indigenous practices and implications for climate-smart agriculture in northern Ghana. *Heliyon*, 9(11), e22162. DOI: 10.1016/j.heliyon.2023.e22162 PMID: 38053880

Ebenezer, O. J. (2024). Indigenous knowledge management practices in subsistence farming: A comprehensive evaluation. *Sustainable Technology and Entrepreneurship*, 3(2), 100058. DOI: 10.1016/j.stae.2023.100058

Ekuri, K. A. Betiang, P. A, Andong, H. A., & Eyam S. O. (2022). The relevance of adult education in a contemporary world: the implications. *LWATI: A Journal of Contemporary Research*, 19(3), 20-31.

Ellen, R., & Harris, H. (1997). Concepts of Indigenous environmental knowledge in scientific and development studies literature: a critical assessment. In *Avenir des Peuples des Forets Tropicales Working Paper* (Vol. 2).

Eyong, C. T. (2007). Indigenous Knowledge and sustainable development in Africa: case study of Central Africa. In *Indigenous Knowledge Systems and sustainable development: relevance for Africa* (pp. 121–139). Kamlla-Raj Enterprises.

Fadhlur, R. A., Suhaiza, Z., & Mastura, R. (2023). A review of the critical gaps in the food security literature: Addressing key issues for sustainable development. *Information Management and Business Review*, 15(2), 35–46.

Faloju J. O. & Sobola E. (2021). The meaning, function and contextual usage of metaphors of women in Russian and Yorùbá. *Lagos Notes and Records: A Journal of the Faculty of Arts*, 27, 160-189.

Faure, E., Herrera, F., Kaddoura, A., Lopes, H., Petrovsky, A. V., & Rahnema, M.. (1972). *Learning to be: the world of education today and tomorrow*. UNESCO.

Futi, M. G. (2019). Enhancing Food Security in Lagos, Nigeria. Restructuring Africa's largest city's food supply through urban farming a policy proposal for Babajide Sanwo-Olu, Governor-Elect of Lagos State. Master's, diss. Harvard Extension School, Policy Analysis Capstone. Retrieved on 10 April 2024 from https://capstone.extension.harvard.edu/files/capstone/files/futimwaku_203040_7923998_futi_policy_capstone.pdf

Galadima, I. A. (2012). The role of adult education in community development. *Knowledge Review*, 26(2).

Hai-Ninh, D., Bich, N. D., & Minh, H. N. (2023). How do constructivism learning environments generate better motivation and learning strategies? The design science approach. *Heliyon*, 9(12), e22862. DOI: 10.1016/j.heliyon.2023.e22862 PMID: 38125439

Harris, D. R., & Fuller, D. Q. (2020). Agriculture: definition and overview. In *Encyclopedia of Global Archaeology*. https://doi.org/DOI: 10.1007/978-3-030-30018-0_64

Heim, T. (2020). The Indigenous origins of regenerative agriculture. National Farmers Union (NFU). Retrieved on 30 April 2024 https://nfu.org/2020/10/12/the-Indigenous -origins-of-regenerative-agriculture/

Jackson, I. I., Ronald, L., Drummond, D. K., & Camara, S. (2007). What is qualitative research? *Qualitative Research Reports in Communication*, 8(1), 21–28. DOI: 10.1080/17459430701617879

Jacobi, P., Drescher, A. W., & Amend, J. (2000). *Urban agriculture - justification and planning guidelines*. Urban Vegetable Promotion Project.

Johnson, M. (1992). *Capturing traditional environmental knowledge*. IDRC.

Johnson, M., & Majewska, D. (2022). *Formal, non-formal, and informal learning: What are they, What are they, and how can we research them*? Cambridge University Press & Assessment Research Report.

Jules, N. P. (1995). Participatory learning for sustainable agriculture. *World Development*, 23(8), 1–17.

Kanosvamhira, T. P. (2024). Sustainable urban agriculture: Unlocking the potential of home gardens in low-income communities. *The Professional Geographer*, 76(5), 587–596. DOI: 10.1080/00330124.2024.2355179

Kapucu, N., Ge, Y., Rott, E., & Isgandar, H. (2024). *Urban resilience: multidimensional perspectives, challenges and prospects for future research*. Urban Governance., DOI: 10.1016/j.ugj.2024.09.003

Kapur, R. (2019). Introduction to adult education. Retrieved on 21 April 2024 from https://www.researchgate.net/publication/335422676_Introduction_to_Adult_Education/citation/download

Kerr, J. & Sanghi, N. (1992). *Indigenous soil and water conservation in India's semi-arid tropics*. International Institute for Environment and Development (IIED).

Kindon, S., Pain, R., & Kesby, M. (2007). *Participatory action research approaches and methods: connecting people, participation and place*. Routledge. DOI: 10.4324/9780203933671

Knoema (2021). Nigeria - urban population as a share of the total population. Retrieved on 1 June 2024 from https://knoema.com/atlas/Nigeria/Urban-population

Kolawole, O. D. (2001). Local knowledge utilization and sustainable rural development in the 21st Century. *Indigenous Knowledge and Development Monitor*, 9(3), 13–15.

Kom, Z., Nethengwe, N. S., Mpandeli, S., & Chikoore, H. (2022). Indigenous knowledge indicators employed by farmers for adaptation to climate change in rural South Africa. *Journal of Environmental Planning and Management*, 66(13), 2778–2793. DOI: 10.1080/09640568.2022.2086854

Lagos State Government. (2019). About Lagos. Retrieved on 1 April 2024 from https://lagosstate.gov.ng/about-lagos/

Lagos State Government. (2022). About Lagos. Retrieved on 1 August 2024 from https://lagosstate.gov.ng/about-lagos/

Lakshmi Reddy, M. V. (2000). *Dynamic role of adult education: marching education towards a learning society*. Aravali Books International (P) Ltd.

Lalonde, A. (1991). African Indigenous knowledge and its relevance to environment and development activities. *The Common Property Conference*. Winnipeg. International Association for the Study of Common Property (IASCP).

Lawal, O. A. (2015). Indigenous languages as tools for effective communication of science and technology for food production in Nigeria. *Theory and Practice in Language Studies*, 5(3), 463–468. DOI: 10.17507/tpls.0503.02

Lincoln, Y., & Guba, E. (1985). Naturalistic inquiry. *Sage (Atlanta, Ga.)*.

Loeng, S. (2020). Self-directed learning: A core concept in adult education, *Hindawi. Education Research International*, 2020, 1–12. Advance online publication. DOI: 10.1155/2020/3816132

Lokot, M., Hartman, E., & Hashmi, I. (2023). Participatory approaches and methods in gender equality and gender-based violence research with refugees and internally displaced populations: A scoping review. *Conflict and Health*, 17(1), 58. DOI: 10.1186/s13031-023-00554-5 PMID: 38066619

Mabhaudhi, T., Chibarabada, T. P., Chimonyo, V. G. P., Murugani, V. G., Pereira, L. M., Sobratee, N., Govender, L., Slotow, R., & Modi, A. T. (2019). Mainstreaming underutilized Indigenous and traditional crops into food systems: A South African perspective. *Sustainability*, 11(1), 172. DOI: 10.3390/su11010172 PMID: 37681213

MacRae, R. J., Hill, S. B., Henning, J., & Mehuys, G. R. (1989). Agricultural science and sustainable agriculture: A review of the existing scientific barriers to sustainable food production and potential solutions. *Biological Agriculture and Horticulture*, 6(3), 173–219. Advance online publication. DOI: 10.1080/01448765.1989.9754518

Magesa, M., Jonathan, J., & Urassa, J. (2023). Digital literacy of smallholder farmers in Tanzania. *Sustainability (Basel)*, 15(17), 13149. DOI: 10.3390/su151713149

Maweu J. M. (2011). Indigenous ecological knowledge and modern western ecological knowledge: complementary, not contradictory. *Thought and Practice: A Journal of the Philosophical Association of Kenya (PAK)*, 3(2), 35-47.

Melchias, G. (2001). *Biodiversity and conservation*. Science Publishers, Inc.

Merriam-Webster. (n.d.): Language. In *Merriam-Webster.com dictionary*. Retrieved on 1 February 2024 from https://www.merriam-webster.com/dictionary/language

Mougeot, L. J. A. (2000). *Urban agriculture: definition, presence, potentials and risks, and policy challenges*. Cities Feeding People Series Report 31. Retrieved on 10 February 2024 from https://idl-bnc-idrc.dspacedirect.

Muin, F., Noortyani, R., Elyani, E. P., & Chaemsaithong, K. (2021). Rethinking arbitrariness of language and its implication in language use. *Cogent Arts & Humanities*, 8(1), 1868687. Advance online publication. DOI: 10.1080/23311983.2020.1868687

Ndlovu-Gatshen, S. J. (2015). Decoloniality as the future of Africa. *History Compass*, 13(10), 485–496. DOI: 10.1111/hic3.12264

Ngugi wa, T. (1986). *Decolonizing the mind: the politics of language in African literature*. James Currey.

Nweke, A. C. (2019). Rural-urban migration in Nigeria, implication on the development of the society: Anambra State as the focus of the study. *Journal of Public Administration and Governance*, 9(2), 209. Advance online publication. DOI: 10.5296/jpag.v9i2.14912

NYSERDA. (2013). *Sustainable urban agriculture: confirming viable scenarios for production*. New York State Energy Research and Development Authority (NYSERDA)

OECD. (1971). *Equal educational opportunities: a statement of the problem with special reference to recurrent education*. Organization for Economic Cooperation and Development (OECD).

Ogar, E., Gretta Pecl, G., & Mustonen, T. (2020). Science must embrace traditional and indigenous knowledge to solve our biodiversity crisis. *One Earth*, 3(2), 162–165. DOI: 10.1016/j.oneear.2020.07.006

Ogwo, B. A., & Oranu, R. N. (2006). *Methodology in formal and non-formal technical/vocational education*. University of Nigeria Press.

Oxford English Dictionary. (1971). *The Oxford English dictionary*. Oxford University Press.

Ozurumba, I. G., Ochiagha, C. D., Echem, G. C. C., & Ugwuoke, A. C. (2022). Contributions of adult literacy programmes towards achieving the empowerment of adults in Enugu State, Nigeria. *Journal of Nigeria National Council for Adult Education*, 7(1), 91–113.

Pearson, L. J., Pearson, L., & Pearson, C. J. (2011). Sustainable urban agriculture: stock take and opportunities. In *Urban agriculture: diverse activities and benefits for city society*. https://doi.org/DOI: 10.3763/ijas.2009.0468

Premium Times. (2023). Updated: Fact-checking claims made by Bola Tinubu at NESG dialogue. Retrieved on 1 April 2024 from https://www.premiumtimesng.com/news/headlines/583799-fact-checking-claims-made-by-bola-tinubu-at-nesg-dialogue.html

Ramirez-Santos, A. G., Ravera, F., Rivera-Ferre, M. G., & Calvet-Nogués, M. (2023). Gendered traditional agroecological knowledge in agro-food systems: A systematic review. *Journal of Ethnobiology and Ethnomedicine*, 19(11), 11. Advance online publication. DOI: 10.1186/s13002-023-00576-6 PMID: 37024903

Reeves, N. P., Sal, V. G., Lutomia, A. N., Medendorp, J. W., Bello-Bravo, J., & Pittendrigh, B. (2024). Agricultural education in Africa using YouTube multilingual animations: A retrospective feasibility study assessing costs to reach language-diverse populations. *PLoS One*, 19(4), e0302136. DOI: 10.1371/journal.pone.0302136 PMID: 38635490

Richards, P. (1985). Indigenous agricultural revolution: ecology and food production in West Africa. *Indigenous agricultural revolution: ecology and food production in West Africa*. https://doi.org/DOI: 10.2307/632952

Ritchie, H., & Roser, M. (2018). Urbanization. Our world in data. Retrieved on 2 April 2024 from https://ourworldindata.org/urbanization

Schneider, P., Rochell, V., Plat, K., & Jaworski, A. (2021). Circular approaches in small-scale food production. *Circular Economy and Sustainability*, 1(4), 1231–1255. DOI: 10.1007/s43615-021-00129-7 PMID: 34888585

Senanayake, S. (2006). Indigenous knowledge as a key to sustainable development. *Journal of Agricultural Sciences*, 2(1), 87–94. DOI: 10.4038/jas.v2i1.8117

Simatimbe, W., Mbozi, E. H., & Hambulo, L. (2018). Integrating Indigenous knowledge practices into agricultural extension education in Luano District of Central Zambia. *International Journal of Advanced Research*, 6(10), 178–190. DOI: 10.21474/IJAR01/7798

Tektigul, Z., Bayadilova-Altybayeva, A., Sadykova, S., Iskindirova, S., Kushkimbayeva, A., & Zhumagul, D. (2023). Language is a symbol system that carries culture. *International Journal of Society. Culture & Language*, 11(1), 203–214.

Tharakan, J. (2017). *Indigenous knowledge systems for appropriate technology development*. InTech., DOI: 10.5772/intechopen.69889

The World Bank. (1998). Indigenous knowledge definitions, concepts and applications. Retrieved on 11 April 2024 from https://chm.cbd.int/api/v2013/documents/4A27922D-31BC-EEFF-7940-DB40D6DB706B/attachments/209070/Hoda%20Yacoub%20-%20IK%20Report%20(1).pdf

The World Bank. (2010). Land area square kilometre - Nigeria. Retrieved on 11 April 2024 from https://data.worldbank.org/indicator/AG.SRF.TOTL.K2?locations=NG

Thomas, A., Menon, A., Boruff, J., Rodriguez, A. M., & Ahmed, S. (2014). Applications of social constructivist learning theories in knowledge translation for healthcare professionals: A scoping review. *Implementation Science : IS*, 6(9), 54. DOI: 10.1186/1748-5908-9-54 PMID: 24885925

Tollefson, J. W., & Tsui, A. B. M. (2014). Language diversity and language policy in educational access and equity. *Review of Research in Education*, 38(1), 189–214. https://www.jstor.org/stable/43284067. DOI: 10.3102/0091732X13506846

Tume, S. J. P., Kimengsi, J. N., & Fogwe, Z. N. (2019). Indigenous Knowledge and Farmer Perceptions of Climate and Ecological Changes in the Bamenda Highlands of Cameroon: Insights from the Bui Plateau. *Climate (Basel)*, 7(12), 138. DOI: 10.3390/cli7120138

Ugbor, O. (2004). Benefits of Indigenous language in the teaching and learning of adult farmers in Delta State, Nigeria. *Journal of Agriculture and Social Research*, 4(2), 54–60.

UNDESA. (2018). 2018 revision of world urbanization prospects. United Nations - Department of Economic and Social Affairs. Retrieved on 22 April 2024 from https://population.un.org/wup/

UNESCO. (1976). Recommendation on the development of adult education, adopted by the General Conference at its nineteenth session. Nairobi: United Nations Educational, Scientific and Cultural Organization (UNESCO). Retrieved on 22 April 2024 from https://uil.unesco.org/fileadmin/keydocuments/AdultEducation/en/declaration-nairob_e.pdf

UNFPA. (2007). *State of world population 2007: unleashing the potential of urban growth*. United Nations Population Fund, Thoraya Ahmed Obaid, Executive Director. Retrieved on 22 April 2024 from https://www.unfpa.org/sites/default/files/pub-pdf/695_filename_sowp2007_eng.pdf

Vanguard (2022). Nigeria not on track to meet many SDGs by 2030 — UN. Retrieved on 22 April 2024 from https://www.vanguardngr.com/2022/09/nigeria-not-on-track-to-meet-many-sdgs-by-2030-un/

Velten, S., Leventon, J., Jager, N., & Newig, J. (2015). What is sustainable agriculture? A systematic review. *Sustainability (Basel)*, 7(6), 7833–7865. Advance online publication. DOI: 10.3390/su7067833

von Kotze, A. (2022). Adult education for "resilience": educating in precarious times. In Evans, K., Lee, W. O., Markowitsch, J., & Zukas, M. (Eds.), *Third international handbook of lifelong learning*. Springer International Handbooks of Education., DOI: 10.1007/978-3-030-67930-9_55-1

Wadumestrige-Dona, C. G., Mohan, G., & Fukushi, K. (2021). Promoting urban agriculture and its opportunities and challenges – a global review. *Sustainability (Basel)*, 13(17), 9609. DOI: 10.3390/su13179609

Warren, D. M. (1991). *Using Indigenous knowledge in agricultural development*. World Bank Discussion Papers, 127. Retrieved on 11 April 2024 from https://documents1.worldbank.org/curated/pt/408731468740976906/pdf/multi-page.pdf

Zidny, R., Sjöström, J., & Eilks, I. A. (2020). Multi-perspective reflection on how Indigenous knowledge and related ideas can improve science education for sustainability. *Science & Education*, 29(1), 145–185. DOI: 10.1007/s11191-019-00100-x

ADDITIONAL READING

Ayeni, A., Soneye, A., & Badru, F. (2014). Adaptation to water stress in Nigeria derived savanna area: The Indigenous knowledge and socio-cultural nexus of management and humanitarian services. *[JMPP]. Journal of Management Policy and Practice*, 15(3), 78–87.

Ayeni, A. O., & Aborisade, A. G. (2022). African indigenous knowledge systems and the world. In Oloruntoba, S. O., & Falola, T. (Eds.), *The Palgrave handbook of Africa and the changing global order*. Palgrave Macmillan., DOI: 10.1007/978-3-030-77481-3_8

Ayeni, A. O., Cho, M. A., Mathieu, R., & Adegoke, J. O. (2016). The local experts' perception of environmental change and its impacts on surface water in Southwestern Nigeria. *Environmental Development*, 17, 33–47. DOI: 10.1016/j.envdev.2015.09.007

Ayeni, A. O., & Olorunfemi, F. B. (2014). Reflections on environmental security, Indigenous knowledge and the implications for sustainable development in Nigeria. *J. Res. Natl. Dev.*, 12(1), 46–57.

KEY TERMS AND DEFINITIONS

Language: A system of symbols and rules used for communication, expressing thoughts, emotions, and ideas between people.

Adult Education: Organized learning aimed at helping adults acquire new skills, knowledge, or qualifications for personal or professional growth.

Indigenous Knowledge: The traditional and/or local knowledge developed by native communities, rooted in their culture, environment, and experiences.

Sustainable Urban Agriculture: Eco-friendly cultivation of food in cities, enhancing food security, biodiversity, and local economies.

Chapter 17
Pedagogy of Ecological Kinships:
Community Teaching-Learning Practices of the Siddis in Gujarat

Sayan Dey
https://orcid.org/0000-0002-7569-5519
Bayan College, Purdue University Northwest, Oman

Niharika Sinha
https://orcid.org/0009-0002-3563-4735
Alliance University, India

ABSTRACT

Ecological kinships serve as one of the several ecologically and emotionally sustainable pedagogical practices of Gujarat's Afro-Indian diaspora communities, popularly known as the Siddis. The Siddis arrived in Gujarat in the 13th century with the Islamic invaders from different parts of eastern Africa as slaves, palace guards, harem keepers, and musicians. They gradually settled in Gujarat and embraced Afro-Sufi cultural practices like Zikr songs, Dhamaal dances, and Urs celebrations. These practices have enabled the Siddis to preserve their ancestral knowledge values as well as curate community-based teaching-learning practices. The musical and dance performances of their knowledge values serve as an effective teaching-learning practice and enable the Siddi communities to remember and re-member their multi-rooted and multi-routed histories. This chapter will discuss how the Afro-Indian cultural practices of the Siddis unfold alternate, radical pedagogies of ecological kinship.

INTRODUCTION: SIDDIS AND THE INDIAN OCEAN

The Indian Ocean is a multifaceted, intertwined, and rhizomatic repository of folklore, histories, sciences, technologies, civilizations, and other knowledge systems. Folklores, histories, and cultures have been archived on the ocean beds as fossils, ship wreckages, banned books, precious stones, sea anemones, phytoplanktons, kelps, coral reefs, seagrasses, mangroves, musk grasses, sargassums, etc. (Hofmeyr, 2021; Dey, 2023). How rhizomatic, multi-perspective, Indigenous, and ecological modes of thinking, learning,

DOI: 10.4018/979-8-3693-8447-3.ch017

and interpreting produce innovative pedagogical practices serve as the central argument of this chapter. Apart from conserving and enhancing the terraqueous environments, the Indian Ocean has also served as a critically diverse (Steyn & Vanyoro, 2023) pedagogical tool to curate and share knowledge-making patterns. The ocean's knowledge-making pattern lies outside the bureaucratic, hierarchized, fortified, exclusionary, and human-centric knowledge-making enclaves like schools, universities, classrooms, and laboratories, which the glorified textbooks of sciences, technologies, cultural studies, histories, and societies have failed to capture to date. For instance, besides the scientific, geological, and geographical perspectives, the evolution of the planet Earth is understood through multiple aquatic myths and folklore, whose ethical and logical credibility is no less significant than the human-centric knowledge systems. According to Japanese folklore, Earth was created by the sea gods Izanami and Izanagi, who churned the primordial ocean with their magic swords to create landmasses (Roorda, 2020). In Chinese folklore, Pangu is regarded as the creator god who segregated the landmass and the ocean with its horns and gave birth to the continents (Wang, 2012). Boshongo folklore in Congo regards Bumba as the creator who created different forms of floral, faunal, and celestial lives on Earth by vomiting. The Boshongo community believes that before Bumba vomited, the planet was completely submerged under the ocean and was lifeless and dark (Scheub, 2002). The Cherokee folklore describes how living beings decided to explore life around the aquatic bodies due to overpopulation in the skies. Dayunishi, the water beetle, dived into the sea to investigate what is beyond water and returned with a handful of mud. This encouraged other creatures to follow in the beetle's footsteps, and the Cherokees believe that this is how aquatic life was born (Francis, 2019). These folklores provoke us to interpret the evolution of the planet Earth and its varied knowledge systems through multi-perspective lenses instead of restricting it to a few academic and institutional narratives that mainly gain validation only after being credentialized by the self-celebratory Euro-North American-centric intellectual bodies. The chapter will explore how, as a porous and fluid space, the Indian Ocean and other ecological spaces have historically facilitated the evolution and mobilization of Indigenous, multirooted, and multi-routed transoceanic communities like the Siddis in India. The Indian Ocean has also created knowledge-producing structures that embrace human and environmental socio-cultural and intellectual perspectives in diverse oral, musical, performative, and kinesthetic narratives like *Dhamaals*, *Zikrs*, and *Salaamis*.

Before delving into these narratives, let us briefly reflect on the historical origin of the Siddi community in India. The terms' Siddis' and 'Habshis' refer to the Afro-Indian community in India. Around the 13th century, the arrival of the Islamic empire in Gujarat through West Asia, Central Asia, and the Persian Gulf brought several Africans from eastern Africa as slaves, merchants, mercenaries, mistresses, weavers, performers, and scholars. Several narratives revolve around the origin of the terms 'Habshi' and 'Siddi.' According to Fitzroy Baptiste, the term 'Habshi' can be traced from the ancient Egyptian word 'Khebtsi,' which was used to refer to the people of Punt (ancient Ethiopia) (Baptiste, 2008, p. 121). Many historians argue that the 'Habshis' refers to the African community that came to India from Ethiopia. Wilfred H. Schoff philosophizes that the ancient Egyptian inscriptions have used the word 'Hbsti' to denote the people from Punt (Schoff, 1995, p. 62). Based on these different perspectives, Schoff also noted that the root for the word 'Hbsti' lies in the Arabic word 'Habash' or 'Habashat,' meaning the "people of Abyssinia or Ethiopia" (Schoff, 1995, p. 63). British historian John Burton-Page has a different narrative to share about the origin and the usage of the term 'Habshi.' In his third edition *of Encyclopedia of Islam* (1971), Burton-Page says: "The majority, at least in the earlier periods, may well have been Abyssinian, but certainly the name was applied indiscriminately to all Africans, and in the days of the Portuguese slave-trade with India, many such 'Habshis' were in fact of the Nilotic and the Bantu races" (Burton-Page, p. 14-15).

Concerning the term 'Siddi,' there are not many perceptions about its origin. According to the ninth edition of the *Oxford English Dictionary* (1933), the term 'Siddi' originated from the Arabic word 'Sayyid,' meaning 'lord' or 'prince' (p. 22). Robert Vane Russell, in his book *The Tribes and Castes of Central Provinces in India* (2015), reflects that 'Siddi' is commonly used to denote the Africans (either Abyssinian or Negro) in India (Russell, p. 145). Baptiste observes that in *Bhagavadgomandal*, an encyclopedia that was compiled under the patronage of Maharaja Bhagvatsinhji of Gondal in the early 20th century, the term 'Siddi' is equated with 'Habshi' and 'Negro' (Baptiste, p. 122). Reginald Enthoven, in his book *The Tribes and Castes of Bombay* (1990), argues the same (Enthoven, p. 332).

Since the arrival of the Siddis in India, their professional and personal identities have always been in a state of fluidity. On the one side, their professions not only changed based on their performances; the Siddis indulged in multiple occupations simultaneously. For instance, an African slave worked as a royal army chief and a royal office attendant at the same time, while an African eunuch functioned as a harem guard and performed the role of the king at the same time; an African woman slave performed the role of a mistress and a queen at the same time, as can be seen in the case of Nawab Wajid Ali Shah's wives (Dey, 2023a). Moreover, under Islamic rule, the African slaves in India hardly faced any form of racial segregation. It was with the arrival of the Portuguese colonizers in India that the African slaves in India were racially dehumanized and compartmentalized. When the Portuguese came to India, they brought with them African slaves from Mozambique, Eritrea, Somalia, and South Africa, which diversified the existing population of Abyssinian slaves in the country (Dey, 2023). Before the arrival of the British colonizers, the Portuguese subjected the African slaves to severe states of existential inhumanity, as they did with the indigenous communities in other parts of the world. Many slaves escaped from the clutches of the Portuguese into the forests of Goa, Maharashtra, Karnataka, and Gujarat. It is in these places that a majority of the Siddi settlements are found today.

The arrival of the British colonizers changed the scenario for the Siddis, with many of them liberated from slavery in return for embracing Christianity. Today, a majority of the Siddis in India are Christians and Muslims; few are Hindus. Though their communities in most parts of India are officially recognized as Scheduled Tribes, they exist in a critical state socially, culturally, politically, and economically (Dey, 2023a). As Scheduled Tribes, they are granted reservations and scholarships at government-run education and work institutions. Due to several race, caste, economic, political, and geographical complexities, most Siddis work as domestic workers, cleaners, untrained laborers, agricultural workers, garage mechanics, tourist guides, and other odd jobs. Only a handful of Siddis go to school, complete their education, and get permanent jobs in the country (Dey, 2021; Dey, 2023a). Despite these challenges, the community members maintain their African ancestral roots through habitual socio-cultural practices and embrace the local native socio-cultural practices of India at the same time. Over the years, this has eventually given birth to pedagogies of ecological kinships (Dey, Maity & Srivastava, 2024).

These diverse narratives about the historical evolution and contemporary situations of the Siddi community unfold how their nomenclatures, identities, celebrations, and crises have generated multitudinous pathways through which their ecological pedagogies and knowledge systems have been established. The aspects of porosities and fluidities across the Indian Ocean naturally lie at the core of Siddi's experiences and practices. As a result, their creative, cross-pollinated, and cross-cultural pedagogies have emerged through "interacting and collaborating with other individuals and systems" (Mavodza & Ngulube, 2012, p. 1). This will be elaborately discussed in the forthcoming sections. Unlike the teaching-learning methods of most mainstream educational institutions in India, fixated visions, enclaved knowledge-building patterns, and sterilized intellectualities do not have a place in the Siddi experience.

RESEARCH METHODS

Indigenous kinships and intersectionality as research methods serve this chapter's theoretical and thematic foundations. Considering the thematic, theoretical, and practical scopes of this chapter, these two research methods have been specifically chosen as they will be able to justify the environmental and pedagogical aspects of the chapter. Both these methods invite us to understand how the creative teaching-learning methods of the Siddi community are deeply intertwined with the spirits, celestial bodies, and the natural environment.

Indigenous kinship is a "theory and practice in its own right" (Weismantel, 2019, p. 687). It is "rooted and framed within theories of interdependence – intertwining actions influencing experiences" (Francis, Bowman & Sutton, 2023, p. 146). This research method enables us to understand the relationships between living beings beyond "genealogical relationships" (Francis, Bowman & Sutton, 2023, p. 145). The method generates the ability to "transverse between Indigenous perspectives and pedagogies to learn and hear Indigenous voices, knowledge, and culture that impact professional disciplines" (Francis, Bowman & Sutton, p. 146) through the practices of "respect, relationship, relevance, reciprocity, responsibility, [and] representation" (Francis, Bowman & Sutton, p. 147). To elaborate further, this method makes us realize Indigenous modes of knowledge production beyond touristic exoticism and capitalistic flattery. In this chapter, the musical, mnemonic, and kinesthetic pedagogical practices of the Siddi community will outline the theoretical and methodological values of Indigenous kinships through ecological, social, cultural, emotional, and intellectual intersectionality. As will be discussed in the following sections, the dance, musical, and spiritual practices of the Siddis function as interdependent and rhizomatic pedagogies, where humans, oceans, forests, mountains, deserts, animals, and plants operate together in co-cared, co-shared, and intersectional manners that celebrate the "continuity of culture … spirituality, family, society, and ancestry" (Sorby, Buchanan & Smith, 2024, p. 4).

The performances not only serve as representations of Siddi histories and cultures but also as teaching-learning methods of "deep reciprocity, respect, non-interference, self-determination, and freedom" (Simpson, 2017, p. 8) about histories, societies, and cultures that are still not familiar to people across India and the world. While sharing our research project with colleagues, we have often encountered systemic ignorance (Steyn, 2012), hierarchies, mockeries, and suggestions about how insignificant it is to work with a community that is hardly known to people.

This chapter will also argue how the intersectional ecological pedagogies of Siddi kinships have established ecologies of intimacies between human, natural, and spiritual communities. Kimberle Crenshaw conceptualized the theory and method of intersectionality to understand how the experiences of gender and sexuality are interrelated to diverse social dynamics and power structures like race, color, political ideologies, and academic systems (Crenshaw, 1991). Eventually, intersectionality has been applied across diverse knowledge disciplines to understand how different cultures, societies, and knowledge-sharing structures function in coalition through agreements, disagreements, resistances, and tensions. The creative pedagogies of the Siddis uphold the practice of intersectionality to create a "radical resurgent present based on deep reciprocity and the gorgeous generative refusal" (Simpson, 2017, p. 9) of the mainstream authoritarian, coerced, and hierarchical academic disciplines in India.

The arguments in this chapter will be supported by varied ethnographic narratives documented through personal conversations with various Siddi men and women in Gujarat between September 2022 and July 2024. The conversations occurred physically in the cities of Ahmedabad, Sasan Gir, and Ratanpur in Gujarat, and the participants ranged between 40 and 85 years. They were approached based on their availability and consent. Contact with the participants was established through Farooq Murima Siddi, who has guided the authors throughout this research project by organizing musical and dance performances, establishing connections with the Siddi community members across Gujarat, and taking them around markets, mosques, gateways, and sidewalks that are connected to Siddi histories and cultures.

The conversations and documentation took place across different spaces, like in the dargahs[1] of Bava Gor and Mai Misra in Ahmedabad and Ratanpur, the house of Hameeda Makwa Siddi (an 85-year-old Siddi woman) in Paththar Kuwa (a locality in Ahmedabad), around Teen Darawaza (three gateways) of Ahmed Shah Badshah, and the adjacent market in Ahmedabad. The conversations took place in Hindi with six participants in semi-structured ways. They were asked open-ended questions. Typically, conversations were started with generic questions: Can you share stories about Siddi cultural and knowledge-making practices? How are they preserved and performed in contemporary times? Based on the responses, the rest of the questions were spontaneously framed. Drawing cues from Lorisia Macleod's "More Than Personal Communication" (2021), where Macleod has devised citation methods to "improve the respectful recognition of Indigenous knowledges and voices" (Macleod, 2021, p. 1), the views of the participants have been cited like any other references. As a part of this acknowledgment and kinship pedagogy, it is also essential to voice out loudly that even though Sayan and Niharika's names appear as authors, the roles of the participants like Hameeda Makwa Siddi, Farooq Murima Siddi, Altaf Siddi, Sikandar Siddi, and Ujma Siddi in the ideation of this chapter need to be equally recognized. This chapter would not have taken shape without sharing their knowledge and experiences.

ECOLOGICAL KINSHIPS AS PEDAGOGICAL TOOLS

In the Siddi community, the pedagogy of ecological kinships is practiced through narrating perspectives of cross-communal existences, symbiotic experiences with plants, and emotional relationalities with the Indian Ocean. In sync with the porous and fluid ancestral pedagogical patterns, the teaching-learning practices of kinships amongst the Siddis are not fixated within fortified spaces, degrees, and awards. They are "continually shifting" in nature (Basu, 2008, p. 64) and are practiced beside seacoasts, inside dargahs, and on thresholds of their homes. Besides specific spiritual occasions, the Siddi welfare organizations like Sidi al-Mubrik Trust, Bava Gor Trust, Sidi Jamaat, and others would organize community gatherings across different locations in Gujarat that have socio-historically related to the Siddis where they would discuss different spiritual and cultural folklore of the Siddis. The gatherings are open to both Siddis and non-Siddis and are widely attended by youngsters and elders.

Communal Kinships

On 11 September 2022, we met Hameeda Makwa Siddi, one of the few surviving Siddi elders, in the Paththar Kuwa locality in Ahmedabad. Paththar Kuwa is one of the many localities that became a residency for the Siddis in the 15th century. After some general conversations about Siddi histories, cultures, and contemporary scenarios, she took us to the threshold of their common housing space, where a group of Siddi children and a few Siddi men and women eagerly awaited her arrival. After greeting each other, we all sat together, and after introducing us to the gathering, she began narrating folktales about the practice of inter-communal kinships amongst Siddis. That day, she narrated the two following stories about the communal and cultural coexistence between the Hindus and the Muslims in Gujarat:

a. Running away from Goddess Luxmi from the Kingdom of Ahmad Shah Badshah: Once, Goddess Luxmi[2] became annoyed with the materialistic extravagance and financial wastages in the kingdom of Ahmad Shah Badshah and decided to leave the kingdom. As she was about to leave the Teen Darwaza (the three gateways), Sidi Suzaad, an African palace guard, stopped her. Sidi Suzaad pled with her not to leave the kingdom and enquired about the reasons behind her annoyance. She said she was not very pleased with the kingdom's materialistic extravagances and economic inequalities, so she wished to leave. Listening to this, Siddi Suzaad asked if she could wait and let him inform the king about her departure. Luxmi agreed to the proposal on the condition that she would wait until Suzaad returned and then take Suzaad with her. Suzaad accepted the proposal and went to the king.

Listening to the story, Ahmad Shah Badshah got confused and suggested that Suzaad decide. Suzaad suggested that if the emperor could kill Suzaad, the goddess would remain in the kingdom since she had promised not to leave without Suzaad. Ahmad Shah rejected this proposal: Suzaad could not be convicted and killed without reason. So, Suzaad killed himself by slashing his throat. This is how Luxmi remained in the kingdom (Hameeda Makwa Siddi, 2022).

Besides oral narrations, this folklore is practiced today by worshipping Luxmi, whose shrine is adjacent to the Teen Darwaza. Folklore that centers on the interaction between Makhaan Devi and Mai Misra also reflects the knowledge values of intercommunal kinship.

b. The curse of Makhaan Devi: In the 14th century, Mai Misra, along with her seven brothers, arrived on the coast of Kuda (a coastal town) in Gujarat. They reached in the evening and decided to rest. While Mai Misra was resting, the seven brothers strolled along the seashores. At that time, Makhaan Devi, the local sea goddess, took one of the brothers as a captive. When the brothers requested Makhaan Devi to release him, she denied them and said she would keep him as her playmate. Concerned, the remaining brothers returned and informed Mai Misra about the incident. Then, Mai Misra met Makhaan Devi and requested that she release their brother. However, she denied it again. Mai Misra proposed an alternative to her. She said that if Makhaan Devi releases their brother, they will bless and heal their devotees together. This is why the coastal shrines of Mai Misra and Makhaan Devi face each other on the coast of Kuda. Also, during the celebration of *Urs*[3] in Bhavnagar (a city near Kuda), the shrines of Mai Misra and Makhaan Devi are worshipped together by the Siddis and the local Hindus (Hameeda Makwa Siddi, 2022).

These narratives display how, historically, the community has valued intercommunal interactions and well-being. This is specifically significant within the communally fractured sociopolitical state of Gujarat. Gujarat, especially after the Godhra riots in 2002, has been perceived as a breeding ground of communal tensions between Hindus and Muslims across India and the world (Patel & Nath, 2013). In such a circumstance, these folklores interrogate and dismantle stereotypes by producing tangential historical narratives that India's mainstream institutions, curriculums, pedagogies, and textbooks are oblivious to.

The connections with the Indian Ocean and eastern Africa have expanded within Gujarat's cultural and communal realms and formed a symbiosis with local cultural tales. The ignorance of the mainstream education institutions toward the intercommunal histories and cultures of the Siddis is what has provoked the community to generate alternate pedagogical archipelagoes[4] on thresholds, such as the one we attended, and lakesides, where they can sensitize their community members about the entangled, rhizomatic, co-habitual, and collaborative pedagogical and knowledge systems of the Siddis.

The archipelagic pedagogy of kinship is not limited to their folklore but can be seen through the organizational and structural patterns of teaching, learning, and sharing. In Hameeda's gatherings, the participants do not think and argue individually and hierarchically but interact in a decentralized and co-shared manner, where everybody's viewpoints, agreements, disagreements, concerns, and criticisms are welcomed irrespective of socioeconomic status, age, class, and gender. Hameeda shared: "Even though, as per the demand of the times, the Siddi children need to go to modern educational institutions to learn modern-day subjects, I try to make sure that they do not forget their ancestral education and histories" (2022). She added, "These daily learnings about their ecological and intercommunal histories and cultures not only make them aware of their ancestral education systems but also make them critical towards modern-day knowledges" (2022). Hameeda's threshold becomes an informal educational space that instills multi-perspective visions within the Siddis and serves to "bind people together" (Sorby, Buchanan & Smith, 2024, p. 2) within spaces, timelines, and cosmologies that are multidirectional, nonlinear, and nonconforming to the cherry-picked, metric-based, degree-based and market-based templates of popular knowledge systems (Mbembe, 2016; Dube & Ngulube, 2012). These stories about the intersections of religion, cultures, geographies, and societies through Goddess Luxmi, Makhaan Devi, Mai Misra, palaces, littorals, and sea coasts have generated an "intricate multilayering" (Ghosh, 2021, p. 71) of teaching-learning methods that are not top-down, unidirectional, dictatorial, and exclusionary, and instead encourage "mutual recognition of distinct territories and overlapping boundaries" (Ghosh, 2021, p. 71).

Environmental Kinships

The kinship-based pedagogies of the Siddis are not restricted to human-human knowledge sharing. They also encompass and embrace human-environmental exchanges of knowledge. Like Hameeda, Ujma Siddi, a 40-year-old woman from Mangrol (a town near Sasan Gir in Gujarat), has curated an intimate space of co-learning and co-caring in the threshold of her home to sensitize the Siddi community about Indigenous medicinal and healing practices inherited from their eastern African ancestors. She organizes weekly and monthly workshops where she shares how to identify naturally grown medicinal plants in the forests of Sasan Gir and use them to treat various health issues. Ujma specializes in infertility and gynecological issues and teaches community members to treat them with seeds, bark, fruits, and animal excreta. According to Ujma, "Civet excreta is dried and mixed with water in a certain proportion and then fed to women who fail to conceive or suffer from consistent abortions" (2024). She also shares the usage of Shivlingi seeds to treat miscarriages and cancer. Shivlingi is an herbaceous plant that belongs to the Cucurbitaceae family. This plant bears fruits that have "a red shell with black seeds" (Ujma Siddi, 2024). The seeds are extracted, dried, and then converted into a powder for curing. She also uses tea tree bark to treat abnormal vaginal discharges, hemorrhoids, and piles. The tree bark is picked up, cleaned, sliced, and then rubbed across infectious regions on the body. The use of these natural cures not only unfolds the rich hygiene and medical practices of the Siddis but also highlights how they have intergenerationally learned from the natural world. Such pedagogies of environmental kinships are a crucial part of Siddi cultural upbringing. The Siddi people around the Sasan Gir forest have followed the footsteps of their elders by going into the forests with them since their childhood days and learning through the sounds, smells, and movements of the natural environment. During our conversations, Ujma shared how the science and geography textbooks in formal schools could not teach them as meticulously about the floral and faunal civilizations as their elders could through hands-on teaching-learning procedures. As she said: "While walking through the forests, as I touched the plants, smelled the flowers, picked up the smell of animals or listened to the rustling of the leaves, I felt they are deeply communicating with me and teaching me knowledges that cannot be textually noted or experimented in laboratories or calculated mathematically" (2024). She added: "After returning from the forests, we would share what we have learned about and from the natural environment, and it can be applied in daily life. Every time we tell a story, we acknowledge our ancestors, elders, brothers, and sisters on the other side of the Indian Ocean. They live within us, we live within them" (Ujma Siddi, 2024). Ujma's initiative is not just scientifically, pedagogically, and medically valuable but also culturally therapeutic. The regular gatherings, discussions, and knowledge sharing encourage the Siddis to be proud of their ancestral knowledge systems, revive them, and introduce them within the mainstream knowledge-producing spaces of India.

It is also essential to note that the traditional medicinal practices of the Siddis are not unique to their community and can also be traced to other traditional medicinal cultures like Ayurveda. Nevertheless, they still need to be highlighted and loudly discussed because, unlike Ayurveda, which is communally labeled and patented as a Hindu medicinal discovery (Rai, 2019), Siddis do not claim communal ownership of their ancestral medicinal cultures. On the contrary, they consistently strive to find avenues to sensitize both Siddis and non-Siddis about natural ways of maintaining health and hygiene. The medicinal knowledge systems of the Siddis have yet to be documented in textbooks and curriculums; they have been carried forward orally across generations. This is how Ujma imbibed the teachings of her mother, who, in turn, learned from her mother. Ujma is getting trained at a local traditional medicine college to formalize her ventures. As Ujma philosophizes: "The workshops not only enable me to sensitize, educate, and share my acquired medical science knowledges with my community members but also allow me to experiment with Ayurvedic and Siddi medicinal systems" (Ujma Siddi, 2024). These workshops are archipelagic grounds of kinship pedagogies, where the Siddi medicinal practices are taught through collaborative, hands-on experiences like learning to identify medicinal plants, examining the edibility of plants, accurately calculating the mixing proportions, and extracting necessary medicinal ingredients, among others.

The learning processes are simultaneously incentivized by stories about the intersectional existence of human, environmental, oceanic, and spiritual ancestors and how every form of living civilization and their knowledge systems have essential roles in shaping their ecologically motivated teaching-learning practices. While workshopping, Ujma and other Siddis share ancestral stories about oceans, deserts, plants, mountains, forests, and spirits that also teach medicinal knowledge. In this way, Siddis embraces, acknowledges, and internalizes cultural and emotional intimacies and memories with the natural environment.

Spiritual Kinships

In Ratanpur, Altaf Siddi's pedagogy of spiritual kinships intersects songs, lyrics, spaces, objects, and memories. Altaf Siddi is a 45-year-old Siddi man from the Ratanpur district of Gujarat, a professional *Zikr* singer and a teacher. For him, *Zikr* music is not just about spirituality and aesthetics but also an oral teaching-learning process to remember his Eastern spiritual ancestors like Bava Gor, Mai Misra, Baba Habas, Siddi Muttaq, Sidi Nabi Sultan, Sidi Sultan Khan, and others. Zikrs are a form of prayer song sung mainly by the Sufi Muslims to revive the memories and blessings of their spiritual ancestors. The stories about the ancestors are documented in the lyrics which narrate stories about how their spiritual ancestors traveled across deserts, mountains, and oceans to bless the Siddis in Gujarat during a time when they were subjected to sociocultural, communal, religious, and racial brutalization by the Islamic invaders. As an instance, let us look into a *Zikr*[5] narrated, explained, and sung by Altaf Siddi:

La Ilala Illaha
Dhamama Bol Nabi
Baba Bol Nabi
Haq Nabi
Dada Haq Nabi
Ya Illaha ya Nabi
Badshah
Aye Dhamama Bol Nabi
Kalma Nabi ka Dhamama bol Nabi
Ya Ilala bol Nabi
Dom Gori Badshah Nabi
Kalma Nabi ka Ya Nabi
Asra Nabi ka ya Nabi
Hey Madina wale ya Nabi
Dhamama bol Nabi
ya Ilala ya Nabi

Meaning: This *Zikr* celebrates Siddi Nabi Sultan's contributions towards the cultural and spiritual upliftment of the Siddi community. The *Zikr* asks the Siddi people to repeat the name of Sidi Nabi wherever they travel because his presence and blessings can be spiritually perceived everywhere. The *Zikr* praises the multiple routes Sidi Nabi has traveled to Gujarat: from the shrines of Mecca and Medina to the waters of the Indian Ocean, the forests of eastern Africa, and the deserts of the Nubian valley. The *Zikr* also praises the different living beings of the natural environment for cooperating with Sidi Nabi Sultan and assisting him to reach the shores of Gujarat safely.

This *Zikr* and several others unpack how the notion of spirituality within the Siddis is not restricted to human figures but encompasses nature deeply. Before singing, Altaf always expresses his gratitude to his parents, grandparents, and ancestors for teaching his community's histories, cultures, and emotions through *Zikrs*. This expression is not just a cultural performance but a pedagogical performance as well. As Altaf said: "When I sing the Zikrs, I receive spiritual satisfaction and feel like I am revising the teaching-learning values I inherited from my community elders. When I sing these *Zikrs*, I learn about historical, geographical, cultural, and spiritual knowledges that no school books could teach me to date" (Altaf Siddi, 2024). Altaf's practice of spiritual kinship is not restricted to self-learning. He shares his knowledge about *Zikr* with his community members, especially the children, through organizing informal gatherings inside the shrine of Bava Gor (a Siddi spiritual ancestor) in Ratanpur. The shrine in Ratanpur is embroidered with forests, stone mines (mostly exhausted), and lakes. Some of these lake shores serve as gathering points where Altaf teaches *Zikrs* to the Siddis. Apart from teaching the rhythms and lyrics, Altaf ensures that the learners are aware of the social, cultural, emotional, and historical contexts of the *Zikrs*. As Altaf shared: "The *Zikrs* are dedicated to different spiritual leaders, and their intent and rhythms are not the same. Therefore, the learners need to understand the contexts of the differences" (2024)). He continued: "The learners should know why some Zikrs focus on mountains, some on deserts, some on the forests, and some on the oceans. They should know that the natural environment is not just an object of materialistic entertainment and aesthetic pleasure. They are ordained with deep spiritual and ecological values that must be respected and inculcated" (2024). Besides nurturing historical, geographical, and cultural knowledge values, the *Zikrs* function as a tool for instilling ecological consciousness within the Siddi community and beyond. To understand further, let us look into another *Zikr* by Altaf Siddi:

Ye bolo Sabaya Hua hey
Ya bolo sabaya hua wey
Hu Sabaya
Salwale Nabi Sultan
Salwale Nabi Sultan
Pyara Muhammad Dastagui
Sachha Muhammad Dastagui
Allah hu
Nabi hu
Allah hu
Nabi hu
Dungar Rangaye Mian
Allah hu
Nabi hu
Allah hu
Nabi hu
Dungar Rangaye Mian
Utho Noori Karo Bandagi
Subaah ka waqt noorani
Andheri ki gaflat meye
Tajawwul noor ka saya hain
Baitho na baitho na
Saccha Pir baitho na
Maulana maulana
Badey Pir maulana
O jungle ke janawar
Baitho Na baitho na

Meaning: This *Zikr* is also sung in praise of Sidi Nabi Sultan, who is believed to have first arrived in Gujarat on the coast of Kuda (located in the Bhavnagar district of Gujarat) from the Nubian Valley (encompasses the geographical regions of Sudan and Egypt). The *Zikr* means that after praising Allah and Bava Gor, it is time to praise Nabi Sultan, and if the blessings of Sidi Nabi Sultan exist, then no evil can befall the Siddis of Gujarat. It also talks about how Bava Gor has brought with him the blessings of Prophet Muhammad for every living being in Gujarat. As Bava Gor has arrived, it is time to decide the way forward for the Siddi community in Gujarat. In order to decide, the mountains, birds, trees, forests, animals, and humans must wake up, sit together with Bava Gor, and seek his spiritual suggestions.

Like the previous one, this *Zikr* also outlines the cultural, emotional, intellectual, and spiritual symbiosis between humans and the natural environment within the Siddi community. These pedagogical practices of human-nature kinships enable the community to teach, learn, and share in archipelagic, fluid, viscous, and porous ways (Ghosh, 2022, p. 2), unlike the ignorant, self-inflated, arrogant, textbook-centric, and decontextualized pedagogical practices of the institutions in India. However, these discourses are still widely limited within the Siddi knowledge systems, and many other communities are unaware of them. Even within the Siddi community, the young generation appears reluctant to imbibe their ancestral knowledges. Thus, the community across Gujarat has been collaborating to establish formal schools for both Siddi and non-Siddi children, where, besides teaching modern-day curriculums, the students are regularly taught about the Siddi dance, songs, music, lyrics, rhythms, stories, histories, rootedness, routed-ness, and spiritualities. Such forms of teaching-learning challenge "pretensions to discrete individuality" and establish a tangential, inconclusive, and non-linear knowledge-making mechanism of "webbed, meshed, and interactive" pedagogies (Ghosh, 2022, p. 2). A few schools have already been established in the Siddi-dominated localities of Jamnagar and Bhavnagar, and more schools are lined up. The general anti-black racist attitude of the state and central governments makes the process of establishing such alternative, archipelagic pedagogical and academic spaces quite challenging, especially in terms of seeking the necessary affiliations and accreditations. Moreover, a significant section of society interprets the Siddi pedagogical practices as uncivilized, unscientific, backdated, and barbaric, discouraging their children and the educational bodies from recognizing their knowledge structures. Despite these challenges, the Siddis, with the support of their welfare organizations, have been able to curate archipelagoes of radical ecological pedagogies that are "visible, … referential, structural, … [and] differential" in nature (Ghosh, 2022, p. 5). The pedagogies will be further discussed in the next section.

DHAMAALS AND SALAAMIS AS PEDAGOGIES OF RADICAL KINSHIPS

Dhamaals are a form of dance and musical practice during which the Siddis sing varieties of *Zikrs* to re-member their social, cultural, ecological, racial, and spiritual roots of eastern Africa on the one side and appreciate the local social and spiritual cultures of Gujarat on the other. *Dhamaals* serve as a pedagogical example of intercommunal symbiosis, where people are welcome to participate without prejudices, regardless of cultural, religious, or national identity. *Dhamaals* are of two types – dance *Dhamaal* and *baithaaki* (sitting) *Dhamaal*. In the dance *Dhamaal*, the *Zikrs* played are loud and frenzied; often, the lyrics are overwhelmed by the musical confluence of hand drums and conch shells. The music is complimented by dance movements that mimic the motion of leaves, plants, animals, and ocean waters, movements that are deeply embedded within the Siddi histories, ancestries, cultures, and folklores.[6]

Apart from being a cultural and spiritual ritual, dance *Dhamaals* function as a kinesthetic pedagogy of preserving, acknowledging, and reminding each other of how their spirits, intellects, and knowledges have been shaped not only by humans but by more-than-human civilizations like oceans, forests, deserts, and cosmic constellations as well. They must be acknowledged in an era of ceaseless environmental depletion and climate catastrophes. The twirls, swings, trips, feet thumping, and acrobatics generate a deep sensory perception of their histories of geographical and cultural mobilizations and their trans-oceanic ancestral connection with Eastern Africa, where similar dance movements are visible amongst Sufi Muslims. For instance, like *Zikrs*, the Sufi Islamic community in Sudan performs *Dhikr* music and dance (Graves, 2019).

In between music and dance, the Siddi performers often pause to explain to the audience the lyrics and dance movements and how they are performed in different formats amongst the Afro-Sufi Muslims in eastern Africa and other parts of the world. In this way, the dance *Dhamaals,* as a kinesthetic pedagogy, produces an "embodied memory of collective enjoyment" (Kabir, 2021, p. 243). Collective enjoyment is produced through an intersectional procedure of "performances, gestures, orality, movement, dance, singing – in short, all those acts usually thought of as ephemeral and nonreproducible knowledge" (Taylor, 2003, p. 20) that cannot be captured and restricted within the popular, mainstream, and tangible knowledge producing spaces like written documents, textbooks, and physical libraries. The performers specifically encourage Siddi children to attend the *Dhamaal* performances so that the legacy of Indigenous Siddi knowledge systems gets carried forward across generations. The impact of *baithaaki Dhamaals* is no different. Farooq Murima Siddi, a 45-year-old Siddi dancer, reflects on *Dhamaals* as a teaching-learning practice: "As a kid, my parents and grandparents took me to various *Dhamaal* performances across Gujarat. When the dancers and singers performed, they explained the intentions, purposes, and significance of the songs, lyrics, and dances. They encouraged me not to forget my transoceanic ancestral roots" (2024). He further observes: "Every time I dance, every movement of my body feels like a manifestation of the cultural teachings of my ancestors" (Farooq Murima Siddi, 2024).

Like the *Dhamaals*, the purpose of performing *Salaami* is deeply immersed in the philosophies of mutual reciprocity, co-caring, co-sharing, respect, togetherness, life, and death. *Salaami* is an offshoot of Zikrs performed at the end of *baithaaki Dhamaals* to acknowledge the spiritual ancestors for blessing them with the knowledge values of community kinships. To further understand, let us look into the lyrics of a *Salaami*:

Part I

Ho kisko pata
Jo amar rahe
Jiyenge to phir milenge (twice)
Ho naat-jamaat
Jo amar rahe (twice)
Ho jiyenge to phir milenge
Ha jiyenge to phir milenge
Ho jiyenge to phir

Part II

Boley ya apney yamey ya
Saya jotey musaley
Ho dada
Saya jotey musaley

Part III

Ho oliya salaam walaikum
Baba mere salaam walaikum (twice)
O Misra peer ko
Walaikum salaam
O salaam waliyon ko
Walekum salaam
Salaam walekum
Walekum salaam (twice)

Part IV

Bava Gor salaam walaikum
Kanga misra salaam walaikum
Mai Misra salaam walaikum
Gori Shah salaam walaikum
Pir Badshah salaam walaikum

This *Salaami* was shared with us by Sikandar Siddi, a 48-year-old Siddi man from Ahmedabad who is a professional *Dhamaal* singer and dancer. As he explained during our conversations, this *Salaami* invokes the blessings of Siddi spiritual leaders like Bava Gor, Mai Misra, and Sidi Gori Shah so that the Siddi community can continue living together in life and the afterlife. The Siddis accept that life and death are normative parts of human life and cannot be avoided. With the community's spiritual, cultural, and historical efforts, the knowledge and values of the Siddis can be intergenerationally preserved, transferred, and practiced. The *Salaami* also says that they should greet each other by saying 'salaam walaikum' (peace be unto you) and share mutual respect whenever the Siddis meet instead of spreading hatred, conflicts, and tensions so that even in the physical absence of each other, the spiritual bond does not cease to exist.

The Salamis function as mnemonic pedagogies in teaching Siddi knowledge systems across generations. As a part of this pedagogy, the ancestral memories function as an archive, with the archives transferred and multiplied across people of the community and beyond. As Sikandar Siddi says: "I have learned all the *Salaamis* from my community elders, and I encourage my children and others to learn them. They do not have to choose a profession with it, but at least they should know about their cultural and ancestral roots. They should know where they come from. They should know their connections to Africa and India" (2024).

The formal and informal Siddi-run schools have started building audiovisual archives to systematize these radical pedagogical practices for future applications. Before Siddis took these initiatives, many researchers and research organizations across India and the world documented Siddi cultural practices for archival in their respective research institutions. It is a novel task to work towards documenting and archiving the tangential and radical sociocultural practices of a community like the Siddis. However, the self-promotional, self-profiting, and myopic intentions of the researchers must be investigated. Stereotypical representations of the Siddi community, either in a hyper-glorified way of always singing, dancing, and worshipping or as a poverty-stricken community, have internally and externally impacted the community (Dey, Maity & Srivastava, 2024). During one of our conversations, Hameeda reflected: "The Siddi community has been historically humiliated, racialized, and marginalized to such an extent that today's children are very reluctant to learn their own cultures and traditions. Many do not want to identify as Siddis" (2022). She added: "They wish to be referred to as Gujaratis because they feel embarrassed to be identified as Afro-Indians and feel that their cultural practices are backdated and filthy" (Hameeda Makwa Siddi, 2022).

This is why it is essential to shift beyond the stereotypical, limited, and appropriative representations of the Siddi cultures to deeply experience their diverse, reciprocating, radical, tangential, and ecological teaching-learning procedures and contextually intertwine them with mainstream knowledge-producing systems of India. Doubtlessly, the process is not straight jacketed, and the community will encounter multiple intellectual, practical, infrastructural, and bureaucratic challenges. As already discussed earlier, the Siddi knowledge systems and the Siddi-run schools are burdened with the humongous task of justifying the authenticity and cultural validity of their pedagogies and knowledge, and they face isolation, ignorance, and marginalization from both within their community and outside. Nonetheless, in pieces and fragments, the Siddi community has already started challenging the popular teaching-learning and knowledge-producing systems, formally and informally, across pockets of Gujarat through their alternate, radical, tangential, and creative pedagogies of storytelling, singing, and dancing.

The Indigenous eco-centered pedagogical approaches of the Siddis also open gateways of research beyond their popular, stereotypical, and racialized representations of Afro-Indian dancers and singers. In a time when the right-wing Bharatiya Janata Party (BJP) regime has communally and culturally fractured the country and has intellectually silenced the Muslims through uncontrollable Hindumania and Islamophobia, the Siddis, as Afro-Sufis, with their symbiosis of eastern African ancestral knowledges with local traditional knowledges shed new rays of cultural and intellectual hope in the country. The researchers who are interested in working with the Siddi community in India in general and Gujarat, in particular, should further explore how their creative teaching-learning methods can contribute towards co-building a diversely contextualized education system that is free from the caste, class, cultural, racial, and communal prejudices and is situated within human as well as more-than-human knowledge constellations.

CONCLUSION: WAYS TOWARDS SYMBIOTIC PEDAGOGICAL FUTURES

How are Siddi Knowledge Systems Integrated Formally Within School Curriculums?

The Siddi-run schools in Gujarat do not teach Siddi and non-Siddi students about their cultural, historical, and ancestral knowledge systems for extracurriculars. It is ensured that these knowledge systems are thoroughly integrated within their curriculum as a part of school assignments. As Hameeda argued: "If we just teach the children for fun and leisure, they do not take the cultural practices seriously. Therefore, our schools make sure that they are thoroughly examined and graded for singing *Zikrs* and dancing *Dhamaals* like any other subject" (2024). Apart from reading mainstream subjects, learning about Siddi cultural practices for the arts and culture classes is compulsory. The students are trained by experienced Siddi men and women, who have not only inherited the musical cultures as a continuity of the Siddi knowledge constellation but have professionally performed across India and the world in dance and music festivals. When such trainers teach the students, they feel encouraged to overcome notions of inferiority about their own cultures and enthusiastically embrace them. They are also made to practically understand the different creative pathways available instead of choosing solely from the usual job market-centered professions. While dancing and singing, trainers and students dance and sing together; they learn with each other and the natural environment, reproducing the philosophies and cultures of plants, animals, and oceans in varied ways. This teaching-learning process fluidifies solidified thinking (Malabou, 1997, p. 115) by training the students in de-hierarchical and enmeshed pedagogies which are not driven by dictating-notetaking philosophies and instead by reciprocating, horizontal, collaborative, and co-curative practices.

What can Mainstream Educational Institutions in India Learn From the Siddi Community?

In their respective teaching-learning contexts, the mainstream educational institutions in India should reformulate their pedagogical practices beyond the anthropocentric lens to encompass the knowledge values of the natural environment. Despite generating conservation policies and awareness programs, knowledge about the natural environment remains concentrated within the curriculums of geography and environmental sciences, which exclusively function in theoretical propositions and fail to engage with practical concerns. The practice of ecological kinships of the Siddis in Gujarat in songs and dances invite educational institutions to generate possible ways of integrating the functional, existential, and knowledge systems of the natural environment within the daily teaching-learning methods. The transoceanic, transcultural, and transborder histories and cultures of the Siddis in Gujarat also invite the teachers and learners from the mainstream educational spaces to engage with knowledge systems that are usually considered intellectually insignificant, redundant, and marginalized.

The pedagogical practices of the Siddis are relevant not only to India but also to education systems worldwide. The musical and story-telling practices of the Siddis can be found within the Afro-Sufis of eastern Africa, like in Zanzibar and Sudan, within the Zanzibari and other eastern African diaspora communities in Gulf countries like Oman, Qatar, Kuwait, and the United Arab Emirates, and other parts of the world. In these countries, the eastern African Afro-Sufi knowledge systems have not yet found a space in the formal education systems but are habitually practiced during festivals and community gatherings. In a time when knowledge-producing systems are becoming more and more surveilled, censored, enclaved, and bubbled within selective market-based and revenue-generative disciplines, the eco-entangled symbiotic pedagogical practices of the Siddis evoke "cross-cutting and tangential" (Ghosh, 2024, p. 4) teaching-learning methodologies that are critical, diverse, praxis-driven, open-ended, rhizomatic, and invitational. In sync with the Siddi pedagogical practices, we (the authors) are inviting the readers not necessarily to agree with us but, even in disagreements, to collectively explore the possibilities of generating nonconforming symbiotic pedagogical futures of human-nature intersections that are not hierarchical, exclusionary, and binarized in structure.

ACKNOWLEDGMENTS

The authors express profound gratitude to Aarthi Sridhar of the Dakshin Foundation for generously funding this project's fieldwork. The authors would also like to thank every Siddi community member in Gujarat who has welcomed, supported, and encouraged them to carry out this project.

REFERENCES

Baptiste, F. A. (2008). From "Invisibility" to "Visibility": Africans in India through the Lens of Some Select Sources from the Late Classical Period to the Late 18th century A.D. In Prasad, K. K., & Angenot, J.-P. (Eds.), *TADIA: The African Diaspora in India*. Jana Jagrati Prakashana.

Basu, H. (2008). History on the Line: Music and the Formation of Sidi Identity in Western India. *History Workshop Journal*, 65(1), 161–178. DOI: 10.1093/hwj/dbm069 PMID: 19618564

Burton-Page, J. (1971). *Encyclopedia of Islam*. Brill.

Crenshaw, K. (1991). Mapping the Margins: Intersectionality, Identity Politics, and Violence against Women of Color. *Stanford Law Review*, 43(6), 1241–1299. DOI: 10.2307/1229039

Dey, S. (2021). "Killing with kindness": Political icons, socio-cultural victims: visual coloniality of the Siddis of Karnataka. In Karam, B., & Mutsvairo, B. (Eds.), *Decolonising Political Communication in Africa: Reframing Ontologies* (pp. 91–108). Routledge. DOI: 10.4324/9781003111962-9

Dey, S. (2023a). *Performing Memories and Weaving Archives: Creolized Cultures Across the Indian Ocean*. Anthem Press. DOI: 10.2307/jj.8973372

Dey, S. (2023b). Baine Maraa-Indu Mama-Siddi Dhamaal: Interwoven Performances of Epistemic Justice and Cognitive Freedom by the Siddis of Karnataka, India. *International Journal of Critical Diversity Studies*, 5(2), 17–32. DOI: 10.13169/intecritdivestud.5.2.0017

Dey, S., Maity, T., & Srivastava, T. (2024). Gender Empowerment in Transoceanic Feminine Folklore and Shrines: A Kin Study of Siddi Women's Participation in Mai Misra Worship in Gujarat, India. *Journal of International Women's Studies*, 26(1), 1–14.

Dube, L., & Ngulube, P. (2012). Knowledge sharing in multicultural environment: Challenges and opportunities. *South African Journal of Library and Information Science*, 78(1), 68–77. DOI: 10.7553/78-1-48

Enthoven, R. E. (1990). *The Tribes and Castes of Bombay*. Asian Educational Services.

Francis, C. D., Bowman, N., & King, S. (2023). Kinship pathways: Nurturing and sustaining resilient, responsible, and respected indigenous evaluators. *New Directions for Evaluation*, 177(177), 145–152. DOI: 10.1002/ev.20537

Francis, J. D. L. (2019). Eastern Cherokee Creation and Subsistence Narratives: A Cherokee and Religious Interpretation. *American Indian Culture and Research Journal*, 43(1), 83–98. DOI: 10.17953/AICRJ.43.1.LOFTIN-FREY

Ghosh, R. (2021). The Plastic Turn. *Diacritics*, 49(1), 64–85. DOI: 10.1353/dia.2021.0003

Ghosh, R. (2022). *The Plastic Turn*. Cornell University Press.

Ghosh, R. (2024). *Plastic Tagore: Thinking After Yesterday*. Oxford University Press. DOI: 10.1093/9780198922995.001.0001

Graves, J. (2019). Filling the Pot: The Remembrance of African Sufi Ancestor-Saints and the Reclamation of African-Historical Heritage in Ahmedabad, Gujarat. *Journal of Africana Religions*, 7(1), 94–104. DOI: 10.5325/jafrireli.7.1.0094

Hofmeyr, I. (2021). *Dockside Reading: Hydrocolonialism and the Custom House*. Wits University Press.

Kabir, A. J. (2020). The Fleeting Taste of Mazaa: From Embodied Philology to an Alegropolitics for South Asia. *South Asia*, 43(2), 243–254. DOI: 10.1080/00856401.2020.1703486

MacLeod, L. (2021). More Than Personal Communication: Templates for Citing Indigenous Elders and Knowledge Keepers. *KULA: Knowledge Creation, Dissemination, and Preservation Studies, 5*(1), 1–5.

Malabou, C. (1997). Who Is Afraid of Hegelian Wolves? In Patton, P. (Ed.), *Deleuze: A Critical Reader* (pp. 114–138). Wiley-Blackwell.

Mavodza, J., & Ngulube, P. (2012). Knowledge management practices at an institution of higher learning. *South African Journal of Information Management*, 14(1), 1–8. DOI: 10.4102/sajim.v14i1.496

Mbembe, A. J. (2016). Decolonizing the university: New directions. *Arts and Humanities in Higher Education*, 15(1), 29–45. DOI: 10.1177/1474022215618513

Oxford English Dictionary. (1933). Clarendon Press.

Patel, N. G., & Nath, S. R. (2013). Where Identity and Trauma Converge: Hindu-Muslim Perceptions of the 2002 Gujarat Riots. *The Journal of Muslim Mental Health*, 7(2), 25–45. DOI: 10.3998/jmmh.10381607.0007.202

Rai, S. K. (2019). Invoking 'Hindu' Ayurveda: Communalisation of the late colonial Ayurvedic discourse. *The Indian Economic and Social History Review*, 56(4), 411–426. DOI: 10.1177/0019464619873820

Roorda, E. (2020). Izanagi and Izanami, Japanese Sea Gods. In Roorda, E. (Ed.), *The Ocean Reader: History, Culture, Politics* (pp. 13–13). Duke University Press. DOI: 10.1215/9781478007456-005

Russell, R. V. (2015). *The Tribes and Castes of the Central Provinces of India*. CreateSpace Independent Publishing Platform.

Scheub, H. (2002). *A Dictionary of African Mythology*. Oxford University Press.

Schoff, W. H. (1995). *The Periplus of the Erythraean Sea: Travel and Trade in the Indian Ocean by a Merchant of the First Century*. Munshiram Manoharlal Publishers Pvt. Ltd.

Siddi, A. (2024). *On the Spiritual Pedagogies of the Siddis*. Personal Conversation.

Siddi, F. U. (2024). *Siddi Dhamaals as a form of Pedagogy*. Personal Conversation.

Siddi, H. M. (2022). *On the Kinship Pedagogies of the Siddis*. Personal Conversation.

Siddi, H. M. (2024). *On teaching-learning Siddi cultures in Siddi-run schools*. Personal Conversation.

Siddi, S. (2024). *Salaamis as Mnemonic Pedagogies*. Personal Conversation.

Siddi, U. (2024). *On the Ecological Pedagogies of the Siddis*. Personal Conversation.

Simpson, L. B. (2017). *As We Have Always Done: Indigenous Freedom Through Radical Resistance*. University of Minnesota Press. DOI: 10.5749/j.ctt1pwt77c

Sorby, J., Buchanan, F., & Smith, A. (2024). Kinship and Cultural Strengths—Learning from an Aboriginal Perspective. *Australian Social Work*, •••, 1–13. DOI: 10.1080/0312407X.2024.2370265

Steyn, M. (2012). The ignorance contract: Recollections of apartheid childhoods and the construction of epistemologies of ignorance. *Identities (Yverdon)*, 19(1), 8–25. DOI: 10.1080/1070289X.2012.672840

Steyn, M., & Vanyoro, K. P. (2023). Critical Diversity Literacy: A framework for multicultural citizenship education. *Education, Citizenship and Social Justice*, 0(0). Advance online publication. DOI: 10.1177/17461979231178520

Taylor, D. (2003). *The Archive and the Repertoire: Performing Cultural Memory in the Americas*. Duke University Press.

Wang, Q. (2012). Phenomenological Interpretation of Chinese Creation Myth of Pangu. *Theoretical Studies in Literature and Art*, 32(3), 134–137.

Weismantel, M. (2019). Making kin: Kinship theory and Zumbagua adoptions. *American Ethnologist*, 22(4), 685–704. DOI: 10.1525/ae.1995.22.4.02a00010

ENDNOTES

1 A shrine of a Muslim saint.

2 According to the Hindu mythology, Luxmi is considered as the goddess of wealth.

3 *Urs* is a Sufi Islamic cult that celebrates the death anniversary of their spiritual leaders.

4 I have referred to the alternate teaching-learning spaces and systems of the Siddis as archipelagoes because, unlike many mainstream and popular pedagogical frameworks, they are not conforming, assimilative, and integrative. Instead, like a group of islands, they mutually respect and reciprocate their differences, disagreements, and diversities and cohabit.

5 To access more *Zikrs* by Altaf Siddi, please go through the following link: https://www.youtube.com/watch?v=Kw6aRakDIC8.

6 To learn more about the dance *Dhamaals*, please go through the following link: https://www.youtube.com/watch?v=df9HWgYYQ6w&t=7s.

Chapter 18
Fostering Cultural and Cognitive Justice Through Inclusive Learning:
Promotion of Indigenous Knowledge Systems for Sustainable Learning

Edward Shizha
Wilfrid Laurier University, Canada

ABSTRACT

This theoretical chapter discusses the relevance and importance of indigenous African knowledges in promoting sustainable learning for indigenous African students. It argues that sustainable learning cannot be achieved when schooling content and pedagogical practices exclude colonized and marginalised indigenous knowledges. Culture involves politics of knowledge construction, and school knowledge that excludes indigenous knowledges lacks cultural justice and cognitive justice— the right of all forms and traditions of knowledge to co-exist in producing meaningful learning for students. Cognitive justice contributes to meaningful education requiring students' capacities and capabilities that are rooted in their social and cultural lived experiences, which in turn, should be the epistemological foundations defining school knowledge. The chapter concludes that to promote cultural justice and cognitive justice that can lead to sustainable learning and sustainable livelihoods indigenous knowledge systems should be the foundation of school knowledge in Africa.

INTRODUCTION

Education and schooling must be meaningful and purposeful to achieve sustainable social and economic development goals. Formal or informal education plays a crucial role in shaping individuals and societies, providing them with knowledge, skills, and values necessary for personal growth and societal progress. Meaningful and purposeful schooling is reflected by the forms of knowledge and pedagogical modes of learning that students encounter and experience in schools and in life. Education encompasses various activities and processes that facilitate learning and promote intellectual, social, emotional, and

DOI: 10.4018/979-8-3693-8447-3.ch018

physical development. At its core, education involves transmitting knowledge, values, and skills from one generation to another, ensuring the continuity and advancement of individuals and society. Without formal and informal education, there would likely be no sustainable development. Personal growth and societal progress are sustained by the process of continuous learning and relearning, buttressed by the construction and re-construction of knowledge.

Schooling is partly a contributing factor in the education process and in advancing learning. Learning, itself, must be sustained for students to achieve their cognitive and social growth and for them to contribute to sustainable societal progress. In both colonial and postcolonial Africa, the education system has failed to provide pathways to sustainable learning and cognitive justice. African education systems continue to marginalize African Indigenous knowledge systems that can provide cultural liberation or justice to African learners and societies. Thus, sustainable learning can be considered when policy makers and curriculum planners design an education system and bodies of knowledge that are desirable, meaningful and purposeful. Additionally, sustainable learning involves teachers who practise effective knowledge construction foundations and inclusive pedagogical practices in their schools and classrooms (Graham, Berman & Bellert, 2015). It involves inclusive practices that deconstruct disembodied learning, which embraces a conceptual transference of knowledge from the teacher- expert into the brains of students who are considered as 'tabula rasa'.

Inclusive pedagogical modes and sustainable learning utilize embodied acquisition of knowledge through the involvement of learners who can use their lens of prior experience to produce insights, knowledge schemas and perceptions that contribute to sustainable and embodied learning. Embodied learning involves learners using their multisensory representations of their environment and symbolic representations (Anzellotti & Young, 2020; Glenberg et al., 2013; Pulvermüller, 2013; Zwaan, 2014) from their social and cultural environment to develop their conceptual knowledge framework. That form of learning creates engaging learning experiences that promote a deeper understanding of concepts and make learning more enjoyable and memorable for students (Hrach, 2021) leading to sustainable learning experiences. As Graham et al (2015, p. 2) argue, "Sustainable learning is learning for all, teaching those matters and learning that lasts. Its foundation is effective classroom practice, and its goal is to provide for the learning needs of all students throughout their school years and into lifelong learning." In Africa, sustainable learning and sustainable knowledge acquisition for sustainable societal growth and personal development cannot be achieved without considering inclusive learning and Indigenous knowledge systems. Inclusive learning, which considers students' social and cultural backgrounds, aims to provide quality learning experiences to all students, and it would be an elusive and elusive enterprise if Indigenous knowledge systems are marginalised and not part of the school curriculum.

This chapter aims to provide a theoretical and conceptual argument for sustainable learning and cognitive justice for Indigenous African students that can be achieved through inclusive knowledge systems that promote Indigenous knowledge systems in African schools. The chapter argues that inclusive learning in Africa requires inclusive knowledge systems for sustainable learning and sustainable personal and social development. The objective is to provide a disruptive Indigenous paradigm that decolonises inherited African education systems that contribute to the miseducation of the African child. The aim is to provide a theoretical and conceptual debate focusing on academic and inclusive schooling knowledge pathways, such as Indigenous knowledge systems, that promote sustainable learning anchored in cultural justice and cognitive justice for Indigenous African students. The chapter provides insights into how policymakers, curriculum planners and educators may transform school knowledge into an inclusive vehicle for sustainable learning and promote academic achievement for Indigenous African students. The chapter is organised under the following sections: contextual background, inclusive learning; Indigenous knowledge systems and inclusive learning; inclusive knowledge systems and education; transformative learning and sustainable learning; cultural justice and school knowledge; cognitive justice and sustainable learning and an exploration of further research directions.

CONTEXTUAL BACKGROUND

Research has shown that students learn best when they can associate the subject matter and the methods of teaching with their lived experiences. African education systems are still embedded in colonial knowledge systems that do not align with the Indigenous African learners' lived experiences. The modern school system across Africa has its origins in colonialism. The dominant features – educational institutions, curriculum knowledge practices and language of instruction in schools – all are directly traced to the colonial institutional structure and the discourses with vilified representations of the African child within a politics of educational exclusion (Adzahlie-Mensah & Dunne, 2019). These colonial structures that include language-in-education policies are incongruent with inclusive learning and sustainable learning. These policies emerged from a history that imposed European languages and knowledge systems on colonial education structures as part of the architecture of colonisation (King et al., 2020). Children have a valuable, innate ability to learn by figuring things out. That is smothered by rote memorisation that is promoted by the colonial education system. The loss is incalculable – to the child and any society that faces challenges and needs people who can figure out answers rather than expecting the solution to come from others. Instead, rote-based schooling is the most colonial lesson of all African education systems (Alyson, 2022). The hallmark of colonial education was to silence the voice of the African learners by excluding their Indigenous knowledges from the education system. Thus, it is vital to note that it is almost impossible to provide effective sustainable education and learning to the Indigenous African child within an inherited colonial structure without taking cognisance of the need to incorporate African Indigenous knowledge systems.

Indigenous knowledge systems give a unique socio-cultural perspective and can benefit students by promoting cultural understandings of knowledge, sustainability in learning, and holistic sustainable learning. Welden, Chausson and Melanidis (2021) observe that when students experience their cultural knowledge and heritage as part of school knowledge, they become connected to their learning, which improves their cognitive development. In addition, Devkota and Timilsena (2023) agree with Welden et al (2021) that learners are shaped by their native languages, oral customs, rituals, ceremonies, and learning by doing, which facilitate cognitive understandings of their world realities and their experiences of knowledge acquisition. A longitudinal study by Teresa McCarty (2009) carried out in American Indian schools reported that children who learned to read first in Navajo and integrated their Indigenous languages and cultural knowledge systems in their education outperformed students in English programmes in academic achievement. Shizha (2005) reports similar findings in Zimbabwe that in primary schools, learners who use their Indigenous languages and bring their home cultural experiences and practices to the school learning settings respond positively and engage actively to their learning in science classes. The language gives students' freedoms and cultural justice to learn in a language they know well. Shizha (2015) concludes that adding Indigenous knowledges and incorporating Indigenous languages to science lessons makes the lessons more exciting and engaging and shows how vital this cultural knowledge is to cognitive development and sustainable learning. Supporting the use of Indigenous perspectives and languages in schools in Tanzania, Rubagumya, Sane and Ndabakurane (2022) praise the practice as a pathway to more sustainable learning and futures.

In South Africa, Cindi (2021) conducted a study on incorporating African Indigenous knowledge systems into the basic education curriculum and found that most teachers agreed that there is value in incorporating the knowledge systems into the school curriculum because of its benefits to students' cognitive understandings of school knowledge and improvement in sustainable learning and academic achievements. Advocating for the incorporation of Indigenous paradigms Barnhardt and Kawagley (2005) carried out a study on Indigenous students' experiences in Alaskan schools in the USA and concluded that the knowledge and skills derived from thousands of years of careful observation, scrutiny and survival in a complex ecosystem readily lends itself to the in-depth study of basic principles of biology, chemistry, physics and mathematics, particularly as they relate to areas such as botany, geology, hydrology, meteorology, astronomy, physiology, anatomy, pharmacology, technology, engineering, ecology, topography, ornithology, fisheries and other applied fields, and contributed to cultural justice and cognitive justice, and to sustainable learning for students. Devkota and Timilsena (2024) argue that incorporating these Indigenous knowledge practices and principles into educational settings can lead to a more holistic and culturally sensitive approach to learning, while also boosting and sustaining students' grasp of school knowledge. School settings that incorporate Indigenous knowledge systems provide a safe space for students to experience cultural justice. Webb and Mashford-Pringle (2022) point out that establishing a culturally safe environment is a complex process that requires participants to critically reflect and challenge personal belief systems to effectively validate the relevance of Indigenous knowledge systems to Indigenous African learners' learning experiences. In support of safe spaces Chilisa (2020) surmises that Indigenous paradigms need their own space so that they can support Indigenous students in their intersection with Western approaches to learning. Indigenous African scholars such as Abdi (2020), Dei (2008, 2020) and Odora-Hoppers (2021) also argue for Indigenous and decolonisation paradigms that culminate in safe spaces for Indigenous African students to benefit from the education they get from African schools.

Efforts to incorporate Indigenous perspectives, histories, and knowledges into the education system aim to address the systemic exclusion and marginalisation of Indigenous cultures and foster a more inclusive and respectful approach to education (Shay & Oliver, 2021) that promotes academic success for Indigenous learners. Such efforts should involve Indigenous communities in co-constructing what is regarded as school knowledge. For example, the Australian integration of the Aboriginal and Torres Strait Islander peoples culture highlights the importance of involving Indigenous communities in the development and implementation of educational strategies by collaborating with diverse Indigenous communities to provide a foundation for a more inclusive education system (Chakma, 2024). The interaction between what is defined as school knowledge and cultural/Indigenous knowledges is promoting sustainable learning, cultural and cognitive justice among the Indigenous people of Australia. A study by Onwu and Mufundirwa (2020) on using cultural contexts in teaching science to high school students in Zimbabwe revealed that integrating Indigenous knowledge and technology as relevant pedagogical contexts with the school science content results in intrinsic motivation for meaningful learning among students. Basing on these findings, Ochwo-Oburu (2020) believes that the absence of Indigenous ways of knowing denies the African child's mind essential tools for the development of thoughts, feelings, emotions, personality and interpretation of reality. In Kenya, Ng'asike (2019) reports on the school failures of Turkana Indigenous pastoral students because of school settings in current Kenyan standard formal education that represent a different world view from that of the nomadic pastoralist children and that modern classrooms undermine the unique cultural experiences of these children and reflect a persistent colonisation of the mind. Decolonising the mind and the education systems in Africa require a radical transformation of the system by indigenising and Africanising school knowledge. Actions currently being taken by Indigenous people in communities throughout the world, to promote their knowledge systems in schools, clearly demonstrate that a significant "paradigm shift" is under way in which Indigenous knowledges and ways of knowing are beginning to be recognised as consisting of complex knowledge systems with an adaptive integrity of their own that contributes to school knowledge and sustainable learning through inclusive learning (Webb & Mashford-Pringle, 2022). Applying Indigenous knowledge systems in African schools provides opportunities for inclusive knowledge and inclusive learning.

WHAT IS INCLUSIVE LEARNING? A CONSTRUCTIVIST APPROACH

Inclusive learning is based on the ideas of diversity, equity, and inclusive education. It is an ongoing process that ensures that schools strive to provide caring, inclusive, safe, and accepting environments that support different knowledge systems and promote the achievement and well-being of every student. Indigenous African students learn knowledge and are surrounded by learning environments that present an externally defined world that has enveloped their existence (Msimanga & Shizha, 2014). The externally prescribed world and its knowledge foundation lack the preparedness that African community knowledge or Indigenous knowledge systems equip the students with, that is, skills for the fulfillment of roles in their family or society. The acquisition of prescribed external knowledge that is phenomenologically foreign and alien to their lived experiences alienates and excludes the students from their community and social experiences that define their existence and sustainable livelihoods. It disengages them from their ontological existence and experiences of the real world from the perspective of their indigeneity and African worldview.

The foreignity of the knowledge and cultural dissonance arising from hegemonic school experiences disrupt inclusive learning possibilities. Inclusive learning possibilities are born out of accepting diversity, culturally responsive and equitable classroom environments, epistemologically inclusive school curriculum and inclusive teaching that promotes diversity in learning and knowledge acquisition. Culturally responsive classrooms and pedagogies are a constructive method that promotes the importance of integrating students' cultural backgrounds and social contexts with the curriculum of the school. Krasnof (2016) argues that because culture strongly influences the attitudes, values, and behaviours that students and teachers bring to the instructional process, teachers who lack inclusivity possibilities are a major factor in disengaging students and creating problems of underachievement. This argument points to the importance of teachers who infuse socio-cultural consciousness in their teaching and classroom climate. Krasnof (2016, p. 2) raises a substantive argument by stating,

> A teacher's own way of thinking, behaving, and being are influenced by race, ethnicity, social class, and language. Prospective teachers must critically examine their own socio-cultural identities and biases in the context of the inequalities culturally diverse segments of society experience.... A teacher's use of strategies that support a constructivist view of knowledge, teaching, and learning assists students in constructing knowledge, building on their personal and cultural strengths, and examining the curriculum from multiple perspectives, thus creating an inclusive classroom environment. (p. 2)

Approaching knowledge acquisition from the students' societal and cultural experiences enhances sustainable learning. Students become active learners who are co-constructors of knowledge.

Culture is an important resource and tool for knowledge construction. Knowledge represents ideas and practices of a particular society and its culture. The politics of knowledge is cultural. For Indigenous African students, the prospects of their cultures and languages being represented in the curriculum and teaching and learning builds self-efficacy. Self-efficacy refers to an individual's belief in their capacity to execute behaviours necessary to produce specific performance attainments (Bandura, 1977, 1986, 1997). Self-efficacy reflects confidence in the ability to exert control over one's own motivation, behaviour, and social environment. Teachers who learn about and support students' past experiences, home and community culture, and world in and out of school help students build and maintain a strong self-efficacy. Inclusive learning experiences contribute to this high level of self-efficacy.

Inclusive learning provides possibilities for sustainable learning and academic achievement for all students. A teacher's contention that all students are capable of learning requires building scaffolding between what students already know (i.e., Indigenous knowledge, cultural knowledge, community knowledge) through their own experiences and what they need to learn (prescribed and predetermined knowledge). Kumar and Gupta (2019) note that the education system now requires strategies emphasising student involvement in learning, focusing on knowledge building instead of knowledge transmission. According to this approach to learning, learning is creating an understanding related to the world (Ahmedi, Kurshumlija, & Ismajli, 2023). This constructivist teaching and learning promote critical thinking, problem solving, collaboration, and the recognition of multiple perspectives. Constructivism is an approach to learning that holds that people actively construct or make their own knowledge and that reality is determined by the experiences of the learner (Von Glasersfeld, 2013). Vygotsky (1978) believes that community plays a central role in the process of making meaning. Hence community knowledge or Indigenous knowledge systems play a vital role on creating school knowledge and in constructing meanings that are important to Indigenous African students' sustainable learning. Constructivism promotes a sense of personal agency as students have ownership of their learning as knowledge is actively constructed by the students based on their experiences. Thus, all teaching and learning is a matter of sharing and negotiating socially constituted knowledge. For example, Vygotsky (1978) states that cognitive development stems from social interactions from guided learning within the zone of proximal development as students and their teachers co-construct knowledge for inclusive learning possibilities.

INDIGENOUS KNOWLEDGE SYSTEMS AND INCLUSIVE LEARNING FOR SUSTAINABILITY

Although globalisation and postmodernism have brought new thinking and ways of behaviour in Africa, Indigenous ways of thinking and practices (Indigenous epistemologies) are still highly recognised and acknowledged. Indigenous knowledge systems are still practised and recognised in many communities, including among the so-called "westernised" Africans. Indigenous knowledge systems cover all aspects of Indigenous African people's lives, and they encompass diverse fields that include agriculture, medicine, ecology, spirituality, and traditional technologies, and therefore provide a rich resource for sustainable development (Olaopa & Ayodele, 2022). According to Shizha (2022a), characteristics of Indigenous epistemologies include social, cultural, and economic ways of life, traditional forms of social organisation, political institutions, customs and laws and historical continuity of lived experiences that are transmitted and shared intergenerationally. As noted by Asamoah and Ngulube (2023), most Africans are predominantly dependent on Indigenous knowledges for their livelihood, which is manifested through informal education, agriculture, traditional medicine, arts and craft, music and dance, natural resources management, weather and climate, chieftaincy, festival, conflict resolution among others. For these forms of knowledge to be sustainable and retainable, they must be acquired and transmitted intergenerationally.

Acquiring Indigenous knowledges requires an ethical and culturally sensitive and responsive approach that recognises the right of Indigenous communities to control their knowledge, resources and practices (Turner, Cuerrier, & Joseph, 2022). The acquisition and dissemination of Indigenous knowledges have become essential in African schools to assist Indigenous African learners to gain learning experiences they can identify with and that are meaningful and sustainable. While there are genuine concerns about the appropriation and uncontrolled documentation of Indigenous knowledges for fear that it will be misused for material gains (Buthelezi, Ocholla, & Dlamini, 2024), Chigwada and Ngulube (2023) suggest that for Indigenous knowledge systems to be sustainable, they need documentation to avoid the danger of them becoming extinct. In addition, Shizha (2017) echoes the same sentiments by pointing out that knowledge codification and cataloguing are important in the preservation of Indigenous knowledge systems. Storing Indigenous knowledge systems through codification and cataloguing makes access to the knowledge possible and makes it easier to integrate them into the school knowledge system and knowledge economy. Integration of Indigenous knowledges into the African school system, plays a key role in the preservation and utilisation of these tacit forms of knowledge.

Preserving Indigenous forms of knowledge through the education system is an emancipation and decolonising process. Colonial education was a colonising process that marginalised and excluded Indigenous knowledge systems, hence alienating Indigenous African learners from their lived experiences. Schools and the academy, in general, were the epicentre of colonial Eurocentric hegemony, indoctrination, and mental colonisation, and the decolonisation process entails a process of reclaiming, rethinking, reconstituting, rewriting, and validating Indigenous knowledge and, by implication, Africa's history and worldview (Shizha, 2015). Asamoah and Ngulube (2023) add that the transformation of knowledge systems is the starting point of knowledge inclusion and integration and the social inclusion of marginalised knowledge, equality, and sustainable development. Rewriting and rethinking the transmission of school knowledge using African cultural resource knowledge is a means of epistemological recuperation and is counter-hegemonic (Dei, 2008; Emeagwali, 2014). Although much has been said about indigenisation, transformation and renaissance in African education, education policymakers have yet to forge a new identity to cope with the pressures of frustration amongst African learners when they indulge in exclusivist and alienating learning practices. In that respect, Ndlazi (2021) asserts that equipping learners with sustainable knowledge requires an Indigenous African-centred approach to learning that focuses on reframing schooling practices and African education, generally promoting the Africanness and African philosophies of indigeneity. Without an African-centred approach devoid of indigenised educational transformation, Indigenous African students will continue to be left out of productive learning experiences and sustainable learning if the education system neglects their traditional knowledge systems.

Unfortunately, schools tend to promote a particular type of knowledge, which, in the case of African schools, is a Eurocentric colonial knowledge system. The knowledge and associated school and learning experiences are devoid of the cultural epistemological foundations of the African child. Indigenous knowledge systems are unique to Indigenous peoples; their traditions, political structures, cultural epistemologies, social organisation, and ways of life are distinct from the dominant Eurocentric knowledge systems invested in African school curricula (Shizha, 2022b; da Silva, Pereira, & Amorim, 2023). While the connotation of 'Indigenous' might entail traditional and old, Indigenous knowledge systems are evolving and constantly changing and transforming as Indigenous peoples' lives transform due to their interconnectedness to the everchanging social, political, cultural and economic realities. This, however, does not mean that external knowledge systems overshadow Indigenous knowledge systems. They continue to be vibrant and pervasive despite colonial hegemony. In this regard, Indigenous knowledge systems, in their diversity and complexity, are often neglected in knowledge production environments, being generally described with words such as 'primitive,' 'backward,' 'savage,' 'rural,' and 'unscientific' (Ezeanya-Esiobu, 2019; Shizha, 2008). Despite evidence that Indigenous knowledge systems provide knowledge pluralism in classrooms, this labelling and stigma usually lead to their neglect in the educational enterprise and schooling for Indigenous African students. Knowledge plurality and diversity is vital to inclusive learning for sustainability. As da Silva et al. (2023) observe, the educational issue resulting from the confrontation between Indigenous knowledge and colonial Western school knowledge is structured in a complex way that imposes obstacles to a possible integration of these epistemologies in schools and/or a possible indigenisation of the school curriculum. For Indigenous African students in African schools, Indigenous knowledge systems promote sustainable learning that is learning for all, teaching that matters and learning that lasts.

INCLUSIVE KNOWLEDGE SYSTEMS AND EDUCATION

Over the years, UNESCO has highlighted the idea of equity in and inclusive education, which has become a global policy vision for education (UNESCO, 2015). For UNESCO, education is its top priority because it is a basic human right and the foundation on which to build equality and drive sustainable development (UNESCO, 2017). A well-planned education policy can influence and support inclusive thinking and practices by establishing the equal rights of every individual to education and by outlining the teaching, support, and leadership forms that lay the foundation for quality education for all (UNESCO, 2015). The issue of equality and inclusive education policies is emphasised by Ainscow (2020), who reminds us that:

> Over the last 30 years there have been major international efforts to encourage inclusive educational developments. In particular, the United Nation's Education for All (EFA) movement has worked to make quality basic education available to all learners. Agreed in 1990, the EFA Declaration sets out an overall vision, which is about being proactive in identifying the barriers some learners encounter in attempting to access educational opportunities. (p. 7)

The EFA vision was reaffirmed by the World Education Forum meeting in Dakar in 2000, which declared that the EFA must take particular account of the needs of disadvantaged children in providing equal access to educational opportunities and experiences. In that vein, inclusive education that involves inclusive knowledge must be realised in contexts where available options for action are restricted by several and often contradictory educational policies on different levels of the education system (Magnússon et al., 2019). According to UNESCO (2012), policies that promote the equitable distribution of diverse forms of knowledge can add value to human experience and progress by enhancing livelihoods and contributing to societies' social, cultural, political and economic development. UNESCO advocates for the building of sustainable and inclusive knowledge societies, which are equitable, open and participatory and are grounded on freedom of expression (should apply to both traditional and contemporary forms of learning), access to quality education for all and respect for cultural and linguistic diversity.

The principle behind inclusive education and inclusive knowledge and learning is the central message that every learner matters and matters equally. According to UNESCO (2017), the complexity of equality arises when we try to put this message into practice. Implementing this message will likely require changes in thinking and practice at every level of the education system, from teachers and others who provide educational experiences directly, to those responsible for national policy. UNESCO (2017) suggests that integrating the principles of equity and inclusion into education policy involves:

- Valuing the presence, participation and achievement of all learners, regardless of their contexts and personal characteristics.
- Recognising benefits of student diversity, and how to live with, and learn from differences.
- Collecting, collating and evaluating evidence on children's barriers to education access, to participation and to achievement, with particular attention to learners who may be most at risk of underachievement, marginalisation or exclusion.
- Building a common understanding that more inclusive and equitable education systems have the potential to promote gender equality, reduce inequalities, develop teacher and system capabilities, and encourage supportive learning environments. These various efforts will, in turn, contribute to overall improvements in educational quality.
- Engaging key education and community stakeholders to foster the conditions for inclusive learning and to foster a broader understanding of the principles of inclusion and equity.

Collectively, the above factors, when applied holistically, will contribute to inclusive knowledge and inclusive learning. Indigenous knowledge systems and how Indigenous people transmit or modes of traditional teaching need to be part of the learners' learning experiences.

Research points to levers for change that can be used to promote equity and inclusion in education (Ainscow, 2020). In particular, research identifies important evidence that concurs that the presence of diverse forms of knowledge encourages student participation and promotes academic achievement and success (Ainscow, 2005; Emeagwali, 2014; Jaimungal & Dei, 2018; Shizha, 2014). A broader understanding of inclusive education aims to meet the social and cultural needs of all pupils and allow them the opportunity to influence their learning capabilities (Göransson & Nilholm, 2014). In African schools, this can be achieved if Indigenous African students are given the opportunity to engage in learning that constitutes their ways of viewing the world, that is, applying their Indigenous knowledge systems and lived experiences. The insistence on Indigenous knowledge systems and pluralisation of school knowledges rests on the conviction that there is no single uniform model of knowledge production (Jaimungal & Dei, 2018; Shizha, 2014) to which all students and societies must conform. Sustainable Development Goal (SDG)-4 on education refers to values of equity, inclusion, diversity, equal opportunity and non-discrimination, positing education as a tool for building a just society (UNESCO 2020). It is concerned with issues of diversity, equity, inclusivity and the building of an equitable society, and this requires including local or student-generated knowledge in classroom practice. Odora Hoppers (2015) argues that plural forms of knowledge bring about social and cultural justice.

TRANSFORMATIVE EDUCATION FOR SUSTAINABLE LEARNING

If inclusive learning has a positive impact on sustainable learning, what does sustainable learning mean? In what ways can students achieve sustainable learning? According to Graham et al (2015), sustainable learning presents a new way of looking at effective inclusive practices for contemporary classrooms. Hays and Reinders (2020) also point out that sustainable learning in education is a fairly new term with very little work or consensus related to its definition, assessment, or application. Since the 1980s, the word "sustainability" has gained traction both academically and colloquially, leading to its current definition being fluid, which often leads to confusion about what it actually means when used (Birdsall, 2014; Lange, 2018). Its use has been largely in terms of futuristic environmental survival. Thus, many academics reference the Brundtland Report (WCED, 1987) for their definition of sustainability, that it is a way to meet current needs without jeopardising the needs of future generations. Analytically, this concept of sustainability incorporates environmental, social and economic perspectives (Lange, 2018). The term can be extended to education and learning with regards to sustainable education and sustainable learning. As Lange (2018, p. 2) shared, "...sustainability education is a profound act of hope in the future." This hope in the future comes from how students are engaged in sustainable learning and inclusive knowledge production and usage.

The capacity for students to develop sustainable skills comes from competency to acquire and grasp knowledge, which could be described as 'sustainability knowledge', which introduces the learner to a broad, general knowledge (Christie et al., 2013; Foley et al., 2017; Mintz & Tal, 2018) acquired for lifelong socio-economic existence. For African students, this sustainability knowledge is their Indigenous knowledge system. Sustainability knowledge is integral to their lived experiences and has been part of their different societal generations. It has been sustained by being passed from one generation to the other. Indigenous knowledges should be reclaimed and reframed to be a major component of the epistemology of the schooling and learning enterprise (Shizha, 2022a). As Potter-Nelson and Meyers (2022) point out, fostering a sustainability literate society will involve education of a different kind. Education of a different kind looks to curtail current unsustainable student learning behaviours that are deeply embedded and rooted in the colonial education system (Shizha, 2022b). Education of a different kind is knowledge transformation that includes but is not limited to the content taught, the pedagogy used by teachers, and the systems of injustice that permeate deep into existing educational structures.

Transformative knowledge requires transformative learning that entails appropriate instructional methods that reflect the students' cultural backgrounds and cultural spaces where their knowledge repertoire comes from. Appropriate instructional methods and knowledge content are likely to result in sustainable learning that produces quality learning experiences that will be available to all students. To accomplish this, all educators need to focus on acquiring the skills underpinning teaching that matters (Graham et al., 2015). By this, Graham et al. (2015) mean teaching that is evidence-based, purposeful, relevant and responsive to students' needs. Purposeful and meaningful learning that is central to sustainable learning is likely to contribute to students who are able to demonstrate that knowledge is mastered and learning maintained to the extent that, as far as possible, they become independent learners throughout their lives (Emeagwali & Shizha, 2016; Kapoor & Shizha, 2010; Ndlovu & Shizha, 2022; Shizha & Abdi, 2014). Independent and active learning is fundamental to sustainable learning and to meeting contemporary challenges. When students apply their cultural, tacit or everyday situational knowledge, they become problem solvers, an essential foundation for developing students' learning ability, especially in understanding the challenges that they and their communities face.

CULTURAL JUSTICE AND SCHOOL KNOWLEDGE

Schools are powerful and compelling cultural validators that have a phenomenal influence on learners. Eurocentric cultural knowledge systems found in African schools, which are colonial systems of knowledge, force Indigenous people and students to adapt to and work within a colonial cultural framework, which neglects Indigenous African knowledges and practices. They create a cultural injustice that alienates the students from productive and sustainable learning experiences. Cultural injustice occurs when people are forced by coercion or persuasion to submit to the burdensome condition of suspending – or permanently surrendering – their culture (Odora Hoppers, 2021). This means that, the subjugated person has no linguistic or cultural 'default drive' – that critical minimum of ways, customs, manners, gestures, and postures that facilitate uninhibited, un-self-conscious action (Kwenda 2003, p, 70). Linguistic or cultural 'default drive' is the ability to utilize one's internalised knowledge schemas that are embedded and grounded in one's epistemological and ontological cultural foundations. For Indigenous African students, their Indigenous knowledge systems provide a cultural 'default drive' that sustains their learning and co-production of school knowledge. This is the foundation of cultural justice in schooling and knowledge production.

Cultural justice means the burden of constant self-consciousness is shared or at the very least recognised, and possibly rewarded (Odora Hoppers, 2021). It, therefore, requires at minimum that the burden of unfamiliar cultural experiences is shared more equitably by people from different cultural backgrounds across society (Kwenda 2003). Sharing this burden is important because it builds a sense of identity and reciprocity among diverse peoples and communities. It is in this sharing that cultural difference is transcended, and cultural arrogance ruptured (Odora Hoppers, 2021). For schools and teachers, it implies that cultural diversity is made the building block for knowledge construction, and that students and community knowledge stakeholders are considered as co-producers of school knowledge. Even more important is translating this recognition into concrete reforms, seeing individual differences not as problems to be fixed, but as opportunities for democratising and enriching learning.

Being co-producers of knowledge awards students and their communities the right to cultural integrity and recognition. For instance, the maintenance of languages, religious practices and customs not only hold value in themselves, but rights which preserve cultural practices reinforce Indigenous peoples' self-determination. In education, integrating ethnomathematics and ethnosciences (Mkhwelie, Satamwe & Shizha, 2022), for instance, encourages sustainable learning and the preservation of Indigenous knowledge systems and cultural justice as it considers Indigenous African students' cultural repertoires. When Indigenous cultures are included in school knowledge, they become preserved knowledge for future generations of learners. Indigenous knowledge preservation can lead to sustainable development and facilitation of the achievement of one of the United Nations' sustainable development goals (Chigwada & Ngulube, 2023). Surprisingly, in many African education systems, not much has been done in terms of integrating African Indigenous knowledge systems into the school curriculum.

Research reveals that schools that incorporate elements that are culturally relevant to students are able to allow students to articulate both home- and school- based values and, thereby, integrate diverse knowledge systems that facilitate cognitive and emotional development as well as social participation (Shang, O'Driscoll & Roche, 2016). Importantly, the preservation of Indigenous cultural traditions cannot be carried out when schools perform an assimilationist role and excludes Indigenous knowledge systems from their curricula. African schools, in their African ecological settings, should be communities through which African traditions can be practised though not exclusively but in partnership with other forms of traditions (Shizha, 2021; Valadez, 2012) that can advance cognitive justice for African students. As Odora Hoppers (2021) aptly points out moving towards cultural justice requires an ethically sound and ecologically constituted way of thinking, the affirmation of the multiplicity of worlds and forms of life; becoming critical explorers of human and societal possibilities; the establishment of new evaluation and appraisal criteria, and the transformation to new futures. Moving towards an equitable and inclusive learning requires bridging the gap between school knowledge and the learners' home life using Indigenous knowledge systems such as ethnoscience or ethnomathematics (Sunzuma, & Maharaj, 2020) in schools that are embedded in ethnoecological settings. This is an African education of the future that needs to build a fraternity between diverse forms of knowledge (Visvanathan 2001) to foster not only cultural justice but also cognitive justice.

COGNITIVE JUSTICE AND SUSTAINABLE LEARNING

A sustainable learning environment should uphold cognitive justice for students. It is an environment where knowledge multiplicity and diverse forms of learning exist. It considers the contribution of students to their learning and the prior and existing knowledge and experiences learners bring into their school and learning engagement. The concept of cognitive justice has its roots in the works of Visvanathan (1977), an Indian anthropologist and human rights researcher (Muchenje, 2017). According to Makoelle (2014), Visvanathan coined this term to encourage the positive acceptance of different forms of knowledge. The concept arose as a reaction to the dominance of Western knowledge forms, which have tended to establish hegemony regarding ways of knowing (Muchenje, 2017). Makoelle (2014) explains that the concept came about because of a concern that Western forms of knowledge and the way knowledge is acquired and validated seemed to perpetuate the dominant hegemonic discourses of Western cultures.

Cognitive justice is unattainable and unsustainable without disruption to knowledge coloniality and Western hegemonic epistemological impositions on African schools and African Indigenous students. The colonial structures of knowledge and power remain an albatross to sustainable learning for Indigenous African students because Indigenous knowledge systems continue to be marginalised. The knowledge coloniality in African schools is reflected in the hegemonic influences of Western assumptions of what knowledge is good and right for Africa and its citizens for sustainable development to be achieved on the continent. Neoliberalism in educational policies, as well as in values and principles of education, which are embedded in coloniality and hegemonic practices (Veintie, 2013), is a hindrance to successful cognitive justice, inclusive learning and cultural justice. For Visvanathan (1997), cognitive justice entails: a) engaging and empowering the historically marginalised voices; b) bringing Indigenous knowledge systems out of the ascribed paradigm of ethnosciences and equally valuing their contributions to collective knowledge production; c) recognising expertise across sectors (i.e. every citizen as a scientist and layperson an expert); and d) bringing seemingly incommensurable knowledges together into a positive heuristic for dialogue. What Visvanathan is pointing out is the decolonisation and decentering of hegemonic Eurocentric knowledge in postcolonial education systems and providing a disruptive scope to reclaim and restore subaltern and submerged forms of knowledge.

Cognitive justice is the right of all forms and traditions of knowledge to co-exist without duress and advances dialogic engagement between the dominant, modernising, western knowledge structures and Indigenous, traditional knowledge formations (Odora Hoppers, 2021, 2015). Cognitive justice is a product of knowledge pluralism and an ethic of respect for diversity, which invites the school and teachers not only to value knowledge differences but also to collaborate with students and communities to actively create practices and platforms that foster reciprocal engagement among diverse peoples, cultures, and worldviews. In Africa, Indigenous knowledge systems provide an avenue for promoting cognitive justice. Indigenous African students are primarily and cognitively a product of their Indigenous upbringing and Indigenous worldviews. According to Odora Hoppers (2015), African education and schooling should function as 'connective tissues' across diverse communities that mediate differences and generate common solutions that are sustainable for the communities. In contemporary Africa, epistemic pluralism, the coexistence of different knowledges and worldviews, especially those emanating from Indigenous and subaltern peoples, should take on greater prominence in educational institutions and societies worldwide (Odora Hoppers, 2021).

FURTHER RESEARCH DIRECTIONS

While some African education policymakers have policy statements regarding the inclusion of Indigenous knowledge systems into their school curricula, there is paucity of data on how these policies have materialised into practical implementations. Educational research is required in the following areas:

a) Examination of policy directives designed by education policymakers on how school knowledge is defined and the place of Indigenous knowledges in African education.
b) Exploratory research to examine the integration of Indigenous knowledge in African schools and whether teachers are prepared during their training to integrate Indigenous knowledges in their pedagogical practices.
c) Curriculum analysis to assess the extent to which Indigenous knowledge systems are included as school knowledge and the nature of the knowledge systems.
d) Cognitive and cultural justice, as well as sustainable learning, arise from using students' experiential knowledge. Therefore, studies are required to explore the implementation of pedagogical practices in schools and how these apply inclusive knowledge systems that include students' social and cultural lived experiences.
e) Studies involving document analysis of texts used in schools should be conducted to systematically review and interpret the textbook content to uncover an understanding of the epistemological foundation of the content to develop culturally inclusive content.

CONCLUSION

Inclusive knowledge, inclusive learning, cultural justice, and cognitive justice are not just buzz phrases. They are processes that affect Indigenous African students and shape social and sustainable livelihoods. Co-created knowledge—its acquisition and utilisation can transform individuals, families and societies. However, the forms of knowledge that are acquired must be meaningful, relevant and capable of transforming communities. In Africa, communities can attain socio-economic transformation if schooling, the education system and teachers' pedagogical practices utilise Indigenous epistemologies that relate to the ontological existentialities of the Indigenous African peoples. This entails that classroom knowledge and pedagogical practices should be inclusive and not exclude and marginalise non-Western knowledge. African education systems rely on colonial, western and Eurocentric hegemonic knowledge systems that lack the lived experiences of Indigenous African learners. Most learners struggle to make sense of the world presented by these alienating curriculum epistemologies devoid of equity, diversity and inclusion. School knowledge and modes of teaching should promote cognitive justice by integrating the cultural and social well-being of Indigenous African students to achieve cognitive justice for learners. Cognitive justice cannot be achieved when Indigenous learners do not identify with the cultural knowledge schemas presented in their school curriculum. All knowledge is cultural. It emanates from the cultures of those who design and define what it is and how it is transmitted. Therefore, curriculum planning in Africa should be based on Indigenous knowledge systems and should be the foundation of school knowledge to promote cultural justice and cognitive justice that can lead to sustainable learning and sustainable livelihoods.

The school or educational epistemology—what can be counted as knowledge, where it is located, and how it is transmitted should be of considerable concern to African policy makers, academics and parents. What and how their children learn and how knowledge is transmitted should focus on building cultural justice and cognitive justice to promote sustainable learning and sustainable knowledge acquisition. What binds the people as a community is their common way of viewing the world or their culture, which forms the basis of their epistemology and, ultimately, school knowledge that can contribute to socio-cultural and sustainable development. Education policymakers, curriculum planners and educators should be at the forefront of disrupting colonial education systems in Africa. They should lead in creating school systems that are inclusive in promoting cultural justice, cognitive justice and sustainable learning by advancing Indigenous perspectives and paradigms in the planning and practice of education.

REFERENCES

Abdi, A. A. (2020). Decolonizing knowledge, education and social development: Africanist perspectives. *Beijing International Education Research*, 2(4), 503–518. DOI: 10.1163/25902539-02040006

Adzahlie-Mensah, V., & Dunne, M. (2019). Continuing in the shadows of colonialism: The educational experiences of the African child in Ghana. *Perspectives in Education*, 36(2), 44–60. DOI: 10.18820/2519593X/pie.v36i2.5

Ahmedi, V., Kurshumlija, A., & Ismajli, H. (2023). Teachers`attitudes towards constructivist approach to improving learning outcomes: The case of Kosovo. *International Journal of Instruction*, 16(1), 441–454. DOI: 10.29333/iji.2023.16124a

Ainscow, M. (2005). Developing inclusive education systems: What are the levers for change? *Journal of Educational Change*, 6(2), 109–124. DOI: 10.1007/s10833-005-1298-4

Ainscow, M. (2020). Promoting inclusion and equity in education: Lessons from international experiences. *Nordic Journal of Studies in Educational Policy*, 6(1), 7–16. DOI: 10.1080/20020317.2020.1729587

Alyson, S. (2022). Colonial education remains alive and well. *Pan African Review*. Retrieved on 10 October 2024 from https://panafricanreview.com/colonial-education-remains-alive-and-well/

Anzellotti, S., & Young, L. L. (2020). The acquisition of personal knowledge. *Annual Review of Psychology*, 71(1), 613–634. DOI: 10.1146/annurev-psych-010419-050844 PMID: 31553673

Asamoah, C., & Ngulube, P. (2023). Exploring models for the management of Indigenous knowledge in academic libraries of Ghana. *Information Development*, 39(4), 787–797. DOI: 10.1177/02666669211052928

Bandura, A. (1977). Self-efficacy: Toward a unifying theory of behavioral change. *Psychological Review*, 84(2), 191–215. DOI: 10.1037/0033-295X.84.2.191 PMID: 847061

Bandura, A. (1986). *Social foundations of thought and action: a social cognitive theory*. Prentice-Hall.

Bandura, A. (1997). Self-Efficacy: The exercise of control. *The Freeman.*

Barnhardt, R., & Kawagley, A. O. (2005). Indigenous knowledge systems and Alaska native ways of knowing. *Anthropology & Education Quarterly*, 36(1), 8–23. DOI: 10.1525/aeq.2005.36.1.008

Birdsall, S. (2014). Measuring student teachers' understandings and self-awareness of sustainability. *Environmental Education Research*, 20(6), 814–836. DOI: 10.1080/13504622.2013.833594

Buthelezi, S. C., Ocholla, D., & Dlamini, P. (2024). Strategies for documenting and disseminating Indigenous knowledge at a South African university. *South African Journal of Information Management*, 26(1), a1648. DOI: 10.4102/sajim.v26i1.1648

Chakma, U. (2024). Indigenous knowledge integration in Australian curriculum: Culturally responsive pedagogy and perspectives. *VNU Journal of Foreign Studies*, 40(1), 1–10. DOI: 10.63023/2525-2445/jfs.ulis.5200

Chigwada, J., & Ngulube, P. (2023). Stakeholders in the acquisition, preservation, and dissemination of Indigenous knowledge projects. *Information Development*, 02666669231192851. Advance online publication. DOI: 10.1177/02666669231192851

Chilisa, B. (2020). *Indigenous research methodologies* (2nd ed.). Sage Publications.

Christie, B. A., Miller, K. K., Cooke, R., & White, J. G. (2013). Environmental sustainability in higher education: How do academics teach? *Environmental Education Research*, 19(3), 385–414. DOI: 10.1080/13504622.2012.698598

Cindi, L. (2021). *Incorporating African Indigenous knowledge systems into the basic education curriculum: experiences from two schools in the Gauteng and KwaZulu-Natal Provinces, South Africa*. M.Phil. diss. University of the Free State, South Africa.

da Silva, C., Pereira, F., & Amorim, J. P. (2023). The integration of Indigenous knowledge in school: A systematic review. *Compare: A Journal of Comparative Education*, 54(7), 1210–1228. DOI: 10.1080/03057925.2023.2184200

Dei, G. J. S. (2008). Indigenous knowledge studies and the next generation: Pedagogical possibilities for anti-colonial education. *Australian Journal of Indigenous Education*, 37(S1), 5–13. DOI: 10.1375/S1326011100000326

Dei, G. J. S. (2020). Elders' cultural knowledges and African indigeneity. In Abidogun, J. M., & Falola, T. (Eds.), *The Palgrave handbook of African education and Indigenous knowledge* (pp. 279–301). Palgrave Macmillan.

Devkota, K. M., & Timilsena, N. P. (2023). Indigenous knowledge in science classroom practices: An explorative study. *Kepes Journal*, 21(3), 847–856.

Emeagwali, G. (2014). Intersections between Africa's Indigenous knowledge systems and history. In Emeagwali, G., & Dei, G. J. S. (Eds.), *African Indigenous knowledge and the disciplines* (pp. 1–17). Brill Publishing. DOI: 10.1007/978-94-6209-770-4_1

Emeagwali, G., & Shizha, E. (2016). *African Indigenous knowledge and the sciences: Journeys into the past and present*. Sense Publishers. DOI: 10.1007/978-94-6300-515-9

Foley, R., Archambault, L., Hale, A., & Dong, H. (2017). Learning outcomes in sustainability education among future elementary school teachers. *Journal of Education for Sustainable Development*, 11(1), 33–51. DOI: 10.1177/0973408217725861

Glenberg, A. M., Witt, J. K., & Metcalfe, J. (2013). From the revolution to embodiment: 25 years of cognitive psychology. *Perspectives on Psychological Science*, 8(5), 573–585. DOI: 10.1177/1745691613498098 PMID: 26173215

Göransson, K., & Nilholm, C. (2014). Conceptual diversities and empirical shortcomings – A critical analysis of research on inclusive education. *European Journal of Special Needs Education*, 29(3), 265–280. DOI: 10.1080/08856257.2014.933545

Graham, L., Berman, J., & Bellert, A. (2015). *Sustainable learning: Inclusive practices for 21st century classrooms*. Cambridge University Press. DOI: 10.1017/CBO9781107280243

Hays, J., & Reinders, H. (2020). Sustainable learning and education: A curriculum for the future. *International Review of Education*, 66(1), 29–52. DOI: 10.1007/s11159-020-09820-7

Hrach, S. (2021). *Minding bodies: How physical space, sensation, and movement affect learning*. West Virginia University Press.

Jaimungal, C., & Dei, G. J. S. (2018). *Indigeneity and decolonial resistance: alternatives to colonial thinking and practice*. Myers Education Press.

Kapoor, D., & Shizha, E. (Eds.). (2010). *Indigenous knowledge and learning in Asia/Pacific and Africa: perspectives on development, education and culture*. Palgrave Macmillan. DOI: 10.1057/9780230111813

King, T. L., Navarro, J., & Smith, A. (Eds.). (2020). *Otherwise worlds: against settler colonialism and anti-Blackness*. Duke University Press.

Krasnof, B. (2016). *Culturally responsive teaching: a guide to evidence-based practices for teaching all students equitably*. Region X, Equity Assistance Center at Education Northwest.

Kumar, R., & Gupta, V. K. (2019). An introduction to cognitive constructivism in education. *Journal of Indian Education*, 35(3), 39–45.

Kwenda, C. V. (2003). Cultural justice: The pathway to reconciliation and social cohesion'. In D. Chidester, W. James & P. Dexter (eds.), *What holds us together: Social cohesion in South Africa* (pp. 67–80). HSRC Press.

Lange, E. A. (2018). Transformative sustainability education: from sustainable to a civilization leap. In Milana, M., Webb, S., Holford, J., Waller, R., & Jarvis, P. (Eds.), *The Palgrave international handbook on adult and lifelong education and learning* (pp. 1–22). Palgrave Macmillan. DOI: 10.1057/978-1-137-55783-4_21

Magnússon, G., Göransson, K., & Lindqvist, G. (2019). Contextualizing inclusive education in educational policy: The case of Sweden. *Nordic Journal of Studies in Educational Policy*, 5(2), 67–77. DOI: 10.1080/20020317.2019.1586512

Makoelle, T. M. (2014). Cognitive justice: Road map for equitable inclusive learning environments. *International Journal of Education and Research*, 2(7), 505–518.

McCarty, T. (2009). *Language planning and policy in Native America: History, theory, praxis*. Bilingual Education & Bilingualism series. Clevedon: Multilingual Matters

Mintz, K., & Tal, T. (2018). The place of content and pedagogy in shaping sustainability learning outcomes in higher education. *Environmental Education Research*, 24(2), 207–229. DOI: 10.1080/13504622.2016.1204986

Mkhwelie, N., Satamwe, N., & Shizha, E. (2022). Integration of Indigenous mathematical knowledge into the mathematics education for sustainable learning. In Ndlovu, C., & Shizha, E. (Eds.), *The dynamics of African Indigenous knowledge systems: A sustainable alternative for livelihoods in Southern Africa* (pp. 17–35). Peter Lang.

Msimanga, A., & Shizha, E. (2014). Indigenous knowledge and science education in South Africa: What messages from the curriculum? In Shizha, E. (Ed.), *Remapping Africa in the global space: Propositions for change* (pp. 137–150). Sense Publisher. DOI: 10.1007/978-94-6209-836-7_10

Muchenje, F. (2017). Cognitive justice and Indigenous knowledge systems in the postcolonial classroom. In Shizha, E., & Makuvaza, N. (Eds.), *Re-thinking postcolonial education in Sub-Saharan Africa in the 21st Century* (pp. 69–83). Sense Publisher. DOI: 10.1007/978-94-6300-962-1_5

Ndlazi, G. (2021). A critical reflection of African philosophy and decolonization of the educational discourse in South Africa. *Journal of Contemporary African Philosophy*, 2, 49–76.

Ndlovu, C., & Shizha, E. (Eds.). (2022). *The dynamics of African Indigenous knowledge systems: A sustainable alternative for livelihoods in Southern Africa*. Peter Lang. DOI: 10.3726/b19639

Ng'asike, J. T. (2019). Indigenous knowledge practices for sustainable lifelong education in pastoralist communities of Kenya. *International Review of Education*, 65(1), 19–46. DOI: 10.1007/s11159-019-09767-4

Ochwo-Oburu, S. (2020). East Africa and Indigenous knowledge: its nature, contents, aims, contemporary structures, and vitality. In Abidogun, J. M., & Falola, T. (Eds.), *The Palgrave handbook of African education and Indigenous knowledge* (pp. 303–318). Palgrave Macmillan. DOI: 10.1007/978-3-030-38277-3_15

Odora Hoppers, C. (2015). Cognitive justice and integration without duress: The future of development education – perspectives from the South. *International Journal of Development Education and Global Learning*, 7(2), 89–106. DOI: 10.18546/IJDEGL.07.2.08

Odora Hoppers, C. (2021). Research on Indigenous knowledge systems: The search for cognitive justice. *International Journal of Lifelong Education*, 40(4), 310–327. DOI: 10.1080/02601370.2021.1966109

Olaopa, O. R., & Ayodele, O. A. (2022). Building on the strengths of African Indigenous knowledge and innovation (AIK&I) for sustainable development in Africa. *African Journal of Science, Technology, Innovation and Development*, 14(5), 1313–1326. DOI: 10.1080/20421338.2021.1950111

Onwu, G. O., & Mufundirwa, C. (2020). A two-eyed seeing context-based approach for incorporating Indigenous knowledge into school science teaching. *African Journal of Research in Mathematics. Science and Technology Education*, 24(2), 229–240.

Potter-Nelson, E., & Meyers, S. (2022). *An Introduction to sustainability education*. Massachusetts Institute of Technology. White paper series.

Pulvermüller, F. (2013). How neurons make meaning: Brain mechanisms for embodied and abstract-symbolic semantics. *Trends in Cognitive Sciences*, 17(9), 458–470. DOI: 10.1016/j.tics.2013.06.004 PMID: 23932069

Rubagumya, C. M., Sane, E., & Ndabakurane, J. J. (2022). Vignette: implementing language supportive pedagogy to support content learning in Tanzania. In Erling, E. J., Clegg, J., Rubagumya, C. M., & Reilly, C. (Eds.), *Multilingual learning and language supportive pedagogies in Sub-Saharan Africa* (pp. 170–174). Routledge.

Shang, S., O'Driscoll, M. P., & Roche, M. (2016). Moderating role of acculturation in a mediation model of work- family conflict among Chinese immigrants in New Zealand. *Stress and Health*, 33(1), 55–68. DOI: 10.1002/smi.2674 PMID: 27018155

Shay, M., & Oliver, R. (Eds.). (2021). *Indigenous education in Australia:lLearning and teaching for deadly futures*. Routledge. DOI: 10.4324/9780429263453

Shizha, E. (2008). Indigenous? What Indigenous knowledge? Beliefs and attitudes of rural primary school teachers towards Indigenous knowledge in the science curriculum in Zimbabwe. *Australian Journal of Indigenous Education*, 37(1), 80–90. DOI: 10.1017/S1326011100016124

Shizha, E. (2014). Indigenous knowledge systems and the curriculum. In Emeagwali, G., & Dei, G. J. S. (Eds.), *African Indigenous knowledge and the disciplines. Anti-colonial educational perspectives for transformative change* (pp. 113–129). Sense Publishers.

Shizha, E. (2015). Reclaiming Indigenous cultures in African education. In Jacobs, W. J., Cheng, S. Y., & Porter, M. K. (Eds.), *Indigenous education: Language, culture, and identity* (pp. 301–317). Springer. DOI: 10.1007/978-94-017-9355-1_15

Shizha, E. (2017). Indigenous knowledges and knowledge codification in the knowledge economy. In Ngulube, P. (Ed.), *Handbook of research on theoretical perspectives on Indigenous knowledge systems in developing countries* (pp. 267–288). IGI Global. DOI: 10.4018/978-1-5225-0833-5.ch012

Shizha, E. (2022a). Mainstreaming and investing in Indigenous knowledge epistemologies for sustainable livelihoods. In Ndlovu, C., & Shizha, E. (Eds.), *The dynamics of African Indigenous knowledge systems: A sustainable alternative for livelihoods in Southern Africa* (pp. 157–176). Peter Lang.

Shizha, E. (2022b). Indigenous epistemologies and decolonized sustainable livelihoods in Africa. In Abdi, A. A., Misiaszek, G. W., & Popoff, J. M. (Eds.), *Palgrave international handbook on critical theories of education* (pp. 465–480). Palgrave Macmillan. DOI: 10.1007/978-3-030-86343-2_26

Shizha, E., & Abdi, A. A. (Eds.). (2014). *Indigenous discourses on knowledge and development in Africa*. Taylor & Francis/Routledge.

Sunzuma, G., & Maharaj, A. (2020). In-service mathematics teachers' knowledge and awareness of ethnomathematics approaches. *International Journal of Mathematical Education in Science and Technology*, 52(7), 1–16.

Turner, N. J., Cuerrier, A., & Joseph, L. (2022). Well grounded: Indigenous peoples' knowledge, ethnobiology and sustainability. *People and Nature*, 4(3), 627–651. DOI: 10.1002/pan3.10321

UNESCO. (2012). *Inclusive knowledge societies for sustainable development*. UNESCO.

UNESCO. (2015). Incheon declaration. [*towards inclusive and equitable quality education for all*. UNESCO.]. *Education*, •••, 2030.

UNESCO. (2017). *A guide for ensuring inclusion and equity in education*. UNESCO.

Valadez, J. M. (2012). Indigenous rights. In Callahan, D., & Singer, P. (Eds.), *Encyclopedia of applied ethics* (pp. 696–703). Academic Press, Elsevier. DOI: 10.1016/B978-0-12-373932-2.00298-2

Veintie, T. (2013). Coloniality and cognitive justice: Reinterpreting formal education for the Indigenous peoples in Ecuador. *International Journal of Multicultural Education*, 15(3), 45–60. DOI: 10.18251/ijme.v15i3.708

Visvanathan, S. (1997). *A carnival for science: Essays on science, technology and development*. Oxford University Press.

Von Glasersfeld, E. (2013). *Radical constructivism* (Vol. 6). Routledge. DOI: 10.4324/9780203454220

Vygotsky, L. S. (1978). *Mind in society: the development of higher psychological processes*. Harvard University Press.

Webb, D., & Mashford-Pringle, A. (2022). Incorporating Indigenous content into K-12 curriculum: Supports for teachers in provincial and territorial policy and post-secondary education spaces. *Canadian Journal of Educational Administration and Policy*, 198(198), 55–73. DOI: 10.7202/1086427ar

Welden, E. A., Chausson, A., & Melanidis, M. S. (2021). Leveraging nature-based solutions for transformation: Reconnecting people and nature. *People and Nature*, 3(5), 966–977. DOI: 10.1002/pan3.10212

World Commission on Environment and Development [WCED]. (1987). *Our common future*. Oxford University Press.

Zwaan, R. A. (2014). Embodiment and language comprehension: Reframing the discussion. *Trends in Cognitive Sciences*, 18(5), 229–234. DOI: 10.1016/j.tics.2014.02.008 PMID: 24630873

ADDITIONAL READING

Ainscow, M. (2020). Promoting inclusion and equity in education: Lessons from international experiences. *Nordic Journal of Studies in Educational Policy*, 6(1), 7–16. DOI: 10.1080/20020317.2020.1729587

Emeagwali, G., & Shizha, E. (2016). *African Indigenous knowledge and the sciences: journeys into the past and present*. Sense Publishers. DOI: 10.1007/978-94-6300-515-9

Ezeanya-Esiobu, C. (2019). *Indigenous knowledge and education in Africa*. Springer. DOI: 10.1007/978-981-13-6635-2

Graham, L., Berman, J., & Bellert, A. (2015). *Sustainable learning: inclusive practices for 21st century classrooms*. Cambridge University Press. DOI: 10.1017/CBO9781107280243

Ndlovu, C., & Shizha, E. (Eds.). (2022). *The dynamics of African Indigenous knowledge systems: a sustainable alternative for livelihoods in Southern Africa*. Peter Lang. DOI: 10.3726/b19639

Odora Hoppers, C. (2021). Research on Indigenous knowledge systems: The search for cognitive justice. *International Journal of Lifelong Education*, 40(4), 310–327. DOI: 10.1080/02601370.2021.1966109

Olaopa, O. R., & Ayodele, O. A. (2022). Building on the strengths of African Indigenous knowledge and innovation (AIK&I) for sustainable development in Africa. *African Journal of Science, Technology, Innovation and Development*, 14(5), 1313–1326. DOI: 10.1080/20421338.2021.1950111

Potter-Nelson, E., & Meyers, S. (2022). *An Introduction to sustainability education.* Massachusetts Institute of Technology. White paper series.

KEY TERMS AND DEFINITIONS

Cognitive Justice: The idea that different forms of knowledge should be able to coexist and that all knowledge is partial and complementary. It's also about the "equality of knowers" and the plurality of knowledge (Visvanathan, 1997).

Cultural Justice: This entails that cultural experiences are shared more equitably by people from different cultural backgrounds across society. It can also be defined as the principle of equitable representation of the values and practices of diverse groups within societal institutions.

Inclusive Knowledge Systems: Involve recognizing and valuing the knowledge of diverse groups to create more sustainable communities and strengthen relationships among communities and their institutional knowledge.

Indigenous African Knowledge: The knowledge systems that were developed by African peoples and communities before colonialism based on the culture and worldview of the people and often used in local decision-making to determine sustainable socio-cultural, political and economic development.

Sustainable Learning: It is learning for all that lasts and is the foundation of effective classroom practice, and its goal is to provide for the learning needs of all students throughout their school years and into lifelong learning (Graham et al., 2015)

Chapter 19
Answering the Call to Integrate Indigenous Education Into the Nursing Curriculum

Muhammad Salman Alam
Shifa Tameer-E-Millat University, Islamabad, Pakistan

Abel Jacobus Pienaar
https://orcid.org/0000-0003-0148-752X
Durban University of Technology, Durban, South Africa

ABSTRACT

This chapter addresses the critical need to integrate Indigenous education into nursing education, especially in low- and middle-income countries. Emphasising the proverb "pawpaw can only bear pawpaw," it advocates revisiting nursing's approach, highlighting the disparity between Western curriculum and Indigenous education in nursing. Employing qualitative research, it utilised educational workshops and individual interviews to explore the understanding of Indigenous education in nursing and recommended ways to facilitate Indigenous education in nursing. Four main categories emerged: local characterisation of Indigenous knowledge, nature of Indigenous knowledge systems, inclusion of cultural, physical, and psychological well-being, and advocacy by nurses and healthcare professionals. It advocates for cultural inclusivity, recommending the recognition of indigenous health systems, incorporating indigenous education in nursing, and evidence-based decision-making. This chapter guides the transformative role that indigenous education can have in nursing, especially in Pakistan.

INTRODUCTION AND BACKGROUND OF THE CHAPTER

There is a current push to integrate Indigenous education into nursing and to oppose the Westernization of Indigenous beliefs (Drummond, 2020). This aforementioned author uses the proverb: *'Pawpaw can only bear pawpaw, it cannot bear another fruit, not mango or coconut.' (Zenadth Kes Proverb).* This proverb emphasises that we cannot inherently change peoples' beliefs. Therefore, professions like nursing should revisit their approach to teaching nursing students to respect Indigenous beliefs. Sadly, health professional education, including nursing, is more often not linked to the Indigenous practices in

DOI: 10.4018/979-8-3693-8447-3.ch019

the respective communities (Drummond, 2020), thus pushing challenges for nurses' practice in Indigenous communities.

On the other hand, it is evident that nursing scholars from an Indigenous community who remained connected to their roots, identity and upbringing attest to serving the community better by applying Indigenous education in health and healthcare practices (Bourque Bearskin et al., 2016). It must also be noted that since time immemorial, Indigenous communities have managed births and healed diseases through a number of culturally originated ways, including religious as well as cultural practices (Narasimhan & Chandanabhumma, 2021). Culturally rooted wisdom, knowledge, and practices tied to specific regions offer a valuable chance for collaboration and knowledge-sharing among peers within the same community (Gainsford & Evans, 2021). Unfortunately, in many countries, cultural practices originating from these communities are prohibited due to colonial health laws, health policies, and healthcare practices. Thus, this has exacerbated the erosion of marginalised cultural identities, driven by an unjust distribution of power and the superimposition of one worldview over another (Matthews, 2017). Furthermore, the marginalisation of Indigenous Knowledge (IK) in healthcare systems has significant real-world consequences. In rural Pakistan, traditional birth attendants (TBAs) use IK, such as herbal remedies and culturally specific birthing positions. However, the dismissal of these practices by Western-trained nurses often creates distrust, leading families to prefer TBAs over hospitals perceived as culturally insensitive. Similarly, in Canada, Indigenous patients frequently encounter barriers when seeking to incorporate traditional healing practices, like smudging ceremonies or herbal supplements, into their treatment plans. This exclusion fosters mistrust and disengagement, as seen in the case of a Cree cancer patient who was denied access to complementary traditional remedies. Adding to the latte, in the United States, Native American communities face a disconnect in mental health care, where holistic approaches involving spiritual and communal elements are often overlooked. Programs like a youth suicide prevention initiative in South Dakota only succeeded after integrating community-led talking circles and spiritual counselling, emphasising the need to bridge these gaps.

Besides curriculum and education, western worldviews disregard the health practices of local communities, especially in low-and-middle-income countries (Van Bewer et al., 2020). Therefore, being ashamed to be associated with one's culture is a serious matter that needs attention by the academician, nurse practitioners and professional regulatory bodies (Ngunyulu et al., 2020).

Consequently, a disparity exists between the community's customs and beliefs regarding healthcare and what students are taught in nursing programs (Van Bewer et al., 2020). A chasm between the theory taught in the Western Nursing Curriculum and where the student nurses normally go for practice ensues because the Community has its own Indigenous health beliefs and practices (Matthews, 2017).

However, the issue arises from the Western curriculum, which fails to encompass the practices of local communities rooted in their culture, environment, and beliefs within their unique context (Van Bewer et al., 2020). This imposition of Western healing practices as the sole solution for addressing health-related issues has sidelined Indigenous ways of knowing. Likewise, the nursing field has yet to fully embrace the Indigenous perspective, which incorporates unique knowledge systems and beliefs within its traditional framework for practice (Van Bewer et al., 2020). Nonetheless, combining Western and Indigenous approaches to knowledge in nursing education and practice offers an innovative opportunity for nursing education (Kennedy et al., 2021).

Background

Indigenous knowledge is unique knowledge, inventions, and practices of local groups produced through experience under specific conditions over time and suited to a given geographical location's local culture and nature (Pienaar, 2016; Ngulube et al., 2015). It is derived from Latin roots meaning "native" or "within," it has evolved to signify original inhabitants, plants, or traditions. However, various regions prefer alternative terms like "Aboriginal" in Australia or "First Nations" in North America, reflecting localised identities and cultural sensitivities. Adding to the previous assertion, Indigenous education in nursing refers to education that integrates the unique cultural beliefs, practices, and values of the Indigenous peoples into nursing education and practice (Van Bewer et al., 2020). These practices may consist of social and economic systems, language, traditional medicine, healing techniques, spiritual beliefs, and artistic and literary works (Mogawane et al., 2015). Indigenous education struggles are related to the ongoing (often European) colonisation of Indigenous peoples, along with the evisceration of their languages and cultures on the social, political, and educational levels (May et al., 2010).

However, as nurses originating from Indigenous communities, we inherently bring our cultural knowledge to our nursing practice, drawing on the wisdom passed down by our families, communities, ancestors, and spiritual beliefs (Bourque Bearskin et al., 2016). As stated in the inception of this chapter, we should respect the beliefs and practices of our Indigenous communities as nurses.

Problem Statement

Nursing education should integrate Indigenous Knowledge (IK) as a respected and distinct system alongside Western methodologies, ensuring culturally inclusive and holistic care. However, the marginalisation of Indigenous Knowledge persists due to the dominance of Eurocentric ideologies in higher education. This devaluation limits nurses' understanding of Indigenous education, leaving them unprepared to address the diverse needs of Indigenous populations (Kennedy et al., 2021). If unaddressed, this gap perpetuates cultural insensitivity, health inequities, and mistrust in healthcare systems. Evidence highlights that culturally appropriate care improves patient satisfaction and outcomes, emphasising the urgent need to include Indigenous education in nursing. Hence, this research first explored Indigenous education in nursing, then capacitated nursing interns about Indigenous education and recommended ways to facilitate the learning of Indigenous education in nursing through educational workshops.

Significance of this Research

This research addressed the complex interplay between cultural, religious, and territorial influences on healthcare beliefs and practices, particularly within the nursing profession. It has the potential to bring about positive change in healthcare education, practice, and policy by promoting inclusivity and cultural sensitivity. Specifically, nursing interns and nursing practitioners can benefit by gaining a deeper understanding of Indigenous education, enabling them to deliver culturally sensitive and holistic care and improving patient-provider relationships. Moreover, healthcare educators and institutions can adopt innovative teaching models that integrate Indigenous and Western knowledge systems, enriching nursing curricula (Blackstock, 2016). Furthermore, policymakers can benefit from data and recommendations to create inclusive policies that address healthcare inequities. However, Indigenous communities can experience improved health outcomes and trust in the healthcare system, while healthcare systems will benefit from enhanced quality of care through holistic, community-centred approaches. Therefore, by bridging the gap between Indigenous and non-Indigenous nursing knowledge, this study fosters innovation in nursing education and practice, contributing to quality nursing education and more meaningful healthcare delivery (Van Bewer et al., 2020). Thus, this study advocated for a broader perspective that incorporates holistic values, spirituality, and community relevance. It recommends ways to enrich nursing education by integrating Indigenous perspectives, contributing to a more inclusive and meaningful educational experience.

Research Objectives

This research achieved the following objectives:

- Explored the understanding of nursing interns regarding Indigenous education in nursing.
- Recommended ways to facilitate the learning of Indigenous education in nursing through educational workshops.

METHODOLOGY OF THIS RESEARCH

This section presents the research methodology to explore the understanding of nursing interns regarding Indigenous education in nursing through educational workshops. The research objectives of the study were to explore the understanding of nursing interns regarding Indigenous education in nursing and to recommend ways to facilitate the learning of Indigenous education in nursing. The section also explains the study approach, design, setting, population, sample and sampling technique, data collection, and data analysis. Furthermore, the study rigour and ethical considerations are described.

The research methodology of the study is summarised in the following table:

Table 1. Research methodology

Approach	Qualitative
Study Design	Educational Workshop
Study area	One Tertiary Care Hospital in Islamabad One Tertiary Care Hospital in Peshawar
Study population	Nursing Interns
Sampling technique	Purposive sampling
Sample size	Achieved through data saturation and data sufficiency (LaDonna et al., 2021)
Inclusion Criteria	Nurse interns having a minimum of five to six months of clinical rotations after graduation
Exclusion Criteria	Nurse Intern who attended workshop or training on Indigenous education
Data collection	Educational Workshops (Inmark, 2010)
Trustworthiness	Credibility, Dependability, Confirmability, Transferability, Authenticity (Lincoln & Guba, 1985)
Data analysis	Open Coding (Creswell & Creswell, 2018)
Ethical considerations	IRB approval, Institutional Permission, APA research guidelines (Beneficence and non-maleficence, Fidelity and Responsibility, Integrity, Justice & Respect) (Berenson, 2017).

Qualitative research involves a method of investigating and comprehending the significance that individuals attach to a social phenomenon (Creswell & Creswell, 2022). As this research intended to explore the understanding of nursing interns regarding Indigenous education in nursing, therefore, the qualitative approach was found to be the most appropriate means to fulfil the study's objectives. The study design used in this research was the educational workshop. The educational workshop method is used as a research design to fulfil the research objectives as well as the learning needs of the study participants (Orngreen & Levinson, 2017). Furthermore, it encourages interaction among participants through group discussions and constructive feedback and facilitates prolonged engagement to explore the phenomenon (Orngreen & Levinsen, 2017). Therefore, this study used an educational workshop as a research design to achieve its objectives.

The study population was nurse interns who completed their four years of bachelor's degree in nursing and were practising under the supervision of registered nurses in the hospital to obtain their license (Shahzeydi et al., 2022). At the time of the study, these nursing interns had 6 months of clinical experience in critical and general nursing through hospital rotation. The research was conducted at two private-sector tertiary care hospitals located in Islamabad and Peshawar. These hospitals were chosen based on the availability of nurse interns and the convenience of conducting the workshops. Both hospitals had 500 beds. One hospital was JCIA-accredited, and the other was ISO-certified. Nurse interns (Pre-Licensure Nurses) have a minimum of five to six months of clinical experience after graduation. This study was conducted from September 2022 to August 2023.

The purposive sampling technique was used to recruit the study participants because purposive sampling provides an opportunity to select individuals who can offer the most valuable insights to explore the research phenomena (Creswell & Creswell, 2018; Etikan, 2017). Moreover, This research relied on data saturation and data sufficiency principles of qualitative research for sample size (Creswell & Creswell, 2018; Rendle et al., 2019). Data saturation is the point at which additional data collection is unlikely to uncover new insights or themes. Therefore, it guides sample size in qualitative studies, including those employing the workshop methodology (Shahzeydi et al., 2022). However, data sufficiency pertains to the degree to which the existing data is ample and thorough, enabling the derivation of comprehensive conclusions and meaningful interpretations (LaDonna et al., 2021).

After the approval from the institutional head, the unit manager was requested to facilitate the recruitment process of study participants. A list of all eligible participants was obtained from the unit in charge of the hospitals. The researcher individually contacted every potential participant to provide a concise overview of the study's objectives, the process of collecting data, and the potential advantages and risks associated with participating in this research. The participants were requested to express their willingness and approval to participate in the study. Those who willingly provided written consent and were fully aware of the study's details were selected for the study. Data collection was done through a workshop process explained by (Inmark, 2010; Orngreen & Levinsen, 2017) consisted of two phases as follows:

To plan the workshops, the researcher collaborated with the supervisory team and decided on the workshop objectives, content, typologies and number of participants in the form of workshop protocol (Attached as Annexure B). Pamphlets and banners were developed and shared one week before the workshop via social media and electronic forums (WhatsApp, emails). For the workshop session, a high-quality icebreaking session thought-provoking short audio-visual Vignette of 4 minutes from the actual case study of the researcher's past clinical experience was developed to portray the day-to-day Indigenous practices in nursing (Script attached in Appendix D). Moreover, a PowerPoint presentation was developed to facilitate the session along with a Google form for pre and post-evaluation of the workshop. Prior to the workshops, a formal meeting was held with nursing managers and the directors of nursing hospitals to share the workshop protocol and to request their facilitation in data collection. Additionally, to establish communication with the participants, a WhatsApp group was made for each workshop where the study participants received instructions regarding the day, venue and time three days prior to the workshop. This workshop utilized a micro group facilitation format (5-10 participants), which is considered suitable for workshops designed for knowledge creation, learning and action planning (Inmark, 2010a). In each workshop, the workshop protocol was followed, and the researcher employed the fishbowl teaching and learning strategy to facilitate group discussion and active participation (Yabarmase, 2013). In the fishbowl strategy, students were divided into inner and outer circles for interactive discussion and to explore a deeper understanding of the phenomenon through group discussion (Yabarmase, 2013). Workshops were conducted in separate conducive rooms along with audio and video recordings. Each session was approximately 160 to 180 mins (Inmark, 2010) and was in English and Urdu language as per participant preference and comfort. During the workshop, the researcher and two observers took field notes, one observer participated remotely, while the other was physically present in each workshop session.

Total of three workshop with 21 participants were conducted and analyzed to reach the data saturation point (Creswell & Creswell, 2018). Moreover, to ensure data sufficiency (LaDonna et al., 2021) Two face-to-face (Two Participants) interviews were also conducted using semi-structured interview guides (Attached in Annexure C). Each interview was of 40 to 50 mins duration audio recorded in English and Urdu language. Aligned with the guideline of Cresswell and Cresswell (2018), this study followed open coding approach for data analysis. Open coding, as a qualitative method of data analysis allows line by line coding of the data and to get comprehensive understanding of the phenomenon (Creswell & Creswell, 2018). Following steps of open coding were followed for data analysis.

In the initial phase of organisation and preparation, the data was transcribed and made ready for analysis. The primary researcher transcribed each audio-recorded workshop and interview. After the transcripts were meticulously organised into separate folders, each labelled with a participant code (e.g., P1, P2, etc.).

The second step involved reading all the data meticulously, where the primary researcher thoroughly reviewed and cross-verified all the transcripts with the corresponding audio recordings. During this reading process, researchers immerse themselves in the data to gain a deep understanding of its content by reviewing the data multiple times, taking notes and making observations. In the coding phase, researchers proceeded with line-by-line coding, a systematic process of dissecting the data into smaller segments known as "codes'. These codes, referred as in-vivo codes, are derived from highlighted segments within the original transcripts. Following this, conceptual codes were assigned to the in-vivo codes.

Rigour in qualitative studies is referred to as trustworthiness, which is parallel to reliability and validity in quantitative studies. The trustworthiness of this study was ensured by following the guidelines of (Lincoln & Guba, 1986) as follows:

Table 2. Trustworthiness

Trustworthiness	Measures to ensure
Credibility	The researcher ensured study credibility through member checking, prolonged engagement, observations, and in-depth data collection procedures. The workshop's intense engagement, akin to prolonged engagement, supports qualitative study credibility (Ahmed & Asraf, 2018). A pilot workshop, subject engagement, and doctoral student cross-checking of transcriptions further enhanced credibility. Furthermore, the study's credibility was strengthened by the supervision of the research committee.
Dependability	To ensure dependability, a comprehensive account of both the data collection and data analysis procedures was presented.
Conformability	Confirmability of the study was achieved by maintaining reflexive journals and conducted weekly investigator meetings. Additionally, two in depended co-coders analyzed the data.
Transferbility	The researcher provided possible explaination of the stud population and setting along woth detailed description of the workshop methodology and data collection process which added to the transferability of the study.
Authenticity	The authenticity of the study findings was ensured by including study timeline, records of participant interviews, transcriptions of audio-video recordings, field notes, and the application of data saturation and sufficiency principles.

Ethical Considerations

This study followed the American Psychological Association 2017 guidelines as ethical considerations and human rights during research work. The study was approved by the institutional review board of Shifa Tameer-e-Millat University, with the IRB NO# 096-23. Moreover, Institution permission was obtained from the head of the institutes before the data collection.

Table 3. Ethical considerations

Ethical Principle	Measures
Beneficence and Nonmaleficence	To uphold this principle, the researcher informed the workshop participants briefly at the beginning of the session about study purpose, benefit and risk and also convey that they had the right to withdraw from the study at any time. Consequently, written informed consent was obtained from each participant before conducting the research (Attached as annexure A).
Fidelity and Responsibility	The researcher took the responsibility for maintaining the professional behavior and moderate the conflict of interest in order to prevent any harm to each of the participants during the workshops.
Integrity	Before obtaining informed consent, participants were provided with an explanation of the study's objectives. They were also informed in advance about the workshop schedule, including the time and location. No ambiguous or unattainable commitments were made, and participants were assured that their responses would use for research purpose only.
Justice	The researcher maintained that principle by ensuring that the benefits and burdens of the research are distributed fairly among the participants and that no particular group or individual is unfairly advantaged or disadvantaged Moreover, the findings were shared with the participants of the study.
Respect for People's Rights and Dignity	To adhere to this principle, the researcher ensured that participants' respect was maintained, regardless of their race, gender, religion, or ethnicity. The researcher upheld participants' rights to choose whether to participate in the study. Additionally, codes were assigned to the participants during the data analysis process to protect their identities.

FINDINGS OF THE RESEARCH

In the previous section, the researcher delved into the methodological aspects of the study. Building on that, this section presents the demographic information of the study participants, and the findings of the open-coding analysis of workshops and face-to-face individual interviews as follows:

Demographic Characteristics

This present study was carried out among nursing interns who had five to six months of clinical rotation in the hospital in various clinical placements such as the Pediatric Ward, Surgical-ICU, and Emergency. Gender is noted as men or women, and a transition to face-to-face interviews is apparent for two participants in Pediatric and Surgical Wards. A total of 23 nursing interns participated in this study, out of which nine were male and 14 were female.

Study Findings

The open coding approach revealed four main categories, including Local Definition of Indigenous Knowledge, Education and Practices (KEP); Nature of Indigenous Knowledge System; Holistic Patient-Centered Care (*Preference, Respect);* Inclusion of Cultural, Physical and Psychological Well-being grounded on Beliefs, Role and Outcome; and Advocacy by Nurses and Health Administrators to Recognize Indigenous Health System. Under the umbrella of these four main categories, there are 13 sub-categories. Each category further consists of sub-categories.

Table 4. Categories and sub-categories of participants' data (n=23)

Categories	Sub- Categories
Local Definition of Indigenous Knowledge, Education and Practices (KEP)	• Ambiguity (Doubts) Regarding the Concept Indigenous • Indigenous Education and Understanding • Comprehensive Nature of Indigenous Practices • Importance of Indigenous Knowledge Transfer • Appreciation for preserving Local Knowledge
Nature of Indigenous Knowledge System: Holistic Patient-Centered Care (Preference, Respect)	• Patient-Centered Care • Holistic Local Approaches
Inclusion of Cultural, Physical and Psychological Well-being grounded on beliefs, Roles and Outcome	• Embracing Cultural Values and Practices in Health and Education • Living With Eastern Indigenous and Western Practices • Intertwining of Physical and Psychological Well-being • Indigenous Healing Practices based on beliefs, Role and Outcome
Advocacy by Nurses and Health Administrators to Recognize Indigenous Health System	• Patient safety • Acts and Omissions of nurses

Detailed descriptions of Categories and Sub-categories are provided below. The discussion is categorized in the description of the themes and sub-themes, quotations from the participants, ending with a clear contextualization of the themes.

Local Definition of Indigenous Knowledge, Education and Practices (KEP)

This initial category was derived from the open coding in data analysis of the data. It encompasses a wide array of knowledge, skills, and values deeply rooted in the heritage, cultural-, and environmental context of Pakistani communities; the ways of knowing of Indigenous Pakistani. This category consists of five subcategories, Ambiguity (Doubts) Regarding the Concept Indigenous; Indigenous Education and Understanding, Comprehensive Nature of Indigenous Practices; Importance of Indigenous Knowledge Transfer; and Inheritance of Local Knowledge. The findings are presented, through a brief explanation, followed by quotes from the participants in cursive.

Ambiguity (Doubts) Regarding the Concept 'Indigenous'

This subcategory "Ambiguity (Doubts) Regarding the Concept Indigenous" pertains to the uncertainties and questions that arise in relation to the term "Indigenous" or "indigeneity." In the context of this research, it signifies a lack of clarity or confusion surrounding what it means for a community or group to be considered Indigenous. Participants indicated that as much as they cannot relate to the word Indigenous, but they know Indigenous 'things. Therefore, there was a level of understanding of Indigenous, though doubt prevailed. Contrary to their ambiguity, participants define 'Indigenous' in their different words providing the same essence of the meaning.

"I had very little understanding of what Indigenous means" (P2)

"I cannot relate the what word Indigenous is but I know actually what is Indigenous things. Indigenous is naturally growing things or natural like naturally occurring of the whatever place or something." (P3)

"I don't know much but Indigenous practices are not just about medicines; they are something important, something essential in my area" (P12)

Clearly most participants perceived that they do not know what Indigenous mean, but their subsequent explanation indicated that a level of comprehension of 'Indigenous' exist among them.

Indigenous Education and Understanding

The sub-category focuses on the education and the comprehension of Indigenous knowledge, practices, and education within Indigenous communities. The findings reveal that Indigenous education is deeply rooted in the local context, taking into account the unique ecological, geographical, and cultural aspects of the community. It emphasizes the importance of local knowledge for sustainable living and resource management. Indigenous education similarly involves the local methods by which Indigenous knowledge is passed down from one generation to the next. This transmission typically occurs within the community, often *"through oral storytelling, hands-on learning, and direct experience"*. Contrary to this, participants have verbalized that they don't understand Indigenous education, but they narrated it in different relevant statements:

"Indigenous education is Local knowledge and the information that people have about their areas" (P8)

"I don't have clear understanding but It's important to consider a patient's spirituality, beliefs, traditions, and everything related to their culture. We can call this Indigenous education" (P1)

Through these statements participants grasp the essence of what they learn from their parents, but struggle to fully comprehend or position it.

Comprehensive Nature of Indigenous Practices

This sub-category revolves around the intricate and all-encompassing characteristics of native practices within Indigenous communities. These practices maintain a deep connection with the natural world, involving ceremonies, rituals, and customs that mirror the community's bond with the land, ecosystems, and natural resources. Furthermore, a significant number of these Indigenous practices hold spiritual or religious importance:

"Local practices are the things we do in our area to treat different illnesses or health problems. This can include spiritual, medical, or home remedies, and it all comes under the term "local practices" when it's related to our health. (P5)

Similarly, the other participant defines the Indigenous practices as *"Local practices are like traditions that have been passed down over time"*.

Additionally, some Indigenous practices, such as local healing practices, prioritize the well-being of community members. They are based upon generations of accumulated knowledge to address various diseases, whether physical, psychosocial or spiritual in nature.

"When someone is sick, they might seek help from a religious leader or a healer". (P2) *Another participant said* "I don't believe it guarantees a cure for diseases. It's more about their faith. By following specific rituals or practices and seeking their assistance, they will be cured" (P4)

To put it concisely, the profound and holistic nature of Indigenous practices is deeply rooted in the natural world and spirituality. Indigenous practices encompass a wide range of remedies and rituals, bridging faith, practice, and the pursuit of healing, although they may not guarantee cures in the conventional sense.

Importance of Indigenous knowledge transfer

Linking to the previous subcategory, this sub-category underscores the significance of passing down traditional knowledge, skills, and wisdom from one generation to the next within Indigenous communities. Transferring this knowledge reinforces a sense of belonging and pride among community members and helps maintain cultural continuity.

"I didn't understand how my father learned things but he has learned about the herb's healing properties from the local knowledge and beliefs in the village" (P8)

However, knowledge transfer encompasses not only practical skills but also the cultural, spiritual, and social aspects of Indigenous life. It equips the younger generation with a comprehensive understanding of their culture.

"I think that I learn my culture from my family and grandfather, who usually told me to do local practices for his health issues". (P10)

Furthermore, the transfer of Indigenous knowledge often occurs within families and communities, strengthening intergenerational bonds and respect for elders. This is very pertinent to our Pakistani culture.

"I belong from Pukhtoon community and we all live together. In our culture it is mandatory to respect our elder." (P11)

In essence, Indigenous knowledge transfer emphasizes the requirement to adapt and advance Indigenous knowledge to effectively tackle the convolutions of the world, while upholding its central values and principles in the specific community.

Appreciation for Preserving Local Knowledge

The mentioned sub-category focuses on the process of how Indigenous knowledge, skills, and traditions across generations within native communities are inherited.

"It's a common practice, especially among the older people in my area. They (foreparents) used to go to a hakeem." (P2)

The influence of acculturation has a substantial impact on the preservation of local medicinal knowledge and practices. Without proper documentation, there's a risk of losing valuable knowledge that could otherwise benefit future generations.

"Sometimes, we become influenced by Western cultures, prioritising them over our own. It would be highly beneficial if our education system incorporated our cultural values and beliefs." (P9)

In short, this inheritance is crucial for ensuring that Indigenous knowledge remains relevant and continues to benefit both the community and the wider world.

Nature of Indigenous Knowledge System: Holistic Patient-Centered Care (Preference, Respect)

This second category refers to the approach and characteristics of Indigenous knowledge systems in the context of healthcare and patient well-being. This category focuses on the holistic and patient-centred nature of Indigenous healthcare practices, with particular attention to the principles of preference and respect for the participants' culture. It has two sub-categories: Patient-centred care and Holistic local approaches.

Patient Centered Care

This sub-category revolves around the approach to healthcare that prioritises the patient's needs, preferences, and well-being at the centre of the care process. It recognises that patients have a unique understanding of their own bodies and health, and their input should be valued in health care.

"We have been advised to respect their practices, even if we don't fully understand them. It's important to respect their beliefs and faith." (P6)

Patient-centered care acknowledges that each patient is distinct, with their own set of values, preferences, and cultural background. Its tailors' treatment plans to align with these individual factors to ensure that care is as effective and relevant as possible.

"By understanding a patient's culture and beliefs. I can provide care that's tailored to their specific needs." (P14)

Ultimately, patient-centered care seeks out to improve patient satisfaction and healthcare outcomes, as including patients in their care decisions and respecting their values is more likely to result in positive treatment experiences and better health results.

"It's really important to consider and respect the patient's perspective." *Another participant also said that* "In our BSN training, we have learned to adapt to the patients' and attendants' perspectives." (P11)

Hence, the importance of patient-centered care, which prioritizes patients' unique needs and values, leads to improved healthcare outcomes and patient satisfaction. Respecting patients' beliefs, cultural backgrounds, and perspectives is fundamental to this approach.

Holistic Local Approaches

The above-mentioned is centered on healthcare and well-being strategies within Indigenous communities that take a comprehensive and interconnected approach to health and healing. Holistic local approaches consider well-being in a comprehensive manner, addressing not only physical health but also emotional, mental, spiritual, and social aspects. They are influenced by traditional practices, belief systems, and customs that have evolved over generations.

"We have discovered various things, including spiritual and cultural practices. We are even implementing some of these practices in the hospital." (P2)

Holistic local approaches often utilise natural resources and remedies available within the local environment. This can include the use of local medicinal plants, traditional healing techniques, and local knowledge about the environment. As verbalised

"Filling a glove with water and gently place it on the patient's hand. This offers human touch, reducing feelings of isolation. It's a unique and almost like a local way of providing comfort and connection in nursing". (P5)

This quote underscores the cultural richness and holistic nature of Indigenous healthcare, integrating local practices and resources into comprehensive well-being strategies.

Inclusion of Cultural, Physical and Psychological Well-Being

The above-mentioned category includes an approach to healthcare and well-being that acknowledges and addresses the interconnected aspects of an individual's life. This category recognises that an individual's cultural identity, physical health, and psychological state are interrelated and should be collectively addressed to promote holistic and effective care. Moreover, it focuses on creating a healthcare environment that is culturally sensitive, physically supportive, and psychologically nurturing, ultimately aiming to improve the overall quality of life and well-being of individuals within diverse communities. The following sub-categories ensure:

Embracing Cultural Values and Practices in Health and Education

This sub-category highlights the integration and promotion of cultural values, traditions, and practices in healthcare and education, emphasising the preservation of Indigenous cultural traditions, values, and beliefs, particularly in the context of healthcare and well-being. Indigenous communities place significant importance on their cultural practices.

"We have been so focused on modernising our practices that we have forgotten about the cultural and local ways of caring for patients." (P8)

In education, it emphasizes the inclusion of cultural teachings and practices in the curriculum to enrich both academic and cultural aspects.

"The focus was primarily on Western education, but Indigenous education will be a valuable addition to nursing curriculum, and a positive development." (P12)

This approach ensures culturally relevant healthcare practices as part of clinical education. The participant stated that *"In nursing education, Indigenous practices and skills are bit like the "Implementation" step in the nursing process."* Participant also said that *"When we actually practice it as an assignment in clinical settings, it becomes a valuable learning experience." (P17)*

Overall, it highlights the significance of cultural respect in both healthcare and education, fostering more effective care and inclusive educational settings.

Living with Eastern Indigenous and Western Practices

The sub-category focuses on the harmonious integration of local Eastern Indigenous healing practices with Western medical approaches. It acknowledges the dual nature of individuals or communities who incorporate both sets of practices into their daily lives. The participants quoted

> "We often prioritized seeking fast-acting medicines over patient Indigenous practices, which offer slower yet enduring effects. Our impatience led to this realization that valuing the knowledge of our elders is essential for well-being" (P17)

It also recognizes the importance of striking a balance between these two approaches, with individuals recognizing the unique value each brings to specific health concerns.

> "What we do is provide guidance and educate them about home remedies and what they should or shouldn't do alongside their medications." "We think about the disease Physically. But no one thinks about emotional and mental health." (P11)

Additionally, this coexistence promotes cultural harmony by enabling individuals to respect their cultural heritage while taking advantage of the progress made in Western medicine.

> "When we ask those who follow these old local practices, their behavior and opinions can influence us, making us think that these practices are universal truths." (P13)

Therefore, it embodies the coexistence, integration, and balance between local Eastern Indigenous practices and Western medical approaches, that allows individuals to harmoniously navigate their health and well-being by embracing the best of both worlds.

Intertwining of Physical and Psychological Well-Being

This sub-category delves into the dynamic interaction and equilibrium between an individual's physical and psychological well-being. It recognizes the interdependence of physical health and mental well-being, with one often influencing the other. The emphasis here lies on achieving a harmonious co-living to preserve overall health.

> "Indigenous not only food and eatables. It includes all natural, local things and it include all things except chemically manufactured." (P15)

The maintenance of physical well-being is pivotal for overall health and can have repercussions on one's mental state. Most of the participants stresses on that factor, as one participant verbalized that:

> "Having a strong willpower is crucial, and good mental health is essential for recovering from a disease. People's belief that they will get better plays a significant role in the healing process through their faith." (P2)
>
> "Mental health and willpower are vital for healing." (P8)

Additionally, in certain instances, cultural and Indigenous practices are integral to the balance of physical and psychological well-being. These practices, rooted in traditional healing methods, rituals, and belief systems, address both physical and mental health concerns.

> "In Gilgit-Baltistan, where I come from, we often rely on local home remedies and herbs for healing mentally and physically." (P5)

Indigenous Healing Practices Based on Beliefs, Role and Outcome

The sub-category explores the profound connection between Indigenous healing practices and the cultural beliefs of Indigenous communities, the pivotal role these practices play in promoting health and well-being, and the hope and confidence they engender. As research participants verbalised that:

> "I may not comprehend the clinical significance of certain beliefs. However, I do understand that due to religious beliefs, people utter "peace be upon him" when they exit a city's boundaries. They hold the belief that it safeguards them from diseases in that region." (P1)

> *Another participant said* "When I asked them what it was, they explained that they have certain practices for this disease in their area. It seems to be related to their faith, spirituality, or some religious beliefs." (P12)

Indigenous healing practices are driven by the goal of reinstating balance and harmony in individuals' lives, ultimately benefiting their overall health and well-being. These practices are deeply rooted in cultural beliefs that detail the guiding principles and healing rituals, highlighting the profound interconnection of various well-being aspects.

> "In my family, there is a strong emphasis on the past and sharing stories about how they used to maintain good health through a healthy diet. Their belief is firmly rooted in their cultural practices, and they don't hold strong beliefs in modern medications and treatments." (P16)

Within these Indigenous practices exists a powerful source of hope and confidence. The rituals, ceremonies, and remedies are administered with a resolute faith in their effectiveness. This unwavering belief cultivates a profound sense of hope, affirming that healing is achievable, and it instils confidence in the power of these practices to foster recovery and well-being.

> "When someone engages in a practice they believe is right, their body naturally responds by releasing certain chemicals associated with that action. This creates a connection between the practice and its effects on the body. However, when people lack understanding of the practice and its impact, they lose hope in recovery." (P19)

It underscores the interplay between Indigenous healing practices, cultural beliefs, their pivotal role in health and well-being, and the hope and confidence they generate. These practices are deeply rooted in cultural heritage, and they play a crucial role in maintaining and restoring the health and well-being of Indigenous individuals and communities.

Advocacy by Nurses and Health Administrators to Recognise Indigenous Health System

The category "Advocacy by Nurses and Health Administrators to Acknowledge Indigenous Health Systems" revolves around the initiatives led by healthcare professionals, particularly nurses and health administrators, to enhance awareness and acknowledgement of Indigenous health systems. Two subcategories, Patient safety and Acts and Omissions of nurses, which is discussed below:

Patient Safety

The sub-category is focused on the critical intersection of patient safety and the recognition of Indigenous health systems within healthcare practices. It emphasises the dual objectives of safeguarding the safety and well-being of Indigenous patients and recognising and respecting their cultural beliefs and traditional healing practices within healthcare.

> "I understand, as nurses, we adhere to a Western-based curriculum for patient care. If we abruptly alter their practices, it may not be effective." (P11)

Patient safety within healthcare involves implementing measures and procedures to prevent harm and errors during healthcare delivery.

> "If Western medicine proves less effective in a particular case, then you willingly consider local practices but with caution." (P9)

Simultaneously, it underscores the significance of acknowledging and incorporating Indigenous health systems into the broader healthcare framework, including the recognition of cultural beliefs, local healing practices, and Indigenous knowledge systems.

> "Patient satisfaction is a crucial aspect of healthcare. If incorporating specific local practices enhances patient satisfaction and comfort, healthcare providers are more inclined to do so." (P13)

The core idea is to promote patient safety by acknowledging and respecting the Indigenous health systems, ensuring that healthcare services are provided in a manner that is culturally sensitive, safe, and aligned with the values and practices of Indigenous populations.

Acts and Omissions of Nurses

In the Pakistani healthcare system, the actions of nurses regularly involve the application of conventional Western knowledge and often neglect local healing practices in their care to the patient.

> "Values and traditions come with Indigenous knowledge. It's about recognising the importance of our culture in our lives. The hospitals did not accept it, but in our homes, these practices are acknowledged and followed." (P16)

Similarly, the omissions of nurses in Western healthcare for recognition of the Indigenous health system are shaped by the need to respect and uphold cultural values and traditional practices. Nurses may refrain from certain medical interventions that conflict with Indigenous beliefs or customs, recognising the importance of preserving the patient's cultural identity and well-being. As it is pertained from the participants' verbalisation:

> "People hold diverse beliefs, often relying on cultural local practices to address their health issues. However, these practices may not be as effective due to our (Nurses) focus on standard hospital western procedures." (P2)

> "Nurses should understand the importance of incorporating local practices and treatments for patients who have believe in them." (P7)

Through these statements, it is clear that nurses should navigate complex decisions to provide holistic care to Indigenous patients, respecting their traditional and local practices.

DISCUSSION TO OPERATIONALISE THE FINDINGS OF THIS RESEARCH IN NURSING

In answering objective one of the discussions, the researcher focuses on the deeper illumination of the previous findings. Objective two will be highlighted by focusing on the realisation of the process of the educational workshops as learning facilitation for Indigenous education in nursing.

Local Definition of Indigenous Knowledge, Education and Practices (KEP)

A Holism, Depicting Theory-Practice Integration

The category "Local Definition of Indigenous Knowledge, Education and Practices (KEP)" in the discussion focuses on the perceptions and comprehension of the study participants regarding Indigenous knowledge, education and practices. It explored the depth of understanding among the participants and how they define and relate to these concepts within their cultural context. The discussion highlights the nuanced understanding of Indigenous education, acknowledging that participants may not have a clear-cut definition but possess an intuitive sense of its importance based on their upbringing and experiences. Noteworthy to state that the participants had doubts about their understanding of Indigenous knowledge, education and practice, but in their verbalisation, they aligned with the actual definition as stated by the literature.

Furthermore, participants did not clearly differentiate between knowledge, education, and practice, but they interchanged these concepts as a holistic entity. In the local definition stated by participants, Pienaar (2016) refers to Indigenous knowledge as special knowledge, skills, and traditions developed by local communities through their experiences in specific places. These understandings are shaped by the local culture and environment of that area over time, whereas Dhungana (et al., 2023) opine that Indigenous knowledge is information, beliefs, and customs designed to uphold, convey, and place into context how Indigenous communities relate to their culture and environment over the time. It is clear that both the previous authors concur with the definition provided by the participants of this research regarding Indigenous knowledge, context and culture, therefore demystifying the doubt that the participants verbalised.

Perusing the definition of Indigenous Education Ogbo and Ndubisi (2017) state that it is the process of imparting and preserving the knowledge, practices, and cultural content of Indigenous communities within educational systems, both formal and informal; and Lutomia and Bravo, (2017) asserts that Indigenous education involves the transfer of knowledge and skills from previous generations, such as parents and the community, to the younger generation through informal means, rather than formal schooling. This includes sharing history, culture, religion, traditions, life skills, and occasionally gender-specific knowledge, preserving the heritage and values of the community. These authors equally confirmed the understanding of the research participants.

These practices become a natural part of their daily life (Mogawane et al., 2015). Furthermore, defined by Owan and Ekereke (2023) on how traditional knowledge, skills, and ways of doing is handed down in a community or culture since time immemorial. These practices are closely connected to the local area and the culture of the people, and they are shaped by how people live and what they believe. Both definitions of practice, as found in the literature, concur with the comprehension of the participants.

Hence, the participants' conception highlights the holistic nature of Indigenous practices, which are deeply intertwined with the natural world, spirituality, and culture. These Indigenous practices include such as local Healing Methods (traditional healers who use local plants, spiritual rituals), The plant- and animal based-mixtures (use of honey and black pepper for cough relieve), Spiritual Healing (prayer, rituals, and seeking blessings from religious leaders to promote physical and mental well-being), and Local Midwives (Indigenous practices often involve the assistance of local midwives and traditional birth attendants during childbirth, relying on their experience and knowledge of natural childbirth practices). While Indigenous practices may not always guarantee conventional cures, the discussion acknowledges their value in maintaining cultural identity and promoting overall well-being.

Nature of Indigenous Knowledge System: Holistic Patient-Centred Care

A Virtue of Respecting Patients' Culture and Preference

The category is the portrayal of the core principles and characteristics of Indigenous knowledge systems in healthcare. It highlights the holistic and patient-centric approach to healthcare within Indigenous communities, shedding light on their unique perspectives and practices.

Indigenous knowledge systems are rooted in a holistic understanding of health and well-being. Similar to Govender (2012), this research findings explained Indigenous knowledge (IK) encompasses a specific type of knowledge, while Indigenous knowledge systems (IKS) describe how IK is conserved and passed down through diverse cultures in various expressions like traditions and customs. Indigenous knowledge systems (IKS) can be understood as a complex reservoir of insights and alternative knowledge perspectives, though it has often been perceived as inferior and primitive by Eurocentric views.

Furthermore, this holistic approach of IKS recognizes the intricate interconnectedness of these dimensions in shaping a person's overall health. Indigenous communities prioritize balance and harmony in all these dimensions for a person to be truly healthy. As Stansfield & Browne (2013) argued Indigenous knowledge is a source of wisdom and unique perspectives that have provided valuable insights for humanity over millennia. However, nursing education largely overlooks it, and the current discussions on healthcare and health research do not adequately incorporate this knowledge. These authors also contended that Indigenous knowledge varies across countries, communities, and cultures due to its deep connection to local environments, groups, and historical contexts (Stansfield & Browne, 2013). Therefore, one must consider the contextual Indigenous knowledge of a specific community.

This approach places the patient at the centre of their healthcare decisions actively involves them in their treatment plans, and respects their preference. The research acknowledges that patients possess a unique understanding of their health and cultural background. Similarly, Gartner et al (2022) assert that patient centricity involves prioritizing the patient by actively and continuously engaging with them in a respectful and compassionate manner to ensure the best health outcomes.

Adding to the previous discussion, unfortunately, Indigenous knowledge is often overlooked in nursing education and healthcare. Therefore, in the light of this research findings advocacy is recommended to acknowledge Indigenous knowledge systems in Pakistan's health care system.

Inclusion of Cultural, Physical and Psychological Well-Being

Inclusive Care and Wellbeing

The category "Inclusion of Cultural, Physical, and Psychological Well-being" pertains to a comprehensive and inclusive approach to promoting well-being that takes into account various aspects of an individual's life. It emphasizes the importance of addressing cultural, physical, and psychological dimensions to achieve holistic health and well-being.

Oishi and Gilbert (2016) presented a useful explanation of cultural well-being, which refers to an individual's feeling of attachment to their cultural identity, ancestry, and customs. They also argued that it encompasses a sense of belonging, pride, and satisfaction in one's cultural heritage (Oishi & Gilbert, 2016). In the context of this research, recognizing cultural well-being highlights the importance of safeguarding and advancing cultural identity and traditions as a vital component of overall well-being. The research finding also aligns with the conclusions from a study conducted by Gott (2020), stating that cultural activities contribute to higher reported well-being in individuals.

As much as physical well-being underlines the critical role of sustaining robust physical health as a fundamental component of overall well-being, Dhanabhakyam et, al (2023), refer psychological well-being to an individual's mental and emotional condition, encompassing factors such as emotional strength, positive interpersonal connections, self-esteem, and overall mental health.

Therefore, recognising the importance of cultural, physical and psychological well-being is a necessity, according to the findings in this research.

Advocacy by Nurses and Health Administrators to Recognize Indigenous Health Systems

Promotion of Co-Living of Indigenous and Western Healthcare Systems

The category "Advocacy by Nurses and Health Administrators to Recognize Indigenous Health Systems" underpins the efforts that should be made by healthcare professionals, specifically nurses and health administrators, to promote awareness and recognition of Indigenous health systems. The fundamental goal of advocacy is to respect and recognise Indigenous healing practices, cultural beliefs, and knowledge in the broader healthcare framework. In line with this finding from the study, Jones et al. (2019) also argued that healthcare workers must be empowered to confront healthcare system barriers. Adding to the latter, the responsibility of healthcare professionals is to advocate for Indigenous rights and dismantle inequitable systems. Moreover, the authors call for leadership roles in shaping public policy and advocating for the integration of these competencies into the educational curricula (Jones et al., 2019).

Furthermore, this advocacy emphasises the importance of cultural sensitivity and the integration of local healing practices within Western healthcare systems. This finding is supported by Shiwarti et al. (2023), who argued that people's culture and religion can affect how they see healthcare and healthcare providers. Advocacy is to provide care that respects patients' cultures and beliefs, as well as understanding patients' backgrounds and local practices.

Ways of Knowing to Facilitate Indigenous Education in Nursing

This part of the discussion recommends ways of knowing to facilitate learning of Indigenous education in nursing through educational workshops. The self-explanatory diagram of Ways of knowing… is illustrated below:

Figure 1. Illustrating the pillars of ways knowing of indigenous education in nursing

Alam, M. S., & Pienaar, A. J. (2023). Ways of Knowing: Four Pillars of Indigenous Education in Nursing. [Diagram] Unpublished.

As stated in the workshop, the first step was an exploration of what was known about the concept to the participants. However, it must be noted that the participants' uncertainty should not be mistaken for a lack of knowledge but rather as an aspect of viewing the concepts as a whole (holism) and interchanging it. During the facilitation, this should be clarified to strengthen the pillar of the theory-practice integration of Indigenous nursing education (*Pillar One*).

Most importantly, participants should be aware that, through their upbringing, they acquired knowledge and understanding about their local ways of knowing. They should be allowed to disclose it through the initial fish-bowl techniques.

Step two highlights the other pillars that emerged through the findings as a facilitation guideline. These pillars focus on the virtues of respecting the culture and preferred care of the patient (*Pillar Two*), hence providing inclusive care based on cultural, physical and psychological aspects (*Pillar Three*) and advocating for both Indigenous and Western health care to co-live (*Pillar Four*) in the health sector of Pakistan.

This discussion emphasises the holistic essence of Indigenous knowledge, the patient-centric healthcare approach within Indigenous communities, the importance of well-being in cultural, physical, and psychological dimensions, and the significant role of healthcare professionals in advocating for the acknowledgement of Indigenous health systems.

RECOMMENDATIONS TO SUSTAIN CULTURALLY APPROPRIATE NURSING IN INDIGENOUS COMMUNITIES

Practice

It is evident from the study that Indigenous health systems exist in our context as well as in the rest of the countries. Therefore, it is recommended that healthcare systems in Pakistan should recognize and incorporate Indigenous knowledge, education, and practices into their nursing and medical frameworks. This should include understanding the holistic nature of these concepts and their significance in the context of local culture and beliefs.

Education

It is further recommended that continuous training initiatives and workshops tailored for nurses be created. These programs should emphasize hands-on experiences that expose nurses to Indigenous knowledge, diverse local practices, and cultural sensitivity in their care, along with the Western healthcare context. Additionally, given the technological advancements in the current era, virtual stimulatory workshops focusing on the significance and incorporation of Indigenous education in nursing could be developed to enhance nurses' understanding and ownership of this knowledge. This area also presents an intriguing avenue for future research. Moreover, this is a strong recommendation from the participants to consider Indigenous education as a noteworthy addition to the nursing curriculum and clinical education. These additions should include the incorporation of local knowledge, practices, and approaches that align with the evolving healthcare landscape and patient needs.

Research and Evidence-Based Decision-Making

Furthermore, based on the findings of this study, it is recommended to conduct further research on local healthcare practices to assess their justifiable evidence for healthcare practitioners to include in their daily practices. That research should incorporate continued learning and innovation of Indigenous knowledge in healthcare, emphasizing the importance of evidence-based care. Additionally, there is a need for further evidence-based research to fill the gaps in our understanding of local practices. Healthcare decision-making should be guided by an evidence-based approach to ensure the best possible patient outcomes.

LIMITATIONS AND STRENGTHS OF THE RESEARCH

The study may not offer specific guidance on how to implement the integration of Indigenous education within nursing education and clinical settings. Hence, the study's recommendations may need to be adapted to various healthcare contexts and may not be universally applicable, as local healthcare practices and education systems vary across regions. Furthermore, a notable limitation of this study lies in its confinement to only two settings and the study grapples with a significant time constraint for research.

On the other hand, this study explores local healthcare practices, emphasizing the need for evidence-based care and ongoing learning and innovation in healthcare. The study offers practical recommendations, such as incorporating evidence-based approaches into healthcare decision-making and bridging the gap between local and Western healthcare systems. Moreover, the study's findings are particularly relevant to nursing education, as they suggest the advocacy for the inclusion of Indigenous knowledge and practices in the nursing curriculum and clinical education to enhance patient care by encouraging a more evidence-based and holistic approach to healthcare.

FURTHER RESEARCH DIRECTIONS

1. Development of Interventions to Promote the Effectiveness of Indigenous Education Integration in Nursing Education

Future research should focus on designing, implementing, and evaluating targeted interventions that facilitate the effective integration of Indigenous education into nursing curricula. These interventions may include culturally sensitive teaching methodologies, workshops, and modules that highlight Indigenous perspectives on health and wellness. Additionally, studies could explore the impact of such interventions on student outcomes, including cultural competence, empathy, and the ability to provide equitable care to Indigenous populations.

Potential areas of investigation include:

- Identifying best practices for incorporating Indigenous knowledge and practices into existing nursing education frameworks.
- Make developing interactive tools and resources, such as case studies or digital platforms, that enhance students' understanding of Indigenous cultures.
- Conducting longitudinal studies to assess the sustained impact of these interventions on nursing graduates' performance in real-world healthcare settings.
- Exploring the perspectives of Indigenous community members to co-create educational content that aligns with their values and traditions.

2. Conceptualizing a Policy Brief Through Participative Educational Research

A critical avenue for future research involves the development of policy briefs aimed at guiding the integration of Indigenous education into nursing curricula. Using participative educational research methods, stakeholders such as educators, nursing students, policymakers, and Indigenous community leaders can collaborate to conceptualize actionable policy frameworks. This approach highlights inclusivity and ensures that the resulting policies are inform by diverse perspectives and grounded in practical realities. Key areas to explore include:

- Engaging Indigenous knowledge holders and nursing educators in co-developing policy recommendations.
- Examining existing policies on cultural education and identifying gaps specific to Indigenous representation in nursing education.
- Proposing policy solutions that address structural barriers, such as lack of funding, limited access to Indigenous educational resources, or institutional resistance.
- Piloting policy recommendations at regional levels to refine and evaluate their effectiveness before scaling to broader contexts.

CONCLUSION

In conclusion, it became clear through this research that Indigenous education in nursing is an important transformation, even in Pakistan. What is unique is the fact that the participants were uncertain about the compartmentalized concepts, but they verbalized the core concepts for Indigenous education in nursing interchangeably and holistically. Furthermore, implementing substantial changes in nursing education and clinical practice may face resistance from established Western systems and individuals who are hesitant to adapt to that neglected inherited approach.

REFERENCES

Aziz, M. A., Adnan, M., Khan, A. H., Shahat, A. A., Al-said, M. S., & Ullah, R. (2018). *Traditional uses of medicinal plants practiced by the Indigenous communities at Mohmand Agency*. FATA., DOI: 10.1186/s13002-017-0204-5

Berenson, E. (2017). Ethical Principles Of Psychologists And Code Of Conduct. American Psychological Association, 1–21. DOI: 10.5430/jnep.v7n1p11

Blackstock, S. Y. (2016). Shifting the academic lens: Development of an interdisciplinary Indigenous health nursing course. *Journal of Nursing Education and Practice*, 7(1). Advance online publication. DOI: 10.5430/jnep.v7n1p11

Bourque Bearskin, R. L., Cameron, B. L., King, M., & Weber Pillwax, C. (2016). Mâmawoh Kamâtowin, "Coming together to help each other in wellness": Honouring Indigenous Nursing Knowledge. *International Journal of Indigenous Health*, 11(1), 18–33. DOI: 10.18357/ijih111201615024

Browne Gott, H. (2020). Exploring the relationship between culture and well-being. In Whaleves, Governemnt Social Research. https://gov.wales/exploring-relationship-between-culture-and-well-being

Creswell, W. J., & Creswell, J. D. (2022). Research Design Qualitative, Quantitative, and Mixed Methods Approaches. In Paper Knowledge. Toward a Media History of Documents (Sixth Edit). SAGE Publications, Inc. https://lccn.loc.gov/2017044644

Dhanabhakyam, & Sarath M. (2023). Psychological Wellbeing: Asystematic Literature Review. International Journal of Advanced Research in Science, Communication and Technology, April, 603–607. DOI: 10.48175/IJARSCT-8345

Drummond, A. (2020). Embodied Indigenous knowledges protecting and privileging Indigenous peoples' ways of knowing, being and doing in undergraduate nursing education. *Australian Journal of Indigenous Education*, 49(2), 127–134. DOI: 10.1017/jie.2020.16

Etikan, I. (2017). Sampling and Sampling Methods. *Biometrics & Biostatistics International Journal*, 5(6), 215–217. DOI: 10.15406/bbij.2017.05.00149

Gainsford, A., & Evans, M. (2021). Integrating andragogical philosophy with Indigenous teaching and learning. *Management Learning*, 52(5), 559–580. DOI: 10.1177/1350507620972528

Gartner, J. B., Abasse, K. S., Bergeron, F., Landa, P., Lemaire, C., & Côté, A. (2022). Definition and conceptualization of the patient-centered care pathway, a proposed integrative framework for consensus: A Concept analysis and systematic review. *BMC Health Services Research*, 22(1), 1–24. DOI: 10.1186/s12913-022-07960-0 PMID: 35473632

Inmark. (2010a). Concept and methodology of Interactive Workshops. PROmotion of an ICT Dialogue between Europe and America Latina – Extension towards Mexico, Colombia, Cuba, Costa Rica, 1–20. http://cordis.europa.eu/docs/projects/cnect/4/246644

Jones, R., Crowshoe, L., Reid, P., Calam, B., Curtis, E., Green, M., Huria, T., Jacklin, K., Kamaka, M., Lacey, C., Milroy, J., Paul, D., Pitama, S., Walker, L., Webb, G., & Ewen, S. (2019). Educating for Indigenous Health Equity: An International Consensus Statement. *Academic Medicine*, 94(4), 512–519. DOI: 10.1097/ACM.0000000000002476 PMID: 30277958

Jones, R., Crowshoe, L., Reid, P., Calam, B., Curtis, E., Green, M., Huria, T., Jacklin, K., Kamaka, M., Lacey, C., Milroy, J., Paul, D., Pitama, S., Walker, L., Webb, G., & Ewen, S. (2019). Educating for Indigenous Health Equity: An International Consensus Statement. In Academic Medicine (Vol. 94, Issue 4). https://doi.org/DOI: 10.1097/ACM.0000

Kennedy, A., Bourque, D. H., Bourque, D. E., Cardinal, S., & Bourque Bearskin, R. L. (2021). Reconciling Taking the "Indian" out of the Nurse. Quality Advancement in Nursing Education - Avancées En Formation Infirmière, 7(1). DOI: 10.17483/2368-6669.1276

LaDonna, K. A., Artino, A. R.Jr, & Balmer, D. F. (2021). Beyond the Guise of Saturation: Rigor and Qualitative Interview Data. *Journal of Graduate Medical Education*, 13(5), 607–611. DOI: 10.4300/JGME-D-21-00752.1 PMID: 34721785

Lincoln, Y. S., & Guba, E. G. (1986). But Is It Rigorous? Trustworthiness and Authenticity in Naturalistic Evaluation. *New Directions for Program Evaluation*, 30(30), 73–84. https://doi.org/https://doi.org/10.1002/ev.1427. DOI: 10.1002/ev.1427

Lutomia, A. N., & Bravo, J. B. (2017). Communities of Practice and Indigenous Knowledge: A Case Study of Empowering Women in Processing Shea Butter Using Scientific Animations. In Handbook of Research on Social, Cultural, and Educational Considerations of Indigenous Knowledge in Developing Countries (pp. 252-270). United State America: Chigago DOI: 10.4018/978-1-5225-0838-0.ch012

Madden, B. (2015). Pedagogical pathways for Indigenous education with / in teacher education. *Teaching and Teacher Education*, 51, 1–15. DOI: 10.1016/j.tate.2015.05.005

Matthews, R. (2017). The cultural erosion of Indigenous people in health care. *Canadian Medical Association Journal*, 189(2), E78–E79. DOI: 10.1503/cmaj.160167 PMID: 27620632

McCleland, A. (2011). Culturally Safe Nursing Research: Exploring the Use of an Indigenous Research Methodology From an Indigenous Researcher's Perspective. *Journal of Transcultural Nursing*, 22(4), 362–367. DOI: 10.1177/1043659611414141 PMID: 21807959

Mogawane, M. A., Mothiba, T. M., & Malema, R. N. (2015). Indigenous practices of pregnant women at Dilokong hospital in Limpopo province, South Africa. *Curationis*, 38(2), 1553. DOI: 10.4102/curationis.v38i2.1553 PMID: 26842096

Narasimhan, S., & Chandanabhumma, P. P. (2021). A Scoping Review of Decolonization in Indigenous-Focused Health Education and Behavior Research. *Health Education & Behavior*, 48(3), 306–319. DOI: 10.1177/10901981211010095 PMID: 34080471

Ngulube, P., Dube, L., & Mhlongo, M. (2015). Towards a cartography of Indigenous knowledge systems in library and information science training and education in Anglophone eastern and southern Africa. *Indilinga*, 14(2), 145–168.

Ngunyulu, R. N., Sepeng, N., Moeta, M., Gambu, S., Mulaudzi, F. M., & Peu, M. D. (2020). The perspectives of nursing students regarding the incorporation of African traditional Indigenous knowledge in the curriculum. *African Journal of Primary Health Care & Family Medicine*, 12(1), 1–8. DOI: 10.4102/phcfm.v12i1.2171 PMID: 32370527

Ogbo, C. V., & Ndubisi, E. J. O. (2017). African Indigenous Education in Contemporary Society. *African Indigenous Knowledge Systems: Problems and Perspectives*, 4, 40–53.

Oishi, S., & Gilbert, E. A. (2016). Current and future directions in culture and happiness research. Current Opinion in Psychology, 8(August 2010), 54–58. DOI: 10.1016/j.copsyc.2015.10.005

Ørngreen, R., & Levinsen, K. (2017). Workshops as a research method. *Electronic Journal of e-Learning*, 15(1), 70–81. https://eric.ed.gov/?id=EJ1140102

Owan, V. J., Owan, M. V., & Ekereke, A. B. (2023). Students' Entrepreneurship Development Through Indigenous Palm Wine Production and Alcohol Distillation: The Case of Yala-Obubra Community. In *Digital Preservation and Documentation of Global Indigenous Knowledge Systems* (pp. 75–101). IGI Global Publisher. DOI: 10.4018/978-1-6684-7024-4.ch003

Pienaar, A. J. (2017). Learning and Asserting an African Indigenous Health Research Framework. In Ngulube, P. (Ed.), *Handbook of Research on Theoretical Perspectives on Indigenous Knowledge Systems in Developing Countries* (pp. 85–99). IGI Global Scientific Publishing., DOI: 10.4018/978-1-5225-0833-5.ch004

Rendle, K. A., Abramson, C. M., Garrett, S. B., Halley, M. C., & Dohan, D. (2019). Beyond exploratory: A tailored framework for designing and assessing qualitative health research. *BMJ Open*, 9(8), e030123. Advance online publication. DOI: 10.1136/bmjopen-2019-030123 PMID: 31462482

Shahzeydi, A., Taleghani, F., Moghimian, M., Farzi, S., Yazdannik, A., & Farzi, K. (2022). Exploring nursing faculty, managers, newly graduated nurses, and students' experiences of nursing internship program implementation in Iran: A descriptive qualitative study. *BMC Nursing*, 21(1), 1–8. DOI: 10.1186/s12912-022-01159-8 PMID: 36572907

Stansfield, D., & Browne, A. J. (2013). The relevance of Indigenous knowledge for nursing curriculum. *International Journal of Nursing Education Scholarship*, 10(1), 1–9. DOI: 10.1515/ijnes-2012-0041 PMID: 23813335

Swihart, D. L., Yarrarapu, S. N. S., & Martin, R. L. (2023). Cultural Religious Competence in Clinical Practice. [Updated 2023 Jul 24] In *StatPearls* [Internet]. StatPearls.

Thurston, J. M., & Mashford-Pringle, A. (2015). Nursing & Indigenous education integration. *Journal of Nursing Education and Practice*, 5(10). Advance online publication. DOI: 10.5430/jnep.v5n10p9

Van Bewer, V., Woodgate, R. L., Martin, D., & Deer, F. (2020). The importance and promise of integrating Indigenous knowledges in nursing education. Witness. *The Canadian Journal of Critical Nursing Discourse*, 2(1), 11–24. DOI: 10.25071/2291-5796.46

Wilson, D., & Barton, P. (2012). Indigenous hospital experiences: A New Zealand case study. *Journal of Clinical Nursing*, 21(15–16), 2316–2326. DOI: 10.1111/j.1365-2702.2011.04042.x PMID: 22642700

Yabarmase, D. (2013). The Fishbowl Strategy: An Effective Way to Improves Students' Speaking Ability. *Indonesian Journal of English Language Teaching*, 9(2), 137–144. 296261880

FURTHER READINGS

AlMehrzi, A. A., Tipu, S. A., & Sarker, A. E. (2024). Determinants, processes, and impacts of Indigenous entrepreneurship: A systematic literature review. *Journal of Enterprising Communities: People and Places in the Global Economy*, 18(3), 646–681. DOI: 10.1108/JEC-10-2022-0155

Ife, J., Mehta, R., & Tusasiirwe, S. (2023). *Re-imagining social work: towards creative practice*. Cambridge University Press. DOI: 10.1017/9781108394949

Moyo, M. (2021). *Teaching our ways of knowing: A cultural imperative for integrating Indigenous languages in Zimbabwe's school curriculum*. Midlands State University Press and Publications.

APPENDIX

A. Informed Consent

I am Muhammad Salman Alam, a student in MSN final year at Shifa College of Nursing, Islamabad. I am doing my research project as partial fulfilment of my degree program. The purpose of this research was to explore the understanding of nursing interns regarding indigenous education in nursing and to recommend ways to facilitate the learning of indigenous education in nursing through educational workshops. Your confidentiality will be maintained. Your name and identities will be masked and will not be shared with anyone at all stages.

Your participation in this research study is completely voluntary. There will be no penalty to you if you choose not to take part. You may choose not to answer specific questions or to stop participating at any time.

Agreement to Participate

Your participation in this study is completely voluntary. If you decide to leave the study at any time, you can leave without any penalty. Would you like to participate in this study?

Yes No

If Yes, Please provide the information:

Name: Signature: Date:

B. Workshop Protocol

Basic Information

Workshop Place: Workshop No:

Date: Time: Number of Participants:

Duration of Workshop:

Workshop program:

- **To begin with:** Welcome the participants, introduce the workshop topic and team members, and take informed consent. Address any queries of the participants.
- **Pre-workshop evaluation**: Fill the Google Form based on open-ended questions*.
- **Opening the workshop:** a short Video clip depicting an ideal and non-ideal way of indigenous education in nursing.
- **Lead the interactive session:** for idea exploration to acknowledge the understanding of participant. Use PowerPoint to facilitate interaction with "Fish Bowl" teaching and learning strategy.
- **Reach the saturation of the responses:** close the discussion when there is no new information is coming out.
- **Post-workshop evaluation**: Fill the Google form based on same open-ended questions.
- **Refreshment**
- **Reflective note:** Request the participants to write a reflection on the workshop.
- **Closing remarks**: Thank the participants for their time and answer any remaining questions.

*Open-ended questions

1. What is the meaning of indigenous education in nursing?

   Other Questions (After a short Video clip depicting an ideal and non-ideal way of indigenous education in nursing)

2. What did you see in the video?

Probing Question in the workshop

a) What is your understanding after watching?
b) Explain the importance for nurses to learn about indigenous education in nursing?
c) After the video: How would you differentiate between indigenous practices, indigenous education, and indigenous knowledge in nursing?

3. How did you experience this workshop regarding the learning of indigenous education in nursing?
4. How can you improve this workshop to learn indigenous education in nursing during your BSN program?

C. Interview Guide

Setting:
Participant's Name:
Interviewer Name:
Demographic Information
Gender of the Participant: Male ☐ Female ☐
Age in years:
Education: Uneducated ☐ Primary ☐ Secondary ☐ Matric ☐ ≥ Intermediate ☐
Perception/ Experiences:

1. Opening Question:

Mr./Ms. [NAME], tell me about yourself, including your age, education, and your social life, etc.

2. Content Questions:

Opening Question 1: what is your understanding of local health practices?

Probing:

Can you explain more?

Opening Question 2. What is your understanding else about local health practices in Pakistan? What is the scope of that? Would people appreciate it or not?

Probing.

Tell me a bit more
Tell me about your local practices?

Closing Question:

Anything else that you want to ask me or share with me?
Thank you very much for your time and for answering questions with concentration.

D. Script for the Audio-Video Vignette Workshop

Greetings, ladies, and gentlemen

Following is an illustration of an indigenous health practice that occurred in Pakistan.

This scenario took place in a tertiary care hospital where a patient was admitted to the ICU under strict observation.

The patient was a child who was in critical condition and required strict compliance with medical and nursing care.

Considering his condition, he required one nurse to take care of him at one time according to ICU protocol. Therefore, I was assigned as his bedside nurse.

His attendants were very sad and worried about his condition and often used to visit him one after another, praying for his well-being.

One day while I was taking care of this patient in the hospital, one of his attendants came to me and requested to bring a black goat to the hospital for the child to put his hand on and afterward, they could Sacrifice this goat.

A Goat

Yes! I was astonished and shocked by this prospect as it was unheard of

Additionally, it was against the hospital standards and infection control precautions to bring an animal into the hospital and endanger all the patients for contamination.

However, the patient's attendant kept insisting, and they wouldn't let go of this proposition for the well-being of the child, hence, we were forced to look for a solution

It was obvious that we could not allow the animal to live in the hospital, so we proposed that the goat should be slaughtered first and the meat be brought in double packing for the patient to be touched and prayed upon and taken back for the completion of their sacred

This is a brief depiction of the indigenous health practices that emerge from time to time in our culture for us to make improvisations in the Western adopted practices.

Thank you!

Chapter 20 Transforming Research in Postgraduate Education: An African Indigenous Perspective

Mishack Thiza Gumbo
https://orcid.org/0000-0001-6760-4341
University of South Africa, South Africa

ABSTRACT

This literature study aims to interrogate the Western approach to postgraduate education and research from an Indigenous perspective to advance a postcolonial Indigenous paradigm with its transformation goal. Postgraduate education heavily depends on Western paradigms and theories to the detriment of indigenous alternatives. There is a need to transform research at this level to add to the Africanisation of the curriculum project. In this chapter, three critical aspects are up for discussion: the need to transform the institutional culture, supervisors who should supervise as change agents, and understanding of the worldview of indigenous students in the supervision. The discussions in the chapter offer insights into the transformation of the academic project at this level, which culminates in the contribution of an Indigenisation of the Postgraduate Research Framework.

INTRODUCTION

This chapter developed from numerous series of seminars that I presented during the master's and doctoral seminars at South African universities. Traditionally, master's and doctoral students' seminars or workshops present research from Western paradigms and lenses, making it difficult for African students to understand and relate research to their Indigenous contexts (Gumbo, Gasa & Knaus, 2022). The institutional culture, academic programmes, and supervisors are deeply immersed in the Western culture (Chilisa, 2012; Smith, 2012), thus disregarding the worldviews of Indigenous students. This moulds a research orientation (Datta, 2018), resulting in the non-contribution of education and research toward Indigenous epistemological sustainability and cognitive injustice. Research-intensive degrees, i.e. postgraduate degrees, follow Western paradigms and methodologies. Thus, this chapter interrogates the Western approach to postgraduate education and research to advance a postcolonial Indigenous par-

DOI: 10.4018/979-8-3693-8447-3.ch020

adigm with its transformational goal – transforming the practice will give equal footing and recognition for Indigenous and Western knowledge systems.

Indigenous people are original inhabitants of a certain geographical area of non-Indigenous peoples. In other contexts, they are called Native people, First Nations, Aborigines, and Subalterns. Owuor (2007) describes Indigenous people according to their approach to education, which is based on knowledge transmission involving progression through age groupings, seniority, and wisdom of Elders. An Elder is a "person older in age, who is a respected keeper of cultural knowledge and traditional ceremonies" (Quinn, 2013, p. 7). Supporting the description of Indigenous people above, my research team visited Dr Esther Mahlangu's Foundation in 2022. We observed education and training in Indigenous art as elderly women taught younger women how to weave, decorate, knit, etc. This sets the tone for the transformation of education and research, i.e. to recognise the contribution of Indigenous knowledge systems in postgraduate education alongside Western ways of knowing.

Recognising Indigenous epistemologies in postgraduate education and research will transform the institutional culture in higher education down to the supervision of postgraduate students, thereby making it relevant to the students' cultural contexts. Indigenous knowledge systems, such as community as an educational institution, technological and scientific practices, art and music can frame students' research. There are also multiple benefits of braiding Indigenous and Western knowledge systems: equity of accorded to both knowledge systems; building trust in research findings; bridging multiple ways of knowing; recognising Botho/Ubuntu by researching with empathy and resonance; engaging in participatory research (Keane & Khupe, 2017; Levac et al., 2018; Mehltretter et al., 2023; Photo & Gumbo, 2024). However, institutions and supervisors are the critical gatekeepers of education and research, designing programmes from Western paradigms and lenses to the disfavour of Indigenous ones. It will, therefore, take institutions to dismantle the colonial powers to nurture a transformative culture that can accommodate Indigenous knowledge and protect and support students who want to pursue postgraduate education in Indigenous knowledge systems. Supervisors should be agents of transformation by promoting research in this field and be prepared to learn along with their students. In this way, students will be more interested in doing their studies in the field. This chapter discusses these three aspects, i.e. institutional transformation, supervisors as agents of change, and Indigenous students' worldview. However, there is a need for a conceptual description of Africanisation, colonisation, decolonisation and Indigenous epistemologies before delving into these aspects. These related concepts are essential as the discussion in this chapter is premised on them.

AFRICANISATION

Africanisation is about making Africa the starting point for education so that African students are not taught away from their cultures and communities but that they become graduates who can contextualise their knowledge and practices within African ethos (Gumbo, 2020a). African students' quest to learn and research Indigenous knowledge systems should be affirmed and promoted. Africanisation is a conscious and deliberate assertion to be an African (Ramose, 1998). African students should bring their identity into research; they should be supported to research as Indigenous scholars who are primarily passionate about advancing Indigenous knowledge systems rather than perpetrating Western knowledge systems. They should be aware that Africanisation aims to experience the true humanity in an African conceptualisation of Botho (Mashabela, 2017). African philosophy is inherently inclusive (Msila, 2017), thus suggesting blending Indigenous and Western knowledge systems instead of advancing the latter. These descriptions of Africanisation are critical in this chapter; African students should position themselves as African scholars to centre their cultural knowledge and exercise their rights by immersing themselves and their research activities within Botho.

COLONISATION

Colonisation, related to Africa, was the Europeans' invasion of and exertion of power on the continent's people, the role-players being Spain, Portugal, Britain and France, among others. In this light, colonisation is about a political and economic relation in which a nation's sovereignty rests on another nation's power, which makes such a nation an empire (Borocz, 2012; Mignolo, 2005a; 2005b). In its historical context, colonisation is a process that culminated in the invasion, conquest and direct administration of Africa (Ndlovu-Gatsheni, 2013; 2016). Borocz (2012) (in Gumbo 2023) lists activities of colonialism – oppressive acts, resource extraction, extractive enterprises, destruction of Indigenous industries, the establishment of modernisationist projects, the enslavement of the oppressed, creation of colonial elites, the massification of education systems, dehumanisation of colonised people, perpetration of inter-colony wars, prevention of colonial subjects to Europe. The two generations of colonisation included conquering physical spaces and bodies of colonised people and oppressing their minds through education, science, economics, law and other means (Le Grange, 2016). In this sense, colonisation robbed Africa of its sovereignty, self-respect, freedom and power, parading it as non-existent in history; it is a symbol of 'castration' that assigned a negative ontological and cognitive status to everything African and Indigenous (Odora Hoppers, 2001). Hence, postgraduate research could immensely contribute toward the decolonisation agenda by freeing Indigenous and non-Indigenous students' minds as they research the issues outlined in these descriptions of colonisation. For example, in Technology Education as a discipline, students could critique designs of colonial establishments and braid them with Indigenous alternatives.

DECOLONISATION

Decolonisation is resistance to colonisation. Self-determination confronts colonisation due to its dehumanising function (Gumbo, 2020a). Decolonisation demands centring Indigenous knowledge (Sium, Desai & Ritskes, 2012). From a research perspective, decolonisation is "a process for conducting research with Indigenous communities that places Indigenous voices and epistemologies in the center of the research process" (Simonds & Christopher, 2013, p. 2185). Decolonisation is the contestation for power that must necessarily push back against colonisation, threatening Indigenous ways of being (Sium et al., 2012). It is a theory of self-ownership that is relational and codified in innate human rights, abilities and arguments (Gumbo, 2020a) reinforced by critical theory (Ndlovu-Gatsheni, 2013). According to Mbembe (2015), decolonisation demythologises whiteness and history; it corrects the myth that whiteness is superior, and indigeneity is inferior and primitive. Smith (1999; 2012) declares that Indigenous people live by a decolonial story; they continue to live under the colonial matrix to a more significant extent (Grosfoguel, 2007). Hence, their story is premised on knowledge as power, as Borocz (2020) exposits, it will free the colonised people. Hence, decolonisation can also be viewed as a resistance against the injustices inflicted by colonialists on Indigenous people (Gumbo, 2020b).). Practically, this was witnessed in South

African higher education through student uprisings about 10 years ago. The #FeesMustFall movement was a clarion call to decolonise the curriculum. They not only demanded free education but profusely rose against unbefitting Western paradigms, theories and methods of Indigenous cultures. Research in postgraduate education can ensure cognitive justice that will resist colonialism.

UNDERSTANDING THE CULTURAL WORLDVIEW OF STUDENTS

Understanding the cultural worldview of students is a starting point worth considering so that supervisors do not impose their worldviews but are prepared to learn along with their students. The Biblical statement is relevant in this context: "Do two walk together unless they have agreed to do so?" (Amos 3:3). The introductory meeting between the supervisor and the student is not just about understanding the idea the student intends to investigate. It is even more than signing an agreement; the supervisor should first understand the student and their cultural background to, in turn, understand the student's proposed research idea. Investing the supervision journey in rapport will ensure good working relationships – an effort must be made to mount the supervisor-student working relationship on Botho.

Several studies (Anderson et al., 2022; Azure, 2016; Cekiso et al., 2019; Martin, 2019; Singh & Major, 2017; Uink, Day & Desmoulins, 2020) have identified a recurrence of supervisor-student working relationships as one of the key factors affecting the success of supervision. In their published article, Singh and Major (2017) reflect on their experiences as Indigenous student and non-Indigenous supervisor. The experiences hang on using the Western theoretical framework of poststructuralism and Indigenous methodologies. They addressed tensions that surfaced during the research design process. As part of an introduction in the article, they introduced themselves. Major declares that she remained critically aware of her privileged status as a non-Indigenous supervisor. Her university study, teaching, travel, and diverse friendships made her learn about culture and identity, and she developed a strong sense of social justice and commitment to equity in her work. This made her adopt the poststructuralist paradigm in her work – "poststructuralism rejects totalizing and essentialist assumptions about the world and reality and questions the aim of research to develop grand theories or metanarratives that seek to explain everything in a complete or universal way" (Singh & Major, 2017, p. 8). Thus, Major viewed her supervision of Singh's study as an opportunity to support and work with an Indigenous student and to use Indigenous methodologies and approaches in research (Singh & Major, 2017).

Singh's study dealt with the concepts of social justice, which attracted the cultural appropriateness to her research. Initially, Singh felt comfortable with Western research traditions' epistemological positions and practices due to her experiences as a learner and teacher in the mainstream Australian education system. However, her experiences in Aboriginal education created a passion for working within cultural and ethical guidelines in conducting her research. Marginalisation of Indigenous Australians because of colonial injustices was the force behind her passion. The authors declared, "we found many points of connection in our supervision relationship, especially our shared ontological stance in relation to identity" (Singh & Major, 2017, p. 7). Singh and Major's (2017) experiences are a compelling case for transforming postgraduate education, especially concerning the colonial undertones in science.

COLONIAL UNDERTONES IN SCIENCE: A REASON FOR TRANSFORMING RESEARCH

In this section, I treat science and knowledge as pillars of understanding students' worldviews through Indigenous epistemologies. The section is divided into two sub-sections, i.e. the historical origins and function of science and Indigenous knowledge (science). The section heavily relies on Aikenhead and Ogawa (2007). This is because "the historical origins and evolution of science within Euro-American cultures naturally cause its practitioners (today's scientists) to embrace certain fundamental worldviews, epistemologies, ideologies, and values, all related to science's origin and evolution" (Aikenhead & Ogawa, 2007, p. 541). This evolution of science ultimately shapes how research is conducted in universities and how postgraduate students are socialised into Western notions of research.

The Historical Origins and Function of Science

In European terms, science's evolution introduced the concept of natural philosophy, which was adopted as a social institution when the Royal Society was founded in 1662 (Aikenhead & Ogawa, 2007). Natural philosophy introduced the social transformation of exercising dominion over nature with the resultant entrepreneurial methodological approach that introduced the Industrial Revolution. The development of the concept was shaped by engineering the socio-economic world to serve the Westerners' interests. Hence, natural philosophy was purposefully and politically renamed science with a drive to professionalise it "during the birth of a new organization in 1831, the British Association for the Advancement of Science" (Aikenhead & Ogawa, 2007, p. 542). Today, the Anglo world privileges an operational meaning of (Eurocentric) science taught in universities (Aikenhead & Ogawa, 2007). Hence, modern science has always been viewed as Western thinking – the strategic training of people by universities to be employed in business, industry, the military, government agencies, and so forth, thus exerting power and privilege associated with research and development (Aikenhead & Ogawa, 2007).

The professional, scientific communities are the order of the day. They are sponsored to do specific strategic projects pre-designed and determined by the funding agencies to ensure dominance aligned with industrial advancement. In this light, Eurocentric science advances Western culture with its related norms, values, beliefs, expectations and conventional actions (Aikenhead & Ogawa, 2007; Datta, 2018). Membership in these professional organisations counts more than methodology and content and characterises an operational legitimacy of Eurocentric sciences (Aikenhead & Ogawa, 2007; Miller et al., 2023). Academic citizenship takes precedence as academics are required to sign up for membership in scientific communities and apply for funding for their research projects with tight conditions of accountability to the funders. It is important to note that "knowledge within a Eurocentric worldview is … separate from the knower" (Aikenhead & Ogawa, 2007, p. 553). This colonial conceptualisation of science denies the world its Indigenous perspectives, calling for researchers' attention to research the issue for transformational purposes.

Indigenous Knowledge (Science)

Indigenous knowledge provides a different perspective to Eurocentric science suggesting the need to bring it into the research debate. It "benchmarks the limitations of Eurocentric theory – its methodology, evidence, and conclusions – reconceptualizes the resilience and self-reliance of Indigenous peoples, and underscores the importance of their philosophies, heritages, and educational processes" (Battiste, 2002, p. 5). Indigenous knowledge and the knower are inseparable. Knowledge is closely tied to the pursuit of wisdom (wisdom-in-action), which considers living harmoniously with nature to ensure the community's survival (Aikenhead & Ogawa, 2007; Gumbo, 2020a). Coming to know differs from the Eurocentric science process of knowing (i.e. discovering). Researchers always claim to have discovered, thus being selfish in acknowledging that they have learned from the participants. One cannot discover, only learn (Datta, 2018). "Indigenous coming to know is a journey toward wisdom . . . , not a destination of discovering knowledge" (Aikenhead & Ogawa, 2007, p. 553). It is experiential, participatory, process-oriented and spiritual; thus, ways of living in nature are actionable (Aikenhead & Ogawa, 2007). "Indigenous ways of living in nature are communicated and learned in the oral tradition through modelling the practices of others, listening to stories, singing songs, reciting prayers, dancing at celebrations, and participating in spiritual ceremonies; all of which are passed from generation to generation" (Aikenhead & Ogawa, 2007, p. 554). Elders perform an important educational role.

Indigenous epistemologies are characterised by an interconnection between the spiritual and physical worlds and their implication on science (Cajete, 2000; Hart, 2010). Knowledge is, therefore, fluid as it is passed from generation to generation (primarily through storytelling). "It cannot be given, accumulated, banked, and assessed by paper and pencil examinations; reading an article is not an adequate experience to learn how to live in nature" (Aikenhead & Ogawa, 2007). Instead, perceptual experiences take precedence through the connection between exteroception and enteroception (i.e., environmental knowledge as it informs one's body). According to Aikenhead and Ogawa (2007, p. 565), "Indigenous ways of knowing share this structure: (1) knowledge of and belief in unseen powers in the ecosystem; (2) knowledge that all things in the ecosystem are dependent on each other; (3) knowledge that reality is structured according to most of the linguistic concepts by which Indigenous people describe it; (4) knowledge that personal relationships reinforce the bond between persons, communities, and ecosystems; (5) knowledge that sacred traditions and persons who know these traditions are given responsibility for this specialised knowledge and its dissemination; and (6) knowledge that an extended kinship passes on teachings and social practices from generation to generation. This leads to mapping characterising Eurocentric science and Indigenous knowledge, as in Table 1. I also offer a braided view of the two, leading to the transformative paradigm in the next section.

Table 1. Characteristics of Eurocentric science and Indigenous knowledge (information from the first two columns was adapted from Aikenhead & Ogawa 2007)

Eurocentric science	Indigenous knowledge (coming to know)	Integrated Eurocentric and Indigenous views
Nature is knowable – conclusive generalisations and mechanistic explanations; eradication of mystery is the key intellectual goal.	The mystery of nature is dynamic and unfolds endlessly; thus, mystery and living harmoniously with it must be celebrated spiritually and physically.	Nature continues to be a wonder. Being enthusiastic about knowing it up to its mysterious intricacies deepens the knower's understanding.
Social goals include curiosity, gaining peer credibility, financial remuneration, competition over research grants, fame and fortune.	All deserve to develop and be celebrated as community members. Remuneration and fame are the honour of the community, not only individuals.	We are better fulfilled and celebrated as community members, balancing an individual's goals with the collective's goals in how one approaches research.
Predictive validity through experiments to test hypotheses.	Systematic research sustained over many generations through holism and spiritual power (dreams, visions, intuitions, etc.) validates the data, especially by content validity, time and the generations' survival.	Interpretive research extends the scope of positivist research, which is laboratory dependent. Data validation can be enhanced by combining these two.
Uniformitarianism – scientific constructs must be applied consistently through time and space.	Constructs are dictated by time and space, considering the ever-changing world in which people live and learn.	Time and space are bounded by nature's dynamism, offering multiple realities that can expand the scope of enquiry.
Rectilinear time – mechanical clock; time quantified.	Cyclical time, as observed by the seasons and natural occurrences. Time is aligned to and interpreted with the unfolding of nature.	Expanding the concept of time to unfolding events and phenomena can transform time into a data-gathering tool, e.g., the events that define the birth of a child at a particular time.
Cartesian dualism – matter and mind are bifurcated, i.e., knowledge of matter is independent of human intuition, spiritual forces or divinities.	Everything in the universe is animate, including animals, plants, humans, rocks, etc. Nature is physical and spiritual; science, art, religion, philosophy or aesthetics, etc are connected.	People are a totality of the intellectual, physical, spiritual, psychic, religious, etc. Hence, their research is influenced by this totality; thus, as scientists, they may not divorce themselves from these elements.
Reductionism – understanding the whole through the structure and function of its parts through measurement of variables, conceptualisation and experimentation.	Holism – everything is relational and has a spirit and knowledge. Particulars can be understood about the whole; one should look at the whole to understand the parts.	Reductionism and holism provide powerful ways to view phenomena; parts can be better understood within the whole and vice versa.
Anthropocentrism – nature is viewed and treated as a servant to humankind, the exertion of power and dominion over nature.	Human beings were created to care after and respect nature; as much as they have dominion over nature, they are stewards thereof – they depend on it.	Human beings, nature and spirituality are intricately connected. Nature is, therefore, an identity, not an object. Man has destroyed nature by treating it as a servant, hence the current problems such as climate change. Solutions can be researched by respecting nature.
Quantification – materialistic reality consists of objective mathematical relationships; science strips an entity of its qualitative, human or spiritual attributes and depersonalises scientists and what they study.	The knower and the known are inextricably interconnected within the metaphysical.	Quantification-qualification – mixing these approaches (mixed methods) offers a deeper understanding of the phenomena because researchers bring their quantitative and qualitative knowledge into research.

continued on following page

Table 1. Continued

Eurocentric science	Indigenous knowledge (coming to know)	Integrated Eurocentric and Indigenous views
Realism – the reality is described as independent of the scientists' act of perceiving.	Scientists receive nature; they experience reality so much that they cannot dissociate themselves from it.	The scientist is part of reality (nature); he brings his/her profundities (intellect, spirit, culture, beliefs, etc) into perceiving reality.
Positivism – adherence to empirical and experimental methodologies (objective, value-free, universal, secure knowledge of nature) defined by logical procedures; one ideal, one norm, and one standard and regards all other knowledge systems and ways of knowing nature as inferior.	Indigenous knowledge is not a fixed body of data; knowledge is passed from generation to generation; individuals observe, compare their experiences with what their teachers have told them, conduct experiments to test the reliability of their knowledge, and exchange their findings with others.	Positivism-interpretivism – merging these approaches invokes multi-methodical approaches (complementary Indigenous and Western methods).

The preceding discussion suggests transforming research such that it is not skewed to Western paradigms but reconciles them with Indigenous paradigms. Students should be exposed to these two worlds of paradigms where the latter is respected just like the former.

TRANSFORMING RESEARCH: ADVOCATING A TWO-EYED PARADIGM

Decolonising research does not mean rejecting all Western methods and theories but giving equal credence to both to ensure mutual respect and social justice (Datta & Starlight, 2024). In light of Table 1, some Western methods may be adapted if found relevant and beneficial to the local community. The third right column in Table 1 suggests two-eyed seeing, which is about using the Indigenous eye (knowledge and ways) and the Western eye (knowledge and ways) and using both together for the benefit of all (Quinn, 2013; Reid et al. (2021). Table 1 illustrates this paradigm, i.e. Column 1 (Western eye) plus Column 2 (Indigenous eye) yield Column 3 (using both Western and Indigenous eyes). There is, therefore, a fervent need for transformative research paradigms, which are decolonial (Chilisa, 2012). Chilisa (2012) and Smith (1999; 2012) think the postcolonial Indigenous paradigm would better serve this transformation. It inspires multi-methods (e.g. orality, experiential, acting, dancing, music) from diverse knowledge systems (Chilisa, 2012; Matowanyika, 2016) (refer to Table 1). Such methods align with transformation, matching theory to non-written texts such as oral traditions, stories, legends, language, and artefacts (Chilisa, 2012).

Conventional methods, which heavily depend on a written text can be enriched with non-written ones; writing and orality are complementary (Quinn, 2022). Other compatible methods include digital storytelling, phenomenology, photovoice, and theatre scripts (Simonds & Christopher, 2013). Sparkes and Piercey (2015) share and describe other multiple frameworks to draw together Indigenous and Western approaches to knowledge creation, such as the Constellations model, Cultural Interface, Insurgent research, etc. In a project I am leading, we collect Indigenous stories and artefacts through pictures from our Indigenous co-researchers (knowledge holders). We view ourselves as learners who learn from them, especially the Elders.

A transformative paradigm advocates a collective engagement with a sense of commitment to others. In southern Africa, collective engagement stems from letsema, a Setswana/Southern Sotho/Nothern Sotho word for working together. According to Twala (2018), letsema relates to black Africans' collective approach to agricultural activities, who grew food crops for subsistence and as a significant socioeconomic activity. Letsema can be applied in research; supervisors usually connect their students with other students and scholars to engage in epistemological and research discussions. These role-players must show an interest in the student's success. Letsema is enshrined within the Botho framework, where values, ethics, culture, and principles guide the organisation of people and their tasks as they respect one another and are protected as a collective (see Table 1). Botho suggests a community-based participatory research (CBPR) model, which is more dialogical and egalitarian (Simonds & Christopher, 2013). Though CBPR was developed from the Western tradition (Simonds & Christopher, 2013), it is culturally sensitive and can be further adapted to suit Indigenous research – it must incorporate Indigenous epistemologies.

Being culturally sensitive calls upon researchers to adopt a non-intrusive treatment of the participants and data and reserve their judgment. They should listen deeply and embark on a reflective non-judgment (Hart, 2010). They should honour what is shared, keep to fidelity, and have a cognitive-emotional connection with the participants (Hart, 2010). This means exercising self-awareness and openness about one's subjectivity. Singh and Major's (2017) case, presented in the earlier parts of this chapter, is an example from which supervisors can learn. Even Indigenous supervisors can learn from this case.

Simonds and Christopher (2013) observe that Western science's postpositivist research tradition promotes grounded theory. Grounded theory is suitable for qualitative research as data is not forced into existing theory (Quinn, 2013). This implicates the extent to which Indigenous participants should be engaged in the research process. Western research paradigms tend to involve participants only during fieldwork activities because that is only when researchers think they need them. Indigenous research can extend this through the continuous engagement of Indigenous people as co-researchers, from the research design to data analysis (Quinn, 2022).

Indigenous knowledge and beliefs can help reshape Western ways of doing research. For instance, the traditional notion that researchers discover knowledge does not apply in Indigenous contexts. Researchers can only learn from Indigenous people instead of discovering them (Datta, 2018; Simonds & Christopher, 2013). They do not own the data (Quinn, 2022). Also, there is sacred knowledge that cannot be easily accessed or that must be treated with care – its access must be earned (Simonds & Christopher, 2013). A critical matter is that the research ethics policy should be flexible to accommodate Indigenous cultural ethics, which can clash with institutional policies. For instance, Quinn (2022), an (Indigenous) researcher, had to offer sacred tobacco to the gatekeeper. The ethics committee at her institution opposed this act, claiming it was a health hazard. Another contentious issue is giving gifts, which is considered bribery and spoils research ethics and trustworthiness in the Western world. On the contrary, this is a gesture of respect among many Indigenous cultures, such as the Lamba people of Zambia (Kalenga, 2015). So, this can bring a paradigm shift that conflates Indigenous and Western worldviews. Since many of my students research with Indigenous people, we may not initially submit some research instruments with the ethics application because we must develop them after consulting our Indigenous participants. Furthermore, traditionally, students are advised to treat their participants' identity confidentially. This should transform into researchers' getting the participants' consent to opt for the concealment or revelation of their identity instead of deciding for them. Quinn (2023) argues that informed consent should be obtained throughout the study to confirm whether the participants are still willing to participate.

A paradigm shift necessitates aligning supervision to (Indigenous) students' ways of living.

ALIGNING SUPERVISION TO (INDIGENOUS) STUDENTS' WAYS OF LIVING

The preceding section raises the need for supervisors to supervise from a culturally responsive and sensitive position. Simonds and Christopher (2013) give an account of a doctoral student from a Crow Indigenous community in the US, who was at the analysis stage of her research. Her advisors, unfamiliar with the Crow community, wanted her to have a theory to guide her interview data analysis. When they did not suggest an Indigenous theory, she thought she had to find a Western theory, which created tension for her study; "she was biased toward finding an appropriate Western theory and assumed that only a Western theory would be acceptable to her advisors" (Simonds & Christopher, 2013, p. 2186). According to these authors, this student struggled to relate Western theories to Indigenous communities; she "learned prescribed ways of conducting research, including choosing from clearly labeled theories, and selecting from clearly outlined methods for collecting and analyzing data" (p. 2188). Social learning and social support theories, as well as the integrated multicultural models, are some of those that can fit within Indigenous culture, considering family and community ties and learning from each other (Simonds & Christopher, 2013; Quinn, 2013). Supervisors should then be critical of the dominance of Western methods as they are not the only actual science, as shown in this chapter.

Science should be conceptualised Indigenously to accommodate Indigenous and non-Indigenous students. Spirituality should form part of such redefinition. Considering the letsema concept, supervisors should push for collaborative research, collective ownership, collective data analysis, collective presentations, acknowledgement of participants or collective publishing, adopting a 'we' in research, and respecting Elders and other knowledge holders. They should inculcate these strategies in their students. Some of these strategies will already be exhibited in the work of Indigenous students, needing to be supported and nurtured.

Approaching research as a collective will prioritise the participants' benefits instead of only the researchers. Supervisors will not supervise for prestige, recognition and institutional rewards. Instead, they should consider learning Indigenous ways along with their students and aim for the students' and community's benefits. They should also encourage students to consult Indigenous scholars' literature instead of relying only on Western scholars – this contributes toward the transformation of the researcher and research. In a study conducted by Datta (2018) with the Dene First Nations community in Sakatchewan, Canada, on *Decolonizing Both Researcher and Research and its Effectiveness in Indigenous Research,* an Elder asked: "Why are you doing research and for whom? How can we benefit from your research?" (Datta, 2018, p. 5). This shows that the participants are aware of the politics of research – research is skewed, designed to serve the researcher's interests more than the participants themselves. In support of the postcolonial Indigenous paradigm, Datta (2018) suggests adopting an Indigenous approach to data collection, analysis, validity, and reliability; this means that the researcher should connect with practical life, people, and culture. Students should be guided appropriately on approaching these critical research aspects Indigenously and non-Indigenously.

Supervisors should develop a sense of the researcher's responsibility, centring the participants' values and following protocol. From an emic perspective, they should listen to participants with empathy and act on their issues to be a voice for the voiceless (Smith, 1999; 2012) regarding transformation. By adopting a research design such as participatory action research, supervisors can guide students in adding the participants' questions to their research (Datta, 2018) and adjust their ethical clearances and notify the ethics committee. They should invest in their initial engagements with the students by considering Indigenous approaches and/or compatible Western approaches towards selecting research topics and guiding them properly in their research proposals by thinking about significant issues important for the community instead of the researcher's objectives or academic research goals (Datta, 2018). Students will be trained to transform their writing style since they co-own research findings with the participants. Thus, they will ensure the exact representation of the participants' voices. The Elders (participants) in Datta's study were opposed to the researchers declaring, "I discovered"/"I found…") (2018, p. 8). Scholars are learners, not discoverers. One Elder stated: "If I share our knowledge with you, you should not say that you discovered it. If you do so, it will be stealing" (Datta, 2018, p. 8). This confirms Semali and Kincheloe's (1999, p. 37) "nothing about us without us". The knowledge holders' voices must be respected.

A transformative/decolonial approach to supervision can free students from the fixed Western and academic mindset. Students will not only focus on completing their degrees and satisfying the institutional requirements, treating the communities as data mines or objects, an unwelcomed attitude described in Datta (2018, p. 9), "researchers are like mosquitos; they suck your blood and leave". The institutional culture comes under the spotlight.

ON A TRANSFORMED UNIVERSITY

Transforming the university is principally a matter of dismantling the hegemonic institutional culture that radiates the Western notions of curricula, epistemologies, and research in favour of a two-eyed seeing. For this to happen, universities in Africa, which resemble Western establishments regarding knowledge advancement should be transformed. Though the scope of discussion in this section is mainly on South African higher education, it is also relevant to other contexts where Indigenous and Western knowledge and ways are polarized, with the latter taking precedence over the former. Heleta (2016) illustrates that South African higher education has reproduced hegemonic identities instead of eliminating them, as seen in Western theorisation that is inconsiderate of locally generated theories. Luvalo (2019) also argues that transforming higher education cannot be completely realised until the institutional culture changes. Limited transformation has been realised in this regard (The Ministerial Oversight Committee on Transformation in South African Public Universities, 2015).

Machingambi (2020) explored the issue of decolonising higher education in South Africa and how a new university can respond to transformation. He interrogated ten senior academics from one new comprehensive university. The findings revealed that the decolonisation of the higher education system was not only urgent but also long overdue. The findings further revealed that academics remain the vanguard of the decolonisation project, given their critical role as teachers, curriculum developers, assessors and researchers. Universities should be centres of epistemic freedom that democratise knowledge by regulating the overrepresentation of Eurocentric thought in education, knowledge, and social theory (Ndlovu-Gatsheni in Machingambi, 2020). "This, among other things calls for the recognition and incorporation of Indigenous knowledge systems into the HE system" (Machingambi, 2020, p. 35). There is a need for knowing students and what matters to them (The Ministerial Oversight Committee on Transformation in South African Public Universities, 2015).

According to Simonds and Christopher (2013), academic institutions should reexamine the teaching of research methods and theories to make students understand the underlying cultural values and assumptions and appreciate non-Western alternatives. A paradigm shift should then see the meshing of complementary methods between Indigenous and Western worldviews.

Indigenisation of Postgraduate Research Framework

Transforming postgraduate education as a research project needs a guiding framework. Thus, this chapter contributes an Indigenisation of the Postgraduate Research Framework (see Figure 1). It starts with reorientating the institutional culture and transforming the research agenda toward recognising Indigenous worldviews and knowledge generation, so they enjoy recognition alongside Western worldviews. This implicates the transformation-centred mindset the university leadership should cultivate in its academic community. Systems of transformation should be in place that will constantly monitor the inclusion of African perspectives in research in the postgraduate programme. Supervisors should be supported and trained to supervise as agents of decolonisation of research. In this way, they will support and nurture the research aspirations of students that are informed by their worldviews.

Supervisors should also take responsibility for transforming research and postgraduate education. They are critical knowledge gatekeepers who will either encourage or discourage students to research in the Indigenous knowledge systems field. They should adopt the letsema and Botho philosophies so that they can enjoy working with students. Considering the experiences of Singh and Major, supervisors can reserve their preconceived ideas about students' ideas and chart a way forward that will make it comfortable for the pair to work together and learn from each other's cultural worldviews. This way, students will come into the research programme positioning themselves well as African Indigenous scholars who have the interest of their communities at heart. They will primarily embark on research relevant to and can uplift or develop their communities. Throughout the research, they will also learn to involve their target participants (Elders and other Indigenous knowledge holders).

Figure 1. Indigenisation of postgraduate research framework

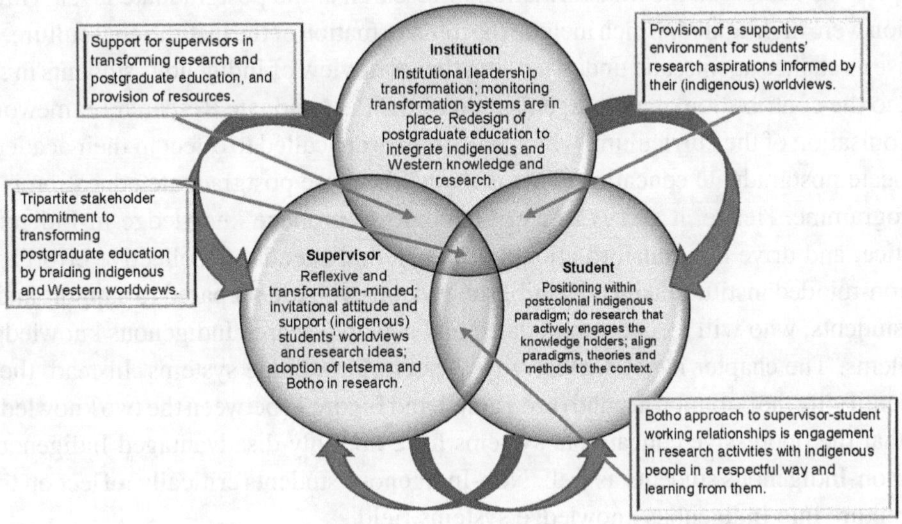

(by Author)

FUTURE RESEARCH DIRECTIONS

The following future research directions should be considered:

- Africanisation and/or decolonisation of education is concentrated on undergraduate education. More research is needed on postgraduate education. Postgraduate students' research should be prioritised as a panacea to transform it. Hence, students must not be buffered from researching in Indigenous knowledge systems field and applying frameworks and methods that can braid Indigenous and Western knowledge systems.
- Research on the monitoring transformation frameworks in postgraduate education is necessary. Research on the design, execution and effectiveness of such frameworks should be done at all institutional leadership levels.
- Transformation may not be realised in the absence of institutional support. Hence, the extent of institutional support should be researched. Managers/leaders of the postgraduate programmes at all levels should be targeted as participants in the study.
- Tracking the students' topics on completed dissertations and theses focusing on Indigenous knowledge systems should be done, which will springboard further research and inspire both Indigenous and non-Indigenous students to research in the field.
- Supervisors should not undermine their knowledge gatekeeping inclination; they influence students' choices and interests in what they want to research. Their (supervisors') attitudes towards Africanisation or decolonisation of education and research necessitate research.

CONCLUSION

This chapter has addressed the transformation of research at the postgraduate level. Three pillars of transformation were highlighted, which include the transformation of the institutional culture, supervisors who supervise as change agents, and understanding the worldview of Indigenous students in supervision. This has led to the contribution of an Indigenisation of the Postgraduate Research Framework in Figure 1. The decolonisation of the curriculum, which universities are called to effect in their academic project, should permeate postgraduate education. This is mainly because postgraduate education is a research-intensive programme. Hence, it serves as a springboard to produce knowledge in Indigenous terms, inform practice, and drive the transformational or decolonial agenda. A collective effort that involves transformation-minded institutional leadership, supervisors who are prepared to support and encourage Indigenous students, who will feel accommodated and safe to research Indigenous knowledge systems-related problems. The chapter is not exclusive to Western knowledge systems. Instead, the discussion advocates identifying the common ground (see Table 1 and Figure 1) between the two knowledge systems. My take is that the oppressive educational systems have not only disadvantaged Indigenous students; they did the non-Indigenous students as well. Non-Indigenous students critically reflect on their cultural lenses and venture into Indigenous knowledge systems field.

REFERENCES

Aikehead, G. S., & Masakata, O. (2007). Indigenous knowledge and science revisited. *Cultural Studies of Science Education*, 2(3), 539–620. DOI: 10.1007/s11422-007-9067-8

Anderson, P., Levon, B., Thu, P., & Melanie, S. (2022). *Higher degree by research: factors for Indigenous student success*. Springer. DOI: 10.1007/978-981-19-5178-7

Atwi, S. K., & Hamza, K. (2015). Qualitative and quantitative research paradigms in business research: A philosophical reflection. *European Journal of Business and Management*, 7(3), 217–225.

Azure, J. A. (2016). Students' perspectives of effective supervision of graduate programmes in Ghana. *American Journal of Educational Research*, 4(2), 163–169.

Battiste, M. (2002). *Indigenous knowledge and pedagogy in First Nations education: a literature review with recommendations.* Indian and Northern Affairs Canada.

Borocz, J. (2012). Colonialism. In Anheier, H. K., Juergenmeyer, M., & Faessell, V. (Eds.), *Encyclopaedia of global studies* (pp. 229–234). Sage.

Borocz, J. (2020). Decolonial theories in comparison. *Journal of World Philosophies*, 5, 43–60.

Cajete, G. (2000). *Native science: natural laws of interdependence.*

Cekiso, M., Baba, T., Rose, M., & Themba, S. (2019). Supervision experiences of postgraduate research students at one South African higher education institution. *South African Journal of Higher Education*, 33(3), 8–25. DOI: 10.20853/33-3-2913

Chilisa, B. (2012). Indigenous research methodologies. *Sage (Atlanta, Ga.).*

Datta, R. (2018). Decolonizing both researcher and research and its effectiveness in Indigenous research. *Research Ethics*, 14(2), 1–24. DOI: 10.1177/1747016117733296

Datta, R., & Starlight, T. (2024). Building a meaningful bridge between Indigenous and Western worldviews: Through decolonial conversation. *International Journal of Qualitative Methods*, 23, 1–9. DOI: 10.1177/16094069241235564

Department of Higher Education and Training. (2023). *The state of transformation in South Africa's public universities: Research Report*. Government Printers.

Desmoulins, L. (2020). Indigenous postgraduate education: intercultural perspectives. *Teachers College Record: The Voices of Scholarship in Education,* 320.

Grosfoguel, R. (2007). The epistemic decolonial turn: Beyond political-economy paradigms. *Cultural Studies*, 21(2-3), 211–223. DOI: 10.1080/09502380601162514

Gumbo, M. T. (Ed.). (2020a). *Decolonization of Technology Education: African Indigenous perspectives.* Peter Lang. DOI: 10.3726/b16660

Gumbo, M. T. (2020b). Indigenising technological pedagogical content knowledge in open distance learning. *Africa Education Review*, 17(4), 72–86. DOI: 10.1080/18146627.2020.1868073

Gumbo, M. T. (2022). The colonial undertones of Queen Elizabeth II's Death: Africa speaks. *Africa Update,* XXX(1). https://www2.ccsu.edu/africaupdate/?issue=152

Gumbo, M. T., Gasa, V., & Kanus, C. B. (2022). Centring African knowledge to decolonise higher education. In Kanus, C. B., Takako, M., & Seroto, J. (Eds.), *Decolonising African higher education: practitioner perspectives from across the continent* (pp. 21–36). Routledge. DOI: 10.4324/9781003158271-2

Hart, M. A. (2010). Indigenous worldviews, knowledge, and research: The development of an Indigenous research paradigm. *Journal of Indigenous Voices in Social Work*, 1(1), 1–16.

Heleta, S. (2016). Decolonisation of higher education: Dismantling epistemic violence and eurocentrism in South Africa. *Transformation in Higher Education*, 1(1), a9. DOI: 10.4102/the.v1i1.9

Kalenga, R. C. (2015). Adjusting Western research techniques to accommodate research in the Indigenous realm. *Indilinga*, 14(1), 1–11.

Keane, M., Khupe, C., & Seehawer, M. (2017). Decolonizing methodology: Who benefits from Indigenous knowledge research? [ERSC]. *International Research for Social Change*, 6(1), 12–24. DOI: 10.17159/2221-4070/2017/v6i1a2

Le Grange, L. (2016). Decolonising the university curriculum. *South African Journal of Higher Education*, 30(2), 1–12.

Letseka, M. (2013). Educating for Ubuntu: Lessons from Basotho Indigenous education. *Open Journal of Philosophy*, 3(2), 337–344. DOI: 10.4236/ojpp.2013.32051

Levac, L., McMurtry, L., Stienstra, D., Baikie, G., Hanson, C., & Mucina, D. (2018). Learning across Indigenous and Western knowledge systems and intersectionality: reconciling social science research approaches. Retrieved on 24 November 2024 from https://www.criaw-icref.ca/wp-content/uploads/2021/04/Learning-Across-Indigenous -and-Western-KnowledgesFINAL.pdf

Luvalo, L. M. (2019). Relationship between transformation and institutional culture. *South African Journal of Higher Education*, 33(1), 184–199. DOI: 10.20853/33-1-2934

Machingambi, S. Analysis of decolonization of higher education in a new South African university. *African Perspectives of Research in Teaching & Learning,* 4(1), 32-45.

Mashabela, J. K. (2017). Africanisation as an agent of theological education in Africa. *Hervormde Teologiese Studies*, 73(3), a4581. DOI: 10.4102/hts.v73i3.4581

Matowanyika, J. Z. Z. (2016). Looking to the other world: Indigenous epistemologies and their relevance in research. A Paper presented during a seminar in the College of Education. Pretoria, South Africa. Unpublished.

Mbembe, A. (2023). Decolonising knowledge and the question of the archive. Retrieved on 24 November 2024 from https://wiser.wits.ac.za/system/files/Achille%20Mbembe%20-%20Decolonizing%20Knowledge%20and%20the%20Question%20of%20the%20Archive.pdf

Mehltretter, S., Longboat, S., Luby, B., & Bradford, A. (2023). Indigenous and Western knowledge: Bringing diverse understandings of water together in practice – Technical report for the Global Commission on the Economics of Water.https://www.waterjusticehub.org/wp-content/uploads/2023/03/Mehltretter.pdf

Mignolo, W. D. (2005). *The idea of Latin America*. Blackwell.

Mignolo, W. D. (2005). *The darker side of Western modernity: Global futures, decolonial options*. Duke University Press.

Miller, M., Warner, S. A., Gumbo, M. T., Alarcón, I. V., & Petrina, S. (2023). Lenses for critiquing and improving the Standards: Design, indigeneity, access and equity, and literacy. In Bartholomew, S. R., Hoepfl, M., & Williams, P. J. (Eds.), *Standards-based Technology and Engineering Education: 63rd Yearbook of the Council on Technology and Engineering Education* (pp. 287–308). Springer., DOI: 10.1007/978-981-99-5704-0_18

Msila, V. (2017). *Decolonizing knowledge for Africa's renewal: Examining African perspectives and philosophies*. KRPublishing.

Naidoo, N. (2011). What is research? A conceptual understanding. *African Journal of Emergency Medicine : Revue Africaine de la Médecine d'Urgence*, 1(1), 47–48. DOI: 10.1016/j.afjem.2011.05.011

Ndlovu-Gatsheni, S. J. (2013). Why decoloniality in the 21st century? *The Thinker: For Thought Leaders*, 48, 10–15.

Ndlovu-Gatsheni, S. J. (2016). Introduction: The coloniality of knowledge – between troubled histories and uncertain futures. In Ndlovu-Gatsheni, S. J., & Zondi, S. (Eds.), *Decolonising the university, knowledge systems and disciplines in Africa* (pp. 3–22). Carolina Academic Press.

Odora Hoppers, C. A. O. (2001). Indigenous knowledge systems and academic institutions in South Africa. *Perspectives in Education*, 19(1), 73–85.

Photo, P., & Gumbo, M. T. (2024). Postgraduate student supervision in open distance learning: towards a Botho supervision framework. In Gumbo, M. T., Gaotlhobogwe, M., Pedzisai, C., Jojo, Z. M. M., & Knaus, C. B. (Eds.), *Global perspectives on decolonizing postgraduate education* (pp. 166–178). IGI Global. DOI: 10.4018/979-8-3693-1289-6.ch011

Quinn, A. L. (2022). Bridging Indigenous and Western methods in social science research. *International Journal of Qualitative Methods*, 21, 1–14. DOI: 10.1177/16094069221080301

Ramose, M. B. (1998). Foreword. In S. Seepe (Ed.). *Black perspectives on tertiary institutional transformation* (pp. iv-viii). Vivilia Publishers & University of Venda.

Ramose, M. B. (2004). In search of an African philosophy of education. *South African Journal of Higher Education*, 18(3), 138–160.

Reid, A. J., Eckert, L. E., Lane, J. F., Young, N., Hinch, S. G., Darimont, C. T., Cooke, S. J., Ban, N. C., & Marshall, A. (2021). Two-eyed seeing: An Indigenous framework to transform fisheries research and management. *Fish and Fisheries*, 22(2), 243–261. DOI: 10.1111/faf.12516

Simonds, V. W., & Christopher, S. (2013). Adapting Western research methods to Indigenous ways of knowing. *American Journal of Public Health*, 103(12), 2185–2192. DOI: 10.2105/AJPH.2012.301157 PMID: 23678897

Singh, M., & Major, J. (2017). Conducting Indigenous research in Western knowledge spaces: Aligning theory and methodology. *Australian Educational Researcher*, 44(1), 5–19. DOI: 10.1007/s13384-017-0233-z

Sium, A., Desai, C., & Ritskes, E. (2012). Towards the 'tangible unknown': Decolinization and the Indigenous future. *Decolonization*, 1(1), I–XIII.

Smith, L. T. (1999). *Decolonizing methodologies: research and Indigenous peoples*. Zed Books.

Smith, L. T. (2012). *Decolonizing methodologies: research and Indigenous peoples* (2nd ed.). University of Otago Press.

Sparkes, L. L., & Piercey, D. W. (2015). What are Indigenous and Western ways of knowing? Retrieved on 24 November 2024 from https://www.criaw-icref.ca/images/userfiles/files/Fact%20Sheet%202%20EN%20FINAL.pdf

The Ministerial Oversight Committee on Transformation in South African Public Universities. (2015). *The transformation of South African higher education*. 2nd Edition. National Higher Education Summit. Retrieved on 24 November 2024 from https://www.dhet.gov.za/summit/Docs/2015Docs/Annex%208_TOC_Transformation%20of%20SA%20HE.pdf

Twala, C. (2018). The 'letsema/ilima' campaign: A smokescreen or essential strategy to deal with the unemployment crisis in South Africa. *Journal of Contemporary History*, 29(1), 184–198.

Uink, B., Day, A., & Martin, G. (2019). Wings to fly: Case study of supporting Indigenous students success through a 'whole-of-university' approach. *Australian Journal of Indigenous Education*, 48(2), 206–206. DOI: 10.1017/jie.2019.11

ADDITIONAL READING

Department of Science and Technology. (2004). *Indigenous Knowledge Systems Policy*. Government Printers.

Kaya, H. O., & Seleti, Y. N. (2013). African Indigenous knowledge systems and relevance of higher education in South Africa. *The International Educational Journal: Comparative Perspectives*, 12(1), 30–44.

Keane, M., Khupe, C., & Muza, B. (2016). It matters who you are: Indigenous knowledge research and researchers. *Education as Change*, 20(2), 163–183. DOI: 10.17159/1947-9417/2016/913

Mosimege, D. M. (2017). Research protocols and ethical considerations in Indigenous knowledge systems. In Ngulube, P. (Ed.), *Handbook of research on theoretical perspectives on Indigenous knowledge systems in developing countries* (pp. 156–172). IGI Global. DOI: 10.4018/978-1-5225-0833-5.ch007

Odora Hoppers, C. (2021). Research on Indigenous knowledge systems: The search for cognitive justice. *International Journal of Lifelong Education*, 40(4), 310–327. DOI: 10.1080/02601370.2021.1966109

KEY TERMS AND DEFINITIONS

Transformation: In higher education, transformation refers to change in form, nature or appearance as it pertains to staffs' and students' experiences of higher education; institutional culture; teaching and learning; curriculum and pedagogy; research content and methodology; and community engagement; production of citizens capable of promoting a democratic, ethical and prosperous society (Department of Higher Education and Training, 2023, p. xii).

Research: Research is wondering about nature and systematically investigating it and society. The aim is to validate and refine existing knowledge and to generate a new one (Naidoo, 2011). Its characteristics include relevance, especially of content and context, its conception in terms of its contribution and significance, research orthodoxies (accommodating marginalised communities, especially through qualitative research methods), theoretical orientations (e.g. critical theories, feminism, eco-systemic trends, etc.), ethical framework (beneficence, non-maleficence and respect for participants' autonomy) (Naidoo, 2011).

Postgraduate Education: This type of study occurs after completing an undergraduate degree. In the context of this chapter, it refers to the master's and doctoral studies.

Indigenous Knowledge: It is a community-based, unique, complex, dynamic, eclectic, non-formal, and passed down from generation to generation in a variety of contexts (including cultural, ecological, economic, ethical, political, social, spiritual, and technological) to assist Indigenous communities in solving problems and making decisions that are critical to their existence, survival, and adaptation (Ngulube, Dube & Mhlongo, 2015, p. 148) .

Paradigm: It is the researcher's philosophical assumptions (worldview) about what constitutes valid research and research method appropriate for the development of knowledge in a given study (Antwi & Hamza, 2015).

Botho: Botho is to be aware of one's own being in relation to another person; it is also a form of human engagement that allows for critical thinking, non-domination and optimal development of human relationships (Letseka, 2013).

Chapter 21
The Body as a Divination Bone:
An Application of the Bone Oracle as a Methodology

Tebogo Victoria Kgope
https://orcid.org/0000-0003-2318-532X
University of South Africa, South Africa

ABSTRACT

This chapter casts the body or people as symbolising divining bones when reimagining the Indigenous African concepts in Indigenous knowledge. With the application of the African ear, the researcher illustrates the conceptual duology of Dani Nabudere and Ngũgĩ wa Thiong o as encompassed with Credo Vusamazulu Mutwa's application of the bone oracle in retelling the indigenous ways of knowing. Drawing from Credo Mutwa's methodology of divining bones, the concepts motlhakola or dispossession, thwagadima as death or danger and mpherefere as chaos and disputes are explored using the conceptual framework of Nabudere and wa Thiong o to comprehend Mutwa's application of the bone oracle. The methodological approach and position are of the fall of the divination bones, applied figuratively as people. The aim is to recognise the state and positionality projected in different people as bodies that characterise ditaola or divining bones especially when inculcating African spiritual perspectives as realised in an African university. Is the body in synch with the dominant theories and concepts used by the academy and society? The chapter calls for a political will for the state and the African universities to appreciate the observed realities of its people.

INTRODUCTION

Mmele wa motho ke taola is a Setswana lingua franca that aptly portends that one's body can be observed as a divination bone in any setting. In this chapter, the African university is encouraged to inculcate African conceptions of thinking. Nabudere, (2003, p.3), related that the African university "must first and foremost be built on a sound spiritual basis that highlights aspects of African spiritual life that has enabled the African people to survive as a human community throughout the centuries". While Mazrui "challenged African academies to strive for political autonomy from their governments" (Mazrui, 2003). Certainly, to date appreciation for conceived indigenous African concepts remains challenged and opposed in other facets as opposed to the suitability and support of concepts of Western modernity.

DOI: 10.4018/979-8-3693-8447-3.ch021

This distinction is held when for example, research critically explores what re-imagining the body or people and complex situational contexts through the lens of indigenous knowledge means. What would be the objectives of contributing to conceptions of theorising *motlhakola, thwagadima and mpherefere*? As against Western modernity concepts such as "state capture". Interestingly to explore this aspect for the sake of explaining the concept of the body as a divining bone we also note that "state capture" has been defined as political and economic corruption, but theoretically "state capture" is a means of social order and control" (Croucamp and Malan (2018;Scott and Pettersson, 2019).

A critical exploration of Indigenous knowledge conception of state capture would relate the concept with *motlhakola* or as dispossession and loss of socio-economic resources of the state and its people, thereby killing institutional structures and the people; and *mpherefere* as chaos and disputes derived from ill-functionalities of those who are mischievous. To bring forward the emphasis of this chapter with its application in the academy is when such indigenous concepts are utilized than notions derived in tandem with Western concepts of modernity (Mignolo, 2011). In other words, state capture can be argued to be *mpherefere*, or *motlhakola* just to indicate what the previous statement suggests, when engaging a mode of delinking showing "that it is not enough to change the content but that the terms of the conversation must be changed" (Mignolo, 2011, p.122). And what is learned is that when the terms of the conversation are changed "hidden geo and bio-graphical politics of knowledge of imperial epistemology are revealed" (Mignolo, 2011, p.119).

The body or *mmele* and people are promoted to re-member and recognise in themselves what approves and aggravates the innate person. The concept of *taola* or divination bone emerges from the philosophy that *ditaola* (plural for *taola*)/ divination bones reprimand, they bestow or impart the law upon those they are divining on. This method of seeing and discerning is entrenched in African indigenous knowledge systems. It is in this position that we submit that the application of bone divination as practised by indigenous healers can be incorporated in any location by seeing people and things as representing divination bones. In other words, this mode is without the actual undertaking of utilising or possessing divining bones. The metaphorical gaze of classifying people or the body and several situation contexts as representing the Indigenous African divination bones when communicating *motlhakola, thwagadima* and *mpherefere* is a traditional standard of analysis that should be practised at the intellectual level in the academy. Therefore, the chapter is a kaleidoscope of re-imaginations of a South African nation that was supposed to be flourishing and thriving in various facets that emanate from the rich resources of natural and potential avenues the country possesses. In this aspect, people are placed in juxtaposition to the sustainable rebirths and developments they would have achieved during the apartheid era in curbing the ills of the past. Other scholars account that "South Africa is the world's most unequal country, that twenty-five years of freedom have failed to bridge the divide" Croucamp and Malan (2018; Scott and Pettersson, 2019). The current timelines and era often ascribed as the new dispensation in South Africa since post-1994 have been dominated by calls for reparations for the painful past and the need to reiterate, and restate, the articulation and promotion of African epistemologies. The demands have been catalysed by "the frustrations experienced by survivors in persuading the government to effect a comprehensive reparations policy" (Makgalemela, 2009). This statement is fundamental to strengthening suitable reparation goals and strategies.

Therefore, strides to re-imagining and realising critical thinking based on African peoples' experiences with reality cannot be defocused by institutions of power such as the government and tertiary institutions. In addition, the state or government as a key and strategic institution that holds sovereign power over its citizens is also signified as a force that should ensure good governance to various societal challenges. Higher education especially African universities is not only reimagined as establishments that are expected to demonstrate political autonomy but also as higher echelons of the highest form of reasoning and intelligence. In this regard "in a democratic South Africa, academic freedom in all higher education institutions is integral to the intellectual life of the country" (Academy of Science of South Africa, 2010). Therefore, we argue that critical analysis of disarrays in society can be resolved by the state and institutions of learning that integrate and link societal challenges with Indigenous epistemologies as part of the solutions. Given the extensive periods of the influence of Western scholarship, it is a known fact that anything that is focused on expands. It is the expansion of African ways of knowing as a principled body of knowledge that will potentially develop and ascend when more emphasis and focus are placed on them.

The defence of this chapter rests in demonstrating and casting the body or people as representing divination bones that seek to address and express the difficulties and complexities that may obstruct the development of indigenous epistemologies in an African university. The researcher proposes that the philosophy emerging from divination bones is inclusive and part of an all-knowing arena of the experiential and ontological realities of Indigenous people. The chapter is in addition conceptualised during the COVID-19 pandemic timeline. This period added more strain and constraints to the already existing challenges that are experienced in post-apartheid South Africa and remain painfully the site of evidence to date. As a useful feature to address, an African university must be intensely symbolic in advancing quality education that eliminates the transgressions that have been persistent to marginalised groups of people and situational contexts that are definitive of atrocities and injustices. How do these bodies as people work or deal with other people or bodies? In other words, how often would an African concept and frame of thinking be applied, and the emphasis of such alignments allowed to be made counter-hegemonic towards the dominant frames of reasoning and seeing? How did the body feel with all the knowledge that it knows in attempts to alleviate the global pandemic of COVID-19? There were more questions than answers at the time. Thus, it is in this manner that the emphasis of the chapter is that of acquaintance with the self, how the body or people can understand themselves as divining bones and analysing the concepts which emanate from an African indigenous knowledge holder's paradigm of examining within the framework of healing.

KEY CONCEPTS IN THE CHAPTER

***Ditaola*/ divination bones:** From *go laya, ditaola di a laya, di a laola, di fa molao*, i.e., divination bones reprimand, they examine and impart law (Kgope 2022, p. xvii).

***Mmele wa motho* as *taola* (the human body as a divination bone):** This is the representation of a person or the body with signs and symptoms that give light and meaning like what divination bones do. An instant example is when one is supposed to catch a bus that is about to leave, and you suddenly have an urgent sensation that you want to answer the call of nature and the bus embarks and leaves you behind, a few hours later you learn that the same bus was involved in a fatal collision. The urge to want to relieve oneself is understood to be the body imparting some knowledge to the individual and driving the individual away from unforeseen danger. To advance the main contribution of this chapter is that our bodies as people offer us the best signals to understand ourselves and our surroundings. In the instalment of this chapter, the body as a divining bone is metaphorically applied to imagine and cast the body or people as they appear holistically in the environments they occupy. It thus explains the interpretation of people's circumstances in relation to the three concepts of the fall of the bones.

***Motlhakola/* dispossession:** The state of being dispossessed of goods, loss of loved ones, the loss and perishing of a nation, *batho bantsho ba felang*, that is, Black people who are perishing in numbers, *go tlhakolega khumo, lefa, bophelo*, i.e., dispossession or loss of wealth leading to various states of impoverishments, loss of knowledge and loss of vitality.

***Thwagadima/* death/ danger:** often implying something that is seen or observed with the naked eye but usually heralding a danger or attack. These would be indicated in the mass murders, killings that afflict men, women, and children especially those of Black and African descent.

***Mpherefere/* chaos:** Signals chaos, confusion, disputes *mpherefere* (Kgoroeadira, 2014, p.53) casts people as having quarrels or disputes amongst each other, never-ending disagreements.

Knowledge: as with how we know things – and the ledge we utilise to attach or secure ourselves to the ledge of that essential imminent knowledge imparted to us through various ways of our understandings.

CONCEPTUAL FRAMEWORK

The assertion by African scholars Ngulube, Mathipa and Gumbo (2015, p. 2) that "theoretical and conceptual frameworks do not come as ready-made in literature waiting for researchers" cannot be adequately emphasised. These Indigenous scholars aptly impart that as arts of conceptions, conceptual frameworks can appear as conflicting because they describe the degree and intensity of the experiential and existential realities as they appear out there. Their positionality on the conceptual framework is that the chosen conceptual framework directs the research problem, the methodology approached and how the knowledge produced is interpreted and analysed. It is in this perspective that the application of the African ear infers from the conceptual duology of Dani Nabudere (2011) and Ngũgĩ wa Thiong'o (2006, 2009) in retelling and remembering the Indigenous ways of knowing. Ngũgĩ wa Thiong'o (2006, 2009) and Dani Nabudere (2011) concepts traverse with African ontological & epistemological realities. African people have specific ways of knowing and understanding themselves alongside the realities in their life world as per their cultures and languages (Nabudere 2011, p. 39). Ngũgĩ wa Thiong'o (2006, 2009, pp. 37, 39) through intellectual decolonisation and the sanitisation of the mind applies the mode of contextualising that which is African to highlight the causalities of empirical African realities and worl-dviews globally, from dismemberment to remembering visions and practices that ought to be supported.

Figure 1. A conceptual overview of the application of Credo Mutwa's bone divination

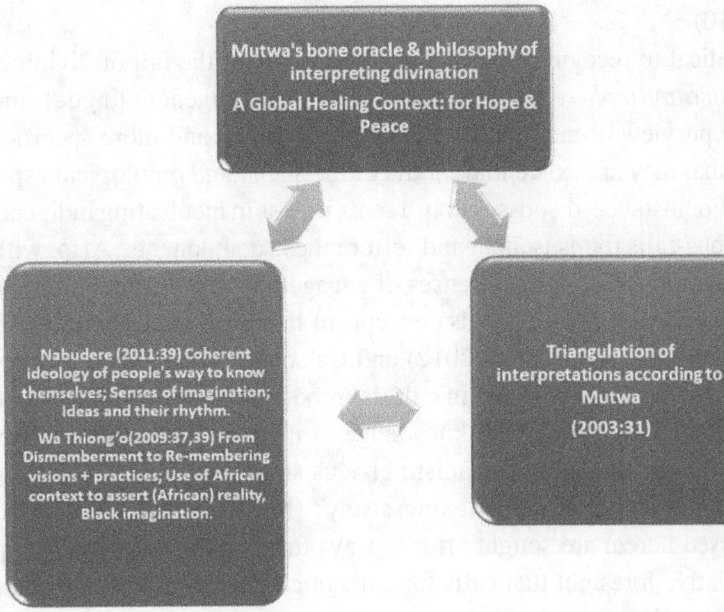

(Kgope, 2022)

The conceptual framework above in Figure 1 is an illustration of the duology of concepts as emanating from Nabudere (2011) in embracing the senses of imagination and ideas with their rhythm and wa Thiong'o's (2006, 2009) assertions from dismembering to remembering visions and practices on how to apply an African context with an emphasis of implementing Black imagination and African scholarship. The combination of Nabudere and wa Thiongo's concepts when fused with Credo Mutwa's bone oracle and ways of interpreting divination bones highlight the contention that theoretical and conceptual frameworks should inform, define and validate the research methodologies and questions that are selected (Ngulube et al. 2015). In the context of this chapter, the question asked is how do people understand themselves in the era or timelines characterised by liberating thoughts of inculcating African thought patterns. It thus becomes appreciative to emphasise the dimension of the duology of concepts from Nabudere (2011) and wa Thiong'o (2006, 2009) with the philosophy and coherence of Credo Mutwa's bone divination. It is this interpretation of the African Indigenous bone divination that informs that African indigenous knowledge systems are a significant domain of knowledge that can generate a clear understanding and become dependable knowledge for a sustainable future. Moreover, there is an aspect of embracing African languages when interpreting divining bones. Cakata (2015, p.iv, 41) emphasises that "there is a need for a more humanising approach to treat indigenous languages with respect". The accordance of this respect which elevates the development of Africa, and its people is in general grounded in culture and language being the cardinal centres that advance African development (Prah 2012, p. 2,9). Actualising wa Thiong'o's (2006, 2009) concept of disengaging from the mode of dismemberment to acts of remembering visions and practices that utilise African contexts to develop African conceptions of thinking is Prah (2012, p.10) also indicated that:

African languages should be used throughout the different stages of the educational system. The argument goes further to suggest that African languages should be used in all areas and dimensions of social life. (p. 10)

Therefore, it is critical to recognise that the three concepts of the fall of divining bones describing *motlhakola, thwagadima and mpherefere* (the Setswana African Indigenous lingua franca) when imagined or engaged under the purview of the South African Black people and more specifically in institutions of higher learning is that they are existential motifs of the social and ontological aspects of people who occupy these spaces. For instance, dispossession, a sense of loss in inculcating indigenous knowledge and languages hence various calls to resuscitate and restore these components. Also, within the framework of acknowledging senses of loss, emerges senses of engagement to capacitate African scholarship and Black imagination. When African indigenous concepts of language are given the respect and humanity they deserve as alluded to by (Prah 2010, 2012) and Cakata (2015) there is an opportunity "to learn from each other and become equal partners in cultural production" Mukoma Wa Ngugi (2018, p. 9). In the promotion and articulation of applying knowledge as it emanates from divination marks that "the theory and practice of divination and shamanism give us some insight into how language and culture fashion people's ways of relating to reality expressively" (Nabudere 2011, p. 40). It is henceforth the three concepts discussed herein are sought after in ways that express the unseen experienced African realities in an appreciative foresight that calls for authentic interpretations and instalments of African or Black scholarship over Western scholarship.

The conceptual framework above shows how Credo Mutwa's bone oracle and divination can be employed to solve rife African problems. Essentially problems that are concentrated with dispossession in its variety, including chaos, disputes, volatile acts of danger and violence as described by the notable concepts of the fall of oracle bones in discussion. Mutwa and Larsen (2003) affirm that:

When the twasa or sangoma casts the bones all aspects of the arrangement are considered carefully, these include which way the image of the bone is facing, the distance between the bones, and any unusual configurations in the pattern." (p. 26)

It is this way of seeing and understanding which this chapter elevates should be defended by African and Black people as gifted people of African spirituality. For example, Nabudere (2011, p. 42) also asserted that in African conditions divination plays a critical position even in contemporary conditions in the lives of the people. Moreover, it is this way of seeing and understanding ourselves from the geometric arrangement of the fall and image of the divination bones, that the body or people with their circumstantial aspects are cast as divination bones that exemplify the fall of bones revealing *motlhakola, thwagadima and mpherefere*. Credo Mutwa on the ability of humans to see spiritual things argues that this way of seeing can be promoted even at the university level by those who are not sangomas or healers hence the position of this chapter that the body is a divining bone. And it is through such a body or people who have not neglected the consciousness of their African indigenous spirituality, that having a sense of knowing themselves, the people or the body become divining bones that examine, diagnose, and treat the self.

According to Makgopa and Koma (2009, p. 52), the throwing of divining bones can be associated with how medical practitioners use the stethoscope to detect and examine the patient. From an African Indigenous perspective, a divining bone and the interpretation of the fall of the bones "reveal the power and the presence of ancestral spirits (Makgopa & Koma, 2009, p. 52). Interestingly, the application of divining bones has also been described in relations to "ritual technology", more like "information technology" (Croucamp, 2013, p. 2) while Ross (2010) likened the art of throwing bones to floor x-ray because the African Indigenous healer can understand the extent of the problem based on the way the bones fall on the floor.

In this chapter, the casting of people or the body as symbolising the fall of the divining bones that reveal the description of the concepts of *motlhakola, mpherefere* and *thwagadima* is reimagined from observation of everyday ordinary people whose lives represent the meaning of these concepts. Similarly, the holistic appearance of people and situations and how they move about their business or daily activities allow an indigenous knowledge expert or holder to understand the extent of the problems faced by these people simply based on patterns they appear to depict. This assertion or relation is different from the out-of-body experiences that may be experienced by some indigenous knowledge healers where they go into a trance-like state that shows the healer or person is in the company of their ancestors or the spirit of their ancestors. Our assertion here is based on the knowledge of the self, alongside the knowledge of ancient African wisdom. But how will this special type of knowledge be gained in the academy if African knowledge systems are not convincingly infused to their advantage and dominance? For example, what wa Thiong'o (2009, p. 39) deems as the act of remembering or elevation of Indigenous concepts of African knowledge systems is not fused even at South African medical schools where there is no incorporation of this knowledge into the curriculum (Chitindingu, George & Gow, 2014, p. 1). For this reason, Kgope (2022, p. 32) problematises the aspect of social and cognitive injustices as a result of the exclusion of indigenous knowledge systems to some extent in higher institutions of learning "where a large majority of the African population obtain their education, an education which is largely Western and disconnected and dismembering to their African reality". Through the method of divination, there is a "special type of knowledge that is shared and gained" and it is this hidden-like knowledge that offers opportunities to make rightful decisions on various matters that affect humanity (Peek, 1991, p. 1).

Ditaola or divination bones are one of the different methods used for divination some include rocks, sunbeams, and trance postures, according to Nabudere (2011, p. 44). Western scientific disciplines regard divination bones as superstitious or fortune telling even though the philosophy of divination bones and the bone oracle communicates a deep understanding of events and influences surrounding a situation or people or a person. It is this philosophy that provides the body or people and situational contexts with an identity. If identities that are explored and revealed by the concepts that emanate from the fall of the divination bones would be adhered to by comparing identities of people, bodies and various situational contexts, levels of complicitness would decline. Therefore, it is important to observe how people understand themselves, and where they go when things go wrong in their lives. From an African perspective or through Western modernity. Nabudere (2011, p. 45) writes that "Western scholarship and epistemologies have 'ring-fenced' themselves from relating to other valuable human experiences that could be of use to their communities". In the sections to follow below such applicable valuable experiences as espoused by Nabudere are underscored as identities that are often observed in everyday ordinary people and their varied circumstances or situations.

Although there exist many studies in the sector of African Indigenous knowledge systems and different types of healers under the category of African indigenous healing practices, the novelty and emphasis of casting people or the human body and relatable situational contexts as divination bones arc of a unique type. Although commensurate when conceptualised as delinking from concepts entrenched in Western conceptions of thought using demonstrating independent thought and freedom (Mignolo, 2011, p. 118), the novelty is in focusing on the body as the knower and one that delivers uninterrupted knowledge about the self and the surrounding circumstances. The latter is expounded on by Kgope (2023) on Mutwa as a fitting characteristic divination bone that:

Mutwa epitomised *taola* or *ditaola* (divination bones). As *taola* and at times talking in the plurality of *ditaola* he examined and diagnosed humanity, and the universe, sharing with humanity dipheko (medicines) in the form of words and statements he articulated in his various books. As a divining bone, he aimed to observe an atmosphere of peace and humanness for all humanity. As many of us saw his physical appearance, but, as one who was imbued with the wisdom of being a Sanusi, as *taola* that sees and reveals the truth he would also characterise all genders in appearance. Symbols of a man, woman, old man or old woman, the transgender group would interchangeably be observed in his dress code, appearance, and how those who guided him would want him like a divining bone to project himself. (p. 7)

The uninterrupted knowledge is incorporated into the awareness of the interweaving nature of when what is embodied applies to the existential and ontological realities of people. These attributes are multifaceted ways which have capabilities to design methodological faculties that produce new thinking and new knowledge.

METHODOLOGY

In this section, the spectacle that knowledge epistemologically or ontologically has been colonised is brought forward. The African conception and ideation of casting the body or people as divination bones is entrenched in self-acquaintance, self-knowledge, and familiarity with indigenous knowledge. Mutwa's approach to any method is that "a person who is not familiar with Africa and its people might find it difficult to understand" (Mutwa, 1964, p. 529). The research methods used are those espoused by Kovach (2010, p. 40; 2019, p. 123); Smith (1999, p. 143,147) and Mutwa (1964) of oral traditions, yarning or *go toloka* in Setswana, and re-membering (Kovach 2010, p. 40; wa Thiong'o 2009, p. 39). Put directly, it implies that Credo Mutwa's approach to research methodology and design is based on the knowledgeability or familiarity of the methods that are related to African people and the knowledge of how people understand themselves. This is in contrast with "Western scholarly approaches that have separated the body from the mind" Nabudere (2011, p. 45). Again, when viewing Mutwa's reproach and submission it reveals that when Africa and its people are not adequately familiar with the Western-centric mode of knowing, both Africa and its people will have difficulties transcending their knowledge systems to familiar competencies. This will prove to be the case, especially in the higher education sector. And similarly, when Africans understand themselves from an African standing and context there are minimal complexities. The implication of this submission of Mutwa is through afterthoughts of asking whether Mutwa applied qualitative or quantitative methods when conducting research activities that are paramount to promoting new knowledge. To respond to this deliberation, we re-member Ramose (2016) and other liberal scholars on breaking the phrase research to research signaling that liberatory pedagogy and emancipation must be rooted with emancipatory methodologies that seek the "quest for truth" in researching.

Reconsidering the methodologies that also encompass the coloniality of knowledge, it is in this regard that not only do we continue to maintain epistemological disobedience or epistemicide, but that the status quo is also that of nullifying this impediment and hegemony while sustaining the theory and praxis of methodologies of coloniality. Research methodologies are also colonially constructed. The body feels out of sync in overarching terms inter alia, "qualitative or quantitative, mixed methods, or sequential etc. In the era of curriculum reform and periods that call for good quality and improved education, scholarly conducts in the wave of redressing the past discriminations continue to support and maintain the rigidity and support of these methodologies in what is often framed that "it is the standard'' (Kgope 2022, p. 76). Should redress comply with the standard or be disruptive of methods we are used to? When redressing epistemicide and methodological decay the process of that rejection leads to research designing methods that privilege African indigenous methodologies. The latter would also imply the reflective meaning of the notion that Eurocentric paradigms are opposed to African paradigms of knowing. Does this notion prove and speak to the body as a divining bone, divining by rejecting, dismembering itself from that which in its nuclearity was colonially constructed? This section also accounts to confirm that a conceptual framework expresses, classifies and confirms the research methodology that is selected (Ngulube et al. 2015). As a result, reimagining, retelling, remembering and redressing the injustices of the past is characteristic of embracing the epistemologies and ontologies of the global South (De Sousa Santos 2016). These are the ways of knowing that have been contested and always frowned upon. Gordon in theory in Black (2010, p. 202), reveals that it is true that if knowledge was colonized, it would be probable that the pacts of research methodologies are also colonised. However, to strike a balance, Gyekye (1995, p. 54) cautions that it is also fundamental that we evaluate the Eurocentric ways and methods and differentiate 'if' and 'how' they apply to the indigenous socio-political life and the impact thereof.

De Sousa Santos (2016, p. 27) asserts that "epistemologies of the South call not just for new methodologies, non-extractive methodologies, but also for new ontologies. This chapter applied non-extractive methodology when privileging alternative and Indigenous knowledge as the lingua franca and dialect often produced by those defined by contested concepts and are met with resistance. Mukoma wa Ngugi (2019) also reminds us that indigenising English privileges Indigenous languages but also helps us understand and define contextually Black or African identities, beings, experiences and the body as *taola* or a divining bone. According to Feyerabend (1978, p. 98) a proponent who was against methodological linearity stated that "every project, every theory, every procedure has to be judged on its own merits, the idea of a universal and stable method is unrealistic". In other words, what we think we know, the terms of language often used and the capacity to trace the etymology of these terms allow for non-linearity in methodological designs. Nabudere (2011, p. 90) established this coherence that "there has to be a return to the divine origin of the word and concept" to improve sustainable indigenous knowledge systems.

Following single-dimension methodologies breeds the same answers which fade the transcendence of othered knowledge when there is a variety. Mamdani (2016) considers vigorous engagement using intellectual decolonization from institutions that constructed and designed how the body or people should be defined especially across colonially constructed disciplines and methodologies. Nabudere (2011, p. 124) asserts that these disciplines were constructed and further broken down into subdisciplines that degraded African ways of knowing. Other scholars like Modiri (2021, p. 47) raised similar concerns that "South African disciplinary knowledge and knowledge about South Africa and intellectual traditions have been constructed almost exclusively based on a Euro-American canon". In this regard, the phenomenon of critiquing methodologies of Western coloniality is not new as it has been extensively explored by many scholars (Sandoval, 2013). Therefore, the methodological approach embraced in this chapter is conceptualised from Mutwa's lens of the bone oracle in observing everyday ordinary people, situational contexts and their circumstances in their interpretation of *motlhakola, thwagadima* and *mpherefere*. Ethical considerations adopted were those of demonstrating currency for and not against indigenous knowledge systems.

RESULTS AND DISCUSSION

The findings in this chapter indicate that observations and conversations with different people bring a realisation that everyday ordinary people all experience dispossession or *motlhakola* or *batlhakodiwa* something, this could be dispossession of their households, knowledge or history to be advanced and dignity; in *thwagadima* as death or impending danger, for example, accidents, violence, conducting conversations on restoring Indigenous knowledge systems but encountering acts of the leftist sense of doing the other; or even some form of chaos or experiences of *mpherefere* in their life world, in instances such as corruption, or arrangements that go wrong. The positionality of casting the body as a divining bone that can complement the bone oracle concepts in addition to *motlhakola, thwagadima* and *mpherefere* supports existing experiential African concepts that could inform sustainable development for impactful Indigenous knowledge. In so doing the academy will be moving away from the "historical mannerisms of marginalising Black people and Black thoughts" (Modiri 2021, p. 46). Hence Kgope (2022, p. 42) posits that despite the challenges and contentions faced by African knowledge systems, emphatically centering "African existentialism proposes that the lifeworld which encapsulates African experiences ought to inform critical pedagogy" as we will illustrate below.

Our methodological approach casts the body or people as representing the interpretation of divination bones or *ditaola*. Mutwa (2003, p. 26) transmits that the healer identifies the fall of the bones on the ground and interprets the fall or lie of the bones and prescribes remedies that are needed. On evaluation and analysis, the fall of the bones describe certain mystic patterns on the ground' (Mutwa 2003, p. 31). A global healing context by Mutwa (2003, p. 31) that excavates possibilities for hope and peace tells us these bones identify people's future, in good faith or bad faith. We relate Mutwa's (2003) description of the fall of bones to people who are also on the ground, in the South African setting. The position of people as bodies on the ground or earth plane represents different interpretations of divining bones.

The Body or People and Situational Contexts as *Motlhakola*

As also attested by Kgoroeadira (1993, p. 55) *motlhakola* approves of death or loss in secret and similarly, Makgopa and Koma (2009, p. 151) correspondingly refer to the fall of divining bones reflecting *motlhakola* to represent "the death of a person or loss of property", in essence, they contend that "the message communicated by this fall is that something has been lost". It is in this regard that people who are dispossessed or robbed of their ways of knowing or evicted from their land; or people who are dismembered, dismembered by death in some cases; homeless people; the unemployed; and people or bodies living in hell or abject poverty, the so-called *The Wretched of the Earth* as Fanon (1963) would say, people who ask themselves "in reality who am I" (Fanon, 1963, p. 250), signalling their loss of self and indignity with the self. All these complex circumstances are descriptive of people or bodies that are in the terrain of *motlhakola* as they appear on the ground or earth plane, they reside in symbolising the fall of the bones. Another case in point is the position the global world found itself in during the COVID-19 pandemic, an era that epitomised all three concepts, *motlhakola*, as many people lost their lives leaving behind large scales of inequality. The aspect of *thwagadima* is observed in the impending doom and danger that was palpably visible to many, particularly during the lockdown periods.

Situational Contexts of *Thwagadima*

Thwagadima as a concept in the fall of divination bones symbolises a danger or something that could be seen and observed with the naked eyes. Kgoroeadira (1993, p. 58) explains it as a "grave danger". People or bodies that represent the fall of divination bone as *thwagadima,* are engulfed by dangerous situations, death or lightning, what we see with our naked eyes. People who are susceptible to accidents, for instance, the road carnage or fatalities that are endless; people or bodies that are prone to various forms of violence, floods including flood victims who experience large magnitudes of loss all represent people marred by *thwagadima*.

People or Bodies With Situational Contexts of *Mpherefere*

According to Makgopa and Koma (2009, p. 152), *mpherefere* indicates that "things are not in order that there is trouble somewhere", Kgoroeadira (1993, p. 53) expanded relating mpherefere to "confusion, disputes or trouble". Everyday observable scenarios entail people or bodies that perpetuate or cause chaos, confusion, troubles, and disputes as realised in families, industries of work; and amongst state institutions. As highlighted previously the term state capture is synonymous with *mpherefere*, alongside the situational contexts of various corrupt activities. Kgoroeadira (1993, p. 54) describes this phenomenon as *mpherefere o matepe,* i.e., cheeky, as it causes infighting. Through divining bones, it can be comprehended that interpretations or the terms or language being used also help define (current) timelines people are living in and extrapolations of existing everyday realities as periods that reflect m*otlhakola* or *thwagadima* and or *mpherefere.* The following section is a summation of all three concepts in what is the spectacle of the chapter.

THE BODY OR PEOPLE AS DIVINING BONES

The African conception of the body or people demonstrating divination bones is explored in the previous sections in more than one dimension, i.e., as positioned in the three concepts discussed. Here the body as the divining bone is a spectacle of this chapter illustrating what characterises the conceptual and schematic thinking of the various examples highlighted in previous sections. It is herein illustrated by Credo Mutwa's philosophy on senses of African imagination and familiarity as verified by Mutwa's triangulation methodology, i.e., assigning "the divination in several different places: indoors, outdoors, at a sacred site or cave or on a mountain top" (Mutwa, 2003, p. 26,27). In this chapter, various examples and scenarios are mentioned as reflecting the three concepts occurring "in several different places". The following cases or scenarios are anecdotal evidence demonstrating the ideology of what Mutwa said about the body or people who are familiar with their surroundings and selves. In a picture captured on 27 February 2020 late in the afternoon Mutwa's innate intelligence of "the need to feel our surroundings, to feel what is happening around us" was re-membered. A description in support of Nabudere's (2011, p. 39) senses of Black imagination and ways of Africans knowing themselves and wa Thingo'o (2009, p. 37,39) on modes of remembering visions and practices that assert African realities with their contexts. Figure 2 below depicts the clouds which are in a wing-like shape dimmed partially by the sun, the individual thought *"gona le motho o mogolo o tsamayang"* meaning there's a prominent figure who will leave or depart this part of the earth. Remarkably on 25 March 2020, Mutwa (1921-2020) departed and negated the lockdown planned from 27 March 2020 in his country South Africa. Advocating for the figure below with Mutwa's philosophy on senses and assertion of an African's way of conducting research is Nabudere (2011, p. 66) augmenting that "there is a lot of scientific and philosophical knowledge hidden in African cosmologies & epistemologies".

Figure 2. The picture of wing-like shaped clouds obscuring the sun.

(Picture by Kgope, 27 February 2020)

According to Mutwa (Mutwa, 2003, p. 27,31) triangulation of research in bone divination deploys bones to be thrown in three or various places. Mutwa (2003, p. 31) asserts that if the findings or results revealed are the same, that gives credibility and validity for the findings to be accepted and recognised as relevant knowledge. Triangulation of the divination concepts used here is imagined as people or bodies positioned in the academy; in (blended) families, and or as different race groups. Mutwa empowers us that we must be able to feel what is happening around us. Mutwa coins this as the 'mother mind'. Urging humanity to be able to feel extra-terrestrially "What could be embedded in a stone or rock feature, what could be the materials that lie in the stone and what could be in the trees"? Ngulube and Ngulube (2017) concur that triangulation is an approach that corroborates information through various methods or multiple sources of evidence.

In another anecdotal evidence depicting that the body is a divining bone, the case is foretold when a 48-year-old man called Bra Rhu[1] who works as a handyman tells a story of his late uncle who visited him twice in a vision instructing him to drink a jar of water every morning before he leaves the house. According to his habits of drinking liquids, he rarely prefers and dislikes drinking water in large quantities. Nonetheless, Bra Rhu chooses to abide by the message his uncle passed on to him in a vision, although with the discomfort of experiencing a urinary bladder that was full while driving on the highway to town. He narrates a chilling ordeal of feeling pressed but with limited chances to drive off the road to relieve himself. He recounts that when he arrives at his place of destination the first thing he does is urgently request directions to where the lavatories are, like a horse, he runs to the loo, and in a few minutes he hears banging sounds of gunshots and later finds out that the very same people he went to consult with and who directed him to the latrine have all been shot dead. Although a shocking and unbelievable relation and account but it is a special type of knowledge (as with how we know things –

and the ledge we utilise to attach ourselves to the ledge of that essential imminent familiar knowledge imparted to us). In this case, the body served as a divining bone through a dream or vision of the late uncle, Bra Rhu relates that after the tragedy, the same uncle came back in another vision where he only smiled at him and abruptly disappeared.

Arguably when depicting African experiences as components of critical pedagogy for the sustainability of Indigenous knowledge and methodologies Mutwa entrusts that the following must justify that:

> The African conducts research on a very different plane from that of the White man, and for this reason, he has made discoveries that the White man has overlooked in his headlong rush to outer space. The Black man possesses tremendous knowledge that could make a great impact, even on the modern world, and he has hidden this knowledge, now for hundreds of years under the cloak of "voodoo" and" black magic". (Mutwa, 1964, p. 657)

The statement is in support of what Mutwa said about the tenets of approaching any method, that it is through the perspective of being familiar with Africa and its people because "a person who is not familiar with Africa and its people might find it difficult to understand" (Mutwa, 1964, p. 529). The question is do African universities, people and bodies understand themselves from the vantage point of Mutwa's familiarity competence to curb the terrains of *motlhakola, thwagadima and mpherefere*?

Legwame a divination bone representing the ancestors or old man (Mutwa 2003, p. 26) is here equated to the late indigenous knowledge holder Credo Vusamazulu Mutwa because what he projected and warned the nation and global communities about is evidenced by the three concepts described. Like *legwame* as a plant, it also flourishes when all spheres in the ecosystems are in harmony.

FURTHER RESEARCH DIRECTIONS

Future studies recommend the need to disseminate and develop indigenous healing methods in association with practitioners in traditional healing, spiritual healing therapies and prophets. More studies are required to expand the critical steps in knowing and motivating initiation into African indigenous knowledge and spiritual healing practices.

CONCLUSION

This chapter has conceptualised and cast the body, people, and various situational contexts as symbolic of the descriptions of the three divination *(ditaola)* bone concepts, *motlhakola, thwagadima and mpherefere*. Using the application of Credo Mutwa's bone oracle and approach as a lens and triangulation in confirming the descriptive nature of the fall of the divination bone concepts as occurring in different settings to people, bodies, and situations it is shown that when African contexts, senses of the African imagination, remembering of visions and practices to follow, the body or people may be acquainted with modes of hope and peace as validated by the sense of their familiarity competence with who they are. The chapter concludes with a sense of imagination that submits that there needs to be a referendum to privilege indigenous knowledge epistemologies in the academy across all disciplines. It is such endeavours that validate how Africans venerate their values, and customs with the responsibilities of leading functional lifestyles that project hope and peace. Put in another vein, this chapter asked how Africans can be true to themselves.

The exploration of the fall of the divination bones describing dispossession or *motlhakola*, danger, death or *thwagadima* and chaos, and disputes as *mpherefere* were discussed. The states are about casting the descriptions of the concepts as symbolising the current timelines in post-apartheid South African democracy as characterised by growing inequality, poverty and unemployment resulting in conditions that are marred by complexities of loss, disputes, and chaos. It remarks that if we truly reflect who we are in terms of African indigenous values systems we will not know dispossession because nothing will be stolen from us. Mutwa's application of the bone oracle is also explained in the context of the academy in the complexities of the twenty-first century. The question is does the body or people and the African universities understand themselves from the vantage point of Mutwa's familiarity competence to project bodies, people and institutions of hope and peace? The scenarios indicated to represent the three concepts discussed suggest that the body or people inclined to the discoursed divining bone concepts are not familiar with who they are hence the overreliance on Western scholarly approaches as signs and symptoms of defeat in challenges that should steer indigenous knowledge epistemologies for sustainable futures. The chapter calls for a political will for the state and the African universities to appreciate the observed realities of its people.

ACKNOWLEDGEMENTS

The financial assistance from the National Institute for the Humanities and Social Sciences (NIHSS), in collaboration with the South African Humanities Deans Association (SAHUDA) towards this research is hereby acknowledged. Opinions expressed and conclusions arrived at are those of the author and are not necessarily to be attributed to the NIHSS and SAHUDA.

REFERENCES

Academy of Science of South Africa. (2010). "Academic Freedom Statement From The Academy Of Science Of South Africa (ASSAf)." DOI: 10.1186/1472-6920-14-40 PMID: 24575843

Cakata, Z. (2015). "In search of the absent voice: The status of indigenous languages in post-apartheid South Africa." PhD diss., University of South Africa.

Chitindingu, E., George, G., & Gow, J. (2014). A review of the integration of traditional, complementary and alternative medicine into the curriculum of South African medical schools. *BMC Medical Education*, 14(1), 40. DOI: 10.1186/1472-6920-14-40 PMID: 24575843

Croucamp, A. (2013). "Traditional African divination systems as information technology." *Mindburst, viewed* 23DOI: 10.1186/1472-6920-14-40 PMID: 24575843

Croucamp, P. A., & Malan, L. (2018). The theory of systemic patronage and state capture: The liberal democratic project and its regime contenders. *African Journal of Public Affairs*, 10(4), 86–105.

de Sousa Santos, B. (2016). "Epistemologies of the South and the future." *From the european south* 1, no. 2016: 17-29. DOI: 10.1186/1472-6920-14-40 PMID: 24575843

Gordon, L. (2010). Theory in black: Teleological suspensions in philosophy of culture. *Qui Parle*, 18(2), 193–214. DOI: 10.5250/quiparle.18.2.193

Gyekye, K. (1995). *An essay on African philosophical thought: The Akan conceptual scheme*. Temple University Press.

Kgope, T. V. (2022). "A Critical Reflection on Complementary, Alternative and Indigenous Knowledge Medicine in Gauteng Province: A Model for Articulation and Promotion." PhD diss., 2022.

Kgope, T. V. (2023). The indigenous knowledge system of Credo Mutwa: A pedagogical challenge in higher education in South Africa. *Curriculum Perspectives*, 43(S1), 1–9. DOI: 10.1007/s41297-023-00182-2

Kgoroeadira, K. O. (2014). *The Praise Poetry of the Bafokeng of Phokeng*. University of Johannesburg.

Kovach, M. (2010). Conversational method in indigenous research. *First Peoples Child & Family Review*, 5(1), 40–48. DOI: 10.7202/1069060ar

Kovach, M. (2019). Conversational method in indigenous research. *First Peoples Child & Family Review*, 14(1), 123–135. DOI: 10.7202/1071291ar

Makgopa, M., & Magaelane, K. (2009). The use of ditaola (divination bones) among indigenous healers in Sekhukhune District, Limpopo Province: Indigenous African healing practices. *Indilinga*, 8(1), 51–58. DOI: 10.4314/indilinga.v8i1.48242

Makhalemele, O. (2009). "Still Not Talking: The South African government's exclusive reparations policy and the impact of the R30,000 financial reparations on survivors." In *Reparations for Victims of Genocide, War Crimes and Crimes against Humanity*, pp. 541-566. Brill Nijhoff. DOI: 10.1163/ej.9789004174498.i-576.139

Mazrui, A. A. (2003). Towards re-Africanizing African universities: Who killed intellectualism in the post colonial era? *Alternatives. Turkish Journal of International Relations*, 2(3&4).

Mignolo, W. (2011). *The darker side of western modernity: Global futures, decolonial options*. Duke University Press.

Modiri, J. (2021). Azanian political thought and the undoing of South African knowledges. *Theoria (Pietermaritzburg)*, 68(168), 42–85. DOI: 10.3167/th.2021.6816804

Mũkoma wa Ngũgĩ. (2019). *The rise of the African novel: politics of language, identity, and ownership.* Pietermaritzburg: University of KwaZulu-Natal Press.

Mutwa, C. (1964). *Indaba, my children: African tribal history, legends, customs and religious beliefs: African tribal history, legends, customs and religious beliefs*. Canongate Books.

Mutwa, C.& Larsen. (2003). *Zulu shaman: dreams, prophecies, and mysteries*. Destiny books.

Nabudere, D. (2011). *Afrikology, philosophy and wholeness. An epistemology: An epistemology*. African Books Collective.

Nabudere, D. W. (2003). Towards the establishment of a pan-African University: A strategic concept paper. *Afr J Polit SCI*, 8, 1–30.

Ngũgĩ wa Thiong'o. (1986). *Decolonising the mind: the politics of language in African literature.* London: James Currey.

Ngũgĩ wa Thiong'o. (2009). *Something torn and new: an African renaissance.* New York: Basic Civitas Books.

Ngulube, P., Mathipa, E. R., & Gumbo, M. T. (2015). "Theoretical and conceptual frameworks in the social and management sciences." *Addressing research challenges: Making headway in developing researchers*, pp. 43-66.

Ngulube, P., & Ngulube, B. (2017). Application and contribution of hermeneutic and eidetic phenomenology to indigenous knowledge research. In *Handbook of research on theoretical perspectives on indigenous knowledge systems in developing countries* (pp. 127–155). IGI Global. DOI: 10.4018/978-1-5225-0833-5.ch006

Peek, P. M. (1991). *African divination systems: Ways of knowing*. Georgetown University Press.

Prah, K. (2010). Multilingualism in urban Africa: Bane or blessing. *Journal of Multicultural Discourses*, 5(2), 169–182. DOI: 10.1080/17447143.2010.491916

Prah, K. (2012). The language of development and the development of language in contemporary Africa. *Applied Linguistics Review*, 3(2), 295–313. DOI: 10.1515/applirev-2012-0014

Ross, E. (2010). Inaugural lecture: African spirituality, ethics and traditional healing–implications for indigenous South African social work education and practice. *South African Journal of Bioethics and Law*, 3(1), 44–51.

Sandoval, C. (2019). *Methodology of the Oppressed*. Vol. 18. U of Minnesota Press, 2013.

Scott, K., and H. Pettersson. (2019). "South Africa is the world's most unequal country. 25 years of freedom have failed to bridge the divide. CNN, 10 May."

Smith, L. T. (1999). *Decolonizing methodologies: research and indigenous peoples*. Zed Books.

ENDNOTES

1 Not his real name

Compilation of References

Abdi, M. (2020). Decolonising higher education. In: *Decolonising education: Let's talk about it*. YouTube channel. Retrieved on 25 July 2023 from: https://www.youtube.com/watch?v=2Y0eBb0UuvQ&feature=youtu.be

Abdi, A. A. (2020). Decolonizing knowledge, education and social development: Africanist perspectives. *Beijing International Education Research*, 2(4), 503–518. DOI: 10.1163/25902539-02040006

Abdullah, H. J., & Hassan, T. K. (2015). The use of Indigenous knowledge in agriculture and its role in sustainable development. *International Journal of Scientific Research*, 4(10), 1312–1317.

Absolon, K. (2010). Indigenous wholistic theory: A knowledge set for practice. *First Peoples Child & Family Review*, 5(2), 74–87. DOI: 10.7202/1068933ar

Academy of Science of South Africa. (2010). "Academic Freedom Statement From The Academy Of Science Of South Africa (ASSAf)." DOI: 10.1186/1472-6920-14-40 PMID: 24575843

Adam, A., Othman, N., & Halim, A. (2021). Indigenous knowledge transfer among Dusuns and Bajaus in Kota blue, Sabah, Malaysia: Approaches and challenges. *International Journal of Academic Research in Business & Social Sciences*, 11(11). Advance online publication. DOI: 10.6007/IJARBSS/v11-i11/11656

Adamo, G. (2020). *Investigating business process elements: A journey from business process management to ontological analysis, and back* (Doctoral dissertation, University of Genoa, Italy).

Addae, D., & McIntyre-Mills, J. J. (2022). Balancing individualism and collectivism: a conversation on Afrocentricity. In Kwapong, O. A. T. F., Addae, D., & Boateng, J. K. (Eds.), *Reimagining development education in Africa* (pp. 55–74). Springer. DOI: 10.1007/978-3-030-96001-8_4

Adeniran, A. (2022). A report on Nigerian Gross Domestic Product. Q1 NBS. Retrieved on 10 June 204 from https://nigerianstat.gov.ng/elibrary/read/1241175

Adjimah, H. P., Atiase, V., & Dzansi, D. Y. (2023). Do government incentives increase Indigenous innovation commercialisation? Empirical evidence from local Ghanaian firms. *International Journal of Entrepreneurial Behaviour & Research*. Advance online publication. DOI: 10.1108/IJEBR-02-2022-0157

Adyanga, F. A., & Romm, N. R. A. (2016). Researching Indigenous science knowledge integration in formal education: Interpreting some perspectives from the field. *International Journal of Educational Development*, 3(1), 1–14. DOI: 10.25159/2312-3540/2696

Adyanga, F. A., & Romm, N. R. A. (2022). Reflections upon our way of invoking an Indigenous paradigm to co-explore community mobilization against irresponsible practices of foreign-owned companies in Nwoya District, Uganda. *The Qualitative Report*, 27(7), 13. Advance online publication. DOI: 10.46743/2160-3715/2022.5147

Adzahlie-Mensah, V., & Dunne, M. (2019). Continuing in the shadows of colonialism: The educational experiences of the African child in Ghana. *Perspectives in Education*, 36(2), 44–60. DOI: 10.18820/2519593X/pie.v36i2.5

Afolabi, O. S. (2020). Globalisation, decoloniality and the question of knowledge production *HEA/RESA*, 18(1), 93-109.

African Journals Online (AJOL) (sa). Welcome to AJOL. Retrieved on 25 June 2023 from: https://www.ajol.info/index.php/ajol

African Union. (2015). *African Agenda 2063 – The Africa we want, Framework document*. Addis Ababa.

Ageh, P. A., & Lall, N. (2019). Biopiracy of plant resources and sustainable traditional knowledge system in Africa. *Global Journal of Comparative Law*, 8(2), 162–181. DOI: 10.1163/2211906X-00802003

Aggarwal, J. C. (1989). *Adult and social education in India: global perspective*. Doaba House.

Agrawal, A. (2004). Indigenous and scientific knowledge: Some critical comments. *Indigenous Knowledge and Development Monitor*, 3(3), 7–8.

Ahmedi, V., Kurshumlija, A., & Ismajli, H. (2023). Teachers`attitudes towards constructivist approach to improving learning outcomes: The case of Kosovo. *International Journal of Instruction*, 16(1), 441–454. DOI: 10.29333/iji.2023.16124a

Aikehead, G. S., & Masakata, O. (2007). Indigenous knowledge and science revisited. *Cultural Studies of Science Education*, 2(3), 539–620. DOI: 10.1007/s11422-007-9067-8

Ainscow, M. (2005). Developing inclusive education systems: What are the levers for change? *Journal of Educational Change*, 6(2), 109–124. DOI: 10.1007/s10833-005-1298-4

Ainscow, M. (2020). Promoting inclusion and equity in education: Lessons from international experiences. *Nordic Journal of Studies in Educational Policy*, 6(1), 7–16. DOI: 10.1080/20020317.2020.1729587

Aiseng, K. (2023). Challenges and opportunities of preserving African Indigenous knowledge using digital technologies: the case of Bogwera. In Masenya, T. (Ed.), *Digital preservation and documentation of global Indigenous knowledge systems* (pp. 123–141). IGI Global., DOI: 10.4018/978-1-6684-7024-4.ch007

Ajani, Y., Oladokun, B., Olarongbe, S., Amaechi, M., Rabiu, N., & Bashorun, M. (2024). revitalizing Indigenous knowledge systems via digital media technologies for sustainability of Indigenous languages. *Preservation. Digital Technology & Culture*, 53(1), 35–44. DOI: 10.1515/pdtc-2023-0051

Ake, C. (2011). *Social science as imperialism: the theory of political development*. Ibadan University Press.

Akinbosede, D. (2022). Acknowledge racist roots in science. *Nature*, 612(7938), 178–179. DOI: 10.1038/d41586-022-04123-3 PMID: 36424502

Akinwale, A. A. (2017). Digitalization of Indigenous knowledge for natural resource management in Africa. *AERN SUMMIT 2012 Conference Paper.* Retrieved on 29 October 2024 from https://api-ir.unilag.edu.ng/server/api/core/bitstreams/f0e2186a-f930-471e-9d01-90056f1d17d9/content

Alidou, H., Boly, A., Brock-Utne, B., Diallo, Y. S., Heugh, K., & Wolff, H. E. (2006). Optimizing learning and education in Africa – the language factor: a stock-taking research on mother tongue and bilingual education in sub-Saharan Africa. Paper presented at the ADEA Biennial Meeting, Libreville, 27-31 March. Retreieved on 10 February 2023 from https://biennale.adeanet.org/2006/doc/document/B3_1_MTBLE_en.pdf

Aliyu, U. S., Ozdeser, H., Çavuşoğlu, B., & Usman, M. A. M. (2021). Food security sustainability: A synthesis of the current concepts and empirical approaches for meeting SDGs. *Sustainability (Basel)*, 13(21), 11728. DOI: 10.3390/su132111728

Allison-Cassin, S., Bateman, R., Bhardwaj, R. K., & Chisita, C. T. (2023). Introduction. *The Serials Librarian*, 84(5–8), 62–65. DOI: 10.1080/0361526X.2024.2313400

Alvesson, M., & Skoldberg, K. (2000). Reflexive methodology: New vistas for qualitative research. *Sage (Atlanta, Ga.).*

Alyson, S. (2022). Colonial education remains alive and well. *Pan African Review*. Retrieved on 10 October 2024 from https://panafricanreview.com/colonial-education-remains-alive-and-well/

Amenyedzi, S. B. (2023). The Afrocentric-Youth Paradigm. *Journal of Youth and Theology*, 1, 1–12.

Anderson, P., Levon, B., Thu, P., & Melanie, S. (2022). *Higher degree by research: factors for Indigenous student success*. Springer. DOI: 10.1007/978-981-19-5178-7

Andreas, S. (2016). Better skills, better jobs, better lives. *Global Comparative Education: Journal of the WCCES*, 1(1), 144–149.

Angus, F. G. (1967). *South Australia. A.H. and A.W.* Reed Publishers.

Ankoma, A. (2020). "Repatriation of African Art." *Arts24*, 9 25 December 2020). Retrieved 10 October 2025 from News24.com/amp/life/arts-and-entertainment/arts/its-time-for-African-countries-to-take-advantage-of-the-moment-and-coordinate-a-decisive-position-on-the-custodianship-of-African-art.

Ann Rahman, E., & Cochrane, T. (2023). Pedagogy and Indigenous knowing and learning. *Oxford Review of Education*, 49(4), 429–445. DOI: 10.1080/03054985.2023.2228640

Anthony, G. B. & Omang, T. N. (2018). Adult literacy education: a pivot for environmental sustainability. *LWATI: A Journal of Contemporary Research*, 15(4), 172-193.

Antonelli, A. (2023). Indigenous knowledge is key to sustainable food systems. *Nature*, 613(7943), 239–242. DOI: 10.1038/d41586-023-00021-4 PMID: 36627421

Anzellotti, S., & Young, L. L. (2020). The acquisition of personal knowledge. *Annual Review of Psychology*, 71(1), 613–634. DOI: 10.1146/annurev-psych-010419-050844 PMID: 31553673

Appadurai, A. (1990). Disjuncture and difference in the global cultural economy. *Theory, Culture & Society*, 7(2-3), 295–310. DOI: 10.1177/026327690007002017

Appiah-Opoku, S. (2005). *The need for environmental impact assessment: the case of Ghana*. The Edwin Mellen Press.

April, W. I., & Itenge, D. I. (2020). Fostering Indigenous entrepreneurship amongst San people: An exploratory case of Tsumkwe. *International Journal of Business and Globalisation*, 24(4), 496–512. DOI: 10.1504/IJBG.2020.106955

Ariyani, F., Putrawan, G. E., Riyanda, A. R., Idris, A., Misliani, L., & Perdana, R. (2022). Technology and minority language: An Android-based dictionary development for the Lampung language maintenance in Indonesia. *Tapuya : Latin American Science, Technology and Society*, 5(1), 2015088. Advance online publication. DOI: 10.1080/25729861.2021.2015088

Arko-Achemfuor, A., & Dzansi, D. Y. (2015). Business doing well by doing good in the community. *The Journal of Commerce*, 7(2), 53–68.

Arko-Achemfuor, A., Romm, N., & Serolong, L. (2019). Academic-practitioner collaboration with communities towards social and ecological transformation. *International Journal of Transformative Research*, 6(10), 1–9. DOI: 10.1515/ijtr-2019-0002

Arowolo, D. (2010). The effects of Western civilisation and culture on Africa. *Afro-Asian Journal of Social Sciences*, 1(1), 1–13.

Asakitikpi, A. (2018). Western versus African medical knowledge systems: A comparative epistemological. *Analysis Synthesis Philosophica*, 33(1), 91–104. DOI: 10.21464/sp33106

Asamoah, C., & Ngulube, P. (2023). Exploring models for the management of Indigenous knowledge in academic libraries of Ghana. *Information Development*, 39(4), 787–797. DOI: 10.1177/02666669211052928

Asante, M. K. (2005). Afronography. *Encyclopaedia Of Black Studies*, 76-77.

Asante, M. K. 2009. *Afrocentricity*. Retrieved on 20 October 2024 from http://www.asante.net/articles/1/afrocentricity

Asante, M. K. (1991). The Afrocentric idea in education. *The Journal of Negro Education*, 60(2), 170–180. DOI: 10.2307/2295608

Asante, M. K. (2003). *Afrocentricity: A theory of social change* (2nd ed.). African American Images.

Asante, M. K. (2007). *An Afrocentric manifesto: toward an African renaissance*. Polity Press.

Asante, M. K. (2007). An Afrocentric manifesto: Toward an African renaissance. *Polity*.

Asante, M. K. (2009). *Afrocentricity*. SUNY Press.

Asante, M. K. (2013). *The African American people: a global history*. Routledge. DOI: 10.4324/9780203145807

Asante, M. K. (2014). *Facing south to Africa*. Lexington Books.

Asante, M. K. (2017). Afrocentricity and knowledge: Adjusting the narrative lens. *The Thinker Magazine*, 74, 24–26.

Athayde, S., Silva-Lugo, J., Schmink, M., & Heckenberger, M. (2017). The same but different: Indigenous knowledge system retention, erosion and innovation in the Brazilian Amazon. *Human Ecology: an Interdisciplinary Journal*, 45(1), 533–544. DOI: 10.1007/s10745-017-9919-0

Atiles-Osoria, J. (2018). Colonial state crimes and the CARICOM mobilization for reparation and justice. *State Crime*, 7(2), 349–368. DOI: 10.13169/statecrime.7.2.0349

Atleo, E. R. (2004). *Tsawalk, a Nuu-chah-nulth worldview*. UBC Press.

Atwi, S. K., & Hamza, K. (2015). Qualitative and quantitative research paradigms in business research: A philosophical reflection. *European Journal of Business and Management*, 7(3), 217–225.

AU (2015). African Union Agenda 2063: The Popular Version @African Union Commission

Auerbach, R., Rundgren, G., & El-Hage Scialabba, N. (2013). (Eds). *African experiences in resilience and sustainability*. Food and Agricultural Organization (FAO) of the UN. https://www.fao.org/4/i3294e/i3294e.pdf

Australian Government. (2024). Respectful use of Indigenous knowledge. Retrieved on 10 October 2024 from https://www.ipaustralia.gov.au/understanding-ip/Indigenous-knowledge/how-to-engage-Indigenous-knowledge

Ayeni, A. O., & Aborisade, A. G. (2022). African Indigenous knowledge systems and the world. In Oloruntoba, S. O., & Falola, T. (Eds.), *The Palgrave handbook of Africa and the changing global order*. Palgrave Macmillan., DOI: 10.1007/978-3-030-77481-3_8

Ayeni, A. O., & Oladeji, K. O. (2023). Pond water as a cultural element: Understanding Aringíyà's household activities and festivities. [JECA]. *Journal of English and Communication in Africa*, 6(3&4), 20–35.

Ayeni, A., Soneye, A., & Badru, F. (2014). Adaptation to water stress in Nigeria derived Savanna area: The Indigenous knowledge and socio-cultural nexus of management and humanitarian services. [JMPP]. *Journal of Management Policy and Practice*, 15(3), 78–87.

Aziz, M. A., Adnan, M., Khan, A. H., Shahat, A. A., Al-said, M. S., & Ullah, R. (2018). *Traditional uses of medicinal plants practiced by the Indigenous communities at Mohmand Agency*. FATA., DOI: 10.1186/s13002-017-0204-5

Azure, J. A. (2016). Students' perspectives of effective supervision of graduate programmes in Ghana. *American Journal of Educational Research*, 4(2), 163–169.

Babagana, U. I., & Kemboi, G. (2024). Digital preservation of Indigenous Knowledge in Ramat Library (University of Maiduguri, Nigeria). *iPRES*. Retrieved on 10 October 2024 from https://ipres2024.pubpub.org/pub/o5shuj0u

Badat, S. 2023. Colonialism shaped modern universities in Africa – how they can become truly African. *The Conversation*. Retrieved on 22 October 2024 from https://theconversation.com/colonialism-shaped-modern-universities-in-africa-how-they-can-become-truly-african-214494 downloaded in 22/10/2024

Badewa, A. S., & Dinbabo, M. F. (2023). Multisectoral intervention on food security in complex emergencies: A discourse on regional resilience praxis in Northeast Nigeria. *GeoJournal*, 88(2), 1231–1250. DOI: 10.1007/s10708-022-10679-4 PMID: 35729954

Balogun, O. K., & Adeseke, A. E. (2022). Advancing Indigenous African values and ethos for film directing and production in Nollywood. [IJCRH]. *International Journal of Current Research in The Humanities*, 26(1), 87–97. DOI: 10.4314/ijcrh.v26i1.6

Balogun, T. (2023). Digitization of Indigenous knowledge systems in Africa: The case of South Africa's National Recorded System (NRS). *Records Management Journal*, 33(1), 88–102. DOI: 10.1108/RMJ-10-2022-0029

Balogun, T., & Kalusopa, T. (2022). Web archiving of Indigenous knowledge systems in South Africa. *Information Development*, 38(4), 658–671. DOI: 10.1177/02666669211005522

Bandura, A. (1977). Self-efficacy: Toward a unifying theory of behavioral change. *Psychological Review*, 84(2), 191–215. DOI: 10.1037/0033-295X.84.2.191 PMID: 847061

Bandura, A. (1986). *Social foundations of thought and action: a social cognitive theory*. Prentice-Hall.

Bandura, A. (1997). Self-Efficacy: The exercise of control. *The Freeman*.

Bansal, S., Sarker, T., Yadav, A., Garg, I., Gupta, M., & Sarvaiya, H. (2024). Indigenous communities and sustainable development: A review and research agenda. *Global Business and Organizational Excellence*, 43(4), 65–87. DOI: 10.1002/joe.22237

Baptiste, F. A. (2008). From "Invisibility" to "Visibility": Africans in India through the Lens of Some Select Sources from the Late Classical Period to the Late 18th century A.D. In Prasad, K. K., & Angenot, J.-P. (Eds.), *TADIA: The African Diaspora in India*. Jana Jagrati Prakashana.

Barkin, D. (2022). Shaping a communitarian ethos in an era of ecological crisis. *Frontiers in Sustainability*, 3, 1–14. DOI: 10.3389/frsus.2022.944252

Barnett, A. (2024). Why I've removed journal titles from the papers on my CV. *Nature*. Advance online publication. DOI: 10.1038/d41586-024-02596-y PMID: 39122980

Barnhardt, R., & Kawagley, A. O. (2005). Indigenous knowledge systems and Alaska native ways of knowing. *Anthropology & Education Quarterly*, 36(1), 8–23. DOI: 10.1525/aeq.2005.36.1.008

Basu, H. (2008). History on the Line: Music and the Formation of Sidi Identity in Western India. *History Workshop Journal*, 65(1), 161–178. DOI: 10.1093/hwj/dbm069 PMID: 19618564

Basu, M., & Dasgupta, R. (2023). Exploring Indigenous water knowledge, values, and practices: insights and examples. In Basu, M., & DasGupta, R. (Eds.), *Indigenous and Local Water Knowledge, Values and Practices*. Springer., DOI: 10.1007/978-981-19-9406-7_1

Battiste, M. (2000). *Reclaiming Indigenous voice and vision*. University of British Columbia Press.

Battiste, M. (2002). *Indigenous knowledge and pedagogy in First Nations education: a literature review with recommendations*. Indian and Northern Affairs Canada.

Baxter, B. (2018). The impact of globalization on communication & education. HIGHVIBE Network. Open Access. Retrieved on 29 October 2024 from https//medium.com.highgate.

Beach, D. N. (1980). *The Shona in Zimbabwe 900-1850*. Mambo Press.

Behailu, B. M., Pietilä, P. E., & Katko, T. S. (2016). Indigenous practices of water management for sustainable services: Case of Borana and Konso, Ethiopia. *SAGE Open*, 6(4), 2158244016682292. Advance online publication. DOI: 10.1177/2158244016682292

Benagr, S., & Ofosu, T. B. (2016). Ghanaian screendance perspectives: The nuance of 'Sankofaism' as emerging aesthetics and rejection of orthodoxy. *The International Journal of Screendance*, 6, 78–93. DOI: 10.18061/ijsd.v6i0.4913

Bennedetti, L. V., de Almeida Sinisgalli, P. A., Ferreira, M. L., & Lemes de Oliveira, F. (2023). Challenges to promote sustainability in urban agriculture models: A review. *International Journal of Environmental Research and Public Health*, 20(3), 2110. DOI: 10.3390/ijerph20032110 PMID: 36767476

Bennett, K. (2015). Towards an epistemological monoculture: mechanisms of epistemicide in European research publication. In Plo Alastrué, R., & Pérez-Llantada, C. (Eds.), *English as a Scientific and Research Language: Debates and DiscoursesEnglish in Europe* (Vol. 2, pp. 9–36). De Gruyter Mouton. DOI: 10.1515/9781614516378-004

Berenson, E. (2017). Ethical Principles Of Psychologists And Code Of Conduct. American Psychological Association, 1–21. DOI: 10.5430/jnep.v7n1p11

Berkes, F. (2004). Rethinking community-based conservation. *Conservation Biology*, 18(3), 621–630. DOI: 10.1111/j.1523-1739.2004.00077.x

Berndt, R. M. (1952). *Djanggawul: An Aboriginal Religious Cult on North Eastern Arnhem Land*. Routledge and Kegan Paul Ltd.

Best, A. C. (1970). General trading in Botswana, 1890–1968. *Economic Geography*, 46(4), 598–611. DOI: 10.2307/142943

Bhaskar, R. (2008). *Dialectic: the pulse of freedom, with new introduction*. Routledge. DOI: 10.4324/9780203892633

Birdsall, S. (2014). Measuring student teachers' understandings and self-awareness of sustainability. *Environmental Education Research*, 20(6), 814–836. DOI: 10.1080/13504622.2013.833594

Boadu, E. S. (2022). Indigenous knowledge and community development in Africa: a framework to explore the approaches, philosophies and practices in Ghana. In: Mbah, M.F., Leal Filho, W., Ajaps, S. (eds) *Indigenous methodologies, research and practices for sustainable development*. World Sustainability Series. Springer. https://doi.org/DOI: 10.1007/978-3-031-12326-9_9

Boamah, E., & Liew, C. L. (2017). Conceptualising the digitisation and preservation of Indigenous knowledge: the importance of attitudes. In: Choemprayong, S., Crestani, F., Cunningham, S. (eds) *Digital libraries: data, information, and knowledge for digital lives*. ICADL 2017. Lecture Notes in Computer Science, vol 10647. Springer. https://doi.org/DOI: 10.1007/978-3-319-70232-2_6

Bodeker, G., & Kariippanon, K. (2020, February 28). Traditional medicine and Indigenous health in Indigenous Hands. *Oxford Research Encyclopedia of Global Public Health.* Retrieved on 10 October 2024 from https://oxfordre.com/publichealth/view/10.1093/acrefore/9780190632366.001.0001/acrefore-9780190632366-e-155

Boezak, W. (2015). *Struggle of an ancient faith – the Khoisan of South Africa.* [Self-published].

Boggs, C. (1976). *Gramsci's Marxism.* Pluto Press.

Bohaker, H. (2006). The significance of Algonquian kinship networks in the Eastern Great Lakes Regions. *The William and Mary Quarterly*, 63(1), 23–52. DOI: 10.2307/3491724

Bola, G., Mabiza, C., Goldin, J., Kujinga, K., Nhapi, I., Makurira, H., & Mashauri, D. (2014). Coping with droughts and floods: A case study of Kanyemba. Mbire district, Zimbabwe. *Physics and Chemistry of the Earth Parts A/B/C*, 67–69(1-2), 180–186. DOI: 10.1016/j.pce.2013.09.019

Bolderston, A. (2008). Writing an effective literature review. *Journal of Medical Imaging and Radiation Sciences*, 39(2), 86–92. DOI: 10.1016/j.jmir.2008.04.009 PMID: 31051808

Bol, G., & van Niekerk, D. (2024). Application of African Indigenous knowledge systems and practices for climate change and disaster risk management for policy formulation. *International Journal of Disaster Risk Reduction*, 111, 104670. DOI: 10.1016/j.ijdrr.2024.104670

Borocz, J. (2012). Colonialism. In Anheier, H. K., Juergenmeyer, M., & Faessell, V. (Eds.), *Encyclopaedia of global studies* (pp. 229–234). Sage.

Borocz, J. (2020). Decolonial theories in comparison. *Journal of World Philosophies*, 5, 43–60.

Bourque Bearskin, R. L., Cameron, B. L., King, M., & Weber Pillwax, C. (2016). Mâmawoh Kamâtowin, "Coming together to help each other in wellness": Honouring Indigenous Nursing Knowledge. *International Journal of Indigenous Health*, 11(1), 18–33. DOI: 10.18357/ijih111201615024

Bramer, W. M., Rethlefsen, M. L., Kleijnen, J., & Franco, O. H. (2017). Optimal database combinations for literature searches in systematic reviews: A prospective exploratory study. *Systematic Reviews*, 6(1), 245. DOI: 10.1186/s13643-017-0644-y PMID: 29208034

Braun, V., & Clarke, V. (2023). Toward good practice in thematic analysis: Avoiding common problems and be (com)ing a knowing researcher. *International Journal of Transgender Health*, 24(1), 1–6. DOI: 10.1080/26895269.2022.2129597 PMID: 36713144

Browne Gott, H. (2020). Exploring the relationship between culture and well-being. In Whaleves, Governemnt Social Research. https://gov.wales/exploring-relationship-between-culture-and-Creswell, W. J., & Creswell, J. D. (2022). Research Design Qualitative, Quantitative, and Mixed Methods Approaches. In Paper Knowledge. Toward a Media History of Documents (Sixth Edit). SAGE Publications, Inc. https://lccn.loc.gov/2017044644

Brown, J., Knol, D., Prevost-Derbecker, S., & Andrushko, K. (2007). Housing for Aboriginal youth in the inner city of Winnipeg. *First Peoples Child & Family Review*, 3(2), 56–64. DOI: 10.7202/1069464ar

Bruce, H. G. (2012). *Northern lights: Fourth World Nations and the geopolitical dance in the Arctic.* SIT Graduate Institute.

Bruchac, M. M. (2020). Indigenous knowledge and traditional knowledge. In *Encyclopedia of Global Archaeology*. https://doi.org/DOI: 10.1007/978-3-030-30018-0_10

Bruchac, M. M. (2014). Indigenous knowledge and traditional Knowledge. In Smith, C. (Ed.), *Encyclopedia of global archaeology*. Springer., DOI: 10.1007/978-1-4419-0465-2_10

Bruton, G. D., Zahra, S. A., Van de Ven, A. H., & Hitt, M. A. (2021). Indigenous theory uses, abuses, and future. *Journal of Management Studies*, •••, 1–17. DOI: 10.1111/joms.12755

Buarqoub, I. A. S. (2019). Language barriers to effective communication. *Utopía y Praxis Latinoamericana*, 24(6), 64–77.

Bujold, R. (2022). *Sharing intergenerational food stories to engage children in indigenous food sovereignty*. Masters diss., Dalhousie University, Halifax, Nova Scotia.

Bukovova, S., & Foltinova, L. (2020). Sustainable development and its impact on the quality of human capital. In *SHS Web of Conferences* (Vol. 83, p. 01005). EDP Sciences. DOI: 10.1051/shsconf/20208301005

Burchardt, H. J., & Dietz, K. (2014). Neo- extractivism: A new challenge for development theory from Latin America. *Third World Quarterly*, 35(3), 468–486. DOI: 10.1080/01436597.2014.893488

Burton-Page, J. (1971). *Encyclopedia of Islam*. Brill.

Buthelezi, N. N., & Hughes, J. C. (2014). Indigenous knowledge systems and agricultural rural development in South Africa: Past and present perspectives. *Indilinga*, 13(2), 231–250.

Buthelezi, S. C., Ocholla, D., & Dlamini, P. (2024). Strategies for documenting and disseminating Indigenous knowledge at a South African university. *South African Journal of Information Management*, 26(1), a1648. DOI: 10.4102/sajim.v26i1.1648

Cajete, G. (2000). *Native Science: Natural Laws of Interdependence. Santa FE.* Clear Light Publishers.

Cajete, G. A. (2000). *Native science: natural laws of interdependence*. Clear Light Books.

Cajete, G. A. (2020). Indigenous science, climate change, and Indigenous community building: A framework of foundational perspectives for Indigenous community resilience and revitalization. *Sustainability (Basel)*, 12(22), 9569. DOI: 10.3390/su12229569

Cakata, Z. (2015). "In search of the absent voice: The status of indigenous languages in post-apartheid South Africa." PhD diss., University of South Africa.

Campbell, P. (1991). *Lessons From the World: A Cross-cultural Guide to Music Teaching and Learning*. Schirmer Books.

Canclini, G. N. (1990). *Culturas híbridas: estrategias para entrar y salir de la modernidad*. Consejo Nacional para la Cultura y las Artes/Grijalbo.

Capra, F. (1975). *The Tao of physics*. Shambala.

Carballo-Garcia, A., & Boté-Vericad, J. J. (2022). Fair data: History and present context. *Central European Journal of Educational Research*, 4(2), 45–53. DOI: 10.37441/cejer/2022/4/2/11379

Carnie, T. (2024). UNDP moves to scrap support for "risky" R165bn Limpopo heavy industry zone. *The Daily Maverick*, 25 February 2024. https://www.dailymaverick.co.za/article/2024-02-25-limpopo-heavy-industry-plan-on-shaky-ground-as-un-agency-rethinks-support/

Carroll, S. R., Garba, I., Figueroa-Rodríguez, O. L., Holbrook, J., Lovett, R., Materechera, S., Parsons, M., Raseroka, K., Rodriguez-Lonebear, D., Rowe, R., Sara, R., Walker, J. D., Anderson, J., & Hudson, M. (2020). The CARE principles for Indigenous data governance. *Data Science Journal*, 19(43), 1–12. DOI: 10.5334/dsj-2020-043

Carroll, S. R., Herczog, E., Hudson, M., Russell, K., & Stall, S. (2021). Operationalizing the CARE and FAIR principles for Indigenous data futures. *Scientific Data*, 8(1), 108. DOI: 10.1038/s41597-021-00892-0 PMID: 33863927

Carroll, S. R., Rodriguez-Lonebear, D., & Martinez, A. (2019). Indigenous Data Governance: Strategies from United States Native Nations. *Data Science Journal*, 18(31), 1–15. DOI: 10.5334/dsj-2019-031 PMID: 34764990

Carter, A., Sisco, S., & Fowler, R. (2023). Since we are, therefore, I am ubuntu and the experiences of black women leadership coaches. *Consulting Psychology Journal*, 75(1), 51–67. DOI: 10.1037/cpb0000227

Cason, T. N., Sheremeta, R. M., & Zhang, J. (2017). Asymmetric and endogenous within-group communication in competitive coordination games. *Experimental Economics*, 20(4), 946–972. DOI: 10.1007/s10683-017-9519-2

Cavino, H. M. (2013). Across the colonial divide: Conversations about evaluation in Indigenous contexts. *The American Journal of Evaluation*, 34(3), 339–355. DOI: 10.1177/1098214013489338

Cekiso, M., Baba, T., Rose, M., & Themba, S. (2019). Supervision experiences of postgraduate research students at one South African higher education institution. *South African Journal of Higher Education*, 33(3), 8–25. DOI: 10.20853/33-3-2913

Celidwen, Y. (2022). Why we need Indigenous wisdom: The rich perspective of Indigenous peoples has a crucial role to play in advancing climate solutions rooted in an understanding of our interdependence. In: *Insights Journey into the heart of contemplative science*. Retrieved June 2023 from: https://www.mindandlife.org/insight/why-we-need-Indigenous-wisdom/

Chakma, U. (2024). Indigenous knowledge integration in Australian curriculum: Culturally responsive pedagogy and perspectives. *VNU Journal of Foreign Studies*, 40(1), 1–10. DOI: 10.63023/2525-2445/jfs.ulis.5200

Chambers, R. ., Pacey, A., & Thrupp, L. A. (1989). *Farmers first*. Intermediate Technology Publications Ltd (ITP).

Chambers, R. (1991). *Rural development: Putting the last first*. John Wiley.

Chandler, J., Cumpston, M., Thomas, J., Higgins, J. P. T., Deeks, J. J., & Clarke, M. J. (2022). Introduction. In: J. P. T. Higgins, J. Thomas, J. Chandler, M. Cumpston, T. Li, M.J. Page, & V. A. Welch (eds.), *J Cochrane Handbook for systematic reviews of interventions version 6.3* (updated February 2022). Cochrane, 2022. Retrieved 9 July 2023 from: www.training.cochrane.org/handbook

Chanza, N., & De Wit, A. (2013). Epistemological and methodological frameworks for Indigenous knowledge in climate science. *Indilinga*, 12(2), 203–216.

Chanza, N., & Gundu-Jakarasi, V. (2020). Deciphering the climate change conundrums in Zimbabwe: an exposition. In Tiefenbacher, J. (Ed.), *Global warming and climate change*. Intech Open. DOI: 10.5772/intechopen.84443

Charatsari, C., Fragkoulis, I., Anagnou, E., & Lioutas, E. D. (2022). Can adult education boost sustainability transitions? Some evidence from farmers and teachers. *Sustainability (Basel)*, 14(16), 9859. DOI: 10.3390/su14169859

Charmaz, K. (2017). The power of constructivist grounded theory for critical inquiry. *Qualitative Inquiry*, 23(1), 34–45. DOI: 10.1177/1077800416657105

Chaudhary, B. R., Erskine, W., & Acciaioli, G. (2022). Hybrid knowledge and climate-resilient agriculture practices of the Tharu in the western Tarai, Nepal. *Frontiers in Political Science*, 4, 969835. DOI: 10.3389/fpos.2022.969835

Chavunduka, M. (1995). *The missing links. keynote address to the workshop on the study and promotion of Indigenous knowledge systems and sustainable natural resources management in Southern Africa.* Midmar [April.]. *KwaZulu-Nata*, 1, 24.

Chawla, D. S. (10 February 2023). Elsevier journal under fire for rejecting papers that didn't cite enough of its old papers. *Chemical & Engineering News*. Retreieved on 10 February 2023 from https://cen.acs.org/policy/publishing/Elsevier-journal-under-fire-for-rejecting-paper/101/web/2023/02

Chemla, K. (2019). Needham and the issue of chinese as a language for science: Taking a linguistic turn materially. *Isis*, 110(1), 109–115. DOI: 10.1086/702915

Chibememe, G., Dhliwayo, M., Gandiwa, E., Mtisi, S., Muboko, N., & Kuipa, O. (2014). *Review of national laws and policies that support or undermine Indigenous peoples and local communities*. Natural Justice and Ford Foundation.

Chifamba, N.. (2020). Climate Change and the Mopane Worm: Implications for rural livelihoods in Zimbabwe. *Zimbabwe Journal of Environmental Studies*, 45(2), 135–148.

Chigangaidze, R., Matanga, A., & Katsuro, T. (2021). Ubuntu philosophy as a humanistic–existential framework for the fight against the covid-19 pandemic. *Journal of Humanistic Psychology*, 62(3), 319–333. DOI: 10.1177/00221678211044554

Chigwada, J., & Ngulube, P. (2023). Stakeholders in the acquisition, preservation, and dissemination of Indigenous knowledge projects. *Information Development*, 0(0), 02666669231192851. Advance online publication. DOI: 10.1177/02666669231192851

Chigwada, J., & Ngulube, P. (2024). Librarians' role in the preservation and dissemination of Indigenous knowledge. *IFLA Journal*, 50(2), 242–256. DOI: 10.1177/03400352231217270

Chikaire, J., Osuagwu, C. O., Ihenacho, R. A., Oguegbuchulam, M. N., Ejiogu-Okereke, N., & Obi, K. U. (2012). Indigenous knowledge system: The need for reform and the way forward. *Journal of Agricultural Science*, 1(8), 201–209.

Chikoko, V. (2021). Re-visiting the decolonising of the South African higher education question: A systematic literature review. *South African Journal of Higher Education*, 35(1), 21–36. DOI: 10.20853/35-1-4420

Chikwanha, O. (2016). Gender dynamics in the Mopane Worm Economy: A case study of Mangwe District. *Gender and Development Studies*, 34(3), 205–218.

Chilisa, B. (2015). A synthesis paper on the "Made in Africa" evaluation concept. African Evaluation Association. Retrieved February 4, 2020, from https://afrea.org/wp-content/uploads/2018/06/MAE2-Final-31st-august.pdf

Chilisa, B. (2024). Philosophical underpinnings of mixed methods: Decolonizing evaluation practice through decolonizing paradigms. In: Y. Shan Y (Ed,), *Philosophical foundations of mixed methods research: Dialogues between researchers and philosophers* (pp. 54-82) Routledge.

Chilisa, B. (2012). Indigenous research methodologies. *Sage (Atlanta, Ga.).*

Chilisa, B. (2017). Decolonizing transdisciplinary research approaches: An African perspective for enhancing knowledge integration in sustainability science. *Sustainability Science*, 12(5), 813–827. https://link.springer.com/article/10.1007/s11625-017-0461-1. DOI: 10.1007/s11625-017-0461-1

Chilisa, B. (2020). *Indigenous research methodologies* (2nd ed.). Sage.

Chilisa, B. (2024). Philosophical underpinnings of mixed methods: decolonizing evaluation practice through decolonizing paradigms. In Shan, Y. (Ed.), *Philosophical foundations of mixed methods research* (pp. 54–82). Routledge.

Chilisa, B., Major, T. E., Gaotlhobogwe, M., & Mokgolodi, H. (2017). Decolonizing and indigenizing evaluation practice in Africa: Toward African relational evaluation approaches. *The Canadian Journal of Program Evaluation*, 30(3), 312–328.

Chilisa, B., Major, T. E., & Khudu-Petersen, K. (2017). Community engagement with a postcolonial, African-based relational paradigm. *Qualitative Research*, 17(3), 326–339. DOI: 10.1177/1468794117696176

Chilisa, B., & Mertens, D. M. (2021). Indigenous made in Africa evaluation frameworks: Addressing epistemic violence and contributing to social transformation. *The American Journal of Evaluation*, 42(2), 241–253. DOI: 10.1177/1098214020948601

Chilisa, B., & Tsheko, G. N. (2014). Mixed methods in Indigenous research: Building relationships for sustainable intervention outcomes. *Journal of Mixed Methods Research*, 8(3), 222–233. DOI: 10.1177/1558689814527878

Chimakonam, J. O. (2019). Decolonisation, Africanisation and transformation: Why we need 'That' African contribution to world history and civilisation. In: *Ezumezu: A System of Logic for African Philosophy and Studies* (pp. 181–205). Springer. DOI: 10.1007/978-3-030-11075-8_11

Chimhundu, H. (2019). UNESCO/ Flanders project on strengthening Sub-regional cooperation and national capacities in seven Southern African countries for implementing the 2002 convention for the safeguarding of intangible cultural heritage. Retrieved on 10 October 2024 from www.saich.org/home/downloads/SAICH-Platform-Narrative-Report-final-draft-amended-291019.pdf

Chimhundu, H. (2019a). SAICH news 2019: on the tracks of living heritage in Southern Africa. Retrieved on 10 October 2024 from https://ich.unesco.org/doc/src/SAICH_NEWS_2019.pdf

Chirimuuta, C., Gudhlanga, E., & Bhukuvani, C. (2012). Indigenous knowledge systems: A panacea in education for development? The place and role of Indigenous knowledge systems in curriculum innovations in Zimbabwe. *International Journal of Social Science Tomorrow*, 4(5), 131–138.

Chisa, K., & Hoskins, R. (2014). The effects of information and communication technologies on Indigenous communities in South Africa: A library and information science perspective. *Mousaion*, 32(3), 49–68. DOI: 10.25159/0027-2639/1676

Chisa, K., & Hoskins, R. (2016). African customary law and the protection of Indigenous cultural heritage: Challenges and issues in the digitization of Indigenous knowledge in South Africa. *Indilinga*, 15(1), 1–15.

Chisita, C. T. (2011). Role of libraries in promoting the dissemination and documentation of Indigenous agricultural information: Case Study of Zimbabwe. In Paper of World Library And Information Congress: 77th IFLA General Conference And Assembly. Retrieved 10 October 2025 from http://conference. ifla.org/past-wlic/2011/78-chisita-en. pdf, diakses.

Chisita, C. T., & Abdullahi, I. (2010, August). The challenges and opportunities of preparing LIS students for orally based communities. In World Library and Information Congress: 76th IFLA General Conference and Assembly (pp. 10-15). Retrieved 10 October 2025 from https://www.ifla.org/past-wlic/2010/152-chisita-en.pdf

Chivaura, V. G. (2006). African Indigenous worldviews and ancient wisdom: a conceptual framework for development in Southern Africa. In Kunnie, J. E., & Goduka, N. I. (Eds.), *Indigenous Peoples' Wisdom and Power: Affirming our Knowledge Through Narratives*. Ashgate Publishing Aldershot.

Chiwanza, K., Musingafi, M. C. C., & Mupa, P. (2013). Challenges in preserving Indigenous knowledge systems: Learning from the past experiences. *Information and Knowledge Management*, 2(23), 19–25.

Chiwome, E. M., Mguni, Z., & Furusa, M. (2000). *Indigenous knowledge and technology in Africa and diaspora communities. Multidisciplinary approaches*. National Council for Black Studies.

Chouinard, J. A., & Cram, F. (2019). Culturally responsive approaches to evaluation. *Sage (Atlanta, Ga.)*.

Chowdhury, J., Wahab, H., Saad, M., Roy, P., Hamidi, M., & Ahmad, M. (2021). *Ubuntu philosophy.*, 361-381. DOI: 10.4018/978-1-7998-7480-5.ch022

Chowdhury, G., McLeod, J., Lihoma, P., Teferra, S., & Wato, R. (2023). Promoting access to Indigenous information in Africa: Challenges and requirements. *Information Development*, 39(3), 611–623. DOI: 10.1177/02666669211048488

Christen, K. (2020). Mukurtu: a digital platform that does more than manage content. CAS in the media, November 13. 2020. Retrieved on 11 October 2024 from https://cas.wsu.edu/news/2020/11/13/mukurtu-a-digital-platform-that-does-more-than-manage-content/

Christen, K., Merrill, A., & Wynne, M. (2017). A community of relations: Mukurtu hubs and spokes. *D-Lib Magazine : the Magazine of the Digital Library Forum*, 23(5/6). Advance online publication. DOI: 10.1045/may2017-christen

Christian Care. (2010). *Vulnerable group feeding end of programme report*. Christian Care.

Christie, B. A., Miller, K. K., Cooke, R., & White, J. G. (2013). Environmental sustainability in higher education: How do academics teach? *Environmental Education Research*, 19(3), 385–414. DOI: 10.1080/13504622.2012.698598

Cindi, L. (2021). *Incorporating African Indigenous knowledge systems into the basic education curriculum: experiences from two schools in the Gauteng and KwaZulu-Natal Provinces, South Africa*. M.Phil. diss. University of the Free State, South Africa.

Coates, S. K., Trudgett, M., & Page, S. (2023). Indigenous institutional theory: A new theoretical framework and methodological tool. *Australian Educational Researcher*, 50(3), 903–920. DOI: 10.1007/s13384-022-00533-4

Cobo, J. R. M. (1982). *Study of the problem of discrimination against Indigenous populations*. Final Report. United Nations.

Collins, C. S., & Stockton, C. M. (2018). The central role of theory in qualitative research. *International Journal of Qualitative Methods*, 17(1), 1609406918797475. Advance online publication. DOI: 10.1177/1609406918797475

Collins, P. H. (2000). *Black feminist thought* (2nd ed.). Routledge.

Collins, P. H. (2009). *Black feminist thought: Knowledge, consciousness, and the politics of empowerment*. Routledge Classics.

Collyer, F., Connell, R., Maia, J., & Morrel, R. (2019). *Knowledge and global power: making new sciences in the South*. Wits University Press.

Córdoba-Pachón, J. R. (2024). *Ritual and systems thinking: managing an initial encounter*. Taylor & Francis.

Cordova, V. F. (2007). *How It Is: The Native American Philosophy of V.F. Cordova*. The University of Arizona Press.

Cortes-Ramirez, E. (2015). Cultural hegemony today: From cultural studies to critical pedagogy. *Postcolonial Directions in Education*, 4(2), 116–139.

Cortina, R., & Earl, A. K. (2020). Advancing professional development for teachers in intercultural education. *Education Sciences*, 10(12), 1–12. DOI: 10.3390/educsci10120360

Cram, F., Tibbitts, K. A., & LaFrance, J. (2018). A steppingstone in Indigenous evaluation. In F. Cram, A. Tibbetts & J. LaFrance (Eds.), *Indigenous Evaluation. New Directions in Evaluation*, 159, pp. 7-16.

Cram, F., Chilisa, B., & Mertens, D. M. (2013). The journey begins. In Mertens, D. M., Cram, F., & Chilisa, B. (Eds.), *Indigenous pathways into social research* (pp. 11–41). Left Coast Press.

Cram, F., & Mertens, D. M. (2015). Transformative and Indigenous frameworks for multimethod and mixed methods research. In Hesse Biber, S., & Johnson, R. B. (Eds.), *The Oxford handbook of multimethod and mixed methods research inquiry* (pp. 91–109). Oxford University Press.

Craven, J. M. (2020). Indigenous approaches to economic development and sustainability. *International Critical Thought*, 10(1), 126–137. DOI: 10.1080/21598282.2020.1742529

Crenshaw, K. (1991). Mapping the Margins: Intersectionality, Identity Politics, and Violence against Women of Color. *Stanford Law Review*, 43(6), 1241–1299. DOI: 10.2307/1229039

Creswell, J. W. (2013). *Qualitative inquiry and research design: Choosing among five approaches* (3rd ed.). Sage Publications.

Creswell, J. W., & Plano Clark, V. L. (2018). *Designing and conducting mixed methods research* (3rd ed.). Sage.

Critchley, W. R. S., Reij, C., & Willcocks, T. J. (1994). Indigenous soil and water conservation: A review of the state of knowledge and prospects for building on traditions. *Land Degradation & Development*, 5(4), 293–314. Advance online publication. DOI: 10.1002/ldr.3400050406

Crone, E. A., Wendelken, C., van Leijenhorst, L., Honomichl, R. D., Christoff, K., & Bunge, S. A. (2009). Neurocognitive development of relational reasoning. *Developmental Science*, 12(1), 55–66. DOI: 10.1111/j.1467-7687.2008.00743.x PMID: 19120413

Cross, J. T. (2010). Property rights and traditional knowledge. [PELJ]. *Potchefstroom Electronic Law Journal*, 13(4), 12–47.

Croucamp, P. A., & Malan, L. (2018). The theory of systemic patronage and state capture: The liberal democratic project and its regime contenders. *African Journal of Public Affairs*, 10(4), 86–105.

Cutshall, C. R. (1990). *Kanyemba/Chapoto Ward: a socio-economic baseline survey of community households*. University of Zimbabwe Press.

Čutura, M. (2020). Consumer ethnocentrism and social identity: theoretical backgrounds and empirical studies overview. Market-Tržište, 32(SI), 131-146. https://doi.org/DOI: 10.22598/mt/2020.32.spec-issue.131

da Silva, C., Pereira, F., & Amorim, J. P. (2023). The integration of Indigenous knowledge in school: A systematic review. *Compare: A Journal of Comparative Education*, 54(7), 1210–1228. DOI: 10.1080/03057925.2023.2184200

Dahlberg, A. C., & Trygger, S. B. (2009). Indigenous medicine and primary health care: The importance of lay knowledge and use of medicinal plants in rural South Africa. *Human Ecology: an Interdisciplinary Journal*, 37(1), 79–94. DOI: 10.1007/s10745-009-9217-6

Dana, L. (2007). A comparison of Indigenous and non-Indigenous enterprise in the Canadian sub-Arctic. *International Journal of Business Performance Management*, 9(3), 278–286. DOI: 10.1504/IJBPM.2007.013308

Dana, L. (2015). Indigenous entrepreneurship: An emerging field of research. *International Journal of Business and Globalisation*, 14(2), 158–169. DOI: 10.1504/IJBG.2015.067433

Dana-Sacco, G. (2010). The Indigenous researcher as individual and collective: Building a research practice ethic within the context of Indigenous languages. *American Indian Quarterly*, 34(1), 61–82. DOI: 10.1353/aiq.2010.a370587

Daniel, R. A., Wilhelm, T. A., Case-Scott, H., Goldman, G., & Hinzman, L. (2022). What is "Indigenous Knowledge" and why does it matter? integrating ancestral wisdom and approaches into federal decision making.". The White House. December, 2. Retrieved 10 October 2025 from https://www.whitehouse.gov/ostp/news-updates/2022/12/02/what-is-indigenous-knowledge-and-why-does-it-matter-integrating-ancestral-wisdom-and-approaches-into-federal-decision-making/

Das, D., & Semaan, B. (2022, April). Collaborative identity decolonization as reclaiming narrative agency: Identity work of Bengali communities on Quora. In *Proceedings of the 2022 CHI Conference on Human Factors in Computing Systems* (pp. 1-23). DOI: 10.1145/3491102.3517600

Data Fairport. (2014). Data FAIRport. Retrieved on 11 October 2024 from https://www.datafairport.org/index.html

DataPhyte. (2022). Can SDGs be achieved in Nigeria before 2030? According to data, not likely. Retrieved on 1 April 2024 from https://www.dataphyte.com/latest-reports/sdgs/can-sdgs-be-achieved-in-nigeria-before-2030-according-to-data-not-likely/

Datta, R. (2018). Decolonizing both researcher and research and its effectiveness in Indigenous research. *Research Ethics*, 14(2), 1–24. DOI: 10.1177/1747016117733296

Datta, R., & Starlight, T. (2024). Building a meaningful bridge between Indigenous and Western worldviews: Through decolonial conversation. *International Journal of Qualitative Methods*, 23, 1–9. DOI: 10.1177/16094069241235564

David, I. N. (2023). *The potential of Namibian education policy to promote education for communitarian ecological democracy*. [Unpublished doctoral thesis]. University of the Free State.

Davis, L. (2023). *A case study of gun violence in Philadelphia: an Afrocentric analysis of cultural factors*. PhD. Temple University.

Davis, K. F., & Olayide, O. E. (2018). Multidimensional framework for achieving sustainable and resilient food systems in Nigeria. In Leal Filho, W. (Ed.), *Handbook of climate change resilience*. Springer., DOI: 10.1007/978-3-319-71025-9_115-1

Davison, R. M., & Díaz Andrade, A. (2018). Promoting Indigenous theory. *Information Systems Journal*, 28(5), 759–764. DOI: 10.1111/isj.12203

de Sousa Santos, B. (Ed.). (2005). *Democratizing democracy: beyond the liberal democratic canon* (Vol. 1). Verso.

De Sousa Santos, B. D. (2014). *Epistemologies of the south: justice against epistemicide* (1st ed.). Routledge.

Dei, G. J. S. (2008). Indigenous knowledge studies and the next generation: Pedagogical possibilities for anti-colonial education. *Australian Journal of Indigenous Education*, 37(S1), 5–13. DOI: 10.1375/S1326011100000326

Dei, G. J. S. (2020). Elders' cultural knowledges and African indigeneity. In Abidogun, J. M., & Falola, T. (Eds.), *The Palgrave handbook of African education and Indigenous knowledge* (pp. 279–301). Palgrave Macmillan.

Dei, G. S. (2012). Indigenous anti-colonial knowledge as 'heritage knowledge' for promoting Black/African education in diasporic contexts. *Decolonization*, 1(1), 102–119.

Deleuze, G., & Guattari, F. (2009). *Anti-Oedipus: capitalism and schizophrenia*. Penguin.

Demchenko, Y., Los, W., & de Laat, C. (2018). Data as economic goods: Definitions, properties, challenges, enabling technologies for future data markets. *ITU Journal : ICT Discoveries*, 2(23). Retrieved October 18, 2024, from https://www.itu.int/en/journal/002/Documents/ITU2018-12.pdf

Denzin, . (2008). *Handbook of Critical and Indigenous Methodologies*. SAGE Publications. DOI: 10.4135/9781483385686

Denzin, N., & Lincoln, Y. (2017). *The SAGE handbook of qualitative research*. Sage.

Department of Higher Education and Training. (2023). *The state of transformation in South Africa's public universities: Research Report*. Government Printers.

Deskin, Z. Y., & Harvey, B. (2023). Critical food systems education in university student-run food initiatives: Learning opportunities for food systems transformation. *Frontiers in Sustainable Food Systems*, 7, 1230787. DOI: 10.3389/fsufs.2023.1230787

Desmoulins, L. (2020). Indigenous postgraduate education: intercultural perspectives. *Teachers College Record: The Voices of Scholarship in Education,* 320.

Devkota, K. M., & Timilsena, N. P. (2023). Indigenous knowledge in science classroom practices: An explorative study. *Kepes Journal*, 21(3), 847–856.

Dey, S. (2020). *Western science v/s humanitarian science: decolonization and Indigenization.* Lectures presented at Yonphula Centenary College.

Dey, S. (2021). "Killing with kindness": Political icons, socio-cultural victims: visual coloniality of the Siddis of Karnataka. In Karam, B., & Mutsvairo, B. (Eds.), *Decolonising Political Communication in Africa: Reframing Ontologies* (pp. 91–108). Routledge. DOI: 10.4324/9781003111962-9

Dey, S. (2022). *Green academia: towards eco-friendly education systems*. Routledge. DOI: 10.4324/9781003301950

Dey, S. (2023a). *Performing Memories and Weaving Archives: Creolized Cultures Across the Indian Ocean*. Anthem Press. DOI: 10.2307/jj.8973372

Dey, S. (2023b). Baine Maraa-Indu Mama-Siddi Dhamaal: Interwoven Performances of Epistemic Justice and Cognitive Freedom by the Siddis of Karnataka, India. *International Journal of Critical Diversity Studies*, 5(2), 17–32. DOI: 10.13169/intecritdivestud.5.2.0017

Dey, S., Maity, T., & Srivastava, T. (2024). Gender Empowerment in Transoceanic Feminine Folklore and Shrines: A Kin Study of Siddi Women's Participation in Mai Misra Worship in Gujarat, India. *Journal of International Women's Studies*, 26(1), 1–14.

Dhanabhakyam, & Sarath M. (2023). Psychological Wellbeing: Asystematic Literature Review. International Journal of Advanced Research in Science, Communication and Technology, April, 603–607. DOI: 10.48175/IJARSCT-8345

Dhir, R. K. (2016). *Sustainable development goals: Indigenous peoples in focus*. International Labour Organization (ILO).

Diep, A. N., Zhu, C., Cocquyt, C., De Greef, M., Vo, M. H., & Vanwing, T. (2019). Adult learners' needs in online and blended learning. *Australian Journal of Adult Learning*, 59(2), 223–253.

Dole, D. (2020). Collaborative, appreciative and experiential pedagogy in educational settings. In McNamee, S., Gergen, M. M., Camargo-Borges, C., & Rasera, E. F. (Eds.), *The Sage handbook of social constructionist practice* (pp. 352–362). Sage. DOI: 10.4135/9781529714326.n34

Đorđević Belić, S. (2024). Communication with the deceased in dreams: Overcoming the boundary between this world and the otherworld or its conceptualisation strategy? *Religions*, 15(7), 828. DOI: 10.3390/rel15070828

Dotson, K. (2015). Inheriting Patricia Hill Collins's *Black Feminist* epistemology. *Ethnic and Racial Studies*, 38(13), 2322–2328. DOI: 10.1080/01419870.2015.1058496

Dovie, D. B. K., Shackleton, C. M., & Witkowski, E. T. F. (2006). The role of mopane worm (imbrasia belina) in the livelihoods of the poor in Southern Africa. *Economic Botany*, 60(3), 301–310.

Dramani, J. M. F., & Godwell, N. (2023). Farmers' choice for Indigenous practices and implications for climate-smart agriculture in northern Ghana. *Heliyon*, 9(11), e22162. DOI: 10.1016/j.heliyon.2023.e22162 PMID: 38053880

Drummond, A. (2020). Embodied Indigenous knowledges protecting and privileging Indigenous peoples' ways of knowing, being and doing in undergraduate nursing education. *Australian Journal of Indigenous Education*, 49(2), 127–134. DOI: 10.1017/jie.2020.16

Du Toit, A. (2000). From autonomy to accountability: Academic freedom under threat in South Africa? *Social Dynamics*, 26(1), 76–133. DOI: 10.1080/02533950008458687

Du Toit, A. (2019). The potential of the South African consumer studies curriculum to foster indigenous knowledge. *Journal of Consumer Sciences*, 47, 27–39.

Dube T (2020), *The Kalanga in Historical Perspectives*, online publication.

Dube, L., & Ngulube, P. (2012). Knowledge sharing in multicultural environment: Challenges and opportunities. *South African Journal of Library and Information Science*, 78(1), 68–77. DOI: 10.7553/78-1-48

Dube, S. S., & Chitarika, M. (2024). Dynamics and socio-environmental Impacts of the Mopane Worm Harvesting in Rural Communities in Zimbabwe: Prospects for climate- smart approaches. *Frontiers in Sustainability*, 1(5), 1–10. DOI: 10.3389/frsus.2024.1414058

Dubin, R. (1969). *Theory building*. 2nd ed. Free Press, Collier-Macmillan

Dul, J., & Hak, T. (2007). *Case study methodology in business research*. Routledge., DOI: 10.4324/9780080552194

Duri, F., & Mapara, J. (2007). Environmental Awareness and Management Strategies in Pre-colonial Zimbabwe. *Zimbabwe Journal of Geographical Research*, 1(2), 98–111.

Dutfield, G. (2005, May). Harnessing traditional knowledge and genetic resources for local development and trade. In Draft Paper Presented at the International Seminar on Intellectual Property and Development. In *Draft Paper Presented at the International Seminar on Intellectual Property and Development*. 2005. Retrieved on 14 June 2023 from https://www.wipo.int/edocs/mdocs/mdocs/en/isipd_05/isipd_05_www_103975.pdf

Dzingirayi, P. (2022). The position of Afrocentric Paradigm in the contemporary globalised professionalised counselling services in Zimbabwe. *The Zimbabwe Journal of Health Sciences (ZJHS)*, 193.

Ebenezer, O. J. (2024). Indigenous knowledge management practices in subsistence farming: A comprehensive evaluation. *Sustainable Technology and Entrepreneurship*, 3(2), 100058. DOI: 10.1016/j.stae.2023.100058

Edwards, R., Barnes, H. M., McGregor, D., & Brannelly, T. (2020). Supporting Indigenous and non-Indigenous research partnerships. *The Qualitative Report*, 25(13), 6–15. DOI: 10.46743/2160-3715/2020.4754

Ekpo, D. (2018). Africa without Africanism: post-Africanism vs. Indigenous knowledge systems, culture/art. In *R. de Lange, I. Stevens, R. Kruger and M. Sirayi, The Arts and Indigenous Knowledge Systems in a Modernized Africa*. Cambridge Scholars Publishing.

Ekuri, K. A. Betiang, P. A, Andong, H. A., & Eyam S. O. (2022). The relevance of adult education in a contemporary world: the implications. *LWATI: A Journal of Contemporary Research*, 19(3), 20-31.

El Aidi, A., & Yechouti, Y. (2017). Antonio Gramsci's theory of cultural hegemony in Edward Saidi's orientation. *International Multidisciplinary Research Journal*, 6(5), 1–23.

Ela, J., Mudgway, M., Harré, N., & Hamill, J. (2022). Planetary health, Indigenous knowledge and healthcare: a scoping review. DOI: 10.1101/2022.11.28.22282853

Elendu, C. (2024). The evolution of ancient healing practices: from shamanism to Hippocratic medicine: a review. *Medicine*, 103(28), e39005. DOI: 10.1097/MD.0000000000039005 PMID: 38996102

Elford, D. (2002). Conservation by Self-Determination in Central America. *Fourth World Journal*, 5(1), 98–149.

Ellen, R., & Harris, H. (1997). Concepts of Indigenous environmental knowledge in scientific and development studies literature: a critical assessment. In *Avenir des Peuples des Forets Tropicales Working Paper* (Vol. 2).

Elo, S., Kääriäinen, M., Kanste, O., Pölkki, T., Utriainen, K., & Kyngäs, H. (2014). Qualitative content analysis: A focus on trustworthiness. *SAGE Open*, 4(1), 2158244014522633. DOI: 10.1177/2158244014522633

Emeagwali, G., & Dei, G. J. S. (eds.). (2014). *African Indigenous knowledge and the disciplines*. Sense Publishers.

Emeagwali, G. (2003). *African Indigenous knowledge systems (AIK): implications for the curriculum*. African World Press.

Emeagwali, G. (2003). African Indigenous knowledge systems (AIK): implications for the curriculum. In Falola, T. (Ed.), *Ghana in Africa and the world: essays in honour of Adu Boahen*. Africa World Press.

Emeagwali, G. (2014). Intersections between Africa's Indigenous knowledge systems and history. In Emeagwali, G., & Dei, G. J. S. (Eds.), *African Indigenous knowledge and the disciplines* (pp. 1–17). Brill Publishing. DOI: 10.1007/978-94-6209-770-4_1

Emeagwali, G., & Shizha, E. (2016). *African Indigenous knowledge and the sciences: Journeys into the past and present*. Sense Publishers. DOI: 10.1007/978-94-6300-515-9

Enthoven, R. E. (1990). *The Tribes and Castes of Bombay*. Asian Educational Services.

Equator Initiative. (2022). Nature-based actions advancing sustainable development. United Nations Development Programme. Retrieved on 20 October 2024 from https://www.equatorinitiative.org/2020/04/24/solution11417/

Escobar, A. (2016). Thinking-feeling with the Earth: Territorial struggles and the ontological dimension of the Epistemologies of the South. *Revista de Antropologia*, 11(1), 11–31.

Estrada, A.. (2020). Global importance of Indigenous Peoples, their lands, and knowledge systems for saving the world's primates from extinction. *Science Advances*, 8, 1–20. https://www.science.org/doi/10.1126/sciadv.abn2927 PMID: 35947670

Etikan, I. (2017). Sampling and Sampling Methods. *Biometrics & Biostatistics International Journal*, 5(6), 215–217. DOI: 10.15406/bbij.2017.05.00149

Evens, T., & Hauttekeete, L. (2011). Challenges of digital preservation for cultural heritage institutions. *Journal of Librarianship and Information Science*, 43(3), 157–165. DOI: 10.1177/0961000611410585

Eyong, C. T. (2007). Indigenous Knowledge and sustainable development in Africa: case study of Central Africa. In *Indigenous Knowledge Systems and sustainable development: relevance for Africa* (pp. 121–139). Kamlla-Raj Enterprises.

Ezeanya-Esioba, C., Oguamanam, C., & Ndungutse, V. (2021). Marginalisation of Indigenous knowledge in Africa: the case of Rwandan traditional medicinal treatment for livestock. Working paper 24. Open AIR (African Innovation Research) Network. Retrieved on 29 October 2024 from https://www.oguamanam.com/publications/2021-2-18/marginalisation-of-indigenous-knowledge-in-african-education-the-case-of-rwandan-traditional-medicinal-treatments-for-livestock

Fabiano, E., Schulz, C., & Branas, M. (2021). Wetland spirits and Indigenous knowledge: Implications for the conservation of wetlands in Peruvian Amazon. *Current Research in Environmental Sustainability*, 3(1), 100107. DOI: 10.1016/j.crsust.2021.100107

Fadhlur, R. A., Suhaiza, Z., & Mastura, R. (2023). A review of the critical gaps in the food security literature: Addressing key issues for sustainable development. *Information Management and Business Review*, 15(2), 35–46.

Fairhead, J., & Leach, M. (1996). *Misreading the African Landscape: Society and Ecology in a forest-savanna mosaic*. Cambridge University Press. DOI: 10.1017/CBO9781139164023

Faloju J. O. & Sobola E. (2021). The meaning, function and contextual usage of metaphors of women in Russian and Yorùbá. *Lagos Notes and Records: A Journal of the Faculty of Arts*, 27, 160-189.

Fanon, F. (2008). *Black skin, white masks*. Grove Press. Global action plan the international decade of languages (IDIL2022-2032) Retrieved on 14 June 2023 from https://en.unesco.org/sites/default/files/global_action_plan_int_decade__languages_en.pdf

Fanon, F. (1964). *Toward the African revolution*. Grove Press.

Farber, T. (2015, December 13). The ugly truth behind the plunder of nature. *Sunday Times*, •••, 15.

Faure, E., Herrera, F., Kaddoura, A., Lopes, H., Petrovsky, A. V., & Rahnema, M.. (1972). *Learning to be: the world of education today and tomorrow*. UNESCO.

Fenstad, J. E., Hoyningen, P., Hu, Q., Kokwaro, J., Salick, J., Shrum, W., & Subbarayappa, B. V. (2002). *Report from the ICSU Study Group on Science and Traditional Knowledge*. International Council for Science.

Ferguson, J. (2013). Declaration of dependence: Labour, personhood and welfare in Southern Africa. *Journal of the Royal Anthropological Institute*, 19(2), 223–242. DOI: 10.1111/1467-9655.12023

Fernaandez-Llamazares, A., & Virtanen, P. (2020). Game masters and Amazonian Indigenous views on sustainability. *Current Opinion in Environmental Sustainability*, 43(1), 21–27. DOI: 10.1016/j.cosust.2020.01.004

Fernández-Llamazares, Á., Lepofsky, D., Lertzman, K., Armstrong, C. G., Brondizio, E. S., Gavin, M. C., Lyver, P. O., Nicholas, G. P., Pascua, P., Reo, N. J., Reyes-García, V., Turner, N. J., Yletyinen, J., Anderson, E. N., Balée, W., Cariño, J., David-Chavez, D. M., Dunn, C. P., Garnett, S. C., & Vaughan, M. B. (2021). Scientists' Warning to humanity on threats to Indigenous and local knowledge systems. *Journal of Ethnobiology*, 41(2), 144–169. DOI: 10.2993/0278-0771-41.2.144

First Nation's Information Governance Centre. 2018. The First Nations principles of OCAP®. Retrieved on 12 October 2024 from https://fnigc.ca/ocap

Flavier, J. M.. (1995). The regional program for the promotion of Indigenous knowledge in Asia. In Warren, D. M., Slikkerveer, L. J., & Brokensha, D. (Eds.), *The cultural dimension of development: Indigenous knowledge systems* (pp. 479–487). Intermediate Technology Publications. DOI: 10.3362/9781780444734.043

Fleischhacker, S., Ries, M., Vu, A., & McPhail, A. (2011). Engaging tribal leaders in an American Indian healthy eating project through modified talking circles. *Family & Community Health*, 34(3), 202–210. DOI: 10.1097/FCH.0b013e31821960bb PMID: 21633212

Flynn, S. (2019). Revisiting hegemony: A Gramscian analysis for contemporary social work. *Irish Journal of Sociology*, 29(1), 77–96. DOI: 10.1177/0791603519884201

Foley, R., Archambault, L., Hale, A., & Dong, H. (2017). Learning outcomes in sustainability education among future elementary school teachers. *Journal of Education for Sustainable Development*, 11(1), 33–51. DOI: 10.1177/0973408217725861

Fonteneau, B., Neamtam, N., Wanyama, F., Pereira Morais, L., & de Poorter, M. (2010). *Social and solidarity economy: Building a common understanding*. ILO Job Creation and Enterprise Development Department Cooperative Programme (EMP/COOP). Retrieved on 20 October 2024 from https://www.ilo.org/publications/social-and-solidarity-economy-building-common-understanding

Fowler, J. (2011). The concept of indigeneity: Can the Declaration on the Rights of Indigenous People be understood within western liberal philosophy? *Queensland Law Student Review*, 4(1), 35–46.

Francis, C. D., Bowman, N., & King, S. (2023). Kinship pathways: Nurturing and sustaining resilient, responsible, and respected indigenous evaluators. *New Directions for Evaluation*, 177(177), 145–152. DOI: 10.1002/ev.20537

Francis, J. D. L. (2019). Eastern Cherokee Creation and Subsistence Narratives: A Cherokee and Religious Interpretation. *American Indian Culture and Research Journal*, 43(1), 83–98. DOI: 10.17953/AICRJ.43.1.LOFTIN-FREY

Francis, K. D., & Liew, C. L. (2010). Digitised Indigenous knowledge in cultural heritage organisations in Australia and New Zealand: An examination of policy and protocols. *Proceedings of the American Society for Information Science and Technology*, 46(1), 1–21. DOI: 10.1002/meet.2009.145046025

Fraser, G., Brady, A., & Wilson, M. S. (2021). "What if I'm not trans enough? What if I'm not man enough?": Transgender young adults' experiences of gender-affirming healthcare readiness assessments in Aotearoa, New Zealand. *International Journal of Transgender Health*, 22(4), 454–467. DOI: 10.1080/26895269.2021.1933669 PMID: 37808530

Frederikse, J. (1982). *None but ourselves: masses vs media in the making of zimbabwe*. Ravan Press.

Friere, P. (2005). *Education for critical consciousness*. Seabury Press.

Frohard-Dourlent, H., MacAulay, M., & Shannon, M. (2020). Experiences of surgery readiness assessments in British Columbia. *International Journal of Transgender Health*, 21(2), 147–162. DOI: 10.1080/26895269.2020.1742842 PMID: 33015666

Futi, M. G. (2019). Enhancing Food Security in Lagos, Nigeria. Restructuring Africa's largest city's food supply through urban farming a policy proposal for Babajide Sanwo-Olu, Governor-Elect of Lagos State. Master's, diss. Harvard Extension School, Policy Analysis Capstone. Retrieved on 10 April 2024 from https://capstone.extension.harvard.edu/files/capstone/files/futimwaku_203040_7923998_futi_policy_capstone.pdf

Gade, C. (2012). What is Ubuntu? different interpretations among South Africans of African descent. *South African Journal of Philosophy*, 31(3), 484–503. DOI: 10.1080/02580136.2012.10751789

Gainsford, A., & Evans, M. (2021). Integrating andragogical philosophy with Indigenous teaching and learning. *Management Learning*, 52(5), 559–580. DOI: 10.1177/1350507620972528

Galadima, I. A. (2012). The role of adult education in community development. *Knowledge Review*, 26(2).

Gallagher, E. (2006). Afrofuturism. Tate. Retrieved on 22 October 2024 from https://www.tate.org.uk/art/art-terms/a/afrofuturism

Garba, I., Sterling, R., Plevel, R., Carson, W., Cordova-Marks, F. M., Cummins, J., Curley, C., David-Chavez, D., Fernandez, A., Hiraldo, D., Hiratsuka, V., Hudson, M., Jäger, M. B., Jennings, L. L., Martinez, A., Yracheta, J., Garrison, N. A., & Carroll, S. R. (2023). Indigenous Peoples and research: Self-determination in research governance. *Frontiers in Research Metrics and Analytics*, 8, 1272318. DOI: 10.3389/frma.2023.1272318 PMID: 38033627

Gartner, J. B., Abasse, K. S., Bergeron, F., Landa, P., Lemaire, C., & Côté, A. (2022). Definition and conceptualization of the patient-centered care pathway, a proposed integrative framework for consensus: A Concept analysis and systematic review. *BMC Health Services Research*, 22(1), 1–24. DOI: 10.1186/s12913-022-07960-0 PMID: 35473632

Gašević, D., Vranješ, M., & Tomašević, D. (2019). The influence of consumer lifestyle in the Republic of Serbia on consumer ethnocentrism. *Skola Biznisa*, (2), 78–94. DOI: 10.5937/skolbiz2-22136

Gebru, A., & Wako, W. (2022). Walking the theories we talk: Utilizing African social work theories in African research. *African Journal of Social Work*, 12(4), 189–198.

Gehl, L. (2012). Debwewin journey: A methodology and model of knowing. *Alternative*, 8(1), 53–65. DOI: 10.1177/117718011200800105

Gentles, S. J., Charles, C., Ploeg, J., & McKibbon, K. A. (2015). Sampling in qualitative research: Insights from an overview of the methods literature. *The Qualitative Report*, 20(11), 1772–1789. DOI: 10.46743/2160-3715/2015.2373

George, J., Reddy, V., & Searle-Chatterjee, M. (1990). Eurocentrism in the social sciences. *Race & Class*, 31(4), 1–15. DOI: 10.1177/030639689003100401

Gergen, K. J. (2022). The relational imperative. Taos Institute. Retrieved on 2 October 2024 from https://www.taosinstitute.net/product/the-relational-imperative-resources-for-a-world-on-edge

Gergen, K. J. (2009). *Relational being: beyond self and community*. Oxford University Press.

Gergen, K. J. (2023). *An invitation to social construction: co-creating the future* (4th ed.). Sage.

Ghosh, R. (2021). The Plastic Turn. *Diacritics*, 49(1), 64–85. DOI: 10.1353/dia.2021.0003

Ghosh, R. (2022). *The Plastic Turn.* Cornell University Press.

Ghosh, R. (2024). *Plastic Tagore: Thinking After Yesterday.* Oxford University Press. DOI: 10.1093/9780198922995.001.0001

Gibson-Graham, J. K., & Dombroski, K. (2020). Introduction to the handbook of diverse economies: Inventory as ethical intervention. In Gibson-Graham, J. K., & Dombroski, K. (Eds.), *Handbook of diverse economies* (pp. 1–24). Edward Elgar. DOI: 10.4337/9781788119962.00008

Gilroy, P. (1996). Route work: the black Atlantic and the politics of exile. In Chambers, I., & Curti, L. (Eds.), *The post-colonial question. common skies, divided horizons* (pp. 17–30). Routledge.

Gintis, H. (2009). Beyond homo economicus. In by C. L. Spash (Ed.), *Ecological economics, Vol. 1* (pp. 150-166). Routledge.

Gipson, P., & Petersen, A. (1994). Music Education in South Africa: In *Proceedings of the 21st World Conference of the International Society for Music Education*, 193-287.

Glaser, B., & Strauss, A. (1967). *The discovery of grounded theory: Strategies for qualitative research.* Sociology Press.

Glenberg, A. M., Witt, J. K., & Metcalfe, J. (2013). From the revolution to embodiment: 25 years of cognitive psychology. *Perspectives on Psychological Science*, 8(5), 573–585. DOI: 10.1177/1745691613498098 PMID: 26173215

Global Affairs Canada. (2022). Indigenous business and federal procurement. Available from: https://www.sac-isc.gc.ca/eng/1100100032802/1610723869356. Accessed on 29 September 2024.

Global Indigenous Data Alliance. (2024). CARE Principles for Indigenous data governance. Retrieved on 18 October 2024 from https://www.gida-global.org/care

Go, F. A. I. R. (2024). FAIR principles. Retrieved on 11 October 2024 from https://www.go-fair.org/fair-principles/

Godfrey-Smith, P. (2003). *Theory and reality.* University of Chicago Press. DOI: 10.7208/chicago/9780226300610.001.0001

Goduka, N. (2012). Re-discovering Indigenous knowledge—Ulwazi Lwemveli for strengthening sustainable livelihood opportunities within rural contexts in the Eastern Cape province. *Indilinga*, 11(1), 1–19.

Goel, V., & Dolan, R. J. (2001). Functional neuroanatomy of three-term relational reasoning. *Neuropsychologia*, 39(9), 901–909. DOI: 10.1016/S0028-3932(01)00024-0 PMID: 11516443

Gonçalves, B. (2023, August 15). Belgium's return of Lumumba's tooth: A new moment for anti-colonial struggles? *EJIL Talk!* Retrieved on 22 October 2024 from https://www.ejiltalk.org/belgiums-return-of-lumumbas-tooth-a-new-moment-for-anti-colonial-struggles/

Gondo R et al (2018), *Integration of Indigenous Knowledge Systems in Natural Resources Management in Hurungwe District, Zimbabwe*, International Journal of African Renaissance Studies - Multi- Inter- and Transdisciplinary 13(4): pp 1-10.

Göransson, K., & Nilholm, C. (2014). Conceptual diversities and empirical shortcomings – A critical analysis of research on inclusive education. *European Journal of Special Needs Education*, 29(3), 265–280. DOI: 10.1080/08856257.2014.933545

Gordon, L. R. (2019). Four kinds of invisibility from Euromodernity. Ted-x Talks. Retrieved on 12 October 2024 from https://www.youtube.com/ watch?reload=9&v=bW_G3-DwtQw&t=189s

Gordon, J. (2002). Towards an African historiography. In Asante, M. K., & Mazama, A. (Eds.), *Egypt vs. Greece and the American academy: the debate over the birth of civilization* (pp. 25–36). African American Images.

Gordon, L. (2010). Theory in black: Teleological suspensions in philosophy of culture. *Qui Parle*, 18(2), 193–214. DOI: 10.5250/quiparle.18.2.193

Gordon, R. (2017). Taming' Bushman farm labour: A villainous era in neo-liberal feudal Namibia? *Anthropology Southern Africa*, 40(1), 261–275. DOI: 10.1080/23323256.2017.1395708

Gough, D., Oliver, S., & Thomas, J. (2017). Introduction to systematic reviews. In Gough, D., Oliver, S., & Thomas, J. (Eds.), *An introduction to systematic reviews* (pp. 1–19). Sage.

Government of Botswana. (2016). [– *Achieving prosperity for all.* Lentswe La Lesedi] [PTY] [LTD.]. *Vision (Basel)*, •••, 2036.

Government of Zimbabwe. (2002). *Rural District Councils Act Chapter 29: 13.* Government Printers.

Government of Zimbabwe. (2023). *Statutory Instrument 14 of 2023. Proclamation of Ward Boundary Description.* Government Printers.

Graham, L., Berman, J., & Bellert, A. (2015). *Sustainable learning: Inclusive practices for 21st century classrooms.* Cambridge University Press. DOI: 10.1017/CBO9781107280243

Gramsci, A. (1971). *Selections from the prison notebook* (Hoare, Q., & Nowell-Smith, G., Eds. & Trans.). Lawrence & Wishart.

Grande, S., & McCarty, T. L. (2018). Indigenous elsewheres: Refusal and re-membering in education research, policy, and praxis. *International Journal of Qualitative Studies in Education : QSE*, 31(3), 165–167. DOI: 10.1080/09518398.2017.1401144

Grange, L. (2020). COVID-19 pandemic and the prospects of education in South Africa. *Prospects*, 51(1-3), 425–436. DOI: 10.1007/s11125-020-09514-w PMID: 33100404

Graves, J. (2019). Filling the Pot: The Remembrance of African Sufi Ancestor-Saints and the Reclamation of African-Historical Heritage in Ahmedabad, Gujarat. *Journal of Africana Religions*, 7(1), 94–104. DOI: 10.5325/jafrireli.7.1.0094

Green, E. (2021). Explaining inter-ethnic marriages in Sub-Saharan Africa. *Journal of International Development*, 33(4), 627–654. DOI: 10.1002/jid.3535

Greene, S. (1999). Family concerns: Gender and ethnicity in pre-colonial West Africa. *International Review of Social History*, 44(1), 15–31. DOI: 10.1017/S0020859000115172

Green-Thompson, L., McInerney, P., & Woollard, B. (2017). The social accountability of doctors: A relationship based framework for understanding emergent community concepts of caring. *BMC Health Services Research*, 17(1), 269. Advance online publication. DOI: 10.1186/s12913-017-2239-7 PMID: 28403860

Greenwood, M., & Lindsay, N. M. (2019). A commentary on land, health, and Indigenous knowledge(s). *Global Health Promotion*, 26(3, suppl), 82–86. DOI: 10.1177/1757975919831262 PMID: 30964411

Grenier, L. (1998). *Working with Indigenous knowledge: A guide for researchers*. International Development Research Centre.

Greyling, E., & Zulu, S. (2009). Content development in an Indigenous digital library: a case study in community participation. *IFLA Congress*, Milan, Italy 23-27 August 2009. https://www.ifla.org/annual-conference/ifla75/index.htm

Greyling, E., & McNulty, N. (2012). How to build an Indigenous digital library through community participation: the case of the Ulwazi programme, Information for Sustainable Development in a Digital Environment. *XXth Standing Conference of Eastern, Central and South Africa Library and Information Associations (SCECSAL).*

Griffin, C. T., Wynd, B. M., Munyikwa, D., Broderick, T. J., Zondo, M., Tolan, S., Langer, M. C., Nesbitt, S. J., & Taruvinga, H. R. (2022). Africa's oldest dinosaurs reveal early suppression of dinosaur distribution. *Nature*, 609(7926), 313–319. DOI: 10.1038/s41586-022-05133-x PMID: 36045297

Griffiths, K. E., Blain, J., Vajdic, C. M., & Jorm, L. (2021). Indigenous and tribal peoples data governance in health research: A systematic review. *International Journal of Environmental Research and Public Health*, 18(19), 10318. DOI: 10.3390/ijerph181910318 PMID: 34639617

Griggs, R. (1992). *Background on the Term "Fourth World"*. DayKeeper Press. Olympia, WA USA, Center for World Indigenous Studies. Occasional Paper #18.

Griggs, R., & Hocknell, P. (1996). *The geography and geopolitics of Europe's Fourth World*. IBRU, Boundary and Security Bulletin.

Griswold, W. (1992). *Cultures and societies in a changing world*. Pine and Forge Press.

Grosfoguel, R. (2003). Introduction: geopolitics of knowledge and coloniality of power- thinking Puerto Rico and Puerto Ricans from the colonial difference. In *Colonial subjects: Puerto Ricans in a global perspective*. University of California Press.

Grosfoguel, R. (2007). The epistemic decolonial turn: Beyond political-economy paradigms. *Cultural Studies*, 21(2-3), 211–223. DOI: 10.1080/09502380601162514

Grosfoguel, R. (2013). The structure of knowledge in Westernized universities: Epistemic racism/sexism and the four genocides/epistemicides of the long 16th century. *Human Architecture*, 9(1), 73–90.

Guba, E., & Lincoln, Y. S. (1989). Fourth generation evaluation. *Sage (Atlanta, Ga.).*

Guba, E., & Lincoln, Y. S. (2005). Paradigmatic controversies, contradictions, and emerging confluences. In Denzin, N. K., & Lincoln, Y. S. (Eds.), *Handbook of qualitative research* (3rd ed., pp. 191–216). Sage.

Guizzardi, G. (2020). Ontology, ontologies, and the "I" of FAIR. *Data Intelligence*, 2(1-2), 181–191. DOI: 10.1162/dint_a_00040

Gumbo, M. T. (2022). The colonial undertones of Queen Elizabeth II's Death: Africa speaks. *Africa Update,* XXX(1). https://www2.ccsu.edu/africaupdate/?issue=152

Gumbo, M. T. (2020b). Indigenising technological pedagogical content knowledge in open distance learning. *Africa Education Review*, 17(4), 72–86. DOI: 10.1080/18146627.2020.1868073

Gumbo, M. T. (Ed.). (2020a). *Decolonization of Technology Education: African Indigenous perspectives*. Peter Lang. DOI: 10.3726/b16660

Gumbo, M. T., Gasa, V., & Kanus, C. B. (2022). Centring African knowledge to decolonise higher education. In Kanus, C. B., Takako, M., & Seroto, J. (Eds.), *Decolonising African higher education: practitioner perspectives from across the continent* (pp. 21–36). Routledge. DOI: 10.4324/9781003158271-2

Gupta, A. (1995). People's knowledge for survival: Grassroots innovations for sustainable natural resource management. Paper presented at the IFADs International Conference on Hunger and Poverty.

Gupta, A. K., Sinha, R., Koradia, D., Patel, R., Parmar, M., Rohit, P., Patel, H., Patel, K., Chand, V. S., James, T. J., Chandan, A., Patel, M., Prakash, T. N., & Vivekanandan, P. (2003). Mobilizing grassroots' technological innovations and traditional knowledge, values and institutions: Articulating social and ethical capital. *Futures*, 35(9), 975–987. DOI: 10.1016/S0016-3287(03)00053-3

Gwala, R. S. (2023). The legalisation of cannabis in South Africa: proposing an economic value chain model for South Africa. In *Rapid innovation and development in the global cannabis market* (pp. 98–122). IGI Global. DOI: 10.4018/978-1-6684-6346-8.ch006

Gwerevende, S. (2020). "Venda Cultural Expressions: An inclusive analysis of tshigombela as a performative enactment of ubuntu." Unpublished master's thesis, Choreomundus.

Gwerevende, S. (2022). "Music Ecosystems. Why Do They Matter? Centre for Music Ecosystems." Accessed January 23, 2023. https://www.centerformusicecosystems.com/music-ecosystems-builders/solomon-gwerevende

Gwerevende, S., & Mthombeni, Z. M. (2023). Safeguarding intangible cultural heritage: Exploring the synergies in the transmission of Indigenous languages, dance and music practices in Southern Africa. *International Journal of Heritage Studies*, 29(5), 398–412. DOI: 10.1080/13527258.2023.2193902

Gyekye, K. (1995). *An essay on African philosophical thought: The Akan conceptual scheme*. Temple University Press.

Hachipola, S. (1998). *A survey in the minority languages of Zimbabwe*. University of Zimbabwe Press.

Haddad, B. M., & Solomon, B. D. (2024). Ecological economics as the science of sustainability and transformation: Integrating entropy, sustainable scale, and justice. *PLOS Sustainability and Transformation*, 3(2), e0000098. DOI: 10.1371/journal.pstr.0000098

Hahndiek, C., & Regina, I. (2018). Leadership and role SAHRA can play to facilitate connectivity with and between museums. *The Heritage Portal*, Retrieved on 14 June 2023 from https://www.theheritageportal.co.za/article/leadership-and-role-sahra-canplayfacilitate-connectivity-and-between-museums

Hai-Ninh, D., Bich, N. D., & Minh, H. N. (2023). How do constructivism learning environments generate better motivation and learning strategies? The design science approach. *Heliyon*, 9(12), e22862. DOI: 10.1016/j.heliyon.2023.e22862 PMID: 38125439

Hall, A. (2003). *The American empire and the Fourth World: the bowl with one spoon*. McGill Queen's University Press.

Hallinger, P. (2013). A conceptual framework for systematic reviews of research in educational leadership and management. *Journal of Educational Administration*, 51(2), 126–149. DOI: 10.1108/09578231311304670

Hall, R., Ansley, L., & Connolly, P. (2023). Decolonising or anti-racism? Exploring the limits of possibility in higher education. *Teaching in Higher Education*. Advance online publication. DOI: 10.1080/13562517.2023.2201676

Harris, D. R., & Fuller, D. Q. (2020). Agriculture: definition and overview. In *Encyclopedia of Global Archaeology*. https://doi.org/DOI: 10.1007/978-3-030-30018-0_64

Harris, R., MacGlaughlin, H., Mertens, D. & Perez, J. (2020). Research Ethics for Sign Language Communities [ASL Translation]. *Deaf Studies Digital Journal*, 6.

Harris, L. D., & Wasilewski, J. (2004). Indigeneity, an alternative worldview. *Systems Research and Behavioral Science*, 21(5), 489–503. DOI: 10.1002/sres.631

Harrison, K. (2020). Indigenous Music Sustainability During Climate Change. *Current Opinion in Environmental Sustainability*, 43, 28–34. DOI: 10.1016/j.cosust.2020.01.003

Hart, M. A. (2010). Indigenous worldviews, knowledge, and research: The development of an Indigenous research paradigm. *Journal of Indigenous Voices in Social Work*, 1(1), 1–16.

Haskell, D. L. (2008). Tarascan kingship the production of heirarcy in the Prehispanic Patzcuaro Basin, Mexico. PhD Doctoral, University of Florida.

Hasler, R. (1996). *Agriculture, foraging and wildlife resource use in Africa*. Kigan Paul Publishers.

Hassan, N. R., Lowry, P. B., & Mathiassen, L. (2022). Useful products in information systems theorizing: A discursive formation perspective. *Journal of the Association for Information Systems*, 23(2), 418–446. DOI: 10.17705/1jais.00730

Hays, J., & Reinders, H. (2020). Sustainable learning and education: A curriculum for the future. *International Review of Education*, 66(1), 29–52. DOI: 10.1007/s11159-020-09820-7

Hayward, A., Sjoblom, E., Sinclair, S., & Cidro, J. (2021). A new era of Indigenous research: Community-based Indigenous research ethics protocols in Canada. *Journal of Empirical Research on Human Research Ethics; JERHRE*, 16(4), 403–417. DOI: 10.1177/15562646211023705 PMID: 34106784

Haywood, L. K., Funke, N., Audouin, M., Musvoto, C., & Nahman, A. (2018). The sustainable development goals in South Africa: Investigating the need for multi-stakeholder partnerships. *Development Southern Africa*, 36(5), 555–569. DOI: 10.1080/0376835X.2018.1461611

Healy, R. (2013). *Understanding Triangulation. Evidence Based Nursing Journal Hoppers, C.O. and Richards, H. (2011). Rethinking thinking*. UNISA Press.

Heater, D. (1990). *Citizenship: the civic ideals in world history, politics and education*. Longman.

Hegel, G. W. F. (1956). *The philosophy of history* (Sibree, J., Trans.).

Heim, T. (2020). The Indigenous origins of regenerative agriculture. National Farmers Union (NFU). Retrieved on 30 April 2024 https://nfu.org/2020/10/12/the-Indigenous-origins-of-regenerative-agriculture/

Held, M. (2019). Decolonising research paradigms in the context of settler colonialism: An unsettling mutual collaborative effort. *International Journal of Qualitative Methods*, 18, 1–16. DOI: 10.1177/1609406918821574

Heleta, S. (2016). Decolonisation of higher education: Dismantling epistemic violence and eurocentrism in South Africa. *Transformation in Higher Education*, 1(1), a9. DOI: 10.4102/the.v1i1.9

Hendrick, A., & Young, S. (2017). Decolonising the curriculum, decolonising ourselves: Experiences of teaching in and from the "third space". *Advances in Social Work and Welfare Education*, 19(2), 9–24.

Herald Newspaper, 18 July (2012). Zimbabwe Newspapers

Herald Newspaper, 2 August (2021). Zimbabwe Newspapers.

Heritage Chairs of Australia and New Zealand. (HCOANZ). (2020). Dhawura Ngilan: a vision for Aboriginal and Torres Strait Islander heritage in Australia and the best practice standards in cultural heritage management and legislation, Australian Government Department of Agriculture, Water and the Environment, Canberra.

Heto, P. P. K., & Mino, T. (2023). (Dis)continuity of African Indigenous knowledge. *Alternative*, 19(1), 71–79. DOI: 10.1177/11771801221138304

Higgs, P., Higgs, L. G., & Venter, E. (2003). Indigenous African knowledge systems and. innovation in higher education in South Africa. *SAJHE/SATHO*, 17(2), 40-45.

Higgs, P. (2012). African philosophy and the decolonisation of education in Africa: Some critical reflections. *Educational Philosophy and Theory*, 44(sup2), 37–55. DOI: 10.1111/j.1469-5812.2011.00794.x

Hikwa, L. (2012). *Indigenous languages as a preserver of Indigenous knowledge in Zimbabwe*. NuSpace Institutional Repository. NUST.

Hindle, K., & Lansdowne, M. (2005). Brave spirits on new paths: Toward a globally relevant paradigm of Indigenous entrepreneurship research. *Journal of Small Business and Entrepreneurship*, 18(2), 131–141. DOI: 10.1080/08276331.2005.10593335

Hipwell, W. T. L. (1997). Industria, The Fourth World, and the question of territory. *Middle States Geographer*, 30, 1–10.

Hirschheim, R., & Klein, H. K. (2003). Crisis in the IS field? A critical reflection on the state of the discipline. *Journal of the Association for Information Systems*, 4(5), 237–293.

Hnatkowska, B., Kozierkiewicz, A., & Pietranik, M. (2020). Semi-automatic definition of attribute semantics for the purpose of ontology integration. *IEEE Access : Practical Innovations, Open Solutions*, 8, 107272–107284. DOI: 10.1109/ACCESS.2020.3000035

Hobane, P. A. (1994), *The Effects of the Commercialisation of the Mopane Worms: Natural Resources Management Project,* USAID Project no. 690 0251; by the same author (1995), '*Amacimbi: the gathering, processing, consumption and trade of edible caterpillars in Bulilimamangwe District*', Centre for Applied Social Studies Occasional Paper 67, Harare, Zimbabwe.

Hofmeyr, I. (2021). *Dockside Reading: Hydrocolonialism and the Custom House*. Wits University Press.

Hoppers, C. O. (2021). Research on Indigenous knowledge systems: The search for cognitive justice. *International Journal of Lifelong Education*, 40(4), 1–18.

Hossain, K., & Ballardini, R. M. (2021). Protecting Indigenous traditional knowledge through a holistic principle-based approach. *Nordic Journal of Human Rights*, 39(1), 51–72. DOI: 10.1080/18918131.2021.1947449

Hossain, M. F. (2021). Anthropocentric approach to the environment: An overview. *Journal of Social Sciences*, 1711, 43.

Hountondji, P.J. (1995). *Second Bashorum MKA Abiola Distinguished Lecture Address.* Reported in Khosa, Meshack M. 1996.

Hove Chenjerai. (1996). *Ancestors*. College Press Publishers.

Hrach, S. (2021). *Minding bodies: How physical space, sensation, and movement affect learning*. West Virginia University Press.

Hudson, M., Carroll, S. R., Anderson, J., Blackwater, D., Cordova-Marks, F. M., Cummins, J., David-Chavez, D., Fernandez, A., Garba, I., Hiraldo, D., Jäger, M. B., Jennings, L. L., Martinez, A., Sterling, R., Walker, J. D., & Rowe, R. K. (2023). Indigenous peoples' rights in data: A contribution toward Indigenous research sovereignty. *Frontiers in Research Metrics and Analytics*, 8, 1173805. DOI: 10.3389/frma.2023.1173805 PMID: 37215248

Hudson, R. (1999). The learning economy, the learning firm and the learning region. *European Urban and Regional Studies*, 6(1), 59–72. DOI: 10.1177/096977649900600105

Hulme, D., & Murphree, M. (2001). *African wildlife and livelihoods: the promise and performance of community conservation*. James Currey.

Hummel, J. E., & Holyoak, K. J. (2005). Relational reasoning in a neurally plausible cognitive architecture an overview of the LISA Project. *Current Directions in Psychological Science*, 14(3), 153–157. DOI: 10.1111/j.0963-7214.2005.00350.x

Hungwe, J. P., & Mkhize, T. M. (2022). The Africanisation of universities in Africa: Reclamation of humanity and rationality. *Critical Studies in Teaching and Learning*, 10(1). Advance online publication. DOI: 10.14426/cristal.v10i1.487

Hunter, M. (2007). The persistent problem of colourism: Skin tone, status, and inequality. *Sociology Compass*, 1(1), 237–254. DOI: 10.1111/j.1751-9020.2007.00006.x

Hutchinson, P., McIlduff, C. D., Legare, M., Keewatin, M., Hagel, M., Chapados, M., & Acharibasam, J. B. (2023). Indigenous knowledge mobilization: Reflection on context, content, and relationship. *Alternative*, 19(4), 902–913. DOI: 10.1177/11771801231198082

Hysi, F. (2005). *Estetike ne tri pikepamje*. Uegen.

Ibrahim, M. G., Adjei, J. K., & Boateng, J. A. (2019). Relevance of indigenous conflict management mechanisms: Evidence from Bunkpurugu-Yunyoo and Central Gonja Districts of Northern Region, Ghana. *Ghana Journal of Development Studies*, 16(1), 22–45. DOI: 10.4314/gjds.v16i1.2

Ikeke, M., & Ukutsemuya, J. (2024). The Afrocentric imperative in tackling African environmental challenges: Afroecosolidarity perspectives. *East African Journal of Environment and Natural Resources*, 7(1), 17–27. DOI: 10.37284/eajenr.7.1.1675

Imam, S. S., Mehdi, S., & Mansuri, A. (2024). A comprehensive review on the pharmacological potential of red ginseng oil, tea tree oil, and hemp seed oil. *European Journal of Biomedical*, 11(4), 167–177.

Imoro, Z. A., Imoro, A. Z., Duwiejuah, A. B., & Abukari, A. (2021). Harnessing Indigenous technologies for sustainable management of land, water, and food resources amidst climate change. *Frontiers in Sustainable Food Systems*, 5, 691603. DOI: 10.3389/fsufs.2021.691603

Indigenous peoples, voices (United Nations Permanent Forum on Issues). Retrieved on 19 June 2023 from https://www.un.org/esa/socdev/unpfii/documents/unpfiibrochure_en07.pdf

Inmark. (2010a). Concept and methodology of Interactive Workshops. PROmotion of an ICT Dialogue between Europe and America Latina – Extension towards Mexico, Colombia, Cuba, Costa Rica, 1–20. http://cordis.europa.eu/docs/projects/cnect/4/246644

Inoue, C. Y. A., Ribeiro, T. L., Goncalves, V. K., Basso, L., Moreira, P. F. (2023). Indigenous and traditional communities' ways of knowing and being in planetary justice. *Environmental Politics*, 1–20 (Online First): . 2023. 22934 37DOI: 10. 1080/ 09644 016

Institute of Museum and Library Services. (2024). Mukurtu software preserves Indigenous digital heritage through technologies of today. Retrieved on 11 October 2024 from https://www.imls.gov/grant-spotlights/mukurtu-software-preserves-Indigenous-digital-heritage-through-technologies-today

Inuit Tapiriit Kanatami. (2018). National Inuit Strategy on Research. Retrieved on 11 October 2024 from https://www.itk.ca/wp-content/uploads/2018/03/National-Inuit-Strategy-on-Research.pdf

Isaacman, A. (2000). *Chikunda transfrontiersmen and transnational migration in pre-colonial South-Central Africa, 1850-1900*. University of Minnesota Press.

Jackson, I. I., Ronald, L., Drummond, D. K., & Camara, S. (2007). What is qualitative research? *Qualitative Research Reports in Communication*, 8(1), 21–28. DOI: 10.1080/17459430701617879

Jacobi, P., Drescher, A. W., & Amend, J. (2000). *Urban agriculture - justification and planning guidelines*. Urban Vegetable Promotion Project.

Jaggar, A. M., & Rothenberg, P. (1993). *Feminist frameworks: Alternative theoretical accounts of the relations between women and men*. McGraw-Hill.

Jaimungal, C., & Dei, G. J. S. (2018). *Indigeneity and decolonial resistance: alternatives to colonial thinking and practice*. Myers Education Press.

Jain, P., & Jibril, L. (2016). *Expanding library services for Indigenous community posterity: a case of selected public libraries in Botswana*. IFLA.

Jessen, T. D., Ban, N. C., Claxton, N. X., & Darimont, C. T. (2020). Contribution of Indigenous knowledge to ecological and evolutionary understanding. *Frontiers in Ecology and the Environment*, 20(2), 93–101. DOI: 10.1002/fee.2435

Jethro, D. (2024). History uploaded: Digital archives after thirty years of democracy. *South African Historical Journal*, •••, 1–24. Advance online publication. DOI: 10.1080/02582473.2024.2351828

Johnson, M., & Majewska, D. (2022). *Formal, non-formal, and informal learning: What are they, What are they, and how can we research them*? Cambridge University Press & Assessment Research Report.

Johnson, M. (1992). *Capturing traditional environmental knowledge*. IDRC.

Jones, R., Crowshoe, L., Reid, P., Calam, B., Curtis, E., Green, M., Huria, T., Jacklin, K., Kamaka, M., Lacey, C., Milroy, J., Paul, D., Pitama, S., Walker, L., Webb, G., & Ewen, S. (2019). Educating for Indigenous Health Equity: An International Consensus Statement. In Academic Medicine (Vol. 94, Issue 4). https://doi.org/DOI: 10.1097/ACM.0000

Jones-Dube, E. (1984). *Indigenous and non-Indigenous entrepreneurs in Botswana: Historical, cultural and educational factors in their emergence*. University of Massachusetts Amherst.

Jones, R., Crowshoe, L., Reid, P., Calam, B., Curtis, E., Green, M., Huria, T., Jacklin, K., Kamaka, M., Lacey, C., Milroy, J., Paul, D., Pitama, S., Walker, L., Webb, G., & Ewen, S. (2019). Educating for Indigenous Health Equity: An International Consensus Statement. *Academic Medicine*, 94(4), 512–519. DOI: 10.1097/ACM.0000000000002476 PMID: 30277958

Joseph, M. (2011). India faces a linguistic truth: English spoken here." New York Times (2011). Retrieved on 14 June 2023 from https://www.nytimes.com/2011/02/17/world/asia/17iht-letter17.html

Josyula, K. L., Nambiar, D., Narayan, V., Sathyanarayana, T. N., Porter, J., & Sheikh, K. (2015). Cultural consonance, constructions of science and co-existence: A review of the integration of traditional, complementary, and alternative medicine in low- and middle-income countries. *Health Policy and Planning*, 30(8), 1067–1077. DOI: 10.1093/heapol/czu096 PMID: 25171821

Joullié, J.-E., & Gould, A. M. (2021). Theory, explanation, and understanding in management research. *BRQ Business Research Quarterly*, 0(0). Advance online publication. DOI: 10.1177/23409444211012414

Jukka, M., Blomqvist, K., Li, P. P., & Gan, C.-M. (2017). Trust-distrust balance: Trust ambivalence in Sino-Western B2B relationships. *Cross Cultural & Strategic Management*, 24(3), 482–507. DOI: 10.1108/CCSM-01-2016-0019

Jules, N. P. (1995). Participatory learning for sustainable agriculture. *World Development*, 23(8), 1–17.

Jurdi, H., & Houjeir, R. (2020). Cultural norms and the marketplace: Favor economies in the arab world. *Journal of Consumer Marketing*, 37(2), 226–235. DOI: 10.1108/JCM-05-2018-2681

Justesen, T., Freyberg, J., & Schultz, A. N. Ø. (2021). Database selection and data gathering methods in systematic reviews of qualitative research regarding diabetes mellitus - an explorative study. *BMC Medical Research Methodology*, 21(1), 94. DOI: 10.1186/s12874-021-01281-2 PMID: 33941105

Kabir, A. J. (2020). The Fleeting Taste of Mazaa: From Embodied Philology to an Alegropolitics for South Asia. *South Asia*, 43(2), 243–254. DOI: 10.1080/00856401.2020.1703486

Kakar, Z. U. H., Rasheed, R., Rashid, A., & Akhter, S. (2023). Criteria for assessing and ensuring the trustworthiness in qualitative research. *International Journal of Business Reflections*, 4(2), 148–173. DOI: 10.56249/ijbr.03.01.44

Kalenga, R. C. (2015). Adjusting Western research techniques to accommodate research in the Indigenous realm. *Indilinga*, 14(1), 1–11.

Kamanga, R. (2008). Protection of Intellectual Property: Musical, Literary and Artistic work in Malawi. Unpublished article presented at the PASMAE Conference in Lilongwe, Malawi.

Kamanga, , Vedeld, P., & Sjaastad, E. (2009). Forest Incomes and Rural Livelihoods in Chiadzulu District, Malawi. *Ecological Economics*, 68(3), 122–131. DOI: 10.1016/j.ecolecon.2008.08.018

Kandiyoti, M.. (2021). The Ecological Impact of Overharvesting Mopane Worms: An analysis of Resource Depletion in Mangwe. *Environmental Sustainability Review*, 12(1), 46–71.

Kanengoni, T. (2017). Doma community faces challenges as nature reserve designation restricts hunting. *Global Press Journal*. Retrieved on 5 November 2024 from https://globalpressjournal.com/africa/zimbabwe/doma-community-faces-challenges-nature-reserve-designation-restricts-hunting-fishing/

Kanosvamhira, T. P. (2024). Sustainable urban agriculture: Unlocking the potential of home gardens in low-income communities. *The Professional Geographer*, 76(5), 587–596. DOI: 10.1080/00330124.2024.2355179

Kaomea, J. (2003). Reading erasures and making the familiar strange: Defamiliarizing methods for research in formerly colonized and historically oppressed communities. *Educational Researcher*, 32(2), 14–23. DOI: 10.3102/0013189X032002014

Kapoor, D., & Shizha, E. (Eds.). (2010). *Indigenous knowledge and learning in Asia/Pacific and Africa: perspectives on development, education and culture*. Palgrave Macmillan. DOI: 10.1057/9780230111813

Kapsdorferova, Z., Kadlecikova, M., & Svitova, E. (2015). Utilization of quality managerial systems in business entities in the Slovak Republic. *International Journal of Qualitative Research*, 9(2), 197.

Kapucu, N., Ge, Y., Rott, E., & Isgandar, H. (2024). *Urban resilience: multidimensional perspectives, challenges and prospects for future research*. Urban Governance., DOI: 10.1016/j.ugj.2024.09.003

Kapur, R. (2019). Introduction to adult education. Retrieved on 21 April 2024 from https://www.researchgate.net/publication/335422676_Introduction_to_Adult_Education/citation/download

Karasimos, A. (2022). The battle of language learning apps: A cross-platform overview. *Research Papers in Language Teaching and Learning*, 12(1), 150–166.

Kaunda, M. K. (2021). Mother earth, mother Africa and African women's role in Indigenous religions. *African Journal of Gender and Religion*, 27(1), 91–117. DOI: 10.14426/ajgr.v27i1.89

Kawulich, B. (2009). The role of theory in research. In Garner, M., Wagner, C., & Kawulich, B. (Eds.), *Teaching research methods in the social sciences* (pp. 37–47). Routledge.

Kaya, H. O., & Seleti, Y. N. (2013). African Indigenous knowledge systems and relevance of higher educaiton in South Africa. *The International Education Journal: Comparative Perspectives*, 12(1), 30–44.

Kaya, O. H. (2013). Integration of African Indigenous knowledge systems into higher education in South Africa: Prospects and challenges. *Alternation (Durban)*, 20(1), 135–153.

Kaya, O. H., & Seleti, Y. N. (2013). African Indigenous knowledge systems and relevance of higher education in South Africa. *The International Educational Journal: Comparative Perspectives*, 12(1), 30–44.

Keane, M., Khupe, C., & Seehawer, M. (2017). Decolonizing methodology: Who benefits from Indigenous knowledge research? [ERSC]. *International Research for Social Change*, 6(1), 12–24. DOI: 10.17159/2221-4070/2017/v6i1a2

Keane, M., Raciti, M., van der Westhuizen, G., Motala, S., Stanton, S., Gilbey, K., Manathunga, C., Qi, J., & Msimango, S. (2023). Indigenous knowledge systems in South Africa and Australia: Transforming doctoral education. *Curriculum Perspectives*, 43(S1), 83–93. Advance online publication. DOI: 10.1007/s41297-023-00183-1

Kelman, I., Mercer, J., & Gaillard, J. (2012). Indigenous Knowledge and Disaster Risk Reduction. *Geography (Sheffield, England)*, 97(1), 12–21. DOI: 10.1080/00167487.2012.12094332

Kennedy, A., Bourque, D. H., Bourque, D. E., Cardinal, S., & Bourque Bearskin, R. L. (2021). Reconciling Taking the "Indian" out of the Nurse. Quality Advancement in Nursing Education - Avancées En Formation Infirmière, 7(1). DOI: 10.17483/2368-6669.1276

Kennedy, A., Sehgal, A., Szabo, J., McGowan, K., Lindstrom, G., Roach, P., Crowshoe, L., & Barnabe, C. (2022). Indigenous strengths-based approaches to healthcare and health professions education – recognising the value of Elders' teachings. *Health Education Journal*, 81(4), 423–438. DOI: 10.1177/00178969221088921 PMID: 35531386

Kennedy, V., Cram, F., Paipa, K., Pipi, K., Baker, M., Porima, L., Sauni, P., & Tuagalu, C. (2015). Beginning a conversation about spirituality in Māori and Pasifika evaluation. In Hood, S., Hopson, R., & Freirson, H. (Eds.), *Continuing the journey to reposition culture and cultural context in evaluation theory and practice* (pp. 151–178). Information Age Publishing.

Kerr, J. & Sanghi, N. (1992). *Indigenous soil and water conservation in India's semi-arid tropics*. International Institute for Environment and Development (IIED).

Kettley, N. (2010). Theory building in educational research. *Continuum*.

Kgope, T. V. (2022). "A Critical Reflection on Complementary, Alternative and Indigenous Knowledge Medicine in Gauteng Province: A Model for Articulation and Promotion." PhD diss., 2022.

Kgope, T. V. (2023). The indigenous knowledge system of Credo Mutwa: A pedagogical challenge in higher education in South Africa. *Curriculum Perspectives*, 43(S1), 1–9. DOI: 10.1007/s41297-023-00182-2

Kgoroeadira, K. O. (2014). *The Praise Poetry of the Bafokeng of Phokeng*. University of Johannesburg.

Khan, M. (2018). Traditional knowledge and the commons: the open movement, listening, and learning. Retrieved on 11 October 2024 from https://creativecommons.org/2018/09/18/traditional-knowledge-and-the-commons-the-open-movement-listening-and-learning/

Khoza, R. J. (2011). *Attuned Leadership: African Humanism as Compass*. Penguins Books.

Khumalo, B. N., & Baloyi, C. (2017). African Indigenous knowledge: an underutilised and neglected resource for development. *Library Philosophy and Practice (e-journal)*. 1663. https://digitalcommons.unl.edu/libphilprac/1663

Kindon, S., Pain, R., & Kesby, M. (2007). *Participatory action research approaches and methods: connecting people, participation and place*. Routledge. DOI: 10.4324/9780203933671

King, T. L., Navarro, J., & Smith, A. (Eds.). (2020). *Otherwise worlds: against settler colonialism and anti-Blackness*. Duke University Press.

Knappert, J. (1990). *The Aquarian Guide to African Mythology*. Aquarian Press.

Knoema (2021). Nigeria - urban population as a share of the total population. Retrieved on 1 June 2024 from https://knoema.com/atlas/Nigeria/Urban-population

Knopf, K. (2015). The turn toward the Indigenous: Knowledge systems and practices in the academy. *Amerikastudien. American Studies (Lawrence, Kan.)*, 60(2/3), 179–200.

Kohler, O. (1959). *The study of the Gobabis district. Union of South Africa*. Department of Bantu Administration.

Kohsaka, R., & Rogel, M. (2019). Traditional and local knowledge for sustainable development: empowering the Indigenous and local communities of the world. In: Leal Filho, W., Azul, A., Brandli, L., Özuyar, P., Wall, T. (eds) *Partnerships for the Goals. Encyclopedia of the UN Sustainable Development Goals*. Springer, https://doi.org/DOI: 10.1007/978-3-319-71067-9_17-1

Kolawole, D. (2022). Is local knowledge peripheral? The future of Indigenous knowledge in research and development. *Alternative*, 18(1), 132–140. DOI: 10.1177/11771801221088667

Kolawole, O. D. (2001). Local knowledge utilization and sustainable rural development in the 21st Century. *Indigenous Knowledge and Development Monitor*, 9(3), 13–15.

Kom, Z., Nethengwe, N. S., Mpandeli, S., & Chikoore, H. (2022). Indigenous knowledge indicators employed by farmers for adaptation to climate change in rural South Africa. *Journal of Environmental Planning and Management*, 66(13), 2778–2793. DOI: 10.1080/09640568.2022.2086854

Koot, S., Robert, H., & Catie, G. (2019). Belonging, indigeneity, land and nature in Southern Africa under neoliberal capitalism: An overview. *Journal of Southern African Studies*, 45(2), 341–355. DOI: 10.1080/03057070.2019.1610243

Korn, L. E. (2023). *Rhythms of recovery: integrative medicine for PTSD and complex trauma*. Routledge. DOI: 10.4324/9781003231455

Korn, L., Logsdon, R. G., Polissar, N. L., Gomez-Beloz, A., Waters, T., & Ryser, R. (2009). A randomized trial of a CAM therapy for stress reduction in American Indian and Alaskan Native family caregivers. *The Gerontologist*, 49(3), 368–377. DOI: 10.1093/geront/gnp032 PMID: 19377083

Kothari, A., Salleh, A., Escobar, A., Demaria, F., & Acosta, A. (2019). Introduction: finding pluriversal paths. In Kothari, A., Salleh, A., Escobar, A., Demaria, F., & Acosta, A. (Eds.), *Pluriverse: A post-development dictionary* (pp. xxi–xl). Tulika Books.

Kovach, M. (2010). Conversational method in indigenous research. *First Peoples Child & Family Review*, 5(1), 40–48. DOI: 10.7202/1069060ar

Kozanayi, W., & Frost, P. (2002). *Marketing of Mopane Worm in Southern Zimbabwe,', Institute of Environmental studies*. University of Zimbabwe.

Kraiwanit, T., Limna, P., & Siripipatthanakul, S. (2023). NVivo for social sciences and management studies: A systematic review. *Advance Knowledge for Executives*, 2(3), 1–11.

Krasnof, B. (2016). *Culturally responsive teaching: a guide to evidence-based practices for teaching all students equitably*. Region X, Equity Assistance Center at Education Northwest.

Kugara., ... (2022), *The role of Indigenous Knowledge Systems (IKS) in Climate Change,* Handbook of Research on Protecting and Managing Global Indigenous Knowledge Systems, IGI-Global

Kugara, S. L., & Sekgothe, M. (2022). Challenges presented by digitisation of VhaVenda oral tradition: An African Indigenous knowledge systems perspective. *Hervormde Teologiese Studies*, 78(3), 1–8. DOI: 10.4102/hts.v78i3.7428

Kuhnlein, V., Erasmus, B., & Sprigelski, D. (2009). *Indigenous people's food systems*. FAO.

Kuhn, T. S. (1970). *The structure of scientific revolutions* (2nd ed.). University of Chicago Press.

Kumar, R., & Gupta, V. K. (2019). An introduction to cognitive constructivism in education. *Journal of Indian Education*, 35(3), 39–45.

Kwenda, C. V. (2003). Cultural justice: The pathway to reconciliation and social cohesion'. In D. Chidester, W. James & P. Dexter (eds.), *What holds us together: Social cohesion in South Africa* (pp. 67–80). HSRC Press.

LaDonna, K. A., Artino, A. R.Jr, & Balmer, D. F. (2021). Beyond the Guise of Saturation: Rigor and Qualitative Interview Data. *Journal of Graduate Medical Education*, 13(5), 607–611. DOI: 10.4300/JGME-D-21-00752.1 PMID: 34721785

Lafrance, J., & Nichols, R. (2010). Reframing evaluation: Defining an Indigenous evaluation framework. *The Canadian Journal of Program Evaluation*, 23(2), 13–31. DOI: 10.3138/cjpe.23.003

Lagos State Government. (2019). About Lagos. Retrieved on 1 April 2024 from https://lagosstate.gov.ng/about-lagos/

Lagos State Government. (2022). About Lagos. Retrieved on 1 August 2024 from https://lagosstate.gov.ng/about-lagos/

Laird, P. (2022). Fits and starts: Anti-racism efforts in the biosciences. *Nature*, 612, 177–178.

Lakshmi Reddy, M. V. (2000). *Dynamic role of adult education: marching education towards a learning society*. Aravali Books International (P) Ltd.

Lalonde, A. (1991). African Indigenous knowledge and its relevance to environment and development activities. *The Common Property Conference*. Winnipeg. International Association for the Study of Common Property (IASCP).

Lan, D. (1985). *Guns and rain: guerrillas and spirit mediums in Zimbabwe*. Zimbabwe Publishing House.

Lander, D. (2000). Eurocentricism and colonialism in Africa. *Nepantla*, 1(2), 510–532.

Lander, E., & Past, M. (2002). Eurocentrism, modern knowledges, and the "natural" order of global capital. *Nepantla*, 3(2), 245–268.

Landy, D., & Goldstone, R. L. (1991). Relational reasoning is in the eyes of the beholder: how global perceptual groups aid and impair algebraic evaluations. *Proceedings of the Annual Meeting of the Cognitive Science Society*, 27, 2509.

Lange, E. A. (2018). Transformative sustainability education: from sustainable to a civilization leap. In Milana, M., Webb, S., Holford, J., Waller, R., & Jarvis, P. (Eds.), *The Palgrave international handbook on adult and lifelong education and learning* (pp. 1–22). Palgrave Macmillan. DOI: 10.1057/978-1-137-55783-4_21

Lankford G.E. (1987). Native American Legends. August House/Little Rock.

Lassiter, L. E. (2021). Collaborative ethnography: trends, developments, and opportunities. In B. Diamond & S. El-Shawan Castelo-Branco (eds), *Transforming ethnomusicology volume i: methodologies, institutional structures, and policies*. New York. DOI: 10.1093/oso/9780197517604.003.0004

Lassiter, L. E. (2005a). *The Chicago guide to collaborative ethnography*. University of Chicago Press. DOI: 10.7208/chicago/9780226467016.001.0001

Lassiter, L. E. (2005b). Collaborative ethnography and public anthropology. *Current Anthropology*, 46(1), 83–106. DOI: 10.1086/425658

Lawal, O. A. (2015). Indigenous languages as tools for effective communication of science and technology for food production in Nigeria. *Theory and Practice in Language Studies*, 5(3), 463–468. DOI: 10.17507/tpls.0503.02

Lawrence, S., & Hylton, K. (2022). Critical race theory, methodology, and semiotics: The analytical utility of a "race" conscious approach for visual qualitative research. *Cultural Studies Critical Methodologies*, 22(3), 255–265. DOI: 10.1177/15327086221081829

Le Grange, L. (2016). Decolonising the university curriculum. *South African Journal of Higher Education*, 30(2), 1–12.

Lebaka, M. E. K. (2019). Understanding Indigenous Knowledge as Indigenous Practice, Skill and Know-How: The Case of Communal Music-Making in Bapedi Culture as a Social Practice. *European Journal of Social Science Education and Research* May - August 2019 6(2), 25. ISSN 2411-9563 (Print) ISSN 2312-8429 (Online)

Lebakeng, L. T., Phalane, M. M., & Nase, D. (2006). Epistemicide, institutional cultures and the imperative for the Africanisation of universities in South Africa. *Alternation (Durban)*, 13(1), 70–87.

Lefike, M., Turpin, M., & Matthee, M. (2023). A systems framework to analyze the impact of corporate social investment projects with an information technology focus. *The Electronic Journal on Information Systems in Developing Countries*, 89(5), e12273. DOI: 10.1002/isd2.12273

Lerner, D. (1958). *The passing of traditional society: modernizing the Middle East*. Free Press.

Lethole, P. V., McIntyre-Mills, J. J., Wirawan, R., & Makaulule, M. (2022). AA!! Venda women and social enterprise: stepwise process to regenerative and sustainable living. In McIntyre-Mills, J. J., & Corcoran-Nantes, Y. (Eds.), *Transformative education for regeneration and wellbeing* (pp. 353–372). Springer. DOI: 10.1007/978-981-19-3258-8_17

Letseka, M. (2013). Educating for Ubuntu: Lessons from Basotho Indigenous education. *Open Journal of Philosophy*, 3(2), 337–344. DOI: 10.4236/ojpp.2013.32051

Levac, L., McMurtry, L., Stienstra, D., Baikie, G., Hanson, C., & Mucina, D. (2018). Learning across Indigenous and Western knowledge systems and intersectionality: reconciling social science research approaches. Retrieved on 24 November 2024 from https://www.criaw-icref.ca/wp-content/uploads/2021/04/Learning-Across-Indigenous -and-Western-KnowledgesFINAL.pdf

Levers, L. L. (2009). A meta-theoretical framework for counselling people of African ancestry: An ecologically oriented response to the special section articles. *Journal of Psychology in Africa*, 19(4), 503–506. DOI: 10.1080/14330237.2009.10820322

Levi-Strauss, C. (1966). *The savage mind*. University of Chicago Press.

Lim, W. M. (2024). What is qualitative research? An overview and guidelines. *Australasian Marketing Journal*, •••, 14413582241264619. DOI: 10.1177/14413582241264619

Lincoln, Y. S., & Guba, E. G. (1986). But Is It Rigorous? Trustworthiness and Authenticity in Naturalistic Evaluation. *New Directions for Program Evaluation*, 30(30), 73–84. https://doi.org/https://doi.org/10.1002/ev.1427. DOI: 10.1002/ev.1427

Lincoln, Y., & Guba, E. (1985). Naturalistic inquiry. *Sage (Atlanta, Ga.)*.

Li, P. P. (2008). Toward a geocentric framework of trust: An application to organizational trust. *Management and Organization Review*, 4(3), 413–439. DOI: 10.1111/j.1740-8784.2008.00120.x

Li, P. P. (2012). Toward an integrative framework of indigenous research: The geocentric implications of Yin-Yang balance. *Asia Pacific Journal of Management*, 29(4), 849–872. DOI: 10.1007/s10490-011-9250-z

Li, P. P. (2016). The global implications of the Indigenous epistemological system from the East: How to apply Yin-Yang balancing to paradox management. *Cross Cultural & Strategic Management*, 23(1), 42–77. DOI: 10.1108/CCSM-10-2015-0137

Li, P. P., Sekiguchi, T., & Zhou, K. (2016). The emerging research on Indigenous management in Asia. *Asia Pacific Journal of Management*, 33(3), 583–594. DOI: 10.1007/s10490-016-9484-x

Lo Liyong, T. (1991). *Culture is Rutan*. Longman.

Loeng, S. (2020). Self-directed learning: A core concept in adult education, *Hindawi. Education Research International*, 2020, 1–12. Advance online publication. DOI: 10.1155/2020/3816132

Lokot, M., Hartman, E., & Hashmi, I. (2023). Participatory approaches and methods in gender equality and gender-based violence research with refugees and internally displaced populations: A scoping review. *Conflict and Health*, 17(1), 58. DOI: 10.1186/s13031-023-00554-5 PMID: 38066619

Loomba, A. (2015) Psychoanalysis and colonial subjects. *Colonialism/Postcolonialism*, 139–153.

Lopez-Deflory, C., Perron, A., & Miro-Bonet, M. (2022). A methodological and practical guide to study peripheral voices in qualitative research. *International Journal of Qualitative Methods*, 21(10), 1–20. DOI: 10.1177/16094069221100639

Loud, M. L. (2021). Conclusion: Transforming evaluation practice for "business unusual". *The Canadian Journal of Program Evaluation*, 36(2), 187–190. DOI: 10.3138/cjpe.72730

Louis, R. P. (2007). Can you hear us now? Voices from the margin: Using Indigenous methodologies. *Geographical Research*, 45(2), 130–139. DOI: 10.1111/j.1745-5871.2007.00443.x

Lucero, J., Wallerstein, N., Duran, B., Alegria, M., Greene-Moton, E., Israel, B., Kastelic, S., Magarati, M., Oetzel, J., Pearson, C., Schulz, A., Villegas, M., & White Hat, E. R. (2018). Development of mixed methods investigation of process and outcomes of community-based participatory research. *Journal of Mixed Methods Research*, 12(1), 55–74. DOI: 10.1177/1558689816633309 PMID: 29230152

Lundquist, B. (1989). Sub-Sahara Africa in Multicultural Perspectives in Music Education. In Anderson, W., & Campbell, S. (eds) Reston, VA:*Music Educators National Conference*, pp. 145-178.

Lundquist, C., Harhash, K. A., Armenteras, D., Chettri, N., Mwamodenyi, J. M., Prydatko, V., & Acebey Quiroga, S. (2016). Building capacity for developing, interpreting and using scenarios and models. In *Methodological assessment of scenarios and models of biodiversity and ecosystem services* (pp. 225–254). Secretariat of the Intergovernmental Platform for Biodiversity and Ecosystem Services.

Lutomia, A. N., & Bravo, J. B. (2017). Communities of Practice and Indigenous Knowledge: A Case Study of Empowering Women in Processing Shea Butter Using Scientific Animations. In Handbook of Research on Social, Cultural, and Educational Considerations of Indigenous Knowledge in Developing Countries (pp. 252-270). United State America: Chigago DOI: 10.4018/978-1-5225-0838-0.ch012

Luutu, B. M. (2015). Restorative Action into Public Policy: An African Perspective. Human and Social Dynamics Research Seminar, HSRC.

Luvalo, L. M. (2019). Relationship between transformation and institutional culture. *South African Journal of Higher Education*, 33(1), 184–199. DOI: 10.20853/33-1-2934

Lyman, R. L., & O'Brien, M. J. (2001). The direct historical approach, analogical reasoning, and theory in Americanist archaeology. *Journal of Archaeological Method and Theory*, 8(4), 303–342. DOI: 10.1023/A:1013736416067

Lythreatis, S., Singh, S. K., & El-Kassar, A. (2022). The digital divide: A review and future research agenda. *Technological Forecasting and Social Change*, 175(121259), 121359. Advance online publication. DOI: 10.1016/j.techfore.2021.121359

Mabhaudhi, T., Chibarabada, T. P., Chimonyo, V. G. P., Murugani, V. G., Pereira, L. M., Sobratee, N., Govender, L., Slotow, R., & Modi, A. T. (2019). Mainstreaming underutilized Indigenous and traditional crops into food systems: A South African perspective. *Sustainability*, 11(1), 172. DOI: 10.3390/su11010172 PMID: 37681213

Mabunda, P. L., & McKay, V. I. (2021). Educational curriculum and multispecies relations. In McIntyre-Mills, J. J., & Corcoran-Nantes, Y. (Eds.), *From polarisation to multispecies relationships* (pp. 369–389). Springer. DOI: 10.1007/978-981-33-6884-2_16

Machado, L. M. O. (2021). Ontologies in knowledge organization. *Encyclopedia*, 1(1), 144–151. DOI: 10.3390/encyclopedia1010015

Machida, S. (2020). Ethnocentrism and support for nuclear armament in Japan. *Asian Journal of Comparative Politics*, 7(1), 95–114. DOI: 10.1177/2057891120939314

Machingambi, S. Analysis of decolonization of higher education in a new South African university. *African Perspectives of Research in Teaching & Learning,* 4(1), 32-45.

Machingura, F. (2018). Market Volatility in the Mopane Worm Trade. *Journal of African Economic Studies*, 56(4), 134–147.

Machiridza, L. H., & Thomas, P. T. (2022). Lost and found property: Evaluating efforts to resurrect and resuscitate 'institutionalised garbage in post-colonial Zimbabwean museums. *Curator (New York, N.Y.)*, 65(2), 451–470. DOI: 10.1111/cura.12462

Machulak, E. (2020). Mukurtu: A digital platform that does more than manage content. *Humanities (Washington)*, 41(4). Retrieved October 11, 2024, from https://www.neh.gov/article/mukurtu-digital-platform-does-more-manage-content

MacLeod, L. (2021). More Than Personal Communication: Templates for Citing Indigenous Elders and Knowledge Keepers. *KULA: Knowledge Creation, Dissemination, and Preservation Studies, 5*(1), 1–5.

MacNeil, M. S. (2008). An epidemiologic study of Aboriginal adolescent risk in Canada: The meaning of suicide. *Journal of Child and Adolescent Psychiatric Nursing*, 21(1), 3–12. DOI: 10.1111/j.1744-6171.2008.00117.x PMID: 18269407

MacRae, R. J., Hill, S. B., Henning, J., & Mehuys, G. R. (1989). Agricultural science and sustainable agriculture: A review of the existing scientific barriers to sustainable food production and potential solutions. *Biological Agriculture and Horticulture*, 6(3), 173–219. Advance online publication. DOI: 10.1080/01448765.1989.9754518

Madden, B. (2015). Pedagogical pathways for Indigenous education with / in teacher education. *Teaching and Teacher Education*, 51, 1–15. DOI: 10.1016/j.tate.2015.05.005

Maddock, K. (1972). *The Australian Aborigines: A portrait of their society*. Allen Lane, Penguin Press.

Madlela, B. (2017). Incorporating Indigenous knowledge systems into the formal natural science curriculum: A case study of schools in Pongola, KwaZulu Natal Province. University of South Africa. Available at https://hdl.handle.net/10500/28656 [Accessed on 28 September 2024].

Madonsela, B., Semenya, K., & Shale, K. (2024). A review of Indigenous knowledge systems and their application in sustainable solid waste management. *WORLD (Oakland, Calif.)*, 5(2), 219–239. DOI: 10.3390/world5020012

Maffi, L. (2005). Linguistic, Cultural, and Biological Diversity. *Annual Review of Anthropology*, 29(1), 599–617. DOI: 10.1146/annurev.anthro.34.081804.120437

Magadza, S. N. (2006). Empowerment. *Mobilisation and Initiation of a Community Driven Project: Women and the Marula*, 7(7), 102–112.

Magesa, M., Jonathan, J., & Urassa, J. (2023). Digital literacy of smallholder farmers in Tanzania. *Sustainability (Basel)*, 15(17), 13149. DOI: 10.3390/su151713149

Magni, G. (2017). Indigenous knowledge and implications for the sustainable development agenda. *European Journal of Education*, 52(4), 437–447. DOI: 10.1111/ejed.12238

Magnússon, G., Göransson, K., & Lindqvist, G. (2019). Contextualizing inclusive education in educational policy: The case of Sweden. *Nordic Journal of Studies in Educational Policy*, 5(2), 67–77. DOI: 10.1080/20020317.2019.1586512

Magro, C., & van den Berg, R. (2021). Of portals and paradigms: evaluation, systems thinking and the pandemic. In R. van den Berg, C. Magro, & M.H. Adrien, (Eds.), *Transformational evaluation,* pp. 371-394. International Development Evaluation Association (IDEAS).

Mahlatsi, K. S., Pienaar, A. J., Nare, N. E., & Mulaudzi, T. M. (2021). A conceptual framework for psychosocial health management grounded in the therapeutic merits of Indigenous Khoisan health dialogues. *Health SA = SA Gesondheid*, 26(0), a1626. DOI: 10.4102/hsag.v26i0.1626 PMID: 34522395

Mahomoodally, M. F. (2013). Traditional medicines in Africa: An appraisal of ten potent African medicinal plants. *Evidence-Based Complementary and Alternative Medicine : eCAM*, 617459, 1–14. DOI: 10.1155/2013/617459 PMID: 24367388

Maiam nayri Wingara. (2018). Indigenous Data Sovereignty Communique. Indigenous Data Sovereignty Summit. 20th June 2018, Canberra: Australian Capital Territory. Retrieved on 11 October 2024 from https://static1.squarespace.com/static/5b3043afb40b9d20411f3512/t/5b6c0f9a0e2e725e9cabf4a6/1533808545167/Communique%2B%2BIndigenous%2BData%2BSovereignty%2BSummit.pdf

Mair, J., Battilana, J., & Cardenas, J. (2012). Organizing for society: A typology of social entrepreneuring models. *Journal of Business Ethics*, 111(3), 353–373. https://link.springer.com/article/10.1007/s10551-012-1414-3. DOI: 10.1007/s10551-012-1414-3

Makananise, F. O. (2024). Digital preservation of Indigenous languages and participatory epistemic knowledge systems of the Global South: insights from South African experiences. In Makananise, F. O., & Madima, S. E. (Eds.), *Digital media and the preservation of Indigenous Languages in Africa: toward a digitalized and sustainable society* (pp. 21–48). Lexington Books.

Makgopa, M., & Magaelane, K. (2009). The use of ditaola (divination bones) among indigenous healers in Sekhukhune District, Limpopo Province: Indigenous African healing practices. *Indilinga*, 8(1), 51–58. DOI: 10.4314/indilinga.v8i1.48242

Makhalemele, O. (2009). "Still Not Talking: The South African government's exclusive reparations policy and the impact of the R30,000 financial reparations on survivors." In *Reparations for Victims of Genocide, War Crimes and Crimes against Humanity*, pp. 541-566. Brill Nijhoff. DOI: 10.1163/ej.9789004174498.i-576.139

Makhanya, T., & Zibane, S. (2020). Students' voices on how indigenous languages are disfavoured in South African higher education. *Language Matters (Pretoria)*, 51(1), 22–37. DOI: 10.1080/10228195.2020.1711533

Makobela, T. (2017). *Perceptions of black men in Katlehong about female " yellow bones ": a case study*. Creative Commons.

Makoelle, T. M. (2014). Cognitive justice: Road map for equitable inclusive learning environments. *International Journal of Education and Research*, 2(7), 505–518.

Makombe, G.. (2019). *Economic and Infrastructure Challenges of the Mopane Worm Economy in Rural Zimbabwe,* Development. *Economic Review (Kansas City, Mo.)*, 25(2), 98–110.

Malabou, C. (1997). Who Is Afraid of Hegelian Wolves? In Patton, P. (Ed.), *Deleuze: A Critical Reader* (pp. 114–138). Wiley-Blackwell.

Maldonado-Torres, N. (2007). On coloniality of being: Contributions to the development of a concept. *Cultural Studies*, 21(2-3), 240–270. DOI: 10.1080/09502380601162548

Maldonado-Torres, N. (2017). Frantz Fanon and the decolonial turn in psychology. Modern / colonial methods to the decolonial attitude. *South African Journal of Psychology. Suid-Afrikaanse Tydskrif vir Sielkunde*, 47(4), 432–441. DOI: 10.1177/0081246317737918

Maldonado-Torres, N., & Cavooris, R. (2017). The decolonial turn. In Poblete, J. (Ed.), *New approaches to Latin American Studies: Culture and power* (1st ed.). Routledge., DOI: 10.4324/9781315158365-8

Maluleka, J. R., & Ngulube, P. (2019). Indigenous knowledge in Africa: A bibliometric analysis of publishing patterns. *Publishing Research Quarterly*, 35(3), 445–462. DOI: 10.1007/s12109-019-09656-6

Mamimine, P. (2000). How far the destination? Decentralisation and devolution in governance of commons under CAMPFIRE. In E. M. Chiwome., Mguni, Z. & Furusa, M. (Eds). *Indigenous knowledge and technology in Africa and diaspora communities. Multidisciplinary approaches*. California, National Council for Black Studies.

Mampilly, Z. (2015). Decolonizing the United States: Lessons from Africa. *Ufahamu. Journal of African Studies*, 38(3), 1–22.

Mandikonza, C. (2019). Integrating indigenous knowledge practices as context and concepts for the learning of curriculum science: A methodological exploration. *Southern African Journal of Environmental Education*, 35(1), 1–16. DOI: 10.4314/sajee.v35i1.13

Manuel, G., & Posluns, M. (1974). *The Fourth World: an Indian reality*. Con Mills, Collier-Macmillan.

Manžuch, Z. (2017). Ethical Issues in digitization of cultural heritage. *Journal of Contemporary Archival Studies*, 4(4). https://elischolar.library.yale.edu/jcas/vol4/iss2/4

Mapara, J. (2009). Indigenous knowledge systems in Zimbabwe: Juxtaposing postcolonial theory. *The Journal of Pan African Studies*, 3, 139–155.

Mapara, J. (2009). Indigenous Knowledge Systems in Zimbabwe: Juxtaposing the postcolonial theory. *The Journal of Pan African Studies*, 3(1), 23–19.

Mapara, J. (2023a). The african university and the urgent need for decoupling from the global north. In Ndofirepi, A., Vurayai, S., & Erima, G. (Eds.), *Unyoking African university knowledge: a pursuit of the decolonial agenda* (pp. 43–63). Brill. DOI: 10.1163/9789004548107_004

Mapara, J. (2023b). Tragic labels, catastrophic consequences: colonial treachery and the Cameroonian calamity. In Mavengano, E., & Mhute, I. (Eds.), *Sub-Saharan political cultures of deceit in language, literature, and the media* (Vol. II). Palgrave Macmillan., DOI: 10.1007/978-3-031-42883-8_13

Mapara, J., & Mutasa, D. E. (2011). Avoiding potential knowledge death: Protecting Indigenous knowledge in a globalising world. *Southern African Journal for Folklore Studies*, 21(2), 12–25.

Maphosa, C. (2020). The Decline of Traditional Knowledge in Zimbabwe: The Case of Mopane Worm Harvesting. *Journal of Rural Sociology*, 22(3), 41–56.

Mapira, J., & Mazambara, P. (2013). Indigenous knowledge systems and their implications for sustainable development in Zimbabwe. *Journal of Sustainable Development in Africa*, 15(5).

Marinković, V. (2017). Effects of the Republic of Serbia's citizens' patriotism and animosity towards the European Union on consumer ethnocentrism. *Ekonomski Horizonti*, 19(1), 3–15. DOI: 10.5937/ekonhor1701003M

Marsden, D. (2005). *Indigenous wholistic theory for health: Enhancing traditional-based Indigenous health services in Vancouver. Unpublished Doctor of Philosophy*. University of British Columbia.

Mart, C. T., & Alpaslan, T. (2010). How did British colonial education in Africa become a reason for decolonization? In *Second International Symposium on sustainable development.*

Martel, R., Shepherd, M., & Goodyear-Smith, F. (2021). He awa whiria – A "braided river": An Indigenous Māori approach to mixed methods research. *Journal of Mixed Methods Research.* Advance online publication. DOI: 10.1177/1558689820984028

Martin, D. H. (2012). Two-eyed seeing: A framework for understanding Indigenous and non-Indigenous approaches to Indigenous health research. *Canadian Journal of Nursing Research*, 44(2), 20–42. PMID: 22894005

Martinez, M. A. (1997). *Study on treaties, agreements and other constructive arrangements between states and Indigenous populations: Final Report.* United Nations.

Martone, M. E. (2015). FORCE11: Building the future for research communications and e-scholarship. *Bioscience*, 65(7), 635. DOI: 10.1093/biosci/biv095

Mary, K. (2018). Use of community cultural practices and beliefs in the conservation of Lake Baringo Ecosystem in Kenya. *International Journal Natural Resources Ecological Management*, 3(32), 14–34.

Masaka, D. (2018). The prospects of ending epistemicide in Africa: Some thoughts. *Journal of Black Studies*, 49(3), 284–301. DOI: 10.1177/0021934717753731

Masango, C. A., & Nyasse, B. (2015). Documenting Indigenous knowledge about Africa's traditional medicine: A myth or a reality? *Indilinga*, 14(1), 52–66.

Masenya, T. M. (2023). Revitalization and digital preservation of Indigenous knowledge systems for sustainable development of Indigenous communities in South Africa. *The Serials Librarian*, 84(5–8), 86–102. DOI: 10.1080/0361526X.2023.2277962

Mashabela, J. K. (2017). Africanisation as an agent of theological education in Africa. *Hervormde Teologiese Studies*, 73(3), a4581. DOI: 10.4102/hts.v73i3.4581

Mashelkar, R. (2001). *The Indian innovation system. Development Outreach.* World Bank.

Masika, D. M., Anyona, G. O., & Kowenje, E. A. (2022). Indigenous knowledge and practices for sustainable water resources management: a case of Luo and Banyala in Kenya. In Kowenje, C. (Eds.), *From Traditional to Modern African Water Management.* Springer., DOI: 10.1007/978-3-031-09663-1_13

Masoga, M. (2005). South African research in Indigenous knowledge systems and challenges of change. *Indilinga*, 4(1), 15–30.

Masoga, M., & Shokane, A. (2019). Viewpoint: Indigenous knowledge systems and environmental social work education: towards environmental sustainability. *Southern African Journal of Environmental Education*, 35(1). Advance online publication. DOI: 10.4314/sajee.v35i1.14

Massola, A. (1971). The Aborigines of South Eastern Australia: As They Were. Australia.

Mataga, J. (2014). Practices of pastness, post wars of the dead, and the power of heritage: museums, monuments and sites in colonial and post-colonial Zimbabwe, 1890. Retrieved on 14 June 2023 from https://open.uct.ac.za/bitstream/handle/11427/12843/thesis_hum_2014_mataga_j.pdf?sequence=1&isAllowed=y

Matema, C. (2019). *Zimbabwe's CAMPFIRE public investments: implications on education, adaption and preferences*. PhD Thesis University of Cape Town.

Matema, S., & Anderson, S. (2015). Why are lions killing us? Human-wildlife conflict and social discontent in Mbire district, northern Zimbabwe. *The Journal of Modern African Studies*, 53(1), 93–120. DOI: 10.1017/S0022278X14000664

Materechera, S. A. (2021). Relevance of Indigenous knowledge systems in sustaining agriculture under smallholder farming in sub-Saharan Africa. In: Leal Filho, W., Pretorius, R., de Sousa, L.O. (eds) *Sustainable development in Africa*. World Sustainability Series. Springer. https://doi.org/DOI: 10.1007/978-3-030-74693-3_3

Matowanyika, J. Z. Z. (2016). Looking to the other world: Indigenous epistemologies and their relevance in research. A Paper presented during a seminar in the College of Education. Pretoria, South Africa. Unpublished.

Matsika, C. (2012). *Traditional African education: Its significance to current education practises with special reference to Zimbabwe*. Mambo Press.

Matthews, R. (2017). The cultural erosion of Indigenous people in health care. *Canadian Medical Association Journal*, 189(2), E78–E79. DOI: 10.1503/cmaj.160167 PMID: 27620632

Maunganidze, L., & Halsall, J. (2016). A moral compass that slipped: Indigenous knowledge systems and rural development in Zimbabwe. *Cogent Social Sciences*, 2(1), 1266749. Advance online publication. DOI: 10.1080/23311886.2016.1266749

Mavodza, J., & Ngulube, P. (2012). Knowledge management practices at an institution of higher learning. *South African Journal of Information Management*, 14(1), 1–8. DOI: 10.4102/sajim.v14i1.496

Mawere, J., Kugara, S. L., & Tshamano, W. (2022). Curriculum transformation in South Africa: An Indigenous knowledge systems perspective. *Journal of African Education*, 3(3), 11–27. https://hdl.handle.net/10520/ejc-aa_jae_v3_n3_a2. DOI: 10.31920/2633-2930/2022/v3n3a1

Mawere, M. (2010). *Indigenous Knowledge Systems' (IKS) Potential for establishing a moral, virtuous society: Lessons from selected IKS in Zimbabwe and Mozambique,* Journal of Sustainable Development in Africa (Volume 12, No.7) Zimbabwe'. *Zimbabwe Journal of Geographical Research*, 1(2), 98–111.

Mawere, M. (2013). A Critical Review of Environmental Conservation in Zimbabwe. *Africa Spectrum*, 2(48), 85–97. DOI: 10.1177/000203971304800205

Mawere, M. (2015). Indigenous knowledge and public education in Sub-Saharan Africa. *Africa Spectrum*, 50(2), 57–71. DOI: 10.1177/000203971505000203

Mawere, M., & Awuah-Nyamekye, S. (2015). *Between Rhetoric and Reality: The State and Use of Indigenous Knowledge in Post-Colonial Africa*. Langa RPCID. DOI: 10.2307/j.ctvh9vwc4

Maweu J. M. (2011). Indigenous ecological knowledge and modern western ecological knowledge: complementary, not contradictory. *Thought and Practice: A Journal of the Philosophical Association of Kenya (PAK),* 3(2), 35-47.

Mazama, A. (2014). *Afrocentricity and the critical question of African agency.* Dyabukam. Retrieved on 22 October 2024 from https://www.dyabukam.com/index.php/en/knowledge/philosophy/item/136-afrocentricity

Mazama, A. (2001). The Afrocentric paradigm: Contours and definitions. *Journal of Black Studies,* 31(4), 387–405. DOI: 10.1177/002193470103100401

Mazodze, C., Mapara, J., & Tsvere, M. (2021). Mainstreaming Indigenous African epistemologies into student development in higher education: A case of case. *East African Journal of Education and Social Sciences*, 2(3), 98–210. DOI: 10.46606/eajess2021v02i03.0118

Mazrui, A. A. (2003). Towards re-Africanizing African universities: Who killed intellectualism in the post colonial era? *Alternatives. Turkish Journal of International Relations*, 2(3&4).

Mazzocchi, F. (2020). A deeper meaning of sustainability: Insights from Indigenous knowledge. *The Anthropocene Review*, 7(1), 77–93. DOI: 10.1177/2053019619898888

Mbah, M. F., & Bailey, M. (2022). Decolonisation of research methodologies for sustainable development in Indigenous settings. In: M. F. Mbah, W. Leal Filho, S. Ajaps. (eds.), *Indigenous methodologies, research and practices for sustainable development.* World Sustainability Series. Springer. DOI: 10.1007/978-3-031-12326-9_2

Mbah, M. F., Bailey, M., & Shingruf, A. (2023). Considerations for relational research methods for use in Indigenous contexts: Implications for sustainable development. *International Journal of Social Research Methodology*, 27(4), 431–446. DOI: 10.1080/13645579.2023.2185345

Mbatha, N. T., Majola, Y. L., & Gumede, Z. S. (2023). Language maintenance: Factors supporting the use and maintenance of isiZulu in Soshanguve. *Literator*, 44(1), a1930. DOI: 10.4102/lit.v44i1.1930

Mbembe, A. (2023). Decolonising knowledge and the question of the archive. Retrieved on 24 November 2024 from https://wiser.wits.ac.za/system/files/Achille%20Mbembe%20-%20Decolonizing%20Knowledge%20and%20the%20Question%20of%20the%20Archive.pdf

Mbembe, A. (2016). Decolonizing the university: New directions. *Arts and Humanities in Higher Education*, 15(1), 29–45. DOI: 10.1177/1474022215618513

McCarty, T. (2009). *Language planning and policy in Native America: History, theory, praxis*. Bilingual Education & Bilingualism series. Clevedon: Multilingual Matters

McCleary, D., & Williams, R. (2009). Sociopolitical and personality correlates of militarism in democratic societies. *Peace and Conflict*, 15(2), 161–187. DOI: 10.1080/10781910902837248

McCleland, A. (2011). Culturally Safe Nursing Research: Exploring the Use of an Indigenous Research Methodology From an Indigenous Researcher's Perspective. *Journal of Transcultural Nursing*, 22(4), 362–367. DOI: 10.1177/1043659611414141 PMID: 21807959

McDonough, K. S. (2010). *Indigenous experience in Mexico: readings in the Nahua intellectual tradition. PhD Doctoral*. University of Minnesota.

McGregor, D. (2014). Lessons for collaboration involving traditional knowledge and environmental governance in Ontario, Canada. *Alternative*, 10(4), 340–353. DOI: 10.1177/117718011401000403

McGregor, D. (2021). Indigenous knowledge systems in environmental governance in Canada. *Kula Knowledge Creation Dissemination and Preservation Studies*, 5(1). Advance online publication. DOI: 10.18357/kula.148

McGuire, P. D. (2010). Exploring resilience and Indigenous ways of knowing. *A Journal of Aboriginal and Indigenous Community Health*, 8(2), 117-131.

McHugh, M. (2012). Interrater reliability: The Kappa statistic. *Biochemia Medica*, 22, 276–282. DOI: 10.11613/BM.2012.031 PMID: 23092060

McIntyre-Mills, J. J., Lethole, P., Makaulule, M., Wirawan, R., Widianingsih, I., & Romm, N. R. A. (2023). Towards eco-systemic living: learning with Indigenous leaders in Africa and Indonesia through a community of practice: implications for climate change and pandemics. *Systems Research and Behavioral Science*, 40(5), 779–786. https://onlinelibrary.wiley.com/doi/full/10.1002/sres.2976. DOI: 10.1002/sres.2976

McNulty, N. (2013). The Ulwazi Programme: bridging the gap with digital skills and local content. Retrieved on 11 October 2024 from https://www.niallmcnulty.com/2013/06/the-ulwazi-programme-bridging-the-gap-with-digital-skills-and-local-content

McNulty, N. (2014). The Ulwazi Programme: a case study in community-focused Indigenous knowledge management. In Osei-Bryson, K. M., Mansingh, G., & Rao, L. (Eds.), *Knowledge management for development. Integrated Series in Information Systems* (Vol. 35). Springer., DOI: 10.1007/978-1-4899-7392-4_13

Mdhluli, T. D., Mokgoatšana, S., Kugara, S. L., & Vuma, L. (2021). Knowledge management: Preserving, managing and sharing Indigenous knowledge through digital library. *Hervormde Teologiese Studies*, 77(2), a6795. DOI: 10.4102/hts.v77i2.6795

Meda, L. (2020). Decolonising the curriculum: Students' perspectives. *Africa Education Review*, 17(2), 88–103. DOI: 10.1080/18146627.2018.1519372

Meharg, S. J. (1999). *Making it, breaking it, and making it again: the importance of identity in the destruction and reconstruction of war-tom societies*. Master of Arts Thesis. Kingston, Ontario: Royal Military College of Canada.

Meharg, S. J. (2011). Restoration and reconstruction for environmental security. In Machlis, G., Hanson, T., Špirić, Z., & McKendry, J. (eds) *Warfare ecology*. NATO Science for Peace and Security Series C: Environmental Security (pp. 177-188). Springer. DOI: 10.1007/978-94-007-1214-0_12

Mehltretter, S., Longboat, S., Luby, B., & Bradford, A. (2023). Indigenous and Western knowledge: Bringing diverse understandings of water together in practice – Technical report for the Global Commission on the Economics of Water.https://www.waterjusticehub.org/wp-content/uploads/2023/03/Mehltretter.pdf

Melchias, G. (2001). *Biodiversity and conservation.* Science Publishers, Inc.

Mellos, K. (1988). *Neo-Malthusian theory: critical essay.* Palgrave MacMillan., DOI: 10.1007/978-1-349-19598-5_2

Menezes, M., & Costa, C. S. (2024). The role of collaborative ethnography in placemaking. *Humans*, 4(3), 284–297. DOI: 10.3390/humans4030018

Merriam-Webster. (n.d.): Language. In *Merriam-Webster.com dictionary.* Retrieved on 1 February 2024 from https://www.merriam-webster.com/dictionary/language

Mertens, D. M. (2017). Transformative research: Personal and societal. *International Journal for Transformative Research*, 4(1), 18–24. DOI: 10.1515/ijtr-2017-0001

Mertens, D. M. (2018). Mixed methods design in evaluation. *Sage (Atlanta, Ga.).*

Mertens, D. M. (2020). Transformation as a goal of mixed methods research in the Caribbean. *Caribbean Journal of Mixed Methods Research*, 1(1), 16–28.

Mertens, D. M. (2023a). *Mixed methods research.* Bloomsbury. DOI: 10.5040/9781350273191

Mertens, D. M. (2023b). The pursuit of social, economic, and environmental justice through evaluation: Learning from Indigenous scholars and the fifth branch of the evaluation theory tree. *Journal of Multidisciplinary Evaluation*, 19(44), 11–23. DOI: 10.56645/jmde.v19i44.749

Mertens, D. M. (2024a). *Research and evaluation in education and psychology* (6th ed.). Sage.

Mertens, D. M. (2024b). Mixed methods research for social betterment and a more just world. *Journal of Mixed Methods Research*, 18(3), 216–224. DOI: 10.1177/15586898241256834

Mertens, D. M., & Cram, F. (2016). Integration tensions and possibilities: Indigenous research and social transformation. *International Journal of Qualitative Research*, 9(2), 185–191. DOI: 10.1525/irqr.2016.9.2.185

Mertens, D. M., Hall, J., & Wilson, A. T. (2025). *Program evaluation theory and practice: a comprehensive guide* (3rd ed.). Guilford Press.

Metis Centre. (2011). Principles of ethical métis research. National Aboriginal Health Organization. Retrieved on 11 October 2024 from https://ruor.uottawa.ca/bitstream/10393/30555/1/2011_04_ethics.pdf

Mgbeoji, I. (2008). The civilised self and the barbaric other: imperial delusions of order and the challenges of human security. In *International Law and the Third World.* Routledge-Cavendish.

Mgbeoji, I. (2011). *Collective insecurity: the liberian crisis, unilateralism, and global order.* UBC Press.

Michailova, S., Piekkari, R., Storgaard, M., & Tienari, J. (2017). Rethinking ethnocentrism in international business research. *Global Strategy Journal*, 7(4), 335–353. DOI: 10.1002/gsj.1159

Mignolo, W. (2011). *The darker side of western modernity: Global futures, decolonial options.* Duke University Press.

Mignolo, W. D. (2005). *The darker side of Western modernity: Global futures, decolonial options*. Duke University Press.

Mignolo, W. D. (2005). *The idea of Latin America*. Blackwell.

Mignolo, W. D. (2009). Epistemic disobedience, independent thought and decolonial freedom. *Theory, Culture & Society*, 26(8), 159–181. DOI: 10.1177/0263276409349275

Mignolo, W. D., & Walsh, C. E. (2018). *On decoloniality: Concepts, analytics, praxis*. Duke University Press.

Miller, M., Warner, S. A., Gumbo, M. T., Alarcón, I. V., & Petrina, S. (2023). Lenses for critiquing and improving the Standards: Design, indigeneity, access and equity, and literacy. In Bartholomew, S. R., Hoepfl, M., & Williams, P. J. (Eds.), *Standards-based Technology and Engineering Education: 63rd Yearbook of the Council on Technology and Engineering Education* (pp. 287–308). Springer., DOI: 10.1007/978-981-99-5704-0_18

Mintz, K., & Tal, T. (2018). The place of content and pedagogy in shaping sustainability learning outcomes in higher education. *Environmental Education Research*, 24(2), 207–229. DOI: 10.1080/13504622.2016.1204986

Mitova, V. (2021). How to decolonise knowledge without too much relativism: Colonial insights into the South African experience. In Kumalo, S. H. (Ed.), *Decolonisation as democratisation* (pp. 24–47). Human Sciences Research Council. DOI: 10.1515/9780796926166-007

Mittermeier, C. (2024). Free, prior and informed consent in context: respecting Indigenous peoples' rights is essential to supporting our mission of helping societies to responsibly and sustainably care for nature. Retrieved on 11 October 2024 from https://www.conservation.org/projects/free-prior-and-informed-consent-in-context

Mji, G. (2013). African Indigenous knowledge and research. *African Journal of Disability*, 2(1), 1–5. PMID: 28729984

Mkabela, Q. (2005). Using the Afrocentric method in researching Indigenous African culture. *The Qualitative Report*, 10(1), 178–189.

Mkhwelie, N., Satamwe, N., & Shizha, E. (2022). Integration of Indigenous mathematical knowledge into the mathematics education for sustainable learning. In Ndlovu, C., & Shizha, E. (Eds.), *The dynamics of African Indigenous knowledge systems: A sustainable alternative for livelihoods in Southern Africa* (pp. 17–35). Peter Lang.

Modiri, J. (2021). Azanian political thought and the undoing of South African knowledges. *Theoria (Pietermaritzburg)*, 68(168), 42–85. DOI: 10.3167/th.2021.6816804

Mogalakwe, M. (2006). From pre-colony to post-colony: Continuities and discontinuities in political power relations and governance in Botswana. *Journal of African Elections*, 5(2), 5–20. DOI: 10.20940/JAE/2006/v5i2a1

Mogawane, M. A., Mothiba, T. M., & Malema, R. N. (2015). Indigenous practices of pregnant women at Dilokong hospital in Limpopo province, South Africa. *Curationis*, 38(2), 1553. DOI: 10.4102/curationis.v38i2.1553 PMID: 26842096

Mohamed, A. M. H., Tlemsani, I., & Mathews, R. (2021). Higher education strategy in digital transformation. *Education and Information Technologies*, 27(3), 3171–3195. DOI: 10.1007/s10639-021-10739-1 PMID: 34539217

Mohamedbhai, G. (2013). Indigenous knowledge must be harvested for development. *University World News*, 262. Retrieved on 30 October 2024 from https://www.universityworldnews.com/

Mohamedbhai, G. (2013, March). Indigenous knowledge must be harvested for development. University World News. Issue 262. In British Council's Going Global 2013 Conference in Dubai (pp. 4–6).

Mohan, A., Aswathi, K. R., & Chitra, K. P. (2021). Indigenous farming systems and global sustainability. In: Leal Filho, W., Marisa Azul, A., Brandli, L., Lange Salvia, A., Gökçin Özuyar, P., Wall, T. (eds) *Reduced inequalities. Encyclopedia of the UN Sustainable Development Goals*. Springer. https://doi.org/ DOI: 10.1007/978-3-319-95882-8_129

Mohan, D. (2024). Wisdom of the ancestors: Harnessing Indigenous knowledge for contemporary challenges. *International Journal of Applied Research*, 10(3), 112–119. DOI: 10.22271/allresearch.2024.v10.i3b.11591

Mohlahlo, A. L., & Ditsele, T. (2022). Exploring multilingualism at the national department levels in South Africa post the Use of Official Languages Act of 2012. *Literator-Journal of Literary Criticism. Comparative Linguistics and Literary Studies*, 43(1), 1819.

Mokhachane, M., Green-Thompson, L., George, A., Wyatt, T., & Kuper, A. (2023). Medical students' views on what professionalism means: An ubuntu perspective. *Advances in Health Sciences Education : Theory and Practice*, 29(3), 841–857. DOI: 10.1007/s10459-023-10280-5 PMID: 37710029

Mokotso, R. I. (2022). Considering the Basotho indigenous education and school system as resources for peace-building education in Lesotho. *Journal for Transdisciplinary Research in Southern Africa*, 18(1), 1–10. DOI: 10.4102/td.v18i1.1213

Molefe, M. (2014). Critical reflections on Gyekye's humanism. In Imafidon, E., & Bewaji, J. A. I. (Eds.), *Ontologized ethics: New essays in African meta-ethics* (pp. 121–132). Rowman and Littlefield.

Molina-Azorin, J. F., & Fetters, M. D. (2019). Building a better world through mixed methods research. *Journal of Mixed Methods Research*, 13(3), 275–281. DOI: 10.1177/1558689819855864

Moloi, L. (2024). *Developing Africa? New Horizons with Afrocentricity*. Anthem Press., DOI: 10.2307/jj.12638978

Molose, T., Goldman, G., & Thomas, P. (2018). Towards a collective-values framework of ubuntu: Implications for workplace commitment. *Entrepreneurial Business and Economics Review*, 6(3), 193–206. DOI: 10.15678/EBER.2018.060312

Mombeshora, S., & Le-Bel, S. (2009). Parks-people conflicts: The case of Gonarezhou National Park and the Chitsa community in south-eastern Zimbabwe. *Biodiversity and Conservation*, 18(1), 2601–2623. DOI: 10.1007/s10531-009-9676-5

Monnet, T. A. S., & Arnsperger, C. (2022). Modernity's temporality. *Text Matters*, 12, 7–31.

Montenegro, M. (2023). Bridging histories through intercultural archiving. *Public History Weekly*, 11(2). Advance online publication. DOI: 10.1515/phw-2023-21185

Moore, J. W. (2019). World-Ecology: A global conversation. *Sociologia Urbana e Rurale*, 120, 9–21. DOI: 10.3280/SUR2019-120002

Moreton-Robinson, A. (2013). Towards an Australian *Indigenous* women's standpoint theory: A methodological tool. *Australian Feminist Studies*, 28(78), 331–347. https://www.tandfonline.com/doi/abs/10.1080/08164649.2013.876664. DOI: 10.1080/08164649.2013.876664

Moreton-Robinson, A. (2016). Race and cultural entrapment: Critical Indigenous studies in the twenty-first century. In Moreton-Robinson, A. (Ed.), *Critical Indigenous studies: engagements in first world locations (Critical Issues in Indigenous Studies)*. University of Arizona Press. DOI: 10.2307/jj.423485.10

Moropa, K. (2018). Depiction of indigenous food in translated isiXhosa folk narratives and biographies. *South African Journal of African Languages*, 38(3), 247–254. DOI: 10.1080/02572117.2018.1518017

Morris, K. (2007). Re-examining issues behind the loss of family and cultural and the impact on Aboriginal youth suicide rates. *First Peoples Child & Family Review*, 3(1), 133–142. DOI: 10.7202/1069532ar

Morton, N., Hill, C., Meiring, D., & De Beer, L. T. (2019). Investigating the factor structure of the South African Personality Inventory-English version. *SA Journal of Industrial Psychology*, 45(1), 1–13. DOI: 10.4102/sajip.v45i0.1556

Mougeot, L. J. A. (2000). *Urban agriculture: definition, presence, potentials and risks, and policy challenges*. Cities Feeding People Series Report 31. Retrieved on 10 February 2024 from https://idl-bnc-idrc.dspacedirect.

Mountford, C. P. (Ed.). (1970). *The Dreamtime Australian Aboriginal Myths*. Angus & Robertson.

Moya, P. M. (2011). Who we are and from where we speak. *Transmodernity: Journal of Peripheral Cultural Production of the Luso-Hispanic World*, 1(2), 5–16. DOI: 10.5070/T412011809

Mpanza, C. D. (2023). The emergence of semilingualism in South African schools–Are the Stipulations of the language in education policy to blame? *African Journal of Development Studies*, 13(1), 179.

Msila, V. (2017). *Decolonizing knowledge for Africa's renewal: Examining African perspectives and philosophies*. KRPublishing.

Msila, V., & Gumbo, M. T. (Eds.). (2017). *African voices on indigenisation of the curriculum: Insights from practice*. Reach Publishers.

Msimanga, A., & Shizha, E. (2014). Indigenous knowledge and science education in South Africa: What messages from the curriculum? In Shizha, E. (Ed.), *Remapping Africa in the global space: Propositions for change* (pp. 137–150). Sense Publisher. DOI: 10.1007/978-94-6209-836-7_10

Mtetwa E et al (2013), *Poverty and Rural Development; Tapping from the Grassroots,* in International Journal of Humanities and Social Science Invention, vol. 2. Issue 5, pp 112-126.

Muchenje, F. (2017). Cognitive justice and Indigenous knowledge systems in the postcolonial classroom. In Shizha, E., & Makuvaza, N. (Eds.), *Re-thinking postcolonial education in Sub-Saharan Africa in the 21st Century* (pp. 69–83). Sense Publisher. DOI: 10.1007/978-94-6300-962-1_5

Mugambiwa, S. (2017). *Knowledge of climate change and the use of Indigenous knowledge system to adapt to climate hazards in Mutoko rural district of Mashonaland East Province, Zimbabwe,* MA Dissertation, University of Limpopo.

Mugambiwa, S. (2020). *Climate governance through IKS for sustainable development in Mutoko district of Mashonaland East Province, Zimbabwe,* PhD Thesis, University of KwaZulu Natal.

Mugambiwa, S., Rankoana, S., & Tirivangasi, H. (2023). Climate governance beyond the government: Indigenous knowledge systems in rural Zimbabwe's climate change adaption. *International Journal of Development and Sustainability*, 12(6), 238–249.

Mugovhani, G. N. (2013). African Renaissance, Indigenous African Music, and globalisation: collusion or collision? Mugovhani, N.G. (2008) a. Research Findings of the University of Venda Indigenous Music and Oral History Project 2003-2008. *Paper presented at the NIKSO EXPO* in Pretoria. Transvaal Museum.

Mugovhani, N. G. (2010). The role of Indigenous African choral music in the search for identity. With special reference to Mzilikazi Khumalo's music. *Muziki Journal for Music Research in Africa* vol 17, issue1 2010.

Muin, F., Noortyani, R., Elyani, E. P., & Chaemsaithong, K. (2021). Rethinking arbitrariness of language and its implication in language use. *Cogent Arts & Humanities*, 8(1), 1868687. Advance online publication. DOI: 10.1080/23311983.2020.1868687

Mũkoma wa Ngũgĩ. (2019). *The rise of the African novel: politics of language, identity, and ownership.* Pietermaritzburg: University of KwaZulu-Natal Press.

Mukurtu. (2024). Welcome to Mukurtu CMS: a safe keeping place. Retrieved on 11 October 2024 from https://mukurtu.org/

Mukwakwami, S., & Moyo, M. (2019). Land Use and its effects on Mopane Worm Populations in Zimbabwe's Southern Regions. *Journal of Environmental Management*, 49(1), 32–45.

Mullins, R., Dettmer, S., Eisenbardt, M., & Ziemba, E. (2020). Using knowledge exchange between prosumers and enterprises to implement circular economy activities in businesses. [OJAKM]. *Online Journal of Applied Knowledge Management*, 8(2), 71–90. DOI: 10.36965/OJAKM.2020.8(2)71-90

Mumpande, I., & Barnes, L. (2019). Revitalisation of the Tonga language in Zimbabwe: The motivational factors. *Language Matters (Pretoria)*, 50(3), 46–69. DOI: 10.1080/10228195.2019.1691634

Mundy, P., & Compton, L. (1991). Indigenous communication and Indigenous knowledge. *Development Communication Report*, 74(3), 1–3.

Mungwini, P. (2017). "African know thyself": Epistemic injustice and the quest for liberative knowledge. *International Journal of African Renaissance Studies*, 12(2), 5–18. DOI: 10.1080/18186874.2017.1392125

Munsaka E & Dube E (2018), *The contribution of Indigenous knowledge to disaster risk reduction activities in Zimbabwe: A big call to practitioners*, Jamba, 26;10(1):493. doi: DOI: 10.410.

Murove, M. F. (2005). *The theory of self-interest in modern economic discourse*. [Unpublished doctoral thesis]. University of South Africa (Unisa).

Murove, F. M. (2018). Indigenous knowledge discourse and inclusivity: An Afrocentric quest for recognition in a globalised world. *Journal for the Study of Religion*, 31(1), 159–176. DOI: 10.17159/2413-3027/2018/v31n1a9

Murwira, S. (2020). Indigenous African knowledge systems and teaching and learning at the Catholic University of Zimbabwe: critical investigation.(Doctoral Dissertation, UNISA, Pretoria). Retrieved on 1 November 2024 from http://hdl.handle.net/10500/26722

Murwira, S., & Higgs, P. (2023). Afrocentric epistemic systems and higher education curriculum. The case of a university in Zimbabwe. *South African Journal of Higher Education*, 37(2), 283–297. DOI: 10.20853/37-2-5334

Mutwa, C.& Larsen. (2003). *Zulu shaman: dreams, prophecies, and mysteries*. Destiny books.

Mutwa, C. (1964). *Indaba, my children: African tribal history, legends, customs and religious beliefs: African tribal history, legends, customs and religious beliefs*. Canongate Books.

Muwanga-Zake, J., & Kibukamusoke, M. (2024). Sustainability of Africa through technological innovations and Indigenous knowledge systems: A discussion of key factors and way forward. *African Journal of Social Work*, 14(2), 59–69. DOI: 10.4314/ajsw.v14i2.2

Mwanza, A. J. (2022). Can digital technology be used for addressing digital literacy in the dissemination of Indigenous Knowledge in rural areas? *South African Journal of Library and Information Science*, 88(1), 1–11.

Mwiti, G. (2014). African Indigenous psychotherapy. *Journal of Psychology and Christianity*, 33(2), 142–153.

Myers, M. (1999). Investigating information systems with ethnographic research. *Communications of the Association for Information Systems*, 2. Advance online publication. DOI: 10.17705/1CAIS.00223

Mzimela, J. H., & Moyo, I. (2024). Indigenous knowledge system and agricultural drought adaptation in the uMkhanyakude District Municipality, KwaZulu-Natal. *Journal of Asian and African Studies*, 0(0), 00219096241275391. Advance online publication. DOI: 10.1177/00219096241275391

Nabudere, D. (2011). *Afrikology, philosophy and wholeness*. AISA.

Nabudere, D. (2011). *Afrikology, philosophy and wholeness. An epistemology: An epistemology*. African Books Collective.

Nabudere, D. W. (2003). Towards the establishment of a pan-African University: A strategic concept paper. *Afr J Polit SCI*, 8, 1–30.

Nabudere, D. W. (2011). *Afrikology, philosophy and wholeness: An epistemology*. Africa Institute of South Africa.

Naidoo, N. (2011). What is research? A conceptual understanding. *African Journal of Emergency Medicine : Revue Africaine de la Médecine d'Urgence*, 1(1), 47–48. DOI: 10.1016/j.afjem.2011.05.011

Naidu, T. (2024). Epistemic disobedience-undoing coloniality in global health research. *PLOS Global Public Health*, 4(4), e0003033. DOI: 10.1371/journal.pgph.0003033 PMID: 38648248

Narasimhan, S., & Chandanabhumma, P. P. (2021). A Scoping Review of Decolonization in Indigenous-Focused Health Education and Behavior Research. *Health Education & Behavior*, 48(3), 306–319. DOI: 10.1177/10901981211010095 PMID: 34080471

National Entrepreneurship Policy. (2019). *National Entrepreneurship Policy of Botswana*. Ministry of Investment, Trade and Industry.

National Humanities Alliance. (2024). Mukurtu CSM: an Indigenous archive and publishing tool. https://humanitiesforall.org/projects/mukurtu-an-Indigenous-archive-and-publishing-tool

Ncube, A., Kunguma, O., Nyahwo, M., & Manombe, S. (2017). Nutritional vulnerability: An assessment of 2010 feeding food programme in Mbire district, Zimbabwe and its impact on pregnant women. *Jàmbá*, 9(1). Advance online publication. DOI: 10.4102/jamba.v9i1.406 PMID: 29955343

Ndhlovu, F., & Masuku, J. (2004). *Mainstreaming African indigenous knowledge systems in Higher and Tertiary Education: The Case of Zimbabwe*. UNISA Press.

Ndhlovu, P. T., Asong, J. A., Omotayo, A. O., Otang-Mbeng, W., & Aremu, A. O. (2023). Ethnobotanical survey of medicinal plants used by Indigenous knowledge holders to manage healthcare needs in children. *PLoS One*, 18(3), e0282113. DOI: 10.1371/journal.pone.0282113 PMID: 36972257

Ndlazi, G. (2021). A critical reflection of African philosophy and decolonization of the educational discourse in South Africa. *Journal of Contemporary African Philosophy*, 2, 49–76.

Ndlovu, J., ... (2024), *Mopane Worms(Imbrasia Belina) Value Chain: Experiences of Rural Women on the Impact of Climate Change and Environmental Sustainability in Mangwe District*, published online: http://hdl.handle.net/

Ndlovu-Gatsheni, S. J. (2013). Why decoloniality in the 21st century? *The Thinker: For Thought Leaders*, 48, 10–15.

Ndlovu-Gatsheni, S. J. (2016). Introduction: The coloniality of knowledge – between troubled histories and uncertain futures. In Ndlovu-Gatsheni, S. J., & Zondi, S. (Eds.), *Decolonising the university, knowledge systems and disciplines in Africa* (pp. 3–22). Carolina Academic Press.

Ndlovu-Gatsheni, S. J. (2017). The emergence of trajectories of struggles for an "African university": The case of unfinished business of African epistemic decolonisation. *Kronos (Bellville, South Africa)*, 43(1), 51–77. DOI: 10.17159/2309-9585/2017/v43a4

Ndlovu-Gatsheni, S. J. (2018). *Epistemic freedom in Africa: deprovincialisation and decolonisation*. Routledge. DOI: 10.4324/9780429492204

Ndlovu-Gatsheni, S. J. (2021). Beyond coloniality of internationalism. *International Relations*, 6(1), 1–8.

Ndlovu-Gatsheni, S. J. (2021). The cognitive empire, politics of knowledge, and African intellectual productions: Reflections on struggles for epistemic freedom and resurgence of decolonisation in the twenty-first century. *Third World Quarterly*, 42(5), 882–901. DOI: 10.1080/01436597.2020.1775487

Ndlovu-Gatsheni, S. J. (2022). On the meaning of decolonization. In Rodney-Gumede, Y., Chasi, C., Jaffer, Z., & Ponono, M. (Eds.), *Decolonizing journalism education in South Africa: critical perspectives* (pp. 9–24). Unisa Press.

Ndlovu-Gatshen, S. J. (2015). Decoloniality as the future of Africa. *History Compass*, 13(10), 485–496. DOI: 10.1111/hic3.12264

Ndlovu, I., Nunu, W. N., Mudonhi, N., Dube, O., & Maviza, A. (2019). *Land use-land cover changes mopane worm harvest in Mangwe District in Plumtree*, Zimbabwe. *Environmental Systems Research*, 8(1), 98–112. DOI: 10.1186/s40068-019-0141-5

Ndlovu, M. (2018). Coloniality of knowledge and the challenge of creating African futures. *Ufahamu. Journal of African Studies*, 40(2), 95–116.

Ndlovu, S. (2023). Indigenous knowledge systems in Africa: Exclusion of students with disabilities in South African higher education. *Curriculum Perspectives*, 43(1), 127–140. DOI: 10.1007/s41297-023-00202-1

Ndlovu, Z.. (2018). Government Support for Mopane Worm Harvesters in Zimbabwe: An overview. *Journal of African Policy and Governance*, 31(4), 185–199.

Ndungu, H., Zaba, P., & Ndlovu, E. (2023). Enhancing resilience to recurrent droughts and floods in semi-nomadic communities. a case of CMDRR amongst the Doma community in Mbire district, Zimbabwe. *International Journal of Research and Innovation in Social Sciences*, V11(1), 1532–1554. DOI: 10.47772/IJRISS.2023.70822

Ndwandwe, J. D. (2009). *Akusiko Kwami Kwebantfu unearthing King Sobhuza II's Philosophy*. Mgangatho Publishers.

Netshandama, V. O., & Nevhudoli, N. D. (2021). Creating intergenerational learning spaces: A collaboration between UNIVEN community *engagement* programme and Dzomo la Mupo. *Critical Studies in Teaching and Learning*, 9(2), 39–63. DOI: 10.14426/cristal.v9i2.462

Neuman, W. L. (2007). *Basics of social research: qualitative and quantitative approaches*. Pearson Education Inc.

Ng'asike, J. T. (2019). Indigenous knowledge practices for sustainable lifelong education in pastoralist communities of Kenya. *International Review of Education*, 65(1), 19–46. DOI: 10.1007/s11159-019-09767-4

Ngakane, B. (2024). *Praxis of Indigenous entrepreneurship in Botswana: A shift towards knowledge based economy*. Poeticblood Publishers.

Ngara, R. (2012). Mukwerera, a Shangwe rainmaking ceremony: Its music, dance and symbols, *MA thesis. Eastern Cape*: University of Fort Hare.

Ngara, C. (2007). African ways of knowing and pedagogy revisited. *Journal of Contemporary Issues in Education*, 2(2), 7–20.

Ngara, R. (2017). Multiple voices, multiple paths. In Ngulube, P. (Ed.), *Handbook of research on theoretical perspectives on Indigenous knowledge systems in developing countries* (pp. 332–358). IGI Global. DOI: 10.4018/978-1-5225-0833-5.ch015

Ngatse-Ipangui, R., & Dassah, M. O. (2019). Impact of social entrepreneurs on community development in the Cape Town metropolitan municipality area, South Africa. *The Journal for Transdisciplinary Research in Southern Africa*, 15(1), 1–10. https://td-sa.net/index.php/td/article/view/474/973. DOI: 10.4102/td.v15i1.474

Ngubane, B. S. (2003). Roll-out of Indigenous Music/Oral History Collection, Promotion and Preservation Project in Three Universities: Government *White Paper on Arts, Culture, Science and Technology*. Pretoria: Department of Arts, Science and Technology.

Ngugi wa Thiong'o. (1986). *Deconolising the mind: the politics of languages in African literature*. Zimbabwe Publishing House.

Ngugi wa Thiong'o. (1992). *Moving the centre: the struggle for cultural freedoms*. James Currey & Heinemann.

Ngugi wa Thiong'o. (1997). *Detained: a writer's prison diary*. Heinemann.

Ngũgĩ wa Thiong'o. (1986). *Decolonising the mind: the politics of language in African literature*. London: James Currey.

Ngugi wa Thiong'o. (1986). *Decolonizing the mind: the politics of language in African literature*. Portsmouth.

Ngugi wa Thiong'o. (2009). *Something torn and new: an African renaissance*. Basic Civitas Books.

Ngũgĩ wa Thiong'o. (2009). *Something torn and new: an African renaissance*. New York: Basic Civitas Books.

Ngugi wa, T. (1986). *Decolonizing the mind: the politics of language in African literature*. James Currey.

Ngulube, P., Mathipa, E. R., & Gumbo, M. T. (2015). "Theoretical and conceptual frameworks in the social and management sciences." *Addressing research challenges: Making headway in developing researchers*, pp. 43-66.

Ngulube, P. (2002). Managing and preserving indigenous knowledge in the knowledge management era: Challenges and opportunities. *Journal of information. Development*, 28(2), 107–116.

Ngulube, P. (2018). Overcoming the difficulties associated with using conceptual and theoretical frameworks in heritage studies. In Ngulube, P. (Ed.), *Handbook of research on heritage management and preservation* (pp. 1–23). IGI Global. DOI: 10.4018/978-1-5225-3137-1.ch001

Ngulube, P. (2020). Theory and theorising in information science scholarship. In Ngulube, P. (Ed.), *Handbook of research on connecting research methods for information science research* (pp. 18–39). IGI Global. DOI: 10.4018/978-1-7998-1471-9.ch002

Ngulube, P. (2022). Using simple and complex mixed methods research designs to understand research in Information Science. In Ngulube, P. (Ed.), *Handbook of research on mixed methods research in information science* (pp. 20–44). IGI Global. DOI: 10.4018/978-1-7998-8844-4.ch002

Ngulube, P. (2023). From Orality to Digital Assets: Managing Indigenous Knowledge in Africa in the Wake of the Open Science Movement. *The Serials Librarian*, 84(5–8), 71–85. DOI: 10.1080/0361526X.2023.2296880

Ngulube, P., Dube, L., & Mhlongo, M. (2015). Towards a cartography of Indigenous knowledge systems in library and information science training and education in Anglophone eastern and southern Africa. *Indilinga*, 14(2), 145–168.

Ngulube, P., Mathipa, E. R., & Gumbo, M. T. (2015). Theoretical and conceptual framework in the social sciences. In Mathipa, E. R., & Gumbo, M. T. (Eds.), *Addressing research challenges: making headway in developing researchers* (pp. 43–66). Mosala-MASEDI Publishers & Booksellers., DOI: 10.13140/RG.2.1.3210.7680

Ngulube, P., & Mosha, N. F. V. (2023). Transparency in the application of theoretical frameworks to the advancement of knowledge in selected library and information science journals: A systematic review. *African Journal of Library Archives and Information Science*, 33(2), 129–146. DOI: 10.4314/ajlais.v33i2.2

Ngulube, P., & Ngulube, B. (2017). Application and contribution of hermeneutic and eidetic phenomenology to indigenous knowledge research. In *Handbook of research on theoretical perspectives on indigenous knowledge systems in developing countries* (pp. 127–155). IGI Global. DOI: 10.4018/978-1-5225-0833-5.ch006

Ngulube, P., & Onyancha, B. O. (2011). What's in a name? Using informetric techniques to conceptualize the knowledge of traditional and Indigenous communities. *Indilinga*, 10(2), 129–152.

Ngunyulu, R. N., Sepeng, N., Moeta, M., Gambu, S., Mulaudzi, F. M., & Peu, M. D. (2020). The perspectives of nursing students regarding the incorporation of African traditional Indigenous knowledge in the curriculum. *African Journal of Primary Health Care & Family Medicine*, 12(1), 1–8. DOI: 10.4102/phcfm.v12i1.2171 PMID: 32370527

Nhemachena, C. (2007). Assessment of the economic impacts of climate change approach. Policy Research Paper. *World Bank Development Research Group.*

Nhira, C. (1989). A socio-economic appraisal study of the Chapoto ward-Guruve district. *University of Zimbabwe Working Paper Number 5.*

Nicholas, G. (2022). Protecting Indigenous heritage objects, places, and values: Challenges, responses, and responsibilities. *International Journal of Heritage Studies*, 28(3), 400–422. DOI: 10.1080/13527258.2021.2009539

Nieto, M. J., & Santamaría, L. (2010). Technological collaboration: Bridging the innovation gap between small and large firms. *Journal of Small Business Management*, 48(1), 44–69. DOI: 10.1111/j.1540-627X.2009.00286.x

Nietschmann, B. Q. (1990). Conservation by conflict in Nicaragua. *Natural History*, (November), 42–49.

Nietschmann, B. Q. (1991). Nietschmann rejoings: Conservation by self-determination. *Research & Exploration*, 7(3), 373.

Nietschmann, B. Q. (1994). The Fourth World: nations versus states. In Wood, D. J. D. A. W. B. (Ed.), *Reordering the world, geopolitical perspectives on the twenty-first century* (pp. 226–242). Westview Press.

Nitah, S. (2021). Indigenous peoples proven to sustain biodiversity and address climate change: Now it's time to recognize and support this leadership. *One Earth*, 4(7), 907–909. DOI: 10.1016/j.oneear.2021.06.015

Nkondo, M. (2012). *Indigenous African knowledge systems in a polyepistemic world: The capabilities approach and the translatability of knowledge systems.* Paper presented at the Southern African Regional Colloquium on Indigenous African knowledge systems: methodologies and epistemologies for research, teaching, learning and community engagement in higher education. Howard College Campus: University of KwaZulu-Natal.

Nkrumah, K. (1965). *Neo-colonialism: the highest stage of imperialism.* Heinemann.

Noble, H., & Smith, J. (2015a). Issues about reliability and validity. *Evidence-Based Nursing Research*, 18(2), 163. PMID: 25653237

Noble, H., & Smith, J. (2015b). Issues of validity and reliability in qualitative research. *Research Made Simple. Evidence-Based Nursing*, 18(2), 34–35. DOI: 10.1136/eb-2015-102054 PMID: 25653237

Nompula, Y. (2010). *Finding Our Voice: A Critique of Eurocentrism. Saarbrucken.* Lambert Academic Publishing.

Norris, G. (2015). *Handprint-based NetPositive assessment.* T.H. Chan Center for Health and the Global Environment.

Noyoo N (2007), *Indigenous Knowledge Systems and their Relevance for Sustainable Development: A Case of Southern Africa*, in K Boon & L Hens, Indigenous Knowledge Systems and Sustainable Development: Relevance for Africa, Kamla-Raj Enterprises

Nsamenang, A. B. (2006). Human ontogenesis: An Indigenous African view on development and intelligence. *International Journal of Psychology*, 41(4), 293–297. DOI: 10.1080/00207590544000077

Nweke, A. C. (2019). Rural-urban migration in Nigeria, implication on the development of the society: Anambra State as the focus of the study. *Journal of Public Administration and Governance*, 9(2), 209. Advance online publication. DOI: 10.5296/jpag.v9i2.14912

Nyamnjoh, F. (2012). Potted plots in greenhouses: A critical reflection on the resilience of colonial education in Africa. *Journal of Asian and African Studies*, 47(2), 129–154. DOI: 10.1177/0021909611417240

Nyathi 1(2022), *Negotiating marginalisation: A socio-economic history of the Kalanga of Mangwe, Zimbabwe, 1940 to 2015,* unpublished PhD thesis, Rhodes University.

NYSERDA. (2013). *Sustainable urban agriculture: confirming viable scenarios for production*. New York State Energy Research and Development Authority (NYSERDA)

Obiero, K. O., Klemet-N'Guessan, S., Migeni, A. Z., & Achieng, A. O. (2023). Bridging Indigenous and non-Indigenous knowledge systems and practices for sustainable management of aquatic resources from East to West Africa. *Journal of Great Lakes Research*, 49(1), S128–S137. DOI: 10.1016/j.jglr.2022.12.001

Odora Hoppers, C. (2015). Cognitive justice and integration without duress: The future of development education – perspectives from the South. *International Journal of Development Education and Global Learning*, 7(2), 89–106. DOI: 10.18546/IJDEGL.07.2.08

Odora Hoppers, C. A. O. (2001). Indigenous knowledge systems and academic institutions in South Africa. *Perspectives in Education*, 19(1), 73–85.

OECD. (1971). *Equal educational opportunities: a statement of the problem with special reference to recurrent education*. Organization for Economic Cooperation and Development (OECD).

Oehrle, E. (1994). The First South African National Music Educators' Conference in Context of World Music Education, in the Proceedings of the First National Music Educators' Conference. Durban, South Africa: SAMES.

Ogar, E., Gretta Pecl, G., & Mustonen, T. (2020). Science must embrace traditional and indigenous knowledge to solve our biodiversity crisis. *One Earth*, 3(2), 162–165. DOI: 10.1016/j.oneear.2020.07.006

Ogbo, C. V., & Ndubisi, E. J. O. (2017). African Indigenous Education in Contemporary Society. *African Indigenous Knowledge Systems: Problems and Perspectives*, 4, 40–53.

Ogden, L. (2012). *Swamplife: people, gators, and mangroves entangled in the everglades*. University of Minnesota Press.

Oguamanam, C. (2020). Indigenous Peoples, Data Sovereignty, and Self-Determination: Current Realities and Imperatives. *The African Journal of Information and Communication*, 26(26), 1–20. DOI: 10.23962/10539/30360

Oguamanam, C. (2023). A critical examination of the African legal framework for Indigenous knowledge. *Journal of African Law*, 16(1), 34–57. DOI: 10.1017/S0021855323000049

Ogwo, B. A., & Oranu, R. N. (2006). *Methodology in formal and non-formal technical/vocational education*. University of Nigeria Press.

Oishi, S., & Gilbert, E. A. (2016). Current and future directions in culture and happiness research. Current Opinion in Psychology, 8(August 2010), 54–58. DOI: 10.1016/j.copsyc.2015.10.005

Olaare, S. (2024). The role of technology in language preservation. *European Journal of Linguistics*, 3(2), 44–56. DOI: 10.47941/ejl.2046

Olaopa, O., & Ayodele, O. A. (2021). Building on the strengths of African Indigenous knowledge and innovation (AIK& I) for sustainable development in Africa. *African Journal of Science, Technology, Innovation and Development*, 14(5), 1–14. DOI: 10.1080/20421338.2021.1950111

Olatunji, S. A. (2010). Thematic changes in postcolonial Africa literature: From colonialism to neocolonialism. *Sino-US English Teaching*, 7(10), 125–134.

Olusoga, D. (2015, September 8). The roots of European racism lie in the slave trade, colonialism—and Edward Long. *The Guardian*. Retrieved on 22 October 2024 from https://www.theguardian.com/commentisfree/2015/sep/08/european-racism-africa-slavery

Omodan, B. I. (2024). *Research paradigms and their methodological alignment in social sciences: A practical guide for researchers*. Taylor & Francis. DOI: 10.4324/9781003484066

Omodan, B. I. (2025). Building reciprocal relationships through decolonial practices in academic research. *Cogent Social Sciences*, 11(1), 2443558. Advance online publication. DOI: 10.1080/23311886.2024.2443558

Omoyajowo, K., Danjin, M., Omoyajowo, K., Odipe, O., Mwadi, B., May, A., & Rabie, M. (2024). Exploring the interplay of environmental conservation within spirituality and multicultural perspective: Insights from a cross-sectional study. *Environment, Development and Sustainability*, 26(7), 16957–16985. DOI: 10.1007/s10668-023-03319-5 PMID: 37363007

Onwuegbuzie, H. N. (2010). Sustainable development strategy using Indigenous knowledge and entrepreneurship. Available at *SSRN* 1841787.

Onwu, G. O., & Mufundirwa, C. (2020). A two-eyed seeing context-based approach for incorporating Indigenous knowledge into school science teaching. *African Journal of Research in Mathematics. Science and Technology Education*, 24(2), 229–240.

Onyancha, O. B., Ngulube, P., Maluleka, J. R., & Mokwatlo, K. I. (2018). Towards a uniform terminology for Indigenous knowledge concepts: Informetrics perspectives. *African Journal of Library Archives and Information Science*, 28(2), 155–167.

Open Data Charter. (2015). Open Data Charter Principles. Retrieved on 11 October 2024 from https://opendatacharter.net/principles/

Ophuls, W., & Boyan, A. (1992). *Ecology and the politics of scarcity revisited.* W. H. Freeman.

Organic Without Boundaries. (2020). *How PGS changed the law on organic agriculture in the Philippines!* Retrieved on 20 October 2024 from https://www.organicwithoutboundaries.bio/2020/06/24/how-pgs-changed-the-law-on-organic-agriculture-in-the-philippines/

Ørngreen, R., & Levinsen, K. (2017). Workshops as a research method. *Electronic Journal of e-Learning*, 15(1), 70–81. https://eric.ed.gov/?id=EJ1140102

Osman, A. (2014). Indigenous knowledge in Africa: Challenges and opportunities. *OAlib*, 2(3), 1–9.

Owan, V. J., Owan, M. V., & Ekereke, A. B. (2023). Students' Entrepreneurship Development Through Indigenous Palm Wine Production and Alcohol Distillation: The Case of Yala-Obubra Community. In *Digital Preservation and Documentation of Global Indigenous Knowledge Systems* (pp. 75–101). IGI Global Publisher. DOI: 10.4018/978-1-6684-7024-4.ch003

Owiny, S. A., Mehta, K., & Maretzki, A. N. (2014). The use of social media technologies to create, preserve, and disseminate Indigenous knowledge and skills to communities in East Africa. *International Journal of Communication*, 8, 234–247.

Owiny, S. A., Mehta, K., & Maretzki, A. N. (2014). The use of social media technologies to create, preserve, and disseminate Indigenous knowledge and skills to communities in East. *The Pennsylvania State University International Journal of Communication*, 8(1), 234–247.

Owolabi, A. K., Ovwasa, D. E., Ajayi, T. B., & Odewale, M. O. (2022). Preservation and use of Indigenous knowledge practices in public libraries in Nigeria. *Public Library Quarterly*, 41(50), 485–502. DOI: 10.1080/01616846.2021.1938897

Owusu-Ansah, F. E. & Mji, G. (2013). African Indigenous knowledge and research. *African Journal of Disability*, 2(1), Art. #30 5 pages. .DOI: 10.4102/ajod.v2i1.30

Oxford English Dictionary. (1933). Clarendon Press.

Oxford English Dictionary. (1971). *The Oxford English dictionary*. Oxford University Press.

Oyebade, B. (1990). African studies and the Afrocentric paradigm: A critique. *Journal of Black Studies*, 21(2), 233–238. DOI: 10.1177/002193479002100208

Oyelude, A. A. (2023). Indigenous knowledge preservation as a sign of respect for culture: Concerns of libraries, archives and museums. *Insights: the UKSG Journal*, 36(1), 1–21. DOI: 10.1629/uksg.628

Ozurumba, I. G., Ochiagha, C. D., Echem, G. C. C., & Ugwuoke, A. C. (2022). Contributions of adult literacy programmes towards achieving the empowerment of adults in Enugu State, Nigeria. *Journal of Nigeria National Council for Adult Education*, 7(1), 91–113.

Page, M. J., McKenzie, J. E., & Bossuyt, P. M.. (2021). The PRISMA 2020 statement: An updated guideline for reporting systematic reviews. *BMJ (Clinical Research Ed.)*, 372(71). PMID: 33782057

Pakendorf, B., & Stoneking, M. (2021). The genomic prehistory of peoples speaking Khoisan languages. *Human Molecular Genetics*, 30(R1), R49–R55. DOI: 10.1093/hmg/ddaa221 PMID: 33075813

Panayiotou, C. (2024). Ontological foundation for representing metaphors as part of LOD. *IEEE Access : Practical Innovations, Open Solutions*, 12, 73810–73838. DOI: 10.1109/ACCESS.2024.3396871

Pandey, D. N. (2002). (in press). Cultural resources for conservation science. *Conservation Biology*.

Paredez, D. M. (1964). *Hunab Ku síntesis del pensamiento filosófico maya*. Orion.

Parker, K. L. (Ed.). (1965). *Australian Legendary Tales*. Angus & Robertson.

Parrinder, G. (1982). *African Mythology*. Hamlyn.

Parson, J. (1984). *Liberal democracy and the labour reserve in southern Africa*.

Patel, N. G., & Nath, S. R. (2013). Where Identity and Trauma Converge: Hindu-Muslim Perceptions of the 2002 Gujarat Riots. *The Journal of Muslim Mental Health*, 7(2), 25–45. DOI: 10.3998/jmmh.10381607.0007.202

Patin, B., Sebastian, M., Yeon, J., Bertolini, D., & Grimm, A. (2021). Interrupting epistemicide: A practical framework for naming, identifying, and ending epistemic injustice in the information professions. *Journal of the Association for Information Science and Technology*, 72(10), 1–13. DOI: 10.1002/asi.24479

Patterson, E. E. (2015). "Oral Transmission: A Marriage of Music, Language, and Tradition." *The Research and Scholarship Symposium* 12. https://digitalcommons.cedarville.edu/research_scholarship_symposium/2015/podium_presentations/12

Pearson, L. J., Pearson, L., & Pearson, C. J. (2011). Sustainable urban agriculture: stock take and opportunities. In *Urban agriculture: diverse activities and benefits for city society*. https://doi.org/DOI: 10.3763/ijas.2009.0468

Peek, P. M. (1991). *African divination systems: Ways of knowing*. Georgetown University Press.

Perez, C., & Smith, C. (2019). Indigenous Knowledge System and conservation of settled territories in the Bolivian Amazon. *Sustainability (Basel)*, 11(1), 6099. Advance online publication. DOI: 10.3390/su11216099

Perry, S. R. (2014). Digitization and digital preservation: A review of the literature. *School of Information Student Research Journal*, 4(1). Advance online publication. DOI: 10.31979/2575-2499.040104

Pete, S., Schneider, B., & O'Reilly, K. (2013). Decolonizing our practice: Indigenizing our teaching. *First Nations Perspectives*, 5(1), 99–115.

Petranker, J. (Ed.). (1997). *Light of Knowledge, Essays on the Interplay of Knowledge, Time and Space*. Dharma Publishing.

Photo, P., & Gumbo, M. T. (2024). Postgraduate student supervision in open distance learning: towards a Botho supervision framework. In Gumbo, M. T., Gaotlhobogwe, M., Pedzisai, C., Jojo, Z. M. M., & Knaus, C. B. (Eds.), *Global perspectives on decolonizing postgraduate education* (pp. 166–178). IGI Global. DOI: 10.4018/979-8-3693-1289-6.ch011

Pidgeon, M., & Riely, T. (2021). Understanding the application and use of indigenous research methodologies in the social sciences by indigenous and non-indigenous scholars. *International Journal of Education Policy and Leadership*, 17(8). Advance online publication. DOI: 10.22230/ijepl.2021v17n8a1065

Pllana, D., & Aliu, H. (2018). The impact of consumer ethnocentrism on purchase intentions: The case of Kosovo. *International Journal of Research and Studies Publishing*, 8(10).

Polanyi, K. (1944). *The great transformation: the political and economic origins of our time*. Farrar & Rinehart.

Poonwassie, A., & Charter, A. (2001). An Aboriginal world view of helping: Empowering approaches. *Canadian Journal of Counselling/Revue. Canadian Counsellor*, 35(1), 63–73.

Popa, F., Guillermin, M., & Dedeurwaerdere, T. (2015). A pragmatist approach to transdisciplinarity in sustainability research: From complex systems theory to reflexive science. *Futures*, 65, 45–56. DOI: 10.1016/j.futures.2014.02.002

Pophiwa, N. & Saidi, U. (2002). Approaches of embedding Indigenous knowledge system in made in Africa Evaluation. *African evaluation Journal,* 10(1).137-154. http://dx.doi.org/DOI: 10.4102/aej.v10i1.62

Popper, K. (1980). *The logic of scientific discovery*. Abingdon-on-Thames: Routledge.

Porten, S., Loë, R., & McGregor, D. (2016). Incorporating Indigenous knowledge systems into collaborative governance for water: Challenges and opportunities. *Journal of Canadian Studies. Revue d'Etudes Canadiennes*, 50(1), 214–243. DOI: 10.3138/jcs.2016.50.1.214

Posey, D. (1999). *Cultural and spiritual values of biodiversity*. UN Environmental Programme.

Potter-Nelson, E., & Meyers, S. (2022). *An Introduction to sustainability education*. Massachusetts Institute of Technology. White paper series.

Prah, K. (2010). Multilingualism in urban Africa: Bane or blessing. *Journal of Multicultural Discourses*, 5(2), 169–182. DOI: 10.1080/17447143.2010.491916

Prah, K. (2012). The language of development and the development of language in contemporary Africa. *Applied Linguistics Review*, 3(2), 295–313. DOI: 10.1515/applirev-2012-0014

Prasad, S. M., Patel, B. D. P., Tiwari, N. N., Sapkota, S., & Shah, B. (2020). Traditional healers of Nepal: Their knowledge, skill, practices, and technology. *Journal of Ayurveda Campus*, 1(1), 7–15. DOI: 10.51648/jac.3

Premium Times. (2023). Updated: Fact-checking claims made by Bola Tinubu at NESG dialogue. Retrieved on 1 April 2024 from https://www.premiumtimesng.com/news/headlines/583799-fact-checking-claims-made-by-bola-tinubu-at-nesg-dialogue.html

Prempeh, C. (2022). Polishing the pearls of Indigenous knowledge for inclusive social education in Ghana. *Social Sciences & Humanities Open*, 5(1), 100248. DOI: 10.1016/j.ssaho.2022.100248

Prigg, M. (2014, December 5). The dwindling South African tribe who number just 100,000 but were once the most common humans on earth. *MailOnline*. Retrieved on 10 October from https://www.dailymail.co.uk/news/article-2868526/The-dwindling-South-African-tribe-number-just-100-000-common-humans-earth.html

Pulvermüller, F. (2013). How neurons make meaning: Brain mechanisms for embodied and abstract-symbolic semantics. *Trends in Cognitive Sciences*, 17(9), 458–470. DOI: 10.1016/j.tics.2013.06.004 PMID: 23932069

Quan-Baffour, K. P. (2008). The wisdom of our forefathers: Sankofaism and its education lessons for today. *Journal of Educational Studies*, 7(2).

Quang, T. D., Noseworthy, W. B., & Paulson, D. (2022). Rising tensions: Heritage-tourism development and the commodification of "Authentic" culture among the Cham community of Vietnam. *Cogent Social Sciences*, 8(1), 2116161. Advance online publication. DOI: 10.1080/23311886.2022.2116161

Quinn, A. L. (2022). Bridging Indigenous and Western methods in social science research. *International Journal of Qualitative Methods*, 21, 1–14. DOI: 10.1177/16094069221080301

Radhakrishnan, S. (1929, 2008). *An idealist view of life*. Aristophanes Press - archINFORM.

Raine, S. C., Kukutai, T., Walter, M., Figueroa-Rodrigues, O. L., & Walker, J.. (2019). Indigenous data sovereignty. In Davies, T., Walker, S., Rubinstein, M., & Perini, F. (Eds.), *State of open data: histories and horizons* (pp. 300–319). African Minds., DOI: 10.5281/zenodo.2677801

Rai, S. K. (2019). Invoking 'Hindu' Ayurveda: Communalisation of the late colonial Ayurvedic discourse. *The Indian Economic and Social History Review*, 56(4), 411–426. DOI: 10.1177/0019464619873820

Ramirez-Santos, A. G., Ravera, F., Rivera-Ferre, M. G., & Calvet-Nogués, M. (2023). Gendered traditional agroecological knowledge in agro-food systems: A systematic review. *Journal of Ethnobiology and Ethnomedicine*, 19(11), 11. Advance online publication. DOI: 10.1186/s13002-023-00576-6 PMID: 37024903

Ramose, M. B. (1998). Foreword. In S. Seepe (Ed.). *Black perspectives on tertiary institutional transformation* (pp. iv-viii). Vivilia Publishers & University of Venda.

Ramose, M. B. (2004). In search of an African philosophy of education. *South African Journal of Higher Education*, 18(3), 138–160.

Ramose, M. B. (2020). *Motho ke motho ka batho*, as African perspective on popular sovereignty and democracy. In Jenco, L. K., Idris, M., & Thomas, M. C. (Eds.), *The Oxford handbook of comparative political theory*. Oxford Press.

Ramoupi, N. L. L. (2014). African research and scholarship: 20 years of lost opportunities to transform higher education in South Africa. *Ufahamu. Journal of African Studies*, 38(1), 269–286.

Rappaport, J. (2008). Beyond participant observation: Collaborative ethnography as theoretical innovation. *Collaborative Anthropologies*, 1(1), 1–31. DOI: 10.1353/cla.0.0014

Rasoreka, K. (2008). Information transformation Africa: Indigenous knowledge securing space in the knowledge society. *The International Information & Library Review*, 40(4), 243–250. DOI: 10.1080/10572317.2008.10762789

Redvers, N., Aubrey, P., Celidwen, Y., & Hill, K. (2023). Indigenous peoples: Traditional knowledges, climate change, and health. *PLOS Global Public Health*, 3(10), e0002474. DOI: 10.1371/journal.pgph.0002474 PMID: 37831713

Reeves, H., & Baden, S. (2000), *Report No. 5,* prepared for the Department of International Development (DFID), United Kingdom.

Reeves, N. P., Sal, V. G., Lutomia, A. N., Medendorp, J. W., Bello-Bravo, J., & Pittendrigh, B. (2024). Agricultural education in Africa using YouTube multilingual animations: A retrospective feasibility study assessing costs to reach language-diverse populations. *PLoS One*, 19(4), e0302136. DOI: 10.1371/journal.pone.0302136 PMID: 38635490

Reid, A. J., Eckert, L. E., Lane, J.-F., Young, N., Hinch, S. G., Darimont, C. T., Cooke, S. J., Ban, N. C., & Marshall, A. (2021). "Two-eyed seeing": An Indigenous framework to transform fisheries research and management. *Fish and Fisheries*, 22(2), 243–261. DOI: 10.1111/faf.12516

Reid, B. (2020). Positionality and research: "two-eyed seeing" with a rural Ktaqmkuk Mi'kmaw Community. *International Journal of Qualitative Methods*, 19, 1609406920910841. Advance online publication. DOI: 10.1177/1609406920910841

Rendle, K. A., Abramson, C. M., Garrett, S. B., Halley, M. C., & Dohan, D. (2019). Beyond exploratory: A tailored framework for designing and assessing qualitative health research. *BMJ Open*, 9(8), e030123. Advance online publication. DOI: 10.1136/bmjopen-2019-030123 PMID: 31462482

Republic of South Africa Intellectual Property Laws Amendment Bill of. 2008. (2017), Retrieved on 14 June 2023 from https://www.publishsa.co.za/file/1446644266ord-ip-laws-amendment-act-2013.pdf

Reviere, R. (2001). Toward an Afrocentric research methodology. *Journal of Black Studies*, 31(6), 709–728. DOI: 10.1177/002193470103100601

Reyes Jalaskoski, R. (2023). Integrating Indigenous entrepreneurs into the global economy: Identifying the main obstacles and opportunities of Indigenous-led ventures in Peru.

Richards, P. (1985). Indigenous agricultural revolution: ecology and food production in West Africa. *Indigenous agricultural revolution: ecology and food production in West Africa.* https://doi.org/DOI: 10.2307/632952

Richards, M., Jones, C., Lardy, J., Godden, A., Godden, W., Bock, S., & Lardy, H. (2024). Warrma Mangarrayi: Co-designing an app for learning Mangarrayi, an Indigenous language of Northern Australia. *International Journal of Human-Computer Interaction*, 0(0), 1–18. DOI: 10.1080/10447318.2024.2381924

Ritchie, H., & Roser, M. (2018). Urbanization. Our world in data. Retrieved on 2 April 2024 from https://ourworldindata.org/urbanization

Ritchie, J. (2015). Counter-colonial research methodologies drawing upon postcolonial critique and Indigenous onto-epistemologies. In Cannella, G. S., Pérez, M. S., & Pasque, P. A. (Eds.), *Critical qualitative inquiry* (pp. 77–90). West Coast Press.

Robertson, A. F. (2001). *Greed: gut feelings, growth, and history*. Blackwell.

Robinson, D. (2010). *Confronting biopiracy: challenges, cases and international debates*. Routledge.

Rogers, N. N. (2023). An Afrocentric examination of Afrocentric schools: status, agency, and liberation. Temple University. DOI: 10.4324/9781849774710

Robinson, C. J., Kong, T., Coates, R., Watson, I., Stokes, C., Pert, P., McConnell, A., & Chen, C. (2021). Caring for Indigenous data to evaluate the benefits of Indigenous environmental programs. *Environmental Management*, 68(2), 160–169. DOI: 10.1007/s00267-021-01485-8 PMID: 34046755

Rodriguez, R. C. (2010). *Amoxtli- The Xodex in Lak Ech-Panche Be and Hunab Ku and the forgotten 1524 debate*. Eagle Feather Research Institute.

Romm, N. R. A. (2018). *Responsible research practice*. Springer https://link.springer.com/book/10.1007/978-3-319-74386-8

Romm, N. R. A. (2015). Reviewing the transformative paradigm: A critical systemic and relational (Indigenous) lens. *Systemic Practice and Action Research*, 28(5), 411–427. https://link.springer.com/article/10.1007/s11213-015-9344-5. DOI: 10.1007/s11213-015-9344-5

Romm, N. R. A. (2018). *Responsible research practice: Revisiting transformative paradigm in social research*. Springer International Publishing. DOI: 10.1007/978-3-319-74386-8

Romm, N. R. A. (2020). Eliciting children's/young people's (group) engagement with scenarios as participatory research practice for exploring and extending responses to climate change. *Participatory Educational Research*, 7(1), i–xiv. DOI: 10.17275/per.20.0.7.1

Romm, N. R. A. (2024). An Indigenous relational approach to systemic thinking and being: Focus on participatory onto-epistemology. *Systemic Practice and Action Research*, 37(6), 1–32. https://journals.sagepub.com/doi/abs/10.1177/1077800419830135. DOI: 10.1007/s11213-024-09672-4

Roorda, E. (2020). Izanagi and Izanami, Japanese Sea Gods. In Roorda, E. (Ed.), *The Ocean Reader: History, Culture, Politics* (pp. 13–13). Duke University Press. DOI: 10.1215/9781478007456-005

Rosiek, J. L., Snyder, J., & Pratt, S. L. (2020). The new materialisms and Indigenous theories of non-human agency: Making the case for respectful anti-colonial engagement. *Qualitative Inquiry*, 26(3-4), 331–346. https://journals.sagepub.com/doi/abs/10.1177/1077800419830135. DOI: 10.1177/1077800419830135

Ross, E. (2010). Inaugural lecture: African spirituality, ethics and traditional healing–implications for indigenous South African social work education and practice. *South African Journal of Bioethics and Law*, 3(1), 44–51.

Roy, K., Zvonkovic, A., Goldberg, A., Sharp, E., & LaRossa, R. (2015). Sampling richness and qualitative integrity: Challenges for research with families. *Journal of Marriage and Family*, 77(1), 243–260. DOI: 10.1111/jomf.12147

Rubagumya, C. M., Sane, E., & Ndabakurane, J. J. (2022). Vignette: implementing language supportive pedagogy to support content learning in Tanzania. In Erling, E. J., Clegg, J., Rubagumya, C. M., & Reilly, C. (Eds.), *Multilingual learning and language supportive pedagogies in Sub-Saharan Africa* (pp. 170–174). Routledge.

Rupp, D. E., Thornton, G. C.III, Bisbey, T. M., Hoover, A. N., Salas, E., & Murphy, K. R. (2024). An epistemology for assessment and development: How do we know what we know? *Industrial and Organizational Psychology: Perspectives on Science and Practice*, 17(2), 252–268. DOI: 10.1017/iop.2024.3

Russell, R. V. (2015). *The Tribes and Castes of the Central Provinces of India*. CreateSpace Independent Publishing Platform.

Rutherford, B. (2017). *Farm labour struggles in Zimbabwe: the grounds of politics*. Indiana University Press.

Rutledge, P. B., & Hogg, J. L. C. (2020). In-depth interviews. *The International Encyclopedia of Media Psychology*, 1-7. Available at https://www.researchgate.net/publication/345737833_In-Depth_Interviews [accessed October 17 2024].

Ryser, R. (2011). Fourth World geopolitics. H. G. Bruce. Unpublished.

Ryser, R. (2022). We are people first. E. Barrett Ristroph (ed.), *Who is Indigenous: perspective on indigeneity, identity, dispossession, and transition*. Ristroph Law, Planning, and Research.

Ryser, R. (1985). Fourth World Wars: Indigenous nationalism and the emerging new international political order. In Boldt, M., & Long, J. A. (Eds.), *The quest for justice: Aboriginal people and Aboriginal rights*. University of Toronto Press. DOI: 10.3138/9781442657762-028

Ryser, R. (1994). *International covenant on the rights of Indigenous nations*.

Ryser, R. (1998). Observations on 'self' and 'knowing. In Wautischer, H. (Ed.), *Tribal epistemologies* (pp. 17–29). Ashgate.

Ryser, R. (2012). *Indigenous and traditional knowledge. ecosystem management and sustainability*. Berkshire Publishing Group.

Ryser, R. (2012). *Indigenous nations and modern states: the political emergence of nations challenging state power*. Routledge. DOI: 10.4324/9780203139882

Ryser, R. (2015). The Anáhuac knowledge system: A dialogue between Toltecs and Descartes. *Fourth World Journal*, 14(1), 31–47.

Sabati, S. (2019). Upholding "colonial unknowing" through the IRB: Reframing institutional research ethics. *Qualitative Inquiry*, 25(9–10), 1056–1064. DOI: 10.1177/1077800418787214

SAICH. (2024). Southern African Intangible Cultural Heritage Cooperation. Retrieved on 11 October 2024 from www.saich.org

Salawu, R. O., Bolatitio, A. S., & Masibo, S. (2023). Theoretical and conceptual frameworks in research: Conceptual clarification. *European Chemical Bulletin*, 12(12), 2103–2117.

Sanchez, G. (2018). *The Zambezi River Basin: water resources management. Energy-food-water nexus approach,* MSc Thesis, Stockholm University.

Sandoval, C. (2019). *Methodology of the Oppressed*. Vol. 18. U of Minnesota Press, 2013.

Sangha, K. K., Le Brocque, A., Costanza, R., & Cadet-James, Y. (2015). Ecosystems and Indigenous well-being: An integrated framework. *Global Ecology and Conservation*, 4, 197–206. DOI: 10.1016/j.gecco.2015.06.008

Santos, B. S. (2016). Epistemologies of the South and the future. *From the European South*, 1, 17-29.

Santos, B. S. (2015). *Epistemologies of the South: justice against epistemicide*. Routledge. DOI: 10.4324/9781315634876

Sarmento, D. A. (2020). *Slowly infecting capitalism with solidarity economy: exploring Members' perceptions in a multiple case study in Urban São Paulo* [Unpublished MSc International Development Studies]. University of Amsterdam.

Sartori, G. (2009). Guidelines for concept analysis. In Collier, D., & Gerring, J. (Eds.), *Concepts and method in social science. The tradition of Giovanni Sartori* (pp. 97–150). Routledge.

Savransky, M. (2021). *Around the day in eighty worlds: politics of the pluriverse*. Duke University Press.

Scheub, H. (2002). *A Dictionary of African Mythology*. Oxford University Press.

Scheurich, J. J. (1997). *Research method in the post-modern*. Falmer Press.

Scheurich, J. J., & Young, D. M. (1997). Coloring methodologies: Are our research epistemologies racially biased? *Educational Researcher*, 26(4), 4–16. DOI: 10.3102/0013189X026004004

Schildermans, H. (2023). *Experiments in decolonizing the university*. Bloomsbury.

Schiller, U., & de Wet, G. (2018). Communication, indigenous culture and participatory decision making amongst foster adolescents. *Qualitative Social Work: Research and Practice*, 17(2), 236–251. DOI: 10.1177/1473325016662329

Schiller, U., & de Wet, G. (2019). Indigenous cultural and social work practices creating inequalities among foster care adolescents in decision-making processes. *International Social Work*, 62(1), 22–34. DOI: 10.1177/0020872817710548

Schippers, H., & Bendrups, D. (2015). Ethnomusicology, Ecology and the Sustainability of Music Cultures. *The World of Music*, 4(1), 9–19.

Schneider, M. A. (2006). *The theory primer: A sociological guide*. Rowman and Littlefield.

Schneider, P., Rochell, V., Plat, K., & Jaworski, A. (2021). Circular approaches in small-scale food production. *Circular Economy and Sustainability*, 1(4), 1231–1255. DOI: 10.1007/s43615-021-00129-7 PMID: 34888585

Schoff, W. H. (1995). *The Periplus of the Erythraean Sea: Travel and Trade in the Indian Ocean by a Merchant of the First Century*. Munshiram Manoharlal Publishers Pvt. Ltd.

Schreiber, F., & Cramer, C. (2022). Towards a conceptual systematic review: Proposing a methodological framework. *Educational Review*, 1–22. Advance online publication. DOI: 10.1080/00131911.2022.2116561

Scoones I (2009), '*Livelihood Perspectives and Rural Development,*' in the Journal of Peasant Studies, 1(36), pp 145-158.

Scott, K., and H. Pettersson. (2019). "South Africa is the world's most unequal country. 25 years of freedom have failed to bridge the divide. CNN, 10 May."

Searle, J. (2018). *Constitutive Rules. Argumenta*. University of Sassari.

Seehawer, M. (2018). South African science teachers' strategies for integrating Indigenous and Western knowledges in their classes: Practical lessons in decolonisation. *Educational Research for Social Change*, 7(special issue), 91–110. DOI: 10.17159/2221-4070/2018/v7i0a7

Seehawer, M., & Breidlid, A. (2021). Dialogue between epistemologies as quality education: Integrating knowledges in Sub-Saharan African classrooms to foster sustainability learning and contextually relevant education. *Social Sciences & Humanities Open*, 4(1), 100200. DOI: 10.1016/j.ssaho.2021.100200

Sefotho, M. M. (2022). Foreword. In Adu, E. O., & Okeke, C. I. O. (Eds.), *Fundamentals of research in humanities, social sciences and science education: A practical step by step approach to a successful research journey* (p. v). Van Schaik.

Selelo, M. E. (2023). Indigenous Entrepreneurship and the Fourth Industrial Revolution in South Africa. *African Journal of Development Studies*, 2023(si1), 129.

Senanayake, S. (2006). Indigenous knowledge as a key to sustainable development. *Journal of Agricultural Sciences*, 2(1), 87–94. DOI: 10.4038/jas.v2i1.8117

Seroto, J. (2018). Dynamics of decoloniality in South Africa: A critique of the history of Swiss mission education for indigenous people. *Studia Historiae Ecclesiasticae*, 44(3), 1-14. *Seroto, J. (2019). The embodiment of the indigenous people by European travel writers at the Cape colony, Southern Africa. *Education as Change*, 23(1), 1–24.

Seton, K. (1999). Fourth World Nations in the era of globalisation an introduction to contemporary theorizing posed by Indigenous nations. *Fourth World Journal*, 4(1), 49–69.

Shackleton, C. M., & Shackleton, S. E. (2007). The contribution of mopane worms to the rural economies of southern Africa.'. *African Journal of Ecology*, 45(1), 62–71.

Shahzeydi, A., Taleghani, F., Moghimian, M., Farzi, S., Yazdannik, A., & Farzi, K. (2022). Exploring nursing faculty, managers, newly graduated nurses, and students' experiences of nursing internship program implementation in Iran: A descriptive qualitative study. *BMC Nursing*, 21(1), 1–8. DOI: 10.1186/s12912-022-01159-8 PMID: 36572907

Shai, K. B. (2023). "Pathetic scholarship" in Gukurahundi and related studies: A silver lining from social media cloud. *[APORTAL]*. *African Perspectives of Research in Teaching & Learning*, 7(2), 378–386.

Shanahan, K. (2019). Mukurtu CMS: differential access for the ethical stewardship in cultural and digital heritage. Retrieved on 11 October 2024 from https://sustainableheritagenetwork.org/system/files/atoms/file/NAMSI-Mukurtu%20History%20and%20Intro.pdf

Shang, S., O'Driscoll, M. P., & Roche, M. (2016). Moderating role of acculturation in a mediation model of work- family conflict among Chinese immigrants in New Zealand. *Stress and Health*, 33(1), 55–68. DOI: 10.1002/smi.2674 PMID: 27018155

Shankarmahesh, M. (2006). Consumer ethnocentrism: An integrative review of its antecedents and consequences. *International Marketing Review*, 23(2), 146–172. DOI: 10.1108/02651330610660065

Shay, M., & Oliver, R. (Eds.). (2021). *Indigenous education in Australia:lLearning and teaching for deadly futures*. Routledge. DOI: 10.4324/9780429263453

Sherr, M. E. (2006). The Afrocentric paradigm: A pragmatic discourse about social work practice with African Americans. *Journal of Human Behavior in the Social Environment*, 13(3), 1–17. DOI: 10.1300/J137v13n03_01

Shiri, A., Howard, D., & Farnel, S. (2021). Indigenous digital storytelling: Digital interfaces supporting cultural heritage preservation and access. *The International Information & Library Review*, 54(2), 93–114. DOI: 10.1080/10572317.2021.1946748

Shiva, V. (2022). *Dr. Vandana Shiva's keynote talk at the* 2022 *People's Food Summit.* Retrieved on 2 October 2024 from https://www.youtube.com/watch?v=cKM1QgOsb4I

Shivani, S. & Mishra, S. (2022). Traditional knowledge: much more than what meets the eyes. *International Journal of Health Sciences*, 6(S5)), 1570–1581.

Shiva, V. (2012). *Making peace with the Earth*. Fernwood.

Shizha, E. (2008). Indigenous? What Indigenous knowledge? Beliefs and attitudes of rural primary school teachers towards Indigenous knowledge in the science curriculum in Zimbabwe. *Australian Journal of Indigenous Education*, 37(1), 80–90. DOI: 10.1017/S1326011100016124

Shizha, E. (2014). Indigenous knowledge systems and the curriculum. In Emeagwali, G., & Dei, G. J. S. (Eds.), *African Indigenous knowledge and the disciplines. Anti-colonial educational perspectives for transformative change* (pp. 113–129). Sense Publishers.

Shizha, E. (2015). Reclaiming Indigenous cultures in Sub-Saharan African education. *Indigenous Education: Language. Culture and Identity*, 2(1), 301–317. DOI: 10.1007/978-94-017-9355-1_15

Shizha, E. (2017). Indigenous knowledges and knowledge codification in the knowledge economy. In Ngulube, P. (Ed.), *Handbook of Research on Theoretical Perspectives on Indigenous Knowledge Systems in Developing Countries*. DOI: 10.4018/978-1-5225-0833-5.ch012

Shizha, E. (2022b). Indigenous epistemologies and decolonized sustainable livelihoods in Africa. In Abdi, A. A., Misiaszek, G. W., & Popoff, J. M. (Eds.), *Palgrave international handbook on critical theories of education* (pp. 465–480). Palgrave Macmillan. DOI: 10.1007/978-3-030-86343-2_26

Shizha, E., & Abdi, A. A. (Eds.). (2014). *Indigenous discourses on knowledge and development in Africa*. Taylor & Francis/Routledge.

Shokane, A. L., & Masoga, M. A. (2018). African indigenous knowledge and social work practice: Towards an Afro-sensed perspective. *Southern African Journal of Social Work and Social Development,* 30(1), 1-|18. https://upjournals.co.za/index.php/SWPR

Siamena, E., Mapara, J., & Chiweshe, M. K. (2024). IDEWS and the building of resilient communities in the eastern highlands of Zimbabwe. *Journal of African History. Archaeology and Tourism*, 2(1), 75–97. DOI: 10.31920/2753-3204/2024/v2n1a4

Sibanda, A. (2013). *Striving for Broad Based Economic empowerment in Zimbabwe: Localisation of Policies and Programmes to Matabeleland and Midlands*, a working paper submitted to Ruzivo Trust (Knowledge for Growth, Prosperity, and Life. Thrupp L.A. (1998). "*Legitimizing Local Knowledge: From Displacement to Empowerment for Third World People. Agriculture and Human Values*, (Summer Issue), 13–24.

Sibandze, G. F., & Dludlu, M. N. (2022). Interfacing Indigenous knowledge with scientific knowledge for improved health outcomes: lessons from Eswatini. In Pullanikkatil, D., & Hughes, K. (Eds.), *Socio-ecological systems and decoloniality*. Springer., DOI: 10.1007/978-3-031-15097-5_3

Siddaway, A. P., Wood, A. M., & Hedges, L. V. (2019). How to do a systematic review: A best practice guide for conducting and reporting narrative reviews, meta-analyses, and meta-syntheses. *Annual Review of Psychology*, 70(1), 747–770. DOI: 10.1146/annurev-psych-010418-102803 PMID: 30089228

Siddi, A. (2024). *On the Spiritual Pedagogies of the Siddis*. Personal Conversation.

Siddi, F. U. (2024). *Siddi Dhamaals as a form of Pedagogy*. Personal Conversation.

Siddi, H. M. (2022). *On the Kinship Pedagogies of the Siddis*. Personal Conversation.

Siddi, H. M. (2024). *On teaching-learning Siddi cultures in Siddi-run schools*. Personal Conversation.

Siddi, S. (2024). *Salaamis as Mnemonic Pedagogies*. Personal Conversation.

Siddi, U. (2024). *On the Ecological Pedagogies of the Siddis*. Personal Conversation.

Sifuna, N. D. (2008). From colonisation to globalisation" *TRANS. Internet-Zeitschrift fur Kulturwissenschaften*, 17/2008. Retrieved on 10 November 2024 from www://http.inst.at/trans/17Nr/9-3sifuna.htm

Simatimbe, W., Mbozi, E. H., & Hambulo, L. (2018). Integrating Indigenous knowledge practices into agricultural extension education in Luano District of Central Zambia. *International Journal of Advanced Research*, 6(10), 178–190. DOI: 10.21474/IJAR01/7798

Simonds, V. W., & Christopher, S. (2013). Adapting Western research methods to Indigenous ways of knowing. *American Journal of Public Health*, 103(12), 2185–2192. DOI: 10.2105/AJPH.2012.301157 PMID: 23678897

Simpson, L. B. (2017). *As We Have Always Done: Indigenous Freedom Through Radical Resistance*. University of Minnesota Press. DOI: 10.5749/j.ctt1pwt77c

Sinclair, N. J. (2013). *Nindoodemag Bagijiganan: a history of Anishinaabeg Narrative. PhD*. The University of British Columbia.

Singh, M., & Major, J. (2017). Conducting Indigenous research in Western knowledge spaces: Aligning theory and methodology. *Australian Educational Researcher*, 44(1), 5–19. DOI: 10.1007/s13384-017-0233-z

Sioui, M. (2020). *Indigenous water and drought management in a changing world*. Elsevier.

Sium, A., Desai, C., & Ritskes, E. (2012). Towards the 'tangible unknown': Decolinization and the Indigenous future. *Decolonization*, 1(1), I–XIII.

Slikkerveer, L. (1999). Changing values and attitudes of social and natural scientists towards Indigenous peoples and their knowledge systems. In D. Warren., L. Slikkerveer. & S. Titilola. (Eds.), *Indigenous Knowledge Systems: Implications for agriculture and international* development (pp. 121-137). Studies in Technology and Social Change. Iowa State University Press.

Smiet, K. (2022). Rethinking or delinking? Said and Mignolo on humanism and the question of the human. *Postcolonial Studies*, 25(1), 73–88. DOI: 10.1080/13688790.2022.2030595

Smith, A. D. (1986, 1987). T*he ethnic orgins of nations*. B. Blackwell.

Smith, C., & Babich, C. (2023). *Theories of individual and collective learning*. Campus Ontario. Retrieved 12 December 2024 from https://ecampusontario.pressbooks.pub/ticl/

Smith, L. T. (1999, 2012). *Decolonizing methodologies: research and indigenous people*. Zed Books.

Smith, A. (2002). Power and hierarchy of knowledge. *Geoforum*, 40(1), 230–248.

Smith, L. H. (2019). Epistemic decoloniality as a pedagogical movement in Social Work: A turn of theory to anti-colonial theorists. *Critical Social Work*, 1(5), 1–20.

Smith, L. T. (1999). *Decolonizing methodologies: research and indigenous peoples*. Zed Books.

Smith, L. T. (1999). *Decolonizing methodologies: research and Indigenous peoples*. Zed Books.

Smith, L. T. (2012). *Decolonizing methodologies: Research and Indigenous peoples*. Zed Books.

Smith, R. W. (1970). *Myths and Legends of the Australian Aboriginals*. George G. Atarrap & Co.

Smylie, J., Martin, C., Kaplan-Myrth, N., Steele, L., Tait, C., & Hogg, W. (2004). Knowledge translation and Indigenous knowledge. *International Journal of Circumpolar Health*, 63(sup2), 139-143.

Snapp, S. (2022). Change the reward structure. *Nature*, 612, 178.

Snowball, J., Collins, A., & Nwauche, E. (2022). Ethics, values and legality in the restoration of cultural artefacts: The case of South Africa. *International Journal of Cultural Policy*, 28(5), 513–530. DOI: 10.1080/10286632.2021.1995377

Snowball, J., Collins, A., & Nwauche, E. (2022). Examining restitution and repatriation options for cultural artefacts: An empirical enquiry in South Africa. *International Journal of Cultural Policy*, 28(5), 531–545. DOI: 10.1080/10286632.2021.1995378

Son, J. Y., Smith, L. B., & Goldstone, R. L. (2010). Connecting instances to promote children's relational reasoning. *Journal of Experimental Child Psychology*, 108(2), 260–277. DOI: 10.1016/j.jecp.2010.08.011 PMID: 20926099

Sorby, J., Buchanan, F., & Smith, A. (2024). Kinship and Cultural Strengths—Learning from an Aboriginal Perspective. *Australian Social Work*, •••, 1–13. DOI: 10.1080/0312407X.2024.2370265

South to South Cooperation. (2018). Addressing asymmetries in global evaluation: elevating the knowledge and influence of the global South. Final proposal. Phase 1: 2018–2021. Retrieved 0n 24 October 2024 from https://afrea.org

Sowa, J. F., & Majumdar, A. K. (2003). Analogical reasoning. *International Conference on Conceptual Structures,* Dresden, Germany. Springer-Verlag.

Sparkes, L. L., & Piercey, D. W. (2015). What are Indigenous and Western ways of knowing? Retrieved on 24 November 2024 from https://www.criaw-icref.ca/images/userfiles/files/Fact%20Sheet%202%20EN%20FINAL.pdf

Srinivas, S., & Sutz, J. (2008). Developing countries and innovation: Searching for a new analytical approach. *Technology in Society*, 30(2), 129–140. DOI: 10.1016/j.techsoc.2007.12.003

Stansfield, D., & Browne, A. J. (2013). The relevance of Indigenous knowledge for nursing curriculum. *International Journal of Nursing Education Scholarship*, 10(1), 1–9. DOI: 10.1515/ijnes-2012-0041 PMID: 23813335

Steenkamp, H., & Rensburg, R. (2018). Utilising an Ubuntu-centred communication management framework to analyse CSR messages on SNSS. *Communicatio*, 44(3), 17–40. DOI: 10.1080/02500167.2018.1541914

Stein, S., & de Andreotti, V. O. (2016). Decolonization and higher education: Introduction. In Peters, M. (Ed.), *Encyclopedia of educational philosophy and theory* (pp. 1–6). Springer. DOI: 10.1007/978-981-287-532-7_479-1

Stenning, K., & Lambalgen, M. V. (2012). *Human reasoning and cognitive science*. MIT Press.

Stephens, A., Taket, A., & Gagliano, M. (2019). Ecological justice for nature in critical systems thinking. *Systems Research and Behavioral Science*, 36(1), 3–19. https://onlinelibrary.wiley.com/doi/abs/10.1002/sres.2532. DOI: 10.1002/sres.2532

Steup, M., & Neta, R. (2020). Epistemology. In Zalta, E. N. (Ed.), *The Stanford Encyclopedia of Philosophy*. Stanford University.

Steyn, M. (2012). The ignorance contract: Recollections of apartheid childhoods and the construction of epistemologies of ignorance. *Identities (Yverdon)*, 19(1), 8–25. DOI: 10.1080/1070289X.2012.672840

Steyn, M., & Vanyoro, K. P. (2023). Critical Diversity Literacy: A framework for multicultural citizenship education. *Education, Citizenship and Social Justice*, 0(0). Advance online publication. DOI: 10.1177/17461979231178520

Stokes, M. (1994). *Ethnicity, Identity and Music - The Musical Construction of Place*. Bridgend: WBC Bookbinders.

Stone, C. L. (2004). *In place of gods and kings: authorship and identity in the relación de Michoacán*. University of Oklahoma Press.

Strochlic, N. (2018). The race to save the world's disappearing languages. *National Geographic*, •••, 16.

Subba Rao, S. (2006). Indigenous knowledge organization: An Indian scenario. *International Journal of Information Management*, 26(3), 224–233. DOI: 10.1016/j.ijinfomgt.2006.02.003

Suchitra, M. R. & Parthasarathy, S. (2020). *International Journal of Research in Pharmaceutical Sciences*.

Sundberg, J. 2009. *International Encyclopedia of Human Geography*. Retrieved on 22 October 2024 from https://www.sciencedirect.com/topics/social-sciences/eurocentrism on 22/10/2024

Sunzuma, G., & Maharaj, A. (2020). In-service mathematics teachers' knowledge and awareness of ethnomathematics approaches. *International Journal of Mathematical Education in Science and Technology*, 52(7), 1–16.

Sutton, R. I., & Staw, B. M. (1995). What theory is not. *Administrative Science Quarterly*, 40(3), 371–384. DOI: 10.2307/2393788

Swakopmund Protocol on the Protection of Traditional Knowledge and Expressions of Folklore Within the Framework of the African Regional Intellectual Property Organization (ARIPO). (2010). Adopted by the Diplomatic Conference of ARIPO at Swakopmund (Namibia) on August 9, 2010. Retrieved on 14 June 2023 from https://wipolex-res.wipo.int/edocs/lexdocs/treaties/en/ap010/trt_ap010.pdf

Swanson, D. (2008). Ubuntu: An African contribution to (re)search for/with a 'humble togetherness'. *Journal of Contemporary Issues in Education*, 2(2). Advance online publication. DOI: 10.20355/C5PP4X

Swihart, D. L., Yarrarapu, S. N. S., & Martin, R. L. (2023). Cultural Religious Competence in Clinical Practice. [Updated 2023 Jul 24] In *StatPearls* [Internet]. StatPearls.

Tagle, Y. R. (2009). The Protection of Indigenous knowledge related to biodiversity: the role of databases. Retrieved on 11 October 2024 from https://www.sylff.org/wp-content/uploads/2009/03/sylff_p131-146.pdf

Takada, A., & Noguchi, T. (2024). Diversity and transition of stepfamilies among the Glui and Gllana. *Hunter Gatherer Research*, •••, 1–11. DOI: 10.3828/hgr.2024.37

Taylor, D. (2003). *The Archive and the Repertoire: Performing Cultural Memory in the Americas*. Duke University Press.

Te Mana Raraunga. (2018). Principles of Māori Data Sovereignty Brief #1. Retrieved on 18 October 2024 from https://static1.squarespace.com/static/5d3799de845604000199cd24/t/5d6f940cc5442b00013e4dab/1567593484780/TMR%2BMa%CC%84ori%2BData%2BSovereignty%2BPrinciples%2BOct%2B2018.pdf

Teffo, L. (2019). Harnessing IKS and technologies in the quest for sustainable rural economic development. *Indilinga*, 18(2), 245–256.

Tektigul, Z., Bayadilova-Altybayeva, A., Sadykova, S., Iskindirova, S., Kushkimbayeva, A., & Zhumagul, D. (2023). Language is a symbol system that carries culture. *International Journal of Society. Culture & Language*, 11(1), 203–214.

Tempels, P. (1952, 1959). *Bantu philosophy*. Présence Africaine, Victor Petrus Keuppens.

Thambinathan, V., & Kinsella, E. A. (2021). Decolonizing methodologies in qualitative research: Creating spaces for transformative praxis. *International Journal of Qualitative Methods*, 20, 16094069211014766. Advance online publication. DOI: 10.1177/16094069211014766

Tharakan, J. (2017). *Indigenous knowledge systems for appropriate technology development*. InTech., DOI: 10.5772/intechopen.69889

The Amended Constitution of the Republic of Zimbabwe. (2013). Retrieved on 14 June 2023 from https://www.constituteproject.org/constitution/Zimbabwe_2013.pdf

The Herald, '*Bulilima hard hit by amacimbi extinction*', 20 December 2016. ZBC news online, '*Mopane worm expected to rack in US$500, 000,*' 7 April 2017.

The Herald, '*Kezi chief warns amacimbi harvesters.*' 13 April 2017.

The Ministerial Oversight Committee on Transformation in South African Public Universities. (2015). *The transformation of South African higher education*. 2nd Edition. National Higher Education Summit. Retrieved on 24 November 2024 from https://www.dhet.gov.za/summit/Docs/2015Docs/Annex%208_TOC_Transformation%20of%20SA%20HE.pdf

The World Bank. (1998). Indigenous knowledge definitions, concepts and applications. Retrieved on 11 April 2024 from https://chm.cbd.int/api/v2013/documents/4A27922D-31BC-EEFF-7940-DB40D6DB706B/attachments/209070/Hoda%20Yacoub%20-%20IK%20Report%20(1).pdf

The World Bank. (2010). Land area square kilometre - Nigeria. Retrieved on 11 April 2024 from https://data.worldbank.org/indicator/AG.SRF.TOTL.K2?locations=NG

Thomas, A., Menon, A., Boruff, J., Rodriguez, A. M., & Ahmed, S. (2014). Applications of social constructivist learning theories in knowledge translation for healthcare professionals: A scoping review. *Implementation Science : IS*, 6(9), 54. DOI: 10.1186/1748-5908-9-54 PMID: 24885925

Thurston, J. M., & Mashford-Pringle, A. (2015). Nursing & Indigenous education integration. *Journal of Nursing Education and Practice*, 5(10). Advance online publication. DOI: 10.5430/jnep.v5n10p9

Tippeconnic, J. W., & Faircloth, S. C. (2010). *The education of Indigenous students*. International Encyclopedia of Education., DOI: 10.1016/B978-0-08-044894-7.00123-8

Titon, J. T. (2009). "Music and Sustainability: An Ecological Viewpoint." *The World of Music* 51 (1): 119–137. https://www.jstor.org/stable/41699866

Tollefson, J. W., & Tsui, A. B. M. (2014). Language diversity and language policy in educational access and equity. *Review of Research in Education*, 38(1), 189–214. https://www.jstor.org/stable/43284067. DOI: 10.3102/0091732X13506846

Tom, M. N., Sumida, H. E., & McCarty, T. L. (2019). Indigenous knowledges as vital contributions to sustainability. *International Review of Education*, 65(1), 1–18. DOI: 10.1007/s11159-019-09770-9

Tottie, G. (1991). *Negation in English Speech Writing: A Study in Variation*. Academic Press.

Tran, D. (2021). *Decolonizing university teaching and learning: An entry model for grappling with complexities*. Bloomsbury Academic. DOI: 10.5040/9781350160040

Tshivhase, M. (2018). Love as the foundation of Ubuntu. *Synthesis Philosophica*, 33(1), 197–208. DOI: 10.21464/sp33112

Tume, S. J. P., Kimengsi, J. N., & Fogwe, Z. N. (2019). Indigenous Knowledge and Farmer Perceptions of Climate and Ecological Changes in the Bamenda Highlands of Cameroon: Insights from the Bui Plateau. *Climate (Basel)*, 7(12), 138. DOI: 10.3390/cli7120138

Turner, N. J., Cuerrier, A., & Joseph, L. (2022). Well grounded: Indigenous peoples' knowledge, ethnobiology and sustainability. *People and Nature*, 4(3), 627–651. DOI: 10.1002/pan3.10321

Tuyen, Q. D. (2023). Staging culture, selling authenticity: the commodification of the Cham community's traditions. In *Heritage conservation and tourism development at Cham sacred sites in Vietnam*. Springer., DOI: 10.1007/978-981-99-3350-1_8

Twala, C. (2018). The 'letsema/ilima' campaign: A smokescreen or essential strategy to deal with the unemployment crisis in South Africa. *Journal of Contemporary History*, 29(1), 184–198.

Ugbor, O. (2004). Benefits of Indigenous language in the teaching and learning of adult farmers in Delta State, Nigeria. *Journal of Agriculture and Social Research*, 4(2), 54–60.

Uink, B., Day, A., & Martin, G. (2019). Wings to fly: Case study of supporting Indigenous students success through a 'whole-of-university' approach. *Australian Journal of Indigenous Education*, 48(2), 206–206. DOI: 10.1017/jie.2019.11

Ukwuoma, U. C. (2016). Educational development of Africa: Changing perspectives on the role of Indigenous knowledge. *Research Pedagogy*, 6(1), 123–145. DOI: 10.17810/2015.28

Ulwazi programme. (2024). Indigenous knowledge important for innovation. Retrieved on 15 October 2024 from https://www.ulwaziprogramme.org

UN Development Group. (2009). *Guidelines on Indigenous people's issues.* UNDG Publications.

UN Environment Programme (UNEP). (2020). What does the UN Environment Programme bring to the table in the zero-budget natural farming debate? UN story, 24 February 2020. Retrieved on 12 October 2024 from https://www.unep.org/news-and-stories/story/what-does-un-environment-programme-bring-table-zero-budget-natural-farming

UN Human Rights, (2013). *Indigenous People's and the UNHR system.* UN.

UN Permanent Forum on Indigenous Peoples. (2008). *Indigenous peoples and the indicators of well-being and development: Preliminary report.* United Nations.

UNDESA. (2018). 2018 revision of world urbanization prospects. United Nations - Department of Economic and Social Affairs. Retrieved on 22 April 2024 from https://population.un.org/wup/

UNDP. (2015). *Sustainable development goals.* United Nations.

UNDP. (2024). Indigenous knowledge is crucial in the fight against climate change – here is why. Retrieved on 18 October 2024 from https://climatepromise.undp.org/news-and-stories/Indigenous-knowledge-crucial-fight-against-climate-change-heres-why#:~:text=Many%20Indigenous%20traditional%20practices%20offer,the%20rest%20of%20the%20world

UNESCO. (1976). Recommendation on the development of adult education, adopted by the General Conference at its nineteenth session. Nairobi: United Nations Educational, Scientific and Cultural Organization (UNESCO). Retrieved on 22 April 2024 from https://uil.unesco.org/fileadmin/keydocuments/AdultEducation/en/declaration-nairob_e.pdf

UNESCO. (2008). Education for Sustainable Development. Accessed May 12, 2023, from http://portal.unesco.org/education/

UNESCO. (2012). *Inclusive knowledge societies for sustainable development.* UNESCO.

UNESCO. (2015). Incheon declaration. [*towards inclusive and equitable quality education for all.* UNESCO.]. *Education*, •••, 2030.

UNESCO. (2017). *A guide for ensuring inclusion and equity in education.* UNESCO.

UNESCO. 1953.1968. The Use of the Vernacular Languages in Education Monographs on Fundamental Education VIll. UNESCO, Paris.

UNFPA. (2007). *State of world population 2007: unleashing the potential of urban growth.* United Nations Population Fund, Thoraya Ahmed Obaid, Executive Director. Retrieved on 22 April 2024 from https://www.unfpa.org/sites/default/files/pub-pdf/695_filename_sowp2007_eng.pdf

UNGA. (2007). United Nations Declaration on the Rights of Indigenous People. GA 61/295 (A/61/I.67 and Add. 1). U. N. G. Assembly. United Nations.

United Nations (UN). (2009). *State of the world's Indigenous peoples*. Department of Economic and Social Affairs: Division for Social Policy and Development Secretariat of the Permanent Forum on Indigenous Issues.

United Nations Permanent Forum on Issues Report on the fifth session (15-26 May 2006) Economic and Social Council Official Records Supplement No. 23. Retrieved on 14 June 2023 from https://documents-dds-ny.un.org/doc/UNDOC/GEN/N06/384/35/PDF/N0638435.pdf?OpenElement

United Nations. (2007). *United Nations declaration on the rights of indigenous peoples*. Author.

United Nations. (2007). United Nations Declaration on the Rights of Indigenous Peoples. Retrieved on 16 October 2024 from https://www.un.org/development/desa/Indigenouspeoples/declaration-on-%20the-rights-of-Indigenous-peoples.html

United Nations. (2019). *Indigenous people's traditional knowledge must be preserved, valued globally*. UN.

United States Indigenous Data Sovereignty Network. (2020). Indigenous data governance. Retrieved on 17 October 2024 from https://www.policylink.org/sites/default/files/AdrianDominguez.pdf

Urbinati, N. (1998). From the periphery to modernity: Antonio Gramsci's theory of subordinating and hegemony. *Political Theory*, 26(3), 370–391. DOI: 10.1177/0090591798026003005

Valadez, J. M. (2012). Indigenous rights. In Callahan, D., & Singer, P. (Eds.), *Encyclopedia of applied ethics* (pp. 696–703). Academic Press, Elsevier. DOI: 10.1016/B978-0-12-373932-2.00298-2

Vambe, M. (1999). 'The Dynamics of the 'Mbende' Dance', *MOTO Magazine*, September 1999, Issue, 2000, pp. 26-27

Van Bewer, V., Woodgate, R. L., Martin, D., & Deer, F. (2020). The importance and promise of integrating Indigenous knowledges in nursing education. Witness. *The Canadian Journal of Critical Nursing Discourse*, 2(1), 11–24. DOI: 10.25071/2291-5796.46

Van de Ven, A. H., Meyer, A. D., & Jing, R. T. (2018). Opportunities and challenges of engaged Indigenous scholarship. *Management and Organization Review*, 14(3), 449–462. DOI: 10.1017/mor.2018.28

Vanguard (2022). Nigeria not on track to meet many SDGs by 2030 — UN. Retrieved on 22 April 2024 from https://www.vanguardngr.com/2022/09/nigeria-not-on-track-to-meet-many-sdgs-by-2030-un/

Vasvanathan, S. (2015). Cognitive Justice in Public Policy: A View from the South. HSRC Human and Social Dynamics Research Seminar, HSRC.

Vedeld, P., Angelsen, A., Bojo, A., Sjastaad, E., & Berg, G. K. (2007). Forest Environmental Incomes and the Rural Poor. *Forest Policy and Economics*, 9(7), 156–172. DOI: 10.1016/j.forpol.2006.05.008

Veintie, T. (2013). Coloniality and cognitive justice: Reinterpreting formal education for the Indigenous peoples in Ecuador. *International Journal of Multicultural Education*, 15(3), 45–60. DOI: 10.18251/ijme.v15i3.708

Vela, R. C. (2015). Fourth World People: Continuity, philosophies and sources. *Fourth World Journal*, 13(2), 21–28.

Velten, S., Leventon, J., Jager, N., & Newig, J. (2015). What is sustainable agriculture? A systematic review. *Sustainability (Basel)*, 7(6), 7833–7865. Advance online publication. DOI: 10.3390/su7067833

Venter, E. (1997). A philosophy of education in a new South Africa. *South African Journal of Higher Education*, 11(1), 57–64.

Ventura, J., & Rueter, A. (2018). *Healthcare workers use language to build trust and break down barriers*. CBC.

Viera, A. J., & Garrett, J. M. (2005). Understanding interobserver agreement: The kappa statistic. *Family Medicine*, 37, 360–363. PMID: 15883903

Vijayan, D., Ludwig, D., Rybak, C., Kaechele, H., Hoffmann, H., Schonfeldt, H. C., Mbwana, H. A., Rivero, C. V., & Lohr, K. (2022). Indigenous knowledge in food system transformations. *Communications Earth & Environment*, 3(1), 213. DOI: 10.1038/s43247-022-00543-1

Vilakazi, B. S., Zengeni, R., & Mafongoya, P. (2022). Integrating Indigenous knowledge systems (IKS) with science-based strategies into climate variability mitigation and adaptation of farming communities in Semi-Arid Southern Africa. *Indilinga*, 21(1), 119–140.

Virk, P. S., Andersson, M. S., Arcos, J., Govindaraj, M., & Pfeiffer, W. H. (2021). Transition from targeted breeding to mainstreaming of biofortification traits in crop improvement programs. *Frontiers in Plant Science*, 12, 703990. DOI: 10.3389/fpls.2021.703990 PMID: 34594348

Visvanathan, S. (1997). *A carnival for science: Essays on science, technology and development*. Oxford University Press.

Von Glasersfeld, E. (2013). *Radical constructivism* (Vol. 6). Routledge. DOI: 10.4324/9780203454220

von Kotze, A. (2022). Adult education for "resilience": educating in precarious times. In Evans, K., Lee, W. O., Markowitsch, J., & Zukas, M. (Eds.), *Third international handbook of lifelong learning*. Springer International Handbooks of Education., DOI: 10.1007/978-3-030-67930-9_55-1

Vygotsky, L. S. (1978). *Mind in society: the development of higher psychological processes*. Harvard University Press.

Wa Thiong'o, N. (1986). *Decolonising the Mind*. James Currey.

Waapalaneexkweew (Bowman, N., Mohican/Lunaape), & Dodge-Francis, C. (2018). Culturally responsive Indigenous evaluation and tribal governments: understanding the relationship. In F. Cram, K.A. Tibbetts, & J. LaFrance (Eds.), *Indigenous evaluation. new directions for evaluation*, 159, (pp. 17–31). Wiley Periodicals, Inc.

Wadumestrige-Dona, C. G., Mohan, G., & Fukushi, K. (2021). Promoting urban agriculture and its opportunities and challenges – a global review. *Sustainability (Basel)*, 13(17), 9609. DOI: 10.3390/su13179609

Wagner, A., & de Clippele, M. S. (2023). Safeguarding cultural heritage in the digital era – a critical challenge. *International Journal for the Semiotics of Law*, 36(5), 1915–1923. DOI: 10.1007/s11196-023-10040-z

Walsh, C. (2018). The decolonial for: resurgences, shifts and movements. In Mignolo, W. & Walsh, C (Eds). *On decoloniality: concepts, analytics, praxis* (pp. 15-32). Duke University Press.

Walter, D. (2002). Colonial and post-colonial discourses in Social Sciences: Agricultural critique of colonialism. *Latin American Research Review*, 28(3), 120–134.

Wang, Q. (2012). Phenomenological Interpretation of Chinese Creation Myth of Pangu. *Theoretical Studies in Literature and Art*, 32(3), 134–137.

Wapmuk, S., & Ali, J. M. (2022). The African Continental Free Trade Area (AFCFTA) and regional economic integration: Prospects and challenges. *Zamfara Journal of Politics and Development*, 3(1), 15–15.

Warner, A. (2021). Google launches app for language preservation. Retrieved on 18 October 2024 from https://multilingual.com/google-launches-app-for-language-preservation/

Warren, D. M. (1991). *Using Indigenous Knowledge for Agricultural Development*, World Bank Discussion Paper 127. Washington, D.C. World Bank, (1997). "*Knowledge and Skills for the Information Age, The First Meeting of the Mediterranean Development Forum*"; Mediterranean Development Forum.

Warren, D. M. (1991). *Using Indigenous knowledge in agricultural development*. World Bank Discussion Papers, 127. Retrieved on 11 April 2024 from https://documents1.worldbank.org/curated/pt/408731468740976906/pdf/multi-page.pdf

Warren, D. M., Slikkerveer, L., & Brokensha, D. (1999). *The cultural dimensions of development: Indigenous knowledge systems*. ITD.

Waters, A. (Ed.). (2004). *American Indian Thought: Philosophical Essays*. Blackwell Publishing.

Weaver, K. (2023). Bridging Indigenous and Western knowledge systems in Amazonian Indigenous communities: A systematic realistic review. *Development Studies Research*, 10(1). Advance online publication. DOI: 10.1080/21665095.2023.2203842

Webb, D., & Mashford-Pringle, A. (2022). Incorporating Indigenous content into K-12 curriculum: Supports for teachers in provincial and territorial policy and post-secondary education spaces. *Canadian Journal of Educational Administration and Policy*, 198(198), 55–73. DOI: 10.7202/1086427ar

Weber-Pillwax, C. (1999). Indigenous research methodology: Exploratory discussion of an elusive subject. *The Journal of Educational Thought*, 33(1), 312–323.

Weick, K. E. (1995). What theory is not, theorizing is. *Administrative Science Quarterly*, 40(3), 385–390. DOI: 10.2307/2393789

Weir, W. I. (2007). *First Nation small business and entrepreneurship in Canada*. National Centre for First Nations Governance.

Weismantel, M. (2019). Making kin: Kinship theory and Zumbagua adoptions. *American Ethnologist*, 22(4), 685–704. DOI: 10.1525/ae.1995.22.4.02a00010

Wei, T. (2020). *Terminology and ontology for cultural heritage: application to Chinese ceramic vessels. Formal Languages and Automata Theory*. Université Grenoble Alpes.

Wei, X., Liu, R., & Chen, W. (2023). Meta theories of technological innovation based on the analysis of classic texts. *Heliyon*, 9(6), e16779. DOI: 10.1016/j.heliyon.2023.e16779 PMID: 37292334

Welden, E. A., Chausson, A., & Melanidis, M. S. (2021). Leveraging nature-based solutions for transformation: Reconnecting people and nature. *People and Nature*, 3(5), 966–977. DOI: 10.1002/pan3.10212

Welsh-Asante, K. (2000a). 'The 'Jerusarema' Dance of Zimbabwe', *Journal of Black Studies*, 15(4), African and African American dance, Music, and Theatre, June, 1985. pp.381–403.

Welsh-Asante, K. (1994). *The African aesthetic: keeper of the traditions*. Praeger.

Welsh-Asante, K. (2000b). *Zimbabwe Dance: Rhythmic Forces, Ancestral Voices: An Aesthetic Analysis*. Africa World Press.

Wheeler, H. C., & Root-Bernstein, M. (2020). Informing decision-making with Indigenous and local knowledge and science. *Journal of Applied Ecology*, 57(9), 1634–1643. DOI: 10.1111/1365-2664.13734

White, M. D., & Marsh, E. E. (2006). Content analysis: A flexible methodology. *Library Trends*, 55(1), 22–45. DOI: 10.1353/lib.2006.0053

Whyte, K., Brewer, J.II, & Johnson, J. (2015). Weaving Indigenous science, protocols and sustainability science. *Sustainability Science*, 11(1), 25–32. DOI: 10.1007/s11625-015-0296-6

Widianingsih, I., & Mertens, D. M. (2019). Transformative research and the sustainable development goals: Challenges and a vision from Bandung, West Java. *International Journal of Transformative Research*, 6(1), 27–35. DOI: 10.1515/ijtr-2019-0005

Wight, C. (2019). Bringing the outside in: The limits of theoretical fragmentation and pluralism in IR theory. *Politics*, 39(1), 64–81. DOI: 10.1177/0263395718815784

Wilkinson, M. D., Dumontier, M., Aalbersberg, I. J., Appleton, G., Axton, M., Baak, A., Blomberg, N., Boiten, J. W., da Silva Santos, L. B., Bourne, P. E., Bouwman, J., Brookes, A. J., Clark, T., Crosas, M., Dillo, I., Dumon, O., Edmunds, S., Evelo, C. T., Finkers, R., & Mons, B. (2016). The FAIR Guiding Principles for scientific data management and stewardship. *Scientific Data*, 3(1), 160018. DOI: 10.1038/sdata.2016.18 PMID: 26978244

Williams, C. (1974). *The Destruction of Black Civilization: Great Issues of A Race from 4500 B.C. to 2000 A.D.* Third World Press.

Williamson, B., Provost, S., & Price, C. (2023). Operationalising Indigenous data sovereignty in environmental research and governance. *Environment and Planning. F, Philosophy, Theory, Models, Methods and Practice*, 2(1-2), 281–304. DOI: 10.1177/26349825221125496

Wilson, D., & Barton, P. (2012). Indigenous hospital experiences: A New Zealand case study. *Journal of Clinical Nursing*, 21(15–16), 2316–2326. DOI: 10.1111/j.1365-2702.2011.04042.x PMID: 22642700

Wilson, S. (2001). What is an Indigenous research methodology? *Canadian Journal of Native Education*, 25(2), 175–179.

Wilson, S. (2008). *Research is ceremony: indigenous research methods*. Fernwood.

Wilson, S. (2008). *Research is ceremony: Indigenous research methods*. Fernwood.

WIPO. (2019). The Protection of traditional knowledge: draft articles facilitators' rev. Retrieved on 18 October 2024 from https://www.wipo.int/edocs/mdocs/tk/en/wipo_grtkf_ic_40/facilitators_text_on_tk.pdf

Wirawan, R., McIntyre-Mills, J. J., Makaulule, M., Lethole, P. V., Pitsoane, E., Arko-Achemfuor, A., & Romm, N. R. A. (2023). Together we can grow: Resourcing the commons through pathways to wellbeing. *Systemic Practice and Action Research*, 36(5), 641–690. DOI: 10.1007/s11213-022-09613-z

World Commission on Environment and Development [WCED]. (1987). *Our common future*. Oxford University Press.

Worthington, E.Jr, & Cowden, R. (2017). The psychology of forgiveness and its importance in South Africa. *South African Journal of Psychology. Suid-Afrikaanse Tydskrif vir Sielkunde*, 47(3), 292–304. DOI: 10.1177/0081246316685074

Wright, A. L., Gabel, C., Ballantyne, M., Jack, S. M., & Wahoush, O. (2019). Using two-eyed seeing in research with Indigenous people: An integrative review. *International Journal of Qualitative Methods*, 18, 1609406919869695. Advance online publication. DOI: 10.1177/1609406919869695

Wright, S. (2014). Quantitative research performing other worlds: Lessons from sustainable agriculture in the Philippines. *The Australian Geographer*, 45(1), 1–18. https://www.tandfonline.com/doi/abs/10.1080/00049182.2014.869293. DOI: 10.1080/00049182.2014.869293

Wynberg, R. (2023). Biopiracy: Crying wolf or a lever for equity and conservation? *Research Policy*, 52(2), 104674. DOI: 10.1016/j.respol.2022.104674

Yabarmase, D. (2013). The Fishbowl Strategy: An Effective Way to Improves Students' Speaking Ability. *Indonesian Journal of English Language Teaching*, 9(2), 137–144. 296261880

Yang, K. S. (2000). Monocultural and cross-cultural Indigenous approaches: The royal road to the development of a balanced global psychology. *Asian Journal of Social Psychology*, 3(3), 241–263. DOI: 10.1111/1467-839X.00067

Yende, N. E. (2020). The quest for decolonisation in higher education in the context of globalisation and internationalisation. *African Journal of Development Studies*, 10(3), 277. DOI: 10.31920/2634-3649/2020/10n3a14

Yeo, A. W., Zaman, T., & Kulathuramaiyer, N. (2013). Indigenous knowledge management in the Kelabit Community in Eastern Malaysia. *International Journal of Sociotechnology and Knowledge Development*, 5(1), 23–36. DOI: 10.4018/jskd.2013010103

Yip, S. Y., & Chakma, U. (2024). The teaching of Indigenous knowledge and perspectives in initial teacher education: A scoping review of empirical studies. *Journal of Further and Higher Education*, 48(3), 287–300. DOI: 10.1080/0309877X.2024.2327029

Young People's Potential, the Key to Africa's Sustainable Development. Retrieved on 14 June 2023 from https://www.un.org/ohrlls/news/young-people%E2%80%99s-potential-key-africa%E2%80%99s-sustainable-development

Zaltman, G., Pinson, C. R. A., & Angelmar, R. (1973). *Metatheory and consumer research.* Holt, Rinehart and Winston, Inc.

Zauditu-Selassie, K. (2007). Women who know things: African epistemologies, ecocriticism, and female spiritual authority in the novels of Toni Morrison. *The Journal of Pan African Studies*, 1(7), 38–57.

Zembylas, M. (2018). Decolonial possibilities in South African higher education: Reconfiguring humanising pedagogies as/with decolonising pedagogies. *South African Journal of Education*, 38(4), 1699. Advance online publication. DOI: 10.15700/saje.v38n4a1699

Zidny, R., Sjöström, J., & Eilks, I. A. (2020). Multi-perspective reflection on how Indigenous knowledge and related ideas can improve science education for sustainability. *Science & Education*, 29(1), 145–185. DOI: 10.1007/s11191-019-00100-x

Zimbabwe Dance https://en.wikipedia.org/wiki/Zimbabwe_Dance Consulted (10/01/2007). Candidature file for the nomination of '*Mbende/Jerusarema* dance as a Masterpiece of Oral

ZimStat. (2022). *Population and Housing Census Report. Population distribution by Ward* (Vol. 2). Government Printers.

Ziphora, M. K. (2017). *Integration of Indigenous knowledge systems in the curriculum for basic education: possible experiences of Canada.* University of South Africa., http://hdl.handle.net/10500/25096

Zulu, Z. (2014). *The Sacred Knowledge of the Desert.* Madisebo University Co.

Zuza, E. J., Lambert, K., Macmillan, T., Chiyemura, F., Araya, Y., Bowskill, V., Oluseye, A., Ng'endo Kanui, M., Keding, G., Butler, G., Breman, E., King, I. O., Matumba, L., Heddenhausen, J., Nordin, S., & Bhagwat, S. (2024). Crop species diversity: A key strategy for sustainable food system transformation and climate resilience. *Food and Energy Security*, 13(3), e558. DOI: 10.1002/fes3.558

Zvobgo, L., Johnston, P., Williams, P. A., Trisos, C. H., & Simpson, N. P. (2022). The role of Indigenous and local knowledge in water sector adaptation to climate change in Africa: A structured assessment. *Sustainability Science*, 17(5), 2077–2092. DOI: 10.1007/s11625-022-01118-x

Zvokuomba, K., Kabonga, I., & Chingozha, P. (2023). The state of the human rights of ethno-cultural minority groups in Zimbabwe: reflections on the Doma people of the Zambezi Valley. *Sverige, Raoul Wallenberg Institute,* 73-86.

Zwaan, R. A. (2014). Embodiment and language comprehension: Reframing the discussion. *Trends in Cognitive Sciences*, 18(5), 229–234. DOI: 10.1016/j.tics.2014.02.008 PMID: 24630873

About the Contributors

Patrick Ngulube is a National Research Foundation of South Africa-rated researcher who works in the field of interdisciplinary research. He has various research interests, including indigenous knowledge systems and African epistemologies. He is extensively published and has edited relevant books such as the Handbook of Research on Theoretical Perspectives on Indigenous Knowledge Systems in Developing Countries. Hershey, PA: IGI Global and the Handbook of Research on Social, Cultural, and Educational Considerations of Indigenous Knowledge in Developing Countries. Hershey, PA: IGI Global. He is passionate about transformative discourse, including inclusion, diversity, and social and epistemic justice.

***

Adeola Funmilola Afonja (PhD) is an Associate Professor in the Department of Adult Education at the University of Lagos, Nigeria. She has over 40 articles in both international and national journals.

Muhammad Salman Alam holds a Master of Science in Nursing (MSN) from Shifa Tameer-E-Millat University, Islamabad, Pakistan, and a Bachelor of Science in Nursing (BScN) from Aga Khan University, Karachi, Pakistan. He is currently serving as an Assistant Professor and Principal at Farabi College of Nursing and Health Sciences, Charsadda. Muhammad Salman Alam is a distinguished educator and nursing expert with extensive experience in critical care, clinical nursing education, and nursing research. His career includes roles such as Registered Nurse Renal transplant Nurse at Aga Khan Hospital Karachi, Clinical Nursing Instructor and Senior Nursing Lecturer at Rehman Medical Institute Peshawar, and Graduate Research Associate at Shifa College of Nursing. His scholarly focus is on advancing nursing education and practice through innovative approaches and a commitment to quality healthcare.

Amidu Owolabi Ayeni (PhD) is an Associate Professor in the Department of Geography at the University of Lagos, Nigeria. His research focus is on environment, resource analysis, climate change adaptation, and geo-information sciences. He has published several articles in reputable journals.

Josiline Chigwada is a postdoctoral research fellow at the University of South Africa (UNISA). She is a librarian with 18 years of experience in academic librarianship. She holds an Information Science doctorate from UNISA and a Postgraduate Diploma in Higher Education. She is a member of the Library and Information Association of South Africa (LIASA), SPARC Africa management committee, Zimbabwe Library Association (ZIMLA), Zimbabwe Young Academy of Science, Zimbabwe Rural Schools Library Initiative (ZRSLI), Association for Information Science and Technology (ASIS&T), International Association of Social Science Information Service and Technology (IASSIST), and the Rotary Club of Msasa, Harare. She has authored on indigenous knowledge, open science, research data management, information and digital literacy, academic librarianship, and contemporary library and information science issues.

Collence Takaingenhamo Chisita (PhD) is a National Research Foundation of South Africa (NRF) C2-rated researcher, lecturer, author and editor of books and several journal articles in high impact journals and book chapters. He is based at the Durban University of Technology in the Department of Information Systems. He is also a Research Fellow at the University of South Africa's College of Human Sciences in the Department of Information Science. Collence Chisita has been working with different local and international organizations in the area of LIS. He has collaborated with other researchers from the different IFLA Special interest groups on issues that affect libraries in the modern era. Chisita has written journal articles, book chapters, edited journals and produced conference papers and participated and shared his knowledge on topical issues on libraries and related institutions and emerging technologies through multimedia platforms.

Sayan Dey is a Bengali. He was born and brought up in Kolkata, and he traces his ancestry to different parts of Bangladesh. Currently, he works as an Assistant Professor in the Department of English Studies at Bayan College (affiliated with Purdue University Northwest), Oman. He is also an Associate Fellow at the Harriet Tubman Institute, York University, Canada, a Critical Research Studies Faculty at The NYI Institute of Cultural, Cognitive and Linguistic Studies, New York, and an Affiliated Member of the Global Posthuman Network. His latest monographs are Green Academia: Towards Eco-friendly Education Systems (Routledge, 2022) and Performing Memories, Weaving Archives: Creolized Cultures across the Indian Ocean (Anthem Press, 2023). His research interests are posthumanism, decolonial studies, environmental studies, critical race studies, culinary epistemologies, and critical diversity literacy. He can be reached at www.sayandey.com

John Olubunmi Faloju (PhD) is a Senior Lecturer in the Department of European Languages and Integration Studies at the University of Lagos, Nigeria. His research areas of interest include the linguistics of the Russian language, with a focus on anthropological linguistics, paremiology, onomastics, and semantics.

Mishack Thiza Gumbo (PhD) is a Research Professor in the Department of Science and Technology Education at the University of South Africa. He is a C1 rated researcher by the National Research Foundation. He leads research projects and mentors emerging academics and a Post-Doctoral Fellow. He has published extensively and supervised master's and doctoral students. He has won numerous awards. He has also examined dissertations and theses for several universities locally and internationally.

Tebogo Victoria Kgope holds a PhD in African Studies from the University of Venda, and a National Diploma in Water Care Technology from Tshwane University of Technology (TUT), in South Africa. As a Quality Assurance Administrator in the College of Graduate Studies at the University of South Africa (UNISA), Kgope is multi-factorially active in research and promotes African indigenous knowledge systems from the lens of the late Sanusi Credo Mutwa. She has published works inculcating African thought patterns on indigenous epistemologies, her work has been presented at national and international conferences. Following her studies at the University of Johannesburg, she is also qualified in homeopathy. Her research interests revolve around grappling with issues reflected in the everyday experiential and ontological realities of ordinary people, where she engages in a decolonial epistemic context in rethinking research.

Khatija Bibi Khan is a full Professor in the Department of Communication Science at the University of South Africa. She received her D Litt et Phil from the Department of English Studies, at UNISA, in the field of popular culture. Her main areas of research interests are the intersections between popular culture, literary studies, Genocide studies and Mainstream media. She has published 2 Books, edited a few books, published articles and chapters in South African and international journals on hip hop, film, genocide and literature.

Rudolph Ryser is a PhD in International Relations from the Union Institute and University. He is a Fulbright Research Scholar. He was Chair of the Board of Directors and Executive Director at the Centre for World Indigenous Studies in Olympia, WA. He was Editor in Chief, Fourth World Journal, and Chair of and member of the Public Service Leadership Faculty at Capella University.

Lehasa Moloi (PhD) is an Afrocentric Scholar grounded in Afrocentricity. He is a senior lecturer in the Department of Development Studies at the University of South Africa. He holds PhD in Development Studies. He is passionate about epistemic decolonization debates and the pursuit for re-humanization in the Global South, in particular Africa is his focus, against the ravages of the colonial past. His intellectual work is grounded in the paradigm of Afrocentricity and decoloniality as liberatory thinking frameworks to oppose epistemicides created by European colonialism. He broadly reflects and write on Africa, Knowledge, Development and Decolonization.

Jacob Mapara teaches modules on indigenous knowledge and sustainable technologies, as well as Zimbabwe's higher education context at Chinhoyi University of Technology. He has researched and published in these areas in addition to onomastics and how as part of intangible cultural heritage, these contribute to the development of robust indigenous epistemologies. Professor Mapara is currently the Director of the Institute of Lifelong Learning and Development Studies in the same University. He holds a PhD is African Languages from University of South Africa. Dr. Mapara also serves as the Chairperson of the Zimbabwe National Intangible Cultural Heritage Advisory Committee.

Tlou Maggie Masenya holds PhD in Information Science from the University of South Africa (UNISA) and a Masters in Information Technology from the University of Pretoria (UP). She has 9 years of working experience as an Associate Professor lecturer in the field of Information Science, Information Systems and Technology. Dr Masenya is currently appointed as Head of the Department of Information Technology at the Durban University of Technology. She is a distinguished researcher with extensive and impressive national and international research standing and has published numerous book chapters and articles in peer-reviewed journals within the broad niche area of ICT4D, Digital preservation and Data Curation, Records and Knowledge Management, Technopreneurship, Digital Entrepreneurship, Cloud Computing, Innovative Pedagogies, Disruptive Technologies and Indigenous Knowledge Systems, which had a positive impact on her scholarly growth. She recently edited two books, and the first book is entitled "Innovative technologies for enhancing knowledge access in academic libraries" published by IGI- Global International Academic Publisher in July 2022. The second book is entitled "Digital preservation and Documentation of Global Indigenous Knowledge Systems" also published by IGI- Global

International Academic Publisher in August 2023. These research publications have consistently pushed the frontiers in Information Technology and related fields and have become highly cited by academics and scholars globally. She is currently reviewing journal articles and book chapters for SAJIM, SAJLIS, Mousaion, JSASA and IGI-Global. Dr Masenya is also a guest editor for two special issues "The Role of Medical Culture in Internet Intervention of Mobile Applications", in IEEE International Journal of Biomedical and Health Informatics (J-BHI) and "New Pedagogies and Technology Driven Innovations in Improving the Quality of Learning and Teaching", Educational Research and Evaluation in Taylor and Francis International Journal.

Crispen Mazodze holds a Doctor of Philosophy degree in indigenous knowledge systems and cultural heritage from Chinhoyi University of Technology. His research interests are anchored on student leadership development, decoloniality, indigenous epistemologies and cultural heritage. Dr. Mazodze's point of departure in student development and student welfare management in the global south should be divorced from a situation where Eurocentric epistemologies, culture, values, languages and identities have remained entrenched in the curriculum, something that has resulted in epistemicides, valuecides, culturecides and identicides. He is the currently the Dean of Students at Bindura University of Science Education and is also the former Director for Off Campus Life and Student Development Programmes at Chinhoyi University of Technology.

Donna Mertens, a professor at Gallaudet University, specializes in transformative research and evaluation methodologies that support social, economic, and environmental justice, seen in her publications: Mixed Methods Research, Program Evaluation Theory and Practice (3rd ed.); Mixed Methods Design in Evaluation; and Research and Evaluation in Education and Psychology (6th ed.). She consults with many international organizations such as U.N. International Fund for Agricultural Development and UN Women. Mertens served as the editor of the Journal of Mixed Methods Research, President of the American Evaluation Association (1998), and founding Board member: International Organization for Cooperation in Evaluation and Mixed Methods International Research Association.

Stanley Murwira (PhD) holds a Doctor of Education Degree (Philosophy) from the University of South Africa, obtained on 5 August 2020. He also has a Master of Education Degree Educational Foundations (Philosophy) attained in 2005 and Bachelor of Education [B.Ed.] in 2001, both obtained from the Zimbabwe. His research interests include Indigenous Knowledge Systems, Technical Education, Unhu/Ubuntu, globalization African Philosophy, digitalization and technological development in Africa, and Philosophy of Education. Dr. Murwira has published articles in journals such as the South African Journal of Higher Education (SAJHE) and the Zimbabwe Journal of Teacher Education [ZJTE]. Dr. Murwira is the full-time Principal in Philosophy at Belvedere Technical Teachers' College, Harare, Zimbabwe.

Mosiuoa Nehemiah Ngaka (PhD holds degrees from four different universities. His qualifications include a BA, BA (Hons), MBA, and a PhD. These academic achievements demonstrate his intellectual curiosity and determination to excel in various fields. In addition to his professional and academic accomplishments, Dr. Ngaka also ventured into the realm of creative writing. In 2021, he published his first short story book, titled "Ngwaaang??? Who Are You???" Dr Ngaka's life story is a testament to the power of perseverance, dedication, and the pursuit of knowledge. His journey from a small town in the Free State to a multifaceted career in education, banking, and public service is an inspiring example of what one can achieve through hard work and a passion for making a difference.

Innocent Nyathi (PhD) is an early career researcher in the social, economic, and political histories of groups existing on the margins of society. Nyathi's main interests lie in local histories, where the voice of the subaltern and their indigenous narratives take paramountcy in the academic presentation of their 'stories' and has over twelve years researching and teaching on the above themes. Currently embarking on independent and collaborative research with a view to publish extensively on matters affecting groups existing on the margins, particularly on resilience and survival strategies using the 'everyday' approach, Nyathi infuses the everyday with emerging themes such as temporality in the study of urban and rural contestations of space among different groups, especially those barely studied.

Abel J Pienaar holds a PhD in Health Sciences (capita selecta Nursing and Indigenous knowledge systems). He also completed an M. Ed, MA Community Psychiatric Nursing and a Comprehensive Nursing Midwifery degree, to discern himself as a clinical specialist and educationist. Prof. Pienaar is a globally renowned IKS researcher, Innovative Educationist and Community development expert. His scholarship is on Inclusive Health and Higher Education, especially to advocate and uphold the human rights of indigenous and marginalized communities.

Norma RA Romm (DLitt et Phil), Professor Extraordinarius at the University of South Africa, is author of: The Methodologies of Positivism and Marxism (1991); Accountability in Social Research (2001); New Racism (2010); Responsible Research Practice (2018); People's Education in Theoretical Perspective (with V McKay, 1992); Diversity Management (with R Flood, 1996); and Assessment of the Impact of HIV and AIDS in the Informal Economy of Zambia (with V McKay, 2006). She has published over 120 research articles on social theorizing, transformative research towards social and ecological regeneration, Indigenous paradigms of knowing, and the facilitation of adult learning. She has worked (in teams) for commissioned research for organizations such as: The International Labor Organization (ILO), The International Organization for Migration (IOM), The Association for the Development of Education in Africa (ADEA), The United Nations Educational, Scientific and Cultural Organization (UNESCO), and The United Nations Democracy Fund (UNDEF).

Edward Shizha is Professor of Children and Youth Studies at Wilfrid Laurier University, Brantford Campus in Canada. He is the series editor of Africa in the Global Space, a book series that is published by Peter Lang. To date, he has published 14 books and several book chapters and journal articles. Some of his recent books are Immigrant Lives: Intersectionality, Transnationality and Global Perspectives (2023) and The dynamics of African indigenous knowledge systems: A sustainable alternative for livelihoods in Southern Africa (2022. Recent book chapters include "Black Lives Matter: Decolonizing whiteness in higher education academic spaces in Canada" (2024); and "Social justice education in Canada" (2023). Edward's areas of expertise are social justice in education, indigenous epistemologies, education of immigrant students, sociology of education, integration of newcomer youth. youth at risk, youth cultures.

Ethel Siamena is a doctoral student at Chinhoyi University of Technology's Centre for Indigenous Knowledge and Living Heritage. Her focus is on indigenous knowledge and living heritage with a particular emphasis on knowledge, management, and interpretation of natural phenomena (climate systems, ecological systems, and peoples' livelihoods). She has executed comprehensive studies on inventorying of ICH elements disturbed by Cyclone IDAI and Indigenous weather disaster early warning systems. Ms Siamena has presented several papers at various conferences, all with a focus on indigenous knowledge.

Niharika R. Sinha is an aspiring researcher and writer who grew up between USA and India. She studied neuroscience and psychology as an undergraduate at New York University, focusing on how we form beliefs, make decisions, and subsequently act. Niharika is currently a Global Leaders' Alliance Fellow for interdisciplinary liberal arts at Alliance University, Bangalore. Her research interests lie in interdisciplinary explorations of well-being and community across cultures.

Kwashirai Zvokuomba (PhD) is a Senior Research Associate in the Department of Sociology at the University of Johannesburg in South Africa. He is a social scientist who deploys sociological and anthropological lenses to interrogate socio-cultural, political, economic, and environmental issues in society. He has years of university service experience in Zimbabwe and the Southern African region. He has published and presented research findings at national, regional, and international platforms, journals, and book chapters.

Index

A

B

C

D

E

F

J

K

S

T

U

W

www.ingramcontent.com/pod-product-compliance
Lightning Source LLC
LaVergne TN
LVHW080619300926
849162LV00005B/1207